Cometary Plasma Processes

Geophysical Monograph Series

Including

IUGG Volumes
Maurice Ewing Volumes
Mineral Physics Volumes

GEOPHYSICAL MONOGRAPH SERIES

Geophysical Monograph Volumes

1 Antarctica in the International Geophysical Year *A. P. Crary, L. M. Gould, E. O. Hulburt, Hugh Odishaw, and Waldo E. Smith (Eds.)*

2 Geophysics and the IGY *Hugh Odishaw and Stanley Ruttenberg (Eds.)*

3 Atmospheric Chemistry of Chlorine and Sulfur Compounds *James P. Lodge, Jr. (Ed.)*

4 Contemporary Geodesy *Charles A. Whitten and Kenneth H. Drummond (Eds.)*

5 Physics of Precipitation *Helmut Weickmann (Ed.)*

6 The Crust of the Pacific Basin *Gordon A. Macdonald and Hisahi Kuno (Eds.)*

7 Antarctica Research: The Matthew Fontaine Maury Memorial Symposium *H. Wexler, M. J. Rubin, and J. E. Caskey, Jr. (Eds.)*

8 Terrestrial Heat Flow *William H. K. Lee (Ed.)*

9 Gravity Anomalies: Unsurveyed Areas *Hyman Orlin (Ed.)*

10 The Earth Beneath the Continents: A Volume of Geophysical Studies in Honor of Merle A. Tuve *John S. Steinhart and T. Jefferson Smith (Eds.)*

11 Isotope Techniques in the Hydrologic Cycle *Glenn E. Stout (Ed.)*

12 The Crust and Upper Mantle of the Pacific Area *Leon Knopoff, Charles L. Drake, and Pembroke J. Hart (Eds.)*

13 The Earth's Crust and Upper Mantle *Pembroke J. Hart (Ed.)*

14 The Structure and Physical Properties of the Earth's Crust *John G. Heacock (Ed.)*

15 The Use of Artificial Satellites for Geodesy *Soren W. Henricksen, Armando Mancini, and Bernard H. Chovitz (Eds.)*

16 Flow and Fracture of Rocks *H. C. Heard, I. Y. Borg, N. L. Carter, and C. B. Raleigh (Eds.)*

17 Man-Made Lakes: Their Problems and Environmental Effects *William C. Ackermann, Gilbert F. White, and E. B. Worthington (Eds.)*

18 The Upper Atmosphere in Motion: A Selection of Papers With Annotation *C. O. Hines and Colleagues*

19 The Geophysics of the Pacific Ocean Basin and Its Margin: A Volume in Honor of George P. Woollard *George H. Sutton, Murli H. Manghnani, and Ralph Moberly (Eds.)*

20 The Earth's Crust: Its Nature and Physical Properties *John G. Heacock (Ed.)*

21 Quantitative Modeling of Magnetospheric Processes *W. P. Olson (Ed.)*

22 Derivation, Meaning, and Use of Geomagnetic Indices *P. N. Mayaud*

23 The Tectonic and Geologic Evolution of Southeast Asian Seas and Islands *Dennis E. Hayes (Ed.)*

24 Mechanical Behavior of Crustal Rocks: The Handin Volume *N. L. Carter, M. Friedman, J. M. Logan, and D. W. Stearns (Eds.)*

25 Physics of Auroral Arc Formation *S.-I. Akasofu and J. R. Kan (Eds.)*

26 Heterogeneous Atmospheric Chemistry *David R. Schryer (Ed.)*

27 The Tectonic and Geologic Evolution of Southeast Asian Seas and Islands: Part 2 *Dennis E. Hayes (Ed.)*

28 Magnetospheric Currents *Thomas A. Potemra (Ed.)*

29 Climate Processes and Climate Sensitivity (Maurice Ewing Volume 5) *James E. Hansen and Taro Takahashi (Eds.)*

30 Magnetic Reconnection in Space and Laboratory Plasmas *Edward W. Hones, Jr. (Ed.)*

31 Point Defects in Minerals (Mineral Physics Volume 1) *Robert N. Schock (Ed.)*

32 The Carbon Cycle and Atmospheric CO_2: Natural Variations Archean to Present *E. T. Sundquist and W. S. Broecker (Eds.)*

33 Greenland Ice Core: Geophysics, Geochemistry, and the Environment *C. C. Langway, Jr., H. Oeschger, and W. Dansgaard (Eds.).*

34 Collisionless Shocks in the Heliosphere: A Tutorial Review *Robert G. Stone and Bruce T. Tsurutani (Eds.)*

35 Collisionless Shocks in the Heliosphere: Reviews of Current Research *Bruce T. Tsurutani and Robert G. Stone (Eds.)*

36 Mineral and Rock Deformation: Laboratory Studies—The Paterson Volume *B. E. Hobbs and H. C. Heard (Eds.)*

37 Earthquake Source Mechanics (Maurice Ewing Volume 6) *Shamita Das, John Boatwright, and Christopher H. Scholz (Eds.)*

38 Ion Acceleration in the Magnetosphere and Ionosphere *Tom Chang (Ed.)*

39 High Pressure Research in Mineral Physics (Mineral Physics Volume 2) *Murli H. Manghnani and Yasuhiko Syono (Eds.)*

40 Gondwana Six: Structure, Tectonics, and Geophysics *Garry D. McKenzie (Ed.)*

41 Gondwana Six: Stratigraphy, Sedimentology, and Paleontology *Garry D. McKenzie (Ed.)*

42 Flow and Transport Through Unsaturated Fractured Rock *Daniel D. Evans and Thomas J. Nicholson (Eds.)*

43 Seamounts, Islands, and Atolls *Barbara H. Keating, Patricia Fryer, Rodey Batiza, and George W. Boehlert (Eds.)*

44 Modeling Magnetospheric Plasma *T. E. Moore and J. H. Waite, Jr. (Eds.)*

45 Perovskite: A Structure of Great Interest to Geophysics and Materials Science *Alexandra Navrotsky and Donald J. Weidner (Eds.)*

46 Structure and Dynamics of Earth's Deep Interior (IUGG Volume 1) *D. E. Smylie and Raymond Hide (Eds.)*

47 Hydrological Regimes and Their Subsurface Thermal Effects (IUGG Volume 2) *Alan E. Beck, Grant Garven, and Lajos Stegena (Eds.)*

48 Origin and Evolution of Sedimentary Basins and Their Energy and Mineral Resources (IUGG Volume 3) *Raymond A. Price (Ed.)*

49 Slow Deformation and Transmission of Stress in the Earth (IUGG Volume 4) *Steven C. Cohen and Petr Vaníček (Eds.)*

50 Deep Structure and Past Kinematics of Accreted Terranes (IUGG Volume 5) *John W. Hillhouse (Ed.)*

51 Properties and Processes of Earth's Lower Crust (IUGG Volume 6) *Robert F. Mereu, Stephan Mueller, and David M. Fountain (Eds.)*

52 Understanding Climate Change (IUGG Volume 7) *Andre L. Berger, Robert E. Dickinson, and J. Kidson (Eds.)*

53 Plasma Waves and Istabilities at Comets and in Magnetospheres *Bruce T. Tsurutani and Hiroshi Oya (Eds.)*

54 Solar System Plasma Physics *J. H. Waite, Jr., J. L. Burch, and R. L. Moore (Eds.)*

55 Aspects of Climate Variability in the Pacific and Western Americas *David H. Peterson (Ed.)*

56 The Brittle-Ductile Transition in Rocks *A. G. Duba, W. B. Durham, J. W. Handin, and H. F. Wang (Eds.)*

57 Evolution of Mid Ocean Ridges (IUGG Volume 8) *John M. Sinton (Ed.)*

58 Physics of Magnetic Flux Ropes *C. T. Russell, E. R. Priest, and L. C. Lee (Eds.)*

59 Variations in Earth Rotation (IUGG Volume 6) *Dennis D. McCarthy and William E. Carter (Eds.)*

60 Quo Vadimus *Geophysics for the Next Generation* (IUGG Volume 10) *George D. Garland and John R. Apel (Eds.)*

Maurice Ewing Volumes

1 Island Arcs, Deep Sea Trenches, and Back-Arc Basins *Manik Talwani and Walter C. Pitman III (Eds.)*

2 Deep Drilling Results in the Atlantic Ocean: Ocean Crust *Manik Talwani, Christopher G. Harrison, and Dennis E. Hayes (Eds.)*

3 Deep Drilling Results in the Atlantic Ocean: Continental Margins and Paleoenvironment *Manik Talwani, William Hay, and William B. F. Ryan (Eds.)*

4 Earthquake Prediction—An International Review *David W. Simpson and Paul G. Richards (Eds.)*

5 Climate Processes and Climate Sensitivity *James E. Hansen and Taro Takahashi (Eds.)*

6 Earthquake Source Mechanics *Shamita Das, John Boatwright, and Christopher H. Scholz (Eds.)*

IUGG Volumes

1 Structure and Dynamics of Earth's Deep Interior *D. E. Smylie and Raymond Hide (Eds.)*

2 Hydrological Regimes and Their Subsurface Thermal Effects *Alan E. Beck, Grant Garven, and Lajos Stegena (Eds.)*

3 Origin and Evolution of Sedimentary Basins and Their Energy and Mineral Resources *Raymond A. Price (Ed.)*

4 Slow Deformation and Transmission of Stress in the Earth *Steven C. Cohen and Petr Vaníček (Eds.)*

5 Deep Structure and Past Kinematics of Accreted Terranes *John W. Hillhouse (Ed.)*

6 Properties and Processes of Earth's Lower Crust *Robert F. Mereu, Stephan Mueller, and David M. Fountain (Eds.)*

7 Understanding Climate Change *Andre L. Berger, Robert E. Dickinson, and J. Kidson (Eds.)*

8 Evolution of Mid Ocean Ridges *John M. Sinton (Ed.)*

9 Variations in Earth Rotation *Dennis D. McCarthy and William E. Carter (Eds.)*

10 Quo Vadimus *Geophysics for the Next Generation* *George D. Garland and John R. Apel (Eds.)*

Mineral Physics Volumes

1 Point Defects in Minerals *Robert N. Schock (Ed.)*

2 High Pressure Research in Mineral Physics *Murli H. Manghnani and Yasuhiko Syono (Eds.)*

Cometary Plasma Processes

Alan D. Johnstone

Editor

American Geophysical Union

Published under the aegis of the AGU Geophysical Monograph Board.

Library of Congress Cataloging in Publication Data

Cometary plasma processes / Alan D. Johnstone.

   p. cm. — (Geophysical monograph : 61)
   ISBN 0-87590-027-5
   1. Comets. 2. Space plasmas. 3. Astrophysics. I. Johnstone,
Alan D. II. Series.
QB721.6.C66 1991   91-11777
523.6—dc20   CIP

ISBN 0-87590-027-5

CONTENTS

Preface
A.D. Johnstone

1. The future of cometary plasma research.
   M. Neugebauer   1

2. The solar wind interaction with Venus and Mars: cometary analogies and contrasts.
   J.G. Luhmann   5

3. Dust-plasma interactions in the cometary environment.
   D.A. Mendis and M. Horanyi   17

4. Collisional processes in cometary plasmas.
   T.E. Cravens   27

5. Theory and simulation of cometary shocks.
   N. Omidi and D. Winske   37

6. MHD models of cometary plasma and comparison with observations.
   H.U. Schmidt and R. Wegmann   49

7. A parametric study of the solar wind interaction with comets.
   C.T. Russell, G. Le, J.G. Luhmann and J.A. Fedder   65

8. Simulation studies of the interaction of a neutral gas and an ambient plasma.
   R. Bingham, R. Bollens, F. Kazeminejad and J.M. Dawson   73

9. Cometary plasma observations between the shock and the contact surface.
   H. Reme   87

10. Permanent and nonstationary plasma phenomena in comet Halley's head.
    K.I. Gringauz and M.I. Verigin   107

11. Friction layer in comet Halley's ionosheath.
    H. Perez-de-Tejada   117

12. Charge exchange regime in the plasma flow as source of cometosheath and Halley's plasma tail.
    M.K. Wallis   175

CONTENTS

13. The plasma parameters during the inbound and outbound legs of the Giotto trajectory.
    E. Amata, V. Formisano, P. Torrente and R. Giovi   131

14. Ions in the coma and in the tail of comets - observations and theory.
    K. Jockers   139

15. The 10 January 1986 disconnection event in comet Halley.
    M.B. Niedner Jr, J C Brandt and Y Yi   153

16. Mirror mode waves at comet Halley
    C.T. Russell, G Le, K. Schwingenschuh, W. Riedler and E.G. Eroshenko   161

17. Spectral structure of ultralow-frequency electromagnetic fields near Comet Halley.
    Yu.M. Mikhailov, O.V. Kapustina, G.A. Mikhailova, E.G. Eroshenko, V.A. Styazhkin, J.G. Trotignon and K. Sauer.   171

18. Electric-field discontinuities at comet Halley during the VEGA encounters.
    J.G. Trotignon, C. Beghin, R. Grard, A. Pedersen, M. Mogilevsky and Yu. Mikhailov   179

19. Comets: a laboratory for plasma waves and instabilities.
    B.T. Tsurutani   189

20. Cometary linear instabilities: from profusion to perspective.
    A.L. Brinca   211

21. Quasilinear theory of the ion cyclotron instability and its application to the cometary plasma
    A .A. Galeev, R.Z. Sagdeev, V.D. Shapiro, V. I. Shevchenko and K. Szego   223

22. Ion pickup in the solar wind via wave-particle interaction.
    P.H. Yoon and C.S. Wu   241

23. The spectrum and energy density of solar wind turbulence of cometary origin.
    A.D. Johnstone, D.E. Huddleston and A.J. Coates   259

24. The magnetic field turbulence at comet Halley observed by VEGA 1 and 2.
    G. Le, C.T. Russell, W. Riedler and K. Schwingenschuh   273

25. Acceleration mechanisms for cometary ions.
   T. Terasawa   277

26. The second order Fermi acceleration of pick-up ions.
   P. Duffy   287

27. Ion pickup, scattering and stochastic acceleration in the cometary environment of P/Giacobini-Zinner.
   D.D. Barbosa   291

28. Observations of the velocity distribution of pickup ions.
   A.J. Coates   301

29. Acceleration of pickup ions in the vicinity of comet Halley.
   M.B. Bavassano-Cattaneo, V. Formisano, E. Amata, R. Giovi and P. Torrente   311

30. Observations of energetic water group ions at comet Giacobini Zinner; an overview.
   S.W.H. Cowley, A. Balogh, R.J. Hynds, K. Staines, T.S. Yates, P.W. Daly, I.G. Richardson, T.R. Sanderson, C. Tranquille and K.P. Wenzel   319

31. Model calculations of oxygen ion fluxes from the dissociation of H_2O, CO and CO_2 at gigameter distances from comet Halley.
   P.W. Daly   341

32. Neutral hydrogen shell structure near comet P/Halley deduced from VEGA-1 and Giotto energetic particle data.
   M.I. Verigin, S. McKenna-Lawlor, A.K. Richter, K. Szego and I.S. Veselovsky   349

   Comment
   M. Neugebauer and A.J. Coates

   Reply
   M.I. Verigin et al

33. Energetic water group ion fluxes ($E_{H_2O} > 60$ keV) in a quasiperpendicular and a quasiparallel shock front as observed during the Giotto-Halley encounter.
   E. Kirsch, S. McKenna-Lawlor, P.W. Daly, F.M. Neubauer, A.J. Coates, A.Thompson, D. O'Sullivan and K-P. Wenzel   357

Cometary Plasma Processes

Preface

Since Explorer 1 discovered the Earth's radiation belts more than thirty years ago, there have been many opportunities to show the value of in-situ observations over remote-sensing when it comes to an understanding of the space plasma environment. When one of the inner solar system's regular visitors was due to make it's once-in-a-lifetime appearance in 1986 the opportunity was too important to be missed. For not only is comet Halley one of the most reliable comets it is also nearly two orders of magnitude larger than any other comet with a known period. Well before there was any visible trace of Halley's comet in the night sky, three of the big four space agencies were banking on that reliability and were preparing five spacecraft to make the journey to intercept the comet. Such activity acted as a spur to the ingenuity of the fourth agency who found a way to redirect one of their long-serving spacecraft and to win the race to be the first to a comet, albeit the smaller, and at the time virtually unknown, Giacobini-Zinner. Although a healthy spirit of competition infused the scientific and engineering teams working on the project at various levels, what mattered in the end was the global cooperation between the agencies and many ground-based observers which for example, enabled Giotto to reach the comet with one-tenth of the targetting error that had originally been predicted.

The same combination of competition and cooperation has characterised the scientific analysis. Three years after the Halley encounters seemed to be a good time to get together to take stock of the progress that had been made; to compare results and to test new ideas. An AGU Chapman Conference was held in Guildford in England in July 1989 to discuss the plasma results from the missions under the title of Cometary Plasma Processes. Rather than being its proceedings, it can be said that the papers in this volume were stimulated by the conference since most of them have been written in the year following and have not been limited to material presented at the meeting.

As usual, progress has been uneven; some topics have proved to be more fruitful than others. This is reflected in the contents which should be regarded as a report on the status of the investigations, not as a textbook. While in some areas, in the editor's view, there is a high degree of convergence in the presentations, in others there is still much controversy. I have not attempted to resolve the conflict in the editing process as I feel the author's have the right to present their views and allow the scientific community to make up their minds on the strength of the arguments and the weight of the evidence. Cometary plasma physics is still an active field with much more work to be done on the excellent data obtained by all the missions.

I would like to thank the referees of the individual papers for their contribution to the quality of the volume. It has been my first experience of the process from an editorial point of view and I have been impressed by the number of letters from authors thanking the referees for the care and thought which they had put into the review. I glad to be able to take this opportunity to add my thanks to theirs. The list of those who provided reports is given below. I would also like to thank Malcolm Niedner for his editorial help in cases where there was a potential conflict of interest.

A.D. Johnstone, November 1990

I would also like to thank the following staff at the Mullard Space Science Laboratory for a considerable amount of help in preparing this volume for publication; Sharon Ashley, Libby Daghorn, Rosalind Rose and Dawn Bedwell for their secretarial skill in typesetting a number of the papers and Derek Hoyle for helping to prepare figures for reproduction.

The Editor thanks the following for their assistance in reviewing one or more of the papers

E. Amata	A. Mendis	J. Brandt	M.B. Niedner
R. Bingham	M. Neugebauer	A.J. Coates	R. Reinhard
S.C. Chapman	A.Pedersen	T.E. Cravens	D.A. Roberts
S.W.H. Cowley	H. Reme	P. Duffy	H.U. Schmidt
P.W. Daly	C.T. Russell	S.A. Fuselier	V. Shapiro
C. d'Uston	S.J. Schwartz	B.E. Goldstein	B.T. Tsurutani
K-H. Glassmeier	T. Terasawa	T.I. Gombosi	A.F. Vinas
M.L. Goldstein	O. Vaisberg	W-H. Ip	D. Winske
K. Gringauz	M. Wallis	K. Jockers	C.S. Wu
P. Isenberg	L.J.C. Wooliscroft	J.G. Luhmann	
M.A. Lee	W.I. Axford	F. Neubauer	

The Future of Cometary Plasma Research

MARCIA NEUGEBAUER

Jet Propulsion Laboratory, California Institute of Technology, Pasadena, CA 91109, USA

There are still many unsolved problems in cometary plasma physics; addressing them requires further analysis of and theoretical studies based on data already in hand together with the acquisition of new types of data. Prospects for new cometary plasma data include ground-based or Earth-satellite observations of active comets, possibly a Giotto Extended Mission, the Comet Rendezvous Asteroid Flyby mission, and perhaps the Japanese Soccer mission.

INTRODUCTION

This conference on cometary plasma physics certainly does not mark an occasion to neatly summarize a well understood subfield of physics before moving on to some greater challenge. Some of our discussions clearly indicate that a lot of very interesting problems in cometary plasma physics remain unsolved. This paper gives my personal view of what the future direction and pace of attacking some of those problems will be.

Although the thrust of this paper is on advances in the acquisition and analysis of cometary data, it is clear that important theoretical work and numerical simulations of cometary plasmas will keep pace with the observations as well as raise interesting questions for future observations and analyses to address. This topic is given scant attention because of my personal incompetence in the field, which should not reflect on its importance.

FURTHER ANALYSIS OF EXISTING DATA

At the time of this meeting in Surrey, data from the ICE flyby of comet Giacobini-Zinner were nearly four years old, while those from the Halley armada were over three. Yet we're hearing about some of the Halley data for the first time, and making plans for future collaborative analyses. There are several reasons why the analysis of the Halley data has been so unusually slow. Most importantly, most of the particle instruments on the Halley spacecraft were new designs, with special features required to cope with the environmental hazards. Since the launch dates were inflexible, many instruments were launched with rather hasty calibrations, some of which have been repeated in greater detail with spare hardware. Another consequence of the unique instrumentation and unique environment was that many groups had little data-analysis software ready to use as soon as the data were acquired.

These recalibrations, generation of new software, computer modeling of instrument responses, and cross-checking of re-sults between different sensors has absorbed the energy of many of the instrument teams ever since the Halley flybys. I think that this effort is finally beginning to pay off. Specific areas in which significant progress can now be made include wave-particle interactions (wave excitation, pitch-angle scattering, particle acceleration, and shocks), the physics of cometosheath structures (their permanence and uniqueness), tail features (disconnection events), and global MHD/chemical models. At this meeting we have seen a convergence of some previously disparate measurements, such as the plasma density and composition measured simultaneously by different sensors on Giotto. A tape containing time-merged data from the Giotto magnetometer, JPA, RPA, and IMS is being constructed. Many of the teams are submitting their data to the International Halley Watch (IHW) archives. We may, thus, finally be on the brink of being able to carry out detailed analyses using the data from several instruments. It is unfortunate that we've reached this point at a time when funds specifically ear-marked for the analysis of Giotto data have been turned off and when many of the investigators on the Halley missions are very busy building instruments for the next generation of missions (e.g., Cluster, Regatta, etc.). Thus I expect the pace of new results from analyses of Halley data to decline while their depth and quality improves.

NEED FOR NEW DATA

There will come a point, however, when further progress and tests of the conclusions drawn from the existing data base will be very difficult without access to new types of data. Verification of models of the inner coma, for example, requires measurement of low energy electrons. There are also no data on plasma waves in the inner coma, inside 8000 km. There are important regions of the comet-solar-wind interaction that have not been explored, such as the subsolar streamline and the region where the field-free cavity connects to the tail. We also need to observe a greater dynamic range of the principal determining parameters and conditions to understand the effects of different levels of cometary activity, different solar wind conditions, and different comet personalities.

Cometary Plasma Processes
Geophysical Monograph 61
©1991 American Geophysical Union

OBSERVATIONS FROM GROUND OR EARTH ORBIT

Some new data can be acquired from the ground or Earth orbit. What we'd like is a bright, active, long-period comet, preferably without much dust. The recently recovered short-period comet Brorson-Metcalf doesn't quite fit this description, but it is an unusually gassy comet with a gas production rate greater than that of Giacobini-Zinner at the time of the ICE flyby. Comet Austin (discovered in the fall of 1989, after the Surrey conference) was well placed for observations from the northern hemisphere in May, 1990, and some good plasma observations were obtained. Additional opportunities may require some prayer on the part of comet astronomers.

Several lessons were learned from the Halley apparition that can be taken advantage of for future ground-based studies. At this meeting, Sue Wyckoff pointed out the necessity for using very narrow-band filters for studying cometary ions. Jack Brandt described the usefulness of well-spaced observer networks, such as that set up by the IHW, to observe comet tails with a frequency sufficient to follow the development and evolution of individual features in rapidly changing ion tails.

As for the Hubble Space Telescope (HST), Jack Brandt has pointed out that the field of view of even its Wide Field Planetary Camera (WFPC) is not wide enough to study plasma structures in cometary tails. HST may, however, be able to do some useful cometary spectroscopy, although comets are difficult targets for HST to track.

GIOTTO EXTENDED MISSION

In February, 1990, the Giotto spacecraft was turned on again and its instruments were checked out over the following few months. Seven of the instruments (including the magnetometer, the Johnstone and Rème plasma analyzers, and the energetic particle analyzer) were found to be fully operational or only marginally degraded. Unfortunately, the Halley Multicolour Camera did not work. At the time of the final preparation of this paper (October, 1990), the European Space Agency (ESA) had not yet determined whether or not it would proceed with a Giotto Extended Mission (GEM). In preparation for GEM, the Giotto spacecraft flew by the Earth in July, 1990, and was retargetted for a flyby of P/Grigg-Skjellerup in July, 1992. The comet flyby speed would be 14 km/s at heliocentric and geocentric distances of 1.01 and 1.43 AU, respectively and an approach phase angle of 79°.

COMET RENDEZVOUS ASTEROID FLYBY MISSION

The US Congress and President have approved the start of the CRAF/Cassini project. CRAF and Cassini are two separate missions that use nearly identical spacecraft with different sets of scientific instruments. Cassini is a joint NASA-ESA mission which will put a spacecraft in orbit around Saturn and deliver a probe (named Huygens) into the atmosphere of its satellite Titan.

The acronym CRAF stands for Comet Rendezvous Asteroid Flyby. The spacecraft trajectory is shown in Figure 1. After launch from Earth in August, 1995, and an Earth-gravity-assist in July, 1997, the spacecraft will fly by the asteroid Hamburga in January, 1998 and finally match orbits with P/Kopff in Au-

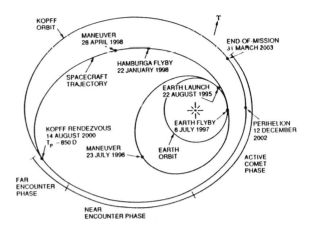

Fig. 1. CRAF trajectory, showing the orbits of Earth and comet Kopff.

gust, 2000. Comet Kopff has a period of 6.4 years and a perihelion distance of 1.58 AU. Its maximum gas production rate is probably somewhat less than 10^{29} s^{-1}. The payload and Interdisciplinary Scientists tentatively selected for the CRAF mission are listed in Table 1; science confirmation is scheduled for the spring of 1991. After launch, there will be an opportunity to propose to work with these instruments as Participating Scientists. Some further description of the CRAF instruments is given in Neugebauer and Weissman [1989].

Tables 2a and 2b give some of the details of the capabilities of the plasma instruments selected for CRAF. The CRAF array of instruments clearly fills many of the gaps in the Giotto measurements of the inner coma; it should be possible to obtain good measurements of both thermal and fast ions, thermal and energetic electrons, and the electric and magnetic components of plasma waves. On the negative side, the maximum measurable ion energy is only 25 keV; unless this gap in coverage is filled with the addition of an energetic particle detector, our knowledge of particle acceleration will be limited to what can be learned from the data from the Giacobini-Zinner and Halley missions and perhaps GEM.

Perhaps the greatest advantage of CRAF over the comet flyby missions is the cometocentric trajectories allowed by a rendezvous. Several mission phases have been marked in Figure 1. The "initial reconnaissance" phase will consist of a series of nucleus flybys. During the "near nucleus science" phase, the spacecraft will orbit the nucleus with a period of ~10 days and an orbit radius of 50-100 km, depending on the mass of the nucleus. There will also be some closer orbits to measure the mass harmonics of the nucleus. During the "perihelion" phase, when the comet is active, the spacecraft will move in and out through the inner coma, ranging between heliocentric distances of 50 and 5000 km at a variety of phase angles and inclinations. At closest approach to the comet, the spacecraft is expected to be well within the field-free cavity, when and if such a structure develops at comet Kopff. After passage through perihelion, it is planned to send the spacecraft on a round-trip excursion ≤50,000 km down the plasma tail.

TABLE 1. Investigations tentatively selected for CRAF

Acronym	Investigation	Principal Investigator (PI) or Team Leader (TL)
ISS	Imaging	J. Veverka, Cornell Univ. (TL)
VIMS	Visual/infrared mapping spectrometer	T. B. McCord, Univ. Hawaii (TL)
TIREX	Thermal infrared radiometer experiment	F. P. J. Valero, NASA Ames Research Center (PI)
CoMA	Cometary matter analyzer	J. Kissel, Max-Planck-Institut für Kernphysik (PI)
CIDEX	Comet ice/dust experiment	G. C. Carle, NASA Ames Research Center (PI)
CODEM	Comet dust environment monitor	W. M. Alexander, Baylor Univ. (PI)
NGIMS	Neutral gas and ion mass spectrometer	H. B. Niemann, NASA Goddard Space Flight Center (PI)
CRIMS	Comet retarding ion mass spectrometer	T. E. Moore, NASA Marshall Space Flight Center (PI)
SPICE	Suprathermal plasma investigation of cometary environments	J. L. Burch, Southwest Research Institute (PI)
MAG	Magnetometer	B. T. Tsurutani, Jet Propulsion Laboratory (PI)
CREWE	Coordinated radio, electrons and waves experiment	J. D. Scudder, NASA Goddard Space Flight Center (PI)
Radio	Radio science	D. K. Yeomans, Jet Propulsion Laboratory (TL)
IDS	Interdisciplinary Scientists	
	Active nucleus	A. H. Delsemme, University of Toledo
	Coma	W. F. Huebner, Southwest Research Institute
	Exobiology	C. McKay, NASA Ames Research Center
	Solar-wind Interaction	D. A. Mendis, Univ. California San Diego
	Asteroid and inactive nucleus	D. Morrison, NASA Ames Research Center

TABLE 2a. Properties of ion detectors on CRAF

Instrument	CRIMS	SPICE	NGIMS
Mass range, amu	1 to >64	1 to 65	1 to 300
Mass resolution	$m/\Delta m = 32$	$\Delta m < 1$ to ~ 2 amu	$\Delta m < 0.5$ amu
Charge sign	Positive	Positive	Positive & negative
n, cm^{-3}	10^{-1} to 10^5	10^{-3} to 10^5	≥ 25
T, K	25 to 10^6	$\geq 10^4$	
v, km/s	0.1 to 40	$(14$ to $2200)/\sqrt{m}$	
E, eV	0 to 150	1 to 25,000	≤ 10
Energy resolution	$\Delta E = 0.0025$ eV	$\Delta E/E \leq 0.04$	
Field of view	140° elevation 300° azimuth	90° elevation 360° azimuth	30° cone
Angular resolution	5° by 15°	10° for energy measurement 2° for mass measurement	None

TABLE 2b. Properties of electron detectors on CRAF

Instrument	SPICE	CREWE/VES	CREWE/Sounder
Energy, eV	0.1 to 32,000	0.5 to 6000	
Temperature, K	$\leq 10^3$ to $\geq 10^8$	$\geq 10^4$	$\geq 10^2$
Energy resolution, $\Delta E/E$	0.05 to 0.32	0.06	
n, cm^{-3}	10^{-3} to 10^4	10^{-2} to 10^5	10^{-2} to 10^5
Field of view	90° by 360°	6 @ 10° by 10°	Omnidirectional
Angular resolution	$\leq 10°$ elevation $\sim 5°$ azimuth	90°	None

COMET SAMPLE RETURN MISSIONS

Two types of comet sample return mission are currently under consideration. Perhaps the first to be launched would be the Japanese Soccer (Sample of Comet Coma and Earth Return) mission. As the name implies, a spacecraft would fly through the coma of an active comet and return collected samples of coma dust and gas to the Earth for detailed analysis. Interest is focussed on the opportunity to collect a sample from comet Kopff in December, 2002, when it is near perihelion and the CRAF spacecraft can act as a navigation beacon for the Soccer spacecraft. If Soccer carried plasma instrumentation as well as sample collectors, it would be possible to obtain simultaneous, two-point measurements using the CRAF and Soccer spacecraft, thereby removing some of the spatial versus temporal variability concerns inherent in single-point measurements.

A comet nucleus sample return mission, named "Rosetta", is one of ESA's four "Cornerstone" missions [European Space Agency, 1984]. The model payload developed by the joint ESA/NASA Science Definition Team [Ahrens et al., 1987] included a plasma analyzer in the category of "highly desirable instruments" (as opposed to "essential instruments"). But it is clear that obtaining a well preserved sample of the comet nucleus is a far more important objective of that mission than is understanding some of the riddles of cometary plasma physics.

CONCLUSIONS

On the whole, I am fairly optimistic about future progress in cometary plasma physics. The interpretation of Halley data is reaching its prime, with quantitative multi-sensor data now available for comparison with evolving numerical models which are steadily increasing in sophistication. In fact, the outlook for substantial progress over the next few years has inspired Karl-Heinz Glassmeier and I to organize another conference on cometary plasma physics to be held in Germany in 1992 or 1993. Beyond that, if we're patient, we should eventually return to an active comet with a set of instruments well designed to address the remaining questions.

Acknowledgments. The research described in this paper was carried out by the Jet Propulsion Laboratory of the California Institute of Technology under contract with the US National Aeronautics and Space Administration.

REFERENCES

Ahrens, T. J., et al., *Rosetta, the Comet Nucleus Sample Return Mission: Report of the Joint ESA/NASA Science Definition Team*, European Space Res. and Technol. Ctr. SCI(87)3, Noordwijk, 1987.

European Space Agency, *European Space Science Horizon 2000, ESA SP-1070*, Paris, 1984.
Neugebauer, M., and P. R. Weissman, CRAF Mission, *Eos, 70,* 633, 1989.

M. Neugebauer, Mail Stop 169-506, Jet Propulsion Laboratory, Pasadena, CA 91109

THE SOLAR WIND INTERACTION WITH VENUS AND MARS:
COMETARY ANALOGIES AND CONTRASTS

J. G. Luhmann

Institute of Geophysics and Planetary Physics
University of California, Los Angeles, CA 90024

Abstract. The weakly magnetized terrestrial planets, Venus and Mars, share some common physical processes with comets because of the "direct" interaction between their atmospheres and the solar wind. However, the importance and outcome of these common processes is generally quite different because of the different system scales and atmospheric properties of these two classes of objects. This paper explores some details of their contrasts and similarities in view of the knowledge that we have obtained from the combination of in-situ data and modeling.

Introduction

"Comparative planetology" is a useful practice in that it often gives us a clearer perspective on the system under study by providing other related data from nature's solar system laboratory of "experiments." In the present case, we are considering the class of bodies for which an intrinsic magnetic field plays no significant role in the solar wind interaction. Both comets and the terrestrial planets, Venus and Mars, share this property. They also share the property of a substantial neutral atmosphere. While the relative scales of these bodies and their atmospheres differ appreciably (Venus and Mars have radii of ~6053 and ~3350 km, respectively, and atmosphere scale heights of 10's to 100's of km, while comets have sizes of a few km and atmospheres that can extend up to millions of km), there are physical processes that occur on both that help us to better understand the other. For example, the process of "mass-loading" of the solar wind by planetary or cometary ions must be governed by common physics. It can even be argued that a terrestrial planet is the limiting case of a very large comet with a very dense, shallow atmosphere. However, before making analogies, it is wise to consider the contrasts and similarities between the two systems as we now know them.

Cometary Plasma Processes
Geophysical Monograph 61
©1991 American Geophysical Union

Comparisons

Since most of our observational information on the characteristics of the environments of weakly magnetized planets come from missions to Venus (e.g., see the reviews by Russell and Vaisberg, 1983 and Luhmann, 1986), it is perhaps most appropriate to compare comets with Venus. Mars, which is currently under study thanks to the recent experiments on PHOBOS 2 (cf. special 1989 NATURE issue), will be specifically included in the discussion in cases where we have some information from the limited past observations or models, but one should be aware that our picture of the Mars-solar wind interaction may change somewhat over the next few years. However, we do know without question that both planets present obstacles to the solar wind that are little more than planet size (e.g. see Slavin et al., 1982, Russell et al., 1985).

Figures 1a and b compare cartoons of the solar wind interaction with comets and Venus. The basic features of a bow shock and region of draped interplanetary magnetic field dominate both of these pictures. Major differences include the draping of the interplanetary field upstream of the cometary bow shock, but not upstream of the Venus shock, and the presence of an "impenetrable" obstacle at Venus that has a size comparable to the size of the entire interaction region. The other major difference is scale size, which is emphasized in Figure 1c where both the comet and Venus systems are roughly scaled on a diagram of the terrestrial magnetosphere. These differences in some sense form the basis of the forthcoming, more detailed discussion. The pre-shock field draping at the comet is a result of the extent of the atmosphere and hence the region of mass-loaded solar wind, while the deflection of the solar wind around Venus and resulting field draping occurs primarily by virtue of the presence of an ionosphere with sufficient thermal pressure to balance the incident solar wind dynamic pressure. The scale size contrast is a result of the low gravity and high neutral gas production rate at the comet compared to Venus. It will be seen that this scale size difference also has other consequences, such as the relative importance of the scale of the pickup ion gyroradius.

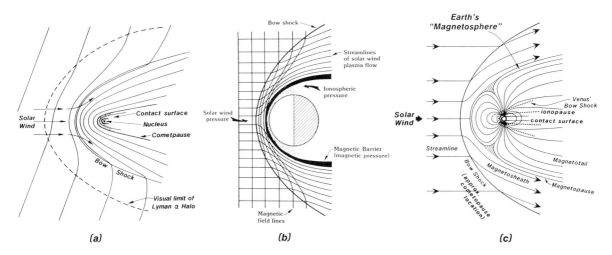

Fig. 1.   (a) Schematic illustration of the main features of a comet's interaction with the solar wind compared with those of the Venus-solar wind interaction (b).   (c) Scaled superposition of the main features of both in comparison to the scale of Earth's magnetosphere.

Global Fluid Models

Fluid models have been used to obtain a first-order, global picture of both the comet and Venus solar wind interactions. In the case of comets, the 3-dimensional MHD equations with a source term have been solved by Fedder et al. (1986), Schmidt and Wegmann (1982), and Ogino et al. (1988) The result resembles the cartoon in Figure 1a. In the case of the comet, the solar wind flow becomes submagnetosonic as a result of the mass addition described by the source term. The cometary bow shock is weak because this mass addition substantially slows the velocity upstream of the shock location. The mass loading slows the flow to stagnation near the nucleus. Overall, the velocity field is such that the frozen-in magnetic field "piles up" and drapes within a large region centered on the nucleus without suffering much lateral deflection. The extent of the "source" term, attributable to the rapidly expanding atmosphere which is subject to ionization, makes the interaction region huge compared to the size of the nucleus. Figure 2a shows streamlines and magnetic field lines obtained for Fedder et al.'s (1986) model of comet Giacobini-Zinner which illustrate these features.

The corresponding models for Venus have been based on a gas dynamic model of the interaction of a supersonic flow with an impenetrable blunt obstacle (cf. Spreiter et al., 1970, Spreiter and Stahara, 1980). In this gas dynamic model, the hydrodynamic equations are solved for the flow properties and then the frozen-in magnetic field is convected through the system. The shock is formed so that the fluid can flow around the obstacle.

The blunt obstacle is the ionosphere, which at ~250 km (subsolar) has sufficient plasma pressure to balance the largely dynamic upstream solar wind pressure (e.g. see Brace et al., 1983). Although it would be highly desirable to have an analogous MHD model with a source distribution based on Venus' atmosphere, analogous to the comet model, we have been able to apply the gas dynamic model for solar wind only with considerable success (e.g., see Luhmann et al., 1986). It appears that at Venus, both mass loading and MHD effects merely perturb the nominal flow around the obstacle (e.g. see Phillips et al., 1986, Belotserkovskii, et al., 1987). Figure 2b shows streamlines and magnetic field lines obtained from Spreiter and Stahara's (1980) gas dynamic model. One can think of this magnetosheath model as resulting from the superposition of the interplanetary field and a perturbation field caused by currents flowing in the ionosphere as illustrated in Figure 3. This "induced magnetosphere," as it is sometimes called, is in its simplest form analogous to the dipole field that is induced in a spherical conductor in a vacuum field, because the impenetrable obstacle surface is an equipotential. Of course, the gas dynamic model, because of its nature ,does not describe either flows or magnetic field inside of the obstacle surface.

The general agreement of these global models with observations is perhaps best illustrated by comparing simulated spacecraft traversals of the model magnetic fields with the corresponding magnetometer data. Figure 4 shows such comparisons for both Comet Halley (from Schwingenschuh et al., 1987) and Venus (from Luhmann et al., 1986). The pile-up of the field close to the object, predicted

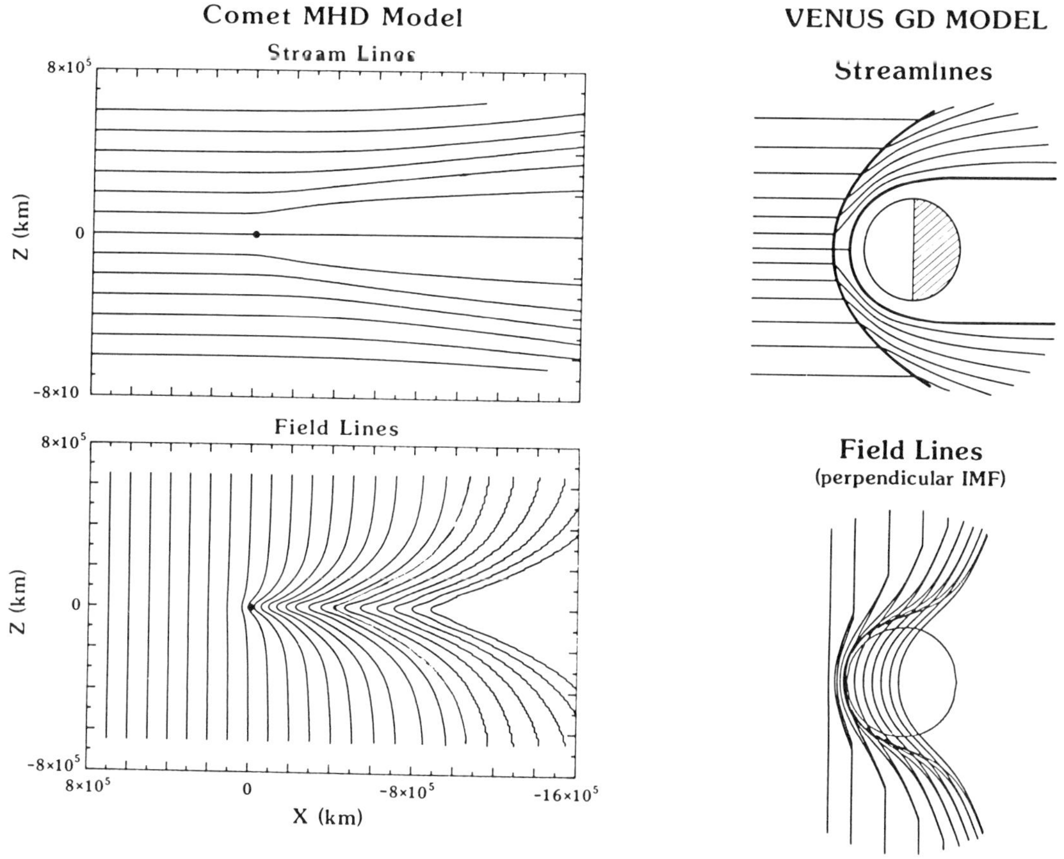

Comet MHD Model

Stream Lines

Field Lines

VENUS GD MODEL

Streamlines

Field Lines
(perpendicular IMF)

Fig. 2.   (a) Plasma streamlines and field lines from the Fedder et al. (1986) MHD computer simulation of comet Giacobini–Zinner.   (b) Streamlines and field lines of a planetary magnetosheath derived from the gas dynamic model of Spreiter and Stahara (1980). In both cases the upstream magnetic field is transverse to the flow.   In the case of Venus, the field lines drape over the obstacle; the view shown is a projection of three dimensional field lines.   For the comet, only field lines in the "equatorial" plane are shown.

by the models, is evident in both cases. Since the comet Halley model includes a simulation of the interplanetary field rotation that occurred during the encounter, a Venus case with a rotation was selected for comparison. It is interesting to note that the signature of the field rotation is seen both inbound and outbound at Halley, but only outbound at Venus. This difference is attributable to the long delay that heavily mass loaded field lines experience in comets like Halley compared to the relatively free "slippage" of fields over the Venus obstacle.

Another feature of interest concerns the dominant pressure components along the stagnation streamlines of these global models. Figure 5 shows how the solar wind pressure starts out as essentially pure dynamic pressure and then evolves in the model of Venus and comet Giacobini–Zinner. At Venus, thermal pressure jumps at the shock but never dominates.   In effect, dynamic pressure is converted to magnetic pressure which actually becomes infinite at the boundary of the gas dynamic

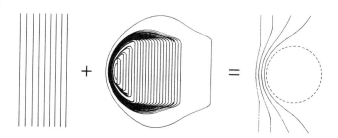

Fig. 3.   Illustration of the field lines one obtains by decomposing the gas dynamic magnetosheath field model into an interplanetary part and a "perturbation" or induced magnetosphere created by currents in the ionosphere.

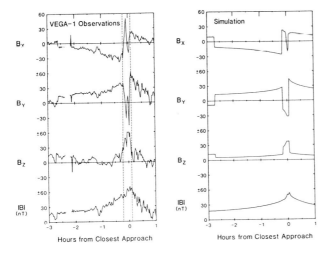

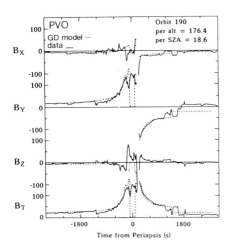

Fig. 4.   (a) Comparison of the time series of comet Halley magnetic fields observed on the VEGA 1 spacecraft and the fields in an MHD simulation (from Schwingenschuh et al., 1986).   (b) A similar comparison between fields observed on the Pioneer Venus Orbiter and the gas dynamic model fields (from Luhmann et al., 1986).

model. In the Venus observations the ionospheric pressure rises to become dominant below the altitude (typically near ~250 km) where ionospheric pressure becomes equal to the solar wind pressure (e.g., see Phillips et al., 1984). In the comet model, the solar wind pressure also falls as the flow stagnates, but the balance is taken up by both magnetic and thermal pressures. The total pressure departs from constancy near the stagnation point as forces related to the curvature of the magnetic field lines become important. (In both of these global models, frictional forces are not explicitly included.) It should be mentioned that in more recent comet models (cf. Schmidt and Wegmann, this volume) an inner surface where the flow is deflected is also found, but for reasons other than

ionospheric pressure balance as will be discussed below.

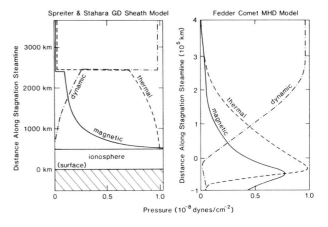

Fig. 5.   Comparison of various pressure components along the stagnation streamlines of the 3-dimensional gas dynamic magnetosheath model (a) and the MHD model of comet Giacobini-Zinner (b). The ionospheric pressure is not explicitly included in the gas dynamic magnetosheath model, but is assumed to provide the impenetrable obstacle around which the model fluid flows.

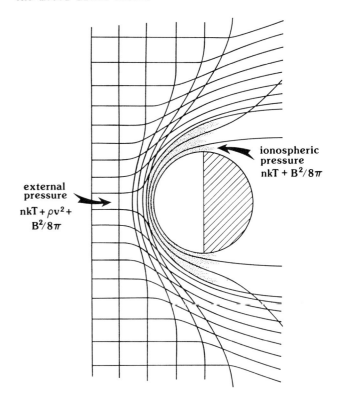

Fig. 6.   Schematic illustration of the magnetization of the ionosphere of Venus by the penetration of magnetosheath fields.

Local Fluid Models

To treat what occurs nearest the objects, investigators have developed one dimensional MHD models which include collisions between plasma and neutral particles, or frictional forces. In particular, for Venus, Shinagawa and Cravens (1987) constructed a model which accurately describes the subsolar ionosphere. Observations at Venus had shown us that the dayside ionosphere can be magnetized by a horizontal field in the direction of the magnetosheath field, as illustrated in Figure 6, if the projected pressure balance level reaches a sufficiently low altitude (cf. a review on this subject by Luhmann and Cravens, 1989). The extremes of this range of behavior are illustrated in Figures 7a and b (from Shinagawa and Cravens,

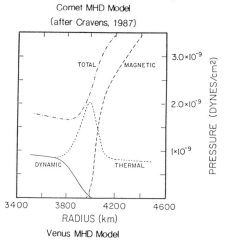

Comet MHD Model
(after Cravens, 1987)

Venus MHD Model
(after Shinagawa and Cravens, 1987)

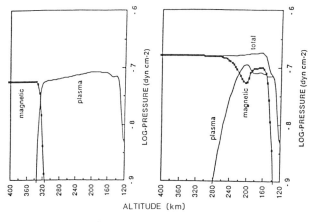

ALTITUDE (km)

Fig. 7. Panels (a) and (b) show altitude profiles of the various components of pressure in the Venus ionosphere according to the one-dimensional MHD model of Shinagawa and Cravens (1987); (a) illustrates the more "usual" case observed at Venus of an unmagnetized ionosphere and (b) shows a magnetized case. For comparison, panel (3) shows a similar display for a one-dimensional MHD model of the inner region along the stagnation streamline of a comet (from Cravens, 1987).

1987). The dominant forces at the top of the model are magnetic and thermal pressure gradient forces, but collisions determine the location of the bottom of the ionospheric magnetic layer and the ionospheric vertical velocity profile determines its shape in between. Essentially the same theory applied to the comet, but with magnetic curvature forces included, produces the result in Figure 7c (from Cravens, 1989). Here it is seen that the magnetic field also comes to a stop, and here again collisions are responsible. In fact, it is important to appreciate that the comparison of the observed magnetic "drop-outs" often observed near periapsis by the Pioneer Venus Orbiter magnetometer at Venus and by the Giotto magnetometer at comet Halley (cf. Neubauer et al., 1986) have different causes. The Venus drop-out shown in Figure 8a (inside the "ionopause") is from ionospheric thermal pressure gradient and magnetic pressure gradient force balance, while the comet drop-out (the "contact surface"), shown in Figure 8b, occurs due to the balance between the magnetic pressure gradient force and the frictional force (cf. Cravens, 1988, Ip and Axford, 1988). Only at the bottom of the magnetized Venus ionosphere do we find the comet-like boundary. This analogy between

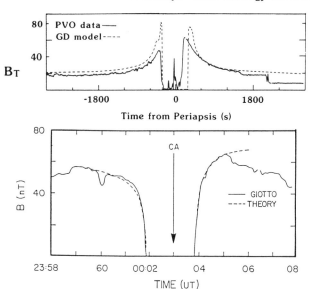

Fig. 8. Comparison of magnetic field "dropouts" often seen at Venus near the periapsis of the Pioneer Venus orbiter and near the Giotto spacecraft's closest approach to comet Halley. As discussed in the text, the physics behind these low altitude dropouts differs.

the magnetized Venus ionosphere case and comets, together with some general scale comparisons, is illustrated schematically in Figure 9 from Breus et al. (1987). An analogous MHD model of the Martian subsolar ionosphere recently been published by Shinagawa and Cravens (1989), shows behavior similar to the Venus magnetized model.

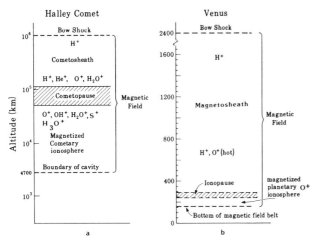

Fig. 9. Schematic illustration of the locations of the various regions and boundaries observed at comet Halley compared to those observed at Venus (from Breus et al., 1987). Note that the altitude scale for the comet is logarithmic.

The Solar Wind–Ionosphere Interface

Unfortunately, missions to Venus did not include sufficiently sophisticated plasma instrumentation to probe the details of the mixing of solar wind and planetary plasmas as they were probed at comet Halley. We know from the available observations that the cold ionospheric plasma largely vanishes above the ionopause boundary, and that within the low altitude magnetosheath both ionospheric and solar wind electrons are present (cf. Spenner et al., 1980), but we know comparatively little about the photoion population at this boundary. In general, we cannot tell whether there is a "cometopause"-like feature where solar wind plasma is replaced by heated or accelerated ionospheric plasma well above the nominal ionopause, although the new Martian data from PHOBOS may show indications of such a transition. On the other hand, we do have observations that tell us about the comet-like fate of planetary ions which originate in the upper reaches of the neutral atmosphere of Venus.

The altitude profiles of neutral density for both Venus and comets are shown in Figure 10. The Venus hot oxygen exosphere (Figure 10a) is produced when O_2^+, the predominant ion in the lower ionosphere, dissociatively recombines to produce two energetic O atoms (cf. Nagy and Cravens, 1988). The scale height of this planetary exosphere is a far cry from the cometary atmosphere scale height (see Figure 10b), but nevertheless it is enough to keep a significant population of neutrals above the ionopause. The consequence is that ionization of these neutrals (by photons, solar wind electron impact or charge exchange) is followed by solar wind "pickup" of the photoions. Some picked up ion observations from the Pioneer Venus Orbiter plasma analyzer (cf. Mihalov and Barnes, 1981, 1982) are shown in Figure 11 which also shows VEGA spacecraft pickup ion data (cf. Gringauz et al., 1986) for comparison. Mars is also expected to have a

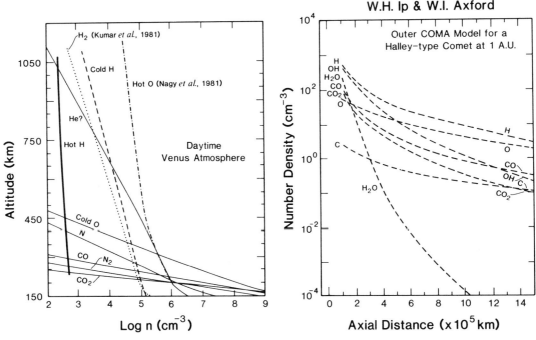

Fig. 10. (a) Altitude profile of the Venus exosphere compared to (b) the profile of a cometary atmosphere. Although the density scales are similar, the distance scales are vastly different.

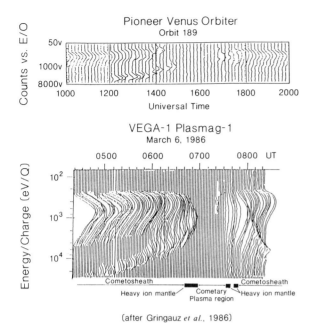

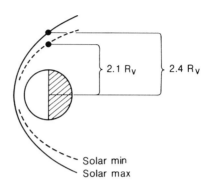

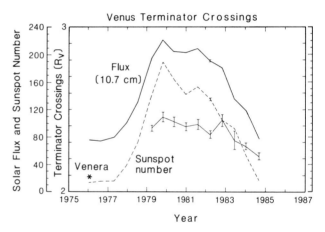

Fig. 11.   Comparison of time series of energy/charge spectra from plasma instruments on the Pioneer Venus Orbiter (while it flew through the magnetosheath and wake) and on VEGA (as it flew by comet Halley).   Both show a high energy peak from the heavy pickup ions. The "dropouts" in the ion fluxes seen here occur in the wake of Venus and just upstream of Halley's nucleus where flow deflections may play a role in causing them.

Fig. 12.   Illustration of the apparent effect of the solar cycle on the position of the bow shock of Venus (from Russell et al., 1988). For purposes of comparison, only the average extrapolated terminator crossing is shown.

similar hot oxygen exosphere, and indeed, preliminary reports from the PHOBOS experimenters describe copious fluxes of oxygen ions in near-Martian space.

Earlier it had been emphasized that we can describe the magnetic field and plasma flow around Venus quite well without considering the effects of mass-loading of the nearby solar wind with picked up ions. Perhaps this is not surprising, since the integrated ion production rate above the ionopause is only ~10^{23}–10^{24} s^{-1} compared to ~10^{28}–10^{30} s^{-1} in the coma of a Halley-like comet. However, now it is appropriate to discuss some of the "perturbations" that this mass-loading causes. Figure 12 (from Russell et al., 1988) shows the observed solar cycle variation in the position of the bow shock which may be at least partially attributable to alteration of the magnetosheath flow by variable mass-loading (both the exosphere density and photoionization rate change with solar EUV). A mass-loaded gas dynamic magnetosheath model developed by Belotserkovskii et al. (1987) suggests that this explanation is viable. Other magnetosheath signatures include enhanced magnetosheath magnetic field draping as described by Phillips et al. (1986) and illustrated by Figure 13. But perhaps the most striking "perturbation" is the existence of a comet-like induced

magnetotail in the wake of the planet. Indeed, the striking similarity between observations behind Venus and those in the central region of the tail of Giacobini-Zinner is evident in Figure 14 from McComas et al., 1986. The new PHOBOS results show a similar structure behind Mars.   The basic interpretation of the Venus observations is that mass-loaded, draped magnetosheath flux tubes sink into the wake as pictured in Figure 15 (from Saunders and Russell, 1986). A self-consistent model of an induced planetary tail has not yet been developed.   However, one can construct an approximate field model of the induced tail of Venus or Mars by configuring a comet model to fit inside of the magnetosheath model "obstacle." Such a model can in turn be used to study the first-order motion of picked up exospheric photoions into the tail (e.g. see Luhmann, 1989a). In particular, the results of test particle calculations in such a model, as illustrated by Figure 16, gives

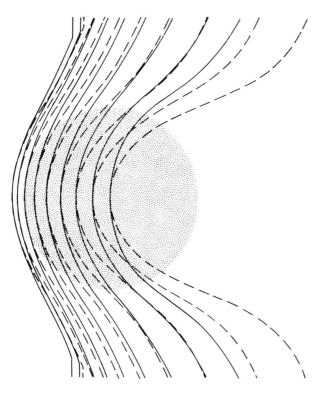

Fig. 13.   Illustration of the magnitude of the Venus magnetosheath field draping enhancement attributed to mass-loading by Phillips et al. (1986).

information about another important aspect of mass-loading at Venus or Mars.

Because the oxygen ions at Venus and Mars generally have gyroradii in the magnetosheath and upstream that are comparable to or exceed the planetary radius, the symmetry of the planetary ion wake is broken as was recognized some time ago (cf. Cloutier et al., 1974). The cometary water ion gyroradii, shown in Figure 16 as they appear when one launches test particles in Fedder et al.'s (1986) model of Giacobini-Zinner, are small enough relative to the scale size of the source region and the system so that they do not have a significant effect on the ion distribution. At Venus and Mars, they are both comparable to the system size and may encounter the absorbing obstacle of the planet and its lower atmosphere in their path. The observations show that there are at least two consequences: the magnetosheath field draping enhancement at Venus, illustrated in Figure 13, is stronger on the side of the magnetosheath where the ions move freely outward (cf. Phillips et al., 1986), and the observations of pickup oxygen ions in both the magnetosheath and tail show a similar asymmetry in flux (cf. Phillips et al., 1986, Slavin et al., 1989, Intriligator, 1989). In fact, it may even be the case that at Mars, where the finite gyroradius of oxygen ions is more significant than at Venus due to the weaker interplanetary field at 1.5 AU and the smaller planetary obstacle, the gyroradius of even solar wind ions cannot be neglected (cf. Moses et al., 1988, Luhmann, 1989b). This would mean that hybrid or kinetic global modeling should replace fluid

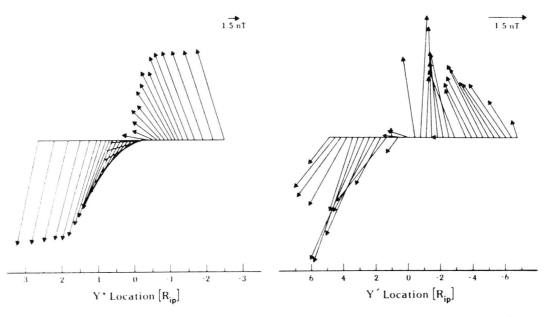

Fig. 14.   Comparison of magnetic field vectors observed behind Venus (averaged) and during the ICE spacecraft flight through the wake of comet Giacobini-Zinner (from McComas et al., 1987). Both show a similar amount of sunward/antisunward field draping in the deep tail lobes.

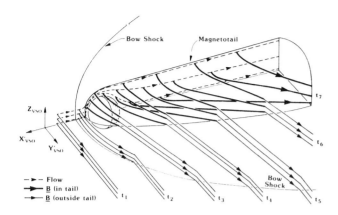

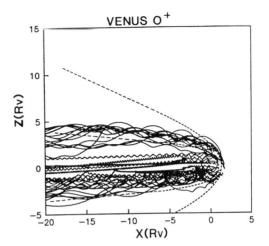

Fig. 15.   Schematic illustration of the manner in which mass-loaded magnetosheath field lines sink, or are pulled, into the wake of Venus (from Saunders and Russell, 1986).

modeling of the solar wind interaction with Mars.

In the same vein, it is important to appreciate the different role that waves and turbulence in the plasma play at Venus as opposed to comets. This is of particular interest here since intense scattering of the particles by field fluctuations could obliterate the finite gyroradius asymmetries and essentially make mass-loading at Venus more symmetric and "fluid"-like. Observations of the magnetic field near Venus tell us that under normal circumstances, the turbulence in the Venus dayside interaction region is determined primarily by the interplanetary field orientation and the bow shock (cf. Luhmann et al., 1983). If the interplanetary field has a significant radial component, the subsolar bow shock is quasiparallel and the magnetosheath and upstream are full of magnetic fluctuations as shown in the right-hand side of Figure 17. If the subsolar shock is quasiperpendicular, the field in the pile-up region is relatively smooth as illustrated on the left-hand side of Figure 17. The interpretation is sketched in Figure 18. In comets, waves generated by plasma instabilities related to the cometary ion pickup have time to develop everywhere and so the cometary environment is always turbulent. In contrast, at Venus, these instabilities evidently do not develop to the extent that we can detect their effects throughout most of the interaction region near the planet. Instead, the level of turbulence is governed by the solar wind plasma parameters and interplanetary field orientation (see Winske (1986) for a description of a numerical simulation of turbulence in the Venus subsolar magnetosheath). The observed magnetosheath turbulence is well described by a picture where plasma convection carries the waves in from the quasiparallel shock. Figure 19 describes the logical consequence, which is that "randomized" ion pickup related to the presence of the shock-generated turbulence (cf. Luhmann et al., 1986) may prevail for fairly radial interplanetary field,

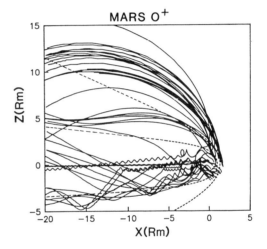

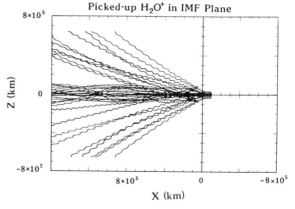

Fig. 16.   Comparison of test particle trajectories for pickup ions at Venus, Mars and comet Giacobini-Zinner. The views shown for Venus and Mars are along the direction of the interplanetary field, while that for the comet is in the plane containing the field. These examples illustrate the size of the pickup ion gyroradius relative to the size of the system.

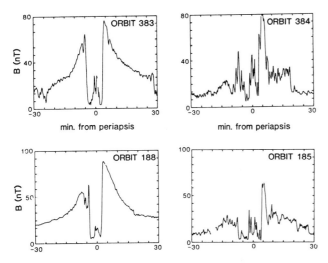

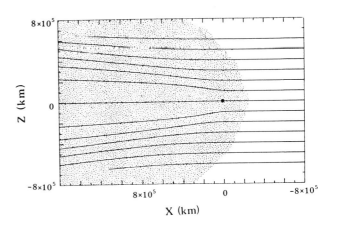

Fig. 17. Several time series of magnetic field magnitude observed as the Pioneer Venus Orbiter passed through the Venus magnetosheath. Periapsis (~150 km altitude) is in the center. The decreases in field magnitude occur inside the cold, dense ionospheric plasma. Cases for largely transverse interplanetary field are shown on the left, and cases for largely radial field are shown on the right. In the latter, turbulence convected from the subsolar quasiparallel shock is seen throughout the magnetosheath.

while the asymmetric pickup described above applies otherwise. It is worth noting here that Omidi et al. (1986) described how cometary ion pickup also works differently for transverse and radial interplanetary field conditions, except that at the comet, this means that the dominant ion-pickup instability changes. In all cases the presence of waves and turbulence seems to be important at comets, thus possibly helping to lend credence to the fluid treatments of the comet-solar wind interaction.

Conclusions

The foregoing discussion has considered some of the observations and models that exist for Venus and Mars, and reviewed them in light of what we now know about comets. The basic reason for expecting some analogies is the occurrence of heavy planetary ion pickup in the solar wind interaction regions at the weakly magnetized terrestrial planets. A secondary, related reason is the existence of induced magnetotails consisting of draped, presumably mass loaded, interplanetary flux tubes at these planets. There is also an important interaction of plasma with neutrals close to the body. However, these analogies do not go much further than the basic phenomenology. Venus, and presumably Mars, are both relatively "hard" obstacles in that their ionospheres appear to stagnate and largely deflect the solar wind. In contrast, comets are "soft" obstacles. The

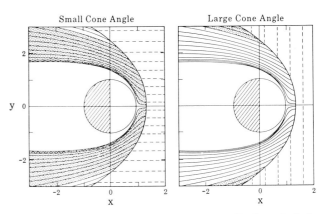

Fig. 18. Schematic illustration of the global distribution of turbulence in the cometary environment (top) and at Venus (bottom). At Venus, the interplanetary field orientation determines the nature of the subsolar shock and hence the distribution of magnetosheath turbulence. In contrast, ion pickup instabilities produce widespread turbulence on the scale of the interaction region at comets.

interpenetration and interaction of cometary and solar wind plasma is what causes the flow stagnation. On the scale of their interaction regions, comets barely deflect the solar wind. The magnetic field draping geometry is determined by the geometry of the stagnation flow rather than by the deflection and compression accompanying stagnation at the subflow point of an "impenetrable" obstacle. It is the ability of the dense, shallow, hemispherical layer of the dayside Venus ionosphere to generate intense local diamagnetic currents that make it appear as an almost impenetrable obstacle to the solar wind. While it is true that heavy ions are picked up at both the planetary and cometary objects, their relative importance in a description of the solar wind interaction is quite different. At comets, because of the scale of the system and the importance of turbulence, it is appropriate to treat the pickup ions as an extended source of

Ion Pickup Processes

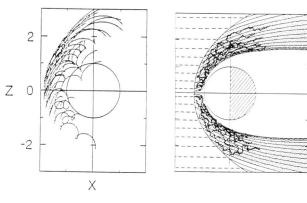

Perpendicular IMF   Parallel IMF

"Large Scale E Field"   "Turbulent"

Fig. 19.   Suggested difference in the mode of planetary ion pickup that occurs at Venus depending on the interplanetary field orientation. When the field is transverse to the upstream flow, the magnetosheath field is smooth and simple $E = -V \times B$ field acceleration occurs. When the interplanetary field is radial, "turbulent" pick-up or acceleration occurs.

heavy "fluid." On the other hand, at Venus, the solar wind and interplanetary field determine the limited role of turbulence. The heavy ion effects are merely perturbations to a description of the solar wind interaction with an impenetrable obstacle, except, perhaps, for their role in the tail formation. Moreover, at the planets the gyroradii of the heavy picked-up ions are so large compared to the interaction region that one must consider their detailed motion. Planetary ion pickup at Venus produces observable global asymmetries in the plasma and field environment that cannot be described by fluid models. The case for a kinetic description of the Mars-solar wind interaction is even stronger. In summary, one can gain some valuable insights into solar wind interactions with atmospheres as a whole by considering Venus and Mars together with comets, but one must always keep in mind that one is dealing with rather different regions of physical parameter space in the planetary and cometary systems.

Acknowledgements. The support of the author by NASA grants NAG2-501 and NAGW-1347 is gratefully acknowledged.

References

Belotserkovskii, O. M., T. K. Breus, A. M. Krymskii, V. Ya. Mitniskii, A. F. Nagy and T. I. Gombosi, The effect of the hot oxygen corona on the interaction of the solar wind with Venus, Geophys. Res. Lett., 14, 503, 1987.

Brace, L. H. et al., The ionosphere of Venus: Observations and their interpretation, in Venus, edited by D. M. Hunten, L. Colin, T. M. Donahue and V. I. Moroz, 779, University of Arizona Press, Tucson, Arizona, 1983.

Breus, T. K., A. M. Krymskii and J. G. Luhmann, Solar wind mass-loading at comet Halley: A lesson from Venus?, Geophys. Res. Lett., 14, 499, 1987.

Cloutier, P. A., R. E. Daniell, Jr. and D. M. Butler, Atmospheric ion wakes of Venus and Mars in the solar wind, Planet. Space Sci., 22, 967, 1974.

Cravens, T. E., Model of the inner coma of comet Halley, J. Geophys. Res., in press, 1989.

Fedder, J. A., J. G. Lyon and J. L. Giuliani, Jr., Numerical simulations of comets: Predictions for comet Giacobini-Zinner, EOS Trans. AGU, 67, 17, 1986.

Gringauz, K. I. et al., First in-situ plasma and neutral gas measurements at comet Halley, Nature, 321, 282, 1986.

Intriligator, D. S., Results of the first statistical study of Pioneer Venus Orbiter plasma observations in the distant Venus tail: Evidence for a hemispheric asymmetry in the pickup of ionospheric ions, Geophys. Res. Lett., 16, 167, 1989.

Ip, W. H. and W. I. Axford, The formation of a magnetic field free cavity at comet Halley, Nature, 325, 418, 1987.

Kimmel, C. D., J. G. Luhmann, J. L. Phillips and J. A. Fedder, Characteristics of cometary picked-up ions in a global model of Giacobini-Zinner, J. Geophys. Res., 92, 8536, 1987.

Luhmann, J. G., M. Tatrallyay, C. T. Russell and D. Winterhalter, Magnetic field fluctuations in the Venus magnetosheath, Geophys. Res. Lett., 10, 655, 1983.

Luhmann, J. G., The solar wind interaction with Venus, Space Sci. Rev., 44, 241, 1986.

Luhmann, J. G., R. J. Warniers, C. T. Russell, J. R. Spreiter and S. S. Stahara, A gas dynamic magnetosheath field model for unsteady interplanetary fields: Application to the solar wind interaction with Venus, J. Geophys. Res., 91, 3001, 1986.

Luhmann, J. G., C. T. Russell, J. L. Phillips and A. Barnes, On the role of the quasi-parallel bow shock in ion pickup: A lesson from Venus?, J. Geophys. Res., 92, 2544, 1987.

Luhmann, J. G. and T. E. Cravens, Magnetic fields in the ionosphere of Venus, Space Science Rev., in press, 1989.

Luhmann, J. G., A model of the ion wake of Mars, submitted to Geophys. Res. Lett., 1989.

Luhmann, J. G., On the breakdown of fluid model of the solar wind interaction with Mars, submitted to Geophys. Res. Lett., 1989.

McComas, D. J., J. T. Gosling, C. T. Russell and J. A. Slavin, Magnetotails at unmagnetized bodies: Comparison of comet Giacobini-Zinner and Venus, J. Geophys. Res., 92, 10111, 1987.

Mihalov, J. D. and A. Barnes, Evidence for the acceleration of ionospheric O^+ in the magnetosheath of Venus, Geophys. Res. Lett., 8, 1277, 1981.

Mihalov, J. D. and A. Barnes, The distant interplanetary wake of Venus: Plasma observations from Pioneer Venus, J. Geophys. Res., 87, 9045, 1982.

Moses, S. L., F. V. Coroniti and F. L. Scarf, Expectations for the microphysics of the Mars-solar wind interaction, Geophys. Res. Lett., 15, 429, 1988.

Nagy, A. F. and T. E. Cravens, Hot oxygen atoms in the upper atmospheres of Venus and Mars, Geophys. Res. Lett., 15, 433, 1988.

Ogino, T., R. J. Walker and M. Ashour-Abdalla, A three-dimensional MHD simulation of the interaction of the solar wind with comet Halley, J. Geophys. Res., 93, 9568, 1988.

Omidi, N. and D. Winske, Simulations of the solar wind interaction with the outer regions of the coma, Geophys. Res. Lett., 13, 397, 1986.

Phillips, J. L., J. G. Luhmann and C. T. Russell, Growth and maintenance of large-scale magnetic fields in the dayside Venus ionosphere, J. Geophys. Res., 89, 10676, 1984.

Phillips, J. L., J. G. Luhmann, C. T. Russell and K. R. Moore, Finite Larmor radius effect on ion pickup at Venus, J. Geophys. Res., 92, 9920, 1987.

Russell, C. T., Planetary bow shock, in Collisionless shocks in the heliosphere: Reviews of current research, Geophysical Monograph 35, American Geophysical Union, p. 109, 1985.

Russell, C. T. and O. Vaisberg, the interaction of the solar wind with Venus, in Venus, edited by D. M. Hunten, L. Colin, T. M. Donahue and V. I. Moroz, University of Arizona Press, p. 873, 1983.

Russell, C. T., E. Chou, J. G. Luhmann, P. Gazis, L. H. Brace and W. R. Hoegy, Solar and interplanetary control of the location of the Venus bow shock, J. Geophys. Res., 93, 5461, 1988.

Saunders, M. A. and C. T. Russell, Average dimension and magnetic structure of the distant Venus magnetotail, J. Geophys. Res., 91, 5589, 1986.

Schmidt, H. U. and R. Wegmann, Plasma flow and magnetic fields in comets, in Comets, edited by L. L. Wilkening, p. 538, University of Arizona Press, Tucson, Arizona, 1982.

Schwingenschuh, K., W. Riedler, Ye. Yeroshenko, J. L. Phillips, C. T. Russell, J. G. Luhmann and J. A. Fedder, Magnetic field draping in the comet Halley coma: Comparison of VEGA observations with computer simulations, Geophys. Res. Lett., 14, 640, 1987.

Shinagawa, H. and T. E. Cravens, A one-dimensional multi-species magnetohydrodynamic model of the d dayside ionosphere of Venus, J. Geophys. Res., 93, 11263, 1988.

Shinagawa, H. and T. E. Cravens, A one-dimensional multispecies magnetohydrodynamic model of the dayside ionosphere of Mars, J. Geophys. Res., 94, 6506, 1989.

Slavin, J. A., R. E. Holzer, J. R. Spreiter, S. S. Stahara and D. S. Chaussee, Solar wind flow about the terrestrial planets 2. Comparison with gas dynamic theory and implications for solar-planetary interactions, J. Geophys. Res., 88, 19, 1983.

Slavin, J. A., D. S. Intriligator and E. J. Smith, Pioneer Venus Orbiter magnetic field and plasma observations in the Venus magnetotail, J. Geophys. Res., 94, 2383, 1989.

Spenner, K., W. C., Knudsen, K. L. Miller, V. Novak, C. T. Russell and R. C. Elphic, Observations of the Venus mantle, the boundary region between solar wind and ionosphere, J. Geophys. Res., 85, 7655, 1980.

Spreiter, J. R., A. L. Summers and A. W. Rizzi, Solar wind flow past nonmagnetic planets — Venus and Mars, Planet. Space Sci., 18, 1281, 1970.

Spreiter, J. R. and S. S. Stahara, A new predictive model for determining solar wind-terrestrial planet interactions, J. Geophys. Res., 85, 6769, 1980.

Winske, D., Origin of large magnetic fluctuations in the magnetosheath of Venus, J. Geophys. Res., 91, 11951, 1986.

DUST-PLASMA INTERACTIONS

IN THE COMETARY ENVIRONMENT

D.A. Mendis

Department of Electrical and Computer Engineering, University of California, San Diego La Jolla, California 92093

M. Horanyi

Lunar and Planetary Laboratory University of Arizona Tucson, Arizona 85721

Abstract. Cometary dust, by virtue of being immersed in a plasma and UV radiative environment are necessarily electrically charged. This charging can lead to both physical and dynamical effects on the dust. On the other hand the dust could have certain effects on the plasma as well as on the interplanetary magnetic field. These processes are discussed and the observations, both remote and in-situ, which pertain to them are critically reviewed.

Introduction

Cometary dust, like all cosmic dust (e.g. intergalactic, interstellar, circumstellar, interplanetary, and circumplanetary) is immersed in a magnetized plasma and UV radiation environment. Consequently the dust and plasma are coupled via the exchange of charge, mass, momentum and energy. This coupling effects both components. For the plasma, it could lead to considerable depletion if the dust density is sufficiently high; it could also lead to heating or cooling depending on the efficiency of photoemission from the dust. Electric currents resulting from systematic relative motion between the charged dust and the surrounding plasma could also lead to the modification of the ambient magnetic field. For the dust, the basic effect is electrostatic charging, which in turn could lead to both physical and dynamical effects.

In this review we will restrict ourselves to the cometary environment, where the existing earth-based observations have more recently been complemented by relevant in-situ observations.

Electrostatic Charging of the Dust

The electrostatic charging of a dust grain in a radiative and plasma environment is given by the equation:

$$\frac{dQ}{dt} = I \qquad (1)$$

where

$$Q = C(\phi - \bar{\phi}) \qquad (2)$$

Cometary Plasma Processes
Geophysical Monograph 61
©1991 American Geophysical Union

Here Q is the charge on the grain, I is the net current on to it, ϕ is its floating potential, $\bar{\phi}$ is the ambient plasma potential, and C is the capacitance of the grain.

If d is the average intergrain distance and λ_D is the Debye shielding distance, given by

$$\lambda_D^2 = \frac{kT_e T_i}{4\pi \, n_e \, (T_e + T_i)} \qquad (3)$$

(where T_e and T_i are the electron and ion temperatures and n_e is the electron density in the "undisturbed" plasma, far away from the grains), then for an isolated spherical grain (i.e. $d \gg \lambda_D$)

$$C = C_{iso} = a(1 + \frac{a}{\lambda_D}) \qquad (4)$$

and $\bar{\phi} = 0$. Here a is the grain radius and all units are in c.g.s.

If this condition is not satisfied (i.e. $d \le \lambda_D$), $\bar{\phi} = \bar{\phi} \, (d, \lambda_D, a) \ne 0$ and $C = C \, (d, \lambda_D, a) > C_{iso}$. If the grain separation becomes small (i.e. $d \ll \lambda_D$) $|\phi - \bar{\phi}|$ becomes very small and thus Q becomes very small despite the increase in C. This follows from the circumstance that while the 'appetite' of the grains for charge has increased, there is not enough charge to satisfy it [e.g. see Goertz and Ip, 1984, Whipple et al., 1985, Goertz, 1989]. In the cometary environment typically, $a \ll \lambda_D < d$ thus giving rise to the case of small "isolated" spheres in a plasma.

The current I depends not only on the physical and electrical properties of the grain and the environmental condition, but also on $\Delta\phi \, (= \phi - \bar{\phi})$ and the relative velocity between the grain and the plasma. Contributions to I come mainly from the collection of thermal electrons and ions, photoemission and secondary emission of electrons due to energetic electron impact. Field emission could be important for very small grains. Smaller contributions result from processes such as thermoionic emission. Expressions for all these currents are rather easily obtained and are given, for example, in Whipple [1981]. For instance the thermal electron and ion currents to the grain in relative motion with respect to the plasma is obtained by integrating the differential flux over a drifting Maxwellian velocity distribution.

The equilibrium potential of an isolated grain, (i.e. the potential achieved when $\frac{dQ}{dt} = 0$) is generally unique, but in certain circumstances, particularly in the presence of strong secondary emission, the current-voltage curve can have triple roots, with the intermediate one being unstable, while the extreme roots are stable [Whipple, 1981]. Which of these the grain assumes would then depend also on history of the charging process.

Using the in-situ dust and plasma data at Halley's comet, it is easy to see that average intergrain distance d (~ 1 m) even for the smallest grains observed, is considerably larger than the Debye shielding distances (~10 cm). Consequently the grains may be considered as isolated, so that we may regard $\bar{\phi} = 0$ and $C = C_{iso} = a \left(1 + \frac{a}{\lambda_D}\right) \simeq a$ (in c.g.s. units) for small ($a \sim 1 \mu m$) grains since $\lambda_D \sim 10$ cm.

A recent calculation [Notni and Tiersch, 1987] of the variation of ϕ_∞ for three different types of grains (Silicate, Graphite and Aluminium) for different plasma compositions (pure H^+ and pure H_2O^+) plasma temperatures and densities are shown in Figure 1. The numbers within

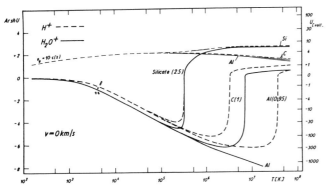

Fig. 1. The equilibrium potentials of grains of different compositions in dense and low density (H^+ and H_2O^+) plasmas as a function of the plasma temperature. The grains are assumed to be stationary with respect to the plasma (from Notni and Tiersch, 1987).

parenthesis are the maximum secondary emission yields for the appropriate material. In this calculation the plasma energy (kT_e) corresponding to the maximum secondary emission yield E_m for silicate, graphite and Aluminium are taken as 420 eV, 250 eV and 420 eV, respectively. Only two values of the electron density are used; very large ($n_e = \infty$) and $n_e = 10 c(x)$, where $c(x)$ is the photoemission yield coefficient.

It is seen that in the region of relatively high plasma densities (where photoemission is unimportant) and relatively low temperatures ($k T_e \leq 30$ eV) where secondary emission is unimportant, the grain potentials are negative and numerically small. Typically $\phi = -2.5 (kT_e/e)$ volts for H^+ plasma and $\phi \simeq -4.0 (kT_e/e)$ volts for H_2O^+ plasma. At higher temperatures as secondary emission becomes important the grain potential is seen to change from negative to positive, particularly for the high yield silicate. Finally it is observed that when photoemission is important (the uppermost curve) the grain potentials are positive.

The rather uncertain secondary emission yield function adopted by these authors and others vary. Consequently the calculated plasma temperature at which the transition from negative to positive potential takes place as well as the maximum positive potential achieved are different. However the basic nature of the potential curves are similar to the ones exhibited in Figure 1.

Inside the dense cometary ionopause where $T_e \leq 10^3 \,^\circ k$ and $n_e \geq 10^3$ cm^{-3}, $\phi_{eq} \leq 0.2V$. In the region of the largely undisturbed solar wind where photoemission is dominant, $\phi_{eq} = 2$ -5V, whereas in the region between the outer shock and the ionopause the grain potentials can assume both positive and negative values. While the potentials of conducting grains (e.g. C) remain negative throughout (-10V − -20V) those of dielectric grains (e.g. Silicates) could switch due to the importance of secondary emission.

The Physical Effects of Electrostatic Charging

When a body is charged, the mutual electrostatic repulsion of the surface charges produces an electrostatic tension. If this electrostatic tension across any section of the body exceeds the tensile strength there, the body will disrupt across that section. In the case of an isolated sphere of radius a (μm) and uniform tensile strength F_t (dynes cm^{-2}), Opik [1956] showed that electrostatic disruption will take place for a given surface potential ϕ (volts) if $a < a_c$ (μm), where

$$a_c \simeq 6.7 \mid \phi \mid F_t^{-1/2} \qquad (5)$$

Cometary grains in common with all cosmic grains are expected to be far from spherical. Consequently, Hill and Mendis [1980a] discussed the disruption of such grains, idealizing them as prolate spheroids, with sufficient surface conductivity to make them equipotentials. The electrostatic tension parallel to the polar (or rotation) axis (i.e. the polar tension P_x at a given distance x along this axis from the center) is obtained using the Maxwell stress tensor, to be

$$P_x(x) = \frac{1}{\pi \rho^2} \int_0^\rho \frac{E^2}{8\pi} \cdot 2\pi \rho \, d\rho \qquad (6)$$

where $\rho = \rho(x)$ is the radius of the section at x.

The variation of $P_x(x)$ (normalized with respect to its value at the center $P_x(0)$) with x, is shown in Figure 2 for various values of λ ($= a/b$,

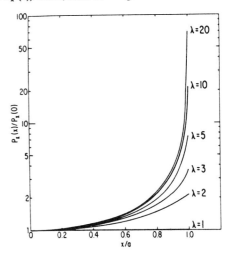

Fig. 2. Variation of the normalized electrostatic tension with polar distance for several prolate spheroids (from Hill and Mendis, 1980a)

where a and b are the semi-major and semi-minor axes). It is seen that while $P_x(x) / P_x(0)$ is constant for a sphere ($\lambda = 1$) as expected, it increased monotomically with x for all prolate spheroids ($\lambda > 1$) reaching a maximum at the extremities. This is due to the increase in the surface charge and therefore, the electric field towards the ends, due to the decrease in radius of curvature.

The consequences of charging up such a grain, of uniform tensile strength, towards it equilibrium potential is obvious from this figure. The tensile strength of the grain will be exceeded by the electrostatic tension first at the ends, if at all. This will result in the ends being "chipped" off. As the potential increases, this chipping process will continue, with the grain becoming more and more spherical, with the final outcome being one of two things. If the tensile strength of the grain is sufficiently large, this end-chipping process will eventually cease while the grain is still prolate. Otherwise this process will continue until the grain becomes more or less spherical, at which point it will explode *en toto*.

It has been argued that certain peculiar features called pseudo-synchronic bands, or striae, observed in the dust tails of several comets are the consequence of the sudden disruption of elongated or chain-like parent grains in the tail [Sekanina and Farrell, 1980]. Hill and Mendis [1980b] argued that the specific mechanism for disruption is the electrostatic one discussed above, when the grains are suddenly charged to high electrostatic potentials by energetic ("auroral") electron beams. They also conclude that the tensile strength of such cometary parent grains lies in the range $10^4 - 10^6$ dynes cm^{-2}.

More recently [Simpson et al., 1987; 1989] have reported the detection of discrete dust "packets" in the environment of comet Halley by the dust detectors on board VEGA-1 and VEGA-2. There packets of dust were detected as 'events' of enhanced flux of fine dust ($m_g \geq 10^{-13}$g) lasting about 10 s, and separating regions of very low flux (see Figure 3).

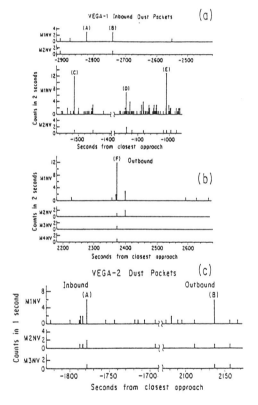

Fig. 3. Time profiles of dust 'events' observed by DUCMA dust detector on board the Vega 1 and 2 spacecrafts, at Halleys comet. The occurrence of discrete dust 'packets' is obvious (from Simpson et al., 1987)

These authors suggest that these dust packets are the result of break-up of larger grains that leave the cometary nucleus. They also suggest that this break up is due to these larger composite grains that leave the nucleus eventually coming "unglued". We are however of the opinion that such a process of "ungluing," presumably due to the sublimation of the volatile icy glue holding the less volatile dust grains, would be a more gradual process than is implied by these short discrete events. Consequently we suggest, as an alternative, that such fragile composite

dust grains could be electrostatically disrupted, in regions where they are seen; i.e. outside the ionopause where substantial electrostatic potentials may be attained by the parent grains. We take the parent grain to be of mass 10^{-12} g and mass density $\rho_d \approx 0.2$ g cm^{-3} leading to a radius a $\approx 1\mu$m. Then if we take the tensile strength $F_t = 10^4$ dynes cm^{-2} for such fragile grains, electrostatic disruption will take place if $|\phi| \geq 16$ V. Electrostatic chipping of sharp edges will take place for a non-spherical grain for even smaller values of $|\phi|$. Values of $|\phi| \sim 10 - 20$ V are typically what is expected in the region between the ionopause and the bow-shock, where the dust packets were observed. Even larger values of $|\phi|$ may be expected if the grains are subjected to sporadic fluxes of "auroral" (keV) electrons [Mendis, 1987], leading to the electrostatic disruption of even larger ($r_g \sim 10\mu$ or $m_g \sim 10^{-9}$g) grains.

The Dynamical Effects of Electrostatic Charging

While the possible importance of electrostatic charging on the dynamics of cometary dust was already recognized by Notni [1966], this question was not pursued until very recently. The study of the dynamics of cometary dust excluding the effects of electrostatic charging of course has a much longer history and has been pursued systematically. In order to provide the background for what follows, we will begin this section with a brief summary of these studies, which incidentally was stimulated by the 1835 apparition of Halley's comet. In an attempt to explain the morphology of the dust tail Bessel [1836] derived the equations of motion of dust particles emitted from the nucleus and driven away from the Sun by some unknown repulsive force, which like solar gravitation varied inversely with heliocentric distance. The physical nature of this postulated repulsive force was recognized to be solar radiation pressure only much later [Arrhenius, 1900; Schwarzschild, 1901].

While significant contributions to dust tail dynamics were subsequently made by a number of authors, it was brought to fruition by Finson and Probstein [1968a, b]. Using a synthetic approach which combined the earlier concepts of syndynes (the locii in the cometocentric frame connecting particles having the same ratio of the solar radiation force to solar gravity) and synchrones (the locii in the cometocentric frame connecting particles emitted from the comet at the same time), they inferred a number of cometary dust parameters, such as the size distribution weighted by the light scattering efficiency of the particles, by comparing calculated and observed isophotes of the tail. This method, with few modification has been the basis of all subsequent studies of cometary dust dynamics. Probstein [1968] was also the first to study in detail the problem of dust-gas coupling in the inner coma. This led to the realization that while the dust-gas coupling was confined to a thin region (≤ 10 R$_n$) surrounding the nucleus, the drag of the dust on the expanding gas made its flow transonic.

Once a dust grain (assumed here to be uncharged) leaves the region of dust-gas interaction, the only two forces acting on it in the heliocentric frame are the solar radiation (pressure) force F_{rad} and solar gravity F_{grav}, since the gravitational force of the cometary nucleus itself is negligible. Since

$$F_{rad} = \frac{Q_{pr}}{c} \left[\frac{L_\odot}{4\pi r^2} \right] \cdot \pi a^2 \qquad (7)$$

and

$$F_{grav} = \frac{GM_\odot}{r^2} \cdot \frac{4}{3} \pi a^3 \rho_d \qquad (8)$$

where $L_\odot$ is the mean solar luminosity ($= 3.90 \times 10^{33}$ ergs^{-1}), Q_{pr} is the scattering efficiency for solar radiation pressure and the other quantities have their usual meaning,

$$\frac{F_{rad}}{F_{grav}} \simeq \frac{5.95 \times 10^{-5}\, Q_{pr}}{\rho_d \cdot a} \qquad (9)$$

Both for dielectrics (e.g. Olivine) and for conductors (e.g. Magnetite) Q_{pr} remains roughly constant for $2 \times 10^{-5}\,cm \leq a \leq 10^{-2}\,cm$ and declines steeply for $a \leq 2 \times 10^{-5}\,cm$, the decline being much steeper for dielectrics. Typically in the range $2 \times 10^{-5}\,cm \leq a \leq 10^{-2}\,cm$, $Q_{pr} \geq 0.5$ for dielectrics, and $Q_{pr} \geq 1.0$ for conductors. Consequently $F_{rad} \simeq F_{grav}$ when $a \approx 0.3\,\mu m$ and $a \approx 0.6\,\mu m$ respectively in the two cases.

For typical times of flight $\leq 5 \times 10^5$s, the cometo-centric frame can be considered an inertial one. In this frame the only force acting on the dust particle is the radiation pressure directed along the Sun-comet line, and which is approximately a constant for a grain of given size. Consequently grains of a given size which are emitted at various angles to the Sun-comet axis with initial speed v_i (= terminal speed acquired due to the gas drag) will move in parabolic orbits, all of which are enveloped within a paraboloid of revolution whose apex distance, $\alpha = v_i^2/2g$, with $g = F_{rad}/m_d$ (see Figure 4).

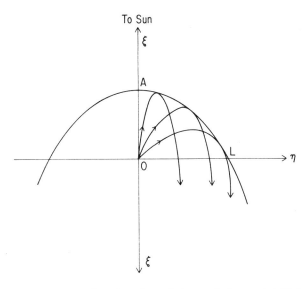

Fig. 4. The parabolic trajectories and the enveloping paraboloid for unchanged dust grains in a plane containing the sun-comet axis.

Using value of v_i calculated by Gombosi and Horanyi [1986] and values of Q_{pr} for Olivine and Magnetite obtained from [Divine, 1984], the values of the apex distance α, for different size grains are shown in Figure 5.

We now proceed to discuss the dynamical effects of charging. Wallis and Hassan [1982] pointed out that the dust grains in the cometary environment, which are necessarily electrically charged as discussed earlier, are subject to electromagnetic forces due to the convectional electric fields both in the undisturbed solar wind as well as in the region of the shocked solar wind outside the ionopause.

The electrostatic acceleration of the grain, $\vec{g}_e$ is given by

$$\vec{g}_e = Q\frac{\vec{E}}{m_d} = -\frac{Q\,\vec{V}_{sw} \times \vec{B}}{c\,m_d} \qquad (10)$$

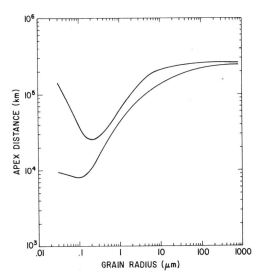

Fig. 5. The apex distance of the bounding paraboloid for unchanged Olivine (upper curve) and Magnetite (lower curve) grains as a function of size.

since $|\vec{V}_d| << |\vec{V}_{sw}|$.

To a good approximation $|\vec{E}|$ is constant outside the ionopause along the Sun-comet axis. Assuming $|\vec{B}| = 5\,\gamma$ and that $\vec{B}$ is inclined at 45° to $\vec{V}_{sw}$ at 1 AU one obtains,

$$X = \frac{|\vec{g}_e|}{|\vec{g}_r|} \simeq \frac{10^{-2}\,|\phi(V)|}{a\,(\mu)\,Q_{pr}} \qquad (11)$$

The values of X for Olivine and Magnetite grains of different sizes are given in Table 1, taking $|\phi\,(V)| = 5$. It is seen that while $X << 1$ when a $= 1\,\mu$, $X \approx 1$ when a $= 0.1\,\mu$ for Olivine. When a $= 0.03\,\mu$, $X > 1$ for Magnetite and $X >> 1$ for Olivine.

TABLE 1. The ratio X for different sized grains.

a (μ)	X	
	Olivine	Magnetite
1.0	0.05	0.04
0.5	0.85	0.06
0.1	1.00	0.28
0.03	16.5	3.35

The grains are now subject to the total acceleration $\vec{g}_r + \vec{g}_e$. Assuming that $\vec{g}_e$ is also constant, it is obvious that dust particles emitted isotropically with initial speed v_i will once again describe parabolic orbits which are enveloped by a paraboloid whose axis is now in the direction of $\vec{g}_r + \vec{g}_e$ as shown in Figure 6. Clearly the axis would be inclined to the solar direction by a large angle for the smaller grains as pointed out by Wallis and Hassan [1982]. For example for a $0.1\,\mu$ Olivine grain ($Q_{pr} \approx 0.5$), if $|\phi| \approx 5V$, the angle θ of the paraboloid axis to the solar direction $\approx 45°$, whereas it is still about 10° even when a $= 0.5\,\mu$ for Olivine.

Of course the smallest grains would not contribute appreciably to the solar radiation scattered by the cometary dust and consequently their bounding envelopes may not be apparent in the visual. This is not the case for grains with radius $\approx 0.5\mu$ and if they are sufficiently charged ($|\phi| \sim 5V$) and are basically dielectric their bounding envelopes should have axis inclined to the Sun-comet line at an angle that is apparent in

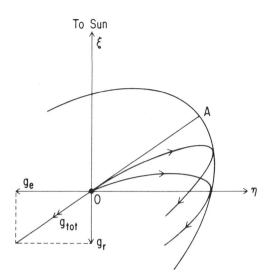

Fig. 6. The parabolic trajectories and the enveloping parabola of charged grains in the plane containing the sun-comet axis and the convectional electric field vector in the cometo-centric frame.

visual observations. Indeed this seems to be the case in some drawings of the sunward envelopes of comet Donati [1958 VI] as seen in Figure 7.

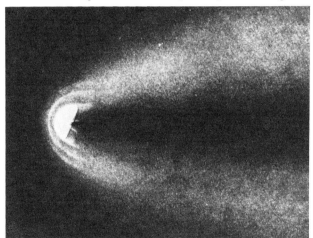

Fig. 7. Drawing based on visual observations of comet Donati 1858 VI by G.P. Bond on Oct. 9, 1858. The Sun comet axis is given by the prolonged tail axis (form Rahe et al., 1969).

So while it is tempting to interpret the skewness of these sunward envelopes as a consequence of the non-radial acceleration of charged dust grains, it must be pointed out that an alternative explanation has been proposed by Sekanina [1987] who interprets the skewness of these sunward envelopes in terms of spiral loci of particles emitted from localized regions in a rotating cometary nucleus.

In this connection it is interesting to point out that the apex distance of the bounding paraboloid of charged grains is given by

$$\alpha = \frac{v_i^2}{2g_{tot}} = \frac{v_i^2}{2\sqrt{g_r^2 + g_e^2}} \qquad (12)$$

So while the grains are even more tightly constrained in the η-direction (see Figure 6) than if they were uncharged, they can move to greater distances from the Sun-comet line in the opposite sense. This could be a possible explanation for the detection of very small $(3 \times 10^{-17} g \leq m_d \leq 10^{-14} g)$ grains by VEGA 1 and VEGA 2 inbound at distances as large as 2.5×10^5 km and 3.2×10^5 km from the nucleus of comet Halley [Vaisberg et. al., 1987]. Of course, optical properties of these very small grains are unknown. However if they are sufficiently absorbing (e.g. "dirty" silicates) as deduced from the infrared thermal emissions from several comets [Hanner, 1980] then the excursions of even $0.05 \, \mu$m grains from the nucleus in the directions of the encounters are likely $\leq 10^5$ km.

The basic equation governing the dynamics of a charged dust grain in the cometocentric frame, regarded as an inertial frame is given by

$$m_d \frac{d\vec{v}_d}{dt} = Q(t) \left[\vec{E} + \frac{\vec{v}_d \times \vec{B}}{c} \right] + \vec{F}_{rad} + \vec{F}_c + \vec{F}_{ig} \qquad (13)$$

where $\vec{E}$ is the convectional electric field $(= -\vec{v} \times \vec{B}/c)$, $\vec{v}$ being the plasma velocity $\vec{F}_{rad}, \vec{F}_c$ and $\vec{F}_{ig}$ are the radiative pressure force the Coulomb drag and the intergrain force respectively and $Q(t)$ is the grain charge which is given by equation (1). In those instances where the cometocentric frame cannot be considered as an inertial frame, the appropriate inertial forces need to be included on the right hand side of equation (13). In situations where the intergrain distance $d > \lambda_D$, which is generally true at comets, F_{ig} may be neglected.

In order to calculate the grain orbits one needs $T_e(r)$, $T_i(r)$, $n_e(r)$, $\vec{v}(r)$ and $\vec{B}(r)$, the last two quantities giving $\vec{E}(r)$. Clearly a detailed model of the plasma flow and the magnetic field in the cometary environment, such as the MHD model of Schmidt and Wegmann [1982] is needed. Horanyi and Mendis [1985] used a simpler approach. They used a simple 'source-in-a uniform stream' model to reproduce the stream lines (see Figure 8) and proceeded to calculate the $\vec{E}$-field noting

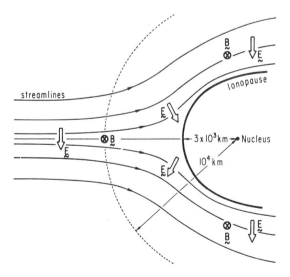

Fig. 8. Schematic of the plasma streamlines and the convectional electric field in the vicinity of the ionopause. The magnetic field is assumed to be into the paper (from Horanyi and Mendis, 1985).

that the stream lines were also electric equipotentials in ideal MHD when $\vec{v}$ and $\vec{B}$ are mutually perpendicular. It must be pointed out that such a source-in uniform-stream model is a gross oversimplification as far as the overall flow is concerned. It however provides a fair approximation to the stream lines obtained by the MHD model. Since one needs only the E field, and not the B field, to calculate grain orbits and since the E-field can be deduced by noting that these stream lines are electric equipotentials, such a model is adequate for our purpose.

Horanyi and Mendis [1986 a,b] subsequently performed numerical simulations to calculate the distribution of charged dust in the environments of comets Halley and Giacobini-Zinner taking into account the orbital motions of the comet. In the latter case, a scatter plot of the distribution of grains of various sizes in a plane normal to the Sun-comet axis at a distance 10^4 km behind the nucleus is shown in Figure 9. In this

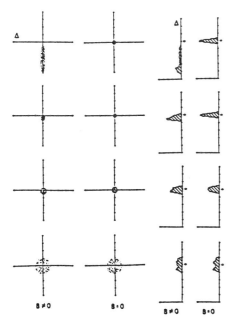

Fig. 9. (Left) Two panels showing the distribution of grains of various sizes (top to bottom, $0.03\,\mu$m, $0.1\,\mu$m, $0.3\,\mu$m, and$1.0\,\mu$m) in a plane normal to the sun-comet axis of comet Giacobini-Zinner at a distance of 10^4 km behind the nucleus. The first column shows the distributions when the effects of the interplanetary magnetic field are included ($\vec{B} \neq \vec{0}$). (Right) two panels showing histograms of the column densities (in arbitrary units) corresponding to the appropriate distributions on the left. The third column corresponds to the case $\vec{B} \neq \vec{0}$, while the fourth column corresponds to the case $\vec{B} = \vec{0}$. The unit length scale (Δ) in each case corresponds to 10^5 km, and the nucleus is always located at the origin of coordinates (from Horanyi and Mendis, 1986b).

case the interplanetary magnetic field is assumed to be in the orbital plane of the comet, so that the convectional electric field is normal to this plane.

The grain sizes increase from $0.03\,\mu$m at the top of the figure to $1.0\,\mu$m at the bottom. The first column shows the distributions when the electromagnetic effects are taken into account ($\vec{B} \neq \vec{0}$), while the second column shows the corresponding distributions when the electromagnetic effects are neglected (e.g., $\vec{B} = \vec{0}$). The two panels on the right are histograms of the column densities (in arbitrary units) corresponding to the distributions on the left. The third column corresponds to the case $\vec{B} \neq \vec{0}$, while the last column corresponds to case $\vec{B} = \vec{0}$. The nucleus in each plot is located at the origin of coordinates. The elongation of the distribution normal to the orbital plane is obvious, particularly for the smallest grains ($0.03\,\mu$m). In that case it is also seen that the grains are concentrated well below the orbital plane. The larger grains have a more symmetrical

distribution about the axis. In the absence of electromagnetic effects (e.g., $\vec{B} = \vec{0}$) it is seen, once more, that the larger ("older") grains concentrate away from the axis, while the smaller ("younger") grains concentrate closer to the axis. These distributions are, of course, axially symmetric in this case.

The NASA-ICE spacecraft intercepted the tail of comet Giacobini-Zinner on September 11, 1985, at a distance about 8×10^3 km from the nucleus moving in a generally south to north direction in the comet's reference frame. Although the spacecraft did not carry a dust detector, the plasma wave instrument detected impulsive signals that were attributed to dust impacts on the spacecraft [Gurnett et al., 1986]. While an asymmetry in the impact rate between the inbound and outbound legs, consistent with the prediction for charged grains was indeed observed, the asymmetry was rather small and may also be consistent with nonisotropic emission of grains from the nucleus as suggested by Gurnett et al., [1986]. However, this observation coupled with the observation that the smaller particles were encountered further away than the larger ones and also had a greater asymmetry between the 2 legs [Gurnett et al., 1986] strongly supports the charged dust model.

In the case of Halley's comet Horanyi and Mendis [1986 a] simulated both the "background" distribution of dust resulting from the uniform emission of dust from the sunward hemisphere as well as the spiral dust features (observed in the 1910 apparition) that would result from an "active" equatorial spot.

Since the effects of electromagnetic forces are important only on the lower end of the dust mass spectrum and are mainly manifested in the projection of the distribution normal to the orbital plane, the dust morphology as seen from the earth in 1910 (which corresponded to larger grains) was shown to be unaffected. However it was shown that the distribution of the smallest particles ($a \leq 0.3\,\mu$m) that would be observed by the various spacecraft, which followed each other 3-4 days apart, were very different due to the rotation of the interplanetary magnetic field.

Wallis and Hassan [1987] discussed the evidence in support of electrical charging in the dust distribution of Halley's comet encountered by the Vega-2 spacecraft. They showed that along the trajectory of Vega-2 the observed magnetic field direction was such that positively charged grains, that were expected in the outer coma, would be deflected towards the inbound flank in conformity with the observation of the smallest grains. For the same reason the negatively charged grains, expected in the inner coma, would be deflected towards the outbound flank, once again in qualitative conformity with the dust jet displacement for the smallest grains. These authors also infer that the required negative potentials in the inner coma require a warm (1 eV) 'halo' in the electron distribution, in addition to the main cool 'core'. As the authors recognize, this is not unlikely, since the photoelectrons could be only partially thermalized in this region.

Horanyi and Mendis [1987] also considered the effect of a sector boundary crossing on the dust tail of a comet. When the grains are electrically charged, the traversal of the sector boundary along the tail leads to a reversal in the electrical acceleration on the grains due to the reversal of the convectional electric field. This leads to a wave-like feature in the distribution of the smaller ($a \leq 0.3\,\mu$m) grains (Figure 10). These authors attribute the peculiar "wavy" morphology of the dust tail of comet Ikeya - Seki [1965f] to such an effect (see Figure 11). Alternatively it had been suggested that these waves are magnetohydrodynamic and arise due to instabilities excited in the plasma tail containing twisted magnetic fields [Krishan and Sivaraman, 1982]. We do not agree with this view because the tail shown in Figure (11) is dominated by the continuous spectrum due to scattering of solar radiation by dust although weak CO^+ emission was observed.

$R_g = 0.3 \, \mu m$

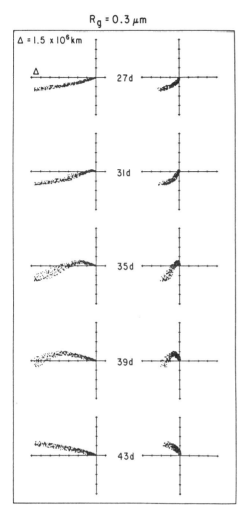

Fig. 10. Time evolution of the (charged) dust tail of a comet on crossing a magnetic sector boundary. All the grains are assumed to be of equal radius $R_g = 0.3 \mu m$. The left panels show the projections of the dust distributions, two days apart, in a plane normal to the orbital plane, and containing the sun-comet axis. The sun is to the right and the sense of orbital motion is into the paper. The panels on the right show the corresponding projections in a plane normal to the sun-comet axis. Here the motion is to the right (from Horanyi and Mendis, 1987).

Effect of Dust on the Plasma and the Magnetic Field

So far we have discussed only the effect of plasma and solar UV radiation on the dust. There are also effects of dust on the plasma, such as depletion and heating or cooling depending on the strength of the UV radiation field. These effects, in the context of dense dusty interstellar molecular clouds has been discussed by several authors [e.g. see Nakano, 1984] for a detailed review. However, they are not important in the cometary context since the dust density is low. There could of course be more subtle effects. These include the modification of plasma waves by the dust, particularly in the low frequency regime (i.e. less than the ion plasma and gyrofrequency). Very little work has been done in this area.

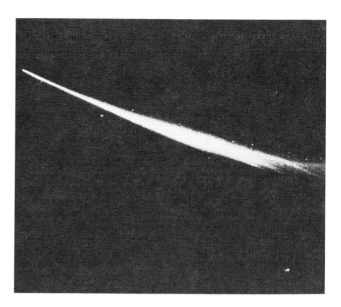

Fig. 11. The dust tail of comet Ikeya-Seki, (1965f). Note the well defined wavy structure far down the tail (Kodaikanal Observatory photograph).

[See Goertz, 1989 for a recent review.] In the cometary context recently de Angelis et al., [1988] investigated the role of ion plasma waves in a dusty plasma to explain the enhanced low frequency $(\omega < \omega_{pi})$ electrostatic noise observed at Halley's comet by probes on the Vega spacecrafts. They obtained a dispersion relation for low frequency electrostatic waves in an inhomogeneous plasma; the inhomogenities being the Debye spheres $(\lambda_D \geq 10 \text{ cm})$ surrounding the distribution of dust particles with an average separation distance $d > \lambda_D$. They argued that these density gradients near the grains could lead to the trapping of such electrostatic waves, with a superposition giving rise to an increase in amplitudes, thereby explaining the enhanced electrostatic noise observed at Halley's comet with $\omega > \omega_{pi}$. This is of course an interesting possibility, but as the above author's recognize there is an alternative interpretation of the above observation simply in terms of dust impacts on the electrical probes [e.g. Gurnett et al., 1986; Mogilevsky, 1987].

The dust can also have an effect on the interplanetary magnetic field. This results from the fact that the relative motion between the charged dust and the surrounding charge neutralizing plasma constitutes an electric current [Hill and Mendis, 1982]. Let us make an estimate of this current in the cometary dust tail, by making the simplifying assumption that all the dust emitted by the cometary nucleus is eventually driven back by solar radiation pressure to form a uniform cylindrical distribution of cross-section r_c. Let us also assume that the dust size distribution at a fixed point down the tail, x, is given by the power law

$$n \, (a) = C \, a^{-\alpha} \qquad (14)$$

The value of α is highly uncertain, and is expected to vary with position as was observed at Halley's comet. However using a typical value of for the cumulative mass index $\beta \approx 1$, one obtains a value of $\beta + 1 = 2$ for the mass index, γ, and a value of $3 \gamma - 2 \approx 4$ for α. Such a value of α would cause the cumulative area of the dust grains to be strongly concentrated at the lower end of the size spectrum. Assuming that the grain acceleration g along the axis is entirely due to radiation pressure, for the purpose of this rough estimate, we have $g(a) = F_{rad} / m_d \approx 7 \times 10^{-5} \, a^{-1} \text{ cm s}^{-2}$, and $v \, (a) = (2g \, (a) \, x)^{1/2} = 1.2 \times 10^3 \, a^{-1/2} \text{ cm s}^{-1}$ taking $x \approx 10^6 \text{km}$. We then have, for the total production rate of dust:

$$\dot{M}_d = \int_{a_0}^{a_1} \frac{4}{3} \pi \ a^3 \ \rho_d \ v(a) \ \pi \ r_c^2 \ n(a) \ da \qquad (15)$$

where a_0 and a_1 are the minimum and maximum radii of the dust grains in the distribution and ρ_d is their bulk density.

The total current carried by these charged grains is then given by

$$I = \int_{a_0}^{a_1} \pi \ r_c^2 \ | \ Q(a) \ | \ (v - v(a)) \ n(a) \ da \qquad (16)$$

where v is the plasma speed (parallel to the axis at x).

Noting that $\dot{M}_d$ for Halley's comet just prior to the Giotto encounter $\simeq 10^7 g \ s^{-1}$ [Hanner et al., 1987], taking $r_c \simeq 10^6$ km and assuming $a_1 >> a_0$, the value of C is obtained from (15) as $C = 2.5 \times 10^{-17} a_0^{1/2}$. Then using this value of C in (16), taking $v \simeq 100$ km s^{-1} and noting that $v >> v(a)$ we get

$$I \ (esu) \simeq 3.10^9 \ | \ \phi \ | \ a_0^{-3/2} \qquad (17)$$

where ϕ is also in e.s.u. (stat-volts). It is unfortunate that I depends so strongly on the highly uncertain value of a_0. Observations [e.g. Vaisberg et al., 1987] indicate the existence of grains with radius $\leq 0.03 \mu$m. However since these very small grains ($a \leq 0.1 \mu$m) are likely to be strongly accelerated normal to the orbital plane of the comet by the convectional electric field in the streaming plasma [Horanyi and Mendis, 1986b] they may not contribute significantly to the axial current that we are considering. Let us therefore take $a_0 \simeq 0.1 \mu$m as being appropriate to our calculation. Also taking $| \ \phi \ | = 5$ volts $= (1/60$ stat-volts) we obtain from (17), $I = 1.5 \times 10^{15}$ esu ($= 5 \times 10^5$A). The toroidal magnetic field due to this current will have a magnitude of

$$B = \frac{2}{c} \ \frac{I}{r} \qquad (18)$$

at a distance r ($\geq r_c$) from the axis. Consequently when r $\simeq 10^6$ km, B $\simeq 10^{-6} \Gamma (= 0.1\gamma)$. This value is only a few percent of the interplanetary magnetic (5-10 γ) and is less than the typical fluctuation ($\geq 10\%$) in it. However due to the large uncertainities in this rough estimate, the question of the role of the cometary tail dust current in modifying the interplanetary magnetic field must remain an open one at the present time.

Conclusions

In this review we have shown that cometary grains, by virtue of being immersed in a plasma and UV radiative environment must be necessarily electrically charged. While the magnitude and polarity of the charge can vary, this charging could lead to both physical and dynamical effects on the dust. It could also lead to certain effects on the plasma as well as on the interplanetary magnetic field. We have discussed a number of observations, both remote and in-situ, which indicate that such effects are indeed present. However, at present alternative non-electrical processes have been proposed to explain many of these observations. So until grain charges are directly measured, as is expected during the Comet Rendezvous Asteroid Fly-by (CRAF) mission to comet Kopff at the beginning of the next century, these questions will remain unresolved. It must however be emphasized that even with very conservative estimates of the electric charge anticipated in the cometary environment the electrodynamic forces experienced by the smallest grains recently observed (in situ) in the dust mass spectrum of comet Halley are larger than or comparable to the radiation pressure force. Consequently these forces cannot be ignored in considering the dynamics of such grains. The reason why we have been able to get by ignoring these forces in the past is that the grains that are strongly effected by electrodynamic forces are too small to be observed optically from the earth.

One area we have left out in this review, since it is only peripheral to the title, is the effect of differential electrostatic charging of the bare cometary nucleus at large heliocentric distances. If loose dust is present on the surface, this could lead to electrostatic levitation and blow-off of fine grains [Mendis et al., 1981]. The large sporadic brightness variations of comet Halley observed at large heliocentric distances (≥ 8 AU) during its last apparition, which also appeared to be modulated by solar wind conditions, has been attributed to this process [Flammer et al., 1986].

Acknowledgements. We wish to acknowledge support from the following grants: (D.A.M) NAGW-1502 of the NASA Planetary Atmospheres program; NRA 88-055A-02 of the NASA Space Science program, (M.H.) NAGW 1656 of the NASA Space Science program.

References

Arrhenius, S.A., *Phys. Zeitschr, 2,* 81, 1900.

Bessel, F.W., *Astron. Nachr., 13,* 185, 1836.

de Angelis, U., Formisano, V. and Giordano, M., *J. Plasma Phys., 40,* 399, 1988.

Divine, N., Status of dust modelling for comet Halley, *JPL interoffice memo;* 5137-84, 1984.

Finson, M. L. and Probstein, R. F., *Astrophys. J., 154,* 327, 1968a.

Finson, M. L. and Probstein, R. F., *Astrophys. J., 154,* 353, 1968b.

Flammer, K.R., Jackson, B.V. and Mendis, D.A., *Earth Moon and Planets, 35,* 203, 1986.

Goertz, C.K., *Rev. Geophys., 27,* 271, 1989.

Goertz, C.K., and Ip, W-H., *Geophys. Res. Letts., 11,* 349, 1984.

Gombosi, T. J. and Horanyi, M., *Astrophys. J., 311,* 491, 1986.

Gurnett, D.A., Averkamp, T.F., Scarf, F.L. and Grun, E., *Geophys. Res. Letts., 13,* 291, 1986.

Hanner, M. S.,in *Solid particles in the Solar System* (Eds. I. Halliday and B. A. McIntosh), Reidel Publ. Co. Holland p. 233, 1980.

Hanner, M.S., Tokunaga, A.T., Golish, W.F., Griep, D.M. and Kaminski, C.D., *Astron. Astrophys., 187,* 653, 1987.

Hill, J.R. and Mendis, D.A., *Can. J. Phys., 59,* 897, 1980a.

Hill, J.R. and Mendis, D.A., *Astrophys. J., 242,* 395, 1980b.

Hill, J.R. and Mendis, D.A., *Geophys. Res. Letts, 9,* 1069, 1982.

Horanyi, M. and Mendis, D.A., *Astrophys. J., 294,* 357, 1985.

Horanyi, M. and Mendis, D.A., *Astrophys. J., 307,* 800, 1986a.

Horanyi, M. and Mendis, D.A. *J. Geophys. Res., 91,* 355, 1986b.

Horanyi, M. and Mendis, D.A., *Earth Moon and Planets, 37,* 71, 1987.

Krishan, V. and Sivaraman, K.P., *The Moon and the Planets, 26,* 209, 1982.

Mendis, D.A., Hill, J.R., Houpis, H.L.F. and Whipple, E.C., *Astrophys. J., 249,* 787, 1981.

Mendis, D.A., *Earth Moon and Planets,* **39,** 17, 1987.

Mendis, D.A., Hill, J.R., Houpis, H.L.F. and Whipple, F.C., *Astrophys. J.,* **249,** 787, 1981.

Mogilevsky, M., et al., *Astron. Astrophys.* **187,** 80, 1987.

Nakano, T., *Fund. Cosmic Phys.* **9,** 139, 1984.

Notni, P., *Nature et Origine des Cometes, Mem. Soc. R. Sci. Liege.,* **12,** (Ser 5), 379, 1966.

Notni, P. and Tiersch, H., *Astron. Astrophys.* **187,** 796, 1987.

Opik, E.J., *Irish Astron. J.,* **4,** 84, 1956.

Probstein, R. F.,in *Problems of hydrodynamics and continuum mechanics, Soc. Industr. Appl. Math,* 568, 1968.

Rahe, J., Donn, B. and Wurm, K., *Atlas of cometary forms,* NASA SP-198 p. 61, 1969.

Schmidt, H.U.and Wegmann, R., (Ed. L.L. Wilkenning), Univ. of Arizona press, p. 538, 1982.

Schwarzchild, L., *Sitz. Bayer. Acad. Wiss. Munchen,* **1901,** 293, 1901.

Sekanina, Z. *Symp. on Diversity and Similarity of Comets,* ESA SP-273, 1987.

Sekanina, Z. and Farrell, J.A.,in *Solid Particles in the Solar System,* (Eds. J. Halliday and B.A. McIntosh), p. 267, 1980.

Simpson, J.A., Rabinowitz, D., Tuzzolino, A.J., Ksanfomality, L.V. and Sagdeev, R.Z., *Astron. Astrophys.,* **187,** 742, 1987.

Simpson, J.A., Tuzzolino, A.J., Ksanfomality, L.V., Sagdeev, L.Z and Vaisberg, O.L., *Adv. Space Res.,* **9,** 259, 1989.

Vaisberg, O.L., Smirnov, V., Omelchenko, A., Gorn L., and Iovlev, M., *Astron. Astrophys.* **187,** 753, 1987.

Wallis, M.K. and Hassan, M.H.A., *Cometary Exploration* Vol. II (Ed. T.I. Gombosi)., Hungarian Acad. Sci., Budapest p. 57, 1982.

Wallis, M.K. and Hassan, M.H.A. ,in *Symp. on Diversity and Similarity of Comets,* **ESA SP-278,** 351, 1987.

Whipple, E.C., *Rep. Prog. Phys.,* **44,** 1197, 1981.

Whipple, E.C., Northrop, T.G. and Mendis, D.A., *J. Geophys. Res.,* **90,** 7405, 1985.

COLLISIONAL PROCESSES IN COMETARY PLASMAS

T. E. Cravens

Department of Physics and Astronomy
University of Kansas
Lawrence, Kansas 66045 USA

Abstract. The interaction of the solar wind with comets is initiated at large distances from the nucleus by the ionization of cometary neutrals. The resulting contamination of the solar wind with cometary ions mass-loads the solar wind flow, causing it to slow down. The plasma-comet interaction is largely collisionless at large cometocentric distances. However, collisional processes become important in the inner coma (within the cometopause). Collisional processes include charge-transfer between solar wind protons and neutrals, ion-neutral friction, electron and ion thermal cooling, and ion-neutral chemistry. For example, the magnetometer on the Giotto spacecraft observed a diamagnetic cavity near closest approach. This cavity is a consequence of the balance between an inward-directed magnetic pressure gradient force and an outward ion-neutral frictional force. Thermalization of the cometary ion distribution function by Coulomb collisions is another important process in the inner coma of an active comet.

1. Introduction

Several plasma populations are known to exist in the inner coma of an active comet, including energetic cometary ions, cold cometary ions, solar wind protons, solar wind electrons, and cold electrons. The plasma-comet interaction is largely collisionless at large cometocentric distances. But the behavior of the plasma in the inner coma of an active comet like comet P/Halley is not just controlled by plasma processes such as magnetohydrodynamics (MHD) but also by collisional processes such as ion-neutral friction, resistivity, electron and ion thermal cooling, and ion-neutral chemical reactions.

This paper will review the collisional processes important for the plasma in the inner coma region (the region inside the cometopause). Some earlier reviews of cometary plasma physics include those by Ip and Axford [1982] and Mendis et al. [1985] prior to the recent spacecraft encounters with comets. More recently, reviews were written by Galeev [1986], Ip and Axford [1989], and Cravens [1990] for cometary plasma processes in general, by Cravens [1987] for the ionosphere, and by Cravens [1989a,b] for cometary plasma boundaries.

The solar wind interaction with a comet starts millions of kilometers from the nucleus with the ionization of a cometary neutral atom or molecule due to photoionization by solar extreme ultraviolet photons, by charge transfer of the neutral with a solar wind proton, or by impact ionization by a solar wind electron. Thus, the interaction begins with a collision of a proton, electron, or photon with a neutral. The cometary neutrals move very slowly (neutral outflow speed $u_n \approx 1$ km/s) in comparison with the solar wind ($u_{sw} \approx 400$ km/s). The newly-born ions are accelerated by the motional electric field of the solar wind, and their motion is a superposition of gyromotion and $E \times B$ drift. The distribution function of these cometary ions in the solar wind reference frame is a ring-beam, which is highly unstable to the growth of low frequency electromagnetic waves. The observed wave amplitudes were as large as the background field near comets Giacobini-Zinner and Halley (c.f. Ip, 1989; Lee, 1989; Galeev et al., 1986). The magnetic fluctuations result in rapid pitch-angle scattering of the cometary pick-up ions, such that the initial ring-beam distribution evolves into a shell distribution with a "radius" equal to the local solar wind speed (c.f. Galeev et al., 1986; Ip, 1989; Terasawa, 1989; Cravens, 1989c). Pick-up oxygen ions have energies of about 20 keV in the solar wind reference frame. The ions also undergo energy diffusion due to scattering by the waves moving along the field, and a small fraction of cometary ions can be accelerated to energies much higher than the initial pick-up energy. The shell distribution thus becomes thicker and more diffuse. The shell also becomes thicker due to the slowing down of the solar wind by the mass-loading resulting from the addition of cometary ions.

A weak bow shock forms at a radial distance of about 350,000 km for comet Halley and 50,000 km for comet Giacobini-Zinner (see reviews listed earlier), and cometary ions picked up downstream of this shock are much less energetic than those picked up upstream, as a consequence of the slower flow speed. A second and lower-energy (≈ 1 keV for O^+) shell thus begins to form in the ion distribution function [Galeev et al., 1985], which can also undergo thickening due to both wave-particle acceleration and the further slowing of the flow. The plasma behavior just described is entirely collisionless except for the initial ionizing collision. However, both the neutral and plasma densities become larger, as the shocked and mass-loaded solar wind approaches closer to the nucleus, and collisional processes start to become important near the cometopause (a cometocentric distance of $\approx 10^5$ km for comet Halley). The charge-exchange process is particularly important and removes the hotter ions from the flow, but leaves the distribution function as a shell distribution. However, very close to the nucleus, Coulomb collisions transform the distribution function into a Maxwellian. The evolution of the cometary ion distribution function takes place in the following sequence:

Ring-beam ---> Shell ---> Thick Shell ---> Maxwellian

A schematic of the cometary plasma environment is given in Figure 1. Three important plasma populations exist upstream of the bow shock: solar wind protons, solar wind electrons, and cometary ions. The cometary ions are hot with a temperature equal to the bulk flow energy (T ≈ 20 keV for O^+). Downstream of the shock the newly picked-up cometary ions are merely warm (T ≈ 1 keV), and most of the ion pressure still comes from the hot ions acquired by the flow upstream of the shock.

Cometary Plasma Processes
Geophysical Monograph 61
©1991 American Geophysical Union

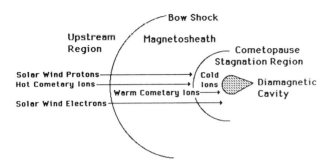

Fig. 1. Schematic of the cometary plasma environment and locations of several plasma populations, including where they originate and where they are removed or significantly modified by collisions.

Charge transfer collisions between the ions and cometary neutrals first become important in the vicinity of the cometopause (see earlier reviews for the many references that now exist on this topic). Solar wind protons and both hot and warm cometary ions are removed from the flow by this process and are replaced by colder ions, whose temperature reflects the relatively slow local flow speed. Inside the cometopause, the flow becomes much slower and the cometary ion density much larger, due to the increasing neutral density and ionization rate. At distances from the nucleus less than a few times 10^4 km, the plasma of comet Halley was observed to be almost stagnant (flow speeds less than a few km/s) and entirely cometary [Balsiger et al., 1986]. Collisional processes, including ion-neutral chemistry and electron collisional cooling, are very important in this region. A diamagnetic cavity was observed by Giotto with a radius of 4500 km [Neubauer et al., 1986]. This cavity is a consequence of the balance between an inward-directed magnetic pressure gradient force and an outward ion-neutral frictional force (i.e., collisions). Collision processes are discussed in the next section.

2. Collision Processes

Many types of collisional processes are important in comets. In this section, the important collision processes for electrons and ions in the cometary environment are listed, and the collisionopause concept is discussed.

2.1 Collision Processes for Ions

The following collision processes are important for ions in the cometary environment:

(1) Ion-neutral charge transfer
$$O^+ + H_2O \longrightarrow O_{fast} + H_2O^+$$

(2) Ion-neutral chemistry
$$H_2O^+ + H_2O \longrightarrow H_3O^+ + OH$$

(3) Ionization
$$O^+ + H_2O \longrightarrow O^+ + H_2O^+ + e$$

(4) Electron removal
$$O + H_2O \longrightarrow O^+ + H_2O + e$$

(5) Electron-ion recombination
$$H_3O^+ + e \longrightarrow H_2O + H$$

(6) Ion-ion Coulomb interaction
$$H_2O^+ + H_2O^+ \longrightarrow H_2O^+ + H_2O^+$$

The importance of each of these processes depends on the region and the particular ion population. For example, collision process (1) -- charge transfer -- first becomes important near the cometopause. Solar wind protons charge transfer with neutrals in this region and are replaced by cold cometary ions. Hot pick-up O^+ ions also charge transfer with neutrals in this region and are replaced by cold ions. Process (2) is very important in the inner coma for colder ions, and results in H_3O^+ rather than H_2O^+ being the major ion species. Dissociative recombination (process (5)) removes the H_3O^+ ions produced by process (2). Processes (3) and (4) are mainly relevant for relatively energetic O^+ and O in the inner coma. Process (6) -- Coulomb collisions -- acts to isotropize and thermalize the cometary ion distribution in the inner coma. Particular emphasis will be given to process (6) in this paper.

2.2 Collision Processes for Electrons

The following collision processes are important for electrons in the cometary environment:

(1) Ionization of neutrals
$$e + H_2O \longrightarrow H_2O^+ + e + e$$

(2) Electronic excitation
$$e + O(^3P) \longrightarrow O(^3S) + e$$

(3) Rotational and vibrational excitation of molecules
$$e + H_2O \longrightarrow H_2O^* + e$$

(4) Electron-ion recombination
$$H_3O^+ + e \longrightarrow H_2O + H$$

(5) Electron-ion Coulomb interaction
$$H_2O^+ + e \longrightarrow H_2O^+ + e$$

(6) Electron-electron Coulomb interaction
$$e + e \longrightarrow e + e$$

Electrons interact with neutrals via elastic or inelastic collisions. The latter includes ionization, electronic excitation, and rotational and vibrational excitation of molecules. Electron impact ionization of neutrals by solar wind electrons is a source of cometary ions (c.f. Cravens et al., 1987). Colder electrons (less than a few eV) can also recombine with ions; this process becomes important in the inner coma, as mentioned in the discussion of ion collision processes. Electrons also undergo elastic Coulomb collisions, which in the inner coma can transfer energy from superthermal electrons (solar wind and photoelectrons with energies of tens of eV) to colder thermal electrons.

2.3 Collisionopause

The collisionopause is defined as the location, or boundary, where collisions first become important (c.f. Mendis et al., 1985; Cravens, 1989a, 1990). The plasma outside the collisionopause can be considered collisionless. There are many types of collision processes and, thus, many collisionopauses are possible, located at different cometocentric distances. The collisionopause for a particular process is located where the transport time, τ_T, is equal to the collision time, τ_c. The transport time is representative of dynamical processes in general and can be approximated as:

$$\tau_T \approx r / u \qquad (1)$$

where r is the cometocentric distance and u is the flow speed. The flow speed near the subsolar region varies with distance roughly as $u(r) = u_2$ (r

/ R_s), where $u_2 \approx 100$ km/s is the post-shock flow speed, and R_s is the distance to the shock [Gombosi, 1987]. Equation (1) then indicates that τ_T is about 3000 s everywhere inside the shock on the subsolar line, although τ_T increases to values of about 10^4 s in the inner coma if more accurate values of the flow speed are used in this region.

The collision time for an ion-neutral interaction process, p, is roughly:

$$\tau_c \approx [<v\,\sigma_p>n_n]^{-1} \qquad (2)$$

where v is the relative ion-neutral speed, σ_p is the relevant ion-neutral interaction cross-section, and n_n is the neutral density. The average is taken over the appropriate distribution function (e.g., shell or Maxwellian). τ_c becomes smaller with decreasing cometocentric distance, mainly due to the increasing neutral density, which for a neutral species, a (e.g., H_2O, CO_2, etc.), is given by:

$$n_{na} = \frac{Q_a}{4\pi u_n r^2}\,e^{-r/(u_n/I_a)} \qquad (3)$$

Q_a is the total gas production rate of species a, u_n is the neutral outflow speed ($u_n \approx 1$ km/s), and $I_a \approx 10^{-6}$ s^{-1} is the ionization frequency of that species.

The cometopause can be considered the collisionopause for charge transfer of solar wind protons. For comet Halley, the cometopause was located at a distance of roughly 10^5 km. The collisionopauses for other processes are located closer to the nucleus [Cravens, 1989a, 1990]. The collisionopause is not a sharp boundary but a fuzzy region across which collisions become increasingly important.

3. Plasma Description

3.1 Boltzmann Equation

A particle population, a, can be described by its single particle distribution function, $f_a(\mathbf{x}, \mathbf{v}, t)$, where $\mathbf{x}$ is the position vector and $\mathbf{v}$ is the velocity vector. This distribution function can be found by solving the Boltzmann equation (c.f. Krall and Trivelpiece, 1973):

$$\frac{Df_a}{Dt} = \frac{\partial f_a}{\partial t} + \mathbf{v}\cdot\nabla f_a + \frac{q}{m_a}(\mathbf{E}+\mathbf{v}\times\mathbf{B})\cdot\nabla_v f_a = \left(\frac{\delta f_a}{\delta t}\right)_{wave}$$
$$+ \left(\frac{\delta f_a}{\delta t}\right)_{collision} + S_a(\mathbf{x},\mathbf{v},t) \qquad (4)$$

where $\mathbf{E}$ and $\mathbf{B}$ are the macroscopic electric and magnetic fields, respectively. The first and second terms on the right-hand side of (4) are the time rate of change of the distribution function due to wave electric and magnetic fields or to collisions, respectively. The last term is a source term. For most space physics applications, the gyrotropic assumption is a good one, in which case the distribution function, $f_a(\mathbf{x},v,\mu,t)$, can be written in terms of the particle speed, v, and the cosine of the pitch-angle, μ. The source term for cometary pickup ions in the solar wind reference frame is given by:

$$\text{Source:} \qquad S_a = I_a\,n_{na}(r)\,\delta(\mu-\mu_0)\,\delta(v-u_{sw}) \qquad (5)$$

$n_{na}(r)$ is the neutral density of species as given by equation (3), I_a is the ionization frequency of species a, and the other factors are delta functions placing the newly-born cometary ions at a speed of the local solar wind flow and at a pitch-angle $\mu_0 = \cos\theta_{vB}$. θ_{vB} is the angle between the solar wind and the interplanetary magnetic field.

The wave-particle term in gyrotropic form is given by:

$$\left(\frac{\delta f_a}{\delta t}\right)_{wave} = \frac{\partial}{\partial\mu}\left[D_{\mu\mu}\frac{\partial f_a}{\partial\mu}\right] + \frac{1}{v^2}\frac{\partial}{\partial v}\left[v^2\,D_{vv}\frac{\partial f_a}{\partial v}\right] \qquad (6)$$

This simplified form of the wave-particle term has omitted cross-terms. The quasi-linear pitch-angle diffusion term for ion cyclotron type waves looks like:

$$D_{\mu\mu} = \frac{\pi e^2}{m_i^2 c^2}\int\frac{dk}{2\pi}\left|B_k^\pm\right|^2(1-\mu^2)\cdot\delta(\omega_k - k\,\mu v \pm\Omega_i) \qquad (7)$$

k is the wave number for the magnetic fluctuations and $|B_k^\pm|^2$ is the power at that wave number, where the $\pm$ refers to the polarization of the waves. In the cometary environment the power spectrum has been measured (e.g., Tsurutani and Smith, 1986). The waves are produced via an ion cyclotron type instability associated with the ring-beam instability [Wu and Davidson, 1972]. The delta function indicates that particles interact with the waves at a resonant wave number. ω_k is the wave frequency, which for simple parallel propagating Alfven waves is just equal to kV_A. Ω_i is the cometary ion gyrofrequency. A detailed discussion of wave-particle interactions does not belong in this paper and has been provided in many review papers [Sagdeev et al., 1986; Galeev et al., 1986; Lee, 1989; Ip and Axford, 1989]. But as mentioned in the introduction, the net effect of wave particle interactions is to first pitch-angle scatter the ions and create a shell distribution, and then to produce energy diffusion and broaden the shell.

The collision term can be written as:

$$\left(\frac{\delta f_a}{\delta t}\right)_{collision} = -\frac{\partial}{\partial\mathbf{v}}\cdot[\mathbf{A}_{ii}f_a] + \frac{\partial}{\partial\mathbf{v}}\cdot\left[\frac{\partial}{\partial\mathbf{v}}\cdot(\mathbf{B}_{ii}f_a)\right]$$
$$\underbrace{\qquad\qquad}_{\substack{\text{Fokker-Planck terms}\\ \text{(Coulomb collisions)}}}$$
$$-\underbrace{\langle\tilde{v}\,\sigma_{ct}\rangle\,n_n f_a + \dots.}_{\text{charge-exchange}} \qquad (8)$$

The last term is the rate of charge-transfer loss of species a at velocity v, and the average < > is taken over the distribution function of neutrals. The velocity in the average is the relative velocity between ion and neutral particles. Other collision terms are possible, including recombination and chemical reactions for cold plasma, but are not shown. In this paper, charge transfer and Coulomb collisions will be emphasized. $\mathbf{A}_{ii}$ and $\mathbf{B}_{ii}$ are the coefficients of dynamic friction and diffusion, respectively, for ion-ion Coulomb collisions. The expressions for these Fokker-Planck coefficients are given in many places including Nicholson [1983], and they include integrals over the target ion distribution function. Rather than reproduce these terms here, the effects of the Fokker-Planck term will be simply summarized.

3.2 Coulomb Collisions

Coulomb collisions between charged particles have two effects on a distribution function. First, these collisions act to isotropize the distribution function, and second, they act to thermalize the distribution [Spitzer, 1962]. The cumulative effect of a large number of distant collisions is more important than the few close strong encounters which take place typically. The thermalization process converts non-Maxwellian distributions to Maxwellian distributions, preserving the overall energy. Thermalization and isotropization occur with approximately the same time constant for collisions of particles within the same species, but thermalization is much slower than isotropization for interactions between particles of two species with significantly different masses. The time constant for thermalization of a species by Coulomb collisions was given by Spitzer [1962]:

$$\tau_{ii} = 1 / \nu_{ii} \cong \frac{11.4\ A_i^{1/2}\ T_i^{3/2}}{n_i\ Z^4\ \ln\Lambda} \tag{9}$$

where the "Coulomb logarithm" depends on the electron density and temperature and is a slowly varying function, approximately equal to $\ln\Lambda \approx 20$ for cometary conditions. T_i is the ion temperature -- or the effective temperature of the cometary shell distribution. A_i is the mass of the ion species in amu, n_i is the ion density, and Z is the charge of the ion. For the inner coma, $A_i \approx 18$ and $Z = 1$. Note that the Coulomb collision time increases with increasing ion temperature, indicating that this type of collision should be most important in the inner coma, where the ions are less energetic.

3.3 Fluid Equations -- Plasma Dynamics

A fluid approach is satisfactory for many treatments of space plasmas. For instance, the dynamics of the solar wind mass-loaded with cometary ions does not depend on the details of the particle distribution function, but does depend on the pressure and density associated with the distribution function. The fluid conservation equations can be found by taking the moments of the distribution function, yielding for each particle species a continuity, a momentum, and an energy equation (and further equations if the procedure is not truncated using a heuristic heat flux). One specialized set of fluid equations is the set of single-fluid magnetohydrodynamic equations. These equations have been very successful in describing the overall flow of plasma near a comet and have been discussed many times before (e.g., Schmidt et al., 1986; see the review papers cited in the introduction). What concerns us in this paper are the collisional aspects of these equations, since they are very important in the inner coma (c.f. Cravens, 1989a). These aspects will now be briefly summarized.

Collisions result in source and sink terms in the continuity equations. Sources for a particular species include ionization processes due to photon or electron collisions with neutrals and sources from ion-neutral chemical reactions. Sinks include loss via ion-neutral reactions and via electron-ion recombination. The most important collisional term in the momentum equation, for comets, is the ion-neutral frictional interaction term. For example, shortly after the encounter of the Giotto spacecraft with comet Halley in 1986, both Ip and Axford [1987] and Cravens [1986, 1987] recognized that the existence of the diamagnetic cavity is a consequence of the balance between an outward ion-neutral friction force on the plasma and an inward-directed magnetic pressure gradient force. The most important term in the ion energy equation is the ion-neutral frictional heating/cooling term, although ion-electron Coulomb energy transfer is also present. Some of these processes will be discussed further in section 5 of this paper.

4. Cometary Ion Distribution Function

4.1 Shell Distribution

The cometary ion distribution function in the magnetosheath is largely isotropized (i.e., a shell distribution function) due to wave-particle interactions. Figure 2 shows a cometary ion distribution function measured just outside the cometopause by the plasma analyzer on the Suisei spacecraft [Mukai et al., 1986a,b]. The distribution is shell-like, indicating significant levels of pitch-angle scattering behind the shock. The radius of the shell is equal to the local flow speed of about 50 km/s. The shell appears to be largely, albeit not completely, hollow, indicating the presence of some energy diffusion. This distribution was measured at a cometocentric distance of 1.8×10^5 km where collisional processes are not yet important, but inside this distance these processes become increasingly important.

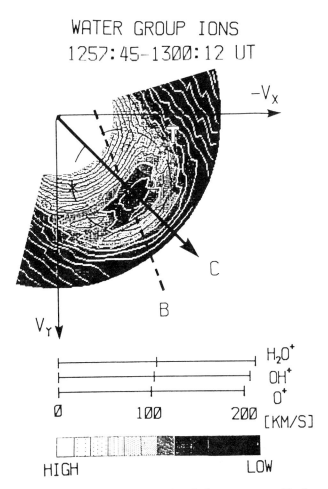

Fig. 2. Distribution of cometary ions in velocity space measured by the electrostatic analyzer onboard the Suisei spacecraft at a distance of 1.8×10^5 km from the nucleus of comet Halley. The distribution is primarily shell-like. From Mukai et al. [1986a].

4.2 Charge Transfer Time Constants

Time constants for charge-transfer of ions with cometary neutrals were calculated as a function of cometocentric distance using equation (2) for the energies shown (Figure 3). The ion was assumed to be O^+ but the 10 keV curve in Figure 3 also gives reasonable charge transfer collision times for solar wind (1 keV) protons. A "generic" charge-transfer cross section of $\sigma = 10^{-15}$ cm^2 was used and is within a factor of 2 of reasonable cross sections for various processes such as charge-transfer of O^+, H_2O^+, and H^+ with H_2O or O. Cross sections for resonant charge-transfer reactions (e.g., O^+ with O) are about a factor of 2 larger than the assumed value and for highly non-resonant reactions (e.g., He^{++} with O), the assumed cross section is too large, especially at lower energies. Time constants for the polarization interaction of cold H_2O^+ or H_3O^+ with H_2O are very similar to the time constants shown for the lowest energy (≈ 1 eV) in Figure 3.

The charge transfer collision time for each energy varies as r^2 due to the r-dependence of the neutral density as given by equation (3). An estimated transport time as a function of r, as given by equation (1), is also shown in Figure 3. For solar wind protons (the 10 keV curve) the charge transfer time becomes less than the transport time near $r \approx 10^5$ km, indicating that the collisionopause for this process (that is, the cometopause) is located

Charge Exchange Times

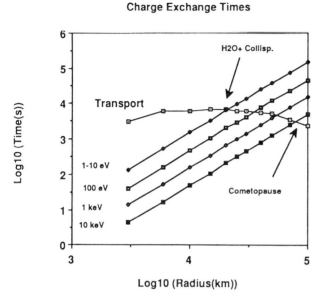

Log10 (Radius(km))

Fig. 3. Collision times versus cometocentric distance for charge-transfer of heavy cometary ions (e.g., O^+ or H_2O^+) with cometary neutrals at comet Halley for several ion energies. The 10 keV curve is also relevant to the charge transfer of solar wind protons ($\approx$ 1 keV energies) with cometary neutrals. An approximate transport time (or time for the plasma to convect through the region of interest) is shown. The location of the cometopause is indicated, as is the H_2O^+ collisionopause which separates the photochemical and transport regions for this ion species. The 1-10 eV curve was calculated for ion-neutral elastic (polarization) collisions rather than for charge-transfer.

near 10^5 km. Gombosi [1987] and Ip [1989] have made careful calculations which show that solar wind proton density decreases relatively rapidly at about this distance; however, the transition does cover $1\text{-}2 \times 10^5$ km, indicating that the cometopause is really a rather wide transition region, rather than a narrow boundary (see discussion in Cravens, 1989a). Cometopause is thus a somewhat misleading term, but it is commonly used now. Ip's calculations were performed along streamlines taken from a global MHD model. Gombosi's calculations were one-dimensional and for the stagnation line, but included the dynamics self-consistently. The first observational evidence for the cometopause came from the PLASMAG instrument on the VEGA spacecraft [Gringauz et al., 1986a,b]. Figure 4 shows data from the ion mass spectrometer (IMS) on Giotto (c.f. Ip and Axford, 1989), which illustrates the disappearance of solar wind protons. The PLASMAG instrument saw the cometopause as a sharp transition (or boundary) in composition, whereas the compositional transition seen in the Giotto data (Figure 4) was quite broad ($\Delta r/r \approx 0.5$). Perhaps transient solar wind structures and/or instrumental observing geometries can account for the difference.

Hot cometary O^+ ions born upstream of the shock also have the time constants indicated by the 10 keV curve, and these ions are removed from the flow near a distance of 10^5 km. This hot population accounts for most of the pressure in the magnetosheath and the charge transfer loss of these ions affects the dynamics by cooling the plasma [Galeev et al., 1985; Wallis and Ong, 1975]. Heavy cometary ions born downstream of the shock have energies approximately 1 keV, with a collisionopause located about 20,000 km inside the nominal cometopause. These warm ions were measured by the HERS spectrometer and are shown in Figure 4. The O^+ ions measured by HIS are either "luke-warm" or cold, and the

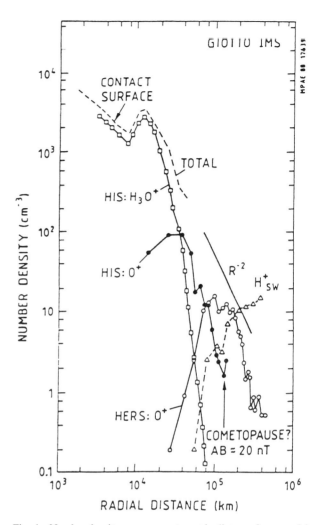

Fig. 4. Number density versus cometocentric distance for several ion species measured at comet Halley by the HIS and HERS sensors on the Giotto ion mass spectrometer [Balsiger et al., 1986; Shelley et al., 1987]. From Ip and Axford [1989].

density of this component starts to increase in the general vicinity of where the other O^+ population is starting to disappear. The charge-transfer time for these ions ($E \approx 100$ eV or less) is ten times larger than for 10 keV ions. H_3O^+ ions make an appearance starting at $r \approx 30,000$ km (Figure 4), which indicates that ion-neutral chemical time constants are becoming small (the 1-10 eV curve in Figure 3) since this ion is only produced by the reaction of H_2O^+ ions with neutral H_2O. Chemical processes are best treated using fluid theory and are discussed briefly in section 5.

4.3 Coulomb Collision Time Constants

The distribution function of cometary ions should remain shell-like, albeit cooler due to charge-exchange, downstream of the cometopause. The transition from a shell distribution to a Maxwellian distribution requires that the Coulomb thermalization time become less than the transport time. Heavy cometary ion thermalization time constants from equation (9) are shown in Figure 5 as a function of distance for several values of ion energy. Ion densities were adopted from the profiles

Coulomb Collision Times

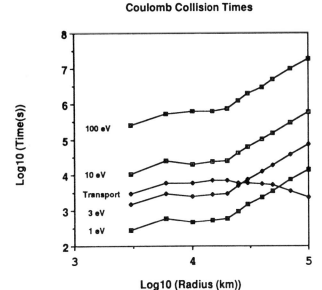

Ion Thermal Energy

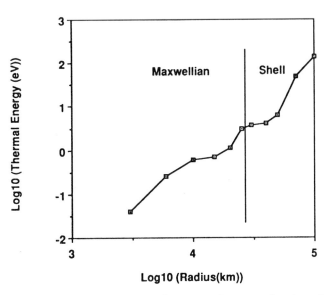

Fig. 5. Relaxation, or thermalization, time constants for Coulomb collisions of cometary ions, as a function of cometocentric distance, for several ion energies at comet Halley. An approximate plasma transport time is also shown.

Fig. 6. The thermal energy of cometary ions as a function of cometocentric distance for comet Halley. These energies were found using measured ion temperatures for smaller distances, and estimated using reasonable flow speeds at larger distances. The approximate transition where the cometary ion distribution evolves from shell-like to Maxwellian is indicated (see the text).

measured by the Giotto ion mass spectrometer [Balsiger et al., 1986; Schwenn et al., 1987]. The thermalization times become larger close to the nucleus due to larger densities, and these times are large for high ion energies due to the energy/temperature dependence of equation (9). 1 eV ions should thermalize near 60,000 km and 3 eV ions near 30,000 km, according to Figure 5. Ions more energetic than about 10 eV have thermalization times which are always larger than the transport time.

The actual evolution of the ion distribution function in the inner coma depends on what remains in the distribution after the charge transfer process. Figure 6 shows the ion thermal energy, which was derived from measured ion temperatures [Balsiger et al., 1986; Schwenn et al., 1987] for small values of r, and was estimated from the local flow speed for large values of r. Comparison of Figures 5 and 6 indicates that for cometocentric distances greater than roughly 3×10^4 km, the cometary ions are too energetic to thermalize in a time less than the transport time; and the distribution remains shell-like. But for distances less than $\approx 3 \times 10^4$ km, the ions are cool enough to have relatively short thermalization times, and the distribution function becomes Maxwellian.

5. The Inner Coma

The MHD equations provide a good description of important aspects of plasma behavior such as the overall density and the flow speed. Collisional terms in these equations are particularly important inside $r \approx 3 \times 10^4$ km, where the cometary ions are largely thermalized. These collision terms were discussed in section 4. Observations and theory of the plasma in the inner coma have been extensively reviewed elsewhere [e.g., Mendis et al., 1985; Ip and Axford, 1989; Cravens, 1986, 1989a,d, 1990]. Only two examples of collisional processes will be included in this section.

5.1 Chemistry

The most abundant cometary ion species observed in the inner coma of comet Halley by the Giotto IMS was H_3O^+, and not H_2O^+ [Balsiger et

al., 1986], even though H_2O is the most abundant neutral constituent, as a consequence of the ion-neutral chemical reaction (2). A large number of other ion species were also observed by the IMS in the inner coma, and the chemistry required to explain the density profiles of these ions is rather elaborate and includes dozens of chemical reactions [Mendis et al., 1985; Allen et al., 1987; Huebner, 1985]. Close to the nucleus, photochemical equilibrium (i.e., local balance between production and loss for a particular ion species) provides a good means of determining ion densities, and transport can be neglected. The collisionopause marking this transition is at a different location for each ion species, and depends on the chemical time constant.

Consider H_2O^+ ions. This ion is mainly produced directly via photoionization of H_2O and is chemically removed by reaction with neutral water. The photochemical equilibrium density profile can be estimated to be $n \approx 500$ cm^{-3} [Cravens, 1986], which should be appropriate inside the H_2O^+ collisionopause. The chemical time constant is $t_c \approx 1/(2 \times 10^{-9} n_n)$, which becomes equal to the transport time at a cometocentric distance of roughly 40,000 km (i.e., the H_2O^+ collisionopause).

Now consider H_3O^+ ions. This is the major ion species in the inner coma and the electron density is approximately equal to the H_3O^+ density Reaction of H_2O^+ with H_2O is the main source and dissociative recombination (reaction (5)) provides the main loss. The photochemical expression for the density varies inversely as r, as was indeed observed by the Giotto IMS out to a distance of about $1-1.5 \times 10^4$ km, where the density appears to be enhanced [Balsiger et al., 1986]. This density enhancement might be a signature of the H_3O^+ collisionopause, although there are also other explanations (c.f. Cravens, 1989a, 1990).

5.2 Ion Energetics

Cometary ions outside the cometopause are energetic and their distribution function is shell-like, but in the inner coma, these ions are removed from the flow by charge exchange collisions and are replaced by colder ions picked-up locally. Collisions are important in determining the nature of the distribution function which evolves into a Maxwellian as discussed earlier.

The ion temperature can be found theoretically in the inner coma by solving a standard energy equation [Banks and Kockarts, 1973]. The ion energetics are dominated by collisional processes in the inner coma. Cravens [1987] and Haerendel [1987] demonstrated that frictional heating by ion-neutral collisions balances cooling by ion-neutral collisions in the inner coma. The frictional heating occurs due to the relative motion of the outflowing neutrals past cometary ions stagnating in the magnetic barrier just outside the boundary of the diamagnetic cavity. Figure 7

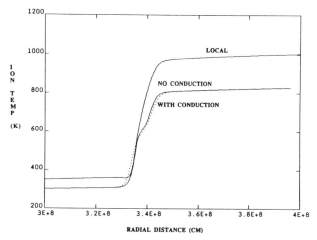

Fig. 7. Ion temperature profiles in the inner coma of comet Halley calculated theoretically (from Cravens, 1989d). Three cases are shown. The "local" case only included local balance between heating and cooling. The other cases were from solutions of the ion energy equation -- with and without heat conduction included. The jump in the ion temperature takes place at the boundary of the diamagnetic cavity, where the plasma flow speed goes from the neutral outflow speed to zero.

shows some calculated ion temperature profiles from the one-dimensional MHD model of Cravens [1989d], in which the continuity, momentum, and ion energy equations plus the magnetic induction equation were solved. Ion temperature profiles for three cases are shown: (1) A simple analytical formula including only collisional heating and cooling terms is used [Cravens 1987; Haerendel, 1987]. (2) The energy equation includes all terms except heat conduction. (3) The energy equation includes all terms including a heat conduction term which incorporates inhibition of the heat flux by the magnetic field. In all these cases, the ion temperature increases sharply at the boundary of the diamagnetic cavity, as observed by Giotto [Schwenn et al., 1987].

5.3 Electron Energetics

Electrons outside the inner coma are basically solar wind electrons [Reme et al., 1986; Gringauz et al., 1986a]. Collisions modify the superthermal electron distribution and a cold electron population forms in the inner coma. The electron temperature, T_e, of the cold electron component of the plasma has been calculated many times for the unmagnetized cometary ionosphere (e.g., Marconi and Mendis, 1986; Korosmezey et al., 1987; see the review by Cravens, 1990), and was calculated for the magnetized inner coma by Gan and Cravens [1990]. Collisional cooling processes were listed in section 2: electron-neutral

elastic collisions, electron-ion Coulomb collisions, rotational excitation of H_2O by electron impact, vibrational excitation by electron impact, and electronic excitation.

Electron cooling rates (per electron and per neutral) from Gan and Cravens are shown in Figure 8. Rotational and vibrational cooling dominate for electron temperatures less than $\approx 3 \times 10^4$ K (≈ 3 eV) and

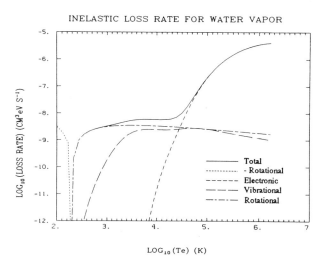

Fig. 8. Electron cooling rates, per molecule and per electron, for water vapor as a function of electron temperature. From Gan and Cravens [1990]. The rotational cooling rate is from Cravens and Korosmezey [1986].

electronic excitation (including ionization) dominates for higher temperatures. The cold electron population is mainly heated by Coulomb collisions with superthermal electrons, which include both solar wind electrons which make it into the inner coma along magnetic field lines and the photoelectrons associated with photoionization. The cold electron heating rate, per neutral and per electron, increases with cometocentric distance. At small values of r, this heating rate is able to be balanced by rotational and vibrational cooling and T_e remains below 10^4 K; however, with increasing distance the heating rate must be balanced by electronic cooling, which requires much higher temperatures. Figure 9 shows calculated electron temperature profiles as a function of distance along the magnetic field for several different field lines. The electron temperature remains low for cometocentric distances less than $\approx 1.5 \times 10^4$ km, as a consequence of collisional cooling. Outside this distance, called the electron thermal collisionopause [Gan and Cravens, 1990; Cravens et al., 1987], T_e exceeds 10^4 K. In fact, at large enough distances, the temperature of the cold component is so large, and the density so low, that these electrons are no longer distinct from the solar wind electron population.

The cometary ion density measured by the Giotto IMS is enhanced in the vicinity of $r \approx 10^4$ km. Ip et al. [1987] suggested several possible explanations for this enhancement, including a localized reduction in the ion loss rate caused by an increase in the electron temperature. Korosmezey et al. [1987] also mentioned this possibility. The rate coefficient for electron-ion dissociative recombination decreases with increasing electron temperature. The sharp gradient of electron temperature at $r \approx 10^4$ km, calculated by Gan and Cravens [1990] and shown here in Figure 9, might provide the explanation for the ion density enhancement seen by Giotto.

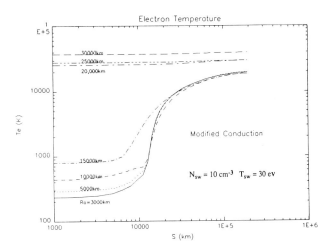

Fig. 9. Theoretically calculated electron temperatures as a function of distance along the magnetic field line, in the inner coma of comet Halley. Electron temperature profiles for several field lines are shown. These field lines were assumed to be parabolic in shape, and the distances of closest approach to the nucleus ranged from r = 5000 km (lowest temperatures) up to 30,000 km (highest temperatures). From Gan and Cravens [1990].

6. Summary

The plasma-comet interaction is largely collisionless at large cometocentric distances.   However, collisional processes become important in the inner coma (within the cometopause). Collisional processes include charge-transfer between solar wind protons and neutrals, ion-neutral friction, electron and ion thermal cooling, and ion-neutral chemistry. Thermalization of the cometary ion distribution function by Coulomb collisions is an important process in the inner coma of an active comet. The distribution function is shell-like for cometocentric distances greater than $\approx$ 30,000 km, but is Maxwellian for smaller distances, where the Coulomb collision time is less than the plasma transport time.

Acknowledgments. This work was supported by NASA grants NAGW 1588 and NAGW 1559, and by NSF grant ATM-8996116.

References

Allen, M., M. Delitsky, W. Huntress, Y. Hung, W.-H. Ip, R. Schwenn, H. Rosenbauer, E. Shelley, H. Balsiger, and J. Geiss, Evidence for methane and ammonia in the coma of comet P/Halley, *Astron. and Astrophys.*, *187*, 502, 1987.

Balsiger, H., K. Altwegg, F. Bühler, J. Geiss, A. G. Ghielmetti, B. E. Goldstein, R. Goldstein, W. T. Huntress, W.-I. Ip, A. J. Lazarus, A. Meier, M. Neugebauer, U. Rettenmudd, H. Rosenbauer, R. Schwenn, R. D. Sharp, E. G. Shelley, E. Ungstrup, and D. T. Young, Ion composition and dynamics at comet Halley, *Nature, 321*, 330, 1986.

Banks, P. M., and G. Kockarts, *Aeronomy*, Academic Press, New York, 1973.

Cravens, T. E., The physics of the cometary contact surface, Proceedings of the 20th ESLAB Symposium on the Exploration of Halley's Comet, edited by B. Battrick, E. J. Rolfe and R. Reinhard, *Eur. Space Agency Spec. Publ. ESA SP-250, 1*, 241, 1986.

Cravens, T. E., Theory and observations of cometary ionospheres, *Adv. Space Res., 7*, 147 1987.

Cravens, T. E., Cometary Plasma Boundaries, *Adv. Space Res., 9*, 3293, 1989a.

Cravens, T. E., The solar wind interaction with non-magnetic bodies and the role of small-scale structures, *Solar System Plasma Physics,* edited by J. H. Waite, Jr., J. L. Burch, and R. L. Moore, Geophysical Monograph 54, AGU, Washington, pp. 353-366, 1989b.

Cravens, T. E., Test particle calculations of pick-up ions in the vicinity of comet Giacobini-Zinner, *Planet. Space Sci.,* *37*, 1169, 1989c.

Cravens, T. E., A one dimensional magnetohydrodynamical model of the inner coma of comet Halley, *J. Geophys. Res., 94*, 15025, 1989d.

Cravens, T. E., Plasma processes in the inner coma, proceedings of IAU Colloquium No. 116 on Comets in the post-Halley Era, International Astronomical Union, in press, 1990.

Cravens, T. E., and A. Korosmezey, Vibrational and rotational excitation cooling of electrons by water vapor, *Planet. Space Sci., 34*, 961, 1986.

Cravens, T. E., J. U. Kozyra, A. F. Nagy, T. I. Gombosi, and M. Kurtz, Electron impact ionization in the vicinity of comets, *J. Geophys. Res., 92*, 7341, 1987.

Galeev, A. A., Theory and observations of solar wind/cometary plasma interaction processes, in Proceedings of the 20th ESLAB Symposium on the Exploration of Halley's Comet, edited by B. Battrick, E. J. Rolfe and R. Reinhard, *Eur. Space Agency Spec. Publ. ESA SP-250, 1*, 3, 1986.

Galeev, A. A., T. E. Cravens, and T. I. Gombosi, Solar wind stagnation near comets, *Astrophys. J.,* *289*, 807, 1985.

Galeev, A. A., K. I. Gringauz, S. I. Klimov, A. P. Remizov, R. Z. Sagdeev, S. P. Savin, A. Yu. Sokolov, M. I. Verigin, and K. Szego, Critical ionization velocity effects in the inner coma of comet Halley: Measurements by Vega-2, *Geophys. Res. Lett., 13*, 845, 1986.

Gan, Lu, and T. E. Cravens, Electron energetics in the inner coma of comet Halley, *J. Geophys. Res., 95*, 6285, 1990.

Gombosi, T. I., Charge exchange avalanche at the cometopause, *Geophys. Res. Lett., 14*, 1174, 1987.

Gringauz, K. I., T. I. Gombosi, A. P. Remizov, I. Apáthy, I. Szemerey, M. I. Verigin, L. I. Denchikova, A. V. Dyachkov, E. Keppler, I. N. Klimenko, A. K. Richter, A. J. Somogyi, K. Szegö, S. Szendrö, M. Tátrallyay, A. Varga, and G. A. Vladimirova, First in situ plasma and neutral gas measurements at comet Halley, *Nature, 321*, 282, 1986a.

Gringauz, K. I., T. I. Gombosi, M. Tatrallyay, M. I. Verigin, A. P. Remizov, A. K. Richter, I. Apathy, I. Szemerey, A. V. Pyachkov, O. V. Balakina, and A. F. Nagy, Detection of a new "chemical" boundary at comet Halley, *Geophys. Res. Lett., 13*, 613, 1986b.

Haerendel, G., Plasma transport near the magnetic cavity surrounding comet Halley, *Geophys. Res. Lett., 14*, 673, 1987.

Huebner, W. F., The photochemistry of comets, in *The Photochemistry of Atmospheres*, Academic Press, New York, p. 437, 1985.

Ip, W.-I., On the charge exchange effect in the vicinity of the cometopause of comet Halley, *Astrophys. J., 343*, 946, 1989.

Ip, W.-H. and W. I. Axford, Theories of physical processes in the cometary comae and ion tails, in *Comets*, edited by L. L. Wilkening, University of Arizona Press, Tucson, p. 588, 1982.

Ip, W.-II., and W. I. Axford, The formation of a magnetic field free cavity at comet Halley, *Nature, 325*, 418, 1987.

Ip, W.-H., and W. I. Axford, Cometary plasma physics, in *Physics of Comets in the Space Age*, edited by W. F. Huebner, Springer-Verlag, 1990.

Ip, W.-H., R. Schwenn, H. Rosenbauer, H. Balsiger, M. Neugebauer, and E. G. Shelley, An interpretation of the ion pile-up region outside the ionospheric contact surface, *Astron. Astrophys., 187*, 132, 1987.

Körösmezey, A., T. E. Cravens, T. I. Gombosi, A. F. Nagy, D. A. Mendis, K. Szegö, B. E. Gribov, R. Z. Sagdeev, V. D. Shapiro, and V. I. Shevchenko, A comprehensive model of cometary ionospheres, *J. Geophys. Res., 92*, 7331, 1987.

Krall, N. A., and A. W. Trivelpiece, *Principles of Plasma Physics*, McGraw-Hill, New York, 1973.

Lee, M. C., Ultra-low frequency waves at comets, in *Plasma Waves and Instabilities at Comets and in Magnetospheres*, edited by B. Tsurutani and H. Oya, Geophysical Monograph 53, AGU, Washington, p. 239, 1989.

Marconi, M. L., and D. A. Mendis, The electron density and temperature in the tail of comet Giacobini-Zinner, *Geophys. Res. Lett.*, *13*, 405, 1986.

Mendis, D. A., H. L. F. Houpis, and M. L. Marconi, The physics of comets, *Fund. Cosmic Phys.*, *10*, 1, 1985.

Mukai, T., W. Miyake, T. Terasawa, M. Kitayama, and K. Kirao, Plasma observation by Suisei of solar wind interaction with comet Halley, *Nature*, *321*, 299, 1986a.

Mukai, T., W. Miyake, T. Terasawa, M. Kitayama, and K. Kirao, Ion dynamics and distribution around comet Halley: Suisei observation, *Geophys. Res. Lett.*, *13*, 829, 1986b.

Neubauer, F. M., K. H. Glassmeier, M. Pohl, J. Raeder, M. H. Acuna, L. F. Burlaga, N. F. Ness, G. Musmann, F. Mariani, M. K. Wallis, E. Ungstrup, and H. U. Schmidt, First results from the Giotto magnetometer experiment at comet Halley, *Nature,* *321*, 352, 1986.

Nicholson, D. R., *Introduction to Plasma Theory,* John Wiley and Sons, New York, 1983.

Reme, H., J. A. Sauvaud, C. D'Uston, F. Cotin, A. Cros, K. A. Anderson, C. W. Carlson, D. W. Curtis, R. P. Lin, D. A. Mendis, A. Korth, and A. K. Richter, Comet Halley--Solar wind interaction from electron measurements aboard Giotto, *Nature*, *321*, 349, 1986.

Sagdeev, R. Z., V. D. Shapiro, V. I. Shevchenko, and K. Szego, MHD turbulence in the solar wind-comet interaction region, *Geophys. Res. Lett.*, *13*, 85, 1986.

Schmidt, H. U., R. Wegmann, and F. M. Neubauer, MHD model for comet Halley, Proceedings of the 20th ESLAB Symposium on the Exploration of Halley's Comet, edited by B. Battrick, E. J. Rolfe and R. Reinhard, *Eur. Space Agency Spec. Publ. ESA SP-250, 1*, 43, 1986.

Schwenn, R., W.-H. Ip, H. Rosenbauer, H. Balsiger, F. Buhler, R. Goldstein, A. Meier, and E. G. Shelley, Ion temperature and flow profiles in comet P/Halley's close environment, *Astron. Astrophys.*, *187*, 160, 1987.

Shelley, E. G., S. A. Fuselier, H. Balsiger, J. F. Drake, J. Geiss, B. E. Goldstein, R. Goldstein, W.-I. Ip, A. J. Lazarus, and M. Neugebauer, Charge exchange of solar wind ions in the coma of comet P/Halley, *Astron. Astrophys.*, *187*, 304, 1987.

Spitzer, L., *Physics of Fully Ionized Gases,* Interscience Publishers/John Wiley, New York, 1962.

Terasawa, J., Particle scattering and acceleration in a turbulent plasma around comets, in *Plasma Waves and Instabilities at Comets and in Magnetospheres*, edited by B. Tsurutani and H. Oya, Geophysical Monograph 53, AGU, Washington, p. 41, 1989.

Tsurutani, B. T., and E. J. Smith, Strong hydromagnetic turbulence associated with comet Giacobini-Zinner, *Geophys. Res. Lett.*, *13*, 259, 1986.

Wallis, M. I., and R. S. B. Ong, Strongly cooled ionizing plasma flows with applications to beams, *Planet. Space Sci.*, *23*, 713, 1975.

Wu, C. S., and R. C. Davidson, Electromagnetic instabilities produced by neutral particles ionization in interplanetary space, *J. Geophys. Res.*, *77*, 5399, 1972.

THEORY AND SIMULATION OF COMETARY SHOCKS

N. Omidi

Department of Electrical and Computer Engineering, University of California, San Diego
La Jolla, California 92093

D. Winske

Applied Theoretical Physics Division, Los Alamos National Laboratory
Los Alamos, New Mexico 87545

Abstract. In this paper recent kinetic simulation studies of cometary bow shocks are reviewed. Cometary shocks are formed due to solar wind mass loading by water group cometary ions. This process is kinetic in nature and varies as a function of the angle between the solar wind flow velocity and the interplanetary magnetic field, as do the properties of cometary shocks. For perpendicular and parallel orientations, quasi-steady shocks with dissipation scales on the order of proton inertial length form. At oblique orientations, no steady shocks are formed; instead, the transition from supersonic to subsonic flow takes place through multiple shocklets (steepened magnetosonic waves) which are generated by the pickup ions via the resonant electromagnetic ion beam instability. This new, time dependent model of cometary bow shocks is further investigated using a large scale kinetic simulation and is compared to the observations at Comet Giacobini-Zinner.

1. Introduction

One of the important aspects of the solar wind interaction with comets is the nature of the bow shock which may form thousands of kilometers away from the nucleus. Early investigations [e.g., Wallis, 1971] had suggested that cometary shocks may be either weak or non-existent, and questions of similar type persist to the present [Cargill et al., 1988]. However, the nature of discussions have to a large degree changed. Earlier studies were concerned with fluid effects and whether they could lead to a smooth transition of solar wind from a supersonic to a subsonic flow [e.g., Wallis, 1971; Schmidt and Wegmann, 1982]. Given the kinetic nature of solar wind mass loading and the resulting waves and turbulence, the focus of discussion has shifted to whether these effects can modify the shock structure or lead to its absence altogether. The remainder of this paper will be concerned with the later issues which determine the nature of cometary shocks. In general, one may expect cometary bow shocks to be different from their planetary counterparts for a number of reasons. For example, the nature of the driving piston is different in that planetary magnetospheres (or ionospheres in the case of planets with small intrinsic magnetic field) act as a hard piston deflecting the solar wind. Comets, however, interact with the solar wind over an extended region resulting in a more gradual deceleration. Similarly, planetary shocks decouple from their deriving piston (i.e. the magnetopause or the ionopause), while the cometary shocks are formed in the mass loading region and as a result stay coupled with the piston. Another consequence is the modification of the usual Rankine-Hugoniot relations due to the presence of source terms (i.e., cometary ion production) in the conservation laws. Given that ion dynamics are known to control the structure of collisionless shocks, it is also reasonable to assume that cometary shocks may have a unique structure due to the presence of multiple ion species. Finally, one may expect cometary shocks to be different because of the presence of large amplitude waves and turbulence due to pickup instabilities. Although in this regard, they may not be as different from the planetary quasi-parallel shocks, which are also turbulent in nature.

With the above possibilities in mind, it is clear that cometary bow shocks are influenced by a variety of nonlinear phenomena which can only be incorporated into a single model through numerical simulations. As in the case of collisionless shocks at planets, it is expected that the structure of cometary shocks is largely determined by ion kinetics, while electrons can be approximated as a charge neutralizing fluid. Thus, one may use an electromagnetic hybrid code with massless electrons and particle ions [Winske and Leroy, 1985] to investigate the nature of cometary shocks. In the recent years a number

Cometary Plasma Processes
Geophysical Monograph 61
©1991 American Geophysical Union

of such studies have been performed [Galeev and Lipatov, 1984; Omidi et al., 1986; Omidi and Winske, 1986, 1987, 1988, 1990] the results of which will be summarized in the next section. As will be shown, these results provide not only an insight into the nature of cometary shocks in general, but also suggest a time dependent multiple shock model that can account for some of the peculiar observations at Comet Giacobini-Zinner. This type of interaction has been further investigated in more detail through large scale, high resolution simulations. The preliminary results from such simulations are presented in section 3. Finally, in section 4 a comparison between these results and the observations at Comet Giacobini-Zinner is made and the outstanding issues regarding cometary shocks and the observations at Comet Halley are discussed.

2. Review

The first kinetic simulation studies related to cometary bow shocks were those by Galeev and Lipatov [1984], and Omidi et al. [1986]. In these studies hybrid simulations were used to investigate the effect of heavy water group ions (O^+) on the structure of quasi-perpendicular shocks. This effect was found to be dependent upon the distribution function of the heavy ions. Namely, a "foot" similar to the ones associated with planetary shocks but scaling with oxygen ion gyro-radius was formed, if the heavy ions formed a beam-ring distribution. On the other hand, O^+ ions with a Maxwellian distribution moving with the solar wind did not greatly modify the shock structure. Although these studies suggested that cometary shock structure may be influenced by the water group ions, they had a major short coming in that mass loading was not included in these studies and the shock was formed using similar techniques in studying planetary shocks (e.g., use of a hard piston).

To include the kinetic nature of solar wind mass loading in the formation of cometary shocks, Omidi and Winske [1986, 1987] performed simulations in which O^+ ions were created throughout the run with zero velocity in a region of simulation box. These ions were then allowed to interact with the flowing solar wind. Figure 1 depicts the model used in Omidi and Winske [1987] where solar wind is continuously injected from the left hand boundary and subsequently interacts with cometary ions whose production rate increases with distance into the box. Both the cometary ions and the solar wind are allowed to leave the box from the right hand boundary. The advantage of such a model lies in the fact that it includes the processes which can affect the nature of cometary shocks, described in the previous section. For example, given that all ions are allowed to exit from the right hand boundary, it is clear that no "hard" piston is present in the model and any shock formed in the system will be entirely due to solar wind mass loading. Of course, one implication of such a model is that while the simulation region consists of only a fraction of the total solar wind coma interaction region (the shaded region in Figure 1), it is nevertheless much larger than the systems used in studying planetary shocks

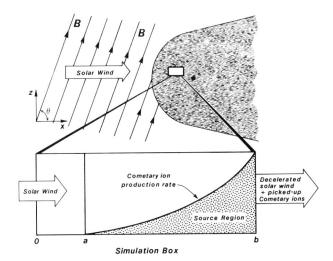

Fig. 1. Schematic of solar wind-comet interaction region (shaded area) and the way that a portion of it is modeled in the simulations. Solar wind protons are injected from the left boundary, cometary ions are created with a production rate that increases from left to right. All particles are allowed to leave the system from the right boundary.

(e.g., 100,000 km as compared to 10,000 km).

Investigation of solar wind mass loading with the model described above has shown that this process can vary with the angle between the solar wind flow direction and the interplanetary magnetic field (θ). At large angles, the motional electric field in the solar wind results in the pick up of cometary ions which in turn causes the solar wind to slow down [see Omidi and Winske, 1987]. When θ is small however, the motional electric field is negligible and the coupling between the solar wind and the cometary ions must take place through microscopic electromagnetic fields generated by plasma instabilities (see e.g., Wu and Davidson, 1972; see also Gary, 1989 for a review of electromagnetic ion beam instabilities). Given that cometary shocks are a direct consequence of solar wind mass loading, it is natural to expect the nature of the shock to also vary with the cone angle θ. Figures $2-4$ show the total magnetic field (B) and the solar wind density N_p for three simulation runs given in Omidi and Winske [1987] with different cone angles θ. In all three runs the simulation box is $1500c/\omega_p$ long where c is the speed of light and ω_p is the proton plasma frequency. Similarly, the simulations are one-dimensional in space and three-dimensional in velocities and electromagnetic fields. As can be seen in Figure 2 ($\theta = 90°$), both the magnetic field and density increase gradually from left to right until $X \sim 610$, where they go through a rapid and large jump which is identified as a shock. The increase in the field and density upstream of the shock is associated with a decrease in the flow velocity and is due to solar wind mass loading. Downstream of the shock, well correlated, large amplitude fluctuations in both the density and magnetic field are seen. It was shown by Omidi and Winske [1987] that

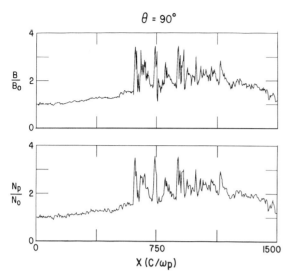

Fig. 2. Plot of total magnetic field and density as a function of distance for $\theta = 90°$. Note the shock at $X \sim 610$.

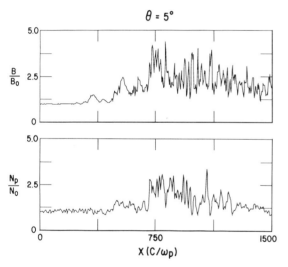

Fig. 3. Plot of total magnetic field and density as a function of distance for $\theta = 5°$. Note the shock at $X \sim 700$.

these fluctuations are due to the decay of transitory, solitary steepened waves that are formed due to macroscopic interactions between the solar wind and the cometary ions. One of the important conclusions to be drawn from this figure is the fact that the shock transition is quite sharp and the dissipation length scales with the proton inertial length. Although there exist no reason to discount the formation of such a shock at comets, observations at Comets Halley and Giacobini-Zinner did not find such a shock structure. As such, we will not pursue the details of this type of interaction here.

Figure 3 shows the density and magnetic field corresponding to a simulation run with a small cone angle ($\theta =$

5°). In this case the density and magnetic field increase in a nonuniform, uncorrelated fashion until $X \sim 700$, where they both undergo a large and sharp increase. This jump in the field and density is associated with a shock whose dissipation length scales with proton inertial length. The density and magnetic field fluctuations downstream of the shock are uncorrelated. As has been shown by Omidi and Winske [1987], the nature of this shock is quite different from the perpendicular case (Figure 2) in that the shock consists of large amplitude electromagnetic waves generated by the ion beam instability and not unlike the parallel shocks studied by Winske [1986] and Quest [1988] for planets. Although a recent study by Neubauer et al., [1990] suggests that Halley's bow shock during the Giotto outbound was quasi-parallel, it is not clear if these observations and our simulation results bear much similarity, especially in regard to the shock thickness (more on this in section 4).

The simulation results at intermediate cone angle ($\theta = 55°$) show yet another type of a transition region. As can be seen in Figure 4, the density and magnetic field show

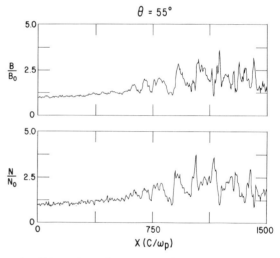

Fig. 4. Plot of total magnetic field and density as a function of distance for $\theta = 55°$. Note that no single shock is present.

a smooth and correlated increase from left to right until $X \sim 500$. Beyond this point, they continue to increase with large amplitude fluctuations superimposed. These fluctuations are correlated at smaller values of X, but become uncorrelated further into the box. The smooth increase in the density and magnetic field is similar to that seen in Figure 2 and is due to macroscopic mass loading of the solar wind. The fluctuations, on the other hand, are due to the electromagnetic waves generated by the microscopic interactions between the solar wind and the cometary ions. One of the striking aspects of this type of interaction is the fact that no single discontinuity can be identified as a shock transition. In other words, although the solar wind is both heated and decelerated as it goes across the simulation box, the transition occurs over a distance much larger than a proton inertial length. This is in contrast to the quasi-perpendicular and the quasi-parallel cases discussed earlier. It may be argued that this difference is related to the presence of large amplitude waves generated by the pick-up ions. However, we note that in the case of $\theta = 5°$ large amplitude waves were also present and yet a shock did form. This suggests that the presence of large amplitude waves alone does not constitute sufficient grounds for the absence of a shock. Instead, the difference in the two cases lies in the nonlinear evolution of the excited waves.

It was shown by Omidi and Winske [1987] that the electromagnetic waves generated during the quasi-parallel run are originally circularly polarized, as expected for parallel and nearly parallel propagating waves. As the waves grow and nonlinear interactions become more dominant, the waveforms become less regular and superimposed with shorter wavelength oscillations. Also the density and magnetic field fluctuations remain uncorrelated. The nonlinear evolution of the waves at larger values of θ is, however, considerably different. These obliquely propagating waves start with elliptical polarization and are compressional in their linear stage. The four panels in Figure 5 show the

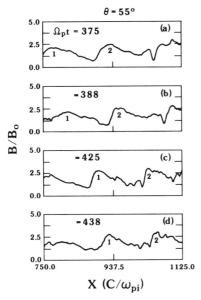

Fig. 5. The four panels show a segment of the total magnetic field at various times during the simulation. As can be seen, the waves steepened during their nonlinear evolution.

total magnetic field in a portion of the simulation box at various times for the $\theta = 55°$ run. The evolution of the waves can be seen from the top to the bottom panel. It is evident that these waves are steepened during their nonlinear phase. Thus, although both the parallel propagating and obliquely propagating waves are on the same dispersion branch and have similar frequencies and wavelengths, their nonlinear evolution is quite different.

As is well known, during the ICE encounter with Comet Giacobini-Zinner a variety of electrostatic and electromagnetic waves were detected over a wide spectrum. Among them are the low frequency (~ 0.01 Hz) electromagnetic waves reported by Tsurutani and Smith [1986] and later identified as magnetosonic wave by Tsurutani et al., [1987]. These waves were shown to be elliptically polarized at large distances from the nucleus where their amplitude was smaller. At closer distances, the waves were found to be steepened and associated with a higher frequency whistler wave train. The polarization was also found to have changed such that the steepened portion of the wave was closer to linear. Given that the obliquely propagating waves in our simulations also steepened during their nonlinear phase, it was suggested by Omidi and Winske [1987] that the type of interaction seen in Figure 4 was similar to that encountered by ICE spacecraft. Despite the similarities, however, a number of questions remained to be answered. One set of issues were related to the nonlinear evolution of the magnetosonic waves, e.g., the origin of the whistler waves (not seen in our simulations) as well as the change in the polarization of the steepened waves was still not clear. Other questions concerned the nature of the interaction itself and how the solar wind could make a transition from supersonic to subsonic flow without a standing bow shock.

To address the nonlinear evolution of the magnetosonic waves in more detail, new simulations were performed by Omidi and Winske [1988, 1990]. These simulations were performed in two steps, where originally a large system was used to generate a series of kinetic magnetosonic waves via the resonant electromagnetic ion beam instability. (We have referred to these waves as kinetic magnetosonic, because their dispersive properties are greatly modified by the beam as compared to the usual fluid mode.) Subsequently, one of these waves was isolated and its nonlinear evolution was studied in a periodic box with much higher spatial and temporal resolutions. Figure 6 shows the total magnetic field and the transverse components of an isolated wave in the periodic system. This

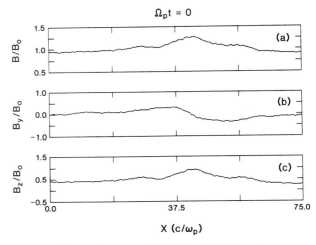

Fig. 6. The total magnetic field and its two transverse components associated with a kinetic magnetosonic wave are shown.

wave was generated by the interaction between the solar wind and newborn protons, as opposed to water group ions. We emphasize, however, that with the exception of the frequency and wavelength the excited waves and their nonlinear evolution are similar in both cases, as will be demonstrated later. Figure 7 shows the total magnetic

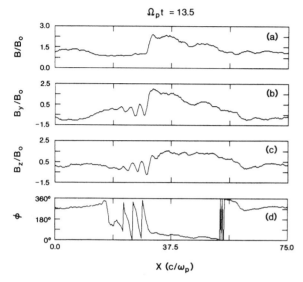

Fig. 7. The same as that in Figure 6 except at a later time. As can be seen the nonlinear evolution of the wave has resulted in its steepening and the generation of a shorter wavelength whistler wave packet. The bottom panel shows the magnetic helicity of the wave, illustrating the circular polarization of the whistlers and the near linear polarization of the steepened part of the magnetosonic wave.

field, its two transverse components, and the magnetic helicity (sense of rotation in space determined by ϕ, the angle between the transverse field vector and the Z-axis) of the wave at a later time in the simulation. As can be seen, the wave has steepened and because of sufficient resolution in the box has also developed a shorter wavelength whistler wavepacket at its steepened edge. The fourth panel in this figure illustrates the fact that the whistler waves are circularly polarized, while the steepened part of the magnetosonic wave is closer to linear. As has been shown by Omidi and Winske [1990], both the presence of the whistlers and the change in polarization are a natural outcome of the nonlinear steepening process. Figure 8 shows a time series plot of the total magnetic field, illustrating how the waveform changes with time and steepens. Note that the original pulse apparently breaks into two parts with the left part spreading in space and steepening, and the right side growing some without steepening. This process is the result of the fact that only half of the original sinusoidal waveform undergoes steepening (see Omidi and Winske, 1990 for more detail).

Thus far, we have discussed the evolution of kinetic magnetosonic waves as a fascinating problem in nonlinear plasma physics. As will be shown in the next section,

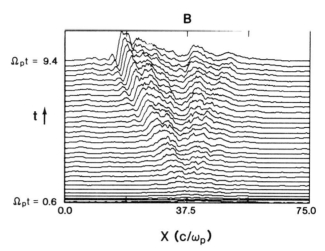

Fig. 8. Time history of the magnetic field shows the steepening of the wave. Note the apparant splitting of the wave.

these waves may have a great influence over the nature of solar wind–comet interaction in the outer parts of the coma to the extent that a single standing bow shock may be replaced by a series of these steepened waves.

3. Time-Dependent Multiple Shock Model

One of the striking aspects of the waveform shown in Figure 7 and the observed steepened waves is their similarity to a shock profile. This similarity led Hoppe et al., [1981] to refer to the steepened magnetosonic waves observed upstream of the Earth's bow shock as shocklets. As it turns out, this name is quite appropriate in that it can be shown that the steepened waves are indeed spatially localized shock waves. Figure 9 shows from top to bottom, the total magnetic field, solar wind velocity (in the wave's frame of reference) and density, density of the proton beam used to generate the wave, and the solar wind proton temperature obtained from the periodic simulation discussed in the last section. The dashed vertical line shows that the increase in the magnetic field is well correlated with a decrease in the solar wind flow velocity and an increase in the solar wind density and temperature. This behavior is, of course, in agreement with what one would expect from a fast magnetosonic shock. The dashed horizontal lines in the figure indicate the expected jumps in the field and plasma properties across a shock based on the Rankine-Hugoniot relations. As can be seen, the agreement between the observed jumps and those predicted is fairly good, further substantiating the shock interpretation. Based on this, Omidi and Winske [1988] identified the shocklets as localized subcritical dispersive shock waves.

Given that the velocity of these waves in the solar wind frame of reference is of the order of 2-3 Alfven speeds (V_A) towards the sun, the shocklets are generally carried away by the solar wind, and therefore, cannot form a standing shock in the frame of the comet. Similarly, while the

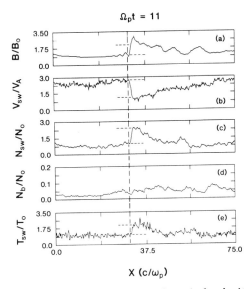

Fig. 9. The magnetic field, the solar wind velocity and density, the beam density, and the solar wind temperature are shown to illustrate the shock-like behavior of the steepened wave.

downstream plasma is subsonic in the frame of the shocklet, it is not necessarily subsonic in the comet's rest frame. This suggests that a parcel of solar wind may go through a number of these shocks before it becomes subsonic in the frame of the comet. With these considerations, the possibility of a new type of cometary shock model emerges, in which the solar wind is not shocked by a single standing bow shock, but instead by a number of moving and localized shock waves (i.e., shocklets). One aspect of this type of interaction is the fact that even for a steady upstream solar wind conditions, no steady state is reached and the system is inherently time-dependent. To further illustrate the multiple shock model, Figure 10 shows a schematic snapshot with the shaded stripes corresponding to regions in space where the solar wind has been locally heated and decelerated by a shocklet. Given the variation of the cometary ion production rate with distance from the nucleus and its influence on the wave growth and steepening rate, one would expect the steepened waves to show up fastest near the sun-comet axis and slower on the flanks (as illustrated in the figure). With time, each stripe is expected to evolve in the following fashion. First, it moves towards the nucleus due to the fact that the shocklets are carried by the solar wind. Second, each stripe grows longer in length by virtue of the fact that waves begin to steepen on the flanks; finally, the central part of the stripe becomes wider because of the relative motion between the solar wind and the shocklet. This later feature along with gradual deceleration of the solar wind will eventually result in merging of the stripes and the formation of a more uniformly heated and decelerated solar wind referred to as the sheath.

Thus far, we have shown that each steepened magnetosonic wave is in fact a localized shock wave and then

tried to construct a more global picture based on this realization. Needless to say, further improvements and refinements of this model are necessary through more theory and observation studies. As far as theoretical work is concerned, the next obvious step is to perform large scale simulations with sufficient time and spatial resolution to allow for the full development of the shocklets in a more realistic (i.e., nonperiodic and inhomogenous) system. We have recently begun to perform such simulations and in the remainder of this section preliminary results from one such run is presented. The simulation model is similar to that used by Omidi and Winske [1987] (see Figure 1) except that a larger box with smaller cell size and time step are used. Specifically, the simulation box is $2000c/\omega_p$ (about 160,000 km), with 2000 cells and a time step $\Delta t = 0.1\Omega_p^{-1}$ where Ω_p is the proton gyrofrequency. In this run the angle $\theta = 30°$ and because of the one-dimensional nature of the simulation, the excited waves all propagate at the same angle with respect to the ambient field. The upstream flow velocity is $4V_A$ and the electron and ion beta are each 0.5. The cometary ions are represented by O^+ and are created through photoionization and charge exchange with the solar wind protons using a similar cross section. Initially, each cell has 100 proton macroparticles, and subsequently O^+ ions are created with the usual spatially varying production rate. By the end of the run, more that 10^6 macroparticles are utilized.

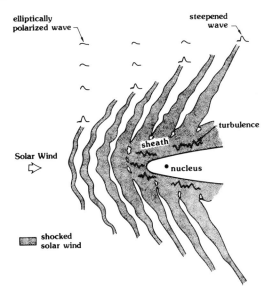

Fig. 10. A schematic of how the dynamic interaction between the solar wind and comets can result in the formation of multiple shocks which eventually lead to a heated and subsonic solar wind (sheath).

Figure 11 shows the time series plot of the total magnetic field for $350 \leq X \leq 1830$ between the times $\Omega_p t = 190$ and 685 in the simulation. Although the details of the interaction are hard to see in this figure, it does provide an overview of how the system evolves with time.

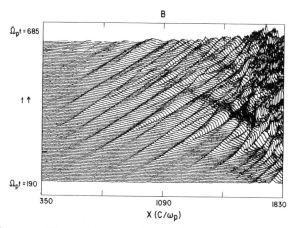

Fig. 11. Time history of the total magnetic field corresponding to a large high resolution simulation. As can be seen, generation and convection of steepened waves leads to formation of a turbulent sheath region on the right hand side.

For example, one can see that at earlier times waves begin to grow closer to the right hand side of the box where ion production rate is larger. These waves are then convected to the right (away from the sun) while at the same time they undergo steepening. Note that the splitting of the wave during steepening shown in Figure 8 can also be seen here, as one would expect. Examination of Figure 11 at later times shows the development of two distinct regions: one closer to the right hand side, where the magnetic field seems turbulent; the other in the upstream of this region, where individual waves (shocklets) can be seen. We note that after some time the extent of each region becomes more or less fixed and stationary in the simulation box. As will be shown shortly, the turbulent region is essentially the sheath drawn in Figure 10, where the plasma is more uniformly heated and decelerated. Finally, a close inspection of Figure 11 shows an aspect in wave evolution · which was absent in our periodic model, namely the interaction between two adjacent waves. As can be seen, there are numerous cases where the leading (i.e., steepening) edge of one wave interacts with the trailing portion of the wave upstream of it. Our initial analysis suggests that this type of interaction can result in considerable modification of the waveforms and probably plays a decisive role in the formation of the turbulence in the sheath region.

In order to see the formation of the shocklets in more detail, Figure 12 shows the total magnetic field at times between 260, and 455 Ω_p^{-1} in the region $1200 \leq X \leq 1750$. As can be seen, in addition to the steepened waves shorter wavelength whistlers have also been generated. An inspection of Figure 12 shows two types of whistlers. A longer wavelength ($\sim 4.5c/\omega_p$) class which is generated as a result of wave steepening, and remains associated with the shocklet (similar to those in Figure 7). The second class has a shorter wavelength ($\sim 3.5c/\omega_p$) and is seemingly associated with the interaction between the leading edge of a shocklet and the trailing portion of the one upstream

of it (see the right hand side of Figure 12). Due to their shorter wavelength (and therefore higher group velocity) these waves can propagate towards the upstream for the flow speeds considered here. While this phenomena seems quite fascinating, we note that the wavelength of these waves is not much larger than the spatial cell size ($1 \ c/\omega_p$) in the present simulations and therefore one must exercise some caution in drawing firm conclusions. If true, however, the results suggest that the interaction between two shocklets can lead to the generation of whistler bursts, which then propagate away from the interaction region.

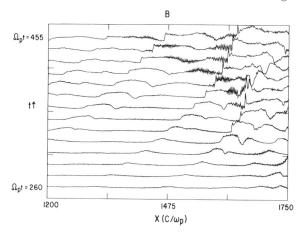

Fig. 12. High resolution plot of the time history of the magnetic field, showing the steepening and interaction of multiple waves and the generation of whistler waves.

These waves could eventually damp and lead to electron heating via Landau resonance [Gary and Mellott, 1985]. Needless to say, a more detailed investigation of this process and its authenticity is necessary in the future.

Whether the interaction between two shocklets results in generation of whistlers or not, there can be little doubt that this interaction modifies their waveform and plays a role in the formation of the turbulence in the sheath region. Figure 13 shows the total magnetic field and its two transverse components at $\Omega_p t = 350$ and $1220 \leq X \leq 1510$. From left to right one can see the evolution of the waveform in space, starting with a growing magnetosonic wave, followed by a steepening one, next by a steepened one, and finally by the remnants of a steepened wave. It is evident from this figure that late in their evolution, steepened magnetosonic waves have complex waveforms that are difficult to characterize. However, an interesting aspect of the evolution of these waves is the change in their helicity. As mentioned earlier, in their linear stage the waves are right hand polarized (in the solar wind frame) which during steepening changes and becomes partially linear. At even later times the polarization changes to a left hand sense. To illustrate this point we show in Figure 14 the magnetic helicity (top) and the transverse components of the field. As can be seen in the left half of the figure where the wave amplitude is relatively small the angle ϕ varies from $360°$ to $0°$ within a wavelength. This corresponds to a helicity which is consistent with magne-

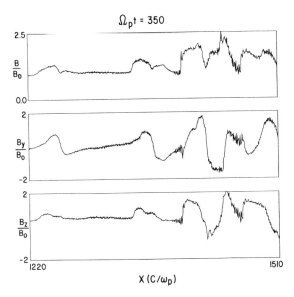

Fig. 13. Plot of total magnetic field and its two transverse components.

tosonic waves. On the right hand of the figure, however, the helicity has changed such that ϕ varies from $0°$ to $360°$. This change in helicity also implies a change in polarization from right to left hand waves. Although the change from right to partial linear polarization is due to steepening alone [see Omidi and Winske, 1990], the change to left hand polarization is most likely due to interaction between the shocklets. The details of this process

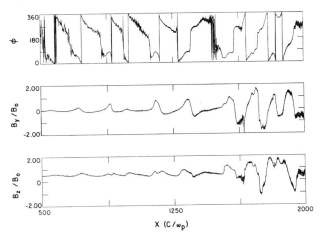

Fig. 14. Magnetic helicity (top) and the transverse components of the magnetic field illustrate the evolution in helicity and wave form from left to right. At small amplitudes the waves are right hand polarized, but after steepening they eventually become left hand polarized.

are unknown and currently under investigation. We note, however, that this change in polarization may explain the observations by Hoppe and Russell [1983] of left handed shocklets upstream of the Earth's bow shock.

To illustrate the behavior of the plasma in this simulation we show in Figure 15 the total magnetic field,

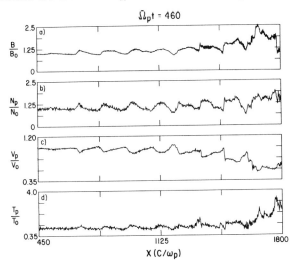

Fig. 15. Total magnetic field, and the solar wind density, velocity, and temperature are shown to illustrate the nature of the interaction seen in the simulation and its agreement with the multiple shock model.

and the solar wind density, velocity and temperature at $\Omega_p t = 460$ and $450 \leq X \leq 1800$. As can be seen in this figure, no single standing bow shock is present in the system. Instead, the magnetosonic waves have resulted in local enhancements of density and temperature and corresponding decreases in flow velocity. It is evident that these changes also lead to an eventual transition to a decelerated and more dense and heated solar wind (sheath) on the right hand side. This, of course, is in agreement with the multiple shock model and provides further evidence that this type of interaction is in fact plausible. An interesting point to note is that some of the peaks in proton temperature occur upstream of the region where the density goes up and velocity decreases (e.g., $X = 1462$). This peak in temperature is due to ions that are reflected by the shocklet. Note that these ions do not necessarily have a negative velocity in the X-direction, due to the fact that shocklets are convected by the solar wind and that reflection takes place in the shocklets' rest frame.

4. Observations and the Outstanding Issues

Given the various types of interactions discussed in the previous sections, an obvious question is how they relate to the observations. As will be shown here, the agreement between theory and the observations is mixed in that the multiple shock model can provide a reasonable explanation for the ICE observations at Comet Giacobini-Zinner. However, the nature of the observations at Comet Halley remains obscure.

During the ICE encounter with Comet GZ, various instruments onboard provided conflicting evidence as to whether a single bow shock was detected or not. For example, plasma waves and energetic particle observations [Scarf et al., 1986; Hynds et al., 1986] detected signatures that suggested the presence of a bow shock. Other instruments such as the magnetometer and the electron detectors, however, showed large amplitude fluctuations without a classic signature of a bow shock [e.g., Tsurutani and Smith, 1986; Gosling et al., 1986]. As a result, while it was clear that Comet GZ did have a sheath region, it was not clear how the solar wind had made this transition. By investigating the properties of the low frequency magnetic field and density fluctuations, Tsurutani et al., [1987] demonstrated the presence of magnetosonic waves upstream and within the transition (to sheath plasma) region. The fact that the steepened magnetosonic waves observed in our simulations bear a close resemblance to the observations provides considerable support for the multiple shock model. As for the plasma data, unfortunately, no low energy proton (i.e. solar wind) observations were available; however, electron measurements were made.

Figure 16 shows typical electron distribution functions

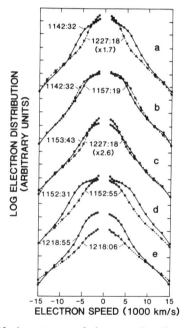

Fig. 16. Various types of electron distribution functions observed during the ICE outbound pass of the transition region at Comet Giacobini-Zinner (courtesy M. F. Thomsen).

obtained during the ICE outbound pass of the transition region first reported by Thomsen et al., [1986]. As discussed by these authors, three types of distribution function can be identified in the figure. One is the solar wind type which shows little or no heating as compared to the upstream plasma, such as the one corresponding to 1227 : 18 UT. The second class has a flattop at lower energies (e.g., at 1142 : 32 UT), similar to the ones observed downstream of weak interplanetary shocks [Feldman et al., 1983]. The increase of the electron temperature asso-

ciated with this type of distribution function is correlated with an increase in density, as one would expect had the heating been due to presence of a shock. An example of the third kind of distribution function found is that at 1218 : 06 UT. As discussed by Thomsen et al., [1986], the temperature associated with this type of distribution function is higher than nominal solar wind temperature, while at the same time the density is lower. Such a distribution could be formed by removing the low energy electrons in the solar wind without altering the higher energy ones. As such, the increase in temperature is not necessarily due to a heating mechanism, but instead could be due to electron removal through an as yet unknown process. By comparing the time of detection and the type of distribution function observed, Thomsen et al., [1986] found an intriguing behavior in that ICE was seemingly encountering shocked (i.e., flattopped) and unshocked solar wind in an intermittent manner (see Figure 16). It was then suggested that this type of behavior could be the result of the cometary shock on the flanks being intermittent. While at the time this interpretation seemed plausible, we now feel that the multiple shock model is a better candidate for explaining this peculiar aspect of the observations at Comet GZ. Given the interpretation of steepened waves as subcritical dispersive shocks, one can draw an analogy between them and low Mach number interplanetary shocks and thereby deduce that shocklets could also lead to the formation of flattop electron distributions. Note that due to the fluid treatment of the electrons in the hybrid code, our simulations do not make any predictions regarding the electrons. But, there seems to be sufficient observational evidence to suggest that the multiple shock model is appropriate for Comet GZ.

Concerning Comet Halley, the Giotto observations during the inbound and outbound suggest quasi-perpendicular and quasi-parallel shock normals [Neubauer et al., 1986, 1990] respectively. However, the entities reported as shocks [e.g., Neugebauer et al., 1987, and Neubauer et al., 1990] do not resemble the types of structures observed in our simulations. For example, Neugebauer et al., [1987] have shown that the so-called shock transition region during the inbound consists of fluctuations in plasma density and magnetic field. While it might be tempting to invoke the multiple shock model for this case, such interpretation is not presently warranted. One difficulty lies with the fact that no shocklets were observed at Comet Halley [see e.g., Glassmeier et al., 1989]. The other is the possibility that the shock structure could have been modified due to the presence of tangential discontinuities in the solar wind. Such discontinuities were observed near the shock region [Neubauer et al., 1988]. Additional simulations will be useful in checking this hypothesis.

As for the Giotto outbound pass, Neubauer et al., [1990] have used the magnetic field data to suggest the presence of a very thick (more than 120,000 km) quasi-parallel shock. It has been shown, however, that the magnetic field orientation in the downstream region is in a direction which is incompatible with the usual Rankine-Hugoniot relations. To resolve this conflict, Neubauer et al., [1990] derived a new set of equations by including source terms in the conservation laws. While addition of sources may address some of the difficulties, it does not explain the large thickness of the shock. As all of our simulations have shown up to now, any kind of a shock (including shocklets) formed in the cometary environment has a dissipative length which scales with proton inertial length. As such, it remains an open question as to why the observed quasi-parallel shock had such a large thickness. It is of course possible that lack of high resolution plasma data during the outbound prevented the detection of a much thinner dissipative structure. We also note that in a recent study of low Mach number parallel and nearly parallel shocks, Omidi et al., [1990] have found that such a shock consists of large amplitude electromagnetic waves generated due to interaction between the incident and (a hard) piston reflected ions. It was found that these waves do not provide the necessary ion dissipation, and as a result the shock remains coupled to the piston and its thickness becomes much larger than a proton inertial length. This result could be relevant to the observations by Neubauer et al. [1990], although simulations similar to that shown in Figure 3 but at smaller Mach numbers are necessary to check this possibility. Recently, Sharma et al., [1988] have suggested the possibility of a broad transition region where generation and subsequent resonant absorption of Alfven waves can result in a so called nonlocal transformation of ordered to random energy. Similarly, Sauer et al., [1989] have used a multifluid formalism to propose a thick quasi-perpendicular cometary shock model. The plausibility of these models remains to be seen through more detailed calculations and comparisons with the observations.

While considerable progress has been made in the area of cometary bow shocks, much more remains to be learned. For example, the nature of the shock observations at Comet Halley remains to be further clarified. Further simulations with perhaps additional physics are needed to understand both the inbound and outbound shock crossings. Similarly, much effort has gone into the analysis of the evolution in velocity space of the pickup ions upstream of and in the shock transition region [e.g., Coates et al., 1989 and references therein]. While a number of simulations have addressed the velocity space development of cometary ions in a local sense [e.g., Gary et al., 1988], the particle information embedded in the simulations similar to the one described in Figures 11 − 15, can address this issue on a more global scale. Such results will be presented in the future.

Acknowledgements. The authors wish to thank M. F. Thomsen for useful discussions and providing them with Figure 16. This research was supported by NASA Grant NAGW-1806. Part of the computations were supported by the San Diego Supercomputer Center which is supported by the National Science Foundation.

References

Cargill, P. J., K. Hizanidis, and K. Papadopoulos, Is the cometary "bow shock" really a shock?, in *Cometary and Solar Plasma Physics*, ed. by B. Buti, World Scientific, 1988.

Coats, A. J., A. D. Johnstone, B. Wilken, K. Jockers, and K. H. Glassmeier, Velocity space diffusion of pick-up ions from the water group at Comet Halley, *J. Geophys. Res.*, *94*, 9983, 1989.

Feldman, W. C., R. C. Anderson, S. J. Bame, J. T. Gosling, R. D. Zwickl, and E. J. Smith, Electron velocity distributions near interplanetary shocks, *J. Geophys. Res.*, *88*, 9949, 1983.

Galeev, A. A., and A. S. Lipatov, Plasma processes in cometary atmospheres, *Adv. Space Res.*, *4*, 229, 1984.

Gary, S. P., Electromagnetic ion/ion instabilities and their consequences in space plasmas: A review, *Space Sci. Rev.*, submitted, 1989.

Gary, S. P., and M. M. Mellott, Whistler damping at oblique propagation: Laminar shock precursors, *J. Geophys. Res.*, *90*, 99, 1985.

Gary, S. P., C. D. Madland, N. Omidi, and D. Winske, Computer simulations of two-pickup-ion instabilities in cometary enviornment, *J. Geophys. Res.*, *93*, 9584, 1988.

Glassmeier, K. H., A. J. Coates, M. H. Acuna, M. L. Goldstein, A. D. Johnstone, F. M. Neubauer, and H. Reme, Spectral characteristics of low-frequency plasma turbulence upstream of Comet P/Halley, *J. Geophys. Res.*, *94*, 37, 1989.

Gosling, J. T., J. R. Asbridge, S. J. Bame, M. F. Thomsen, and R. D. Zwickl, Large amplitude low frequency plasma fluctuations at Comet Giacobini-Zinner, *Geophys. Res. Lett.*, *13*, 267, 1986.

Hoppe, M. M., C. T. Russell, L. A. Frank, T. E. Eastman, and E. W. Greenstadt, Upstream hydromagnetic waves and their association with backstreaming ion populations: ISEE 1 and 2 observations, *J. Geophys. Res.*, *86*, 4471, 1981.

Hoppe, M. M., and C. T. Russell, Plasma rest frame frequencies and polarizatios of the low-frequency upstream waves: ISEE 1 and 2 observations, *J. Geophys. Res.*, *88*, 2021, 1983.

Hynds, R. J., S. W. H. Cowley, T. R. Sanderson, K. P. Wenzel, and J. J. Van Rooijen, Observations of energetic ions from Comet Giacobini-Zinner, *Science*, *232*, 361, 1986.

Neubauer F. M., et al., First results from the Giotto magnetometer experiment at Comet Halley, *Nature*, *321*, 352, 1986.

Neubauer F. M., et al., Magnetic field structure of the inbound bow shock of Comet P/Halley observed by Giotto, *EOS*, *69*, 396, 1988.

Neubauer F. M., et al., Giotto magnetic-field observations at the outbound quasi-parallel bow shock of Comet Halley, *Annal. Geophy.*, *8*, 463, 1990.

Neugebauer M., et al., The variation of protons, alpha particles, and the magnetic field across the bow shock of Comet Halley, *Geophys. Res. Lett.*, *14*, 995, 1987.

Omidi, N., and D. Winske, Simulation of solar wind interaction with the outer region of coma, *Geophys. Res. Lett.*, *13*, 397, 1986.

Omidi, N., D. Winske, and C. S. Wu, The effect of heavy ions on the formation and structure of cometary bow shocks, *Icarus*, *66*, 165, 1986.

Omidi, N., and D. Winske, A kinetic study of solar wind mass loading and cometary bow shocks, *J. Geophys. Res.*, *92*, 13409, 1987.

Omidi, N., and D. Winske, Subcritical dispersive shock waves upstream of planetary bow shocks and at Comet Giacobini-Zinner, *Geophys. Res. Lett.*, *15*, 1303, 1988.

Omidi, N., and D. Winske, Steepening of kinetic magnetosonic waves: Simulations and consequences for planetary shocks and comets, *J. Geophys. Res.*, *95*, 2281, 1990.

Omidi, N., K. B. Quest, and D. Winske, Low Mach number parallel and quasi-parallel shocks, *J. Geophys. Res.*, *in press*, 1990.

Quest, K. B., Theory and simulation of collisionless parallel shocks, *J. Geophys. Res.*, *93*, 9649, 1988.

Sauer, K., U. Motschmann, and T. Roatsch, Plasma boundaries at comet Halley, preprint, Akademie der Wissenschaften, 1989.

Scarf, F. L., F. V. Coroniti, C. F. Kennel, D. A. Gurnett, W. H. Ip, and E. J. Smith, Initial report on plasma wave observations at Comet Giacobini-Zinner, *Science*, *232*, 377, 1986.

Schmidt, H. U., and R. Wegmann, Plasma flow and magnetic fields in comets, in *Comets*, ed. by L. L. Wilkening, p. 538, Univ. of Arizona Press, Tucson, 1982.

Sharma, A. S., P. J. Cargill, and K. Papadopoulos, Resonance absorption of Alfven waves at comet-solar wind interaction regions, *Geophys. Res. Lett.*, *15*, 740, 1988.

Thomsen, M. F., S. J. Bame, W. C. Feldman, J. T. Gosling, D.J. McComas, and D.T. Young, The comet/solar wind transition region at Giacobini-Zinner, *Geophys. Res. Lett.*, *13*, 393, 1986.

Tsurutani, B. T., and E. J. Smith, Hydromagnetic waves and instabilities associated with cometary ion pickup: ICE observations, *Geophys. Res. Lett.*, *13*, 263, 1986.

Tsurutani, B. T., R. M. Thorne, E. J. Smith, J. T. Gosling, and H. Matsumoto, Steepened magnetosonic waves at Comet Giacobini-Zinner, *J. Geophys. Res.*, *92*, 11074, 1987.

Wallis, M. K., Shock-free deceleration of the solar wind, *Nature*, *233*, 23, 1971.

Winske, D., and M. M. Leroy, Hybrid simulation techniques applied to the earth's bow shock, in *Computer Simulations of Space Plasmas—Selected Lectures at the First ISSS*, ed. by H. Matsumoto and T. Sato, D. Ridel, Hingham, Mass., 1985.

Winske, D., Origin of large magnetic fluctuations in the magnetosheath of Venus, *J. Geophys. Res.*, *91*, 11951, 1986.

Wu, C. S., and R. C. Davidson, Electromagnetic instabilities produced by neutral particle ionization in interplanetary space, *J. Geophys. Res.*, *77*, 5399, 1972.

AN MHD MODEL OF COMETARY PLASMA AND COMPARISON WITH OBSERVATIONS

H.U.Schmidt and R.Wegmann

Max-Planck-Institut für Astrophysik, Karl-Schwarzschildstr.1, D-8046 Garching, FRG

Abstract. Cometary plasma flow was sampled by spacecraft along inclined trajectories through comets Giacobini-Zinner and Halley. The interpretation of these observations needs extensive modeling of the whole comet. Here we review some global simulations of cometary plasma flow based mainly on MHD. We discuss features from the first deceleration of the solar wind outside the bow shock to the radial expansion of the purely cometary plasma inside the contact surface and to the cometary tail downwind. For the interaction of solar UV, solar wind, and cometary matter many processes must be considered: In the outermost regions mainly photoionisation of neutrals as well as pitch angle anisotropy and scattering. Towards the nucleus collisional processes become important: charge exchange, ionisation and dissociation by electron impact, possibly critical velocity effects, recombination processes, and momentum exchange with the counterstreaming neutrals. Chemical kinetics of plasma and neutral gases evaporating from nucleus and dust needs accounting. Lorentz forces necessitate 3D simulation of the magnetic pile up.

1.Introduction

MHD modeling of cometary plasma flow began in the early sixties. It predicted first the existence of a detached bow shock when and where the solar wind is loaded with about one percent of cometary ions. It predicted secondly the existence of a contact surface which separates the mixture of solar wind protons and pick-up ions from a purely cometary plasma in the neighbourhood of the nucleus. Calculations started with a one-dimensional model [Biermann et al., 1967] and developed via two-dimensional hydrodynamic models [Brosowski and Wegmann, 1973] to three-dimensional magnetohydrodynamic (MHD) models [Schmidt and Wegmann, 1980, 1982]. The physics included first only a single ion species generated by photoionization. The need for theoretical interpretation of the incoming data from the cometary missions stimulated more sophisticated models which now include realistic

microphysics which determines the plasma flow in a comet [Wegmann et al., 1987, Schmidt et al., 1988]. In order to resolve the flow on the relevant scales which range over three orders of magnitude we calculate on a curvilinear adapted grid.

Time-dependent MHD calculations on cartesian grids have been performed by Fedder et al.[1984, 1986], Ogino et al. [1988], and Schmidt-Voigt [1988, 1989]. A parametric study of stationary MHD models is given by Russell et al. [1990].

In this paper we present some new results calculated with a refined version of our MHD code for cometary plasma flow. We partly review the existing literature in this field.

Our most recent and up to now best model from which most of the data and figures in this paper are chosen is based on conditions which are representative for comet Halley at Giotto encounter. The production rate is 6.9 10^{29} molecules s^{-1}. The solar wind has velocity $u_\infty = 380$ km s^{-1}, a density of $N_\infty = 7$ protons and 7 electrons cm^{-3}, an ion temperature of $T_{i\infty} = 10^5$ K and an electron temperature of $T_{e\infty} = 2.5$ 10^5 K. It carries a magnetic field of $B_\infty = 7$ nT which is inclined with respect to the velocity vector by the gardenhose angle of 56°.

The plasma is described as an ideal gas with $\gamma = 5/3$. We allow for separate temperatures for ions and electrons. The neutral coma is taken from a spherically symmetric model. We assume that it is unaffected by the interaction with the plasma. But it yields new pick-up ions to the plasma by ionization and dissociation by hot electrons and solar UV-photons. Charge exchange replaces hot fast ions by cold slow ions. The plasma looses ions by recombination. The electrons are cooled by inelastic collisions with neutral molecules and atoms. The neutral flow exerts a drag on the plasma by elastic collisions.

These effects are taken into account in a parametrized way which is derived from a detailed chemical calculation as described in Schmidt et al. [1988]. Recently, we have improved the numerical method and have refined the computational grid by a factor of two over the models pu-

Cometary Plasma Processes
Geophysical Monograph 61
©1991 American Geophysical Union

blished earlier. Furthermore, errors in the calculation of the temperatures have been corrected.

We start off in sections 2 and 3 with the description of the adapted grid and the boundary conditions, and a discussion of the validity and applications of MHD models. In sections 4, 5 and 6 we discuss the large scale flow pattern and the deformation of the interplanetary magnetic field, the profile of the subsolar streamline and the distribution of sources and sinks of mass, momentum and energy along this line. We briefly assess the critical velocity effect in section 7.

The MHD model is compared with the observed data along the Giotto trajectory in section 8 followed by a discussion of the stagnation flow and the shape of the contact surface in sections 9 and 10. The tail features are the subject of section 11 and conclusions are drawn in section 12.

2.Grid and boundary conditions

This adapted grid (which is in principle like that of figure 1 in Schmidt and Wegmann, [1980]) is outlined by three families of nearly orthogonal surfaces: a) 61 nearly hyperboloidal surfaces formed by spherical caps on the subsolar side which extend into cones in the tailward direction. They are spaced on the sun-comet axis logarithmically with distance from the nucleus between 1600 and $2 \cdot 10^6$ km, b) 75 flat cones pointed in the nucleus or on the tailward axis and opening towards the sun at angles between 0 and 90 degrees, and c) 41 meridional planes equidistant in azimuth between 0 and 180 degrees with respect to the interplanetary magnetic field.

The outermost of the hyperboloidal surfaces with an asymptotic opening of 30 degrees is placed completely in the unperturbed supersonic solar wind with all characteristics running inward so that the solar wind completely and exactly fixes the boundary values. The outermost of the cones b) starts on the axis $6 \cdot 10^6$ km behind the nucleus and fully engulfes the subsonic region. Therefore, it is again placed completely in supersonic flow with all characteristics running outward so that free extrapolation provides the appropriate boundary condition here. The innermost of the hyperboloidal surfaces a) on the subsolar side is an almost complete half sphere around the nucleus of radius 1600 km which turns into a flat cone whose radius at $6 \cdot 10^6$ km has grown only to 3000 km.Therefore, only a minute region around nucleus and tailward axis is excluded. On this surface we assume completely radial outflow of neutral gas and plasma from the nucleus as calculated selfconsistently in the spherically symmetric hydrodynamical model with complete chemistry by Huebner and coworkers [1980, 1983, 1985], and Giguere and Huebner [1978]. This grid allows sufficient resolution to describe bow shock, subsolar flow, stagnation, and the contact surface.

3. Validity and applications of MHD models

The strength of MHD modelling lies in its capability to describe consistently the global plasma flow throughout the comet over many length scales from outside the bow shock to inside the contact surface, and to get a really three-dimensional picture.

MHD means that one uses a closed finite set of moments of the Boltzmann equations for the velocity distribution of the ionized particles. This presupposes the existence of some processes which can establish some minimum degree of isotropy in the peculiar velocities in timescales which can at least compete with the timescales of advection and other changes so that the effects of free cross streaming of nonlocal particles are limited.

Therefore, each specific MHD model gives useful results for the macroscopic flow on scales which are larger than the minimum mean free path of the scattering processes which establish the specific isotropy which is assumed in the model. In a comet such processes are gyration, pitch angle scattering at magnetic fluctuations induced by the pick-up of cometary ions, growth of two stream instabilities in the outer regions, and collisions among the constituents of the plasma and ion-neutral collisions in the dense inner coma.

So it is not surprising that the early MHD predictions turned out to be successful despite the fact the interplanetary mean free path for two particle collisions is of the order of 1 a.u. The basic reason that a one fluid MHD simulation is so successful in predicting the right distance of the bow shock from gas production, outflow velocity, chemical composition, ionization rates and solar wind parameters only, is fast pitch angle scattering even in front of the bow shock. The pick-up protons form a ring in velocity space which for inclined fields is not centered on the solar wind velocity. Therefore, relative to the solar wind they lag behind and drift along the field. But as shown for these outer regions by Neugebauer et al. [1989] the protons in this shifted ring are gradually scattered over the complete shell of equal energy in the frame moving with the solar wind.

To describe field particle interactions and waves, as they are important e.g. for the local fine structure of the shock one has to use selfconsistent particle and field simulations or a suitable multifluid approach. Such models have been calculated for the shock by Omidi and Winske [1986, 1987, 1989], for one-dimensional flow on the subsolar streamline by Sauer et al. [1990], and for a two-dimensional flow by Raeder [1990].

Gombosi [1988], using a MHD fluid description of the solar wind and treating implanted protons and oxygen ions with coupled Fokker Planck equations with velocity diffusion only, was able to compute spectra for the Fermi acceleration of the implanted ions in the foreshock

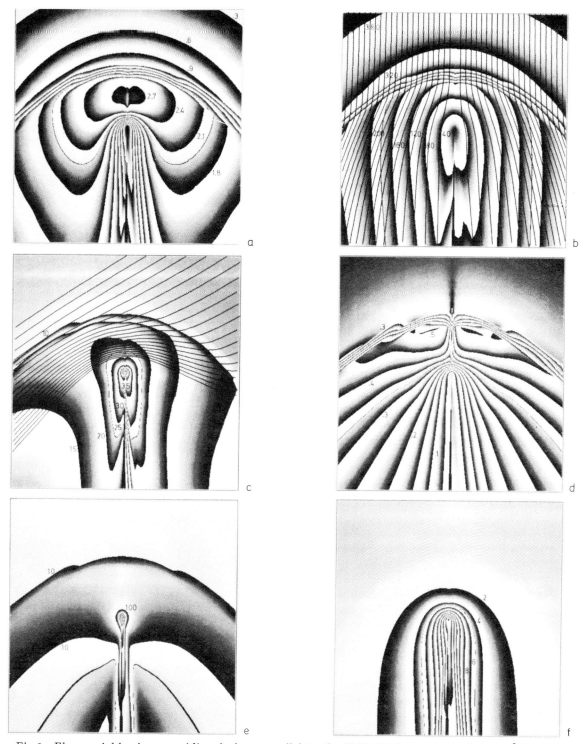

Fig.1. Flow variables in a meridional plane parallel to the IMF. Each map extends to 10^6 km on each side of the nucleus. The sun is above. The view is from north. Right-left asymmetries are caused by the inclined IMF (see c). a) Ion temperature [10^6K], b) velocity [km s^{-1}] with streamlines, c) magnetic field strength [nT] with fieldlines. The field lines follow each other in time intervals of 300 s. d) Electron temperature [10^6K], e) number density of the electrons [cm^{-3}] on a logarithmic scale, f) mean molecular weight of the ions (including solar wind protons).

field disturbances. He neglected the fast lateral diffusion. These spectra would fit the Vega 1 spectra at the bow shock if they were taken at the subsolar position. The spectra calculated for the actual Vega 1 position are too low, and lateral diffusion cannot balance the difference.

Other examples of applications of MHD models can be found in Schwingenschuh et al. [1986, 1987], Huebner et al. [1989], Russell et al. [1990].

4.Large scale flow pattern

In Figure 1 we present contour maps of the cometary large scale flow pattern from our updated stationary MHD simulation. The maps refer to a meridional plane containing the solar wind velocity and the interplanetary magnetic field. The asymmetries with respect to the axis are caused by the finite inclination of the IMF at 56 degrees from the velocity vector. The presentation of the ion temperature (Figure 1a) is corrected and we thank N. Meyer-Vernet (personal communication), who kindly made us aware of an inconsistency in the temperature data presented in Schmidt et al. [1988]. These data contained erroneously the mean molecular weight as a general factor. Fortunately, this error did not falsify any other results.

The most conspicuous features in a comet on scales of the order of several 10^5 km are the shock front, the draping of the magnetic field and the tail.

Considering the solar wind flow as one-dimensional parallel MHD flow with source terms, Biermann et al. [1967] noticed that this flow cannot be extended beyond a certain critical point. Assuming that mass injection is the dominant process the flow would run into a self-reversal when the percentage of added mass is $1/(\gamma^2 - 1)$. Before this point is reached the velocity must get an appreciable lateral component to cope with the added ions.

By the pick-up of slow ions the velocity of the plasma decreases and the temperature rises. Therefore, the Mach number of the originally supersonic solar wind decreases and would be equal to 1 at the critical point. In this situation a shock must form which separates the essentially parallel flow of the solar wind from the diverging flow of the post-shock region. This flow pattern can be visualized by the streamlines shown in Figure 1b.

The comet acts like an obstacle in the solar wind. As a consequence the frozen-in magnetic field lines hang at the comet and are draped around it like a curtain (Figure 1c). The motion of an originally plane sheet of magnetic field relative to the Giotto spacecraft was calculated by Schmidt et al. [1986] and the same for the Vega spacecraft by Schwingenschuh et al. [1987]. Close to the comet the advected and deformed magnetic field vector retains approximately its original azimuth. Sheets of originally

parallel field are deformed to paraboloidal surfaces bent around the nucleus and nested like onion shells.

Using this picture one can, with a stationary model, also describe approximately the situation where the solar wind magnetic field changes suddenly direction. This results in field shells around the nucleus with different azimuthal field orientations.

Raeder et al. [1987] identified in the Giotto magnetic field data pairs of field discontinuities observed before and after closest approach as being related, i.e., apparently corresponding to the same discontinuity in the interplanetary magnetic field (IMF). Assuming that this was generated by an originally plane discontinuity in the IMF, this must fit the topology described above and one can derive from our model a time table for the second encounter with a field structure which Giotto has observed before closest approach. This has been worked out in Huebner et al.[1989]. The observations missed this time table only by about 1 minute and so confirm rather well the global flow pattern of the model.

Though successful, this description can only be an approximation since a time dependent non uniformly oriented magnetic field has dynamic influence in the stagnation region [Schmidt-Voigt, 1988, 1989], which is not taken into account in this kinematic picture.

A simple topological argument shows that Giotto must have passed a field discontinuity either once or thrice. In the latter case it had the unique opportunity to probe a field structure over a spatially extended region. The fact that the third encounter with the discontinuities could not be identified [Neubauer, personal communication] puts a limit of about several 10^5 km on the size of these coherent field structures in the IMF.

5.Values along the axis

The transition of the flow to various zones with different characteristics can best be studied by looking at the situation along the axis which is described on Figure 2 in a logarithmic scale. This is also useful for comparison with one-dimensional models [Galeev et al. 1986, 1987, 1988, 1990, Gombosi 1988, Gombosi et al. 1989, Cravens 1986, 1987, 1989], which must concentrate on the sun - comet line.

The shock is clearly marked by the discontinuity in almost all variables at a distance of 4.5 10^5 km. All conserved quantities jump by an amount determined by the Rankine-Hugoniot relations.

The ions are heated by the subsolar shock to 2.6 10^6 K. Behind the shock the braking by pick-up ions transforms kinetic energy into thermal energy so that the temperature rises to almost 3 10^6 K. A separate calculation of the balance of the protons shows that by charge exchange most fast protons are replaced by cold ions inside 2 10^4

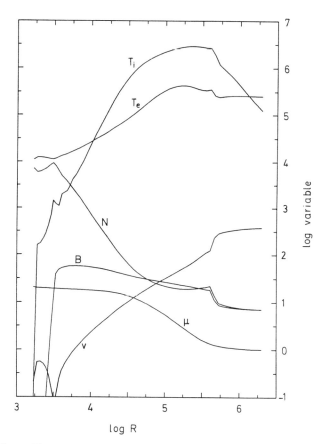

Fig.2. Flow variables along the comet-sun axis. T_i = ion temperature [K], T_e = electron temperature [K], v = velocity [km s^{-1}], N = number density [cm^{-3}], B = magnetic field strength [nT], μ= mean molecular weight, R = distance from nucleus [km].

km. This reduces the velocity dispersion and the temperature drops rapidly.

Behind the shock the number density decreases slightly. This indicates the divergence of the flow. The heavy pick-up ions add to the mass density but little to the number density. The solar protons are depleted by charge exchange at distances less than 3 10^4 km. The density of the plasma now consisting mainly of pickup ions rises somewhat more steeply than R^{-2}. The velocity decreases nearly linearly $\sim R$. Therefore, the particle flux $Nu \sim R^{-1}$ is nearly proportional to the integrated source of pick-up particles which is naturally proportional to the density of the neutrals, which varies like R^{-2} with distance. The gradual rise of the mean molecular weight from the solar wind value of 1 to 20 indicates the change in chemical composition from protons to a mixture of ions, mainly H$_2$O$^+$, OH$^+$, and CO$^+$. The magnetic field of originally 7 nT is also compressed in the shock and then rises steadily with a power of about R$^{-1/4}$ with distance R. The magnetic stress behaves like R$^{-1/2}$.

A noticeable exception from the well known shock relations is the electron pressure p_e which in our model is described by the equation [Schmidt-Voigt 1988, 1989]

$$\frac{1}{\gamma - 1}\frac{\partial p_e}{\partial t} + \text{div}\left(\frac{1}{\gamma - 1}p_e\mathbf{u}\right) + p_e\,\text{div}\,\mathbf{u} = \dot{E}^e \quad (1)$$

with the velocity $\mathbf{u}$ and a source term $\dot{E}^e$. We see from (1) that p_e, in addition to the normal shock compression which is described by the first two terms, gains from the "source term"

$$-(\gamma - 1)\,p_e\,\text{div}\,\mathbf{u}. \quad (2)$$

In order to estimate this effect we assume that the shock is smeared over a region of thickness δx in which the velocity is reduced from u_1 to u_2. Then div $\mathbf{u} \approx (u_2 - u_1)/\delta x$. The time needed for the transition of this shock layer is $\approx 2\delta x/(u_1 + u_2)$. This is the time during which the additional source (2) is active. We calculate from this the total gain $\delta p_e \approx 2(\gamma-1)p_e(u_1-u_2)/(u_2+u_1)$ which does not depend on the shock thickness but only on the ratio u_2/u_1 which is determined by the Rankine-Hugoniot relations. We get finally

$$\frac{p_{e2}}{p_{e1}} \approx \exp\left(2(\gamma - 1)\frac{u_1 - u_2}{u_2 + u_1}\right)\frac{u_1}{u_2}. \quad (3)$$

Since the jump in the density is given by normal shock compression the electron temperature is also discontinuous. The values in front and behind the shock are connected by

$$\frac{T_{e2}}{T_{e1}} \approx \exp\left(2(\gamma - 1)\frac{u_1 - u_2}{u_2 + u_1}\right). \quad (4)$$

For $\gamma = 5/3$ and a Mach number of 2 we get an amplification by a factor of about 1.8.

The electrons are further heated by the surplus energy in photo ionization processes. But further inward the cooling by elastic collisions with neutrals prevails and reduces the temperature to a value, which in this slow moving flow is close to the equilibrium temperature determined by the balance of heating and cooling.

6. Sources and sinks

The flow is determined by the sources which add mass and energy to the plasma and exert forces on it.

Ions are added to the plasma by photo-processes which act throughout the comet with a yield proportional to the neutral coma density which varies approximately as R^{-2} with distance R from the nucleus. The efficiency of electron impact ionization depends on the electron temperature. Therefore, it becomes unimportant in the inner coma where the electrons are cold.

Charge exchange is important in the solar wind where fast protons yield cool heavy ions. But as soon as the mean molecular weight of the plasma comes close to that of the neutral coma there is no more gain in mass. Photo-ionization preferentially picks the heavier molecules. Therefore, the ions generated by photo-ionisation are on the average heavier than the average neutrals. Consequently, in the stagnation region charge exchange replaces heavy ions by less heavy ones which means a loss of mass for the plasma (see Figure 3a). In any case charge exchange replaces fast moving ions by slow ones and in this way exerts a volume force on the plasma.

Electrons are heated by the surplus energy of photo-processes which is of the order 10 - 20 eV. They are cooled by inelastic collisions with the neutrals. The heating is proportional to the neutral density N_n, the cooling to the product $N_e N_n$ of neutral and electron densities with a factor which depends on T_e (see Figure 3b). Therefore, in the inner dense coma the cooling dominates and keeps T_e low. This low T_e as well as the high densities favour recombination which near to the contact surface consumes almost all ions which are produced by other processes. Therefore, the source of ions which form the tail comes mainly from the region outside 10^4 km.

The shock produces a high ion and electron pressure. This is used to exert a force on the plasma in tailward direction. It is counteracted by the momentum loss by charge exchange and the drag exerted by the collisions with the neutrals. The magnetic field plays only a passive role in the outer region, but it becomes dominant in the stagnation region. Inside 10^5 km the radius of curvature R_K of the magnetic field lines is $\sim 1.5R$ (see also Figure 7). In this region the curvature forces B^2/R_K are dominant until near stagnation the rapid drop off of the magnetic field causes an even larger inward directed force (Figure 3c).

The gain in energy is the work done by the aforementioned forces with the noticeable exception of the magnetic field. The gain in magnetic field energy is much larger than the work done by the Lorentz-forces in the coma. This accounts for the energy which flows from the by streaming fast solar wind to the slow central parts.

7. Critical velocity effect

The critical velocity effect has been investigated for comets by Galeev et al. [1986a] and for Venus and Mars by Luhmann and Russell [1990].

Ionisation by the critical velocity effect can be effective if two conditions hold:

The relative velocity v_{rel} of the plasma flow relative to the neutrals must exceed a critical velocity v_{cr} determined by the threshold energy χ for electron impact ionization, i.e., the kinetic energy of a neutral with mass m_i must satisfy

$$m_i v_{rel}^2/2 > |\chi|. \tag{5}$$

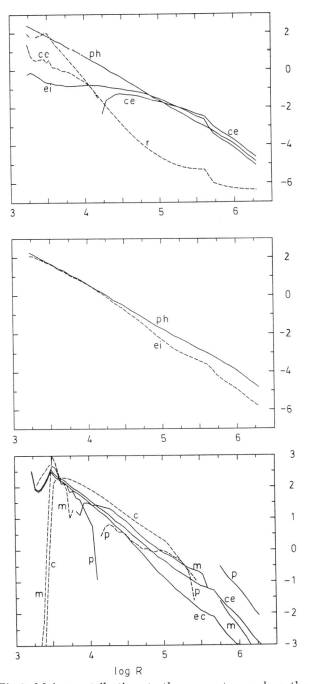

Fig.3. Major contributions to the source terms along the axis. a) Sources and sinks of ion mass [amu cm^{-3}s^{-1}], b) Heating and cooling of the electrons [eV cm^{-3}s^{-1}], c) Forces on the plasma [10^5 amu cm^{-2}s^{-2}]. The symbols have the following meaning: photoprocesses (ph), charge exchange (ce), recombination (r), elastic collisions (ec), ion and electron pressure forces (p), magnetic pressure forces (m), curvature forces (c). Negative contributions are indicated by dashed lines. The effect of lateral outflow or inflow is not shown.

With the velocity taken from the model of Wegmann et al [1987] this is drawn in the bottom panel of Figure 4

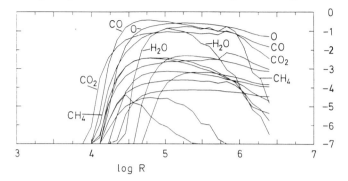

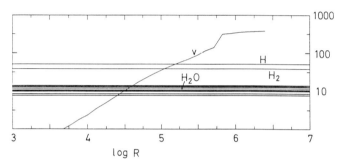

Fig.4. Bottom panel: velocity (v) in the model and critical velocities [km s^{-1}] for several species (horizontal lines); top panel: the logarithm of the Townsend ratio (see text formula (7)) for several different species calculated with model densities and electron temperatures.

The critical velocity is highest for the light species H and H_2. Then follow the water group molecules and all others close together around 10 km s^{-1}.

In order that the process keeps going the Townsend condition must be satisfied. This means that the electron generated in a critical velocity ionisation has a chance for ionising another molecule before leaving the volume. The probability π_{ei} (s^{-1}) for this second electron impact ionization is proportional to the density N_{ni} of the neutral species and a factor κ_i dependent on the energy E_e of the electron

$$\pi_{ei} = N_{ni}\kappa_i(E_e). \qquad (6)$$

For an estimate we make the extreme most favourable assumptions that the new electron gets the full surplus energy from the new pick-up ion, and that the time which the electron spends in the supercritical range at the distance R is $\tau = R/v$. The probability p_{2i} for a second ionisation is then

$$\frac{p_{2i}}{1 - p_{2i}} < \tau\pi_{ei} = N_{ni}\kappa_i\left(m_i\frac{v^2}{2}\right)\frac{R}{v}. \qquad (7)$$

This quantity on the right hand side is drawn for our Halley model data in the upper panel of Figure 4 for several species. At large distances from the comet it falls off with $N_{ni}R \sim R^{-1}$, at small distances due to subcritical velocity. The estimate assumes perfect efficiency and is therefore very optimistic since even efficiency of order 10^{-2} cannot be reached [Goertz et al., 1990]. Even though, for all species this probability is $\ll 1$ and the Townsend condition is not satisfied.

We believe that this result holds for comets in general. We write $\kappa_i = Q_{ei}(m_i/m_e)^{\frac{1}{2}}v$ as the product of a cross-section Q_{ei} and the velocity and observe that

$$N_{ni} = \frac{G_i}{4\pi R^2 w} \qquad (8)$$

with the total gas production G, the production G_i of species i, and the outflow velocity w of the neutrals. Then the right hand side in (7) reduces to

$$\tau\pi_{ei} = \frac{G_i Q_{ei}(m_i/m_e)^{\frac{1}{2}}}{4\pi Rw}. \qquad (9)$$

For a crude estimate we use the scaling law of cometary plasma flow

$$\frac{m_C G\sigma}{4\pi w\rho_\infty u_\infty R} = C_0 \qquad (10)$$

with a universal constant C_0 [Schmidt and Wegmann, 1982, p. 548], and find

$$\begin{aligned}
\frac{p_{2i}}{1 - p_{2i}} &< C_0 \frac{Q_{ei}(m_i/m_e)^{\frac{1}{2}}\rho_\infty u_\infty}{\sigma m_C}\frac{G_i}{G} \\
&\leq C_0 \frac{Q_{ei}(m_i/m_e)^{\frac{1}{2}}\rho_\infty u_\infty}{\sigma m_C}
\end{aligned} \qquad (11)$$

where ρ_∞ and u_∞ are the mass density and the velocity of the solar wind, σ is the average rate of photoionisation and m_C the mass of an average cometary ion. We see that in this approximation the Townsend ratio does not depend on the gas production nor on the distance from the sun but solely on the chemical composition and on the ratio of solar wind flux to the solar UV flux.

8.Values along the Giotto trajectory

For comparison with the data from the Giotto encounter with comet Halley we give in Figure 5 the values of some flow variables along the Giotto trajectory as calculated from our updated model.

The model shock standoff distances of 1.3 10^6 km inbound and and .9 10^6 km outbound agree fairly well with the measured distances. Here we emphasize that besides the relative chemical abundances we do not fit in our model any cometary parameters to Giotto observations.

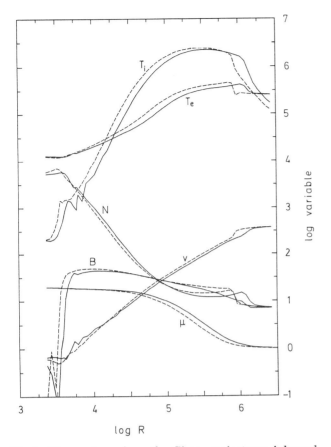

Fig.5. Flow values along the Giotto trajectory: inbound (solid), outbound (dashed), R = distance from closest approach [km], all other variables as in Fig. 2.

The solar wind data are taken from the upstream Giotto measurements before encounter. The main free parameter is the production rate. This is chosen in accordance with the IUE measurements [Festou et al., 1986].

Easily interpretable data are the magnetic field measurements of Neubauer et al. [1986]. The noise in these data is partly due to turbulence but partly also to advected large scale disturbances in the solar wind. Using the method of nesting rotated pieces of our stationary model we have tried to approximate this nonstationary situation with our stationary model. Surprisingly we do not only get a good fit in the y- and z- components of the field but also in the x-component. (Our coordinate system is centered at the comet so that the x-axis points towards the sun and the x-y-plane is parallel to the IMF.) This shows that the approximation is valid to a large extent [Huebner et al., 1989].

The Giotto magnetometer confirmed for the first time the existence of the theoretically predicted magnetic cavity. We can insert in the figure of Neubauer et al. [1986] our magnetic field predictions from the model (Figure 6).

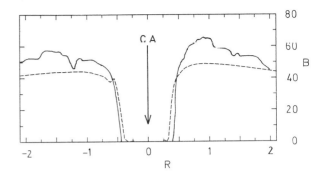

Fig.6. Magnetic field strength B [nT] around closest approach (CA) as observed by Neubauer et al. [1986] (solid line), and obtained in the model (dashed line). R = distance from CA in 10^4 km.

We find good agreement in the position and the shape of the drop of the field to zero. We get also the higher field strength outbound compared to inbound. But the model field strength is systematically to low. In view of this we conjectured that an interplanetary magnetic field slowly rotating around the solar wind velocity vector interferes strongly with the flow pattern (described in the next section) needed for transporting the field off the stagnation region. Therefore, it is amplified more than a stationary field. This effect has been confirmed by the time-dependent calculations of Schmidt-Voigt [1988, 1989] who got an amplification of the field strength by a factor of up to 1.56 depending on the rotation frequency.

For the ion temperature there seem to exist discrepancies between our model and the observations. Meyer-Vernet (private communication) finds the model temperatures systematically too low by a sizable factor in an intermediate region between shock and contact surface. For more ion temperature measurements see Vaisberg et al. [1989]. We conjecture that in this region the drift of the cometary ions relative to the solar protons restricts the applicability of the MHD approach which assumes isotropic distribution in a common frame of reference.

9.Stagnation flow

Behind the shock the flow remains nearly axisymmetric. By an argument first used by Lees [1964] it can be shown that in a rotationally symmetric flow the ratio

$$B^2 u/N_\odot \qquad (12)$$

remains constant on the axis of symmetry where $N_\odot$ is any conserved quantity satisfying div $N_\odot \mathbf{u} = 0$. This implies that in the stagnation region where u tends to zero the magnetic field is amplified till it becomes strong enough to govern the flow and to organize its own transportation. A bundle of field lines is hung up which finally exerts a pressure gradient which enforces lateral expansion.

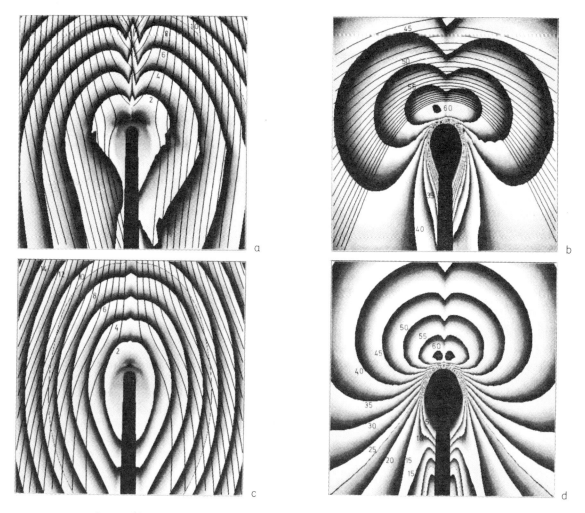

Fig.7. Velocity [km s^{-1}] and magnetic field [nT] in a meridional plane parallel to the IMF (a, b) and a perpendicular meridional plane (c, d) with streamlines and fieldlines. Each panel extends 30 000 km on each side of the nucleus. The fieldlines follow each other in time intervals of 300 s.

The magnetic field can only evade the obstacle of the inner coma by advection in the direction perpendicular to the field. This implies that the flow in the magnetically dominated region will have a tendency to be stronger in planes perpendicular to the field. Since the supply of mass, momentum and energy is axisymmetric, this planar flow together with recombination processes generates a low pressure region near the axis which is filled up by a flow along the field lines directed towards the axis. This flow pattern is shown by the streamlines in planes perpendicular and parallel to the field (Figure 7). It can also be visualized by the projection of the velocity vector in a cut perpendicular to the axis 10 000 km in front of the nucleus (Figure 8) which shows that in a wide region the azimuthal component of the flow is almost aligned with the y-axis. The flow towards the axis is concentrated to the meridional plane parallel to the IMF.

We can get an estimate of the position of this magnetically dominated region in the following way: In the shock the stagnation pressure $\rho_\infty v_\infty^2$ of the solar wind is transformed into thermal gas pressure, which is steadily reduced by the work done against the forces due to the exchange of momentum with the neutral gas by charge exchange and elastic collisions. We see in Figure 3c that the effect of charge exchange is dominant. The force is [Schmidt et al. 1988]

$$\dot{q} = Q_{ce} N_n N_i v_{rel} \mu v \qquad (13)$$

where Q_{ce} is the cross-section for charge exchange which we assume to be 2 10^{-15} cm^2. The density of the neutrals N_n is given by (8). The velocity v_{rel} of ions relative to the neutrals includes the thermal velocity and is therefore always of the order of the solar wind velocity v_∞. The

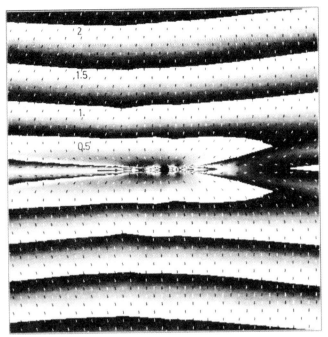

Fig.8. Projected velocity vectors in a plane cut perpendicular to the sun-comet axis, 10 000 km in front of the nucleus. The projected IMF would be horizontal. Contourlines show magnitude of azimuthal velocity [km s^{-1}]. The panel extends 10 000 km on each side of the nucleus.

mass flux $\mu N_i v$ in the regions where the charge exchange force is most efficient is approximately the same as in the solar wind. We equate the force exerted by charge exchange integrated from infinity up to a distance R_S to the stagnation pressure of the solar wind and use the aforementioned approximations. We obtain

$$\int_{R_S}^{\infty} \dot{q} \, dR = Q_{ce} \cdot \frac{G}{4\pi w R_S} v_\infty^2 \rho_\infty = v_\infty^2 \rho_\infty. \quad (14)$$

All solar wind data cancel and we get the estimate

$$R_S \approx Q_{ce} \cdot \frac{G}{4\pi w}. \quad (15)$$

We emphasize the remarkable result that according to this estimate this distance does not depend on the solar wind. With the data for the Halley model already mentioned complemented by the neutral velocity $w = 1$ km s^{-1} we get the estimate $R_S = 10^4$ km in good agreement with the numerical results.

Using arguments of Ip and Axford [1986] we can get from this an estimate of the standoff distance R_K of the contact surface. The forces from charge exchange and elastic collisions generate a current j which reduces the field

B. Neglecting curvature forces we get the simple pressure balance formula

$$\frac{d}{dR} \frac{B^2}{8\pi} = \dot{q} \quad (16)$$

with the force $\dot{q}$ of (13). Using the same approximations as before we get the equation

$$\frac{B^2}{8\pi} + Q_{ce} \cdot \frac{G}{4\pi w R} v_\infty^2 \rho_\infty = \text{const} \quad (17)$$

which yields B as a function of R. The square root drop off of B near stagnation has been found to fit the measurements well [Ip, Axford, 1986, Cravens, 1986].

We assume that at the distance R_S where the thermal pressure is exhausted the magnetic field is piled up to a value such that the magnetic pressure is equal to the stagnation pressure of the solar wind. This gives the value of the constant in (17). The standoff distance R_K of the boundary of the magnetic cavity is the R for which formula (17) yields $B = 0$. Simple algebra gives the estimate

$$R_K \approx R_S/2 \approx Q_{ce} \cdot \frac{G}{8\pi w} \quad (18)$$

which is independent of the solar wind parameters. For the Halley parameters we get $R_K = 5000$ km. Cravens [1986] took into account the balance of dissociative recombination and photo-ionisation and derived a more specific scaling law which which implies that R_K is proportional to $G^{3/4}$. In his high resolution numerical model [Cravens, 1989] he found a drop in plasma density by about a factor of 3 within 50 km distance.

Neubauer [1988] found that the final drop from 20 nT to zero was very steep and concludes that this rest field is balanced by thermal (electron) pressure from within the contact surface. We find in the model (Figure 9) on the contact surface a peak of the gas pressure which brakes the remaining momentum of the flow onstreaming from both sides.

10. Contact surface

The real shape of the contact surface cannot be rotationally symmetric as the flow is partially controlled by the asymmetric Lorentz forces of the magnetic field in the pile up. Nevertheless, a rotationally symmetric approximation for the shape has been deduced by Cravens [1987], Ip and Axford [1986], and by Wu [1987] based on a local theory which neglects the gas pressure of the plasma altogether and balances the volume force, exerted from the expanding neutral gas via elastic collisions, by the local Lorentz force of a draped magnetic field only. This local theory fixes only the local radial gradient of the magnetic field. It has still to assume a value of the field strength from observation, a plausible dependence on the local in-

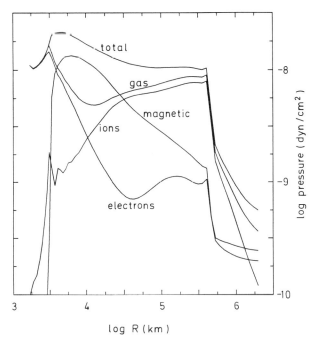

Fig.9. Pressure profiles in [dyn cm^{-2}] along the sun-comet axis. (gas = electrons + ions)

clination of the contact surface relative to the incident solar wind, and independence on the azimuth around the solar wind.

In our three-dimensional model we find the contact surface nearly rotationally symmetric on the subsolar side. We find that the shape still expands on the wings behind the nucleus. But the pressure of the pile-up field line bundle presses a dent into this paraboloid which becomes finally more wedge like. In the tail the contact surface becomes more and more ill defined and its shape as deduced from the numerical model depends on what one takes as defining property. In Figure 10 we have drawn the surface of equal field strength B = 3 nT as an approximation of the boundary of the magnetic cavity. We see that this surface extends into the tail about 15 000 km and then continues into a tail like structure which is the neutral sheet between the two regions of opposite field orientation. Note the offset of this neutral sheet which is caused by the inclination of the IMF (see also next section).

11. The tail

The christening feature of a comet is its tail. From its shape and properties, in particular its direction, Biermann [1951] deduced the existence of the solar wind. For numerical modeling it poses severe problems, since in the tail region all errors in the numerical calculations and/or in the physical assumptions accumulate along the very extended streamlines which makes it almost impossible to achieve reliable quantitative results.

It was noted already in the first calculations [Schmidt and Wegmann, 1980, 1982], that the magnetic field imposes an asymmetry in the density distribution of the ions. This makes the comet look differently if it is observed

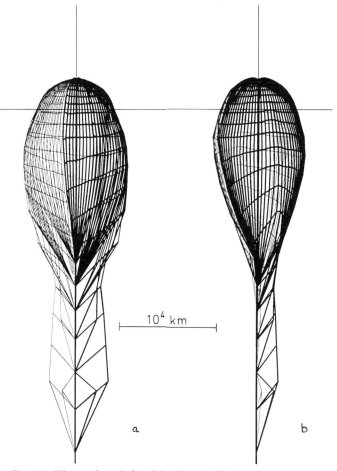

Fig.10. The surface defined by $B = 3$ nT as an approximation to the boundary of the magnetic cavity in projections perpendicular to the sun-comet axis which form with the meridional plane $z = 0$ parallel to the IMF angles of 20 (a) and 70 degrees (b). Thick lines represent the part $z > 0$ and thin lines the symmetric part $z < 0$.

in the direction along the IMF or in a direction perpendicular to the IMF. This has been used by Schmidt and Wegmann [1982] to explain streamers as the image of magnetic field rotations by about 90 degrees.

The different appearance in different directions has been confirmed by Fedder et al. [1984] and Schmidt-Voigt [1988, 1989], and recently by Russell et al. [1990]. Despite some differences in the details of the results obtained by different methods, all workers agree that the tail looks more slender when viewed perpendicular to the field than when looked parallel. Schmidt-Voigt [1988, 1989] made a time-dependent calculation with a tangential field discontinuity representing a rotation of 90^0. This had the effect of a tail condensation which might explain some observed features. A tail disconnection event which according to Niedner and Brandt [1978] should be due to a rotation of the field by 180^0 has up to now not been found in numerical simulations despite some effort of Schmidt-Voigt.

One of the most exciting results in the first calculations of Fedder et al. [1984] was the drastic flattening of the tail far behind the nucleus. The tail apparently is more like a ribbon. In fact, also on some photographic plates the tail has a ribbon like appearance, which sometimes is viewed edge on and then looks like a narrow spike. The interpretation of the ICE measurements in the tail of Giacobini-Zinner often assumes a tail structure of a plasma sheet between the two lobes of magnetic field of opposite polarity. Figure 11 shows the ion densities in cuts perpendicular to the tail axis in distances of 10^5 and $3\ 10^6$ km behind the nucleus. In this Halley model we should find at 10^5 km conditions which are similar to those of comet Giacobini-Zinner at ICE encounter. In both cuts we see the concentration in a sheet which is perpendicular to the projected IMF direction which is horizontal in all panels of Figure 11.

How is this sheet formed? We find that in cuts perpendicular to the tail axis the total pressure is very uniform. Therefore, it cannot exert any forces on the flow which is basically an expansion by a constant factor which leaves structures unaltered. The plasma sheet must be generated much earlier, namely to a certain extent already in the stagnation region. We have seen in section 9 that here the flow expands in the direction perpendicular to the field and contracts in the parallel direction. This flow pattern in a region of high ion production is suitable to generate this ribbon structure.

In our model we use a spherically symmetric neutral gas coma, which is generated by evaporation from the dayside of the nucleus with instantaneous equidistribution over the solid angle of 4π. The concentration of activity on the dayside of the nucleus has been proven by the Halley camera. It is up to now unclear which mechanism distributes the gas so evenly spherically symmetric around the nucleus. To test the hypothesis of the mechanism for the tail flattening, we have separated the ions which are generated in front of the nucleus from those which are generated behind the nucleus. In Figure 11 we have separated the contributions of day and nightside ions to the total ion density. The dayside ions exhibit the marked ribbon like condensation more clearly than the latter not only at the smaller distance but even in the far tail since they are exposed for a longer time to the asymmetric forces of the field. This gives also some credibility to the appearance of the cut through the ribbon. The hole in the center is a shadow of the contact surface, and not

(or not only) due to the effect of the excluded volume in the computational grid.

We should also like to mention the offset of about 1^0 which is reminiscent of the original inclination of the IMF.

For comparison with other work [e.g. Russell et al., 1990] we include plots of the velocity and the magnetic field strength in the cut at a distance of $3\ 10^6$ km on a larger scale. The crescent in the magnetic field contour plot marks the shock which acts on the inclined field differently as it is a quasiparallel shock on the right side and a quasiperpendicular one on the left. In the center is the current sheet with low magnetic field associated with the tail ribbon, between two lobes of magnetic field of opposite orientation.

In the velocity we see the deceleration by the shock and farther inward the asymmetric deceleration caused by the hung up magnetic field.

12. CONCLUSIONS

Global MHD modeling of cometary plasma flow using an appropriately adapted three-dimensional curvilinear grid covering more than three orders of magnitude in distance from the nucleus has been able to resolve and describe the main discontinuities of stationary cometary plasma flow such as the bow shock and the contact surface. In this way the limitations of cartesian grids, which quite naturally have to be rather coarse, are overcome to a large extent.

For future work several distinct improvements and extensions of the global cometary MHD model seem to be conceivable and called for. First there is a need to allow for a systematic drift along the magnetic field between the onstreaming solar wind protons and the implanted cometary ions caused by the inertia of the latter and controlled by mutual interaction via pitch angle scattering, two stream instabilities and collisional processes.

Another challenge is posed by the transient discontinuities in the solar wind, especially in the direction of the IMF, which are advected into and converted by the comet. To resolve and properly describe them a moveable grid seems to be necessary. By the anisotropy of the Lorentz forces these directional discontinuities bring about close contact between plasma elements with a rather diverse history of chemical and photochemical processing and interaction of neutral background and plasma. Therefore, the appropriate evaluation of in situ measurements by probes necessitates the detailed calculation of the chemical kinetics for the individual plasma elements along their individual three-dimensional lines of flow instead of the hitherto used description which uses more or less sophisticated constant one- or two-dimensional parametrizations in the state variables of the plasma only.

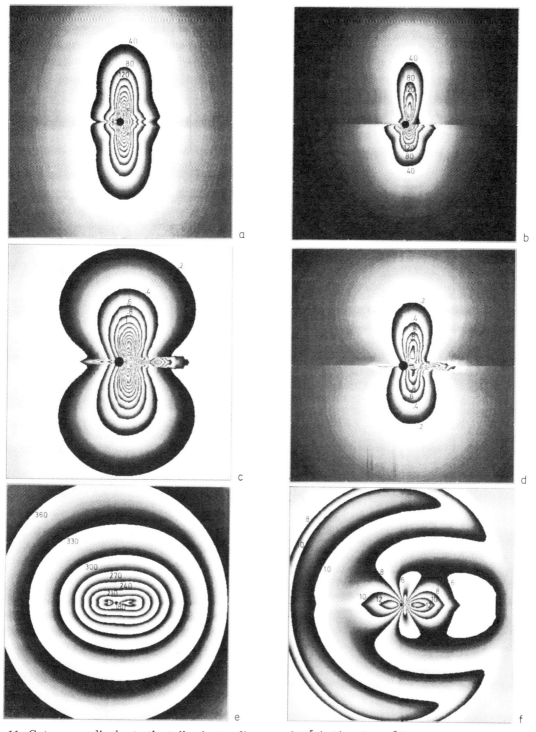

Fig.11. Cuts perpendicular to the tail axis at a distance of 10^5 (a,b) and $3\ 10^6$ km behind the nucleus. The plots extend 10^5 (a,b), 10^6 km (c,d) and $3\ 10^6$ km on each side of the axis. The projected IMF is horizontal. Right-left asymmetries are caused by the inclined IMF (see fig 1c). The orientation of the horizontal axis (the y-axis) is the same as in fig. 1. a) and c) Cometary ion density $[\mathrm{cm}^{-3}]$, b) and d) density $[\mathrm{cm}^{-3}]$ of the cometary ions generated on the dayside (upper half) and on the nightside (lower half), e) velocity $[\mathrm{km\ s}^{-1}]$, f) magnetic field strength $[\mathrm{nT}]$.

References

Biermann L., Kometenschweife und solare Korpuskular-strahlung, Z. Astrophys., 29, 274 -286, 1951.

Biermann L., B. Brosowski, and H. U. Schmidt, The interaction of the solar wind with a comet, Solar Phys., 1, 254 - 284, 1967.

Brosowski B., and R. Wegmann, Numerische Behandlung eines Kometenmodells, Meth. Verf. Math. Phys., 8, 125-145, 1973.

Cravens T. E., Physics of the cometary contact surface ESA SP-250, 1, 241 - 246, 1986.

Cravens T. E., Theory and observations of cometary ionosphere, XXVI COSPAR Meeting, Toulouse, France, 1987.

Cravens T. E., A magnetohydrodynamical model of the inner coma of comet Halley, J.Geophys. Res., 94, 15025, 1989.

Fedder J. A., S. A. Brecht, and J. G. Lyon, MHD simulation of a comet magnetosphere, NRL Memorandum report, 5306, Naval Research Laboratory, Washington, D.C., 1984

Fedder J.A., J. G. Lyon, and J. L. Giuliani, Numerical simulations of comets: Predictions for comet Giacobini-Zinner, EOS, 67, 17 -18, 1986.

Festou M. C., P. D. Feldman, M. F. A'Hearn, C. Arpigny, C. B. Cosmovici, A. C.Danks, L. A. McFadden, R. Gilmozzi, P. Patriarchi, G. P. Tozzi, M. K. Wallis, and H. A. Weaver, IUE observations of comet Halley during the Vega and Giotto encounters, Nature, 321, 361 - 363, 1986.

Galeev A. A., Plasma processes in the outer corona, in Comets in the Post-Halley Era, eds. R.L.Newburn jr, M. Neugebauer, and J.Rahe, D.Reidel, Dordrecht, 1990, to appear.

Galeev A. A., R. Z. Sagdeev, V. D. Shapiro, V. I. Shevchenko, and K. Szegö, Mass loading and MHD turbulence in the solar wind/comet interaction region, in Proc. Varenna-Abastumani International School and Workshop on Plasma Astrophysics, Sukhumi, USSR, 307, 1986.

Galeev A. A., K. I. Gringauz, S. I. Klimov, A. P. Remizov, R. Z. Sagdeev, S. P. Savin, A. Yu. Sokolov, M. I. Verigin, and K. Szegö, Critical ionization velocity effects in the inner coma of comet Halley: Measurements by Vega-2 Geophys. Res.Lett., 13, 845 - 848, 1986a

Galeev A. A., A. S. Lipatov, and R. Z. Sagdeev, Two-dimensional simulation of the relaxation of cometary ions and MHD turbulence in the flow of the solar wind around cometary atmosphere, Sov. J. Plasma Phys., 13, 323, 1987.

Galeev A. A., and R. Z. Sagdeev, Alfvén waves in space plasma and its role in the solar wind interaction with comets, Astrophys.J., 144, 427, 1988.

Giguere P.T. and W. F. Huebner, A model of comet comae I. Gas-phase chemistry in one dimension, Astrophys. J., 223, 638-654, 1978.

Goertz C.K., S. Machida, and G. Lu, On the theory of CIV, Adv. Space Res., 10, (7)33- (7)45, 1990.

Gombosi T. I., Preshock region acceleration of implanted H^+ and O^+, J. Geophys. Res., 93, 35, 1988

Gombosi T. I., Lorencz, and J. R. Jokipii, Combined first- and second-order Fermi acceleration in cometary environments, J. Geophys. Res., 94, 15 011, 1989.

Huebner W. F., and P. T. Giguere, A model of comet comae II. Effects of solar photodissociative ionization, Astrophys. J., 238, 753-762, 1980

Huebner W. F., and J. J. Keady, Energy balance and photochemical processes in the inner coma, in Cometary Exploration, ed. T. J. Gombosi, Budapest, 165-183, 1983.

Huebner W. F., The photochemistry of comets, in The Photochemistry of Atmospheres, ed. J.S. Levine, Academic press, New York, 437-481, 1985.

Huebner W. F., D. C. Boice, H. U. Schmidt, M. Schmidt-Voigt, R. Wegmann, F. M. Neubauer, and J. A. Slavin, Time-dependent study of magnetic fields in comets Giacobini-Zinner and Halley, Adv. Space Res., 9, 385 - 388, 1989.

Ip W.-H., and I. W. Axford, The formation of a magnetic-field-free cavity at comet Halley, Nature, 325, 418 - 419, 1986

Lees L., Interaction between the solar plasma and the geomagnetic cavity, Amer. Inst. Aero. Astronaut. J, 2, 1576 - 1582, 1964

Luhmann J.G., and C. T. Russell, An assessment of the conditions for critical velocity ionization at the weakly magnetized planets, Adv. Space Res., 10, (7)71-(7)76, 1990

Neubauer F.M., K. H. Glassmeier, M. Pohl, J. Raeder, M. H. Acuna, L. F. Burlaga, N. F. Ness, G. Musmann, F. Mariani, M. K. Wallis, E. Ungstrup, and H. U. Schmidt, First results from the Giotto magnetometer experiment at comet Halley, Nature, 321, 352 - 355, 1986.

Neubauer F.M., The ionopause transition and boundary layers at comet Halley from Giotto magnetic field observations, J.Geophys.Res., 93, 7272 -7281, 1988.

Neugebauer M., A. J. Lazarus, H. Balsiger, S. A. Fuselier, F. M. Neubauer, and H. Rosenbauer, The velocity distributions of cometary protons pickup by the solar wind, J. Geophys. Res., 94, 5227 - 5239, 1989.

Niedner M.B., and J. C. Brandt, Interplanetary gas XXIII. Plasma tail disconnection events in comets: Evidence for magnetic field line reconnection at interplanetary sector boundaries? Astrophys. J., 223, 655 - 670, 1978.

Ogino T., R. J. Walker, and M. Ashour-Abdalla, A three-dimensional MHD simulation of the interaction of the solar wind with comet Halley, J.Geophys.Res., 93, 9568 - 9576, 1988.

Omidi N., and D. Winske, Simulation of the solar wind interaction with the outer regions of the coma, Geophys. Res.Lett., 13 , 397, 1986

Omidi N., and D. Winske, A kinetic study of solar wind mass loading and cometary bow shocks, J.Geophys.Res., 92, 13409 -13426, 1987.

Omidi N., D. Winske, and K. B. Quest, The effect of ion/neutral collisions on the structure of electrostatic shocks: Application to cometary inner shocks, EOS, 70, 384, 1989.

Raeder J., F. M. Neubauer, N. F. Ness, and L. F. Burlaga, Macroscopic perturbations of the IMF by P/Halley as seen by the Giotto magnetometer, Astron. Astrophys., 187, 61, 1987.

Raeder J., Ein zweidimensionales, anisotropes Multiionenmodell der Plasmawechselwirkung einer Kometenatmosphäre mit dem Sonnenwind, Mitt. Inst. Geophys. Meteor. Univ. Köln, 69, 1990.

Russell C. T., G. Le, J. G. Luhmann, and J. A. Fedder, A parametric study of the solar wind interaction with comets, this volume, 1990

Sauer K., U. Motschmann, and Th. Roatsch, Plasma boundaries at comet Halley, Annales Geophysicae (in press), 1990.

Schmidt H. U., and R. Wegmann, MHD calculations for cometary plasmas, Comput. Phys. Commun., 19, 309-326, 1980.

Schmidt H. U., and R. Wegmann, Plasma flow and magnetic fields in comets, in: Comets , ed. L.L.Wilkening, University of Arizona Press, Tucson, 538 - 560, 1982.

Schmidt H. U., R. Wegmann, and F. M. Neubauer, MHD-Model for comet Halley ESA SP-250, 43 - 46, 1986

Schmidt H. U., R Wegmann, W. F. Huebner, and D. C. Boice, Cometary gas and plasma flow with detailed chemistry, Comput. Phys. Commun., 49, 17 - 59, 1988.

Schmidt- Voigt M., Zeitabhängige magnetohydrodynamische Modellrechnung für kometare Plasmen, thesis, Ludwigs-Maximilians-Universität München, 1988

Schmidt- Voigt M., Time dependent MHD simulations for cometary plasmas, Astron. Astrophys., 210, 433-454, 1989.

Schwingenschuh K., W. Riedler, G. Schelch, Ye. G. Yeroshenko, V. A. Styashkin, J. G. Luhmann, C. T. Russell, and J. A. Fedder, Cometary boundaries: VEGA observations at Halley, Adv. Space Res., 6, 217, 1986.

Schwingenschuh K., W. Riedler, Ye. G. Yeroshenko, J. L. Phillips, C. T. Russell, J. G. Luhmann, and J. A. Fedder, Magnetic field draping in the comet Halley, coma: Comparison of VEGA observations with computer simulations, Geophys. Res. Lett., 14, 640 - 643, 1987.

Vaisberg O.L., C. T. Russell, J. G. Luhmann, and K. Schwingenschuh, Small scale irregularities in comet Halley's plasma mantle: An attempt at self-consistent analysis of plasma and magnetic field data, Geophys. Res. Lett., 16, 5 - 8, 1989.

Wegmann R., H. U. Schmidt, W. F. Huebner, and D. C. Boice, Cometary MHD and chemistry, Astron. Astrophys., 187, 339 -350, 1987.

Wu Z.-J., Calculation of the shape of the contact surface at comet Halley, in: Diversity and similarity of comets, Brussels, Belgium ESA SP-278, 69 - 73, 1987.

A PARAMETRIC STUDY OF THE SOLAR WIND INTERACTION WITH COMETS

C. T. Russell, G. Le, J. G. Luhmann

Institute of Geophysics and Planetary Physics, University of California, Los Angeles, California 90024-1567

J. A. Fedder

Naval Research Laboratory, Washington, D. C. 20375

Abstract. The Naval Research Laboratory's magnetohydrodynamic simulation code is used to simulate the solar wind interaction with comet Halley for two different outgassing rates and several different solar wind states. The magnetic field is more strongly draped for fast solar wind conditions than slow. For higher mass loading rates, the tail becomes wider and contains more magnetic flux. The visual appearance of the comet differs for the case in which the interplanetary magnetic field lies in the plane of the sky from the case when it lies along the line of sight. The ion tail appears shorter in the latter case. Thus variation in the IMF direction can cause significant changes in the appearance of comets. The comet also creates a large momentum flux deficit in the solar wind with a narrow enhanced region within it corresponding to the ion tail.

Introduction

To date in-situ measurements are available from only two comets. Much has been learned from these encounters but nevertheless the sampling has been limited in terms of spatial coverage and the variety of solar wind conditions and cometary outgassing rates studied. Despite the paucity of in-situ data, it is important to investigate the behavior of cometary tails as best we can. First, cometary ion tails are well-studied optically from the Earth, and second, the Earth itself may have once passed through a cometary tail while relevant terrestrial data were being recorded [cf. Russell et al., 1988].

Comets encounter a variety of solar wind conditions during an apparition, both because of their varying heliocentric distance and because of the temporal variations in the solar wind. Furthermore, as they pass through the inner solar system their outgassing rate increases and then decreases. There are many cometary phenomena that appear to be controlled by the solar wind or interplanetary magnetic field [cf. Niedner and Brandt, 1978; Jockers, 1985]. The only well understood response is the change in tail direction due to a solar wind directional change. The causes of most other phenomena are merely conjectures.

An important advance in our ability to understand comets was the development of magnetohydrodynamic (MHD) simulations [Schmidt and Wegmann, 1980;1982]. These codes do not simulate all of the physical (or chemical) processes occurring in the comet but they do reproduce at least semi-quantitatively the deceleration of the solar wind and the pile up of the magnetic field. Furthermore, they produce a simulated comet that can be probed throughout a volume of space rather than solely along a spacecraft trajectory. It is the purpose of this paper to examine one such MHD simulation, that performed by Fedder et al. [1986], as a function of varying cometary and solar wind parameters to elicit which parameters control what features of a comet. The simulation results used will be in steady state or quasi-steady state conditions. Nevertheless, it is anticipated that by examining these steady state conditions there will be clues as to the causes of dynamic phenomena in comets as transitions between these steady states. It is also hoped that these simulations will aid cometary researchers in determining what parameters should be examined when studying the observed response of comets to changes in the solar wind.

The MHD simulation used here was developed at the Naval Research Laboratory and has been described by Fedder et al. [1986]. The comet is taken to be a spherically symmetric outgassing body releasing gas of atomic weight 20 amu which expands at 1 km/sec. The ionization rate is taken to be 10^{-6} sec^{-1} and due solely to photoionization. The equations of continuity, momentum and pressure, and Faraday's, Ampere's and Ohm's law are then solved on a mesh of dimension 33 x 29 x 29 extending 2.4 million km upstream and 7.6 million km downstream

Cometary Plasma Processes
Geophysical Monograph 61
©1991 American Geophysical Union

Table 1.   Parameters Used in Simulations

Name of run	VEGA	Slow	Normal	Fast
Production Rate (1.0E30/molecules/sec	1.3	0.6	0.6	0.6
Solar Wind Velocity (km/sec)	450	330	450	700
Solar Wind Density (amu/cc)	9	12	9	4
Solar Wind Momentum Flux (nPa)	3.04	2.18	3.04	3.27
IMF (nT)	7	7	7	7
Te (ev)	10	10	10	10
Tp (ev)	10	3	10	20

and 1.2 million km to the north, south, east and west of the nucleus. The numerical algorithm used is a partial donor cell method. The code uses free slip boundary conditions on the top and bottom and sides in which the tangential component of velocity has zero derivative at the boundary. The back surface has a slight drop (10%) in pressure across it. The interplanetary magnetic field is perpendicular to the solar wind flow in this model.

The observed magnetic field profile has been compared with the comet simulation by Schwingenschuh et al. [1986; 1987]. The observed magnetic field profile differs significantly from the predicted profile only close to the nucleus where the distance to the nucleus approaches the mesh spacing of the simulation. The simulation also predicts convection time delays similar to those observed [Schwingenschuh et al., 1987].

We examine herein the results of four simulation runs. The first run is for the conditions most typical of the solar wind at the location of the VEGA and GIOTTO encounters, which we term normal solar wind, [cf. Feldman et al., 1977] and for a mass loading rate close to that observed during the GIOTTO encounter 0.6×10^{30} molecules per second (1.2×10^{31} amu s^{-1}), [Reinhard, 1986]. The second simulation is for conditions typical of a slow solar wind; the third for a fast solar wind. The fourth simulation keeps the conditions for a typical solar wind but increases the outgassing rate to a value close to that observed during the VEGA encounters, 1.3×10^{30} molecules per second (2.6×10^{31} amu s^{-1}), [Gringauz et al., 1986]. The parameters used in these runs are given in Table 1. We note that the solar wind momentum flux does not vary much between these runs even though the other parameters do. This behavior is an observed property of the solar wind. The standard deviation of the momentum flux in the solar wind is only about 8% of its mean value [Feldman et al., 1977].

The plan of the paper is as follows. To introduce the simulation and its parametric dependences we will examine the draped magnetic field pattern for the various runs. Then we examine the integrated line of sight density to determine how the comet might look to an observer under these varying conditions. A preliminary report on these results has been presented elsewhere [Russell et al., 1989] but these results are repeated here so that the results in the later sections of the paper may be properly understood. Next we examine the variation of the magnetic field direction, density, velocity and momentum flux for different models and distances down the tail using contour plots on cross-sections of the tail.

Magnetic Field Line Draping

Figure 1 shows the magnetic field lines in the plane containing the solar wind flow, the upstream magnetic field and the cometary nucleus for the slow, normal and fast solar wind runs at the GIOTTO mass loading rate and for normal solar wind conditions at the VEGA rate. The choice of lines to be drawn was governed by a desire to provide a representation of the field strength while not saturating the plot with field lines in the regions of maximum field. Since the magnetic field strength is proportional to the number of magnetic field lines crossing a unit area perpendicular to the field and since this area is determined by two dimensions (one parallel to the solar wind flow and one into the page) we spaced field lines along the x-axis through the comet inversely proportional to the square root of the field strength and traced the field lines upward and downward from the x-axis. Hence field lines are not equispaced in time in this display. All four runs show the strongly draped field configuration expected for a comet of the size of Halley. As a measure of the amount of draping we take the intersection of the field line

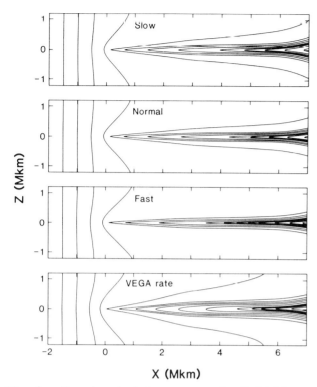

Fig. 1. From top to bottom, magnetic field lines in the plane containing the nucleus for slow solar wind, normal solar wind and fast solar wind for a mass loading rate of 0.6 x 10^{30} molecules per second. The bottom panel shows the magnetic field lines for normal solar wind conditions and a mass loading rate of 1.3 x 10^{30} molecules per second. [Russell et al., 1989].

passing through the nucleus with the sides of the simulation box. The closer is this field line to the x-axis, the more strongly draped is the tail. The draping is stronger for the faster flowing solar wind which produces a more tightly confined tail. The increased mass loading rate field line configuration is most like that of the slow solar wind, strongly draped but with a somewhat wider tail than those of the normal and fast solar wind runs.

Figure 1 also reveals some of the limitations of this MHD model. We would expect for instance that the tail plasma would accelerate with distance from the nucleus. This would cause field lines to spread apart in x. This occurs in Figure 1 to distances of 4 to 5×10^6 km. However after this point the field lines begin to approach each other representing an increasing field with distance from the nucleus. Similarly we see an increasing flare beyond about 6×10^6 km. These artifacts are most probably respectively due to the inability of the finite cell-size to treat the narrow current layer in the distant tail and the growing influence of the rear simulation boundary over the sides as the flow approaches the end of the simulation space.

While the reader should be aware of these imperfections, we feel the accuracy of the simulation is sufficient to provide the lessons concerning cometary behavior for which this paper is intended.

Line-of-Sight Integrated Densities

In order to determine how the comet should appear visually, we must first separate the solar wind ions from the cometary ions and then integrate the number of cometary ions along the line-of-sight. To separate the solar wind ions we note that the solar wind mass flux is divergenceless, i.e.,

$$\nabla \cdot \rho_{sw} \underline{V}_{sw} = 0 \qquad (1)$$

This can be determined from the MHD solution by taking the numerical divergence appropriate for the Partial Donor Cell method used in the computations:

$$\Sigma_\alpha \nabla_\alpha (S_\alpha V_\alpha M_\alpha^+ (\rho)) = 0 \ (\alpha = X, Y, Z) \qquad (2)$$

$$\text{where } \nabla f_j = f_j - f_{j-1}$$

$$M^+ f_j = 1/2 (f_j + f_{j+1})$$

where S is the area of the face of a cube in the computational grid and α denotes the direction of each component or operator in the Cartesian grid. This equation is used to determine the downstream value of the mass flux from the previous upstream value, thus allowing the determination of the mass flux added by the comet (which is not divergence-less).

The favorable comparison between the calculated magnetic field profile and that observed by the VEGA spacecraft [Schwingenschuh et al., 1986; 1987] gives us confidence in the model's applicability. The observed magnetic field profile differs significantly from the predicted profile only close to the nucleus where the distance to the nucleus approaches the grid spacing of the simulation. The simulation also predicts convection times similar to those observed.

The integrated density of cometary ions shown in Figures 2-5 has been calculated along the line-of-sight for two orientations of the interplanetary magnetic field with the observer at infinity perpendicular to the solar wind flow. In the top panel of each display the magnetic field is perpendicular to the line-of-sight, and in the bottom panel along it. Thus in the top panel the observer is looking along the thickest part of the ion tail and in the lower panel across the thinnest part of the sheet. Figure 2 shows contour plots for the normal solar wind case. As one would expect the tail visually extends further from the comet when the magnetic field is perpendicular to the line-of-sight than when it is along it. Thus simple rotations of the IMF about the solar wind flow vector can cause the cometary ion tail to appear to lengthen or shorten without any real change in the comet.

Normal Solar Wind

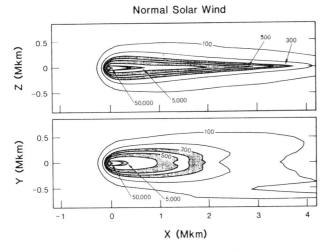

Slow Solar Wind

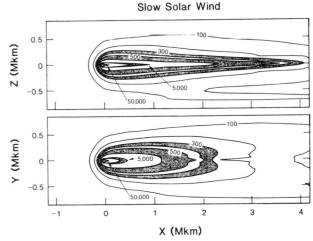

Fig. 2. Line-of-sight integrations of the cometary ion density for the normal solar wind run with a mass-loading rate of 0.6×10^{30} molecules s^{-1}. Top panel shows the projection into the plane containing the interplanetary magnetic field. Bottom panel shows the projection into the orthogonal plane. Integrated densities are given in units of 10^{10} ions cm^{-2}. [Russell et al., 1989].

Fig. 4. Line-of-sight integrated cometary ion densities similar to Figure 3 but for slow solar wind. [Russell et al., 1989].

up to higher values when the flow is slower. Since the ion sheet here is thin and ribbon-like the difference in the orthogonal views is quite marked. Figure 5 shows the integrated cometary ion density

Fast Solar Wind

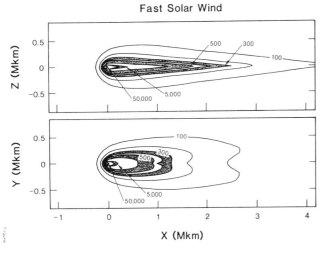

Fig. 3. Line-of-sight integrated cometary ion densities similar to Figure 3 but for fast solar wind. [Russell et al., 1989].

Figure 3 shows the analogous displays for the fast solar wind. Here the difference between the two orthogonal views is less because the ion sheet is not so ribbon-like. The ion tail, perhaps counter-intuitively is shorter. This occurs because the fast solar wind carries cometary ions away faster so that the density does not build up to as high a value. Figure 4 shows the integrated cometary ion density for the slow solar wind. As expected from the comparison of the normal and fast solar wind cases, the slow solar wind causes a much longer visible tail because the ion density builds

VEGA Mass Loading Rate

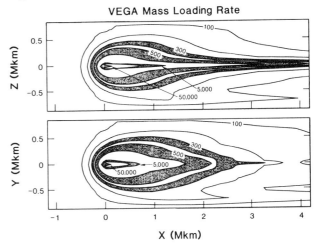

Fig. 5. Line-of-sight integrated cometary ion densities similar to Figure 3 but for outgassing rates similar to those encountered by VEGA. [Russell et al., 1989].

for the higher mass loading rate appropriate to the VEGA encounters. The difference in the two views is again quite marked, even more so than at lower mass loading rates. As in all of these cases, the tail is longer when viewed orthogonal to the magnetic field and shorter when viewed along it. Closer to the nucleus we find that the cometary ion density is greater at all orientations of the field for this higher mass loading rate and extends further upstream.

The cometary magnetic tail and its ion tail are not coincident. The greatest ion density lies at the sharp kink in the magnetic field and extends in a ribbon whose thinnest dimension is in the vertical direction in Figure 1 and whose width is in the direction into the page. Thus, the integrated density along the line-of-sight will vary depending on whether the observer is sighting along the thickest or thinnest dimension of the ion tail

Cross-sections of the Cometary Wake

One of the dominant visual features of a comet is its plasma tail. Surrounding this plasma, or ion tail, is an unseen magnetic tail with enhanced magnetic field and reduced momentum flux. In order to characterize this region which must play an important role in the dynamics of cometary tails we have examined the value of various parameters across the cross-section of the tail at two distances downstream from the nucleus, 1×10^6 km and 6×10^6 km. Figure 6 shows contour maps of the mass density for these two distances for the three models: normal solar wind, the VEGA mass loading rate and the fast solar wind cases. We do not present contour maps of the slow solar wind case because they are very similar to those of the VEGA mass loading case as can be inferred from a comparison of Figures 4 and 5. No symmetry has been assumed in these simulations so that the observed symmetry and deviations from it are some indication of the level of numerical artifacts and transient effects. In each panel the upstream magnetic field is horizontal.

The middle panel shows the mass densities for the normal solar wind. At 1×10^6 km there is a moderately narrow ribbon of high densities near the center of the tail elongated in a direction perpendicular to the magnetic field. The maximum density here is 1000 amu cm^{-3}. Further from the tail axis are two lunes of low densities beyond which the density rises. At 6 million km the ribbon of high density has become pinched into 2 narrower ribbons but the density has fallen by 2 orders of magnitude to 11 amu cm^{-3}. The density minima to the sides are still present but their shape has evolved from lunes to tear drops. The minimum density at 6×10^6 km is about 1/4 that at 1×10^6 km.

The top two panels show the fast solar wind model. Here the density is less than in the normal solar wind case by a factor of about 2 to 3 at both 1 and 6×10^6 km distances. Generally the contours have the same shape as in the normal solar wind case. The high mass loading rate model shown in the bottom two panels shows a similar pattern. The densities in the ion tail are a factor of about 3 greater here than in the normal solar wind case with slightly over a factor of 2 smaller mass loading rate. There are no lobes of minimum density to the sides of the ion tail at 1×10^6 km but such lobes have developed, albeit weakly, at 6×10^6 km.

Figure 7 shows the corresponding contours for the magnetic field strength. Examining the normal

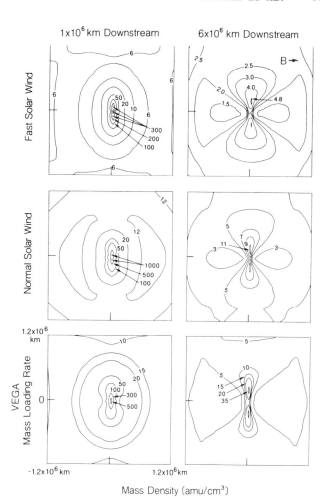

Fig. 6. Contour maps in planes perpendicular to the solar wind flow of total mass density at distances of 1 million and 6 million km downstream for the nucleus for the fast solar wind simulation (top), the normal solar wind simulation (middle) and the simulation using the VEGA outgassing rate (bottom). The upstream magnetic field is horizontal in these diagrams.

solar wind case in the center two panels, we see that there are two magnetic lobes of enhanced field strength at both distances separated by a narrow ribbon of reduced field strength corresponding roughly to the ion density ribbon seen in Figure 6. The minimum field strength in the center of the ion tail falls with distance down the tail in the model. The model is limited in its ability to resolve the narrowest minima in field strength or maxima in density. Nevertheless, the qualitative trends should be correct.

The fast solar wind cases are shown in the top 2 panels. They are very similar to the normal solar wind cases except that the field is about 2 nT stronger at large distances from the tail axis. The higher mass loading rate in the bottom two panels also makes very little difference in the

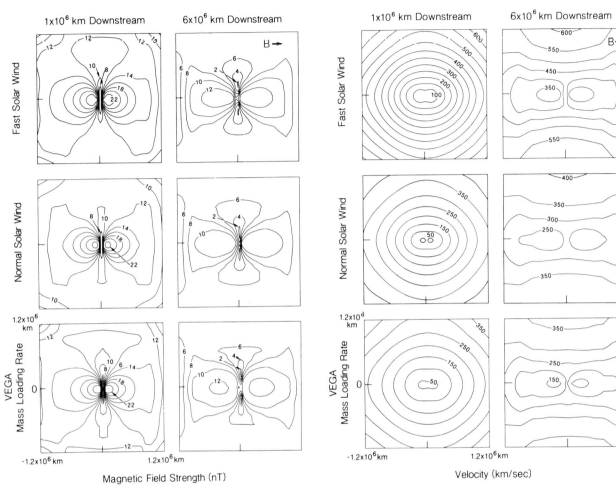

Fig. 7. Similar to Figure 6 but for the magnetic field strength.

Fig. 8. Similar to Figure 6 but for the anti-solar component of the ion velocity.

field strength. At 10^6 km the field strength is about 2 nT greater but at 6×10^6 km it does not differ much from the normal solar wind case. As with the ion density contours the ribbons of field minima are slightly thinner, more ribbon-like and deeper. The lobes of maximum field strength are somewhat higher and larger in dimension.

Figure 8 shows contours of the anti-solar directed component of velocity for the three models. The behavior of all the models is similar. The anti-solar velocity is lowest near the nucleus and in the inner magnetic lobes. The velocity contours are stretched out along the magnetic field direction so that field lines that pass near the nucleus are moving more slowly over their entire length (at least over the interval contoured) than field lines that pass farther from the nucleus. The principal differences between these cases is that the fast solar wind case has the highest velocities and the high mass loading rate case has the lowest velocities. At 6×10^6 km the velocities in the high density ribbon is higher than that in the adjacent plasma. This is in

accord with the acceleration of the ion tail in the anti-solar direction by the Lorentz force and the vanishing contribution of this force in the magnetic lobes. The field lines through the low velocity lobes cross the tail closer to the nucleus where the plasma in the ion tail is moving more slowly.

The solar wind interaction with a comet exerts a drag on the solar wind which decelerates it at the same time the cometary ions are accelerated. The region of solar wind deceleration is quite large compared to the visible ion tail. One way to study this acceleration and deceleration process is to examine the momentum flux moving down the tail. Figure 9 shows these contours in the same format as before. In the bottom 2 models the undisturbed momentum flux in the solar wind is 3.04 nPa and in the top 3.27 nPa. Thus at 1.0 million km downstream in the normal solar wind model the principal effect on the solar wind is a decrease in momentum flux over most of the region contoured with an enhancement in momentum flux in the ribbon of the ion tail. The transverse transfer of momentum flux

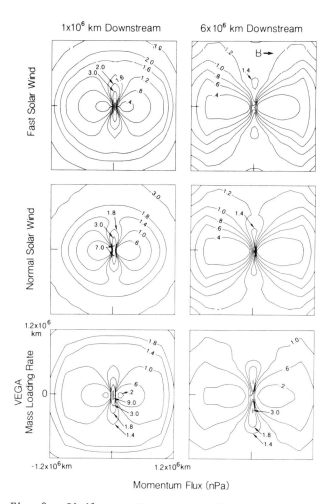

1x10⁶ km Downstream   6x10⁶ km Downstream

Fast Solar Wind

Normal Solar Wind

1.2x10⁶ km

0

VEGA Mass Loading Rate

-1.2x10⁶km   1.2x10⁶km

Momentum Flux (nPa)

Fig. 9. Similar to Figure 6 but for the momentum flux.

from the neighboring solar wind streamlines to the tail proper is affected by the ability of the magnetic stresses to transport momentum along field lines. [cf. Schmidt and Wegmann, 1982]. At 6.0 million km the region of reduced momentum flux is larger and deeper and the momentum flux enhancement in the ion tail ribbon is reduced, presumably due to the spreading of the streamlines. There is very little difference here between the results for the fast and the normal solar winds. In the high mass loading rate model the region of reduced momentum flux is larger and the minima deeper. The momentum flux in the ion tail ribbon is similar to that of the normal solar wind at 1.0 million km and at 6.0 million km.

Discussion and Conclusions

Computer simulations provide an important adjunct to space observations even when the models do not duplicate the entire plethora of physical processes that are taking place because they allow the entire volume of the interaction to be probed

and because they permit parametric studies to be undertaken. Our investigations with an MHD model [Fedder et al., 1986] have shown that magnetic draping and the pile-up of magnetic field lines varies as expected as the solar wind velocity varies and as the mass loading rate changes.

The visual appearance of a comet also depends on whether the magnetic field is in the plane of the sky or is orthogonal to it. When the comet is viewed along the interplanetary magnetic field direction the visible tail appears to be shorter. Hence rotations of the IMF should lead to apparent tail lengthenings and shortenings.

Finally, the comet both blocks the solar wind momentum flux over a wide area and accelerates the mass in the ion tail so that there is a region of reduced momentum flux surrounding a region of enhanced flux. It is possible that when the Earth passed through Halley's tail in 1910 such a region was indeed encountered [Russell et al., 1987, 1988].

Acknowledgments. The work at UCLA was supported by the National Aeronautics and Space Administration under grant NAGW-717.

References

Fedder, J. A., J. G. Lyon and J. L. Giuliani, Jr., Numerical simulations of comets: Predictions for comet Giacobini-Zinner, EOS, 67, 17-18, 1986.

Feldman, W. C., J. R. Asbridge, S. J. Bame and J. T. Gosling, Plasma and magnetic fields from the Sun, in The Solar Output and its Variation, ed. O. R. White, pp. 351-382, Colorado Assoc. Univ. Press, Boulder, Colorado, 1977.

Gringauz, K. I., et al., First in-situ plasma and neutral gas measurements at comet Halley, Nature, 321, 282, 1986.

Jockers, K., The ion tail of comet Kohoutek 1973 XII during 17 days of solar wind gusts, Astron. Astrophys. Suppl. Ser., 62, 791-838, 1985.

Niedner, M. B., Jr. and J. Brandt, Interplanetary gas XXIII. Plasma tail disconnection events in comets: Evidence for magnetic field reconnection at interplanetary sector boundaries, Astrophys. J., 223, 655, 1978.

Reinhard, R., The Giotto encounter with comet Halley, Nature, 321, 313, 1986.

Russell, C. T., M. A. Saunders, J. L. Phillips and J. A. Fedder, Near-tail reconnection as the cause of cometary tail disconnections, J. Geophys. Res., 91, 1417-1423, 1986.

Russell, C. T., J. L. Phillips, J. A. Fedder, J. H. Allen, L. Morris and R. A. Craig, Effect of possible passage through Halley's magnetic tail on geomagnetic activity, J. Geophys. Res., 92, 11195-11200, 1987.

Russell, C. T., M. von Dornum, R. L. McPherron, J. H. Allen and L. Morris, Geomagnetic activity during the passage of the Earth through Halley's tail   in 1910, Nature, 333, 338-340, 1988.

Russell, C. T., G. Le, J. G. Luhmann and J. A. Fedder, The visual appearance of comets under varying solar wind conditions, Adv. Space Res., 9, (3)393-(3)396, 1989.

Schmidt, H. U. and R. Wegmann, MHD calculations for cometary plasmas, Comp. Phys. Com., 19, 309-326, 1980.

Schmidt, H. U. and R. Wegmann, Plasma flow and magnetic fields in comets, in Comets, edited by L. L. Wilkening, University of Arizona Press, Tucson, 1982.

Schwingenschuh, K., W. Riedler, G. Schelch, Ye. G. Yeroshenko, V. A. Styashkin, J. G. Luhmann, C. T. Russell and J. A. Fedder, Cometary boundaries: VEGA observations at Halley, Adv. Space Res., 6, 217, 1986.

Schwingenschuh, K., W. Riedler, Ye. G. Yeroshenko, J. L. Phillips, C. T. Russell, J. G. Luhmann and J. A. Fedder, Magnetic field draping in the comet Halley coma: Comparison of VEGA observations with computer simulations, Geophys. Res. Lett., 14, 640-643, 1987.

SIMULATION STUDIES OF THE INTERACTION OF A NEUTRAL GAS AND A FLOWING PLASMA

R Bingham

Rutherford Appleton Laboratory, Chilton, Didcot,Oxon, OX11 0QX

R Bollens, F Kazeminejad and J M Dawson

Department of Physics, UCLA, Los Angeles, CA.

Abstract The interaction of neutral gas and an ambient flowing plasma have been simulated using 2- and 3-D hybrid codes with kinetic ions and massless fluid electrons. The codes are generalized to include the production of plasma by a gradually ionizing gas in a flowing plasma, allowing important problems in space science, such as the pick up of plasma (comet gas) by the solar wind and artificially created comets produced by the AMPTE barium releases in the solar wind, to be studied. In the simulations of the AMPTE artificial comet, we have been able to demonstrate the generation of a diamagnetic cavity, which slows and deflects the solar wind protons, comet particle acceleration and the sideways deflection of the comet head and density ripples appearing on one side of the comet head which are explained in terms of the Rayleigh Taylor instability. The structure of the cavity as well as the magnetic field surrounding it is highly asymmetric.

Introduction

One of the main experiments of the Active Magnetospheric Particle Explorers (AMPTE) satellite mission was the release of neutral atoms in the solar wind creating for the first time two man-made artificial "comets", the results of which provide an invaluable data set to test both the theoretical and computational models of collisionless coupling at high Mach number ($M_A \gg 1$) [Haerendel et al. 1986]. It is now possible for the first time to develop computational codes which can be vigorously tested using such data sets. This problem is relevant to a number of astrophysical situations such as the pick-up of cometary ions [Ip and Axford 1982], the study of planetary ion exospheres interacting with the solar wind [Hartle et al. 1973] and the pick-up of helium ions of interstellar or planetary origin [Wu et al. 1973]. The experiments are particularly important in trying to understand the collisionless coupling processes occurring in the interaction of the solar wind with the released "cometary" plasma and will provide valuable insight into momentum and energy coupling processes between the solar wind and the plasma of real comets. In this paper we shall concentrate mainly on the modelling of the release which took place on the 27th December 1984 [Haerendel et al. 1986, Valenzuela et al. 1986, Rees et al. 1986, Lühr et al. 1986, Rodgers et al. 1986,

Cometary Plasma Processes
Geophysical Monograph 61
©1991 American Geophysical Union

Woolliscroft et al. 1986] using 2-D and 3-D hybrid codes with kinetic ions and massless fluid electrons. In the experiment a barium cloud was released into the solar wind its ionization process was similar to the case of gases produced by a real comet ie. a gradual ionization by the solar ultraviolet radiation. The cloud resembled those of real comets in that a head and tail were clearly visible [Haerendel et al. 1986]. However, the size of the released cloud being smaller than the ion gyroradius of a solar wind ion or the release ions provides an opportunity of studying small scale effects where finite Larmor radius effects dominate, such effects are sometimes not always observable in planetary magnetospheres or comets.

The artificial comet head was observed to perform a transverse movement with respect to the incoming solar wind flow ie., a sideways deflection. The in situ AMPTE spacecraft were able to measure the plasma parameters and wave signatures as they moved through the cloud. Recording important effects such as the generation of a diamagnetic cavity within the volume enclosed by the expanding cloud and also the generation of a shock like region at the interface of the solar wind and the cometary plasma boundary. Intense bursts of electrostatic [Gurnett et al. 1986] and the magnetic [Klöcker et al. 1988] wave activity together with energization of electrons are also observed on the upstream side of the ion clouds with the electrostatic wave signature similar to observations at the bow shock [Gurnett et al. 1986].

The modelling of the barium release was done using 2 and 3 dimensional hybrid codes [Kazeminejad et al. 1989a,1989b] where the electrons are treated as a massless fluid and the ions as particles. Due to the much smaller inertia of such an artificial comet, electromagnetic forces play a much more significant role than in the case of a real comet and thus rather sophisticated plasma models need to be used to describe the results. The spatial dimension of the comet was of the order of the ion Larmor radius, therefore, prohibiting the use of an MHD description. To model finite Larmor radius effects properly it is important to use a hybrid code. To properly model the observations the codes had to include

1. The gradual ionization of the barium gas

2. Multiple ion species

3. The streaming of the solar wind through the cometary plasma

4. The collective behaviour of the plasma particles in their self-consistent electromagnetic fields

5. Kinetic effects on the ions, by treating them as Vlasov ions.

Using such hybrid codes we can study the ion dynamics and the structure of the diamagnetic cavity together with magnetohydrodynamic waves. Most of the plasma effects mentioned in the AMPTE experimental papers were qualitatively observed in the simulations especially the sideways deflection with the exception of the electron energization. Other features such as the slowing and southward deflection of the solar wind ions, northward acceleration of Ba^+ ions and formation of density ripples on the northern flank are also easily studied using such codes. A notable feature, that of the sideways deflection, required the kinetic nature of the codes which cannot be studied using MHD codes. Other features which require a kinetic description are the upward acceleration of the cometary particles and the density ripples seen on the top boundary giving rise to an asymmetric cloud, MHD gives a symmetric structure due to the very sharp density gradient.

Previous simulations, by various research groups, of the collisionless coupling between the flowing plasma and the released particles using a variety of 1-D, 2-D and 3-D codes have already been carried out. In particular one dimensional simulations using hybrid codes have been used by Lui et al., [1986] and Chapman and Schwartz [1987] to study the early time of the interaction, demonstrating the slowing down of the solar wind due to mass loading and acceleration of the tail particles. Two and Three dimensional results by Bingham et al., [1988] and Brecht and Thomas [1987] obtained from hybrid and fluid codes show the deflection as well as slowing down of the solar wind ions. The deflection allows the solar wind particles to flow around the object as well as magnetic field draping around the object. With regard to the side-ways deflection of the artificial comet it is worth mentioning that there is so far six mechanisms proposed to explain the observation. A complete review of the results of these simulations and mechanisms including aspects of the simulations reported in this paper can be found in the article by Lui [1990], the present paper will only report the results of the authors simulations.

The paper will concentrate on the results of two and three dimensional simulations. Particular attention will be placed on modelling as closely as possible, some of the main features observed in the experiment by the two spacecraft, this includes the formation of a diamagnetic cavity and structure of the magnetic field surrounding the cloud, the behaviour of the solar wind ions and the released ions, as well as the electric fields generated. Although we try to simulate the actual experiment it is impossible to use the exact plasma parameters due to computational limitations the emphasis is on the qualitative similarities with the experiment rather than quantitative.

In the next section a brief description of the release experiment is given, section 3 describes the analytical treatment and section 4 describes the simulation model and results and compares it to observations.

Experimental Observations

On December 27, 1984 in the solar wind and upstream of the earth's bow shock at a distance of approximately 17 Earth radii the IRM (Ion Release Module) spacecraft created an expanding cloud of barium atoms which were rapidly ionized by solar UV with a photoionization time of about 30 sec. The solar wind was flowing at a speed of 540 km/sec, it's density was about $5cm^{-3}$ and the ambient magnetic field before the release was measured to have a strength of about $10nT$ with its main component perpendicular to the solar wind flow in the y direction.

The interaction of the solar wind, with the expanding barium plasma, created a diamagnetic cavity in the solar wind as shown in Figure 1.

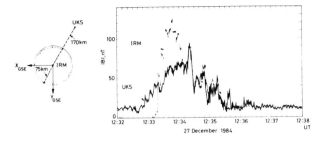

AMPTE Ba (comet) release in solar wind

Fig 1. Diamagnetic cavity created by barium-ion release at 12:32 UT, 27th December. The cavity is evident from the depression to zero of the magnetic field measured by the IRM at the centre of the cavity. Enhanced magnetic fields are recorded by the UKS outside the cavity and the IRM when it emerges. In the Geocentric Solar Ecliptic coordinate system employed above, x is directed towards the sun and y towards the ecliptic pole. On this occasion B remained approximately in the $x - y$ plane. (Courtesy H Lühr and D J Southwood).

The cavity created attained a radius of about 75km after 60 sec. Measurements of the proton velocity [Rodgers et al. 1986] reveal that the solar wind flow is slowed down and deflected around the barium clouds, the velocity component in the direction of flow being brought almost to zero. The UKS and IRM spacecraft were separated by about 170 km with the UKS spacecraft positioned downstream of the IRM with respect to the solar wind flow direction and just outside the cavity. The UKS spacecraft never entered the cavity ie., the region of zero magnetic field, it measured magnetic field line draping as well as energetic particle fluxes of both electrons and ions [Rodgers et al. 1986] generated during the release and also plasma waves activity [Woolliscroft et al. 1986]. As well as the in situ measurements made by the IRM and the UKS spacecraft, ground based and airborne optical data were also obtained using a highly sensitive low-light-level television system [Haerendel et al. 1986, Rees et al. 1986] and the UCL Doppler imaging system [Rees et al. 1986]. It was from these observations that produced one of the greatest surprises. Rather than moving in the flow direction of the solar wind, the comet head is displaced sideways during the initial $4\frac{1}{2}$ min, the speed of the deflection reached about 6km/sec. Only after the density and brightness had decayed substantially did the comet exhibit the expected motion in the solar wind direction. Figure 2 taken from reference 5 depicts the transverse motion of the comet head during the first few minutes and also the subsequent motion in the solar wind direction seen by the low-light-level television camera onboard the aircraft. The comet head is seen to expand and move in a direction opposite to the convective electric field $E(= -\mathbf{v}_{sw} \times \mathbf{B})$ where $\mathbf{v}_{sw}$ is the solar wind velocity and B is the ambient magnetic field before the release.

Similar observations of the comet taken from the South Pacific and from Boulder, Colorado [Valenzuela et al. 1986], figure 1 in the paper of Valenzuela et al. [1986], reveal an asymmetry in the density structure of the comet. The northern flank has very definite protrusions pointing in the direction of the convective electric field while the southern flank appears smooth with a more tightly confined boundary and steeper density gradients as shown by the contours in figure 1. taken from Valenzuela et al. [1986]. Using the 3-D ion energy per charge spectrometer on UKS, [Coates et al. 1988] were able to observe the momentum flow of the observed barium ions and also the solar wind protons as they traversed the cavity. The velocity compo-

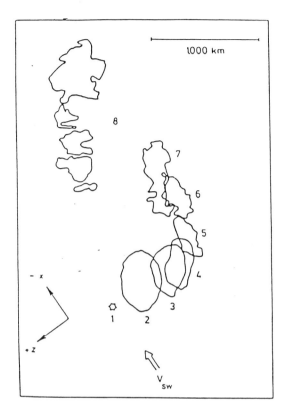

Fig 2. Real experimental observations of the movement of the AMPTE cloud. Rather than moving in the flow direction of the solar wind, the comet head is displaced sideways during the initial 4.5 min. The numbers 1-7 refer to the outlines of the comet head as a function of time.

$1 + 4s; 2, +181s; 3, +242s; 4, +256s; 5, +280s; 6, +288s; 7, +294s;$
$8, +306s$ (Taken from reference 5).

nents in GSE coordinates are shown in figure 3 (taken from reference 14). It is clear from the measurement made by the 3-D ion analyser that the effect of the release was to slow and divert the solar wind, however, owing to its initial high speed the net effect was a deflection in the southward direction as it approached the comet. There was no evidence of reflected ions from the shock like structure. Further studies by [*Coates et al.* 1988] have shown that the solar wind protons are deflected southwards whereas there was a component of barium ions accelerated northwards ie. in the direction of the convective electric field.

The magnetic field showed a very sharp increase from zero in the cavity to a maximum of around $120nT$, which is a 12 fold increase over the ambient field strength at the front, creating a pile up region upstream of the magnetic cavity characterized by an enhanced level of wave activity. This region is followed by a series of magnetic pulses with a period of about 10sec, *Klöcker et al.* [1988] have suggested that these wave pulses could be the magnetoacoustic mode.

Simulation model and results

The simulations were done using particle in cell codes which treat the electrons as a massless fluid and the ions as fully kinetic and known as hybrid Vlasov-fluid Code or hybrid code for short [*Kazeminejad et al.* 1989b]. A hybrid code is intermediate between a full particle and an MHD code. It can resolve disturbances on the scale

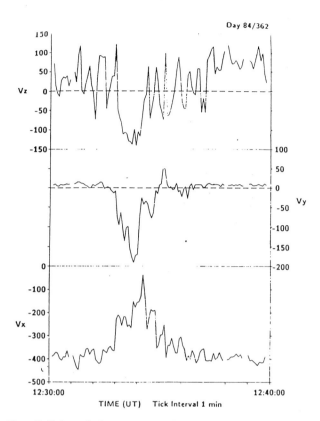

Fig 3. Bulk ion velocity components during the release, assuming all ions are protons (taken from reference 12)

of the ion gyroradius, its basic time scale is the ion gyroperiod.

The equations describing each ion are

$$\frac{d\mathbf{x}_i}{dt} = \mathbf{v}_i \tag{1}$$

$$\frac{d\mathbf{v}_i}{dt} = \frac{e}{M_i}\left(\mathbf{E} + \frac{\mathbf{v}_i \times \mathbf{B}}{c}\right) \tag{2}$$

where subscript i refers to each ion.

The massless electrons however satisfy the following equation:

$$\mathbf{E} + \frac{\mathbf{v}_e \times \mathbf{B}}{c} = 0 \tag{3}$$

where $\mathbf{v}_e$ is the electron fluid velocity. In writing equation 3 we assume perfect conductivity along $\mathbf{B}$; eg.:

$$\mathbf{E}.\mathbf{B} = 0 \tag{4}$$

Quasineutrality is also assumed

$$n_e = n_i = \int f_i d\mathbf{v} = n \tag{5}$$

n_e, n_i are the electron and ion densities and f_i is the ion distribution function.

The current $\mathbf{j}$ is therefore:

$$\mathbf{j} = -en\mathbf{v}_e + e\int \mathbf{v}f_i d\mathbf{v} \tag{6}$$

This current $\mathbf{j}$, according to equation 5 (quasineutrality assumption) must be divergence free. Thus:

$$\underline{\nabla} \cdot \mathbf{j} = 0 \qquad (7)$$

This is equivalent to:

$$\mathbf{j}_L = 0 \qquad (8)$$

where $\mathbf{j}_L$ is the longitudinal current ($\mathbf{j}_L \parallel \mathbf{k}$). The relevant Maxwell equations are Faraday's law and Ampere's law without the displacement current; ie.

$$\underline{\nabla} \times \mathbf{E} = -\frac{1}{c}\frac{\partial \mathbf{B}}{\partial t} \qquad (9)$$

$$\underline{\nabla} \times \mathbf{B} = \frac{4\pi}{c}\mathbf{j} \qquad (10)$$

Substituting 6 into 10 gives:

$$\mathbf{j} = \frac{c}{4\pi}\underline{\nabla} \times \mathbf{B} = -en\mathbf{v}_e + e\int f_i \mathbf{v} d\mathbf{v} \qquad (11)$$

Which then implies:

$$\mathbf{v}_e = \frac{e\int \mathbf{v} f_i d\mathbf{v} - \frac{c}{4\pi}\underline{\nabla}\times\mathbf{B}}{ne} \qquad (12)$$

As a result using 3 and 12 we have:

$$\mathbf{E} = -\frac{\mathbf{v}_e \times \mathbf{B}}{c} = -\frac{\mathbf{B}\times(\underline{\nabla}\times\mathbf{B})}{4\pi ne} + \frac{\mathbf{B}\times\int\mathbf{v}f_i d\mathbf{v}}{nc} \qquad (13)$$

Finally using 9 and 13 we get:

$$\frac{\partial \mathbf{B}}{\partial t} = c\underline{\nabla}\times\left\{\frac{\mathbf{B}\times(\underline{\nabla}\times\mathbf{B})}{4\pi ne} - \frac{\mathbf{B}\times\int\mathbf{v}f_i d\mathbf{v}}{nc}\right\} \qquad (14)$$

Using $\mathbf{v}' \left(= \int \mathbf{v}f_i d\mathbf{v}/n_e\right)$ as the average ion velocity in a given cell equation 13 can be written as

$$\mathbf{E} = -\frac{\mathbf{B}\times(\underline{\nabla}\times\mathbf{B})}{4\pi ne} + \frac{\mathbf{B}\times\mathbf{v}'}{c} \qquad (15)$$

and using equations 2 and 15 we can obtain

$$\frac{d\mathbf{v}_i}{dt} = \frac{e}{M_i}\frac{(\mathbf{v}_i - \mathbf{v}')\times\mathbf{B}}{c} - \frac{e}{M_i}\frac{\mathbf{B}\times(\underline{\nabla}\times\mathbf{B})}{4\pi ne} \qquad (16)$$

and from 14 we get the expression for the magnetic field

$$\frac{\partial \mathbf{B}}{\partial t} = c\underline{\nabla}\times\left\{\frac{\mathbf{B}\times(\underline{\nabla}\times\mathbf{B})}{4\pi ne} - \frac{\mathbf{B}\times\mathbf{v}'}{c}\right\} \qquad (17)$$

Equations 16 and 17 form the analytic basis of our model. They involve only $\mathbf{v}_i$, $\mathbf{v}'$ and $\mathbf{B}$($\mathbf{E}$ does not appear explicitly).

In the simulations, initially the solar wind particles were positioned on a lattice for the 2-D runs and a cell for the 3-D runs in a regular array moving to the right in the simulation box. The system size was 256×256 grids for the 2-D case and $256 \times 256 \times 256$ for the 3-D case, each grid corresponding to one ion skin-depth c/ω_{pi}, where ω_{pi} is the ion plasma frequency and time is measured in proton gyro-periods. New solar wind particles were continually introduced at the left while solar wind particles leaving the box at the right were removed. The comet particles on the other hand, occupied a circular area centred at the grid point 70-128 see figure 4. with a diameter of $8c/\omega_{pi}$ in the 2-D simulations. The comet ions initial speed was in the radial direction. A number of differences existed between the simulations and the experiment such as the exact values of velocity, number density, mass and magnetic field strength. However, for the most important effects studied such as the formation of a cavity, solar wind deflec-

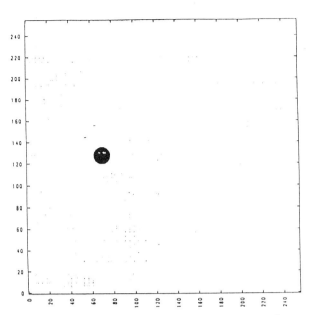

Fig 4. Initialization of the particles in configuration space, the magnetic field is in the plane perpendicular to the figure. Ther barium atoms are centred at the grid point (70-128) with a diameter of 8 grid spacings.

tion, barium acceleration and sideways deflection of the comet head, will have the same qualitative behaviour. The comet particles are initially neutral and they gradually ionize over a time of one proton gyroperiod. Note that distances are measured in c/ω_{pi} ie. ion skin depth and time in ω_{ci}^{-1} where c is the velocity of light, ω_{pi} is the ion (proton) plasma frequency and ω_{ci} is the ion (proton) gyro-frequency. The solar wind ions were given an initial speed of $v_{sw} = 1.5c_s$ with c_s being the ion-acoustic speed; in addition they were given a small thermal jitter. Due to the computational limitations set by system size and time scales the ratio of the cometary gas initial expansion speed to the solar wind speed is 0.2 which is not the same as in the real experiment ie. 1.35/500. For the computer runs different comet to proton mass ratios were used most of the 2-D calculations used a mass ratio of 6 whereas in the 3-D simulations mass ratios of 6,12,24 and 137 were used. One other disadvantage using a hybrid code with zero mass electrons is that length and time scales less than c/ω_{pi} and ω_{ci}^{-1} respectively are not treated properly. All velocities in the model are normalized to the ion acoustic speed while the magnetic field is normalized to the Alfven speed divided by the ion acoustic speed.

Also in the 2-D model particles are cylinders of infinite length which are only allowed to move in the two dimensional plane of the computations. Therefore for our case with $\mathbf{B}$ being almost along the z-axis, the 2-D model lacks draping of the field lines around the cloud but contains its compression against it. As a result most of the magnetic pressure is in the x-y plane, as opposed to the real experiment in which it is more isotropically distributed in space around the cloud. In contrast, however, the 3-D model does include the magnetic field line draping.

Since the scale size of the release is less than the solar wind ion gyroradius this eliminates the possibility of using an MHD code. In the usual MHD treatment, the ions are treated as a fluid and the ion-electron slip ie the Hall term which is the first term on the right hand side of equation 17, is left out of the equation for the fluid acceleration. The fluid particles are accelerated by the Maxwell stress tensor as well as the gradient of pressure. However, it is possible to include the Hall term in the MHD treatment. Two versions of the code were used to investigate the difference between the MHD model including the Hall term and the hybrid model. Figure 5a,b shows the

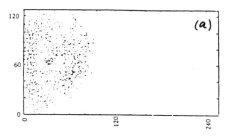

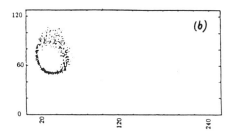

Fig 5a,b. (a) The initial comet particles after one proton gyroperiod using the MHD version of the model.
(b) the same as (a) but using the kinetic version of the model this contains finite ion Larmor radius effects. Magnetic field is ⊥ to figures.

results from an MHD code including the Hall term and a hybrid code in 2-D. As can be clearly seen the hybrid code shows the upward acceleration of the comet particles in the convective electric field figure 5b, while for the MHD case figure 5a just shows a uniform expansion of the comet particles in all directions. The MHD case even with the inclusion of the Hall term cannot reproduce the observations. The upward acceleration of the cometary particles, which is observed experimentally *Coates et al.* [1988], is a purely kinetic effect and can only be reproduced using the hybrid code. The structure shown at the bottom boundary of figure 5b is tightly constrained showing a higher concentration and steeper density gradient than at the top also agreeing with observational images see figure 1 of reference 5. The upward acceleration of the barium ions due to the convective electric field also agrees with the observations made by *Coates et al.* [1988]. Figure 6a,b represent the magnetic field and density contour plots in the x-z plane taken from the 2-D simulations. The magnetic field is taken to be in the y-direction pointing out of the plane (A GSE co-ordinate system is used). The solar wind is flowing from left to right in the x direction. These figures show the large enhancement of the magnetic field and concentration of the barium ions. The fingers at the top of figure 6b are very similar to the experimental observations in particular the contour plot of figure 1 of *Valenzula et al.* [1986]. These fingers or density ripples are a consequence of the Rayleigh Taylor instability caused by the downward acceleration of the dense cometary plasma which is produced as a reaction to the upward acceleration to high velocity, of the low density cometary ions that expand into the solar wind. From the asymmetry of the magnetic field structure shown in figure 6a there develops a strong confining force due to the magnetic pressure at the bottom of the cloud this together with the convective electric field contains the barium ions at the bottom of the comet. Only those ions which escape from the top can be further accelerated by the convective electric field which points

along the z axis. As these low density fast moving cometary ions are ejected by the $E = v_{sw} \times B$ field the dense part recoils in the opposite direction to conserve momentum. An analysis by *Kazeminejad*

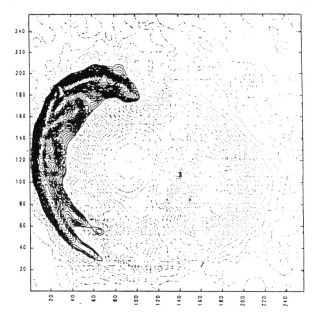

Fig 6a,b. (a) Magnetic field contour plot (the z component) after two proton gyroperiods. Note that in the simulations B is taken to be in the positive z direction and the solar wind flow in the negative x direction. The coordinate system used in experiments is the GSE system.

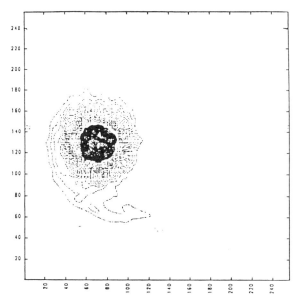

(b) Average density contour plot after one and a half proton gyroperiods showing formations of head and tail.

et al. [1990] has shown that the ripples can be explained in terms of the Rayleigh Taylor instability. A comparison of the Rayleigh Taylor growth rate $\gamma = \sqrt{2\pi a/\lambda}$, where a is the acceleration of the dense fluid and λ is the wavelength of the ripples, with the growth rate obtained from the simulations show very good agreement *Kazeminejad et al.* [1990]. While the top fingers in the head can be attributed to the upward acceleration of a relatively small population of barium ions, the tail fingers or ripples are due to both the solar wind and cometary ions. The dense part that recoils is just a result of momentum balance between the high velocity low density barium ions being accelerated out of the top of the cloud. The movement of the barium cloud was analysed by successively computing the position of the low energy dense portion of the cloud or the centre of mass of the barium ions the result is plotted in figure 7. Initially the position of the cloud

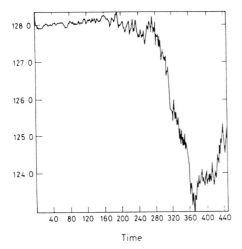

Fig 7. The time evolution of the centre of mass of the comet from 2-D simulation results.

remains fixed and then slowly begins to accelerate in the $-z$ direction in agreement with observations [*Haerendel et al.* 1986].

Figure 8 depicts the trajectory of the solar wind protons as they interact with the barium cloud. The interaction is mainly produced

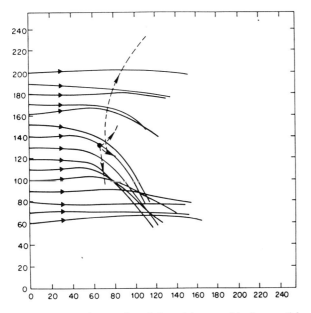

Fig 8. Trajectory of some solar wind particles — and barium particles - - - chosen at random after two and a half proton gyroperiods.

by the ambipolar electric field associated with the magnetic field gradient. The obstacle slows and deflects the protons in a downward direction in agreement with the observations [*Coates et al.* 1988] while the barium ions are extracted from the top of the cloud as shown. The trajectory of the ejected barium ions is cycloidal but due to its very large Larmor radius only part of the cycloidal trajectory is depicted. The slowing down of the solar wind protons is shown in figure 9, here

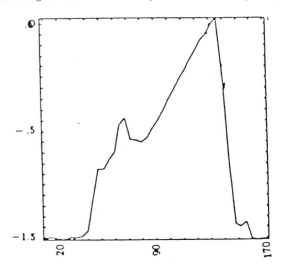

Fig 9. Velocity vector v_x of solar wind protons as a function of distance horizontal axis, after one gyroperiod showing the slowing down.

we plot the x component of the solar wind velocity as it penetrated the barium cloud. The velocity values plotted are measured at points with fixed z on the mesh and with the x values starting at the right hand boundary of the simulation box and progressing to its left hand boundary. This corresponds to a satellite trajectory moving along the x direction (in GSE co-ords) through the cloud. In reality the cloud moved in the $-z$ direction with respect to the satellites therefore the true trajectory would have had a z component as well. In the simulations we see that the solar wind is indeed decelerated as it traversed the cavity in agreement with the observation made by IRM and UKS spacecraft. The 2-D hybrid code describing the above simulation results always assumed that the magnetic field was perpendicular to the mesh plane, thus the effect of magnetic field line "draping" could not be studied. A 3-D hybrid code was used to fully understand the complicated dynamics of the interaction between the barium cloud and the solar wind flow. Figure 10 represents the density structure of the barium ions. In the figure the solar wind protons which are not represented are flowing from left to right and the magnetic field is initially in the plane perpendicular to the page. Important features to notice is the asymmetry of the structure, the top of the cloud is highly structured with density clumps extending upwards into the solar wind while the bottom part of the cloud is extremely smooth and tightly contained. The density structure at the top, a result of the Rayleigh Taylor instability [*Kazeminejad et al.* 1990] is very similar in appearance to the observations made by *Valenzuela et al.* [1986] we can also clearly see the development of the tail. The ripples on the surface of the AMPTE cloud are consistent with the theory of un-magnetized ion Rayleigh-Taylor instability proposed by Hassam and Huba [1988]. Figure 11 depicts the magnetic field structure, it is quite evident that there is indeed magnetic field compression at the front as well as draping of the field around the object. Finite resistivity was included in Amperes equation which can allow for magnetic field diffusion.

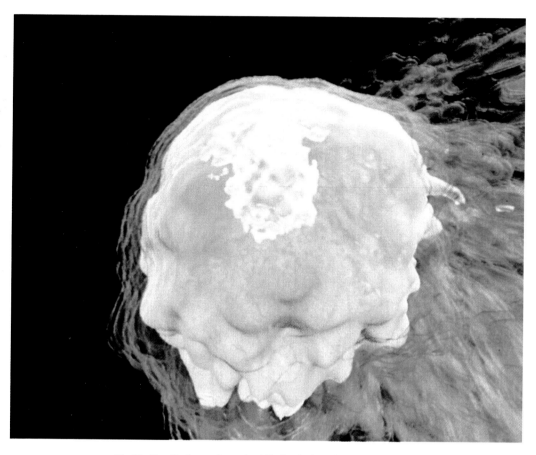

Fig 10. Density image from the 3-D simulations showing the density ripples at the top, attributed to the Rayleigh Taylow instability, formation of tail and smooth profile on the bottom. The solar wind flows from left to right and the magnetic field is perpendicular to plane of figure.

Figures 12a,b,c represent cross sections of the density of the solar wind protons and the total integrated density of the release ions (Figures 12d,e,f) in the different planes. It can be clearly seen that the solar wind particles pile up at the front of the magnetically compressed region creating a region of enhanced solar wind density around the cloud. The figures show in the early stages a very symmetric ion cloud producing the obstacle to the solar wind flow which piles up at the front and flows around the object. The density cuts of the released particles also reveal the presence of structure at the top similar to results of the 2-D simulations and observations. This confirms the result obtained in the 2-D simulations that barium ions are accelerated upwards and the solar wind particles are deflected downwards in the opposite direction.

We can describe theoretically the interaction of the solar wind particles and the cavity using a two fluid model for the plasma and a constant profile structure for the magnetic field which only varies in the x-direction. Such a model was proposed by *Bingham et al.* [1988] to try and understand the slowing down of the solar wind particles. Papadopolous and *Lui* [1986] proposed earlier a momentum coupling force F to produce the downward motion through the $\mathbf{F} \times \mathbf{B}$ drift. The analysis treats the ions as unmagnetized therefore, the solar wind ions do not respond directly to the charge in the magnetic field created by the cavity, only the electrons, because of their small Larmor

orbit respond to this field change which slows them down, the ions continue to penetrate causing a charge separation resulting in an ambipolar field. This ambipolar field is responsible for retarding the ion flow. This field is depicted in figure 13 it is oppositely directed to the magnetic field gradients. Within the cloud the ambipolar field, however, has the opposite directions to that at the front region, here it opposes the expansion of the barium ions.

Using the steady state two fluid model and assuming a constant profile for the magnetic field structure varying in the x-direction the fluid equations become

$$\frac{\partial}{\partial x}(n_\alpha \mathbf{v}_\alpha) = 0 \tag{18}$$

$$v_{\alpha x}\frac{\partial}{\partial x}\mathbf{v}_\alpha = \frac{q_\alpha}{m_\alpha}\left(\mathbf{E} + \frac{\mathbf{v}_\alpha \times \mathbf{B}}{c}\right) \tag{19}$$

where α is the species index.

Adding the y components of equation (19) for electrons and ions and using the boundary condition $v_{ix} = v_{ex} = v_x$ at $x = -\infty$ (ie. in the solar wind electron and ion fluids have same velocity) we obtain

$$v_{iy} = -\frac{m_e}{m_i}v_{ey} \tag{20}$$

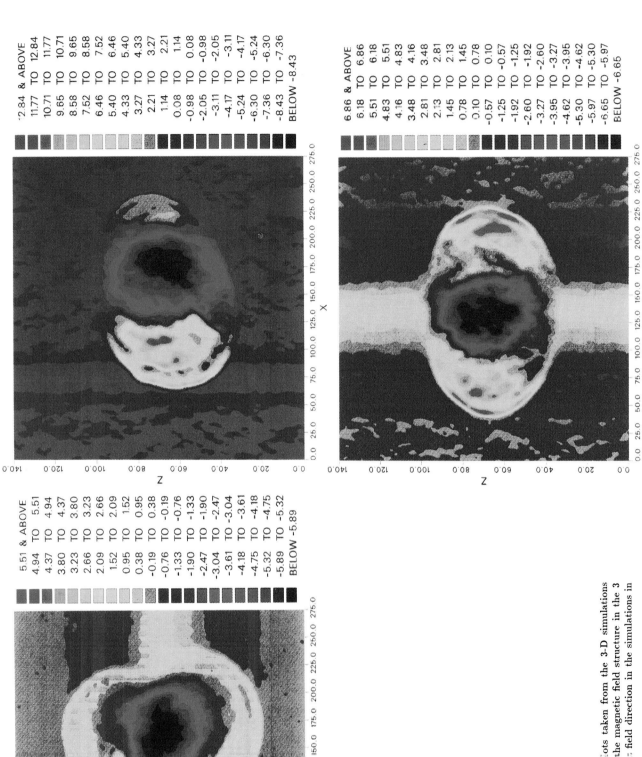

Fig 11a,b,c. Magnetic field plots taken from the 3-D simulations showing the B_z component of the magnetic field structure in the 3 planes (note that the dominant field direction in the simulations in the z direction).

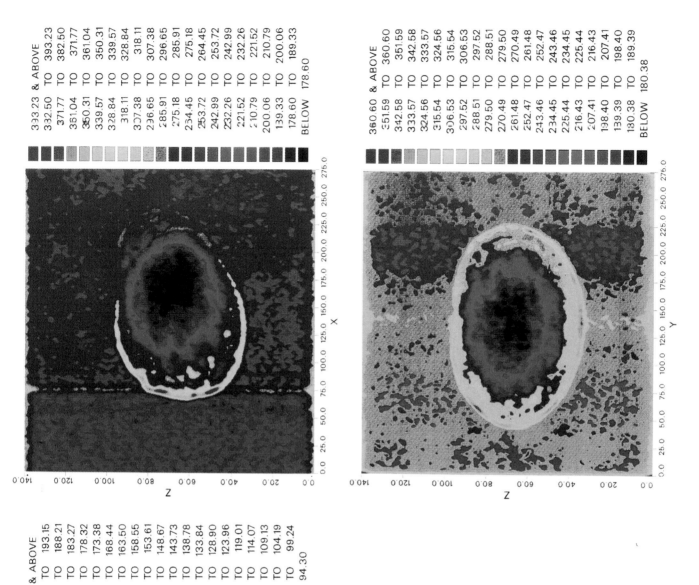

Fig 12a-f. (a,b,c) Cross-section of the density of solar wind protons in each plane.

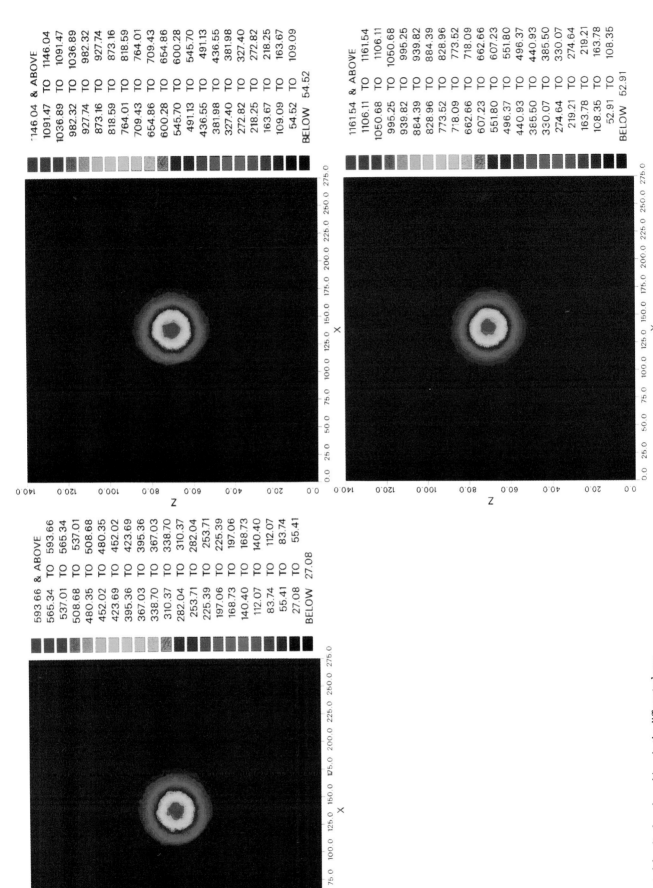

(d,e,f) Total integrated density for released ions in the different planes.

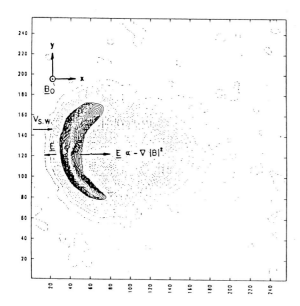

Fig 13. Magnetic field structure from the simulations after one gyroperiod. The arrows indicate the directions of the solar wind flow and electric fields.

Clearly $v_{iy} \ll v_{ey}$, therefore the electron velocity produces most of the J_y current which is responsible for the perturbed B_z field. Taking the x-component of equation (18) for electrons and ions and subtracting one from the other we get, using $\frac{m_e}{m_i} \to 0$

$$E_x = -\frac{v_{ey}B_z}{c} \qquad (21)$$

which simply means that the electrons are in force balance in the x direction.

From the y component of Ampere's law equation (10) together with equation (20) we get

$$v_{ey} = \frac{c}{4\pi e n \left(1 + \frac{m_e}{m_i}\right)}\frac{dB_z}{dz}$$

$$\simeq \frac{c}{4\pi e n}\frac{dB_z}{dx} \qquad (22)$$

Substituting equation (22) into equation (21) results in the ambipolar field in the x direction

$$E_x = -\frac{1}{8\pi e n}\frac{dB_z^2}{dx} \qquad (23)$$

The electric field along the solar wind direction points in the direction opposite to the gradient of the magnetic pressure ie. this component of the field which is the ambipolar field, is proportional to the Maxwell stress tensor. This supports our earlier assertion that the electric field points one way in the compressed field region so as to slow down the incoming solar wind particles and points the opposite way in the cloud region so as to push back on the cloud confining the released plasma in the compressed region.

From the $y-$ components of equation (19) and using $v_{iy} = -\frac{m_e}{m_i}v_{ey}$ get

$$E_y = \frac{v_x B_z}{c} \qquad (24)$$

This simply means that in the rest frame of the cloud, there exists a convective electric field given by $E = -\frac{1}{c}\mathbf{v}_{sw} \times \mathbf{B}$.

The force on an incoming solar wind ion due to the electric field E_x is given by

$$F = eE_x$$

$$= -\frac{1}{8\pi n}\frac{dB_z^2}{dx} \qquad (25)$$

The force required to reduce the velocity of such an ion from u_x to v_x in a distance Δx can be written as

$$F = \frac{m_i(v^2 - u^2)}{2\Delta x} \qquad (26)$$

balancing these forces we obtain the expression

$$u^2 - v^2 \simeq \frac{1}{8\pi m_i n}\Delta B_z^2 \qquad (27)$$

where ΔB_z^2 is the charge in B_z^2 in a distance Δx. The force exerted by E_x is sufficient to account for the slowing down of the solar wind.

The electric field produced by the released ions and solar wind interaction is shown in Figure 14a,b. Figure 14a show the electric field seen by the incoming solar wind particle showing the slowing down field and confining field, and figure 14b shows the field seen by a particle at rest with respect to the solar wind.

The sideways deflection can easily be explained in terms of a $\mathbf{E} \times \mathbf{B}$ drift given by

$$v_D = c\frac{\mathbf{E} \times \mathbf{B}}{B^2} \qquad (28)$$

and shown in figure 15. Here the electric field is due to the magnetic field gradient $\mathbf{E} \propto -\underline{\nabla}|B|^2$ shown in figure 14a this is the confining field seen by the barium ions. This field couples with the diffusing magnetic field to give a substantial drift in the down-ward direction at the same time barium ions at the top are extracted by the convective electric field.

The downward motion does not happen immediately this is due to the fact that the magnetic field has to diffuse back into the cloud before any significant $\mathbf{E} \times \mathbf{B}$ coupling can occur. There is a delay in moving sideways as shown in figure 7, this is due to the finite time for a significant diffusion of the magnetic field. In the code diffusion can occur due to finite resistivity in Ampere's equation. The resistivity in the experiment is from wave-particle interactions.

Conclusions

The two and three dimensional simulations presented of the interaction of released gas in a flowing plasma with an intrinsic magnetic field have provided excellent qualitative agreement with the 27th December Ampte release experiment. We have demonstrated that the codes can produce diamagnetic cavities with a compressed field region on the upstream side of the cloud.

The compressed magnetic field and particle flow show a strong asymmetry which was responsible for the subsequent sideways deflection of the released plasma. The dynamics of the interaction were determined by finite ion Larmor radius effects such effects operate in regions where the scale lengths of interest are smaller than or the order of the ion Larmor radius. It is found that the finite Larmor radius effects help to stabilize the cloud this effect is well known in the stabilization of z-pinch plasmas, such stabilization effects are not possible in the MHD description of the problem.

The asymmetric structure of the release cloud resembled closely the images taken of the actual Ampte release. In both cases density structures or ripples were observed on the top of the clouded while the bottom had a very clean edge these structures are thought to be the result of a Rayleigh Taylor instability set up by the extraction of released ions from the top of the cloud.

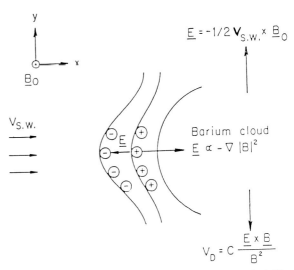

Fig 15. Schematic representation of solar wind flow electric fields set up in the frame of the solar wind and the rest frame of the released particles showing the resultant drift velocity v_D.

References

Bingham, R., D. A. Bryant., D. S. Hall., J. M. Dawson., F. Kazem-inejad., and J. J. Su., Ampte observations and simulation results. *Comp. Phys. Comm.* **49**, 257, 1988.

Brecht S. H., and V. A. Thomas, Three-Dimensional Simulation of an Active Magnetospheric Release, *J. Geophys. Res.*, **92**, 2289, 1987.

Chapman, S. C., and S. J. Schwartz, 1-D hybrid simulations of boundary layer processes in the AMPTE solar wind lithium releases. *J. Geophys. Res.*, **92**, 11059, 1987.

Coates, A. J., D. J. Rodgers., A. D. Johnstone., M. F. Smith., and J. W. Heath., Development of the first artificial comet: UKS ion measurements. *Adv. Space Res.*, **8**, 15, 1988.

Gurnett, D. A., R. R. Anderson., T. Z. Ma., G. Haerendel., G. Paschmann., O. H. Bauer., R. A. Trumann., H. C. Koons., R. H. Holyworth., and H. Lühe., Waves and Electric Fields Associated With the First Ampte Artificial Comet. *J.Geophys.Res*, **91**, 10,013, 1986.

Haerendel, G., G. Paschmann, W. Baumjohann, and C. W. Carlson., Dynamics of the Ampte Artificial Comet. *Nature*, **320**, 21, 1986.

Hartle, T. E., K. W. Ogilvie., and C. S. Wu., Neutral and ion-exospheres in the solar wind with applications to Mercury, *Planet Space Science*, **21**, 2181, 1973.

Hassam, A. B., and J. D. Huba, Magnetohydrodynamic equations for systems with large Larmor radius. *Phys. Fluids.* **31**, 318, 1988.

Ip, W. I., and W. I. Axford., *Comets*, L L Wilkering, (University of Arizona, Tuscon), 588, 1982.

Kazeminejad, F., J. M. Dawson., J-N. Lebeouf., and F. Brunel., MHD Hall term model.*J.Comp. Phys*, (Submitted).

Kazeminejad, F., J. M. Dawson., J-N. Lebeouf, R. Sydora., and D. Holland., Vlasov ion fluid electron model for plasma simulations. *J.Comp. Phys*, (Submitted).

Kazeminejad, F., R. Bingham., and J. M. Dawson., Simulations and qualitative analysis of the AMPTE experiments. *Phys. Rev. Lett.* (Submitted) 1990.

Klöcker, N., H. Lühr., D. J. Southwood., and M. H. Acuna., Magnetic ULF fluctuations in the compressional zone of Amptes artificial comets. *Adv. Space Res.*, **8**, 23, 1983.

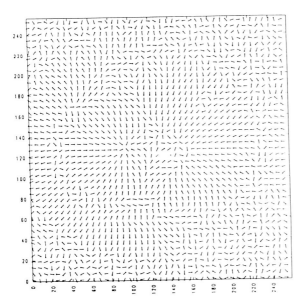

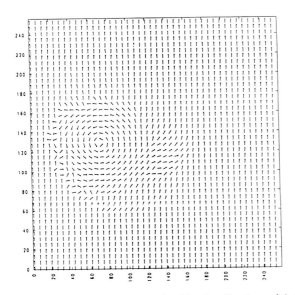

Fig 14a,b. The electric field arrow plot (showing direction only), in the laboratory frame after one and a half proton gyroperiods (a) the frame at rest with respect to the solar wind (b) in the solar wind frame. The solar wind flows from left to right and the solar wind magnetic field points out of plane.

Lühr, H., D. J. Southwood., N. Klöcker., M. W. Dunlop., W. Meir-Jederzejowicy. , R. Rijnbeck., M. Stix., B. Häusler and M. Acuna., In situ magnetic field observations of the Ampte artificial comet. *Nature*, **320**, 708, 1986.

Lui, A. T. Y., Collisionless coupling processes in Ampte releases. *Physics of Space Plasmas*, SPI conf. Proc. **9**, ed. T S Chang. 1990.

Lui, A. T. Y., C. C. Goodrich., A. Mankofsky., and K. Papadopolous., Early Time Interaction of Lithium Ions with the Solar Wind in the AMPTE Mission. *J. Geophys. Res.*, **91**, 133, 1986.

Papadopolous, K., and A. T. Y. Lui., On the initial motion of artificial comets in the Ampte releases. *Geophys. Res. Lett* **13**, 925, 1986.

Rees, D., T. J. Hallinan., H. C. Stenbaek-Nielson., M. Mendillo., and J. Baumgardner. Optical observations of the Ampte artificial comet from the northern hemisphere. *Nature*, **320**, 704, 1986.

Rodgers, D. J., A. J. Coates., A. D. Johnstone., M. F. Smith., D. A. Bryant., D. S. Hall and C. P. Chaloner., UKS plasma measurements near the AMPTE artificial comet. *Nature* **320**, 712, 1986.

Valenzuela, A., G. Haerendel., H. Föpple., F. Melzner., H. Neuss., E. Rieger., J. Stocker., O. Bauer., H. Höfner., and J. Loidl., The Ampte artificial experiment. *Nature*, **320**, 700, 1986.

Woolliscroft, L. J. C., M. P. Gough., P. J. Christiansen., A. G. Darbyshire., H. G. F. Gough., D. S. Hall., D. Jones., S. R. Jones., and A. J. Norris., Plasma waves and wave-particle interactions seen at the UKS spacecraft during the AMPTE artificial comet experiment. *Nature* **320**, 716, 1986.

Wu, C. S., R. E. Hartle., and K. Ogilvie., Interaction of singly charged interstellar helium ions with the solar wind. *J. Geophys Res.* **78**, 306, 1973.

COMETARY PLASMA OBSERVATIONS BETWEEN THE SHOCK
AND THE CONTACT SURFACE

Henri Rème

Centre d'Etude Spatiale des Rayonnements, CNRS/Université Paul Sabatier,
Toulouse, France

Abstract. The ICE passage through the tail of Comet Giacobini-Zinner on September 11, 1985 and the flyby of Comet Halley in March 1986 by a fleet of spacecraft have provided for the first time an *in situ* study of the interaction of the solar wind with a cometary plasma under various solar and cometary conditions.

Several sharp boundaries separating fairly well defined regions and transitions from one flow state to another were observed by the particle and field instruments during the cometary flybys. Near 1 A.U., the cometary particle and field environment is highly complex. Although the main features like the bow shock, the pile-up magnetic region and the contact surface (or ionopause) are relatively close to the theoretical predictions, the *in situ* measurements have revealed many more structures between the shock and the ionopause. These structures were not anticipated and still remain ill understood, but they are the signs of the plasma processes taking place in the coma of a comet. One important region is the "mystery region" where significant and highly variable fluxes of keV electrons were detected by Giotto. This region is limited by the "mysterious transition" which is seen by all the spacecraft about halfway between the shock and the comet nucleus and seems to be an unpredicted permanent structure of the solar wind - comet interaction at least around 1 A.U.. This transition can be called the **cometary transition**. The magnetic pile-up region seen by Vega-1 and Vega-2 was gradual but showed a strong gradient on the inbound leg of Giotto: the Magnetic Pile-up Boundary (MPB), may possibly be a convected solar wind effect. On the contrary the Vega-2 "cometopause" is not a plasma boundary and so is not a "pause".

Introduction

For the first time, *in situ* studies of the interaction of the solar wind with cometary plasmas have taken place thanks to the pass of ICE (International Cometary Explorer) through the tail of Comet Giacobini-Zinner, on September 11, 1985 and the flybys of Comet Halley by 5 spacecraft in March 1986. Table 1 summarizes the main features of these 6 spacecraft-comet encounters.

Before these *in situ* measurements, models of interactions of the solar wind with a negligible gravity, no magnetic field body

Cometary Plasma Processes
Geophysical Monograph 61
©1991 American Geophysical Union

were based upon ground-based cometary measurements and theoretical calculations. Due to the negligible gravity of the small cometary nucleus, the sublimation gases expand supersonically and interact with the solar wind over a length scale that is typically, at about 1 AU, 5-6 orders of magnitude larger than the size of the nucleus, through charge exchange, photoionization and impact ionization, with increases as the cometocentric distance decreases. The newly created ions "mass load" the solar wind and slow it down. Theoretical models predict a gradual slowing and heating of the solar wind until the Mach number of the solar wind flow is reduced to a value of M ~ 2 [Biermann et al., 1967; Wallis, 1973; Brandt et Mendis, 1979]. At that point, a weak bow shock forms which further slows, heats and deflects the flow. Inside the shock, mass loading must continue as the solar wind penetrates closer to the nucleus. A strong deceleration of the solar wind is expected to begin at such a distance where the momentum-transfer collision mean-free path of a solar wind ion is of the order of its radial distance from the nucleus [Mendis et al., 1986]. This distance has been termed the "collisionopause". Inside this boundary, the solar wind would decelerate rapidly and also cool due to exchange of energetic ions formed upstream with new, less energetic ions, continuously being formed in the decelerating flow [Wallis and Ong, 1975; Galeev et al., 1985]. This would lead to the formation of a magnetic barrier. At the inner edge of this barrier there is the ionopause, a tangential discontinuity between the inflowing contaminated solar wind and the purely cometary plasma. The slowing down of the magnetized plasma near the comet also leads to the draping of the interplanetary magnetic field around the comet to form a magnetotail with 2 lobes separated by a cross-tail current sheet. So several boundaries were expected to be found *in situ* in the cometary environments as a consequence of the solar wind-comet interaction.

The spacecraft results show that the reality is much more complicated and far from being fully understoood. This review is devoted to the plasma properties between the shock and the contact surface (ionopause). This region can be called the cometosheath. Upstream, shock, ionospheric and tail phenomena will not be discussed.

1. Main Features of the Cometosheath Plasma and Magnetic Field Observations

The particle and field environment of Comet Halley encountered by spacecraft in March 1986 displays a much more complex structure than expected [Balsiger et al., 1986; Galeev, 1986; Gringauz et al., 1986a; Johnstone et al., 1986; Mukaï et al.,

Spacecraft	Comet	Encounter Geometry	Date at closest approach	Sun Comet distance (AU) at closest approach	Distance to comet nucleus at closest approach (km)	Flyby speed (km/s)	Solar wind speed (km/s)	Comet production
ICE (NASA)	Giacobini-Zinner	tailward side	11 Sept. 1985 11 02 UT	1.05	7800	21	between 400 and 500	Between 2×10^{28} and 5×10^{28} water molecules s^{-1} [Stewart et al., 1985]
VEGA 1 (Inter-cosmos)	Halley	sunward side	6 March 1986 07 20 UT	0.79	8890	79	510	Total gas production rate: 1.3×10^{30} molecules s^{-1} [Gringauz et al., 1986a]
VEGA 2 (Inter-cosmos)	Halley	sunward side	9 March 1986 07 20 UT	0.83	8030	77	620	OH gas production rate : $\sim 9 \times 10^{29}$ molecules s^{-1} [(Moreels et al., 1986] OH production rate : $\sim 2 \times 10^{30}$ molecules s^{-1} [Krasnopolsky et al., 1986] Water production : 5.6×10^{29} molecules s^{-1} [Festou et al., 1986]
SUISEI (ISAS, Japan)	Halley	sunward side	8 March 1986 13 06 UT	0.81	151,000	73	~500	
SAKIGAKE (ISAS, Japan)	Halley	sunward side	11 March 1986 04 18 UT	0.86	6.99×10^6	75	~450	
GIOTTO (ESA)	Halley	sunward side	14 March 1986 00 03 UT	0.89	600	68	~350-400	Total gas production: 6.9×10^{29} molecules s^{-1} Water production : 5.5×10^{29} molecules s^{-1} [Krankowski et al., 1986b] Water production: 5.2×10^{29} with nucleus near a minimum of activity [Festou et al., 1986]

TABLE 1. Main Features of the Various Spacecraft-Comet Encounters

1986; Neubauer et al., 1986; Rème et al., 1986]. In the midst of rapid fluctuations, fairly well defined regions separated by sharp transitions were also distinguished. To determine if these transitions between the shock and the ionopause are permanent or temporary is very difficult due to the problems in comparing the results :
- There is a fast and unique cometary flyby for each spacecraft.
- The solar wind conditions are different for different encounters.
- Two comets (Giacobini-Zinner and Halley) were encountered.
- It is difficult to separate temporal and spatial effects.
But the main problems are linked to the instruments :
- The plasma instruments are very different on these spacecraft.
- There is no full coverage in energy and/or in mass and/or in angle.
- There are no wave measurements aboard Giotto and Suisei.
- There are no cross-calibrations and/or no precise calibrations of the sensors.
- There are effects due to dust (mainly near the comet) and artefacts.
For example, in the Giotto payload various instruments were dedicated to ion composition measurements : the High Energy Range Spectrometer (HERS) and the High Intensity Spectrometer (HIS) [Balsiger et al., 1987], the Implanted Ion Sensor (IIS) [Wilken et al., 1987a], and the Positive Ion Cluster Analyzer (RPA2-PICCA) [Korth et al., 1987a]. This shows the importance attached to composition determination in this mission. However none of these 4 sensors could cover the wide range of phase space

necessary for the cometosheath ions. So HERS and IIS could not really detect cold cometary ions while HIS and PICCA looked only into the ram direction. The Plasma package also included another ion sensor [Johnstone et al., 1987], an electron spectrometer [Rème et al., 1987a] and a neutral mass spectrometer with ion measurements [Krankowsky et al., 1986a].
On Vega-1 and -2 the plasma package included five different sensors: Two of them CRA (Cometary Ram Analyzer) and SDA (Solar Direction Analyzer) measured the E/Q spectra of ions with very limited fields of view and without mass discrimination, RFC (Ram Faraday Cup) was a Faraday Cup recording the solar wind ion fluxes and the electron sensor EA (Electron electrostatic Analyzer) was, with a field of view of 7°x7° oriented perpendiculary to the ecliptic plane, unable to measure the full electron distribution.
During the encounters of Vega-1 and -2 with Comet Halley the following sensors were in operation [Balebanov et al., 1987] (Table 2).

TABLE 2.

Sensor	SDA	CRA	EA	SDFC	RFC
VEGA-1	X			X	X
VEGA-2	X	X	X		X

Furthermore, the results of EA on Vega-2 were not completely reliable. Taking into account the fact that a small peak above 1 keV detected near the encounter remained at the same energy even two days after the encounter, Gringauz et al. [1986c] suggest that this peak could be regarded at least partially as an

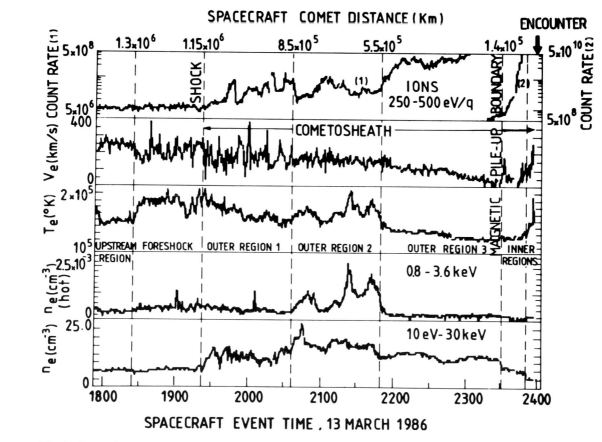

Fig. 1. Some plasma electron data and count rate of the 250-500 eV/q ions in the ram direction during the last 6.2 hours before the Giotto encounter with Comet Halley. Note a change of scale from count rate (1) to count rate (2) at time ~ 2330 SCET in the top panel; from Rème [1990].

indication of an instrumental effect.

Because the Giotto spacecraft was the only one going through the entire cometosheath including the inner cometosheath and the ionopause, with a full coverage of all the solar wind - cometary interaction region, the main features of the cometosheath will be described on the basis of the Giotto plasma and magnetic field results. In each region, comparisons with other spacecraft results will be made, where data are available.

Figure 1 [Rème, 1990] summarizes some plasma data from the Giotto RPA-COPERNIC experiment during the last 6.2 hours before the encounter with comet Halley when the spacecraft went from the upstream region to the closest approach to the comet: n_e is the density of 10eV - 30 keV electrons; n_e (hot) the density of 0.8 - 3.6 keV electrons; T_e the electron temperature estimated from the slope of the energy spectra near 45 eV; and V_e is the electron bulk velocity calculated from the electron distribution. In the top panel the RPA2-PICCA count rate in the ram direction is shown for 250 - 500 eV/q ions. Two seconds before closest approach the spacecraft telemetry signal was lost. When it returned about 30 min later, it was apparent that the RPA-COPERNIC experiment had received some damage, presumably from dust impact. Data from this experiment from the post-encounter period will not be presented here.

From these data the cometosheath:
- began near 1.15 x 10⁶ km from the comet (bow shock

position),
- had an outer part, between the bow shock and the Magnetic Pile-up Boundary (MPB), with 3 regions: the second one, between 8.5 x 10⁵ and 5.5 x 10⁵ km from the comet was characterized by the presence of keV electrons and is called the "mystery region", due to these unexpected energetic electrons; it terminated in a remarkable boundary, the mysterious transition,
- had an inner part between the MPB, found at ~ 1.35 x 10⁵ km, and the ionopause, with several discontinuities,
- ended at the ionopause, 4700 km from the comet nucleus.

Figure 2, from Balsiger et al. [1986], gives the plasma parameters (density, velocity and thermal speed) for the M/Q = 2 ions (mainly solar wind alpha particles at least until ~ 2300 SCET; see Fuselier et al. [1988] during Giotto encounter. Also given is the elevation angle which demonstrates how these ions change direction relative to the sun-comet line as they have to move around the cometary obstacle. The same discontinuities as in Figure 1 are seen for the outer cometosheath between the shock (BS) and the MPB (PB) with 3 regions, and a strong maximum in number density in the mystery region. Note that in Figure 2, time is Earth or Ground Received Time (GRT), 8 minutes delayed from Spacecraft Event Time (SCET).

An overview of the Giotto magnetic field measurements [Neubauer, 1987] is shown in Figure 3 including pre- and post-

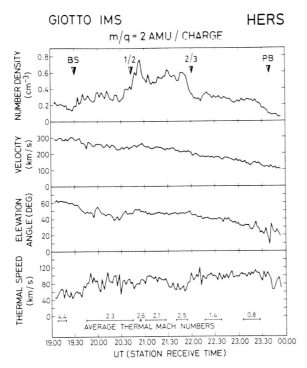

Fig. 2. Plasma parameters for the M/Q = 2 ions during the last 6 hours before Giotto encounter; from Balsiger et al. [1986].

encounter data. This figure gives the magnitude of the field and its direction in Halley-centered solar ecliptic coordinates. The positions of the bow shock (BS) and MPB (PB) are shown, in close agreement with the plasma measurements (on the inbound leg). In this figure the outer cometosheath is called the sheath and the inner cometosheath the pile-up region (PU). On the inbound leg the direction of the field both inside and outside the bow shock was quite variable although it was close to the spiral angle of the undisturbed interplanetary field (PHI = 130°) for some of the time. Because Giotto was close to the heliospheric current sheet during comet encounter it was thought at first sight from the magnetic field data alone that Giotto crossed it several times with a resulting magnetic reversal [Raeder et al., 1987]. The comparison of heat flux electron and magnetic field data clearly shows that there was no traversal of the heliospheric current sheet during the Giotto comet encounter but that due to curving and draping of the field, the magnetic geometry was very complicated. Between the shock and the MPB the magnitude of the field did not increase significantly but at the MPB, 135,000 km from the comet nucleus, there was a dramatic increase associated with strong effects in the plasma instruments. The same feature was searched for on the outbound leg and Neubauer [1987] has placed the MPB at 188,000 km after 0100 UT (SCET). However it must be noted that the corresponding change at this place in magnetic field intensity was not as noticeable as on the inbound leg at this time and that the magnetic field direction was undisturbed. Another possibility exists, namely placing the MPB closer to the nucleus (see Figure 3); it is also possible to consider the absence of a strong MPB on the outer leg leading to a time dependent MPB. Note that this fact is strongly supported by the Vega-1 and -2 magnetic field measurements where no MPB was

detected and where the increase inside the cometosheath was much more steady [Riedler et al.,1986].

Figure 3 shows that inside the MPB the magnitude of the magnetic field continued to increase, reaching a maximum of about 60 nT at a distance of 16,500 km from the nucleus. Then the field dropped to zero after the ionopause, at ~ 4700 km. The ionopause (or contact surface) is the boundary between a region of pure cometary plasma, with no magnetic field (because the comet nucleus has no intrinsic magnetic field) and a region of mixed solar wind and cometary plasmas; the magnetic flux, carried by the solar wind, drapes and piles up around this boundary. On the outbound leg the ionopause was found at 3900 km from the nucleus. The magnetic field draping around the ionopause is due to the fact that the interplanetary magnetic field is frozen into the plasma flow and slows down with it as they approach the comet [Alfven, 1957], while the distant field continues at the undisturbed solar wind speed. Thus the magnetic field lines are draped around the comet and stretched downstream to form the tail.

On the outbound leg, the bow shock was a very broad transition apparently occuring between 0230 and 0305 SCET (between 6 x 10^5 and 7.5 x 10^5 km) from the comet.

2. The Outer Regions of the Cometosheath

The solar wind speed abruptly drops at the bow shock and begins to deflect around the nucleus. The outer cometosheath between the bow shock and the MPB was clearly dominated by solar wind plasma, as can be seen in Figure 4. This Figure shows, between 1800 SCET and the encounter, from top to bottom: the ion flux between 250 - 500 eV/q in the ram direction from RPA-2-PICCA, the density of M/Q = 2 ions (mainly the second most abundant solar wind ion He^{++}) from HERS, and the total electron density between 10 eV and 30 keV from RPA-1. The electron density and the M/Q = 2 ion density are well correlated, showing that a large majority of these electrons are of solar wind origin. The anticorrelation with the ions in the ram direction, mainly of cometary origin, is also very clear.

At the bow shock the pick-up ions and solar wind ions are heated, and the cometary protons can no longer be distinguished from the solar wind protons [Balsiger, 1990]. Pick-up oxygen coming from the dissociation of cometary water was observed by IIS [Wilken et al., 1987b] even outside the bow shock. After an initial density jump at the shock by a factor of 2, its absolute density steadily increased further and, of course, its abundance relative to the solar wind protons went up from the 1% value measured at the shock. The variable slopes of the oxygen radial profile [Wilken et al., 1987b] and complex energy spectra with peaks clearly below the energy which is expected for local pick-up ions [Neugebauer et al., 1987] indicate that the simple pick-up mechanism will probably not explain the features observed behind the shock and in the cometosheath. The results of Coates et al. [1987] also well illustrate the rather variable nature of the bow shock structure.

a. Outer Region 1: The Turbulent Region

Outer region 1 extends from the shock to 8.5 x 10^5 km from the comet. Throughout this turbulent region (Figure 1) there was a large increase in the electron density after the shock with rapid and strong fluctuations on time scales of the order of one minute. The electron temperature and bulk velocity also displayed fluctuations. The average temperature decreased toward the comet

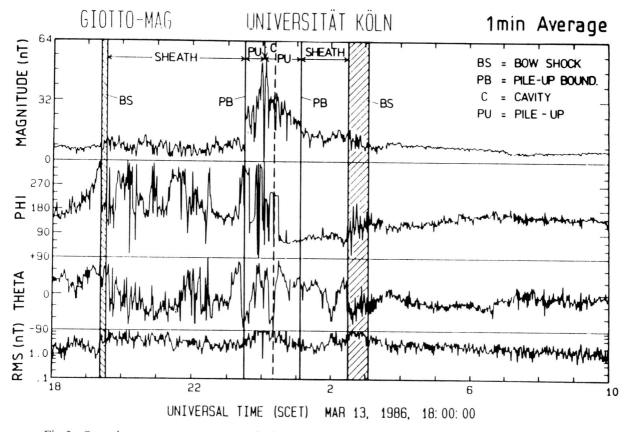

Fig. 3. One minute average vector magnetic field observations in Halley centered solar ecliptic coordinates. Theta is the elevation angle with respect to the X,Y plane and Phi the azimuth of the projection with Phi = 0 towards the sun. RMS is the Pythagorean mean of the component rms values (from Neubauer [1987]). A suggested possible position of MPB on the outbound leg is added (dashed line).

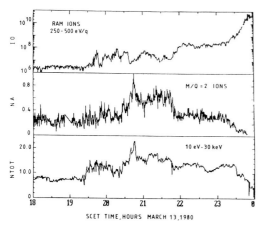

Fig. 4. Total (10eV - 30 keV) electron density, M/Q = 2 ion density and total flux of the 250-500 eV/q ions in the ram direction during the last 6 hours before Giotto encounter.

whereas the average velocity was almost constant but with very large fluctuations. In this region, the electron distribution function may fluctuate rapidly from quiet solar wind-like to nearly isotropic, which explains the variations of the heat flux ratio between 0.45 and 1. Some keV electrons were detected but their fluxes decreased toward the comet. Inside this region, low energy ions (protons) were detected in the ram direction but their fluxes seem to be anticorrelated with the electron bulk velocity. The more the electrons are decelerated, the more the solar wind protons are disturbed and observed in the ram direction.

The data obtained by IMS [Balsiger et al., 1986] and JPA [Johnstone et al., 1986] show that, in this region, there is a large increase in the proton density and temperature, the M/Q = 2 ion density and temperature, and the pick-up ion density, and a decrease of the proton velocity and the M/Q = 2 ion velocity, but with large fluctuations. A Fourier analysis of the electron, proton and He^{++} densities shows that identical periods of the order of one minute can be found for these 3 quantities in spite of the turbulence in this region [Mazelle, 1990].

But the main feature of this region is perhaps the large hydrodynamic fluctuations of both the magnetic field magnitude and direction [Glassmeier et al., 1986] mainly between 1930 and 1952 SCET. Very few quasi-coherent wave trains with at least 3 full periods can be found. On the contrary, the spectrum of transverse or longitudinal magnetic fluctuations exhibits no special peak, but rather a power law with a slope of about -1.72, i.e. similar to a Kolmogoroff type of spectrum and in accordance with the observations of Tsurutani and Smith [1986] at Giacobini-Zinner. This turbulence in the cometosheath, near the shock, was also measured by the two Vegas [Galeev, 1986] and by Suisei [Mukai et al., 1986]. Thus this region can be called the **turbulent region**.

b. Outer Region 2: The Mystery Region

Outer region 2 is crossed at distances between 8.5×10^5 and 5.5×10^5 km from the comet. This region is still turbulent and on Giotto, the magnitude of B was on the average, almost constant. It is also characterized by larger electron, proton and alpha densities (e.g. there is an increase of the alpha particle density by a factor of 3), and by larger variations of electron density and temperature that occur on a slower time scale. The average electron bulk velocity is almost constant. The fluxes of the ions detected in the ram direction are anticorrelated with the electron density and temperature [Rème et al., 1987b]. But the main feature of this region is probably the detection of significant fluxes of keV electrons as seen in the plot of n_e(hot) in Figure 1. These fluxes of 0.8-3.6 keV electrons detected in outer region 2 are not present after it, and are much smaller before it. The 3.9-30 keV electron density (not shown) stays at the background level. Except for the highest energy level, all the densities are a maximum in this region. Due to these unexpected and presently not well understood high energy electron fluxes this region is called the **"mystery region"** [Rème et al., 1987b].

The behaviour of the solar wind ion plasma here can be deduced from Figure 2. Because of the weak contamination expected from cometary H_2^+ particles in the comet's outer plasma environment [Fuselier et al., 1988] the profiles for M/Q = 2 ions reflect the behaviour of solar wind alpha particles. Consequently, the variations indicated in that figure may be used to describe changes that affect the solar wind plasma in that region. The most important effect here is the large increase in density and the small decrease in bulk speed.

The PLASMAG experiment on Vega-2 was turned on only after the first boundary found by Giotto at 8.5×10^5 km from the comet but it identifies this region as a plateau in the ion velocity distribution [Sagdeev et al., 1986].

c. The Mysterious Transition

At about 5.5×10^5 km from the comet, Giotto penetrated outer region 3. The boundary between regions 2 and 3, or the mysterious transition, described in detail in d'Uston et al. [1987], occured on a relatively short time scale, with a remarkable change in the plasma properties separating two different plasma regions without a special magnetic field signature except a slow rotation of the field between 2140 and 2152 SCET (Figure 3).

This discontinuity separates the zone containing energetic electrons from outer region 3. Figure 5 [d'Uston et al., 1987] shows, from bottom to top, the electron bulk velocity, the electron temperature, the 40-73 eV electron density, the 0.8-3.6 keV electron density and the 250-500 eV/q flux of ram ions ions during the traversal of this boundary around 2150 SCET. This transition is characterized by a sudden decrease in electron bulk speed, temperature and density and the beginning of a sustained increase in the flux of ram ions. The energetic electrons (0.8 < E < 3.6 keV) returned to the level they had in the upstream solar wind, having been up to 10 times as intense in the mystery region. The parameters displayed in Figure 5 allow a determination of the width of the transition. It took place between 2146:14 to

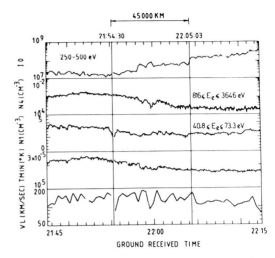

Fig. 5. From bottom to top: electron flow speed, electron temperature, 40-73 eV electron density, 800-3600 eV electron density and 250-500 eV/q ram ion flux at the comet Halley mysterious transition crossing, from d'Uston et al. [1987].

2156:47 SCET, giving a maximum width of 45,000 km. However, a minimum width of about 17,000 km can be estimated from Figure 6. This figure shows, from bottom to top, the 0.8-3.6 keV electron density, the proton density from IMS and the 250-500 eV/q flux of ram ions in a more extended time scale than that of Figure 5. The observations of Berthelier et al. [1986] indicate that the large ion velocity drop, accompanied by a substantial temperature increase, is also detected across a relatively thin (~ 20,000 km) region.

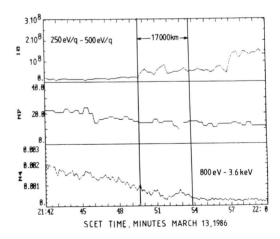

Fig. 6. Variations of 0.8-3.6 keV electron density, proton density and total ram ion fluxes though the mysterious transition.

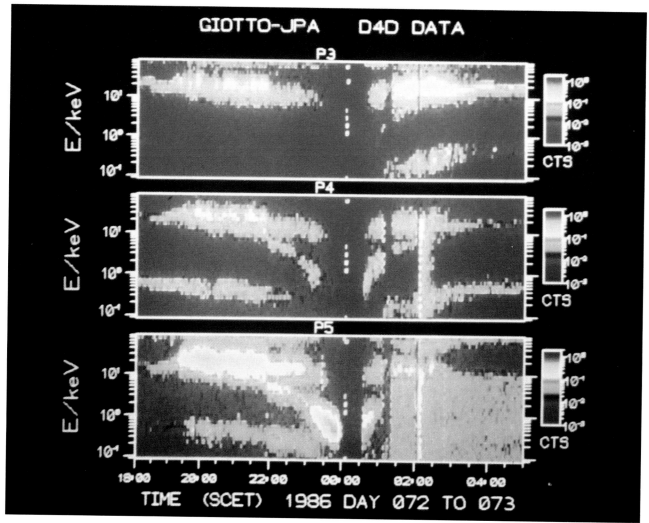

Fig. 7. An energy-time spectrogram giving an overview of positive ion distributions during Giotto's encounter with Comet Halley. Each panel shows data from one of the individual sensors in the array which make up the IIS. They are placed at different angles with respect to the spin axis of the spacecraft. In each case the data are averaged over one spin of the spacecraft. At closest approach the instrument is turned off by electrical noise associated with the dust bombardment and is not restored for 30 min. After that, one sensor is permanently noisy (bottom panel); from Johnstone [1988].

At this mysterious transition there were also (see Figure 2, 2/3 time) a sudden decrease in the bulk velocity and density of M/Q = 2 ions but an increase in ion temperature. Thus there was here a cooling of the electrons and a heating of the ions, but no obvious effect in the magnetic field (see Figure 3).

This transition was also marked by a sudden change in the energy spectrum of the heavy ion population [Johnstone et al., 1986; Wilken et al., 1987b]. Figure 7 shows data from the Implanted Ion Sensor (IIS) of the JPA instrument during which the spacecraft traveled from 1.5 million km before to 1.2 million km after closest approach. Each panel is an energy spectrogram, covering the energy range 90 eV/q to 90 keV/q, for one of the individual sensors within IIS placed at different angles to the spin axis. There are two main lines in the energy spectrum; the upper one is the spectrum of cometary ions and the lower one is the spectrum of the solar wind protons. At the mysterious transition near 2150 SCET it appears that the cometary ion distribution suddenly split into two lines; the upper branch continued at the same energy, while the lower branch decreased rapidly in energy as the nucleus was approached. In fact the lower energy peak can also be seen before the transition showing that the splitting is actually gradual inside the mysterious region.

The explanation of this split is as follows [Johnstone, 1988; Thomsen et al., 1987]. The cometary ion spectrum is made up of ions produced at all positions from the immediate vicinity of the spacecraft to the far upstream region tracking back along the trajectory of the ions. The energy of the ions depends on the flow velocity and the magnetic field direction at their point of creation and on the flow velocity at the point of observation. If the flow field had been smoothly varying then the spectrum would have been a broad feature with a single peak. The double-peaked structure arises because there was a jump in the velocity variation at the bow shock. Very few ions are created at velocities within the range of the jump and this leads to a gap in the spectrum and hence a double peak. But why is there a sudden transition at 2150 with a sudden decrease in velocity by approximately 90 km/s [Johnstone, 1988]? The nature of this velocity change is not yet understood. Figure 7 shows that this phenomenon is also identified on the outbound leg after 0100 SCET (the time accuracy is not good due to data gaps).

Moreover, Amata et al. [1991] have recently shown that the mysterious transition corresponds to the place where the water group ion nucleonic density becomes greater than the solar wind proton density and Fuselier et al. [1988] have found that H_2^+ ions begin to be detected as far back as this transition.

So, in summary, Giotto detected a strong plasma transition, on the inbound and outbound legs, with a width of about 20,000 km, located roughly halfway between the shock position and the comet nucleus. By considering the results of the other spacecraft it is possible to draw the very important conclusion that this transition is detected by all of them: Suisei, Vega-1 and -2 and ICE.

Suisei observations. Figure 8 shows the results of plasma

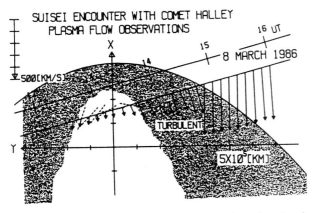

Fig. 8. Plasma flow vectors obtained during the Susei encounter with comet Halley. The flow vectors and angles are represented in the Cometocentric Solar Ecliptic Coordinate system. Dashed lines represent the estimated directions of magnetic field; from Mukai et al. [1986].

flow observations during the Suisei encounter with comet Halley, where the observed flow pattern is shown by arrows starting from the Suisei position at each epoch, using the cometocentric solar ecliptic (CSE) coordinate system [Mukai et al., 1986]. From 1232 UT, when the plasma instrument was turned on at only 2×10^5 km from the comet until 1443 UT, Suisei was in the cometosheath. Inside the cometosheath the flow pattern was roughly symmetric with respect to a direction slightly shifted (4-6° left) from the direction of the sun. This is the expected direction for the solar wind velocity at 400-500 km/s in the CSE frame.

Mukai et al. [1986] estimated the magnetic field directions from the symmetry axes of the pick-up shells. These directions however can be estimated only within ~ 2×10^5 km from the nucleus where the flow is laminar. Outside this region, the fluctuations in the flow direction (of periods ≤ 2 min) become too large to perform this estimate. Flow turbulence is deduced from fluctuations of the anisotropy directions. The turbulent region is shown by hatching in Figure 8. It corresponds to the outer region 1 and 2 as seen by Giotto. The separation between the laminar

flow and the turbulent flow as seen on the outbound leg by Suisei (the plasma experiment was turned on too late to detect it on the inbound leg) suggests the existence of a boundary comparable to the mysterious transition seen by Giotto).

Vega observations. The electron detector did not work on Vega-1 and on Vega-2 was turned on at about 8.5×10^5 km from the comet nucleus. The electron measurements [Gringauz et al., 1986c] show (Figure 9) that the electron temperature was high,

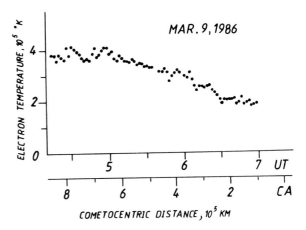

Fig. 9. Distribution of the electron temperature in the cometosheath of comet Halley between 8×10^5 and 1.6×10^5 km. CA = closest approach; from Gringauz et al. [1986c].

with maxima, and ressembled the Giotto electron results inside the mystery region up to 5.5×10^5 km from the nucleus where there is a clear decrease of the electron temperature, as in the Giotto mysterious transition.

Grard et al. [1989] have measured electron parameters with Langmuir probes. They think that the considerable scattering in the electron mean kinetic energy does not represent a real variation; rather it must be linked with the fact that the ambient plasma parameters can vary during a voltage sweep. However for Vega-1 and Vega-2 the level of scattering changed around 6×10^5 km from the comet, and for Vega-2 this change was associated with a decrease in the electron mean kinetic energy (Figure 10). On Vega-1 between 8.7 and 6.5×10^5 km from the comet, Trotignon et al. [1991] have found a region characterized by strong fluctuations in the electron density and in the electric- and magnetic-field measurements. Trotignon et al. [1991] think this region looks like the mystery region identified on board Giotto.

ICE observations. During the encounter with comet Giacobini-Zinner, ICE passed through several distinct plasma regimes, downstream of the shock [Bame et al., 1986]. Afterwards ICE went through the tail of Comet G/Z and those results will not be discussed here.

Bame et al. [1986] identified two different broad regions behind the bow wave and referred to them as the transition region (TR) and the sheath (S) as shown in Figure 11. These two regions are seen on the inbound and on the outbound leg. Bame et al. [1986] did not find evidence for a conventional bow shock but rather the beginning of a transition region inside which the incident solar wind and cometary plasma strongly interact to compress, heat, and slow the solar wind. It is characterized by gradual increases of a factor of 2 to 3 in electron density, by a factor of 1.5 to 2 in electron temperature, and by a nearly linear decrease to roughly half its upstream value in flow speed (Figure

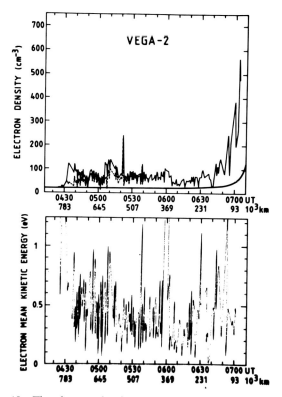

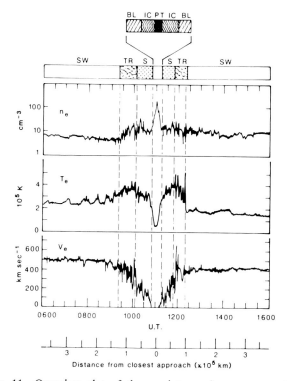

Fig. 10.   The electron density and mean kinetic energy of the comet Halley environment during the Vega-2 flyby. The dashed line represents the electron density after removal of the contribution associated with the N2 gas release. The thick line gives an estimate of the density caused by electron emission from the spacecraft surface; from Grard et al. [1989].

Fig. 11.   Overview plot of three-point running averages of electron n, T, and V, measured during the encounter of ICE with comet Giacobini-Zinner. Plasma regions are identified at the top where SW identifies the solar wind, TR the transition region, S the sheath, BL the boundary layer, IC the cold inner tail, and PT the plasma tail; from Bame et al. [1986].

11). There was a high variability in the electron distributions, with some distributions upstream-like and others sheath-like. The upstream-like distributions were more frequent near the upstream edge, while the sheathlike distributions were more frequent near the inner edge, but there was no monotonic evolution from one to the other.

The sheath was the region in which the electron density fell slightly from its peak value at the inner edge of the transition region, the electron temperature decreased to near its upstream value, and the flow speed continued to fall, reaching values below 10 % of its upstream value. The change between the transition region and the sheath was abrupt and can be compared to the Giotto mysterious transition.

All these observations (Giotto, Suisei, Vega-1 and -2, ICE) suggest that the mysterious transition seems to be a feature characteristic of the cometosheath, at least around 1 a.u., in comets Halley and Giacobini-Zinner and that is not coincidental but reflects the effects of similar phenomena in all cases.

To compare the sizes of the comet Halley and comet Giacobini-Zinner solar wind interactions, Figure 12 shows a single curve which represents the locations of the bow shock in the case of Halley (from Giotto measurements on the inbound leg) and the bow wave corresponding to the entry of the Giacobini-Zinner transition region with different distance scales. It is assumed that these two boundaries are described by a symmetrical paraboloid equation form with the same flaring ratio

of 2 [Mendis et al., 1986; Rème et al., 1986]. Thus, the stand-off distance of the bow wave for Giacobini-Zinner was ~ 6 x 10^4 km deduced by the ICE bow wave-nucleus distance of 1.3 x 10^5 km. The linear factor relating the solar wind-comet interaction regions of the two comets at the time each was observed is thus in the range 6-7, this difference being mainly due to a larger neutral production rate from the Halley nucleus. On the same figure the flyby geometries of each spacecraft are indicated : e.g. the Giotto trajectory was at an angle of ~ 107° to the Sun-Comet Halley line. To compare the data of all the spacecraft the inbound and outbound positions of the shock (when available) are included in Figure 12 for each spacecraft.

Because of the similar behaviour observed by the spacecraft in the cometosheath of the two comets it is possible to compare the position of the mysterious boundary as seen by Giotto with the other spacecraft measurements, including the transition region - sheath boundary identified at comet Giacobini-Zinner. To do this the linear scaling factor of 6.7 determined from the bow shock wave crossing locations at the two comets is used. The Giotto mysterious transition and the G/Z transition region - sheath boundary are constructed by fitting a paraboloid to the two inbound boundary crossing points. The position of the mysterious boundary lies remarkably close to the transition-sheath surface inferred from the ICE data at Giacobini-Zinner. In addition the positions found by Suisei and Vega are included in Figure 12, showing that the measurements of all the spacecraft are surprisingly consistent, as for the shock position, given the limits of the fluctuations due to the solar wind variations. It is possible to conclude that the mysterious transition, about half way between the shock and the comet nucleus, is a permanent global feature of the active comet's plasma environment near 1 a.u. from the Sun.

Pérez-de-Tejada [1989] has noted that the contrast between the lower electron temperatures and the higher ion temperatures that were detected at the mysterious transition, is reminiscent of a similar anticorrelation that exists across the intermediate transition of the Venus ionosheath. In that case the lower electron temperature seen downstream from that transition is interpreted as resulting from contaminant electrons of ionospheric origin that populate the region between that boundary and the Venus ionopause [Spenner et al., 1980]. The same effect with the increase of the contamination by cold cometary electrons could explain the electron temperature decrease.

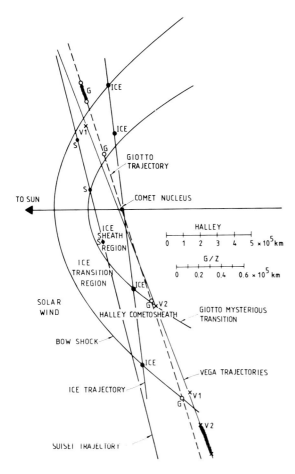

Fig. 12.   Comparison of the size of the comet Halley and comet Giacobini-Zinner solar-wind interaction regions. The bow shock, the Halley mysterious transition and the G/Z transition-sheath boundary, and the distance scale appropriate to each comet are shown. The plane of projection is defined by the spacecraft trajectory and the sun-comet direction. For Halley, inbound Giotto results were taken to put the boundary positions. Positions from the other spacecraft are added when available (G = Giotto, S = Susei, V1 = Vega-1, V2 = Vega-2).

In conclusion, these observations suggest that the mysterious transition is a feature characteristic of the ionosheath exterior to an ionospheric obstacle (be it of planetary or cometary origin), and that its common identification in Venus and in comets Halley and Giacobini-Zinner reflects the effects of similar phenomena in all cases. This boundary, which has not be included in the solar wind-comet interaction models up to now, urgently requires more theoretical work to explain it.

d. Outer Region 3: The Quiet Region

Outer region 3 extends from 5.5×10^5 to 1.35×10^5 km (MPB) from the comet. In this region the average velocity of the plasma flow decreased regularly and the electrons and water group ions cooled continuously (because the pickup ion velocity decreased); the average 10 eV - 30 keV electron density was smaller than before and relatively constant while the electron temperature and bulk velocity decreased smoothly. The heat flux ratio was everywhere greater than 0.57, showing that the electron distribution functions were more isotropic than in the other outer regions. In fact the electron distributions became stable and cooled monotonically. Here the high energy electron density was small (smaller than in all the regions encountered before, see Figure 1). On the other hand, low energy ions were increasingly detected in the ram direction, as seen in Figure 1, and the peak energy of the new low energy component of cometary ions rapidly decreased as the spacecraft moved towards the comet (Figure 7). From around 2.5×10^5 km, IIS, PICCA, HERS and NMS quite consistently measured ion intensities which strongly increased with decreasing distance from the comet [Amata et al., 1986; Rème et al., 1987b; Balsiger et al., 1986; Hodges et al., 1986]. The radial dependence may be estimated to be $1/r^x$ with $x \geq 2$.

In outer region 3, charge exchange of the solar wind ions with the outflowing cometary gas has been recognized to become increasingly important. Figure 13 is from a study of Shelley et al. [1987] and demonstrates how the charge-exchange, as expected,

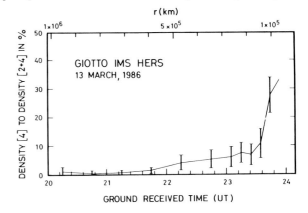

Fig. 13.   Ratio of the density of M/Q=4 ions to the combined densities of M/Q=2 and 4 versus distance to the nucleus on the inbound leg with a strong increase of charge-exchange around the MPB; from Shelley et al. [1987].

becomes more and more important when the comet is approached. $^4He^+$ began to rise relative to the total helium (M/Q = 2 + 4) around the inner mystery region (mysterious) boundary (2/3), and the slope of the curve became very steep at the MPB. Whereas the qualitative picture of the increasing importance of charge exchange towards the comet is as expected, the observed

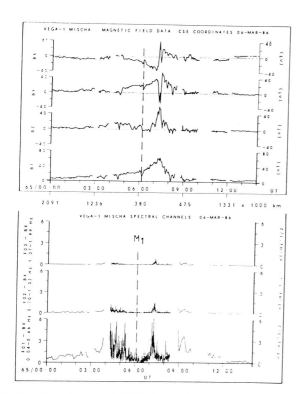

Fig. 14. Vega-1 magnetic field observations during the Halley encounter. The lower panels show three spectral channels of the Vega-1 magnetometer in the frequency range 0.04 to 2 Hz (B_x component). The M1 boundary has a distance to the nucleus of about 330000 km; from Schwingenschuh et al. [1987].

quantity of He^{++} being transformed into He^+ at distances less than ~ 10^5 km was much higher than expected (by a factor of 5 to 10).

Due to the much smaller fluctuations of the plasma properties in this region, it can be called a "**quiet region**". Here the cometary - solar wind interaction may become so important that many things are "frozen". The magnitude of the magnetic field was still almost constant in this region (Figure 3).

The M1 boundary. The magnetic field experiments on board Vega-1 (Figure 14) and Vega-2 detected a sudden increase in the slope of total B during the inbound phase of the Halley encounters [Schwingenschuh et al., 1987]. This M1 boundary was observed at a distance of 330,000 km (Vega-1) and 370,000 km (Vega-2). A magnetic field depression with a length of about 20,000 km separated the two regions with different slopes. The M1 boundary was also associated with a broad minimum of wave activity in the frequency range 250 - 750 Hz [Oberc et al., 1987], and a similar minimum in the lower frequency channels (0.04 - 1.3 Hz) of the magnetometers. At this boundary the population of the solar wind protons became comparable to that of the cometary ions [Grin-gauz et al., 1986a], and the velocity of the plasma started to decrease more rapidly. In the same region the neutral gas density also displayed a discontinuity. This boundary is quite probably caused by an increased massloading of the solar wind. A boundary at the same position was not observed by the Giotto magneto-meter. This indicates that the M1 boundary has a lifetime of only several days and can therefore be regarded as a kind of quasis-tationary feature of Halley's coma.

3. The Inner Regions of the Cometosheath

a. The Giotto Magnetic Pile-up Boundary

After the traversal of the outer regions, the Giotto spacecraft entered the magnetic pile-up region, as indicated by the magnetometer data (Figure 3). Giotto has identified an unpre-dicted sharp boundary separating regions controlled by the solar wind dominated hot plasma from the region where cold cometary ions dominate. This boundary is called the magnetic pile-up boundary (MPB). Outside this MPB, a large region is characte-rized by an anti-correlation between the electron plasma density and the magnetic field strength [Glassmeier, private communi-cation]. The study of the electron pressure tensor and the anisotropy of the distribution function and the intercorrelation between the magnetic field oscillations and the electron density and pressure tensor lead to the identification of these modes as drift mirror modes probably generated by the pressure anisotropy due to the cometary implanted ions [Mazelle, 1990].

Indeed, at 1.35×10^5 km from the nucleus, the Giotto magnetometer detected an abrupt increase in the magnetic field intensity, from ~ 8 γ to ~ 30 γ [Neubauer et al., 1986]. The JPA experiment indicated a sudden sharp decrease in the solar wind proton density at this time; the same strong gradient in the proton density was seen by HERS (Figure 15, from Goldstein et al.

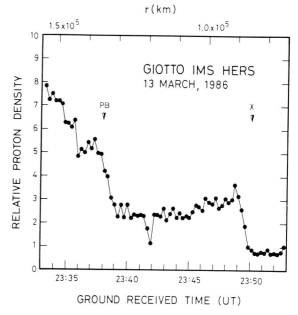

Fig. 15. Radial profile of the relative proton density (arbitrary units). The proton density strongly decreases at about the MPB and at "discontinuity" X; from Goldstein et al. [1987].

[1987], giving the radial profile of the relative proton density around the MPB), while the RPA-COPERNIC electron detector displayed a rapid change in the characteristics of the electron plasma parameters [d'Uston et al., 1989] and the electron bulk velocity tended to its minimum value, i.e. decreased almost down to stagnation (Figure 1). No dramatic discontinuity for the heavy ions was observed at this boundary, in contrast to the situation of the protons [Goldstein et al., 1987] and electrons. Thus the cometary ions became dominant over solar wind protons and

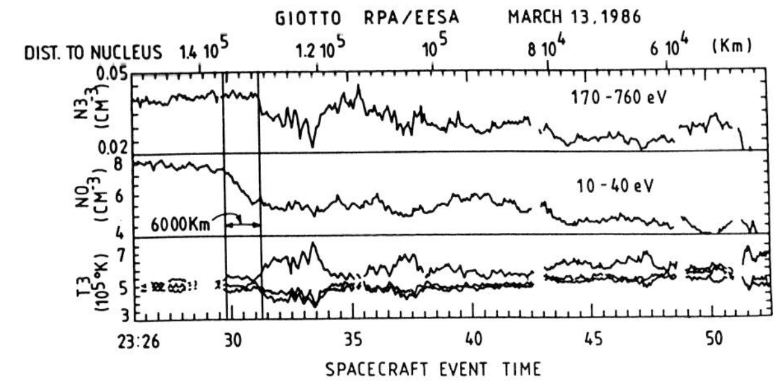

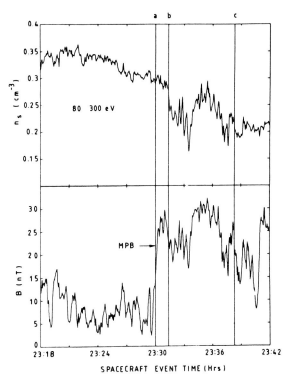

Fig. 16. From bottom to top, 3 electron temperatures given by the diagonal pressure tensor, 10-40 eV electron density and 170-760 eV electron density around the MPB located at 2330 SCET; from Rème [1990].

alpha particles, mainly because of a steep decrease of the protons at this location. At this time the magnetic field pile-up region began.

For example, Figure 16 shows, from bottom to top, the 3 electron temperatures given by the diagonal pressure tensor, the 10-40 eV electron density and the 170-760 eV electron density during the MPB traversal period [Rème, 1990]. Before the MPB, the three temperatures were tightly linked together, meaning that the electron distribution function was nearly isotropic. After the MPB, the parallel temperature increased while the perpendicular one decreased. This indicates that inside the MPB the electron distribution is cigar shaped with the long axis along the magnetic field direction. During the same time the ratio of minimum to maximum heat flux (not shown) transported by the electrons parallel and antiparallel to the magnetic field, close to 1 before the MPB, dropped suddenly to about 0.5 at this boundary. This indicates that the electrons leak along the field lines in a preferential direction [d'Uston et al., 1989]. The 10-40 eV electron density began to decrease around 2329:45 SCET and this decrease continued until the abrupt decrease around 2331:20 SCET of the 170-760 eV electron density. The width of the boundary deduced from these electron measurements was ~ 6000 km, or more than 10 cometary ion gyroradii in a 30 γ field. After 2332 SCET, the suprathermal electron density, the 170-760 eV electron density, and the T3 temperatures underwent a series of low frequency oscillations, as shown in Figure 16. The oscillation frequency was in the range of the gyrofrequency for 18 amu ions in a 30 γ magnetic field (~ 2.5 x 10^{-2} Hz).

On Figure 17, the upper panel shows the 80-300 eV suprathermal electron density and the lower panel the magnitude of the 2 second average magnetic field vector around the MPB [Mazelle et al., 1989]. Around 2330 SCET (discontinuity labelled 'a' on the plot), the magnetic field strength increased by a factor of about 3 over a distance of less than 2 x 10^3 km, and reached a value around 30 nT. At 2331:20 SCET, the 80 - 300 eV suprathermal electron density ns dropped sharply by about 25 %, whereas it was gradually decreasing before. Figure 17 shows that from this time ('b'), ns and B reveal structures which are very well correlated, up to about 2338 SCET ('c'). Moreover, large amplitude quasi-coherent wave events were observed, particularly between 2332 and 2334 SCET. The simultaneous oscillations of the magnetic field appeared at the same frequency but affected the magnitude of the field only, hence showing the compressive nature of these waves. In addition, at 2330 SCET, the very time when the magnetic field began to rise, the 10 - 40 eV electron

Fig. 17. Time evolution of the density of the 80-300 eV suprathermal electron (upper panel) and the magnitude of the magnetic field (lower panel) around the MPB located at 2330 SCET; from Mazelle et al. [1989].

density began to drop (see Figure 15), but decreased smoothly by about 30 % over a distance of 10^4 km. Thus the coincidence between the magnetic field and the electron discontinuity is quite clear but the effect of the B field jump on the electrons seems to be energy-dependent. Afterwards, the electron density above 10 eV never recovered to the level it had prior to the MPB [d'Uston et al., 1989].

Figure 18 presents a comparison between the parallel and perpendicular suprathermal electron pressures and the magnetic pressure. The top panel of this Figure shows that there is no evidence for the MPB in the parallel pressure p// of the suprathermal electrons. On the contrary, the middle panel illustrates significant changes in the perpendicular pressure P with the same evolution as the density ns in Figure 17 : after a gradual decrease it also suddenly droped at time 'b' as did the magnetic pressure shown on the lowest panel. Following this, a close correlation between the variations of these two pressures appeared again over a distance of 3 x 10^4 km ('b' to 'c').

The region between the increase of the field strength at 2330 SCET ('a') and the drop in the suprathermal electron density and perpendicular pressure at 2331:20 SCET ('b') may be interpreted as a transition layer of the MPB also with a width of about 6000 km. Neubauer [1987], taking into account that B_n (the normal field component) was nearly zero at the MPB, proposed that this discontinuity could be either a tangential discontinuity or a slowly propagating rotational discontinuity with very different plasma anisotropies on both sides. The fact that the magnetic moment changed rapidly at the MPB is also consistent with a tangential type of discontinuity.

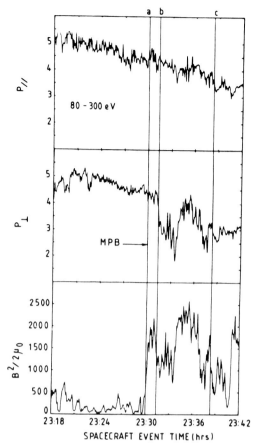

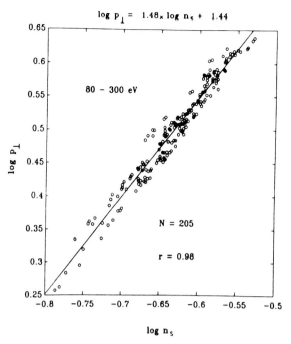

Fig. 19. Plot of log p versus log ns for the 80-300 eV suprathermal electrons in the plasma layer between 2331:30 SCET and 2338:17 SCET. The linear regression leads to a correlation coefficient of 0.98 and the slope is equal to 1.48 ± 0.06 (for a 99% confidence level from a Student test); from Mazelle et al. [1989].

Fig. 18. Time evolution of the parallel pressure (uppermost panel), the perpendicular pressure (middle panel) of the suprathermal electrons and the magnetic pressure (lowest panel), around the MPB located at 2330 SCET; from Mazelle et al. [1989].

The quasi monochromatic waves seen on Figure 18 between b and c were compressive, since they affect both the magnetic field magnitude and the electron density and pressures. Thus this region, between b and c, is called the wavy sheet [Mazelle et al., 1989]. The mean magnetic moment of suprathermal electrons, which had very large fluctuations before, was very constant here. B and n_S (Figure 18) were strongly correlated because the waves, when propagating, cause alternating compressions and rarefactions of both the lines of force and the conducting fluid. This is quite consistent with fast magnetosonic waves if we consider a propagation perpendicular to the lines of force [Glassmeier, private communication]. In this wavy sheet, Figure 19 shows that the suprathermal electron density and perpendicular pressure were linked by a very nice equation of state with a polytropic form:

$$\frac{P}{n_S^{\gamma_\perp}} = constant$$

with $\gamma_\perp$ = 1.48 ± 0.06. Because double adiabatic theory predicts γ = 2 for compression perpendicular to the B field, a multifluid approach may be necessary to explain the departure from an adiabatic law.

In summary, the MPB separating 2 very different plasma regimes seen by Giotto on the inbound leg is a tangential discontinuity at the boundary followed by a wavy sheet with fast magnetosonic waves and a very steady polytropic equation of state. If we consider an ellipsoidal MPB geometry, as for the shock and the mysterious transition (Figure 12), the MPB would have a distance of ~ 5 x 10^4 km from the nucleus in the solar direction

Is there also a MPB on the Giotto outbound leg ? Johnstone [1988] found a similar change in the solar wind protons at 0049 SCET and Neubauer [1987] considered the magnetic field discontinuity at 0107 SCET to be the best candidate for the outbound MPB (see Figure 3) noting that the boundary could be called the isodyne for magnetic field values around 20 nT. However this conclusion is not very strong due to the magnetic asymmetry between inbound and outbound measurements.

In fact, if we compare inbound and outbound MAG data the identification of the MPB is not easy. I suggest a possible candidate near 0021 SCET, taking into account the phi and theta variations of B. The IIS sensor was not on at this time but an increase in the intensity of energetic cometary ions was found by FIS. The time 0042 SCET is also more in agreement with the geometry of the Giotto encounter if the MPB is a "relatively" stable boundary. But in any case, the outbound MPB, if it exists, is very different from the inbound one.

Vega-1 and -2 results. Both Vega spacecraft measured a magnetic pile-up region with a peak field strength of 70-80 nT and observed draping of magnetic field lines around the cometary obstacle [Riedler et al., 1986]. An unexpected rotation of the magnetic field vector was observed, which may reflect either penetration of magnetic field lines into a diffuse layer related to the contact surface separating the solar wind and cometary plasma, or the persistence of pre-existing interplanetary field structures.

However neither Vega-1 nor Vega-2 magnetometers detected a MPB. Why ? In fact Vega-1 and -2 detected a B field increasing starting at least 350,000 km from the comet (Figure 14) while B is nearly constant between the shock and the MPB at 135,000 km from the comet, in the case of Giotto. So even if the Giotto MPB was a very sharp boundary, it was a temporary effect and probably a discontinuity due to a solar wind structure convected into the cometary environment. The differences over a few days, between the Giotto and Vega-1 and -2 results, confirm the high degree of variability of the cometary environment. This is partly confirmed by the differences between inbound and outbound leg results for the same spacecraft.

Several models predicted the formation of a magnetic barrier starting at a collisionopause where the strong deceleration of the solar wind is expected to begin, due to momentum transfer collisions with the outflowing neutrals. Mendis et al. [1989] have therefore identified the MPB with the so-called collisionopause but the measurements have revealed that this transition is probably too sharp, of the order of 6 x 10^3 km.

The Vega-2 "cometopause". Gringauz et al. [1986b] believe they have identified a "sharp, chemical" boundary separating region controlled by the solar wind flow from the region where slowly moving cometary ions dominated near 1.7 x 10^5 km from the nucleus (Figure 20). This boundary appears to be associated

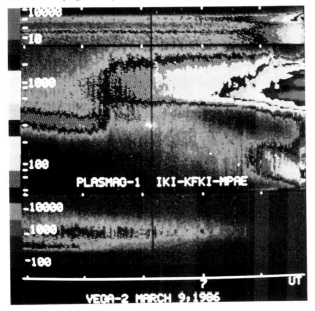

Fig. 20. Color coded summary representation of the VEGA-2 plasma measurements between 2.3x10^5 km and 1.4x10^4 km. Time runs from left to right (there are ten minutes tickmarks in the Figure). Electron energy spectra are shown in the upper panel, while CRA and SDA spectra are presented in the middle and lower panels, respectively. Energy increases upwards in each panel. The color coding varies from dark blue representing the lowest fluxes to red corresponding to the highest intensities (see the colour bar at the left of the figure); from Gringauz et al. [1986b].

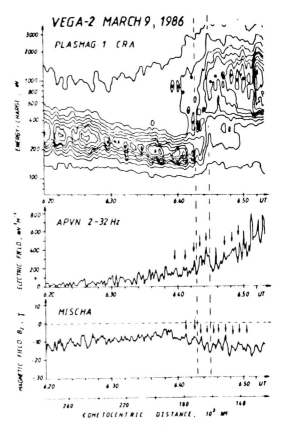

Fig. 21. The top panel shows the ion spectrogram measured by the CRA on Vega-2. The growth rate between the adjacent isolines is 440 sec^{-1} and the outermost isoline corresponds to a counting rate of 10^3 sec^{-1}. Dots on the spectrogram mark the local maxima of ion fluxes observed in an interval of 10 minutes around the "cometopause". Are also shown ion flux fluctuations, electric field oscillations in the lower hybrid frequency range and the amplitude of the B_z component of the magnetic field; from Sagdeev et al. [1987].

with some small effects (Figure 21) in the 2-32 Hz electric field oscillations and in the B_z component of the magnetic field with the same characteristic period (T ~ 1 min). However several questions come up concerning this interpretation:

- colour or grey plots can make it difficult to see a "pause",
- there was no magnetic field increase (while for the Giotto MPB the increase was remarkably strong),
- the apertures of the sensors were small and covered a very limited range of solid angles,
- the solar wind protons were seen again, closer to the nucleus (Figure 20),
- the cometopause crossings, at 0647 UT and 0645 UT for Vega-1 and Vega-2 respectively, cannot be identified in the electron density profiles [Grard et al., 1989],
- how is the Vega-1 cometopause identified if the CRA sensor was not operating on Vega-1 ?

In addition the Giotto ion results are in contradiction with a thin chemical boundary (~ 10^4 km) but show a rather gradual increase of ion density even in the magnetic pile-up boundary.

Therefore my conclusion is that the Giotto MPB and the Vega-2 "cometopause" are totally different. If the MPB is a sharp but unsteady plasma boundary, the cometopause is not a real plasma boundary and is not a pause. It is just a density cross-over point, where the implanted cometary ion density rapidly exceeds the solar wind number density in the ram direction.

b.   The Inner Cometosheath between the MPB and the Contact Surface: the Magnetic Pile-up Region

The MPB is the outer boundary of the magnetic pile-up region due to draping of the magnetic field lines around the conducting obstacle of the comet ionosphere. This region did not show a steady increase of the field magnitude when the spacecraft approached the comet. Several decreases were measured; they could be diamagnetic cavities associated with pressure increases linked to the ray formation seen in the cometary tail [Neubauer et al., 1986]. In addition plasma densities are much more variable when displayed with high time resolution.

However, on the average, the intensity of the magnetic field increased and reached a maximum of ~ 60 γ, around 2359 SCET (Figure 3), i.e. at about 16,000 km from the comet [Neubauer et al., 1986].

Ion results. As Giotto moved inward from ~ 10^5 km the plasma environment became less turbulent, and solar wind protons and alphas became less and less important while there was a fast increase in the cometary ion fluxes detected in the ram direction.

A new strong gradient in the proton density was observed after the MPB around 2341 SCET (2349 GRT), at the location marked X on Figure 15, at 8.6 x 10^4 km from the nucleus, where Johnstone et al. [1986] have noted the gradual loss of energetic solar wind ions, probably due to charge exchange. This location could be important because if the solar wind dominates the regions between the shock and the MPB, with a contamination by pick-up cometary ions, there is a transition between the MPB and X and after X cometary ions clearly dominate. As expected, the cometary ion densities continuously increased when approaching the comet and the most abundant ions were water and ions resulting from the dissociation of water, OH^+, O^+ and H^+ [Balsiger, 1990; Korth et al., 1987b]. From the HIS measurements, the dependence of the main ions on radial distance outside ~ 10^4 km went roughly as r^{-2}. Outside 6 x 10^4 km the dissociation products of water OH^+ and O^+ were more abundant than the ion of the mother product H_2O^+. However Korth et al. [1987b] found with RPA2-PICCA a radial dependence between r^{-2} and r^{-3} for masses below 50 amu and steeper than r^{-3} above 50 amu (Figure 22).

Below 2 x 10^4 km the H_3O^+ ion became more abundant than H_2O^+ demonstrating the importance of ion-molecule reactions at high total densities. Between here ("discontinuity" Z) and the ionopause the plasma was found to be stagnant (Figure 23). Indeed, the velocity (Figure 23) was more than 4 km/s at 3 x 10^4 km but decreased and remained very low between ~ 20,000 km and the ionopause.

This stagnant plasma had a local density peak A at around 10^4 km from the comet (Figure 24), termed the **plasma pile-up region** [Balsiger et al., 1986]. Actually, both particles and magnetic field [Neubauer, 1987] displayed a maximum here. The density fell off as r^{-1} from the inner part of the cometary ionosphere, through the edge of the magnetic cavity (C) out to a distance of 8000 km (B). After the ion pile-up the density fell as r^{-2} as one might expect. A similar maximum like A was seen by

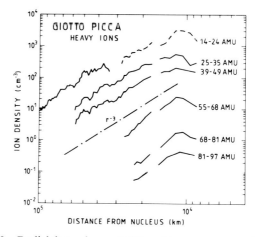

Fig. 22.   Radial dependence of different ion mass groups; from Korth et al. [1987b].

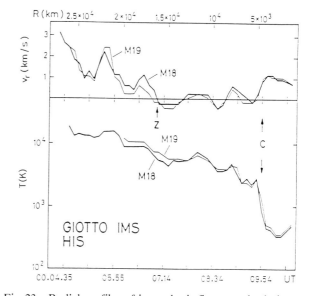

Fig. 23.   Radial profiles of ionospheric flow speed relative to the comet and temperature of ions with mass/charge ratios of 18 and 19 amu/charge. The ionopause(c) and the discontinuity Z are marked with arrows; from Balsiger et al. [1986].

Vega-1 [Vaisberg et al., 1988] and can be deduced from the H_2O^+ ion brightness profile as seen from Earth on March 15, 1986 [Ip et al., 1988].

Whereas the stagnation of the plasma in front of the ionopause was predicted by a model [Ip and Axford, 1982] this observed ion density profile with a maximum at 10^4 km from the comet and a minimum just outside the ionopause was quite unexpected (see figure 24, points A and B). The conclusion of Ip et al. [1987] is that the "pile-up" is in fact a rapid increase of the ion recombination rate as a result of a drop in electron temperature, just inside the A peak. Another possible interpretation could be the non-stationary process of plasma envelope formation as seen in some bright comets.

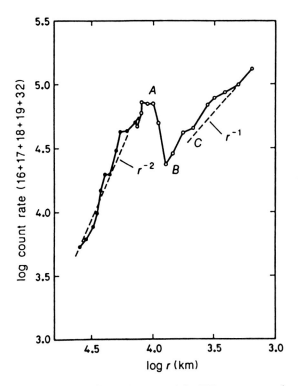

Fig. 24. Radial profile of the sum of the HIS count rates for masses 16, 17, 18, 19 and 32. Inside the "contact surface" (C, at r = 4700 km) the total count rate tends to follow a r^{-1} dependence while in the outer part (r > 16000 km) it tends to follow a r^{-2} dependence; from Balsiger et al. [1986].

Balsiger [1990] has noted that in the stagnation region the plasma was decoupled from the neutral gas. The outflow of neutral water vapour was even accelerated here. The ion temperature, on the other hand, (Figure 23) increased from the ionopause to reach about 20,000°K at 2.7×10^4 km (discontinuity Y) while the gas temperature remained below 500 K over this range [Lämmerzahl et al., 1987; Schwenn et al., 1987].

Electron results. From the MPB, the electron density decreased with several strong gradients (Figure 25). One of them, near 2342 SCET, coincided with the position X. At 4.5×10^4 km from the comet another sharp electron density decrease was detected. Afterwards, the 10 eV - 30 keV electron density decreased to a value as low as 2 cm^{-3}, while the ions in the ram direction were cold, with a high density, and appeared in the form of peaks in the E/q spectra. It is interesting to note that all along the cometosheath, the density of the ram (cold) ions as measured by RPA2-PICCA did not vary in the same way and was almost anticorrelated with the 10 eV-30 keV electron density and, increased strongly when Giotto approached the comet. Due to the neutrality of the plasma, it can be concluded that in the inner coma most of the electrons have an energy much lower than the threshold of RPA1-EESA, well below 10 eV [Rème et al., 1987a]. Grard et al. [1989] give an average value for the mean kinetic energy of 0.53 eV inside the cometosheath on Vega-1 and 0.49 eV on Vega-2. Fitting the RPA1-EESA electron data in the inner coma aboard Giotto confirms these results. In the case of G/Z ICE measurements, Meyer-Vernet et al. [1986] measured by thermal noise spectroscopy, a high density of cold electrons

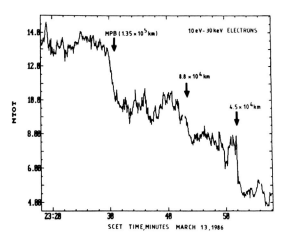

Fig. 25. Total 10 eV-30 keV electron density showing 3 important discontinuities: MPB, X (at 8.8×10^4 km), and one at 4.5×10^4 km from the comet as measured on Giotto.

having a temperature slightly above 1 eV near the tail axis. Thus we can conclude that the electron temperature in the inner coma of the comets was ≤ 1 eV.

The ionopause. After its maximum the magnetic field decreased very rapidly to about 0 at 4700 km from the nucleus when Giotto crossed the ionopause (contact surface) on the inbound pass of Halley's comet [Neubauer et al., 1986; Balsiger et al., 1986; Krankowsky et al., 1986b; Chaizy et al., 1990]. This very clear discontinuity, marked C in Figure 22, was characterized by a steep ion temperature drop, ~ 1,700°K (the temperatures are very low, of the order of 300°K), by an increase of the ion outflow velocity from 0 to about 800 km/s which is the same order as the outward flow velocity of the neutral particles, and by a drop of the magnetic field to zero [Balsiger et al., 1986; Schwenn et al, 1987; Lämmerzahl et al, 1987; Neubauer et al., 1986]. Inside C, the Giotto spacecraft was basically immersed in a cold plasma of purely cometary origin, Halley's ionosphere. Giotto crossed the contact surface on the outbound pass at 3900 km from the nucleus. This "contact surface" was in fact a tangential discontinuity where pressure equilibrium is obtained except for a minor contribution from the frictional forces due to the ions [Neubauer, 1988].

Two important measurements made by the RPA experiment in the regions near the comet on both sides of the ionopause also deserve to be mentioned:

- very complex ions were detected by RPA1-PICCA. Mitchell et al. [1989] think that the ordered series of broad mass groups observed in the thermal ion mass spectra (Figure 26) are characteristic of molecules rich in hydrogen, carbon, nitrogen, and oxygen, associated with the abundant CHON dust particles.

- negative ions were detected with a relatively high density at distances smaller than 40,000 km from the comet. They are organized into 3 well defined peaks centered around 17 ± 2 amu, 30 ± 10 amu and 100 ± 10 amu [Chaizy et al., 1990]. Their radial profile is described by a law in $r^{-2.8}$ outside the ionosphere and in r^{-4} inside. These negative ions had densities about 2 orders of magnitude higher than the proposed theoretical ones.

Conclusion

The global nature of the solar wind - comet interaction has some features which are close to the predictions made before

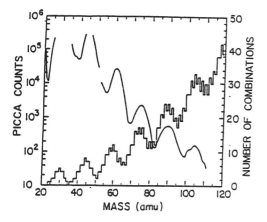

Fig. 26. The total number of combinations of hydrogen, carbon, nitrogen, and oxygen which have enough chemical bonds to form a molecule as a function of mass (histogram). This is compared to a mass spectrum of cold thermal ions in the inner coma region (10,000 to 15,000 km, solid line). Breaks in the spectrum indicate instrument saturation (~ 30 and ~ 45 amu) or change in instrument mode (at 50 amu). The apparent structure in the mass group centred at 105 amu is due to poor counting statistics; from Mitchell et al. [1989].

spacecraft comet encounters. However at 1 a.u. the cometary medium is highly complex, much more so than theoretical predictions. With fast flybys, in different solar wind and cometary conditions and with different instruments, it is very difficult to deduce what is fundamental and what is time dependent.

As predicted the cometosheath is between a weak (~ Mach 2) bow shock which has a structure thickened by heavy pick-up ions and a thin ionopause separating the mass-loaded solar wind and the interplanetary field from the field-free, nearly pure ionospheric cometary plasma (but only Giotto was able to pass through this boundary), with a draping of the interplanetary magnetic field around the comet. There are, however, between the shock and the ionopause, several sudden changes in the plasma properties whose origins, whether temporal or spatial, are still not understood, although the neutral interaction with inflowing contaminated solar wind plasma may play a role in all these transitions:

- The mysterious transition, seen by all the spacecraft, about halfway between the shock and the nucleus seems to be an **unpredicted permanent structure** of the solar wind - comet interaction, at least around 1 a.u. It can be called the **cometary transition**.

- The magnetic pile-up region can be gradual (Vega-1 and -2) or have a strong gradient (Giotto Magnetic Pile-up Boundary), the Giotto MPB being possibly a convected solar wind effect. Inbound and outbound passes are asymmetric. The Vega-2 "cometopause" is not a real plasma boundary like the MPB but just a density crossover point where the cometary ion density exceeds the solar wind number density in the ram direction. It is not a pause, nor a strong, steady or chemical boundary. Thus the Giotto MPB and Vega-2 "cometopause" are totally different.

Many other structures exist, and probably a large majority of them are related to the dynamics of the solar wind and of the comet outgassing.

Contrary to what was sometimes thought, the magnetic field reversals seen by Giotto are not the sign of a traversal of the heliospheric current sheet but, as it was shown by the electron pitch angle distributions, the sign of a very complicated magnetic field topology inside the cometosheath.

Much theoretical work is therefore still needed to understand the structure of the solar wind - comet interaction.

Acknowledgements. Thank all the RPA-Copernic Co-Is, as well as P. Chaizy, S. Fuselier, D. Larson, C. Mazelle and D.L. Mitchell for many helpful discussions.

References

Alfvén, H., *On the Origin of the Solar System*, Oxford Clarenton Press, 1954.

Amata, E., V. Formisano, P. Cerulli-Irelli, P. Torrente, A.D. Johnstone, A. Coates, B. Wilken, K. Jockers, J.D. Winningham, D. Bryant, H. Borg, and M. Thomsen, The cometopause region at comet Halley, *ESA SP-250, 1*, 213, 1986.

Amata, E., P. Torrente, V. Formisano, R. Giovi, Proton and water group ion plasma parameters at Comet P/Halley, *Cometary Plasma Processes, this issue*, 1991.

Balebanov, V.M., K.I. Gringauz, M.I. Verigin, Plasma Phenomena in the vicitiny of the closest approach of Vega-1, -2 spacecraft to the Comet Halley nucleus, *ESA SP-278*, 119, 1987.

Balsiger, H., Measurements of ion species within the coma of comet Halley from Giotto, *Comet Halley: Investigations, Results, Interpretations, vol. 1*, Ed. J.W. Mason, 1990.

Balsiger, H., K. Altwegg, F. Bülher, J. Geiss, A.G. Ghielmetti, B.E. Goldstein, R. Goldstein, W.T. Huntress, W.H. Ip, A.J. Lazarus, A. Meier, M. Neugebauer, Y. Rettenmund, H. Rosenbauer, R. Schwenn, R.D. Sharp, E.G. Shelley, E. Ungstrup, D.T. Young, Ion composition and dynamics at Comet Halley, *Nature, 321*, 330, 1986.

Balsiger, H., K. Altwegg, J. Benson, F. Bühler, J. Fischer, J. Geiss, B.E. Goldstein, R. Goldstein, P. Hemmerich, G. Kulzer, A.J. Lazarus, A. Meier, M. Neugebauer, U. Rettenmund, H. Rosenbauer, K. Säger, T. Sanders, R. Schwenn, E.G. Shelley, D. Simpson, and D.T. Young, The ion mass spectrometer on Giotto, *J. Phys. E: Sci. Instrum.*, 20, 759, 1987.

Bame S.J., R.C. Anderson, J.R. Asbridge, D.N. Baker, W.C. Feldman, S.A. Fuselier, J.T. Gosling, D.J. McComas, M.F. Thomsen, D.T. Young, R.D. Zwickl, The Comet Giacobini-Zinner: Plasma description, *Science, 232*, 356, 1986.

Berthelier, J.J., J.M. Illiano, R.R. Hodges, D. Krankowski, P. Eberhardt, P. Lämmerzahl, J.H. Hoffman, I. Herrwerth, J. Woweries, U. Dolder, W. Schulte, Angular and energy distribution of low energy cometary ions measured in the outer coma of Comet Halley, *ESA SP-250, 1*, 175, 1986.

Biermann, L., B. Brosowski and H.U. Schmidt, The interaction of the solar wind with a comet, *Solar Phys., 1*, 254, 1967.

Brandt, J.C. and D.A. Mendis, The interaction of the solar wind with comets, *Solar System Plasma Physics*, ed. by North-Holland Publishing Company, II, 253, 1979.

Chaizy, P., H. Rème, J.A. Sauvaud, C. d'Uston, R.P. Lin, D.E. Larson, D.L. Mitchell, K.A. Anderson, C.W. Carlson, A. Korth, D.A. Mendis, Detection of negative ions in the coma of Comet P/Halley, *Nature*, in press, 1990.

Coates, A.J., A.D. Johnstone, M.F. Thomsen, V. Formisano, E. Amata, B. Wilken, K. Jockers, J.D. Winningham, H. Borg, and D.A. Bryant, Solar wind flow through the comet P/Halley bow shock, *Astron. Astrophys., 187*, 55, 1987.

Festou, M.C., P.D. Feldman, M.F. A'Hearn, C. Arpigny, C.B. Cosmovici, A.C. Danks, L.A. McFadden, R. Gilmozzi, P. Patriarchi, G.P. Tozzi, M.K. Wallis, H.A. Weaver, IUE observations of comet Halley during the Vega and Giotto encounters, *Nature, 321*, 361, 1986.

Fuselier, S.A., E.G. Shelley, H. Balsiger, J. Geiss, B.E. Goldstein, R. Goldstein, W.-H. Ip, Cometary H_2^+ and solar wind He^{2+} dynamics across the Halley cometopause, *Geophys. Res. Lett., 15*, 549, 1988.

Galeev, A.A., Theory and observations of solar wind/cometary plasma interaction processes, *ESA SP-250, 1*, 3, 1986.

Galeev, A.A., T.E. Cravens and T.I. Gombosi, Solar wind stagnation near comets, *Astrophys. J., 289*, 807, 1985.

Glassmeier, K.H., F.M. Neubauer, M.H. Acuna, F. Mariani, Strong hydromagnetic fluctuations in the comet P/Halley magnetosphere observed by the Giotto magnetic field experiment, *ESA SP-250, 3*, 167, 1986.

Goldstein, B.E., M. Neugebauer, H. Balsiger, J. Drake, S.A. Fuselier, R. Goldstein, W.-H. Ip, U. Rettenmund, H. Rosenbauer, R. Schwenn, E.G. Shelley, Giotto-IMS observations of ion flow velocities and temperatures outside the contact surface of comet Halley, *Astron. Astrophys., 187*, 174, 1987.

Grard R., H. Laakso, A. Pedersen, J.G. Trotignon, Y. Mikhailov, Observations of the plasma environment of Comet Halley during Vega flybys, *Ann. Geophys., 7*, 141, 1989.

Gringauz, K.I., T.I. Gombosi, A.P. Remizov, I. Apathy, I. Szemerey, M.I. Verigin, L.I. Denchikova, A.V. Dyachkov, E. Keppler, I.N. Klimenko, A.K. Richter, A.J. Somogyi, K. Szegö, M. Tatrallyay, A. Varga, G.A. Vladimirova, First in situ plasma and neutral gas measurements at Comet Halley, *Nature, 321*, 282, 1986a.

Gringauz, K.I., T.I. Gombosi, M. Tatrallyay, M.I. Verigin, A.P. Remizov, A.K. Richter, I. Apathy, I. Szeremey, A.V. Dyachkov, O.V. Balakina, A.F. Nagy, Detection of a new "chemical" boundary at Comet Halley, *Geophys. Res. Lett., 13*, 613, 1986b.

Gringauz, K.I., A.P. Remizov, M.I. Verigin, A.K. Richter, M. Tatrallyay, K. Szegö, I.N. Klimenko, I. Apathy, T.I. Gombosi, and T. Szemerey, Electron component of the plasma around Halley's comet measured by the electrostatic electron analyzer of Plasmag-1 on board Vega-2, *ESA SP-250, 1*, 195, 1986c.

Hodges, R.R., J.M. Illiano, J.J. Berthelier, D. Krankowsky, P. Lämmerzahl, J. Woweries, U. Stubbemann, J.H. Hoffman, P. Eberhardt, U. Dolder, and W. Schulte, Measurements of thermal ion energy spectra from the Giotto encounter with Comet Halley, *ESA SP-250, 3*, 415, 1986.

Ip, W.-H. and W.I. Axford, Theories of physical processes in the cometary comae and ion tails, in *Comets*, ed. L.L. Wilkening, 588, 1982.

Ip, W.-H., R. Schwenn, H. Rosenbauer, H. Balsiger, M. Neugebauer, and E.G. Shelley, An interpretation of the ion pile-up region outside the ionospheric contact surface, *Astron. Astrophys., 187*, 132, 1987.

Ip, W.-H., H. Spinrad, and P. McCarthy, A CCD observation of the water ion distribution in the coma of comet P/Halley near the Giotto encounter, *Astron. Astrophys., 206*, 129, 1988.

Johnstone, A.D., Observations of the interaction between Comet Halley and the solar wind by the Giotto spacecraft, in *Cometary and Solar Plasma Physics*, ed. by B. Buti, 110-140, 1988.

Johnstone, A.D., A. Coates, S. Kellock, B. Wilken, K. Jockers, H. Rosenbauer, W. Studemann, W. Weiss, V. Formisano, E. Amata, R. Cerulli-Irelli, N. Dobrowolny, R. Terenzi, A. Egidi, H. Borg, B. Hultqvist, J. Winningham, C. Gurgiolo, D. Bryant, T. Edwards, W. Feldman, M. Thomsen, M.K. Wallis, L. Biermann, H. Schmidt, R. Lust, G. Haerendel, G. Paschmann, Ion flow at comet Halley, *Nature, 321*, 344, 1986.

Johnstone, A.D., J.A. Bowles, A.J. Coates, A.J. Coker, S.J. Kellock, J. Raymont, B. Wilken, W. Studemann, W. Weiss, R. Cerulli Irelli, V. Formisano, E. de Giorgi, P. Perani, M. de Bernardi, H. Borg, S. Olsen, J.D. Winningham, D.A. Bryant, The Giotto three-dimensional positive ion analyser, *J. Phys. E: Sci. Instrum., 20*, 795, 1987.

Korth A., A.K. Richter, A. Loidl, W. Güttler, K.A. Anderson, C.W. Carlson, D.W. Curtis, R.P. Lin, H. Rème, F. Cotin, A. Cros, J.L. Médale, J.A. Sauvaud, C. d'Uston, D.A. Mendis, The heavy ion analyzer Picca for the Comet Halley flyby with Giotto, *J. Phys. E: Sci. Instrum., 20*, 787, 1987a.

Korth, A., A.K. Richter, D.A. Mendis, K.A. Anderson, C.W. Carlson, D.W. Curtis, R.P. Lin, D.L. Mitchell, H. Rème, J.A. Sauvaud, and C. d'Uston, The composition and radial dependence of cometary ions in the coma of comet P/Halley, *Astron. Astrophys., 187*, 149, 1987b.

Krankowsky, D., P. Lämmerzahl, D. Dörflinger, I. Herrwerth, U. Stubbemann, J. Woweries, P. Eberhardt, U. Dolder, J. Fischer, U. Hermann, H. Hofstetter, M. Jungck, F.O. Meier, W. Schulte, J.J. Berthelier, J.M. Illiano, M. Godefroy, G. Gogly, P. Thévenet, J.H. Hoffman, R.R. Hodges, W.W. Wright, The Giotto neutral mass spectrometer, *ESA SP-1077*, 109, 1986a.

Krankowsky, D., P. Lämmerzahl, I. Herrwerth, J. Woweries, P. Eberhardt, U. Dolder, U. Herrmann, W. Schulte, J.J. Berthelier, J.M. Illiano, R.R. Hodges, J.H. Hoffman, In situ gas and ion measurements at Comet Halley, *Nature, 321*, 326, 1986b.

Krasnopolsky, V.A., M. Gogoshev, G. Moreels, V.I. Moroz, A.A. Krysko, Ts. Gogosheva, K. Palazov, S. Sargoichev, J. Clairemidi, M. Vincent, J.L. Bertaux, J.E. Blamont, V.S. Troshin, B. Valnicek, Spectroscopy study of Comet Halley by the Vega-2 three-channel spectrometer, *Nature, 321*, 269, 1986.

Lämmerzahl, P., D. Krankowsky, R.R. Hodges, U. Stubbemann, J. Woweries, I. Herrwerth, J.J. Berthelier, J.M. Illiano, P. Eberhardt, U. Dolder, W. Schulte, and J.H. Hoffman, Expansion velocity and temperatures of gas and ions measured in the coma of Comet Halley, *Astron. Astrophys., 187*, 169, 1987.

Mazelle, C., Etude de l'interaction du vent solaire et de la comète de Halley - Analyse des propriétés du plasma dans la région d'empilement magnétique, Thesis, Toulouse Univ., 1990.

Mazelle, C., H. Rème, J.A. Sauvaud, C. d'Uston, C.W. Carlson, K.A. Anderson, D.W. Curtis, R.P. Lin, A. Korth, D.A. Mendis, F.M. Neubauer, K.H. Glassmeier, J. Raeder, Analysis of suprathermal electron properties at the magnetic pile-up boundary of Comet P/Halley, *Geophys. Res. Lett, 16*, 1035, 1989.

Mendis, D.A., E.J. Smith, B.T. Tsurutani, J.A. Slavin, D.E. Jones, G.L. Siscoe, Comet-solar wind interaction: Dynamical length scales and comets, *Geophys. Res. Lett., 13*, 239, 1986.

Mendis, D.A., K.R. Flammer, H. Rème, J.A. Sauvaud, C. d'Uston, F. Cotin, A. Cros, K.A. Anderson, C.W. Carlson, D.W. Curtis, D.E. Larson, R.P. Lin, D.L. Mitchell, A. Korth, and A.K. Richter, On the global nature of the solar wind interaction with Comet Halley, *Ann. Geophys., 7*, 99, 1989.

Meyer-Vernet N., P. Couturier, S. Hoang, C. Perche, J.L. Steinberg, Physical parameters for hot and cold electron populations in Comet Giacobini-Zinner with the ICE Radio experiment, *Geophys. Res. Lett., 13*, 279, 1986.

Mitchell, D.L., R.P. Lin, K.A. Anderson, C.W. Carlson, D.W. Curtis, A. Korth, H. Rème, J.A. Sauvaud, C. d'Uston, and D.A. Mendis, Complex organic ions in the atmosphere of Comet Halley, *Adv. Space Res., 9*, 35, 1989.

Moreels, G., M. Gogoshev, V.A. Krasnopolsky, J. Clairemidi, M. Vincent, J.P. Parisot, J.L. Bertaux, J.E. Blamont, M.C. Festou, Ts. Gogosheva, S. Sargoichev, K. Palasov, V.I. Moroz, A.A. Krysko, V. Vanysek, Near ultra-violet and visible spectrometry of Comet Halley from Vega-2, *Nature, 321*, 271, 1986.

Mukai T., W. Miyake, T. Terasawa, M. Kitayama and K. Hirao, Ion dynamics and distribution around Comet Halley: Suisei observations, *Geophys. Res. Lett., 13*, 829, 1986.

Neubauer, F.M., Giotto magnetic field results on the boundaries of the pile-up region and the magnetic cavity, *Astron. Astrophys., 187*, 73, 1987.

Neubauer, F.M., The ionopause transition and boundary layers at comet Halley from Giotto magnetic field observations, *J. Geophys. Res., 93*, 7272, 1988.

Neubauer, F.M., K.H. Glassmeier, M. Pohl, J. Roeder, M.H. Acuna, L.F. Burlaga, N.F. Ness, G. Mussmann, F. Mariani, M.K. Wallis, E. Ungstrup, H.U. Schmidt, First results from the Giotto magnetometer experiment at Comet Halley, *Nature, 321*, 352, 1986.

Neugebauer, M., A.J. Lazarus, K. Altwegg, H. Balsiger, B.E. Goldstein, R. Goldstein, F.M. Neubauer, H. Rosenbauer, R. Schwenn, E.G. Shelley, and E. Ungstrup, The pick-up of cometary protons by the solar wind, *Astron. Astrophys., 187*, 21, 1987.

Oberc, P., D. Orlowski, R. Wronowski, S. Klimov, S. Savin, Plasma waves in the Halley's inner coma as measured by the APV-N experiment during Vega mission, *ESA SP-250, 1*, 89, 1986.

Pérez-de-Tejada, H., Viscous flow interpretation of Comet Halley's "mystery" transition, *J. Geophys. Res., 94*, 10131, 1989.

Raeder, J., F.M. Neubauer, N.F. Ness, and L.F. Burlaga, Macroscopic perturbations of the IMF by P/Halley as seen by the Giotto magnetometer, *Astron. Astrophys., 187*, 61, 1987.

Rème, H., Regions of interaction between the solar wind plasma and the plasma environment of comets, in *Comet Halley: Investigations, Results, Interpretations, Vol. 1*, ed. J.W. Mason, 1990.

Rème, H., J.A. Sauvaud, C. d'Uston, F. Cotin, A. Cros, K.A. Anderson, C.W. Carlson, D.W. Curtis, R.P. Lin, D.A. Mendis, A. Korth, A.K. Richter, Comet Halley - solar wind interaction from electron measurements aboard Giotto, *Nature, 321*, 349, 1986.

Rème, H., F. Cotin, A. Cros, J.L. Médale, J.A. Sauvaud, C. d'Uston, K.A. Anderson, C.W. Carlson, D.W. Curtis, R.P. Lin, A. Korth, A.K. Richter, A. Loidl, D.A. Mendis, The Giotto electron plasma experiment, *J. Phys. E:Sci. Instrum., 20*, 721, 1987a.

Rème, H., J.A. Sauvaud, C. d'Uston, A. Cros, K.A. Anderson, C.W. Carlson, D.W. Curtis, R.P. Lin, A. Korth, A.K. Richter, D.A. Mendis, General features of the Comet Halley - solar wind interaction from plasma measurements, *Astron. Astrophys., 187*, 33, 1987b.

Riedler, W., K. Schwingenschuh, Ye. G. Yeroshenko, V.A. Styashkin, C.T. Russell, Magnetic field observations in Comet Halley's coma, *Nature, 321*, 288, 1986.

Sagdeev, R.Z., V.D. Shapiro, V.I. Shevchenko and K. Szegö, Plasma phenomena around comets: interaction with the solar wind, Invited topical lecture on the XVIII International Conference on Phenomena in Ionized Gases, Swansea, 1987.

Schwenn, R., W.-H. Ip, H. Rosenbauer, H. Balsiger, F. Bühler, R. Goldstein, A. Meier, and E.G. Shelley, Ion temperature and flow profiles in comet Halley's close environment, *Astron. Astrophys., 187*, 160, 1987.

Schwingenschuh, K., W. Riedler, H.I.M. Lichtenegger, J.L. Phillips, J.G. Luhman, C.T. Russell, J.A. Fedder, A. Somogyi, Ye. Yeroshenko, Variability of comet Halley's coma: Vega-1 and Vega-2 magnetic field observations, *ESA SP-278*, 63, 1987.

Shelley, E.G., S.A. Fuselier, H. Balsiger, J.F. Drake, J. Geiss, B.E. Goldstein, R. Goldstein, W.-H. Ip., A.J. Lazarus, and M. Neugebauer, Charge exchange of solar wind ions in the comet Halley coma, *Astron. Astrophys., 187*, 304, 1987.

Spenner, K., W.C. Knudsen, K.L. Miller, V. Novak, C.T. Russell and R.C. Elphic, Observation of the Venus mantle, the boundary region between the solar wind and ionosphere, *J. Geophys. Res., 85*, 7655, 1980.

Stewart, A.I.F., M.R. Combi, W.H. Smyth, *Bull. Am. Astron. Soc., 17*, 686, 1985.

Thomsen, M.F., W.C. Feldman, B. Wilken, K. Jockers, W. Stüdemann, A.D. Johnstone, A. Coates, V. Formisano, E. Amata, J.D. Winningham, H. Borg, D. Bryant, and M. Wallis, In situ observations of a bi-modal ion distribution in the outer coma of comet Halley, *Astron. Astrophys., 187*, 141, 1987.

Trotignon, J.G., C. Béghin, R. Grard, A. Pedersen, M. Mogilevsky, and Y. Mikhailov, Electric field discontinuities at Comet Halley during the Vega encounters, *Cometary Plasma Processes, this issue*, 1991.

Tsurutani, B.T. and E.J. Smith, Strong hydromagnetic turbulence associated with Comet Giacobini-Zinner, *Geophys. Res. Lett., 13*, 259, 1986.

d'Uston, C., H. Rème, J.A. Sauvaud, A. Cros, K.A. Anderson, C.W. Carlson, D.W. Curtis, R.P. Lin, A. Korth, A.K. Richter, D.A. Mendis, Description of the main boundaries seen by the Giotto electron experiment inside comet P/Halley - Solar wind interaction region, *Astron. Astrophys., 187*, 137, 1987.

d'Uston, C., H. Rème, J.A. Sauvaud, C.W. Carlson, K.A. Anderson, D.W. Curtis, R.P. Lin, A. Korth, A.K. Richter, D.A. Mendis, Properties of plasma electrons in magnetic pile-up region of comet Halley, *Ann. Geophys., 7*, 91, 1989.

Vaisberg, O.L., G.N. Zastenker, V.N. Smirnov, B.I. Khazanov, D.S. Zakharov, Distribution of the concentration of heavy ions in the head of Comet Halley, *Kosm. Issled., 26*, 9, 1988.

Wallis, M.K., Weakly-shocked flows of the solar wind plasma through atmospheres of comets and planets, *Planet. Space Sci., 21*, 1647, 1973.

Wallis, M.K., and R.S.B. Ong, Strongly cooled ionizing plasma flows with application to beams, *Planet. Space Sci., 23*, 713, 1975.

Wilken, B., W. Weiss, W. Stüdemann, and N. Hasebe, The Giotto implanted ion spectrometer (IIS): physics and technique of detection, *J. Phys. E: Sci. Instrum., 20*, 778, 1987a.

Wilken, B., A. Johnstone, A. Coates, H. Borg, E. Amata, V. Formisano, K. Jockers, H. Rosenbauer, W. Stüdemann, M.F. Thomson, J.D. Winningham, Pick-up ions at comet Halley's bow shock: Observations with the IIS spectrometer on Giotto, *Astron. Astrophys., 187*, 153, 1987b.

THE PLASMA PARAMETERS
DURING THE INBOUND AND OUTBOUND LEGS OF THE GIOTTO TRAJECTORY

E. Amata, V. Formisano, P. Torrente, R. Giovi

Istituto di Fisica dello Spazio Interplanetario
Consiglio Nazionale delle Ricerche, Frascati, Italy

Abstract. New bulk speed, density and temperature values are presented for protons and water group ions based on the data of the two ion sensors of the Giotto JPA experiment. Two water group ion populations, implanted locally and further upstream, evolve with different density and temperature during the fly-by. In the inbound leg of the Giotto trajectory the centre of mass switches from the protons to the water group ions between 5.2×10^5 km and 4.8×10^5 km from closest approach and to the locally picked up ions between 4.7×10^5 km and 3.9×10^5 km; in the outbound leg the centre of mass switches to the water group ions between 3.3×10^5 km and 3×10^5 km and to the locally implanted ions around 2.9×10^5 km from the nucleus.

Introduction

The interaction of the solar wind with comet P/Halley has been investigated by many authors using the Giotto, VEGA and Suisei data [e.g. see the special issue of *Astron. Astrophys., 187, N. 1/2*, 1987]. As the comet has no intrinsic magnetic field, the fundamental process of this interaction is the mass loading of the solar wind plasma by the cometary ions, by which the solar wind is slowed down in front of the comet. As H_2O is the most abundant molecule emitted by the comet [*Balsiger et al.*, 1987], the mass loading is essentially due to the pick up of water group ions by the solar wind; therefore, the central section of this paper will present the plasma parameters of the two main ion species observed during the Giotto encounter: protons and water group ions. New revised parameters are presented for the inbound leg of the encounter trajectory; moreover, the outbound parameters are also presented and discussed for the first time. The existence of two groups of water ions down-

Cometary Plasma Processes
Geophysical Monograph 61
©1991 American Geophysical Union

stream of the shock was predicted theoretically prior to the comet Halley encounters [*Galeev et al.*, 1985]. Actually it has been found experimentally that the water group ion population has a higher energy component due to old pick up ions and a lower energy component due to locally implanted ions [e.g. *Johnstone et al.*, 1986, and *Thomsen et al.*, 1987]. Plasma parameters for the two components are presented in the third section, both for the inbound and outbound legs of the trajectory, in order to gain more insight in the interaction of the solar wind with the cometary ions. The last section of the paper contains the summary and discussion.

The data set and its reduction

The JPA instrument, which is fully described in *Johnstone et al.*, [1986], has two parts: FIS, which measures positive ions from 10 eV/q to 20 keV/q without mass discrimination and with a 35° blind cone about the ram direction, and IIS, which measures positive ions in thirty two energy channels of $dE/E = 0.07$ from 76 eV/q to 86 keV/q with mass discrimination (in this paper only protons and water group ions are considered). IIS is actually made of five equal sensors, each with 10° aperture, centred at 20°, 55°, 90°, 125° and 160° from the ram direction (i.e. IIS does not cover a 15° ram cone); each sensor collects particles in sixteen equal and contiguous azimuthal sectors (eight azimuthal sectors from 2240 SCET on 13 March to 0102 SCET on 14 March).

The FIS data used in the analysis described hereafter are organised in two matrixes: the HAR (High Angular Resolution) matrix, with an intrinsic time resolution of 12 s and thirty logarithmically spaced energy channels covering the entire energy range and the FTR (Fast Time Resolution) matrix, with an intrinsic time resolution of 4 s and fifteen logarithmically spaced energy channels. The HAR matrix covers the polar range from 20° to 180° from the ram direction with six contiguous sectors (the first sector really starting at 35°); the azimuthal range is divided in sixteen 22.5° sectors for the polar range from 72° to 124° and in eight 45° sectors for the polar ranges from 20° to 72° and from 124° to 180°. The FTR matrix has three polar sectors and eight equal azimuthal sectors. The HAR data were used until 2240 SCET inbound and from 0102 SCET onwards outbound; the FTR data were used closer to the nucleus when the HAR matrix was not available because of the reduced telemetry rate.

The proton plasma parameters discussed in the following sections are obtained by calculating the appropriate moments of the proton distribution function provided by the FIS HAR (or FTR) matrix and by the IIS sensor close to the RAM direction (IIS polar sector 5). The use of the IIS data becomes more and more relevant as the spacecraft approaches the nucleus because of the large FIS blind cone about the ram direction. Because of the different time resolution of the two sensors, the HAR and FTR distribution functions have been averaged over 128 s. The use of FIS and IIS data together has been possible after performing a careful intercalibration of the two sensors [see *Formisano at al.*, 1990]. The calibration factors used in this paper differ from those used in a previous paper [*Amata et al.*, 1988] which were found recently to be wrong. Moreover, in that paper the water group plasma parameters were calculated using a wrong value of the IIS geometric factor. These two corrections account for the differences between the plasma parameters presented hereafter for the inbound leg of the Giotto trajectory and those found in *Amata et al.* [1988]; the detailed discussion of the differences and their explanation, together with the description of the methods followed in the data reduction, is given by *Formisano et al.* [1990]. The use of the IIS proton data to complement the HAR and FTR distribution ceases to be effective around 2315 SCET, when the peak of the proton distribution probably moves below the lower energy limit of IIS; after that time the proton data are not reliable any more. As the FIS sensors did not survive closest approach, the outbound proton data are obtained only from the IIS sensor. Because of increased noise in the measured distributions after closest approach, both the proton and the water group ion outbound parameters should be regarded as preliminary. The calculation of the plasma parameters from the HAR (FTR) and IIS data does not yield reliable results for the solar wind proton temperature which would require the use of the FIS solar wind mode data (not described here); therefore, the proton temperature in the solar wind is not discussed hereafter.

The following section contains a description of the plasma parameters of two water group ion populations. The two populations have been separated by visual inspection of the individual distribution functions. Figure 1 shows a typical water group ion distribution function,

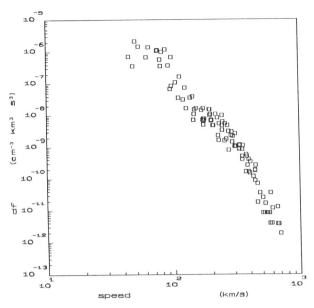

Fig. 1. Distribution function values for the water group ions at 2258 SCET on 13 March 1986. The data refer to all the azimuthal sectors of IIS polar sector 5. The data are plotted against speed in the plasma centre of mass reference system.

taken at 2258 SCET on 13 March 1986. The plotted values refer to all the azimuthal sectors pertaining to the IIS polar sector 5 (close to the ram direction) and to energies in the range 1.5 - 55 keV. The speed is given in the frame of reference of the plasma centre of mass. The plotted distribution function shows two clear components, at higher and lower energies, which by visual inspection can be separated at 150 km/s. As the separation energy depends on the time, all the distribution functions have been visually inspected to allow the calculation of separate momenta for the two components.

All the plasma parameters shown in the following section are given in the Halley cometocentric coordinate system.

The observations

Figure 2 shows the macroscopic parameters of protons (thin line) and water group ions (thick line) from 1900 SCET to 2400 SCET on 13 march 1986. The two vertical dashed lines mark the region, between 5.7×10^5 and 5.2×10^5 km, identified by *Rème* [1990] as the "mystery transition". The proton number density (bottom panel) in front of the shock wave is 6 cm^{-3} (averaged between 1905 and 1914 SCET); downstream it is 17 cm^{-3} (averaged from 1930 to 1935 SCET). After the shock the proton number density does not vary monotonically with

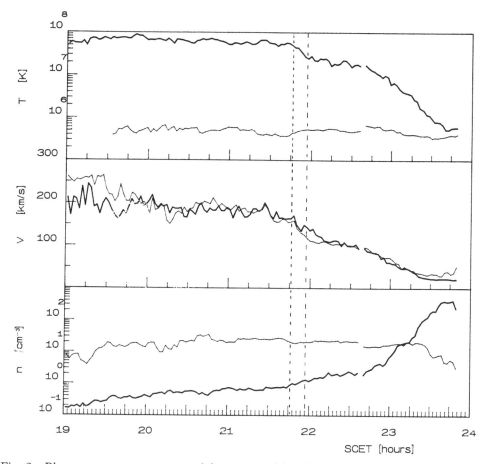

Fig. 2. Plasma parameters measured for protons (thin line) and water group ions (thick line) from 1900 to 2400 SCET on 13 March 1986, during the inbound leg of the Giotto fly-by. From top to bottom: temperature, bulk speed, number density. The two vertical dashed lines mark the "mystery transition" [Rème, 1990].

time; however it reaches a maximum of 32 cm^{-3} around 2045 SCET followed by several decreases, among which a decrease from 28 cm^{-3} to 20 cm^{-3} between 2138 and 2148 SCET, i.e. during the first part of the "mystery transition". The last large decrease after 2320 SCET may be due to the fact that most of the proton population is moving out of the field of view of the two sensors. The water group ion number density (bottom panel) in the solar wind is n$_{wg} \sim 0.2$ cm^{-3}, increasing smoothly across the shock to 0.35 cm^{-3} (averaged from 1930 to 1940 SCET); after the shock it increases steadily reaching a maximum value of 400 cm^{-3} at 2345 SCET. After 2330 SCET the errors on the density values probably increase as the peak of the water group distribution function moves to the instrumental blind cone about the RAM direction. At 1900 SCET the proton bulk speed (central panel of Figure 2) is 250 km/s, lower than the undisturbed solar wind speed, as a result of the mass loading process [e.g. *Coates et al.*,

1987]. Across the shock crossing, starting at 1.6×10^6 km from closest approach, the proton speed decreases from 250 km/s to 200 km/s. The slow down continues with some oscillations until 2148 SCET when the proton speed decreases from 160 to 105 km/s over 15 min, corresponding roughly to the second part of the "mystery transition". After this jump the decrease goes on steadily without fluctuations until the proton data start loosing their reliability around 2315 SCET. Before the shock the water group ion speed is lower than the proton speed while soon after the shock v$_{wg} \sim$ v$_p \sim 200$ km/s. Later the water group ion speed decreases continuously following closely the proton speed (however, the decrease starting at 2148 SCET is less abrupt). The temperature of the water group ions (top panel of Figure 2) increases from 5×10^7 to 7×10^7 K at the shock wave (as explained in the preceeding section the proton temperature in the solar wind has not been plotted). At the "mystery transi-

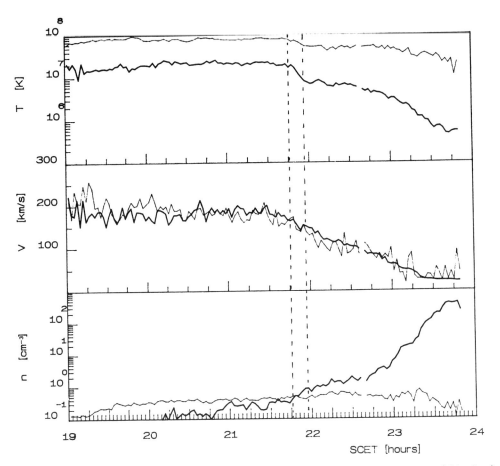

Fig. 3. Plasma parameters for the lower energy (thick line) and higher energy (thin line) water group ions from 1900 to 2400 SCET on 13 March 1986 in the same format of Figure 2. The two vertical dashed lines mark the "mystery transition" [*Rème*, 1990].

tion, between 2144 and 2200 SCET, the water ion temperature decreases from 6×10^7 to 2.3×10^7 K, while the proton temperature increases slightly between 2142 and 2150 SCET. Further inwards the proton temperature decreases slowly reaching 4×10^5 K at 2330 SCET; the water group ion temperature decreases much faster to 6×10^5 K at 2340 SCET.

The splitting of the water group ion population occurs already at the bow shock but becomes evident only at 2150 SCET. Figure 3 shows the moments computed for the high energy (thin line) and the low energy water group ions (thick line). As in Figure 2 the two vertical dashed lines mark the "mystery transition". The separation between the two populations has been performed by inspection of the water group distribution function in the centre of mass reference system, as described in the preceeding section. The higher energy ion density increases from the shock, where $n_h = 0.3$ cm^{-3}, to 2220 SCET, where $n_h = 0.65$ cm^{-3}. After a small decrease around

2300 SCET, n_h goes through a maximum of 0.8 cm^{-3} at 2320 SCET. The lower energy ion number density (plotted only from 2009 SCET onwards) increases from 0.3 cm^{-3} to 1.0 cm^{-3} between 2145 and 2200 SCET, i.e. at the "mystry transition"; then, it keeps increasing showing two large jumps from 3 to 14 cm^{-3} between 2255 and 2305 SCET and from 20 to 100 cm^{-3} between 2314 and 2324 SCET; the density levels off at 400 cm^{-3} at 2338 SCET. The speeds of the higher and lower energy water group ions, central panel of Figure 3, are rather similar. The top panel of Figure 3 displays the temperature of the two water group populations. The lower energy ion temperature shows a clear decrease, from 2×10^7 K to 7×10^6 K between 2148 and 2200 SCET, during the second half of the "mystery transition", and falls to 5×10^5 K by 2345 SCET; the higher energy ion temperature decreases from 8×10^7 K to 5×10^7 K between 2148 and 2200 SCET and remains roughly constant thereafter, were it not for a small decrease between 2310 SCET and 2345 SCET.

PERMANENT AND NONSTATIONARY PLASMA PHENOMENA IN COMET HALLEY'S HEAD

K.I. Gringauz and M.I. Verigin

Space Research Institute, Profsoyuznaya 84/32, 117810 Moscow GSP-7, USSR.

Abstract The characteristics of various plasma phenomena observed near comet Halley in 1986 are studied to determine whether or not they are permanent features of the comet. Taking as the criteria for permanence that they should be observed by all spacecraft or be physically explicable, the permanent features include the near-cometary bow shock, the cometosheath, with its unique energy distribution of ions, the systematic cooling of electrons in this region, the cometopause, and the tangential discontinuity near the cometary nucleus. Among nonstationary events observed there are the unusual burst of ions with energies 100 -1000 eV recorded in a direction from the Sun in the region of cometary ions at $r \sim (1 - 2) \times 10^4$ km, the magnetic field pile-up boundary (in the region of the cometopause), the mystery region, and the precipitation of energetic electrons with ~ 1 keV at $r \sim (1.5 - 2.5) \times 10^3$ km.

1. Introduction

The in situ plasma and magnetic measurements made near Halley's comet nucleus in March 1986 detected a considerable number of phenomena. Some of them had been anticipated before 1986. They are the near-cometary bow shock, the tangential discontinuity near the nucleus (not very adequately called the contact surface or ionopause), and auroral phenomena in the head of the comet due to the events in its tail. The other phenomena were not mentioned in the literature prior to 1986 and received the new names: "cometosheath", "cometopause", [Gringauz et al., 1986], "pile-up boundary" [Neubauer, 1987], and "mystery region" [Reme et al., 1987].

Some of these phenomena were observed from all three spacecraft which came close to the comet nucleus (VEGA-1, 2 and Giotto), while the other ones were detected by only one of these spacecraft. Among them is the tangential discontinuity near the comet nucleus. This phenomenon was observed by the Giotto spacecraft only, because the other spacecraft passed by too far from the nucleus.

Some phenomena always occur in a comet head at a distance of 1 a.u. from the Sun. Along with this it is obvious that the structure of the plasma (and of the magnetic field) near the comet nucleus should change with time since it is a result of the interaction of two opposite flows of particles highly unsteady with time - a solar wind flow (with unsteady interplanetary magnetic field), and a neutral gas flow

evaporating from the nucleus and subjected to ionization. So, one would expect (and it was expected previously to 1986 [Mendis, Houpis and Marconi, 1985]) that the plasma and magnetic field near the comet nucleus will be unsteady. In particular, the changes in the characteristics of the solar wind plasma and interplanetary magnetic field should lead to variations in the characteristics and maybe in the structure, of the near-cometary plasma and magnetic field. For this reason it is clear that not all plasma phenomena near the comet nucleus are permanent.

The spacecraft passed by the nucleus under different interplanetary conditions. In spite of this some phenomena were observed during all three flybys, and so they can probably be considered as permanent. We can only consider phenomena observed from one spacecraft as permanent in the case where there is a clear understanding of its physical nature and of the reason why it could not be observed by other spacecraft.

If we only had the data from a single spacecraft, there would be a risk of believing that all the detected features of the near-cometary plasma are permanent. However, with the information from three spacecraft, obtained at different times and in different conditions we can make an attempt to identify the near-cometary plasma formations and peculiarities created by the changes in interplanetary space and in characteristics of gas flow evaporating from the nucleus. It is this attempt that is the objective of this paper.

In comparing results obtained from all spacecraft, due consideration should be given to the different characteristics of the instruments, particularly to their energy ranges and fields of view. Thus, for example, the lower velocity ions in the cometocentric system were well recorded by the instruments whose fields of view covered the direction of the relative velocity vector, but they could not be recorded by those instruments whose fields of view did not include the above-noted direction.

2. Permanent Phenomena.

2.1 Bow shock

Figure 1 gives the near-cometary trajectory portions of the VEGA-1, 2, Giotto and Suisei spacecraft. The crosses show the bow shock positions determined from plasma velocity changes, from plasma heating, and from magnetic field jumps at the bow shock; the plasma velocity vector projections to the plane of the figure are denoted by the arrows [Suisei data, Mukai et al., 1986]. Each crossing had its individual features (for example, sometimes plasma heating started earlier than

Cometary Plasma Processes
Geophysical Monograph 61
©1991 American Geophysical Union

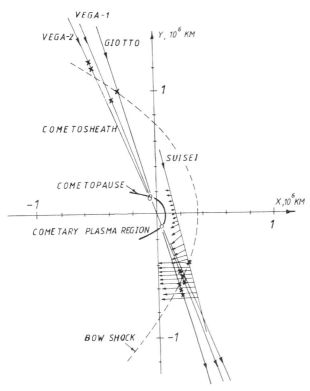

Fig 1. General overview on the in- and outbound locations of the bow shock and of the cometopause as well as of the cometosheath and of the cometary plasma region as identified from VEGA-1,2, Giotto and Suisei observations during their encounters with comet Halley.

the plasma velocity decreased and turned) nonetheless in all cases the observed bow shock positions fit each other well. The bow shock subsolar point was located about 3.5×10^5 km from the comet nucleus.

The observed bow shock positions were close to the smooth empirical shock surface (dashed line in Figure 1), despite the fact that the solar wind velocity during the flyby of Giotto was two times lower than during the flybys of VEGA-1 and VEGA-2. Apparently this was associated with the fact that the neutral gas production by the comet on March 13th, 1986 was lower than that on March 6th and 9th, 1986, perhaps due to the different orientation of the asymmetric nucleus relative to the Sun. The cometary bow shock formation is not caused by solar wind compression and heating due to interaction of the supersonic plasma flow with a sufficiently rigid obstacle (as in the case of solar wind flow around the near-Earth magnetic obstacle or around the non-magnetized ionospheric plasma confined by the strong gravitational field of Venus). The bow shock forms due to mass-loading of the solar wind by picked-up ions of cometary origin [Galeev, 1987]. It is beyond any doubt that the near-cometary bow shock is a permanent feature of Halley's comet at heliocentric distances of ~ 1 a.u.

2.2 Cometosheath

It was proposed that the plasma transition region downstream of the near-cometary bow shock be called "cometosheath" [Gringauz et al., 1986] since the energy distribution of ions in this region is unique compared with similar regions near the solar system planets, for example, the magnetosheath near Earth or the ionosheath near Venus. One of the differences is that three different branches of ions are present in the ion energy distribution; the ratio of intensities of these branches changes with the cometocentric distance. This feature of the cometosheath is associated with the above-noted principle difference in the bow shock formation process near planets and comets.

Figure 2 gives the results of measurements made with the JPA instrument (ion energy-mass analyzer) aboard the Giotto spacecraft in Halley's comet head [Johnstone et al., 1986]. The spectrogram of the ion fluxes shown in Figure 2 presents the energy spectra of ions recorded in the sector of the field of view containing the solar direction (the instrument had a fan-type field of view). Two upper branches (1, 2) in this panel correspond to the ion mass interval which included water group ions; the lowest branch (p) is an instrumental "ghost" of the energy distribution of the protons.

The upper branch of the distribution (1) is formed by water group ions picked-up by the solar wind upstream of the bow shock [Thompson et al., 1987] and coming to the spacecraft with the velocity twice that of the solar wind. Their energy should be greater than that of solar wind protons by a factor of about 4M times where M is the mass of the cometary ions. According to Figure 2, that corresponds to the results of the observations.

The branch (2) is formed by water group ions picked-up in the cometosheath where the velocity of the solar origin plasma is reduced compared with the velocity in interplanetary space. For this reason their energy is also lower. According to Thompson et al. [1987] energy splitting of water ions in the cometosheath in branches (1) and (2) can be explained by the stepwise decrease of plasma velocity on the bow shock. Along with this, the observation that the energy of the ions in branch (2) rapidly decreases and becomes close to that of the protons (also observed by VEGA-1 and VEGA-2 (Figure 3, Gringauz et al.,1986))still remains to be studied.

However, the fact that the cometosheath, with its unique energy ion distribution, is a permanent feature of comet Halley's head at ~ 1 a.u., is beyond doubt.

2.3. Cometopause

The VEGA-2 spacecraft recorded a sharp (~ 10^4 km along trajectory) change of the ion distribution function at a distance of ~ 1.6×10^5 km from the nucleus [Figure 4, Galeev et al, 1988]. This change corresponds to the boundary between two regions: in one, the solar wind protons are predominant, in the other heavy ions of cometary origin dominate. There was no simultaneous change of the distribution function of electrons, so the plasma number density and electron temperature had no discontinuity. This "chemical discontinuity" in plasma was called the "cometopause" [Gringauz et al., 1986; Gringauz et al., 1986a].

According to the VEGA-2 data the magnetic field absolute value actually did not change near the cometopause [Riedler et al., 1986], however the amplitude of electric field oscillations rapidly grew (during ~ 2 min) in the lower-hybrid frequency range (8 -14 Hz). Plasma flux oscillations were recorded in the whistler wave range, and the frequency at which the intensity of these waves was a maximum increased from ~

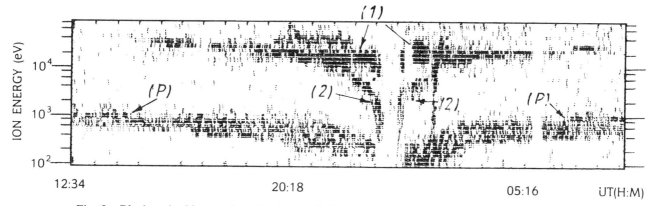

Fig. 2. Black-and-white version of colour coded spectrogram of IIS time-of-flight sensor of JPA instrument onboard Giotto spacecraft. Branch (P) corresponds to protons of solar wind origin, branches (1,2) - to cometary ions of mass 12-22 AMU.

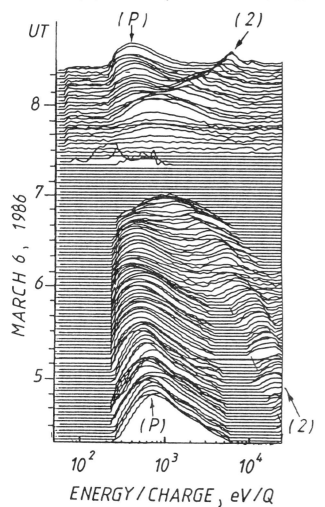

Fig. 3. Sequence of energy-per-charge spectra observed by the solar direction ion analyser onboard of VEGA-1. Branches (P) and (2) are similar to those in Figure 2. Near VEGA-1 closest approach (~ 7.20 UT) short non-stationary spike of accelerated ions was registered.

250 Hz to ~ 900 Hz while the s/c approached the cometopause [Galeev et al., 1988].

The cometopause thickness of $(1 - 2) \times 10^4$ km was determined using the data from the Giotto/JPA instrument [Amata et al., 1986]. According to data from the HERS sensor of the IMS instruments aboard Giotto, the transfer from the light ion region to the heavy cometary ion region seems to more gradual [Balsiger et al., 1987].

The cometopause separated the cometosheath region where protons are dominant (and picked-up water group ions can be considered as minor components) from the region where cometary ions are dominant. There is no doubt now that the fast transition from one region to another is accompanied by the rapid growth of the charge exchange rate between protons and cometary neutrals. In other words, there is a growing number of collisions of protons with neutrals, although the study of the physical processes that can lead to the formation of a sharp cometopause boundary is not completed. The decrease of the proton fluxes recorded by the CRA analyzer onboard VEGA-2 can be partially due to their collisionless isotropization.

It has been noted by Galeev et al. [1988] that, in the vicinity of the cometopause, conditions are met for the firehose instability if one assumes that the involvement of cometary ions in the motion along the magnetic field is not effective.

The fast isotropization of protons in the vicinity of the cometopause promotes the acceleration of the process of charge exchange which decreases their concentration by several times. The characteristic time of proton charge exchange τ is related to their total velocity:

$$\tau \sim (\sigma v n)^{-1} \sim 5 \times 10^3 s,$$

where $\sigma \sim 2 \times 10^{-15}$ cm^2 is the charge exchange cross section; n is the neutral concentration equal to about 5×10^3 cm^{-3} [Remizov et al., 1986], v ~ 200 km/s is the velocity of the motion of protons in front of the cometopause which is of the order of magnitude of the velocity of their gyration on the cometopause (and downstream of it due to pitch-angle scattering by the ion - excited oscillations).

The time estimated above is comparable with the characteristic time of the plasma flow interaction with the

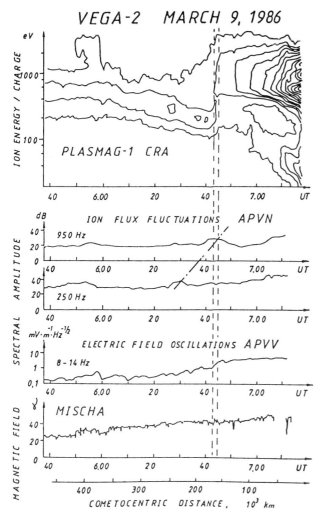

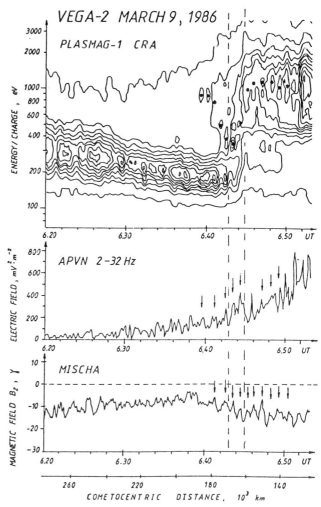

Fig. 4. Plasma and field data collected by four different instruments during last 100 min before VEGA-2 closest approach. From top to bottom: spectrogram of ion flux in ram direction, plasma wave activity in three different frequency ranges, total magnetic field. The cometopause indicated by dashed lines. The outermost isolines in top panel correspond to ram ion analyzer of PLASMAG-1 instrument count rate of $10^3 s^{-1}$, and the ratio between count rates represented by adjacent isolines is equal to 2.

Fig. 5. Fluctuations of ion flux, electric field and B_z component (pointing towards the north pole of ecliptic) of the magnetic field around the cometopause (dashed lines). Maxima are shown by dots and arrows. Here the difference between count rates of PLASMAG-1/VEGA-2 ram ion analyzer (top panel) represented by adjacent isolines is $440 s^{-1}$, and the outermost isolines corresponds to a count rate of $10^3 s^{-1}$.

cometary neutrals at the cometopause which is equal to about $2r/v_t \sim 5 \times 10^3$ s where $v_t \sim 60$ km/s is the flow velocity downstream of the cometopause. This indicates that the charge exchange is effective in this region. However, the characteristic scale of this process exceeds much of the cometopause width.

Modelling of the cometopause within the frame of the two-fluid hydrodynamic model [Gombosi, 1987] cannot completely explain, in some respects, the "too sharp" boundary which was observed by VEGA-2. The effect of collective interactions in the plasma should be incorporated in a complete model. Figure 5 [Galeev et al., 1988] illustrates

the existence of intense plasma wave processes near the cometopause; one can see synchronous oscillations of ion fluxes and electric and magnetic fields near this boundary.

Processes occurring on the cometopause depend on the properties of the plasma flow moving towards the nucleus and connected with the variable solar wind, as well as on parameters of the flow of the neutral gas, which is also varying in time. Therefore, we should expect that the position and, maybe, the width of the cometopause can and should vary in time. However, the existence of the cometopause as a permanent feature in the head of Halley's comet is in no doubt.

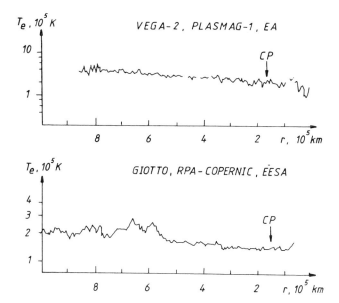

Fig. 6. Cometocentric profiles of electron temperature estimated by VEGA-2 electron electrostatic analyzer EA of PLASMAG-1 instrument data, and Giotto EESA sensor of RPA instrument data.

2.4. Cooling of electrons with the decrease of cometocentric distance

Using Figure 6 one can compare the results of measurements of the plasma's electron component from VEGA-2 [Gringauz et al., 1987], and Giotto [Reme et al., 1987] spacecraft. The top panel show the values of T_e calculated from the data of the EA electrostatic analyzer on board VEGA-2, while the bottom panel is from the data of the EESA instrument onboard Giotto. Within the cometocentric distance range from $r \sim 8 \times 10^5$ km to the cometopause CP, the electron temperature decreases by $(1 - 2) \times 10^5$ K. This cooling of electrons by 10 - 20 eV can be explained by losses of electron energy during their inelastic collisions with the cometary neutral gas.

Indeed, at $r \sim 1.6 \times 10^5$ km, the density of neutral particles n is $\sim 5 \times 10^3$ cm-3 [Remizov et al., 1986]. The electron energy loss due to inelastic collisions during electron motion through water vapour of such density is nL $\sim 2 \times 10^{-11}$ eV/cm, where L $\sim 4 \times 10^{-15}$ cm2 eV is the electron energy loss function value at E ~ 40 eV. If the plasma flow velocity is v ~ 200 km/s the characteristic time of the flow-around is $\sim 2r/v \sim 1.5 \times 10^3$ s (with due account of the fact that the spacecraft is approaching the nucleus at an angle of 110° with the direction to the sun). During this time, an electron moving with a velocity of 4×10^3 km/s can cover a distance of 6×10^6 km, and on this path it can lose an energy of ~ 12 eV. This value is comparable with the observed systematic cooling of the electron component.

Hence, with allowance for the fact that the results of the VEGA-2 observations coincide with Giotto measurements and that the physical process caused this effect is clearly understandable, it should be considered as a permanent

feature of the cometosheath within the interval of cometocentric distances discussed above.

2.5. The tangential discontinuity (ionopause, the contact surface)

The existence of a sharp boundary of the cavity adjacent to the nucleus where there is no magnetic field but where there is plasma had been predicted prior to missions to Halley's comet.

This surface was only detected by Giotto at a distance of $\sim$ 5000 km from the nucleus; the other spacecraft had flight trajectories too far from the nucleus. The main characteristics of this region are well known and we will not discuss them here.

The plasma turned out to be very cold so that the force balance at the cavity boundary was due to the friction between the neutrals and ion balanced by the magnetic field pressure [Cravens, 1986; Ip and Axford, 1987] rather than the equality of the cometary ionosphere plasma and external magnetic field pressures. The physical processes which create such a surface are now well understood. Therefore, this surface can also be considered as a permanent feature of the head of Halley's comet at $\sim$ 1 a.u. regardless of the fact that it has been observed only once.

3. Non-Stationary Phenomena

3.1 Discontinuities of the magnetic field in the cometary plasma region.

It should be again noted that the VEGA-1, VEGA-2 and Giotto flybys in the head of Halley's comet were performed under essentially different conditions in the interplanetary plasma. Interplanetary magnetic field (IMF) conditions were also very different. At the time of the VEGA-1 flyby, one IMF sector boundary passed through the cometary plasma region.

Arrows on Figure 7a show the results of magnetic field measurements with the MISHA magnetometer along the VEGA-1 trajectory [Riedler et al., 1986]. It is seen from this figure that, as a whole, the magnetic field behaves as if it was draped around the cometary nucleus, and was directed sunward at the outbound leg of the trajectory. However, between 7.11 UT and 7.24 UT, the direction of the magnetic field was reversed. Several hours prior to this VEGA-1 detected a change of IMF direction also. Therefore, the magnetometer investigations supposed that part of the magnetic field which had been measured near the closest approach point was the remnant of the previous direction of the interplanetary magnetic field slowly moving towards the nucleus, frozen-in the decelerated plasma flow and hence, approaching the nucleus with a delay in time. Later, 3D MHD-modeling of the solar wind interaction with the comet confirmed this supposition [Schwingenschuh et al., 1987].

Figure 7b illustrates the results of the Giotto magnetic field measurements [Raeder et al., 1987]. It can be seen from this figure that the magnetic field direction changed many times during the Giotto flyby. According to our knowledge, a detailed analysis of the passage of these multiple discontinuities in the IMF through the cometary plasma region has not been made yet. However, we have no doubt that discontinuities in the IMF on the magnetic field influence the cometary plasma region as in the results of the VEGA-1 spacecraft.

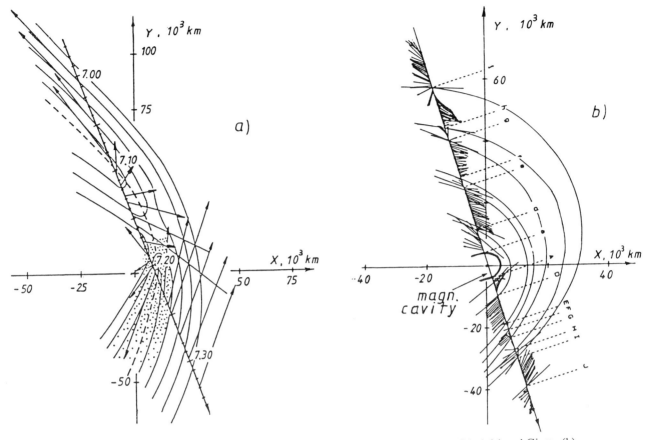

Fig. 7. Overall topology of magnetic field around the closest approaches of VEGA-1 (a) and Giotto (b). The dotted area represents the region in which the burst of accelerated ions (see Figure 3) was observed.

3.2 Unusual fluxes of ions with energies 100-1000 ev in the cometary plasma region.

In the region downstream of the cometopause (in the cometary plasma region), there were no observations of significant fluxes of ions in the direction from the sun except for the VEGA-1 measurements which observed rather intense ion fluxes (F ~ (5 - 8) x 10^9 cm^{-2} s^{-1} by the Sun-oriented electrostatic ion analyzer and Faraday cup) with energies 100 - 1000 eV for five minutes (from 7.19 UT to 7.24 UT, see Figure 3) soon after closest approach.

A group of scientists from the Space Research Institute, USSR Academy of Sciences, and the Max-Planck Institut fur Aeronomie have analyzed this event [Verigin et al., 1987]. Omitting their arguments we should only note that the analysis showed the following: at cometocentric distances of ~ 10^4 km the ion fluxes were observed in the vicinity of the surface which separates the regions with oppositely-directed magnetic fields (the dotted region in Figure 7a). They could have been accelerated in the process of reconnection of magnetic fields, and thus would leave the region of reconnection (around the x-point) with the velocity v directed along the separatrix surface. It was shown that in order to appear in the FOV of the ion sensors oriented towards the Sun (with due account of the relative velocity of the spacecraft

and the cometary nucleus), it is necessary that v should be > 35 km/s. In this case, the energy of the detected water-group ions should be > 200 eV in the spacecraft-fixed system of coordinates; this is in agreement with measurement results. Then from ion flux measurements, the concentration of accelerated ions can be estimated as (1 - 2) x 10^3 cm^{-3}, which also corresponds to the estimates made from the data of the Faraday cup oriented to the ram direction.

Thus, from the independent but self-consistent results of measurements of the magnetometer, the electrostatic analyzer, and the Faraday cup, the conclusion was reached that the unusual five-minute burst of the cometary ion flux observed near the spacecraft closest approach to the cometary nucleus is caused by the directed motion of water group ions accelerated up to a velocity of several tens of km/s. The acceleration of these ions could be due to the reconnection of magnetic fields with opposite polarity retarded by the cometary plasma.

3.3. The "mystery region" in the electron component of plasma detected form the Giotto spacecraft.

Figures 8a and 8c refer to the VEGA-2 electron EA analyzer [Gringauz et al., 1986b] and Figure 8b and 8d - to the Giotto EESA instrument data [Reme et al., 1987]. Comparing the VEGA-2 and Giotto data, for the portions of the trajectories in

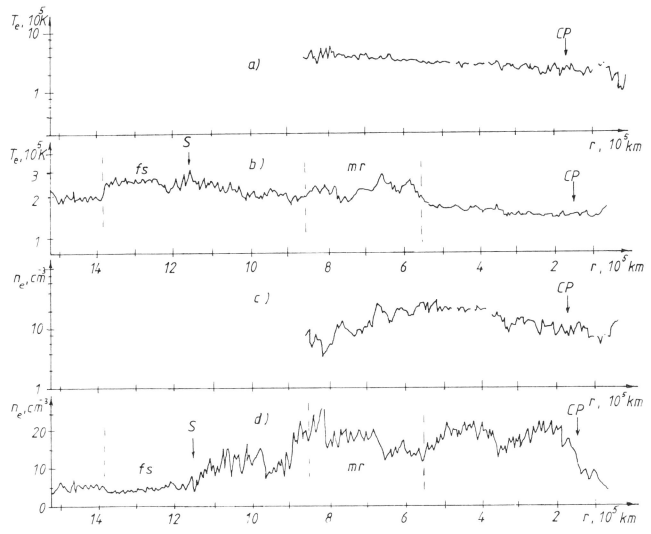

Fig. 8. Comparison of cometocentric profiles of elecron temperature (a, c) and density (b, d) estimated from VEGA-2 PLASMAG-1/EA data (a, b), and from Giotto EESA/RPA data (c, d).

the region called "mystery region" ("mr") by the authors of the Giotto experiment, one can see that in this region the electron temperature does not decrease with decreasing r and even increases (is varying) from time to time, according to the Giotto data, but is decreasing almost monotonically according to the VEGA-2 data. We can suggest that the different peculiarities of the electron temperature variations seen in these two cases are associated with the difference of conditions in interplanetary space. Any other explanations for the behavior of the plasma's electron component in the "mystery region" require an explanation for the absence of such behavior during the fly-by of VEGA-2. It seems to us that the "mystery region" peculiarities should be regarded as a result of nonstationary effects.

3.4 The magnetic field "pile-up boundary"

Figure 9 shows the results of measurements of the magnetic field absolute value made from VEGA-1 (a), VEGA-2 (b)

[Riedler et al., 1986] and Giotto (c) [Neubauer et al., 1987]. At r ~ 1.35 x 10^5 km, at the inbound leg of the trajectory, there is a jump in the Giotto data called the "pile-up boundary" (PB) by the authors of the experiment. Approximately at the same distance the cometopause (CP) is observed in the VEGA-1, 2 data however, any dramatic feature like the "pile-up boundary" has not been detected in the VEGA magnetic data.

The curve (c) illustrates that the pile-up boundary was not observed for the case of Giotto on the outbound leg of the trajectory, so the arrow designated PB was drawn arbitrarily at this position of the trajectory.

The difference between the VEGA-1, 2 and Giotto magnetic data within the interval (1 - 2) x 10^5 km, and the absence of the pile-up boundary signature at the time when Giotto was receding from the nucleus, give grounds to the suggestion that this boundary should also be referred to as a non-stationary event.

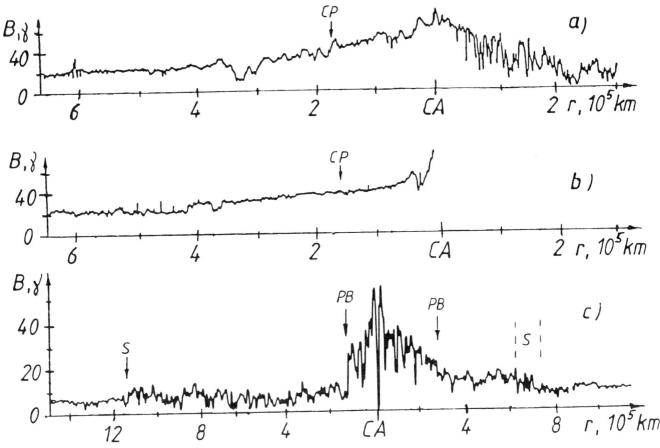

Fig. 9. Cometocentric profiles of magnetic field absolute values as measured by MISHA magnetometer onboard VEGA-1 (a) and VEGA-2 (b), and onboard Giotto (c).

3.5 Precipitation of electrons with energies ~ 1 kev near r ~ 10^4 km.

At a distance of (1.5 - 2.5) x 10^4 km from the nucleus, VEGA-2 recorded electron fluxes with energies of about 1 keV (note that the electron analyzer was oriented perpendicular to the ecliptic plane). Similar electron fluxes were not observed during the Giotto flyby near the cometary nucleus. Thus, it is also a nonstationary event.

Figure 10a illustrates the growth of the peak in the spectrum of electrons at E ~ 1 KeV (VEGA-2). Figure 11 (top spectrum) shows the energy spectrum of electrons detected on March 9th, 1986 from VEGA-2 at a distance of 1.5 x 10^4 km from the nucleus (the bottom spectrum was also recorded by the same instrument two days after the encounter with the comet on March 11th, 1986). The appearance of auroral electrons in the upper atmosphere of the Earth is also a typical non-stationary event. Substorms in cometary magnetospheres (Figure 10b) were predicted by Ip and Mendis [1976], and Ip and Axford [1982].

During the VEGA-2 flyby near the nucleus, the comet was not observed simultaneously in the optical and UV bands. However, the presence of sporadic precipitation of electrons in the atmosphere of Halley's comet is confirmed by non-simultaneous remote observations of this comet in the UV range. The IUE satellite observations made on March 18-19th, 1986 showed that for 37 minutes between measurements of two spectra, the CO^+ ion line intensity decreased by about 4 times whereas the OH line brightness remained practically the same [Feldman et al., 1986]. The authors explain this effect by additional ionization by sporadic fluxes of electrons similar to those observed from VEGA-2.

The 1536 Å line was observed in the UV spectrum recorded on February 26th, 1986 during the rocket experiment. This oxygen line cannot be excited by solar radiation but can be caused by impact of energetic electrons in the inner region of coma [Woods et al., 1986].

Using the data of electron measurements made from VEGA-2 and Giotto, the authors of rocket UV observations performed on February 26th, 1986 and March 13th, 1986 indicate that the presence of impact ionization by electrons can solve the so-called "carbon puzzle in the inner coma" (the excess of atomic carbon at r < 10^5 km which cannot be explained by photodissociation of carbon-bearing molecules). At r > 3 x 10^5 km the amount of carbon is satisfactorily explained by photodissociation of carbon-bearing molecules [Woods et al., 1987].

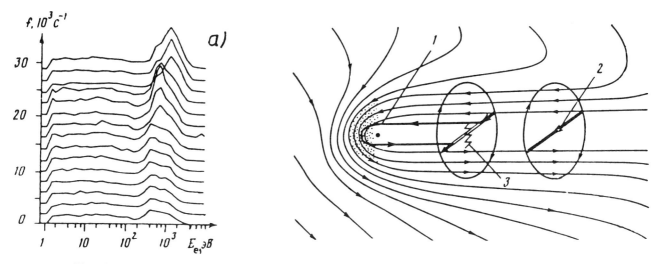

Fig. 10. Results of electron spectra measurements in cometary plasma region (r ~ (24.7 - 14.5) x 10³ km by VEGA-2/EA analyzer (a), and schematics of "cometary substorm" processes (b). 1 - tail aligned current discharging into the coma, 2 - cross tail current, and 3 - partial interruption of the cross-tail current.

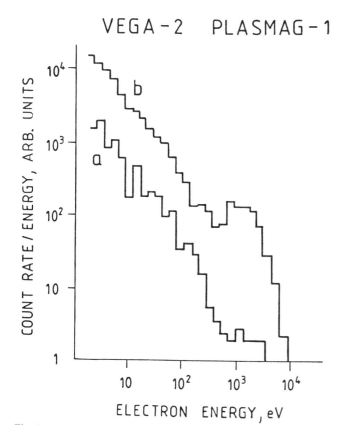

Fig 11. Two electron spectra as measured by the VEGA-2 electron analyzer in solar wind (a, 7.03 UT, March 11th) and near the closest approach (b, 7.16 UT, March 9th). Notice the occurrence of kev electrons close to the nucleus.

4. Conclusion

An attempt has been made above to separate the non-stationary events (not always occurring at ~ 1 a.u.) from the permanent events which had been observed in the head of Halley's comet on three spacecraft - VEGA-1, VEGA-2 and Giotto, which encountered the comet during March 1986. Permanent events should always take place when comet Halley passes at distance of 1 a.u. from the sun, regardless of conditions in the interplanetary medium or variations in gas production by the comet due to its nuclear rotation.

The above analysis employed two criteria: (i) the repetitive character of results obtained from all spacecraft flying near the cometary nucleus, and/or (ii) the existence of a clear physical understanding of the observational results.

The permanent events include the near-cometary bow shock, the cometosheath, with its unique energy distribution of ions, the systematic cooling of electrons in this region, the cometopause, and the tangential discontinuity near the cometary nucleus.

Among nonstationary events observed (boundaries, discontinuities, and so on) there are the unusual burst of ions with energies 100-1000 eV recorded in a direction from the Sun in the region of cometary ions at r ~ (1 - 2) x 10⁴ km, the magnetic field pile-up boundary (in the region of the cometopause), the mystery region, and the precipitation of energetic electrons with ~ 1 keV at r ~ (1.5 - 2.5) x 10³ km.

References

Amata, E., V. Formisano, R. Cerulli-Irelli, P. Torrente, A.D. Johnstone, A. Coates, B. Wilken, K. Jockers, J.D. Winningham, D. Bryant, H. Borg, and M. Thomsen, The cometopause region at comet Halley. *Exploration of Halley's Comet*, ESA SP-250, 1, 213-218, 1986.

Balsiger, H., K. Altwegg, F. Buhler, S.A. Fuselier, J. Geiss, B.E. Goldstein, R. Goldstein, W.T. Huntress, W. H. Ip, A.J. Lazarus, A. Meier, M. Neugebauer, U. Rettermund, H. Rosenbaur, R. Schwenn, E.G. Shelley, E. Ungstrup, and D.T. Young, The composition and dynamics of cometary ions in the outer coma of comet P/Halley. *Astron. Astrophys.* 187, 163-168, 1987.

Cravens, T.E., The physics of the cometary contact surface. *Exploration of Halley's Comet.* ESA SP-250, 1, 241-245, 1986.

Feldman, P.D., M.F. A'Hearn, M.C. Festou, L.A. McFadden, H.A. Weaver and T.N.Woods, Is CO_2 responsible for the outburst of comet Halley? *Nature.* 324, 433-436, 1986.

Galeev, A.A., Encounter with comets: discoveries and puzzles in cometary plasma physics. *Astron. Astrophys.*, 187, 12-20. 1987.

Galeev, A.A., K.I. Gringauz, S.I. Klimov, A.P. Remizov, R.Z. Sagdeev, S.P. Savin, A.Yu. Sokolov, M.I. Verigin, K. Szego, M. Tatrallyay, R. Grard, Ye.G. Eroshenko, M.J. Mogilevsky, W. Riedler and K. Schwingenschuh, Physical processes in the vicinity of the cometopause interpreted on the basis of plasma, magnetic field and plasma wave data measured on board the Vega-2 Spacecraft. *J. Geophys. Res.*, 93, 7527-7531, 1988.

Gombosi, T.I., Charge exchange avalanche at the cometopause. *Geophys. Res. Lett.*, 14, 1174-1177, 1987.

Gringauz, K.I., T.I. Gombosi, A.P. Remizov, I. Apathy, I. Szemerey, M.I. Verigin, L.I. Denchicova, A.V. Dyachkov, E. Keppler, I.N. Klimenko, A.K. Richter, A.J. Somogyi, K. Szego, S. Szendro, M. Tatrallyay, A. Varga and G.A. Vladimirova, First in situ plasma and neutral gas measurements at comet Halley. *Nature*, 321, 282-285, 1986.

Gringauz, K.I., T.I. Gombosi, M. Tatrallyay, M.I. Verigin, A.P. Remizov, A.K. Richter, I. Apathy, I. Szemerey, A.V. Dyachkov, O.V. Balakina and A.F. Nagy, Detection of a new "chemical" boundary at comet Halley, *Geophys. Res. Lett.*, 13, 613-616, 1986a.

Gringauz, K.I., A.P. Remizov, M.I. Verigin, A.K. Richter, M. Tatrallyay, K. Szego, I.N. Klimenko, I. Apathy, T.I. Gombosi and I. Szemerey, Analysis of electron measurements from Plasmag-1 Experiment on board Vega-2 in the vicinity of comet P/Halley. *Astron. Astrophys.*, 187, 287-289, 1987.

Ip W.-H and D.A. Mendis, The generation of magnetic fields and electric currents in the cometary plasma tails. *Icarus,* 29, 147-151, 1976.

Ip W. -H and W.I. Axford, Theories of physical processes in the cometary comae and in tails. *Comets* (ed. L.L. Wilkening). Univ. of Arizona Press. Tucson. Arizona. 588-634, 1982.

Ip W.-H. and W.I. Axford, The formation of a magnetic-field-free cavity at comet Halley. *Nature,* 325, 418-419, 1987.

Johnstone, A., A. Coates, S. Kellock, B. Wilken, K. Jockers, H. Rosenbauer, W. Studeman, W. Weiss, V. Formizano, E. Amata, R. Cerulli-Irelli, M. Dobrowolny, R. Terenzi, A. Egidi, H. Borg, B. Hultquist, J. Winningham, C. Gurgiolo, D. Bryant, T. Edwards, W. Feldman, M. Thomsen, M.K. Wallis, L. Biermann, H. Schmidt, R. Lust, G. Haerendel and G. Paschmann, Ion flow at comet Halley. *Nature,* 321, 344-347, 1986.

Korth, A., A.K. Richter, K.A. Anderson, C.W. Carlson, D.A. Curtis, R.P. Lin, H. Reme, J.A. Sauvaud, K. d'Uston, F. Cotin, A. Cros and D.A. Mendis, Cometary ion observations at and within the cometopause region of comet Halley. *Adv. Space Res.*, 5, No. 12, 221-225, 1987a.

Korth, A., A.K. Richter, D.A. Mendis, K.A. Anderson, C.W. Carlson, D.W. Curtis, R.P. Lin, D.L. Mitchell, H. Reme, J.A. Sauvaud and C. d'Uston, The composition and radial dependence of cometary ions in the coma of comet P/Halley. *Astron. Astrophys.*, 187, 149-152, 1987b.

Mendis, D.A., H.L.F. Houpis and M.L. Marconi, The physics of comets. *Fund. Cosmic Phys.*, 10, 1-380, 1985.

Mukai, T., W. Miyake, T. Terasawa, M. Kitayama and K. Hirao, Plasma observations by Suisei of solar-wind interaction with comet Halley. *Nature*, 321, 299-303, 1986.

Neubauer, F.M., Giotto magnetic-field results on the boundaries of the pile-up region and the magnetic cavity. *Astron. Astrophys.*, 187, 73-79, 1987.

Raeder, J., F.M. Neubauer, N.F. Ness and L.F. Burlaga, Macroscopic perturbations of the IMF by P/Halley as seen by the Giotto Magnetometer. *Astron. Astrophys.*, 187, 61-64, 1987.

Reme, H., J.A. Sauvaud, C. d'Uston, A. Cros, K.A. Anderson, C.W. Carlson, D.W. Curtis, R.P. Lin, A. Korth, A.K. Richter and D.A. Mendis, General features of comet P/Halley: solar wind interaction from plasma measurements. *Astron. Astrophys.*, 187, 33-38, 1987.

Remizov, A.P., M.I. Verigin, K.I. Gringauz, I. Apathy, I. Szemerey, I. Gombosi and A.K. Richter, Measurements of neutral particle density in the vicinity of comet Halley by Plasmag-1 on board Vega-1 and Vega-2. *Exploration of Halley's Comet.* ESA SP-250, 1, 387-390, 1986.

Riedler, W., K. Schwingenschuh, Ye.G. Eroshenko, V.A. Styaskin and C.T. Russell, Magnetic field observations in comet Halley's coma. *Nature*, 321, 288-289, 1986.

Schwingenschuh, K., W. Riedler, Ye.G. Eroshenko, J.L. Phillips, C.T. Russell, J.G. Luhmann and J.A..Fedder, Magnetic field draping in the comet Halley coma: comparison of Vega observations with computer stimulations. *Geophys. Res. Lett.*, 14, 640-643, 1987.

Thomson, M.F., W.C. Feldman, B.Wilken, K.Jockers, W. Studeman, A.D. Johnstone, A. Coates, V. Formisano, E. Amata, J.D. Winningham, H. Borg, D. Bryant and M.K. Wallis, Observations of a bi-modal ion distribution in the outer coma of comet P/Halley. *Astron. Astrophys.*, 187, 141-148, 1987.

Verigin, M.I., W.I. Axford, K.I. Gringauz and A.K. Richter, Acceleration of cometary plasma in the vicinity of comet Halley associated with an interplanetary magnetic field polarity change. *Geophys. Res. Lett.*, 14, 987-990, 1987.

Woods, T.N., P.D. Feldman, K.F. Dymond and D.J. Sahnow, Rocket ultraviolet spectroscopy of comet Halley and abundance of carbon monoxide and carbon. *Nature*, 324, 436-438, 1986.

Woods, T.N., P.D. Feldman and K.F. Dymond, The atomic carbon distribution in the coma of comet P/Halley. *Astron. Astrophys.*, 187, 380-384, 1987.

FRICTION LAYER IN COMET HALLEY'S IONOSHEATH

H. Pérez–de–Tejada

Instituto de Geofísica, Universidad Nacional Autónoma de México
Ensenada, Baja California, México

Abstract. The properties of the mass–loaded shocked solar wind in the region between the intermediate transition and the cometopause in comet Halley's outer plasma environment are examined. It is argued that the flow behavior within that region is consistent with the effects of friction phenomena between the external plasma flow and the bulk of cometary material at the cometopause. In particular, the low–energy branch of cometary ions detected at, and downstream from, the intermediate transition represents a population of particles initially dragged–off from the cometopause and gradually accelerated viscously into the streaming flow. It is argued that the local convective electric field of the mass–loaded shocked solar wind is not sufficient to propel that population of cometary particles, and that a non–local source of momentum is required to explain their motion. The scattering of the particles' momentum, induced by their interaction with turbulent wave fields in the presence of a velocity shear, is assumed to be the underlying process responsible for the momentum exchange among them. The overall character of the plasma between the intermediate transition and the cometopause can be described in terms of a friction layer formed along the flanks of the cometopause. It is further noted that the kinematic viscosity coefficient derived from the geometric aspect of this layer implies that the momentum exchange mean free path of the cometary ions is of the order of their gyroradius.

Introduction

The in–situ measurements conducted with the spacecraft that probed comet Halley's plasma environment in 1986 revealed flow properties whose interpretation is currently the subject of intense theoretical research. In a previous report [Pérez–de–Tejada, 1989] it was proposed that the intermediate ("mystery") transition detected ~ 5 10^5 km from the nucleus along the Giotto trajectory [Johnstone et al., 1986; Reme et al., 1986, 1987; d'Uston et al., 1987; Thomsen et al., 1987] represents the outer boundary of a region dominated by the effects of a viscous–like interaction between the mass–loaded shocked solar wind and the bulk of cometary material at the cometopause. That interpretation is based on the observation that the solar wind ion population decelerates, thermalizes, and expands abruptly at the intermediate transition [Balsiger et al., 1986; Goldstein et al., 1986]. The sudden changes seen at this transition suggest that mass–loading processes and charge exchange collisions may not be sufficient to describe the

plasma behavior in that region of space and that other processes may also be involved.

Within the context of the viscous–like interaction the solar wind flow decelerates abruptly because of the momentum transferred to the cometary plasma located at and downstream from the cometopause (< 1.5 10^5 km from the nucleus along the Giotto trajectory), its temperature is increased because of the heat released through the dissipation associated with the momentum transfer process, and its density decreases because of the ensuing expansion that the heated plasma experiences so that its pressure matches that of the external (inviscid) flow. The intermediate transition marks, in this view, the outer extent of the region affected by frictional effects which, in a supersonic flow around an obstacle, are constrained to a well–defined layer formed around it (see, for example, Mikhailov et al., [1971]). While this description provides a qualitative account of the observations, the underlying physical processes responsible for that behavior remain to be identified. This question has been raised in the past in connection with the suggestion that similar frictional processes are also operative in the Venus ionosheath. In that case there is ample evidence for the existence of a plasma transition, located between the planet's bow shock and its ionopause, with signatures analogous to those seen at comet Halley's intermediate transition [Bridge et al., 1967; Sheffer et al., 1979; Pérez–de–Tejada et al., 1984]. The possible common development of frictional phenomena in the interaction of the solar wind with a local ionosphere, be it of planetary or cometary origin, emphasizes the need to clarify the source of that behavior and, at the same time, allows the possibility of cross–correlating the information obtained in both cases.

These issues are further addressed in the present study by examining some general ideas that help to identify the origin of the viscous phenomenon. Fundamental to these ideas is the observed behavior of the cometary heavy ion component in the region bounded by the intermediate transition and the cometopause in comet Halley's ionosheath. Such a component was not discussed in the Perez–de–Tejada [1989] report, which dealt primarily with the remarkable changes experienced by the solar wind ion population in that region of space. Here we will show that equally important information can be derived from the examination of the measurements of cometary ions moving with the solar wind downstream from the intermediate transition, and that their thermodynamic properties also seem to be consistent with the development of a viscous–like interaction at the cometopause. The first section reviews experimental evidence of cometary ions in the ionosheath moving with speeds smaller than the local solar wind speed. The second section addresses the limitations of views based on the con-

Cometary Plasma Processes
Geophysical Monograph 61

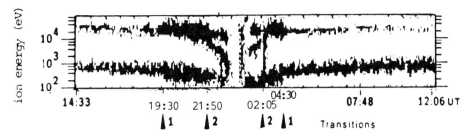

Fig. 1. Energy-vs-time tracks of the ion populations detected near comet Halley with the Implanted Ion Sensor of the Johnstone Plasma Analyzer (IIS-JPA) onboard the Giotto spacecraft on March 14, 1986. The upper (high energy) track, which splits into two components cometward from the transition labeled 2, corresponds to heavy cometary ions. The lower track results from signals received from the proton population (from Johnstone et al., [1986]).

ventional convective electric field–pickup to explain these observations. This discussion is followed by the description of a non-collisional transport of momentum which seems to be necessary to account for the observed plasma behavior. The implications of the erosion that the mass-loaded solar wind produces through such transport of momentum on the cometary plasma at the cometopause are outlined in the last section.

Plasma Measurements

Perhaps the most dramatic evidence of the presence of the intermediate transition in comet Halley's plasma environment is the observation, at that position, of the splitting of the cometary heavy ion population into two branches of clearly distinct energies. Measurements conducted with the Johnstone Plasma Analyzer (JPA) onboard the Giotto spacecraft show that, in addition to the main population of pickup ions moving with nearly the local solar wind speed there is, at and downstream from that transition, a new component of cometary ions whose energies rapidly decrease with decreasing distance from the comet. The local solar wind speed is interpreted here as that associated with the light ion component identified by the low–energy track in the energy-vs-time diagram reproduced in Figure 1. The upper (high energy) track corresponds to the (cometary) heavy ion population which at ~ 21:50 UT (inbound) and at ~ 02:05 UT (outbound) is seen to split into two branches. The position where this occurs defines the intermediate transition and is identified by the label 2 at the bottom of the panel (the comet's bow shock, traversed at ~ 19:30 UT (inbound) and at ~ 04:30 UT (outbound), is identified by the label 1).

Undoubtedly the most significant aspect of the heavy cometary ion population in the lower branch of the high energy track of Figure 1 is the energy deficiency of that component with respect to the main (upper branch) pickup cometary population. That energy deficiency implies that, barring a different chemical composition, the particles in the lower branch move with speeds smaller than those of the particles in the upper branch. Since the latter ions exhibit energies commensurable with their having been accelerated to nearly the local solar wind speed [Thomsen et al., 1987] the above circumstance implies that the cometary ions in the lower branch of the high energy track of Figure 1 move with speeds below the local solar wind value and thus that they have not been assimilated by the local flow. The tendency of the magnetic field vector to co–align with the velocity vector near the cometopause does not seem sufficient to explain this peculiarity since the strong oscillations of the magnetic field vector that are seen in that re-

gion [Neubauer, 1986] should result in conditions in which the pickup ions are accelerated to the local solar wind speed. Cometary ions moving with speeds substantially smaller than the solar wind value are not observed upstream from the intermediate transition where the pick up process nearly immediately accelerates them to the local solar wind speed.

Equally remarkable is the simultaneous onset of other plasma phenomena at the time when the splitting of the heavy cometary population is observed. As noted earlier, that separation coincides with the sudden changes in the plasma properties of the solar wind populations that mark the intermediate transition. Their simultaneous observation suggests that they all are manifestations of a common phenomenon whose effects on each plasma component depend on their own characteristics. For example, unlike the temperature increase exhibited by the solar wind protons and the alpha particles of the solar wind at the intermediate transition [Goldstein et al., 1986], the temperature of the assimilated heavy cometary ion population (upper branch of the high energy track in Figure 1) drops substantially at that transition [Johnstone et al., 1986; Amata et al., 1989]. This latter behavior can be appreciated from the narrower width that the main (upper branch) cometary energy track acquires at that time, and is in the same sense as that of the electron population reported by Reme et al. [1986] from measurements conducted with the electron spectrometer onboard the Giotto spacecraft. As argued below, both results are fundamental to understand the effects that the unassimilated (lower branch) cometary ion population has on the plasma as a whole.

A related issue is the marked energy decrease of the unassimilated (lower branch) cometary ions of the high energy track of Figure 1 with decreasing distance from the comet. While the cometary ions in the upper branch (together with the proton population) exhibit a gradual energy decrease in that direction, the one exhibited by the lower branch heavy cometary ions is more accentuated and implies that the velocity difference between both cometary ion components is larger in the inner regions of the cometary ionosheath. Thus, near the cometo pause the unassimilated cometary ion population moves with speeds significantly smaller than the local solar wind value. A quantitative estimate of the velocity deficiency of that cometary ion population with respect to the local solar wind speed is available from the results of the High Energy Range Sensor of the Ion Mass Spectrometer (HERS-IMS) data analysis reported by Fuselier et al., [1989]. That instrument measures positive ions in the 10 ev – 4.50 kev range which is adequate to register cometary ions within the energy range of the unassimilated (lower branch) population near the cometopause. The bulk speed of

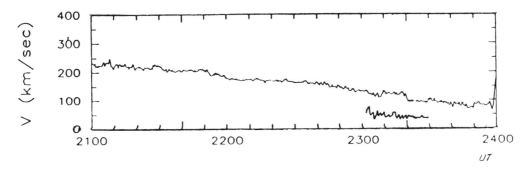

Fig. 2. Bulk speeds of the (long trace) proton and (short trace) water group ion cometary populations measured near comet Halley with the High Energy Range Sensor of the Ion Mass Spectrometer (HERS-IMS) onboard the Giotto spacecraft on March 14, 1986 (Fuselier S., Personal communication, [1989]).

this component (measured in the 23:00 UT – 23:30 UT interval), together with that of the proton population (measured across the entire inbound pass through comet Halley's ionosheath) are reproduced in Figure 2. The difference between the ~ 40 km/s speed of the heavy cometary ions and the ~ 100 km/s of the lighter component (formed mostly by solar wind protons) represents clear evidence that the plasma conditions in the inner regions of comet Halley's ionosheath allow the existence of an important particle population moving with speeds significantly smaller than the local solar wind speed.

Results of data analyses of independent measurements in the plasma environments of comets Giacobini–Zinner and Halley also lead to similar conclusions. Isemberg [1987] noted, in particular, that the distribution of data points in velocity space obtained from measurements of heavy ions in comet Giacobini–Zinner [Gloeckler et al., 1986] contains an important component at energies below the local pickup value (the latter is defined by the convective electric field of the solar wind which in turn is set by the magnitude and relative local orientation of the velocity and magnetic field vectors). Phase space densities with similar properties have also been reported by Coates et al., [1989, Figure 5] from the analysis of the inbound JPA plasma data in comet Halley's ionosheath. In this latter case there is clear evidence that the above peculiarity is basically a post shock phenomenon which becomes particularly strong after ~ 22:00 UT; that is, downstream from the intermediate transition. The observation of particles moving with speeds below the local solar wind value implies that not all contaminant ions are immediately accelerated to the local solar wind speed, but that there are conditions in which the pick up process operates in a more gradual manner.

Non-local Pick-up

In order to properly assess the significance of the properties of the cometary ions of the lower branch of the high energy track shown in Figure 1 it is fitting to mention that measurements conducted with both the JPA and the HERS–IMS instruments onboard the Giotto spacecraft indicate that the density of those ions increases substantially with decreasing distance from the comet, and that near the cometopause (between 23:00 UT and 23:30 UT) it reaches values even larger than the local proton density. For example, the density profile presented by Balsiger et al., [1986, Figure 4] shows that beginning ~ 3.5 10^5 km from the nucleus the density of water group ions (which, as noted earlier, are detected in an energy range consistent with that of the lower branch of the high energy track of Figure 1) rises first gradually from

very low (< 1 cm^{-3}) values to 3–5 cm^{-3} and then more rapidly near 2.5 10^5 km up to ~ 10 cm^{-3} before the inbound crossing of the cometopause (about 1.5 10^5 km from the nucleus). The proton density, on the other hand, decreases gradually from ~ 8 cm^{-3} at the intermediate transition (~ 5 10^5 km from the nucleus) down to ~ 2 cm^{-3} near the cometopause thus becoming, in the vicinity of this latter boundary, significantly smaller than the local cometary ion density. These numbers are in good agreement with the density values of electrons with energies larger than 10 ev reported by Reme et al., [1986, Figure 1]. According to these authors the density of electrons with energies above that value remains nearly uniform at ~ 10 cm^{-3} across the region between the intermediate transition and the cometopause, and drops substantially downstream from this latter boundary.

The larger densities of the cometary heavy ions over those of the proton population that are seen near the cometopause lead to conditions vastly different from those encountered in less contaminated regions of the cometary plasma environment. Since, as noted above, the cometary ions in the lower branch of the high energy track of Figure 1 move near the cometopause with speeds of the order of 40 km/s, while the proton population maintains speeds of about 100 km/s, the density values indicated above reveal that the kinetic energy density of the first component (assuming that it is formed by water group ions) can be up to 10 times larger than that of the proton population. A similar argument also holds in regard to the assimilated (upper branch) component of cometary ions. In this case the estimated number density is < 1 cm^{-3} even near the cometopause [Amata et al., 1989, Figure 2], and thus its kinetic energy density is, at most, comparable to that of the proton population. These results are fundamental to examine the manner in which the (lower branch) cometary ions are set into motion in that region of space. To this effect we should point out, first, that these energy density ratios impose conditions which are not consistent with standard views of the pick-up of cometary ions through the convective electric field of the solar wind. In particular, it is not evident how the solar wind, and its assimilated cometary ions, can locally propel the unassimilated (lower branch) cometary ions whose kinetic energy density near the cometopause is much larger than that of the incident flow. The effect of the pick-up process is to accelerate fresh material at the expense of the momentum of the solar wind; a response that is evident in the gradual deceleration exhibited by the proton population even far upstream from the comet's bow shock [Coates et al., 1987]. However, it should be clear that within the context of an acceleration produced locally through the convective electric field the momentum delivered to the cometary ions is limited to that of the oncoming flow; in other words, the pickup ions cannot end up having a momentum flux larger than that of the incident plasma.

A viable solution to this problem can be prepared in terms of shear forces transferring momentum across large sections of the solar wind around the comet. In this view the large number of (slowly moving) cometary ions near the cometopause are set into motion by the momentum supplied non-locally from the solar wind moving far away from that region. The underlying concept here is that despite the collisionless character of the solar wind, momentum can be exchanged among the ions in regions in which there is a velocity shear. As it has been proposed elsewhere [Pérez-de-Tejada, 1989] it is possible that conditions producing this effect result from interactions of the particles with turbulent wave fields. With the information now available regarding the assimilation of contaminant ions in the solar wind [Wu and Davidson, 1972; Gaffey et al.,1988] we can advance the following ideas to prepare an heuristic description of the manner in which momentum can be exchanged among different ion populations. It is now known that the predominant wave mode in cometary plasma environments is that associated with the Larmor gyration of the water group cometary ions around the local magnetic field [Glassmeier et al., 1989]. This hydromagnetic mode (with frequencies in the 10^{-2} – 10^{-3} Hz range) results from instabilities of the ring distribution that such ions form upon acceleration by the convective electric field of the solar wind. The effect of these waves, and of their harmonics, is to produce an effective scattering of both the pitch angle and the energy of the cometary ions [Wu et al., 1986; Gaffey et al., 1988]. When plotted in the reference frame of the solar wind the measured phase space densities follow a distribution roughly consistent with these concepts even though important discrepancies still await resolution.

An important feature of these results is that the phase space distribution of the cometary ions depends strongly on the local conditions. Neugebauer et al., [1989] have shown this dependance for different relative orientations of the magnetic field and solar wind velocity vectors. Coates et al., [1989] report, on the other hand, phase space density distributions which differ significantly with position along the Giotto trajectory. These variations reveal that the pitch angle and energy scattering of the particles depend strongly on their velocity, on the level of turbulence, and on the degree of contamination, and thus expose the intricate complexities of the scattering process. All these considerations are important because they imply that a cometary particle, which due to its large gyroradius experiences different scattering patterns along its trajectory, may be subject to conditions in which it will be forced to adjust to different phase space distributions. What this means is that cometary ions moving into a region of a velocity shear in which the local speed differs from that of the region where they were first picked up, will be scattered by the wave field in such a way as to adjust to the flow conditions appropriate to that location. The end result of this interaction is that the momentum difference between the ions and the local flow will be ultimately transferred among them by the wave field, provided the scattering of momentum is sufficiently fast.

These various concepts can be formulated by examining the geometry presented in Figure 3. Following a procedure similar to that suggested by Parker [1958] consider a cometary ion of mass M_i executing a Larmor gyration of radius ρ_1 in a region of space in which there is a velocity shear $\partial u/\partial y$. Since the drift velocity u changes in the direction of the shear the scattering interactions that it will experience with the local wave field will result in a net transfer of momentum ($M_i \Delta u$)

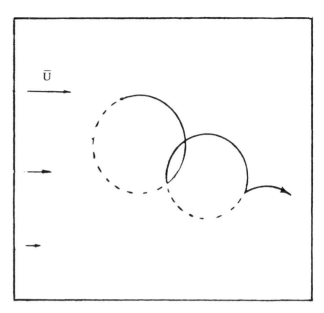

Fig. 3. Schematic diagram of the trajectory of a cometary ion moving in a velocity shear projected on the plane perpendicular to an external magnetic field (the velocity shear is indicated by the different lengths of the velocity vectors in the left hand side). Momentum scattering interactions with local wave fields force the ions to drift in different directions and with different speeds.

per unit time given by: $M_i \omega \rho_1 \partial u/\partial y$ where ω is the frequency of the scattering interactions. We can calculate the rate of momentum transferred per unit area in that region of space by multiplying the above quantity with the total number of cometary ions $n_i \rho_1$ subject to such conditions (n_i being their number density). This is: $\Phi = n_i M_i \rho_1^2 \omega \partial u/\partial y$, and defines an effective kinematic viscosity coefficient ν given by:

$$\nu \sim \rho_1^2 u/\lambda \qquad (1)$$

where we have introduced $\lambda = u/\omega$ as the equivalent mean free path traveled by the cometary ions between the momentum scattering interactions. An empirical comparison between λ and ρ_1 can be obtained from this relationship by using values for ν derived from independent considerations. For example, with $\nu = 3\ 10^5$ km^2/s, as implied by the relative position of the intermediate transition with respect to that of the cometopause along the Giotto trajectory [Pérez-de-Tejada, 1989], and by taking u = 100 km/s and $\rho_1 = 8\ 10^3$ km, as is suitable for the conditions present in the region exterior to comet Halley's cometopause where the magnetic field intensity is ~ 8γ [Neubauer, 1986], we find that $\lambda \sim 2\ 10^4$ km. This value implies that the distance traveled by the cometary ions between the momentum scattering interactions is comparable to their own gyroradius, and hence that they may not complete, on the average, whole unperturbed gyrations. This result is congruent with the basic assumption that the cometary ions can exchange momentum as they move about their Larmor trajectories and, at the same time, justifies the concept introduced earlier in the sense that momentum can be efficiently transported across the velocity shear in comet Halley's ionosheath.

Despite the crude nature of the calculation involved in these estimates it is interesting that they should lead to a picture not different from that obtained in the numerical model studies of Wu et al., [1986]; and Gaffey et al., [1988]. By following the time evolution of the velocity distribution function of newborn cometary ions in the presence of strong hydromagnetic turbulence (as reported, for example, by Neubauer et al. [1986]) these authors conclude that the diffusion times associated with the energy and pitch angle scattering of these particles are comparable to their gyroperiod. Under such conditions it is possible to argue that as they move about the magnetic field lines the particles may experience important changes in both the direction and magnitude of their velocity vector. The fact that the kinematic viscosity coefficient used above leads to similar conclusions supports both, the relevance of the friction process and the stochastic character of the trajectories of the plasma particles that is implied from those conclusions.

A final comment on this issue should be made in regard to the order-of-magnitude value of the kinematic viscosity coefficient used above. That value is supported by the fact that it satisfies an alternative form of (1) relating the magnitude of the velocity fluctuations δu of the plasma and the characteristic frequency Ω of the oscillations responsible for the velocity fluctuations. This relationship is [Saffman, 1970; Korzhov et al., 1984]:

$$\nu \sim (\delta u)^2 / \Omega \qquad (2)$$

and leads to $\sim 10^5$ km^2/s for $\delta u \sim 30$ km/s [Johnstone et al., 1989, Figure 2] and $\Omega = 10^{-2}$ Hz (corresponding to waves in resonance with the gyration motion of the cometary ions, Glassmeier et al., [1989]).

Based on these various concepts we can now address the issue raised before regarding the source of the momentum of the cometary ions that form the lower branch of the high energy track shown in Figure 1. We will show that the integrated momentum flux of that component represents only a fraction of the integrated deficiency of momentum flux of the solar wind population that is implied by its velocity profile downstream from the intermediate transition. By assuming axial symmetry around the Sun-comet line the first quantity can be estimated from:

$$\phi = <n_i> M_i <u_i>^2 \pi (R_L^2 - R_c^2) \qquad (3)$$

where $<n_i> = 10$ cm^{-3} and $<u_i> = 40$ km/s are, respectively, average values for the density and bulk velocity of the unassimilated (lower branch) cometary ion population measured in the vicinity of the cometopause (between $R_c = 1.5\ 10^5$ km and $R_L = 2.5\ 10^5$ km from the nucleus). As noted earlier, this is the region where the density of the cometary ions is comparable or even larger than the proton density. The second quantity, on the other hand, can be estimated from:

$$\psi = <n_p> m_p [u_0^2 - <u>^2] \pi (R_I^2 - R_c^2) \qquad (4)$$

where $<n_p> = 5$ cm^{-3} is the average density of the proton population between the intermediate transition and the cometopause (at $R_I = 5\ 10^5$ km and $R_c = 1.5\ 10^5$ km from the nucleus, respectively). u_0 denotes, in turn, the speed of the proton component just upstream from the intermediate transition, and $<u>$ its average speed between that transition and the cometopause. From Figure 2 we can take $u_0 = 220$ km/s and $<u> = 150$ km/s, so that with $M_i/m_p = 18$ for a cometary population formed by water group ions the above numbers lead to: $\phi/\psi = 0.3$. Thus, the total momentum flux carried by the unassimilated (lower branch) cometary ions can be easily accounted for in terms

of the total momentum flux available from the observed deceleration of the solar wind downstream from the intermediate transition. The implication of this result is that the difficulties mentioned earlier regarding the limitations of the E x B pickup process to explain the large momentum flux of the cometary ion flow near the cometopause can be resolved in terms of cross-flow momentum transport. It should be noted, however, that the concept advanced here does not require that the reduction of momentum flux of the (solar wind) proton population is solely due to momentum transfer. A fraction of that reduction is undoubtedly associated with the local assimilation of cometary ions through the convective electric field of the solar wind, and an additional fraction can also be ascribed to the intrinsic variations of the flow properties of the shocked solar wind as it expands and accelerates with the downstream distance. In the absence of adequate estimates of the magnitude of these two effects the argument presented above should be understood as indicating that there is enough momentum flux in the overall balance of that quantity within the ionosheath to account for that of the unassimilated (lower branch) cometary ions of the high energy track shown in Figure 1.

Discussion

In addition to the issues treated thus far there are other aspects of the observations which seem also congruent with the overall picture derived from the suggested cross-flow transport of momentum downstream from the intermediate transition. Most important is the origin of the unassimilated (lower branch) heavy cometary ions detected in the vicinity of the cometopause. The available density profile of that component along the Giotto trajectory [Balsiger et al., 1986] reveals that the sharp ($\sim 2\ 10^4$ km) chemical transition, identified from the Vega data [Gringauz et al., 1986] as the cometopause region, may also be characterized by large density gradients across a much wider distance. In fact, we noted earlier that the density of heavy cometary ions detected with the HERS-IMS instrument onboard Giotto is seen to first increase gradually with decreasing distance from the comet and then more rapidly as the spacecraft approached the cometopause. This variation is consistent with what would be expected from a gradual and continuous erosion process exerted by the streaming mass-loaded solar wind on the cometopause region proper. The dynamic scavenging of cometary material from that region of space, and its subsequent distribution at increasingly far away distances as the particles are carried downstream by the flow, provides a viable interpretation of the observations on the basis of the non-local transport of momentum discussed above. Accordingly, we can argue that cometary particles initially nearly at rest with respect to the comet in the cometopause are gradually brought into motion through momentum scattering interactions with the local wave field. As a result of this process the cometopause particles are gradually carried downstream by the local flow and, at the same time, distributed over increasingly larger distances away from that boundary. A convenient analogue to view the proposed behavior is that of wind streaming over a sandy surface. In this case the sand grains are continuously removed from the surface and displaced to heights above the surface that depend, together with the grains speed, on the distance traveled. Analogously, the speed of the particles dragged from the cometopause will increase with the downstream distance from the point where they were first removed.

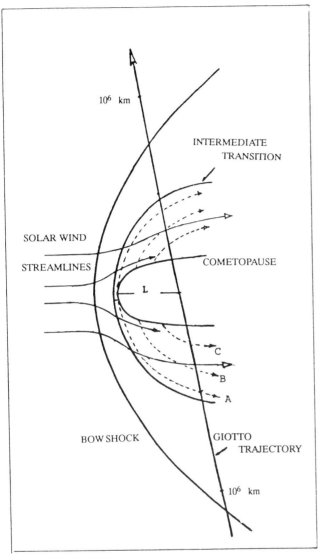

Fig. 4. Schematic diagram of the flow geometry in comet Halley's ionosheath. The solid lines indicate streamlines of the mass-loaded solar wind, and the dashed lines indicate streamlines of cometary ions dragged-off from the cometopause (a cross flow density gradient of such ions prevents their diffusion back to the cometopause), (from Pérez-de-Tejada, [1989]).

A schematic of the flow geometry exterior to the cometopause is depicted in Figure 4. The solid lines indicate streamlines of the solar wind and of its assimilated (upper branch) heavy cometary ions as they move around the cometopause and erode the cometary plasma at this latter boundary. A fraction of both populations may also find its way across the cometopause as a result of turbulent mixing in that region. The dashed lines indicate, on the other hand, streamlines of unassimilated (lower branch) heavy cometary ions as they are removed from the cometopause and gradually accelerated by the oncoming flow. The overall shape of the bow shock, intermediate transition, and cometopause is that calculated in the Pérez-de-Tejada [1989] study.

The most notable implication of the interpretation of the Giotto data given here is that the observed time-vs-energy dependence of the unassimilated (lower branch) heavy cometary ion population shown in Figure 1 can be viewed as reflecting the effects of its gradual acceleration with respect to the distance traveled. For example, the cometary ions that contribute to the innermost tip of the lower branch of the high energy track of that figure (at ~ 23:00 UT inbound, and at ~ 01:30 UT outbound) are particles that were removed from locations on the cometopause not too far upstream from the point of measurement (their low bulk energies are a manifestation of this circumstance) and their position is indicated in Figure 4 by the streamline labeled C. On the other hand, the cometary ions of the same lower branch of the high energy track of Figure 1 that were measured just downstream from the intermediate transition (at ~ 22:00 UT inbound and at ~ 02:00 UT outbound) are particles that were dragged from the cometopause at locations far upstream from where they were detected. Such particles (whose trajectories are labeled A in Figure 4) have traveled distances that are sufficiently long for them to nearly reach the local solar wind speed and may have been removed from the nose of the cometopause .

The relevance of the proposed origin of the cometary ions that form the lower branch of the high energy track shown in Figure 1 finds additional support in its implicit explanation of the peculiar anticorrelation of the temperature change of the various plasma components that is seen at the intermediate transition. We noted earlier that contrary to the temperature increase experienced by the proton and alpha particle populations at that transition, the component of assimilated heavy ion cometary ions that forms the upper branch of the high energy track of Figure 1 exhibits a net temperature decrease. Since the first two components are mostly of solar wind origin there is a marked contrast between what the solar wind flow experiences at the intermediate transition and the manner in which the cometary material responds to the changes that take place at that boundary. Most likely such a difference is simply due to the enhanced addition of cometary material initially removed from the cometopause by the flow, and subsequently distributed throughout the region between this boundary and the intermediate transition. This material represents, in effect, an intense source of cool cometary ions whose eventual assimilation into the main (upper branch of the high energy track of Figure 1) cometary population should result (within the context of a stochastic description) in a net decrease of its overall effective temperature. A similar argument can also be advanced to explain the concurrent decrease of the electron temperature at the intermediate transition. Since the material removed from the cometopause contains an electron component with temperatures appropriate to its origin, the effective values measured downstream from the intermediate transition should also reflect the contamination that this component produces on the solar wind electron temperature. However, although all these arguments suggest variations in the same sense as those observed, quantitative calculations are clearly necessary to establish to what extent they can explain the measurements. Among the issues that should also be addressed stand out those concerning the energetics of the scattering process. In particular, it is necessary to clarify the manner in which the wave field administrates the energy delivered across the velocity shear. The balance of that energy must include the different forms it can be converted into (kinetic and thermal energy of the particles as well as the fraction that is used to feed the wave field). Concepts similar to those advanced by Hizanidis et al., [1988] and by Sharma et al., [1988] regarding the viscous heating of the solar wind protons at the cometary bow shock may be relevant to describe the heating process within the velocity shear exterior to the cometopause. Equally important is the necessity to determine the manner in which the proton and the heavy ion components contribute to produce the viscous phenomenon. These basic questions serve merely to illustrate the complexities of the problems that await examination.

Summary and Conclusions

The behavior and the properties of the mass loaded shocked solar wind in the region between the intermediate transition and the cometopause in comet Halley's plasma environment have been interpreted in this study in terms of friction phenomena. The underlying mechanism responsible for the transfer of momentum among the various ion populations is assumed to be related to momentum scattering processes produced by local wave fields. In particular, the momentum acquired by cometary ions born in the solar wind is transferred, near the cometopause, through such scattering interactions. A calculation of the momentum transfer expected leads to an effective mean free path of the cometary ions of the order of their gyroradius. The role of these ions in transferring the momentum of the incident solar wind to the material at and downstream from the cometopause was also emphasized. The relevance of these concepts is based on the fact that they provide a flow description consistent with the multiple peculiarities of the changes exhibited by the different components of the plasma in that region of space.

For the solar wind ion population the model addresses the following features of the observations: a) the abrupt velocity drop seen at the intermediate transition, followed by a steep velocity decrease with decreasing distance from the comet (this behavior can be accounted for as reflecting changes forced on the solar wind flow by the transfer of a fraction of its bulk momentum to the cometary material at the cometopause); b) the simultaneous sudden temperature increase seen at the intermediate transition, and its subsequent continued rise in the direction of the cometopause (dissipated heat associated with the friction process is expected to result in an effective enhancement of the flow temperature); c) the accompanying large density decrease detected at the intermediate transition, with even lower values measured downstream from it (the expansion of the heated plasma should follow the above changes so that the local pressure can match that of the external inviscid flow); d) the sudden outward deflection of the flow that occurs at the intermediate transition (the lower momentum flux of the incident flow within the friction layer necessarily implies that the flow be diverted as if forced around an obstacle).

For the cometary ion population the model addresses the following features: a) the dramatic splitting of the heavy ion population at the intermediate transition into two branches of clearly distinct energy (cometary material dragged viscously from the cometopause provides, at and downstream from the intermediate transition, a separate component of contaminant ions); b) the gradual energy decrease with decreasing distance from the comet that the assimilated (upper branch) cometary ion population exhibits downstream from the intermediate transition (this variation is analogous to that of the solar wind population and should also result from a net transfer of momentum to fresh cometary material removed from the cometopause); c) the steep energy decrease with decreasing distance from the comet that the unassimilated (lower branch) cometary population exhibits near the cometopause (the gradual acceleration of particles removed viscously from the cometopause should result in bulk speeds that increase gradually with their distance traveled); d) the larger kinetic en-

ergy density of the unassimilated (lower branch) ion population over that of the assimilated (upper branch) component and the solar wind proton population near the cometopause (unlike the convective E x B pick-up processes the transfer of momentum across the friction layer predicts momentum flux values consistent with the observations); e) the temperature decrease of the assimilated (upper branch) population at the intermediate transition (the denser unassimilated cometary material is expected to produce an effective cooling of the cometary population as a whole).

Finally, for the electron population the model addresses the following features: a) the change in the slope of the velocity profile at the intermediate transition, and the large velocity decrease seen downstream from it (this variation is comparable with that exhibited by the solar wind ion population, and is also compatible with the effects of momentum transfer processes within the friction layer); b) the abrupt density drop encountered at the intermediate transition (the local expansion of the mass loaded solar wind ion components should have this effect on the electron population in order to maintain charge neutrality); c) the simultaneous temperature decrease seen at the intermediate transition (this variation is in the same sense as that of the heavy cometary ion populations and, as in that case, should result from contaminant material removed viscously from the cometopause).

While all these properties of the plasma within the region between the intermediate transition and the cometopause can, in principle, be embodied within the context of effects associated with the onset of friction phenomena in that region, it is clear that detailed calculations of the magnitude of the changes expected are still necessary in order to properly substantiate this interpretation. Thus, estimates of the amount of energy dissipated, either in the form of thermal motion of the various particle populations, or to sustain the intense wave field that allows the transport of statistical properties among the particles, should be properly carried out. In this respect it should be stressed that the adequate understanding of the manner in which such properties (mass, momentum and energy) are transferred by the wave field in the collisionless solar wind plasma is a major goal which will require of extended analyses of the experimental data as well as the continued development of suitable theoretical models.

Acknowledgements. I wish to thank A. Johnstone and A. Coates for their kind hospitality and scientific discussions conducted during my visit to the Mullard Space Science Laboratory in Guildford, England, where this study was planned. I am also indebted to S. Fuselier for providing the proton and the heavy cometary ion velocity profiles used in this study. My appreciation is also due to P. Byrdseye and G. Zenteno for technical assistance. Funding provided by the Mexican Academy of Sciences and the British Royal Society is acknowledged.

References

Amata, E., V. Formisano, P. Torrente, M. B. Bavassano-Cattaneo, A. D. Johnstone, and B. Wilken, Experimental plasma parameters at comet Halley, Adv. Space Res., 9, (3)313, 1989.

Balsiger, H., K. Altwegg, F. Buhler, S. A. Fuselier, J. Geiss, R. Goldstein, A. Lazarus, A. Meier, M. Neugebauer, U. Rettenmund, H. Rosenbauer, R. Schwenn, E. Shelley, E. Ungstrup and D. Young, Ion composition and dynamics of cometary ions in the outer coma of Halley, Proc. Heidelberg Symposium on the Exploration of Comet Halley, ESA SP-250, 99, December 1986.

Bridge, H. S., A. J. Lazarus, C. W. Snyder, E. J. Smith, L. Davies, P.L. Coleman and D. E. Jones, Plasma and magnetic fields observed near Venus, Science, 158, 1669, 1967.

Coates, A. J., A. D. Johnstone, M. F. Thomsen, V. Formisano, E. Amata, B. Wilken, K. Jockers, J. D. Winningham, H. Borg, and D. A. Bryant, Solar wind flow through the comet P/Halley bow shock, Astron. Astrophys., 187, 55, 1987.

Coates, A. J., A. D. Johnstone, B. Wilken, K. Jockers, and K. H. Glassmeier, Velocity space diffusion of pickup ions from the water group at comet Halley, J. Geophys. Res., 94, 9983, 1989.

d'Uston, C., H. Reme, J. A. Sauvaud, A. Cross, K. Anderson, C. W. Carlson, D. Curtis, R. P. Lin, A. Korth, A. K. Richter and A. Mendis, Description of the main boundaries seen by the Giotto electron experiment in the comet P/Halley-solar wind interaction region, Astron. Astrophys. 187, 137, 1987.

Fuselier, S., Observations of charge exchange of solar wind ions in comet Halley's coma, Cometary Plasma Processes Conf., Univ. of Surrey, Guildford, England, 1989.

Gaffey, Jr. J. D., D. Winske, and C. S. Wu., Time scales for formation and spreading of velocity shells of picked up ions in the solar wind, J. Geophys. Res., 93, 5470, 1988.

Glassmeier, K. H., A. J. Coates, M. H. Acuna, M. L. Goldstein, A. D. Johnstone, F. M. Neubauer, and R. Reme, Spectral characteristics of low frequency plasma turbulence upstream of comet P/Halley, J. Geophys. Res., 94, 37, 1989.

Gloeckler, G., D. Hovestadt, F. Ipavich, M. Scholer, B. Klecker, and A. Galvin, Cometary pickup ions observed near Comet Giacobini Zinner, Geophys. Res. Lett., 13, 251, 1986.

Goldstein, R. E., M. Neubauer, H. Balsiger, J. Drake, S. Fuselier, R. Goldstein, W. H.-Ip, U. Rettemund, H. Rosenbauer, R. Schwenn, and E. Shelley, Giotto-IMS observations of ion flow velocities and temperatures outside the contact surface of comet Halley, Heidelberg Symposium on the Exploration of Comet Halley, ESA SP-250, 229, 1986.

Gringauz, K. I., T. I. Gombosi, M. Tatrallyay, M. I. Verigin, A. P. Remizov, A. K. Richter, I. Apathy, I. Szemerey, A. V. Dyachkov, O. V. Balakina, and A. F. Nagy, Detection of a new chemical boundary at comet Halley, Geophys. Res. Lett., 13, 613, 1986.

Hizanidis, K., P. J. Cargill, and K. Papadopoulos, Lower hybrid waves upstream of comets and their implications for the comet Halley bow wave, J. Geophys. Res., 93, 9577, 1988.

Isenberg, P. A., Energy diffusion of pickup ions upstream of comets, J. Geophys. Res., 92, 8795, 1987.

Johnstone, A., A. Coates, S. Kellok, B. Wilken, K. Jockers, H. Rosenbauer, W. Studemann, W. Weiss, V. Formisano, E. Amata, R. Cerulli-Irelli, M. Dobrowolny, R. Terenzi, A. Egidi, H. Borg, B. Hultquist, J. Winningham, C. Gurgiolo, D. Bryant, T. Edwards, W. Feldman, M. Thomsen, M. Wallis, L. Biermann, H. Schmidt, R. Lust, G. Haerendel, and G. Paschmann, Ion flow at comet Halley, Nature, 321, 344, 1986.

Johnstone, A., K. H. Glassmeier, M. Acuña, H. Borg, D. Bryant, A. Coates, V. Formisano, J. Heath, F. Mariani, G. Musmann, F. Neubauer, M. Thomsen, B. Wilken, and J. Winningham, Waves in the magnetic field and solar wind flow outside the bow shock at comet P/Halley, Astron. Astrophys., 187, 47, 1989.

Korzhov, N. P., V. V. Mishin, and V. M. Tomozov, On the role of plasma parameters and the Kelvin Helmholtz instability in a viscous interaction of solar wind streams, Planet. Space Sci., 32, 1169, 1984.

Mikhailov, V. V., V. Ya. Neiland, and V. V. Sychev, The theory of viscous hypersonic flow, Annual Reviews of Fluid Mechanics, 3, 371, 1971.

Neubauer, F. M., Giotto magnetic field results on the magnetic field pileup region and the cavity boundaries, Proc. Heidelberg Symposium on the Exploration of Comet Halley, ESA SP-250, 35, December 1986.

Neubauer, F. M., K. H. Glassmeier, M. Phol, J. Raeder, M. H. Acuna, L. F. Burlaga, N. F. Ness, G. Musmann, F. Mariani, M. K. Wallis, E. Ungstrup, and H. U. Schmidt, First results from the Giotto magnetometer experiment at comet Halley, Nature, 321, 352, 1986.

Neugebauer, M., A. J. Lazarus, H. Balsiger, S. A. Fuselier, F. M. Neubauer, and H. Rosenbauer, The velocity distributions of cometary protons picked up by the solar wind, J, Geophys. Res., 94, 5227, 1989.

Parker, E. N., Interaction of the solar wind with the geomagnetic field, Phys. Fluids., 1, 171, 1958.

Perez-de-Tejada, H., D. Intriligator, and F. Scarf, Plasma and electric field measurements of the PVO in the Venus ionosheath, Geophys. Res. Lett., 11, 31, 1984.

Perez-de-Tejada, H., Viscous flow interpretation of comet Halley's mystery transition, J. Geophys. Res., 94, 10131, 1989.

Reme, H., J. Sauvaud, C. d'Uston, A. Cross, K. Anderson, C. Carlson, D. Curtis, R. Lin, A. Korth, A. Richter and A. Mendis, General features of the comet Halley-Solar wind interaction from plasma measurements, Proc. Heidelberg Symposium on the Exploration of Comet Halley, ESA SP-250, 29, December 1986.

Reme, H., J. A. Sauvaud, C. d'Uston, A. Cross, K. Anderson, C. W. Carlson, D. W. Curtis, R. P. Lin, A. Korth, A. K. Richter, and A. Mendis, General features of comet p/Halley-Solar wind interaction from plasma measurements, Astron. Astrophys. 187, 33, 1987.

Saffman, P. G., A model for inhomogeneous turbulent flows, Proc. Royal Soc. London, A317, 417, 1970.

Sharma, A. S., P. J. Cargill, K. Papadopoulos, Resonance absorption of Alfven waves at comet-solar wind interaction regions, Geophys. Res. Lett., 15, 740, 1988.

Sheffer, R., A. Lazarus, and H. Bridge, A re-examination of plasma measurements from the Mariner 5 Venus encounter, J. Geophys. Res., 84, 2109, 1979.

Thomsen, M. F., W. Feldman, B. Wilken, K. Jockers, W. Studemann, A. D. Johnstone, A. Coates, V. Formisano, E. Amata, J. D. Winningham, H. Borg, D. Bryant, and M. K. Wallis, In-situ observations of a bimodal ion distribution in the outer coma of comet P/Halley, Astron. Astrophys. 187, 141, 1987.

Wu, C. S., and R. C. Davidson, Electromagnetic instabilities produced by neutral particle ionization in interplanetary space. J. Geophys. Res., 77, 5399, 1972.

Wu, C. S., D. Winske and J. Gaffey Jr., Rapid pick-up of cometary ions due to strong magnetic turbulence, Geophys. Res. Lett., 13, 865, 1986.

CHARGE EXCHANGE REGIME IN THE PLASMA FLOW
AS SOURCE OF THE COMETOSHEATH AND HALLEY'S PLASMA TAIL

Max K. Wallis

School of Mathematics, University of Wales, Cardiff CF2 4AG, Wales, U.K.

Abstract. The depletion by charge exchange of energetic cometary ions picked up in the solar wind flow into Halley's coma is calculated to coincide with the upstream extension of its cometopause, detected by the comet probes. It relates to the cometary plasma structures observed telescopically. The field–free ionospheric cavity and the cometosheath regions of Halley's plasma must in some way connect to the comet's tail. The origin of the cometosheath–envelope is argued to be due to charge–exchange cooling, this mechanism determining the source and variability of the main tail rays. The ionosphere connects to a denser tail core. The sunward part of the charge–exchange dominated cometosheath is identified on an IPD image of Halley by depletion in CO^+ ions. Where traversed by the comet probes, however, the cometosheath was twice as far out. Pressure–driven expansion occurs there and the sheath's main signature is a change in the flow direction.

1. Introduction

The nature of the major plasma structures detected by the *Vega* and *Giotto* comet–probes is still unclear theoretically. The plasma characteristics of Halley's "cometopause" and "cometosheath" differ in the two instances, and the relation to large–scale tail structures is unexplained. Moreover, the longstanding mystery of what mechanism governs the generation of plasma envelopes in the comet head, as studied by Eddington (1910) for comet Morehouse, remains unclarified. It's evident that the inner magnetic field–free cavity discovered by *Giotto* is of much smaller scale (10^4 km) than the main tail. But the question remains – is a tail core to which it presumably connected, as for comet Giacobini–Zinner, significant for MHD stability (Rees et al. 1987).

While the "magnetic pile–up" boundary at 1.4×10^5 km (Neubauer 1987) is compatible in position with the "cometopause" detected by the two *Vegas*, the magnetic field did not increase in the latter cases (Galeev et al. 1987). Though the cometopause appeared impressively sharp ($\sim 10^4$ km wide) in terms of appearance of cometary ions and disappearance of protons, the failure to see it in the Giotto plasma data indicates an episodic structure.

Indeed Giotto saw sharp changes in direction, which could well give appearance and disappearance from field–of–view of the *Vegas'* ion instruments. On *Giotto's* outbound leg, the "magnetic pile–up" boundary was placed at 2.6×10^5 km, twice as far as analogy would predict, but as it was not sharp it may be that the "pile–up" in field on the inbound pass was anomalous. So the magnetic field would generally satisfy the prediction (Wallis and Johnstone 1983) of enhancement via convection at a radial distance roughly 10% of the bow shock scale.

Discontinuities in the solar wind flow through Halley's coma were surprisingly few and weak. No distinct bow shock was evident on several passes. The "cometopause" and "magnetic pile–up" features showed no strong jumps in plasma parameters u, n_i, n_p, T, nor were they clearly identified on outbound passes. Instead, strong fluctuations were evident, superposed on gradual changes of all parameters. However, the change at 1.5×10^5 km on Giotto's inbound path does characterize a transition between physical regimes. Fluctuations in speed and direction decrease while the flow direction inclines as from a source region ahead of the comet. The magnetic field aligns with the flow and relative streaming builds up between cometary ions and protons of solar wind origin. While the ion speeds decrease in conformity with pick–up ideas, the protons maintain around 100 km/s speeds. Ahead of the comet, at $0.5 - 1 \times 10^5$ km (sect. 3), charge exchanges are resulting in losses of the energetic pick–up ions but few of the protons.

With the drop in plasma pressure, the convected magnetic field becomes strong enough (50 nT) to divert the flow. One infers from the Giotto velocity data that protons accelerate laterally along B, down the pressure and magnetic field gradients, apparently unimpeded by plasma beam or firehose instabilities. Deeper in the cometosheath, the protons themselves deplete via charge exchange, but the strong cooling enables speeds to stay high until the new slow cometary ions fully dominate. Though subsonic, the overall flow is strongly compressible; with the strong lateral motions, the idea that rarefaction waves could propagate outwards and weaken the bow shock (Wallis and Dryer 1985) may well be invalid.

The discontinuity with magnetic field dropping essentially to zero was discovered by *Giotto* at some 5000 km (Neubauer 1987). The interior is composed of purely cometary plasma – but so is the exterior for all practical purposes, as solar wind protons and α–particles disappear

Cometary Plasma Processes
Geophysical Monograph 61
©1991 American Geophysical Union

via charge exchange at over ten times that distance (Fuselier et al. 1987). While the cavity boundary is a contact discontinuity, it is not at a position of plasma pressure balance (Biermann et al. 1967) but within the collisional atmosphere where the solar wind pressure is balanced by a contribution from gas pressure (Wallis and Ong 1976). The magnetic field, in enforcing a short effective path for the ionized component of order of the gyroradius, determines the boundary's structure. Maxwell stresses are balanced on the short scale by ion–neutral friction, as described by Cravens (1986) or Ip and Axford (1987).

This kind of magnetic structure was expected to be highly unstable (Wallis and Dryer, 1976; Ershkovich and Mendis, 1986). It appears, however, that Giotto did see a stable boundary, with detailed directional structure being similar on both inbound and outbound passes (Neubauer 1987). The tailward extension of the cavity has been identified with the long and comparatively dark tail cores observed in some comets. These consist mainly of cool ionospheric plasma, not the commonly recognized H_2O^+ and CO^+ but less visible H_3O^+, as traversed by the ICE–probe of comet Giacobini–Zinner (Rees et al. 1987).

Telescopic images potentially give us global information on the plasma distribution to link up with the 1–D snapshot from the comet–probes. But limited spatial and time resolution, together with heavy contamination from dust and molecular emissions, have proved significant obstacles. Account has to be taken too of the multiplicity of ion species, mentioned above. The IPD images (Rees et al. 1987) reported here are of higher quality than others (Ip et al. 1988) and our conclusions differ.

2. Cometosheath

For modelling the solar plasma flow into a comet, through the bow shock into the stagnating cometosheath, the comet–probes have proven that thermalized plasma of a single species (Biermann et al. 1967) constitutes a poor description. It is important to treat the picked–up cometary ions separately as a highly supathermal species of plasma (Wallis and Ong, 1976). Much of the energy of the keV solar wind is transferred to these ions, and then lost via charge exchange in the dilute atmosphere. Fuselier et al. (1988) have studied the α–particle and H_2^+ data from Giotto, taking the first as solar wind and the second as cometary indicators and finding different density behaviour and systematic velocity deviation. Ip (1989) compares the velocities with the thermalised flow model (unpublished calculations of Vedder) but including charge–exchange losses of suprathermal cometaries. The model fails to reproduce the relatively sharp cometopause, nor does it give strong enough lateral diversion of the solar wind α–particles (Fig. 1). Gombosi (1987) incorporates charge–exchange cooling into the hydrodynamics, but 1–D only, and confirms exponentially fast compression (Wallis and Ong 1976) at a finite radial distance. His computation thus substantiates the hypothesis that the cooling facilitates formation of envelope structures sunward of the nucleus.

These fluid models suffer, however, from inability to reproduce the strong deviation in mean velocity of the suprathermal ions. Deviations of individual test cometary

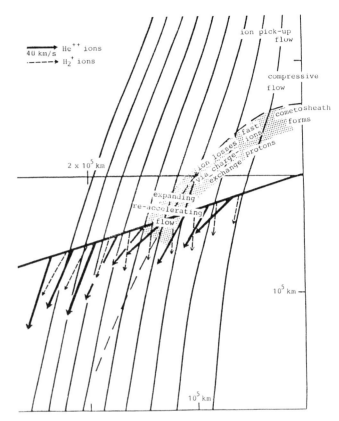

Fig. 1. Plasma velocities measured by IMS (Fuselier et al. 1988) along Giotto's trajectory superposed on the flow field calculated on the thermalized flow model (adapted from Ip 1989). The cometosheath or plasma envelope structure and flow regions (compressive and expansive) inferred in the present study are indicated.

ions have been explored as drifts in realistically draped, large–scale magnetic fields (Wallis and Johnstone, 1983). Such drifts explain why the cometary H_2^+ tend to move across the field and the α–particle flow, as evidenced in the data (Fuselier et al. 1987; Fig. 1). At present, we can make estimates only of the various scales in the dynamics of the complex multi–species plasma, checked against the Halley data. We use the linear dependence of flow speed on radial distance outside the cometosheath found approximately in gasdynamical models

$$u = Ur/Z(\theta), \quad U = 450 \text{ km/s}, \quad Z(0) = 9 \times 10^5 \text{ km} \qquad (1)$$

and confirmed by Gombosi (1987) on the upstream axis with the parameters given in (1). Here r denotes radial distance sunwards and θ the azimuthal angle. $Z(\theta)$ defined in this was does not change much with θ in the Munich model (Schmidt and Wegmann 1982), increasing to 1.1×10^5 km at $\theta = 107^\circ$, appropriate to Giotto's trajectory. The distribution f(v,u) for cometaries ionized at z_* and picked up with suprathermal speeds $v \simeq u(z_*)$, being convected at speed u and lost via charge exchange in the atmosphere $N \propto r^{-2}$, satisfies

$$u \, df/ds = -\sigma_{ex} N v f. \qquad (2)$$

The solution on the upstream axis $ds = -dz = -dr$ follows using (1) as

$$f \propto \exp - Z_1^2/z^2 \text{ with } Z_1 = (\tfrac{1}{2}R_{ex}Z(0)v/U)^{\frac{1}{2}} \tag{3}$$

if R_{ex} is the charge exchange scale defined by $\sigma_{ex} N(R_{ex}) R_{ex} = 1$. Numerically for $\sigma_{ex} = 2 \times 10^{-15} \text{ cm}^2$, $Z_1 \simeq 1 \times 10^5$ km but varies a little with energy as $\sqrt{<\sigma_{ex}(v')v'>}$, allowing for u and averaging over phase angle. By position in the axial flow $Z = Z_1/\sqrt{3} = 6 \times 10^4$ km, some 5% remain, in agreement with Gombosi's (1987) charge exchange "avalanche".

Where the flow is closer to tangential we can estimate depletion with $r = $ const., $ds = rd\theta$ in (2) to give

$$f \propto \exp - 2R_1^2 \Delta\theta/r^2 \text{ with } R_1 = Z_1 \{\overline{Z(\theta)}/Z(0)\}^{\frac{1}{2}} \tag{4}$$

which shows strong (95%) depletion for $\Delta\theta \simeq \pi/2$ at $r = R_1 \simeq Z_1$. From such estimates, it's clear that an envelope structure can condense at $0.6Z_1$ apex distance, but its plasma can only flow laterally if it is strongly flattened to pass outside Z_1 before reaching $\theta = \pi/2$. A pressure gradient is required to drive the flow against atmospheric friction (momentum loss via charge exchange). So with $Z_1 = 1 \times 10^5$ km such an envelope is compatible with the *Vega* cometopause at 1.6×10^5 km and the *Giotto* magnetic pile–up at 1.4×10^5 km. Though the cometo–pause apex is in the charge exchange region, the region where the probes traversed it is pressure–dominated. Several factors check qualitatively with this picture; the solar wind speed declines more slowly than (1) within 2×10^5 km; α–particle velocities tend to align along the B–fields; the depletion of the α–particles implies they must have passed some 3 times closer to the comet (Fuselier et al. 1987). Further confirmation comes from the B–fields being relatively weak. The decrease in B from an estimated 25 nT at the apex to the 8 nT outside the cometopause provides sufficient pressure to accelerate ions ($100/\text{cm}^3$) to typical values $u \simeq 50$ km/s.

3.   Sunward structure in telescopic images

Envelope structures as occasionally observed in cometary heads were difficult to detect in Halley because of high dust and molecular brightnesses. Ip et al. (1988) report results with CCD slit spectroscopy. A preliminary search has been made of University College London's series of narrow band images in CO^+, taken from the Table Mountain Facility in California through the encounter period (Rees et al. 1986, 1987). Because of observational constraints (low on horizon, near dawn) the usual use of separate exposures for sky background gives error. However, some images as that selected for presentation here have sunward margins essentially free of CO^+ emissions. Note that these narrow band images secured with an Imaging Photon Detector (Fabry–Pérot system) are of much higher quality in isolating the CO^+ and in avoiding problems of time–varying comet and sky conditions, because of brief 2 or 3 min. exposures. The 2–D image has advantage over slit spectrometry in the faint outer regions, as we enhance signal:noise by integrating counts over the sunward oriented $45°$ sector on

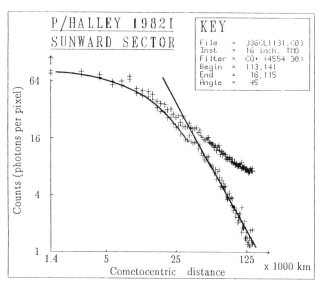

Fig. 2.   Profile of sunward intensity in CO^+ ions from an IPD image (Rees et al. 1987) of 21 March 1986, on logarithmic scales, averaged over a $45°$ sector. The upper data points are prior to sky subtraction, the lower ones after subtraction of intensity at the sunward edge of the field. The straight line shows R^{-2} dependence; the curved line smoothly connects the inner (corrected) data points. The shoulder–discontinuity between the two lines is centred on 35000 km.

the nucleus. Our range of eg. 1.3 to 130×10^3 km is far greater than the 10–50×10^3 km of Ip et al. (1988).

The intensity plotted in Fig. 2 from an image of 21 March, 1986 falls steeply as $\sim R^{-2}$ beyond 4×10^4 km, but is flat within 5000 km. This agrees semi–quantitatively with the model calculations of Ip (1980) for CO^+ density in comet Halley (results at 1AU). Though that model assumed half the C–containing molecules to be CO_2, rather than the complex organics more probably present, at the present state of knowledge that model remains a reasonable comparator. Line–of–sight profiles cannot be readily compared with the density calculations for the sunward axis, but it seems that the latter with $n(CO^+) \sim r^{-3.5}$ inside 1×10^5 km (so that $\int n dz \sim R^{-2.5}$ at offset distance R) are a little steeper than observed. The $R^{-2.5}$ or $r^{-3.5}$ steepness reflects the CO density $N \sim r^{-2}$ combined with sweeping up of the CO^+ in decelerating flow with $u \sim r$, as Equ. (1). The flatness at a few 1000 km reflects conversion to H_2O^+ and HCO^+ (2:1 weighting), which dominates in slowly–moving ionospheric plasma as evident in Ip's calculations.

So what of the shoulder in the CO^+ profile (Fig. 2) at 3–4×10^4 km? It suggests that swept–up ions do not penetrate inside this position; they are lost via charge exchange and lateral flow. Note that the loss time scale for ion–molecule reaction with H_2O (density $2 \times 10^4 \text{ cm}^{-3}$) is some 10^4s, many times the acceleration time for ions picked–up in $\underline{E}$ and $\underline{B}$ fields ($\Omega_i^{-1} = 10$–20s), so this process is effective only in the stagnating flow. With the depletion in CO^+ extending in as far as 2.5×10^4 km (Fig. 2) this value would be the inner limit of the stagnating flow plasma–sheath.

Ion profiles in H_2O^+ have been discussed previously. Our own from images of comet Giacobini–Zinner (Rees et al. 1987) have too short a length–scale for identification of the sheath, the gas source being over 10 times weaker than comet Halley. Ip et al. (1988) present a slit profile from Halley on 14 March (shortly after *Giotto's* encounter), that covers the middle of our distance range. It shows an oscillatory structure in $1-1.6 \times 10^4$ km (presumably an instrumental or data problem) along with a decrease in H_2O^+ intensity, in disagreement with ionosphere models. The profile stops short of 6×10^4 km, the scaled distance of the plasmasheath. The authors discuss other structure at $1.8-2.2 \times 10^4$ km. Our image contains possibly analogous structure at 0.8×10^4 km (see profile of Fig. 2) but of limited angular extent. It may be that such features are quasi–stationary structures as Ip et al. (1988) suggest. But they would be small–scale and not linked spatially to the ion structure encountered on the flank by *Giotto* at 1×10^4 km (rather than at $3-4 \times 10^4$ km expected on the flow pattern of Fig. 1).

4. Discussion

For comparing the telescopic observations with *Giotto's*, the scale radius has to be adjusted according to the H_2O source strength. While the 21 March observation lies between IUE observations of 18/19 and 23/24 March, the strength would appear low, being 0.7 of the *Giotto* value 5×10^{29}/s on 19 March and 0.3 times it on 23 March (Feldman et al. 1987), just prior to a 2.5 times upsurge (McFadden et al. 1987). The heliocentric distance is larger at 1.01 AU rather than 0.9 AU, but the variation in solar UV by the inverse square distance would be relatively unimportant. The charge exchange scale radius for incoming plasma flow scales proportional to source strength, so the 3×10^4 km identification of a plasmasheath boundary (Sect. 3) is compatible with the $Z = 6 \times 10^4$ km cometosheath scale of Sect. 2.

Just as the signature of the cometosheath is a *reduction* in CO^+ in the stagnating flow, a reduction in H_2O^+ is similarly expected due to conversion to H_3O^+ at a very similar rate (2×10^{-9} cm^3/s) to the CO^+ reactions. The electrons are too hot to allow much recombination (e.g. Wallis and Ong 1976). The data of Ip et al. extend only to 5×10^4 km, where there is a hint of the sheath's "shoulder". On the flanks, the cometosheath is an expanding flow and the prime signature is a change in the flow direction. The magnetic field increase on *Giotto's* inbound path is anomalous, probably reflecting an inter-planetary discontinuity (Ip 1988), and the cometopause occurs a little outside it. Where the comet–probes crossed it, the cometosheath is dominated by pressure gradients, driving the plasma out laterally. This differs from the cometosheath's origin in the stagnating flow region at some 6×10^4 ahead of the comet, where charge exchange dominates. The thermalised flow models that do not incorporate the strong exchange cooling, do not show this strong lateral deviation of the flow and give poor agreement with the observed ion densities (Fuselier et al. 1987, Ip 1989). The single–fluid descriptions also, of course, fail to represent the relative drifts of solar wind and cometary ion species, important inside 2×10^5 km.

Ip et al. (1988) suggested that a structure at 2×10^4 km sunwards of the comet seen in an H_2O^+ profile was linked to the ion peak crossed by *Giotto* at 1×10^4 km. But the velocity vectors and the flow pattern inferred here (Fig. 1) make this unlikely; 2×10^4 links to 5×10^4 km on *Giotto's* inbound path. While Ip's (1980) pre–Halley ion density computations appear qualitatively good, he simply "interpolated" across the cometosheath stagnation region, so did not represent this structure.

The flow field with an embedded directional change is unlikely to be stationary, even if the solar wind and gas production are steady. Indeed relative motion at 20 km/s gives a 3–hour time scale, corresponding to changes in envelopes observed at comet Morehouse (Eddington, 1910). The cometosheath–envelope has long been considered to constitute the source of tail structures (Wallis 1968), observed to be inhomogeneous and time–varying. Its formation depends primarily on cooling via charge exchange (Wallis and Ong, 1976) which permits compression and condensation. A plasma condensation instability is unlikely, if the electrons stay hot (10–20 eV) and contribute significant pressure, as *Giotto* found. But the drag instability based on interaction with the neutral atmosphere (Wallis 1968, Wallis and Dryer 1976) would operate. On the other hand, the drag instability is apparently suppressed at the inner ionospheric cavity boundary, where plasma recombination is stronger than plasma flow.

In conclusion, consistancy between the telescopic images of the global plasma structures and the comet–probes' trajectory data has been established. The "cometosheath" is identified with the plasma envelopes that have long been hypothesised to be the source of the major ion tail rays. The sunward part of the envelope is dominated by charge–exchange cooling, but the flanks have pressure–driven expansion.

References

Biermann, L., Brosowski, B. and Schmidt, H.U., Solar Phys., 1, 254–284, 1967.

Cravens, T.E., Exploration of Halley's comet, 20th ESLAB Symp. ESA SP–250, 1, 241–246, 1986.

Eddington, A.S., Mon. Not. Roy. Astr. Soc., 70, 442–470, 1910.

Ershkovich, A.I. and Mendis, D.A., Astrophys. J., 302, 849–852, 1986.

Feldman, P.D. and 15 authors, Astron. Astrophys., 187, 325–328, 1987.

Fuselier, S.A., Shelley, E.G., Balsiger, H., Geiss, J., Goldstein, B.E., Goldstein, R. and Ip, W.H., Geophys. Res. Lett., 15, 549–552, 1988.

Galeev, A.A., Astron. Astrophys., 187, 12–20, 1987.

Galeev, A.A. and 14 authors, Diversity and Similarity of Comets, ESA SP–278, 83–87, 1987.

Gombosi, T.I., Geophys. Res. Lett., 14, 1174–1177, 1987.

Ip, W–H., Astron. Astrophys., 92, 95–100, 1980.

Ip, W–H., Astrophys. J., 343, 946–952, 1989.

Ip, W–H. and Axford, W.I., Nature, 325, 418–419, 1987.

Ip, W–H., Spinrad, H. and McCarthy, P., Astron Astrophys., 206, 129–132, 1988.

McFadden, L.A., A'Hearn, M.F., Feldman, P.D., Roettger, E.E., Edsall, D.M. and Butterworth, P.S., Astron. Astrophys., 187, 333–338, 1987.

Neubauer, F.M., Astron. Astrophys., 187, 73–79, 1987.

Rees, D., Meredith, N.P. and Wallis, M.K., Adv. Space Sci., 5 (12), 225–261, 1986.

Rees, D., Meredith, N.P. and Wallis, M.K., Planet. Space Sci., 35, 299–311, 1987.

Schmidt, H.U. and Wegmann, R., Comets, ed. L. Wilkening, p.538–560, U. Ariz. Press, 1982.

Wallis, M.K., Planet. Space Sci., 16, 1221–1248, 1968.

Wallis, M.K. and Dryer, M., Astrophys. J., 205, 895, 1976.

Wallis, M.K. and Dryer, M., Nature, 318, 646–647, 1985.

Wallis, M.K. and Johnstone, A.D., Cometary Exploration, ed. T.I. Gombosi, CRIP Budapest, 1, 307–311, 1983.

Wallis, M.K. and Ong, R.S.B., The Study of Comets, NASA SP–393, 856–876, 1976.

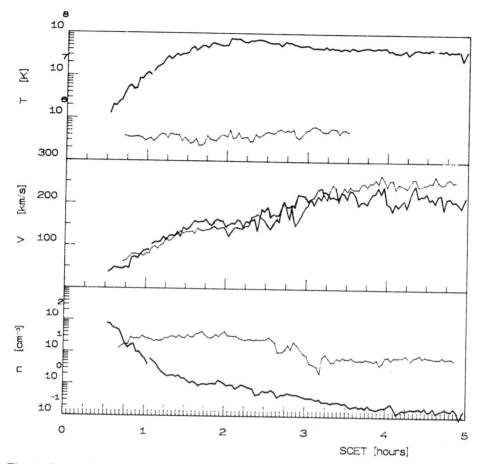

Fig. 4. Proton (thin line) and water group ion (thick line) plasma parameters from 0000 to 0500 SCET on 14 March 1986, during the outbound leg of the Giotto fly-by, in the same format of Figure 2.

Figures 4 and 5 show the proton and the water group ion parameters for the outbound leg of the encounter trajectory, from 0000 to 0500 SCET of day 74, 1986, in the same format used for Figures 2 and 3. As the data are very noisy soon after closest approach, the moments are plotted starting from 0035 SCET for the water group ions and starting from 0044 SCET for the protons. As for the inbound data the solar wind proton temperature has not been plotted. In the following the observations are described by inspecting the plots from the right to the left, i.e. from the solar wind through the bow shock to closest approach. The proton number density (thin line in the bottom panel of Figure 4) in front of the shock wave is 6.5 cm^{-3} (averaged between 0315 and 0320 SCET); downstream it is 15 cm^{-3} (averaged from 0240 to 0250 SCET). After the shock the proton number density does not vary monotonically with time: it displays a plateau around 25 cm^{-3} between 0210 and 0230 SCET, reaches a peak of 43 cm^{-3} at 0140 SCET, and oscillates around 30 cm^{-3} between 0048 and 0130 SCET. At 0048 SCET it

drops from 28 cm^{-3} to 14 cm^{-3}. The water group number density (thick line) in the solar wind is $n_{wg} \sim 0.2$ cm^{-3}, increasing smoothly across the shock to 0.55 cm^{-3} (averaged from 0240 to 0250 SCET); after the shock it increases steadily. However, the values observed in the inbound leg are not reached because the data are plotted starting only at 0040 SCET. The proton bulk speed (thin line, central panel of Figure 4) is 250 km/s at 0400 SCET. Moving toward closest approach the speed falls from 220 km/s to 150 km/s through the shock region, then decreases with some oscillations but with no further clear discontinuity as observed inbound. As in the inbound leg, in the solar wind the water group ion speed is lower than the proton speed; inside the shock $v_{wg} \sim v_p$. The water group ion temperature (thick line, top panel of Figure 4) increases from 5×10^7 to 7×10^7 K at the shock wave and reaches a maximum of 8×10^7 K at 0200 SCET. At that time it displays a sudden jump to 6×10^7 K followed by a continuous decrease, similarly to the inbound leg. The proton temperature does not show any signifi-

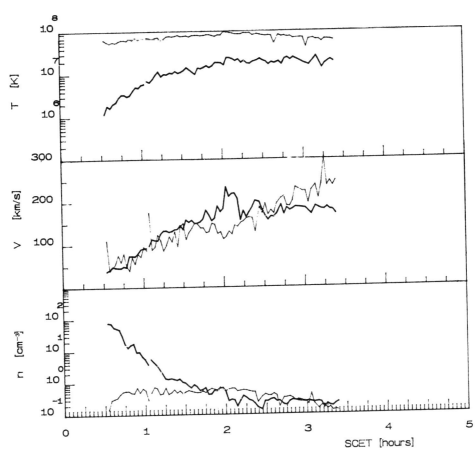

Fig. 5. Plasma parameters for the lower energy (thick line) and higher energy (thin line) water group ions from 0000 to 0500 SCET on 14 March 1986 with the same format of Figure 2.

cant feature, apart from a slow decrease from the shock region to 0140 SCET and a small increase between 0140 and 0120 SCET.

Figure 5 shows the moments computed for the higher energy (thin line) and the lower energy (thick line) water group ions in the outbound leg. The higher energy ion density increases from $n_h = 0.2$ cm^{-3}, at the shock to 0.7 cm^{-3} around 0200 SCET and is roughly constant at 0.6 cm^{-3} between 0045 and 0145 SCET; closer to the nucleus it falls off rapidly. The lower energy ion number density jumps from 0.3 to 0.7 cm^{-3} at 0200 SCET and increases stedily moving toward closest approach as in the inbound leg. The speeds of the two water group populations are rather similar except in the shock region, where the higher energy ion speed is larger, and around 0200 SCET, where the lower energy ion speed is larger. The lower energy ion temperature is roughly constant through the shock and behind it; at 0202 SCET it shows a small jump, followed by a continous decrease moving toward closest approach. The higher energy ion temperature is

roughly constant, going through a maximum of 10^8 K at 0200 SCET.

In order to better compare the observations made by the JPA sensors in the inbound and outbound legs of the trajectory, the density in nucleons cm^{-3} has been plotted in Figure 6 (in a log-log scale) as a function of the distance from closest approach. The three curves refer to the protons (thin line), the higher energy water group ions (dotted line) and the lower energy water group ions (thick line). The density of the higher energy water group ions and that of the protons have a similar behaviour in the inbound and in the outbound legs of the trajectory. The two horizontal bars mark the regions where the total water group ion density (not plotted in Figure 6) equals the proton density (i.e. where the centre of mass switches from the protons to the water group ions). In the inbound leg this region extends from 5.2×10^5 km to 4.8×10^5 km, where the proton speed and density and the water group ion temperature show a sudden decrease and the proton temperature increases (cfr. the discussion

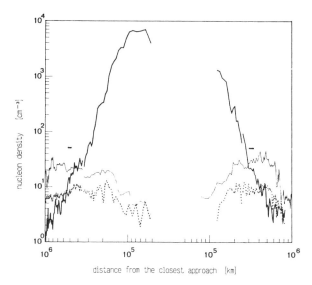

Fig. 6. Density (in nucleons/cm³) of protons (thin line), higher energy water group ions (dotted line) and lower energy water group ions (thick line) plotted on a log-log scale against distance from closest approach. Two horizontal bars mark the regions where the total (not plotted) water group density equals the proton density. The distance log scale is marked until 5×10^4 km both inbound and outbound; the region closer to the nucleus is cut off from the plot.

of Figure 2). This region corresponds to the inner edge of the "mystery transition" described by *Rème* [1990], and occurs just after the "mystery region" identified by *Rème et al.* [1987, 1989], which is characterised by the presence of significant fluxes of keV electrons. Further inwards, between 4.7×10^5 km and 3.9×10^5 km, the centre of mass switches to the lower energy water group ions. In the outbound leg the region marked with the horizontal bar extends from 3.3×10^5 to 3×10^5 km, corresponding to the time interval from 0131 to 0115 SCET, when the water group ion density and the proton temperature increase (as the distance from the nucleus decreases) and the ion water group ion temperature, the proton density and both speeds decrease (see Figure 4). This signature is consistent with that observed for the corresponding inbound region (see Figure 2), but its features are not as abrupt and well defined. Outbound the centre of mass switch to the lower energy ions occurs more abruptly, around 2.9×10^5 km. The inbound and outbound regions of centre of mass switching are both observed roughly halfway between the shock and the closest approach.

Summary and discussion

Revised plasma parameters have been shown for the inbound leg of the Giotto trajectory. Moreover, the out-

bound leg data have been presented for the first time. The following conclusions can be drawn from the data.

a) The outbound plasma parameters show an overall agreement with the inbound parameters. However, the outbound structures are closer to the nucleus than the inbound ones: the bow shock is observed at 7.4×10^5 km, compared to 1.16×10^6 km, and the centre of mass switch is observed between 3.3×10^5 km and 3×10^5 km, compared to 5.2×10^5 km and 4.8×10^5 km. This is an effect of the Giotto trajectory: in the inbound leg the observations are made on a flank of the comet, while in the outbound leg they are made more towards the sun, where the interaction region is expected to be more compressed.

b) In the inbound leg of the Giotto trajectory the region where the centre of mass switches from the protons to the water group ions is observed roughly halfway between the bow shock and closest approach and corresponds to the inner edge of the "mystery transition", which shows up quite clearly in the inbound plasma parameters (increase of the proton temperature and water group ion density and decrease of the speed, proton density and water group ion temperature). The corresponding outbound region is also observed roughly halfway between the bow shock and closest approach and is accompanied by changes of the plasma parameters consistent with the inbound observations, although not as clear. Tentatively we interpret the "mystery transition" as a permanent structure of the solar wind interaction with comet Halley at 1 AU [crf. *Rème*, 1990], characterised by the centre of mass switching from the solar wind protons to the cometary water group ions.

c) Outbound the proton number density is generally higher than inbound, reaching a maximum of 46 cm⁻³ compared to 35 cm⁻³. This may also be an effect of the compression.

d) In the outbound leg the bow shock is less clear than inbound and extends over a much larger region, 9×10^4 km, corresponding to 14 water group gyroradii. This has brought *Neubauer et al.* [1989] to suggest that the outbound shock is a new type of structure, which he called "draping shock".

e) The analysis of the water group ion distribution function leads, both inbound and outbound, to the separation of two distinct populations, which roughly form two rings in velocity space about the solar wind bulk velocity; when the moments of the distribution function are calculated, the radius of the ring yields the thermal speed. The behaviour of the two populations is similar inbound and outbound, with some differences. Inbound, the radius of the velocity space ring ranges from 260 km/s close to the "mistery transition" to 200 km/s at 2330 SCET for the higher energy water group ions, i.e. it is of the order of the bulk speed upstream from the shock or right behind it; on the other hand, the same quantity for the lower energy population ranges from 90 km/s at the "mistery transition" to 45 km/s at 2330 SCET, both values being close to the local bulk speed. Similar values are observed also outbound, confirming the interpretation of the water group ion temperature in terms of the radius of the velocity space ring.

f) The study of the two water group ion populations leads to a further consideration. The lower energy water group ion density shows a clear increase through the inbound "mystery transition" from 0.3 cm^{-3} to 1.0 cm^{-3} (see the discussion of Figure 3). Outbound the lower energy water group ions display a moderate density increase through the region of centre of mass switching, followed by a steeper increase at its inner edge around 2.9×10^5 km (see Figure 6). If both the described density increases are interpreted as due to a local enhancement of the water group ion production rate, this would be a further characteristics of the "mystery transition" which waits for an explanation and deserves further investigation.

Acknowledgements. This research was supported by the Italian Space Agency (ASI).

References

Amata, E., V. Formisano, P. Torrente, M.B. Bavassano Cattaneo, A.D. Johnstone and B. Wilken, Experimental plasma parameters at comet Halley, *Adv. Space Res.*, *9*, 313, 1989.

Balsiger, H., K. Altwegg, F. Bühler, S. A. Fuselier, J. Geiss, B. E. Goldstein, R. Goldstein, W. T. Huntress, W.-H. Ip, A. J. Lazarus, A. Meier, M. Neugebauer, U. Rettenmund, H. Rosenbauer, R. Schwenn, E. G. Shelley, E. Ungstrup, D. T. Young, The composition and dynamics of cometary ions in the outer coma of comet P/Halley, *Astron. Astrophys.*, *187*, 163, 1987.

Coates, A. J., A. D. Johnstone, M. F. Thomsen, V. Formisano, E. Amata, B. Wilken, K. Jockers, J. D. Winningham, H. Borg, D. A. Bryant, Solar wind flow through the comet P/Halley bow shock, *Astron. Astrophys*, *187*, 55, 1987.

Formisano, V., E. Amata, M.B. Bavassano Cattaneo, P. Torrente, A. Coates, A. Johnstone, B. Wilken, K. Jockers, M. Thomsen, D. Winningham,, H. Borg, Plasma flow inside comet P/Halley, *Astron. Astrophys.*, 1990 (in press).

Galeev, A. A., T. E. Cravens, T. I. Gombosi, Solar wind stagnation near comets, *Astropys. J.*, *289*, 807, 1985.

Johnstone, A. D., J. A. Bowles, A. J. Coates, A. J. Coker, S. J. Kellock, J. Raymont, B. Wilken, W. Stüdemann, W. Weiss, R. Cerulli Irelli, V. Formisano, E. de Giorgi, P. Perani, M. de Bernardi, H. Borg, S. Olsen, J. D. Winningham, D. A. Bryant, The Giotto three-dimensional positive ion analyser, *ESA SP-1077*, 1986.

Neubauer, F. M., K. H. Glassmeier, M. Pohl, J. Raeder, M. H. Acuna, L. F. Burlaga, N. F. Ness, G. Musmann, F. Mariani, M. K. Wallis, E. Ungstrup, H. U. Schmidt, First results from the Giotto magnetometer experiment at Comet Halley, *Nature 321*, 352, 1986.

Neubauer, F. M., K. H. Glassmeier, M. H. Acuna, F. Mariani, G. Musmann, N. F. Ness, A. J. Coates, Giotto magnetic field observations at the outbound quasi-parallel shock of comet Halley, *Ann. Geophys. 8*, 463, 1990.

Rème, H., Cometary plasma observations between the shock and the contact surface, AGU Chapman Conference, Guilford, UK, 1989.

Rème, H., J. A. Sauvaud, C. d'Uston, A. Cros, K. A. Anderson, C. W. Carlson, D. W. Curtis, R. P. Lin, A. Korth, A. K. Richter, D. A. Mendis, Gemeral features of the Comet Halley - solar wind interaction from plasma measuRèments, *Astron. Astrophys.*, *187*, 33, 1987.

Rème, H., C. d'Uston, C. Mazelle, J. A. Sauvaud, K. A. Anderson, C. W. Carlson, R. P. Lin, A. Korth, D. A. Mendis, A. J. Coates, K. H. Glassmeier, S. A. Fuselier, Properties of the "mystery" region and the "mystery" transition at Comet Halley, AGU Chapman Conference, Guilford, UK, 1989.

Thomsen, M. F., W. C. Feldman, B. Wilken, K. Jockers, W. Stüdemann, A. D. Johnstone, A. Coates, V. Formisano, E. Amata, J. D. Winningham, H. Borg, D. Bryant and M. K. Wallis, In-situ observations of a bi-modal ion distribution in the outer coma of comet P/Halley, *Astron. Astrophys.*, *187*, 141, 1987.

IONS IN THE COMA AND IN THE TAIL OF COMETS - OBSERVATIONS AND THEORY.

K. Jockers Max-Planck-Institute for Aeronomy, D-3411 Katlenburg-Lindau, FRG

Abstract   This paper summarizes our present knowledge about the behaviour of cometary plasma from the viewpoint of the ground-based observer. Starting with an overview about the dimensions and ion column densities of comet tails the tail velocities and accelerations are studied. The in-situ measurements of the cometary magnetic field are briefly mentioned as far as they concern the optical observations. The problem of ion origin in the cometary coma and the production of ions and its relation to the production of neutrals is discussed. The last section is concerned with comet tail activity, i.e. tail kinks and the associated Rayleigh-Taylor instability, tail rays and the so-called tail disconnection events.

Introduction

The in-situ measurements of cometary plasma in the comets Giacobini-Zinner and Halley have led to a renewed interest in the ground-based study of cometary ions. In this paper an attempt is made to review the status of this field. The discussion concentrates mostly on the observations of cometary plasma and refers to the theoretical models only when interpreting observations.

In the cometary tail the ions are spatially separated from the neutral gas and, in many cases, also from the dust tail. Therefore cometary ions are most easily observed in the cometary ion tails and it is here where their ionic nature was discovered in the beginning of the century. Most of this paper will deal with the ion tail. On the other hand, cometary ions originate from the neutral coma gas, so the effort to understand cometary plasma processes must concentrate in the cometary coma. Observations of the cometary plasma in the coma, where its emissions must be separated from the strong emissions of the neutral radicals and from the dust continuum, have only become possible very recently with the advent of modern panoramic detectors. Therefore, despite the importance of such observations, our knowledge is still very limited. Overviews on the topic of this paper have been given e.g. by Ip [1987] and by Jockers [1985] in his study of the ion tail of comet Kohoutek 1973 XII.

Cometary Plasma Processes
Geophysical Monograph 61
©1991 American Geophysical Union

General properties of cometary ion tails

Chemical composition of ion tails

The cometary plasma tail is most easily photographed on unsensitized blue-sensitive plates. In this spectral range most of the emission comes from the CO^+ ion. The so-called comet-tail band of CO^+ was already identified in the beginning of the century. The water ion H_2O^+ was identified only recently [Wehinger et al. 1974]. Because its emissions are located in the red spectral region, where plates are not as sensitive as in the blue spectral range, H_2O^+ is not as easily photographed as CO^+ but red plasma tails have been photographed long before the ion was identified [Miller 1980]. Other important ions are OH^+ and CO_2^+. The chemical composition of the ion tail was reviewed at the conference by Wyckoff (see Wyckoff and Tegler [1989]).

Dimensions of ion tails

As comets move in the inner solar system, which, because of the radiative and corpuscular output of the sun, is an HII region, all neutral molecules released by a comet must end up as ions. Therefore, as these ions will be swept away by the solar wind, one may say that every comet with a non-zero gas production rate must have an ion tail. Only the stronger tails can be observed, however. First, the column density of the tail must exceed the limit of detectability, and second, the tail must stand out against the background of the night sky. Since the plasma tail has a molecular band spectrum not too much can be gained from observing in narrow wavelength bands. When comet Kohoutek 1973 XII was within the earth's orbit its CO^+ tail could be easily observed over a length of 10^7 km and had an average width of 10^5 km. These dimensions can be considered typical for a sizable comet. CO^+ is a very stable molecular ion and its lifetime against destruction by solar radiation or charge exchange with the solar wind is likely to exceed the few days of travel time along the tail out to a distance of 10^7 km. The H_2O^+ ion, however, may be less stable and it is therefore of interest if the H_2O^+ tail is significantly shorter than the CO^+ tail. No published data are available but there exist plates of

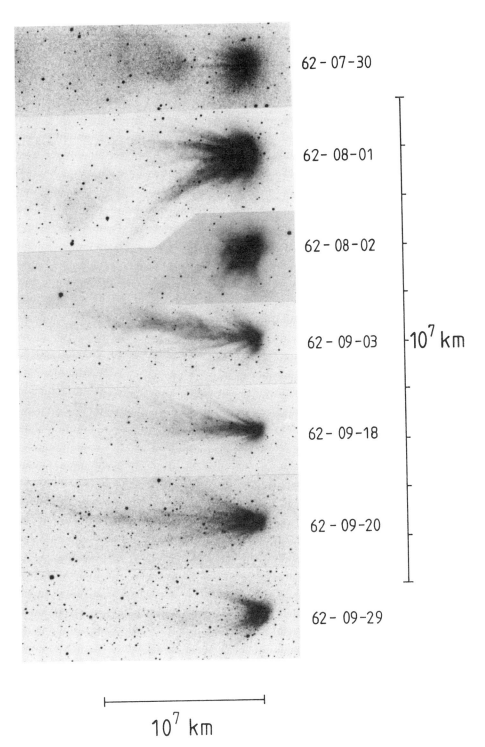

Fig. 1. Comet Humason's ion tail. Boyden Observatory Schmidt telescope. Courtesy E.H. Geyer. The numbers denote the observation date. The sun-comet direction is horizontal. Two scales are shown. The vertical scale gives the distances at the comet perpendicular to the line of sight. The horizontal scale measures the distance along the cometary tail.

comets with long H_2O^+ tails, very similar to the CO^+ tails (see e.g. Miller [1980]). There is an extraordinary type of carbon-rich comets which produce spectacular CO^+ tails even at large solar distances of 2 AU and more. These comets not only have extended CO^+ tails but also the coma and dust emissions are unusually faint. Fig. 1 shows a sequence of images of comet Humason 1962 VIII taken by E. Geyer and coworkers at Boyden observatory to give an example of the tail of such an unusual comet and its activity.

Tail column densities. Bright comets are photographed widely, because they are considered as targets of opportunity and look impressive on large-scale plates. Within about a hundred years of comet tail photography, however, very little photographic surface photometry of comets has been done. Only recently measurements of column densities of individual ions have become available. Fig. 2 shows an image of comet P/Halley 1986

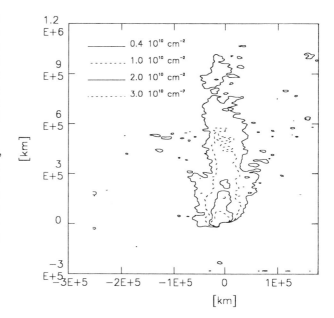

Fig. 3. H_2O^+ column densities in comet Liller 1988V [Rauer and Jockers 1990]. The values of the isocontours are given at the left side of the figure. The anti-solar direction is vertical. The horizontal scale has been stretched to better show the column density contours in the tail. Distances along the tail as in previous figure.

of Hoher List Observatory, again with a focal reducer and in the same wavelength range, is presented in Fig. 3 [Rauer and Jockers 1990]. The image was taken after perihelion on May 13/14 1988 when the comet was at a perihelion distance of 1.16 AU. The column densities of this comparatively small comet are about two orders of magnitude less than in comet Halley. Despite this the tail widths of the two comets are comparable.

Two spectra of comet Humason 1962 VIII were taken at the prime focus of the 200-inch telescope of Mount Palomar Observatory by Greenstein [1962] on August 1 and 2 1962 (see Fig. 1) when the comet was at a heliocentric distance of 2.6 AU. Arpigny [1965] has derived from these spectra a value of the CO^+ column density of 10^{13} ions cm^{-2} at a distance of 10^4km from the cometary nucleus. This number, when compared with the water ion densities quoted above, demonstrates the exceptional nature of this unusual comet.

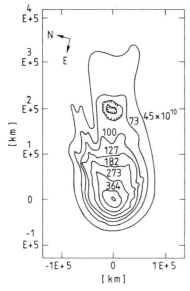

Fig. 2. H_2O^+ column densities in comet P/Halley [after DiSanti et al. 1990]. The labels of the isocontours denote column densities in units of 10^{10} ions cm^{-2}. The anti-solar direction is vertical. As in Fig. 1, distances perpendicular to the tail axis refer to distances at the comet perpendicular to the line of sight and distances along the tail are corrected for the inclination of the tail with respect to the sky plane. The negative values at about 200000 km distance in the tail are an artefact, presumably caused by light backscattered from the CCD and reflected at the interference filter.

III obtained with a focal reducer at the Catalina 154 cm telescope of the University of Arizona Observatories six hours after the Vega 1 encounter on March 06 1986 in the light of the 0–8–0 band of H_2O^+ [DiSanti et al. 1990]. A continuum image has been subtracted so the emission should be entirely caused by the water ion but, as indicated by the authors, some residual C_2 coma may be present. The contours indicate column densities ranging between 40 to 400×10^{10} ions cm^{-2}. A similar image of comet Liller 1988 V, taken at the 1m telescope

Tail velocities and accelerations

Cometary plasma tails are nonuniform. The temporal behaviour of tail structures can be followed with cinematography. To obtain interesting results a long time span is required. Apart from the rare cases of circumpolar comets, or comets which are sufficiently bright in opposition, one must combine observations of different observatories distributed in longitude around the world. Photographs of comet Kohoutek 1973 XII have been collected and analysed by Jockers [1985]. A similar

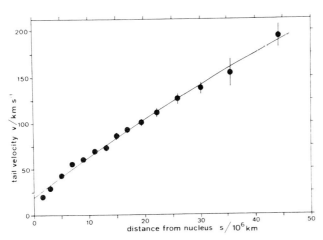

Fig. 4. Mean velocity of condensations (averaged over all observations) and their standard deviations plotted as a function of distance along the plasma tail [Celnik and Schmidt-Kaler 1987].

collection of comet Halley images will be published in the IHW archive.

By tracing the motion of tail structures in different cometary images one can derive tail speeds and accelerations. Fig. 4 shows mean velocities of tail structures observed with this technique in comet Halley by Celnik and Schmidt-Kaler [1987]. The average speeds measured in the tail of comet Halley range from 20 to 200 km s^{-1}. Differentiation of this curve gives tail accelerations ranging between 15 and 75 cm s^{-2}. These numbers give a good idea of the velocities and accelerations to be expected in a comet. It must be understood, however, that there are systematic trends not expressed in the averaged data. An increased solar wind momentum flux will tend to increase the overall acceleration in the cometary tail so the variance of the acceleration will be higher than expected from differentiating the mean velocities. Ahnert [1943] and Hoffmeister [1944] have traced clouds in comet Whipple-Fedke-Tevzadze 1942 I and find for individual clouds accelerations from 30 to 230 cm s^{-1}. The velocities go up to values exceeding 400 km s^{-1}. Comet 1943 I seems to have been carbon-rich like comet Humason 1962 VIII.

The method of tracing plasma tail clouds is about as old as plasma tail photography. The fundamental assumption with this method is that the clouds represent material motion and do not resemble waves. Even if this is true the clouds are denser than the average plasma tail and it is therefore not unlikely that they move more slowly than the rest of the tail. This would introduce a bias towards low tail velocities. The simple fact that fast moving clouds in plasma tails tend to become smeared out during the exposure time of the plate and, hence, will not be traceable in a sequence of exposures, also favours measurement of small tail speeds. This second effect can be checked in principle if during the observations the telescope, instead of tracking the nucleus motion, is set to track tail clouds of an average speed.

A better method, which has become available only very recently, is direct measurement of the Doppler shift in individual rotational lines. In this case the line-of-sight velocity component is measured but, with the exception of cometary phase angles near 90°, one can derive the tail speed rather accurately from the assumption that the velocity vector points in the antisolar direction. Scherb et al [1990] have observed comet Halley in the spin doublet of H_2O^+ at 6158.64 and 6158.85 Ångstrom with a scanning Fabry-Perot having a field of view of 3–5 arcmin corresponding to 1–2 × 10^5 km. Spectral scans were made at several places in the tail with distances up to 3x10^6 km from the nucleus. The velocities range from about 10 to 80 km s^{-1} and usually increase with increasing distance from the nucleus. Accelerations obtained from comparison of measurements of one night along the tail vary between 37 and 300 cm s^{-2}. These are somewhat higher values than indicated from the results of Celnik and Schmidt-Kaler [1987]. High accelerations were found at times of high tail activity, i.e. during so-called tail disconnection events (see below section 3.3). Also these measurements are not completely free from a bias towards low flow speeds as dense clouds will contribute more than average to the brightness signal.

Tail magnetic fields

Magnetic fields up to about 60 nT have been measured during the spacecraft encounters with comets [Neubauer et al. 1986]. Unfortunately such values are too small to produce observable effects in the cometary radiation. Therefore we must rely on the spacecraft measurements and on the theory of solar wind interaction with comets. In the interplanetary and cometary plasma the plasma pressure normally dominates. Nevertheless, as already shown by Lees [1964] for the case of the earth magnetosphere, at the "nose" of the comet there must be a magnetic pile-up region where in a small area the magnetic field dominates and deflects the flow with its embedded magnetic field sideways around the comet. Alfven's [1957] picture of the magnetic field draped around the comet has become very familiar. A more elaborate model must, however, consider the three-dimensional characteristics of the flow. As schematically shown in Fig. 5 we must distinguish the magnetic meridional plane of the magnetic pile-up and the magnetic equatorial plane where the ion pressure dominates and most of the ion flow occurs. The models show that the magnetic pressure is enhanced in the magnetic meridional plane as compared to the equatorial plane. Interestingly, the sum of plasma and magnetic pressure is nearly axisymmetric with respect to the comet-sun axis [Schmidt and Wegmann 1982]. Correlated with this behaviour is a flattening of the ion tail toward the magnetic equatorial plane [Fedder et al. 1984, Schmidt-Voigt 1989]. The situation is further complicated by the nonstationarity of the solar wind and its embedded magnetic field. Discontinuities of tangential or rotational type [Lepping and Behannon 1986] will rotate the magnetic field component perpendicular to the sun-comet line and, consequently, the magnetic meridional and equatorial

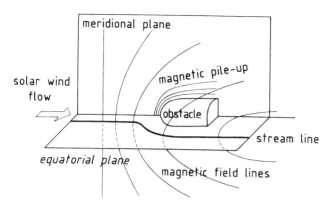

Fig. 5. Non-axisymmetric magnetoplasma flow around a conducting obstacle (comet).

planes. As the solar wind flow speed is reduced in the vicinity of the comet, magnetic layers with different magnetic field orientation will be accumulated in the coma by the draping process.

The predicted draping of the magnetic field around comets has been confirmed by the space probe encounters. Magnetic field data from the Vega-2 encounter show a single reversal of the magnetic field component parallel to the comet-sun line around closest approach, consistent with magnetic field draping around the nucleus [Schwingenschuh et al. 1986]. At the Vega-1 encounter two additional reversals of the draped magnetic field were observed at places nearly symmetric to the comet-sun axis. This indicates that a reversal of the interplanetary field, which passed the comet some time before the encounter, was preserved in the comet in the course of the draping process. The Giotto probe found a very complicated pattern of magnetic shells draped around the comet. The ICE probe intersected the tail of comet Giacobini-Zinner about 8000 km on the tailward side. The measurements indicated a tail current sheet permeated by magnetic field [McComas et al. 1987].

The Giotto space probe confirmed the existence of an inner region in comet Halley free of magnetic field, the so-called magnetic cavity or cometary ionosphere. Its diameter was about 8000 km. Its extent into the tail and its possible role in the tail formation process was reviewed by Neubauer on this conference (see Neubauer [1988]).

Origin of tail plasma

The tail plasma originates from ionization of the neutral coma gases. The main ionization processes are photoionization, charge exchange with solar wind protons and, in the inner coma, collisional ionization and perhaps anomalous ionization. Because of the large cross-sections of ion-neutral interactions the ions play an important role in cometary chemistry in the inner coma. On the other hand, when neutral molecules or radicals have escaped the inner coma region they can move away from the comet up to distances of several million km, because

the charge exchange and photoionisation rates are low. The transport of the ions from their source into the tail is determined by the solar wind-comet interaction and, consequently, by the nonstationary nature of the solar wind. Because of the complexity of the interplay of all these processes and its importance for the understanding of cometary production rates, chemistry and tail activity it is of interest to obtain observations of the source region of the cometary plasma. Plasma structures in the inner coma are, however, very difficult to observe from the ground. Not only are the ion emissions in the coma mixed with radiation from neutral molecules and dust, but the emission of the neutral molecules and dust increases much more strongly towards the nucleus than that of the ions. One reason for this is that the ions are lost in the inner coma by dissociative recombination [Ip 1987]. Images of unusual comets without visible coma (see Fig. 1) and narrow-band images of comet Halley in the light of the CO_2^+ ion in April 1986 [Jockers et al. 1987] show a tapered inner ion tail. This appearance is enhanced by perspective if the comet as seen from the earth is close to opposition. This is the case with the images of Fig. 1. The cometary plasma on the solar side of the comet seems to be particularly nonstationary. Plasma envelopes, as first described by Eddington [1910] in comet Morehouse 1908 III, which also was a carbon-rich comet, are sometimes observed. Examples can be seen in Fig. 1, in the frames of Sept. 3 and 18. In these cases the envelope appears asymmetric on one side of the comet. The narrow-band images of comet Halley, mentioned above, also show indications of such envelopes.

McCarthy et al. [1986] have analyzed spectra of comet Halley with the slit centered at the nucleus and aligned along the tail in order to deduce the H_2O^+ content along the tail axis. They find this content and its distribution highly variable. Often the maximum plasma emission is displaced from the centre of the gas and dust coma, mostly in tail direction. Narrow-band images of comet P/Brorsen-Metcalf 1989o obtained by the author and his coworkers with a Fabry-Perot of 4 Ångstrom bandwidth at the Bulgarian National Observatory seem to confirm this result. More observations are needed to clarify the behaviour of cometary ions in the inner coma.

Ion production rates

Many cometary ions which can be observed from the ground are ions of presumed mother molecules. It is therefore of prime importance to derive ion production rates or at least ratios of the production of pairs of ions. The simplest way to derive ion production rates is to integrate the column density across the tail and to multiply the integral with an average tail speed. The measurements of Scherb el al [1990] allow for the first time to derive such production rates without assumptions. I will focus in the following on their measurements of January 1986. At that time their measuring diaphragm had, when projected on comet Halley, a diameter of about 2×10^5 km. This seems large enough to cover the full width of the tail. Fig. 6 shows

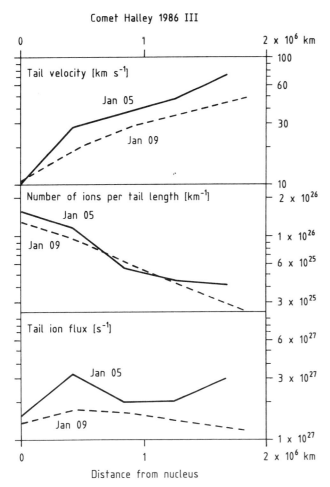

Comet Halley 1986 III

Fig. 6. Tail velocities (upper panel), H_2O^+ ion densities (centre panel), and tail ion flux (bottom panel) in comet Halley, derived from the measurements of Scherb et al. [1990]. The solid lines refer to observations of 05 Jan. 1986 and the dotted lines to observations of 09 Jan. 1986. Note that the ion flux varies from day to day and also within hours, as indicated by the flux variation along the tail.

their measurements of Jan. 5 (full lines) and Jan. 9 (dashed lines). The upper panel displays the measured velocities, rising from 10 km s^{-1} with the diaphragm centered at the nucleus to 73 km s^{-1} at 1.7 million km. The centre panel gives the integrated number density of tail ions per km tail length. It was obtained by dividing Scherb et al.'s numbers of ions within the diaphragm by the diaphragm diameter, projected to the cometary tail, i.e. taking into account that the tail is not perpendicular to the line of sight. The densities per tail length decrease by about a factor of 8 within the measurement range. The lower panel shows the ion fluxes, i.e. the product of the two other panels. As pointed out by Scherb et al. [1990] the flux is not constant, neither on the two nights of observation nor within one night along the tail. The peak in the ion flux at 420000 km from the nucleus (Fig. 6) exceeds the value mea-

sured at the nucleus by more than a factor of two which seems to be out of the uncertainties of the measurements. The rise of the production rate as compared with the nucleus value could perhaps be caused by mass addition into the tail from particles which were ionized on the antisolar side of the nucleus. The decay of the production rate at larger distances is unlikely to be due to destruction of the H_2O^+ ions but seems to indicate temporal changes in the cometary ion production either caused by a real variation of the cometary gas production or by a non-stationary solar wind effect.

From the CCD image shown in Fig. 2 DiSanti et al. [1990] have derived the number density of H_2O^+ ions per tail length in much the same way as shown in the centre panel of Fig. 6. In the absence of a velocity measurement they have assumed that the total ion flux is constant. The above discussion indicates that this assumption is acceptable if nothing better is known, but will not always be true.

It is interesting to compare the H_2O^+ flux in comet Halley, as determined by Scherb et al. [1990] with the flux of a smaller comet. Fig. 7, arranged in a similar way as Fig. 6, gives values for comet Liller 1988 V [Rauer and Jockers 1990, Rauer, *private communication*]. Here the ion content per tail length (centre panel) was determined from integration of the CCD filter images across the tail. The velocities were obtained the classical way, i.e. by tracing tail features in a sequence of frames and making a linear fit (there is no measurable acceleration within the few hours of observations) to the positions. While the velocities are slightly higher in comet Liller than in comet Halley, the H_2O^+ fluxes are almost an order of magnitude less.

The H_2O^+ fluxes found in the two comets are compared with the production of neutral water in Table 1. In case of comet Halley the production rate of neutral water was derived from Feldman et al [1987]. Comet Liller was observed with the

TABLE 1. Comparison of comet Halley 1986 III (Jan. 5, 1986) and comet Liller 1988 V (May 13/14 1988).

	Comet Halley	Comet Liller
Heliocentric distance at observation:	0.95 AU	1.16 AU
Mean flux of H_2O^+ ions in the tail [s-1]:	2.4×10^{27}	3.9×10^{26}
Production rate of neutral H_2O [s-1]:	5×10^{29}	6×10^{28}
Ratio H_2O/H_2O^+:	210	150

Nançay radiointerferometer in the OH 18 cm lines from March 17 to 24, 1988, when the comet was at a heliocentric distance of 0.86 AU [Bockelée-Morvan, Crovisier, Gérard, *private communication*]. The water production rate obtained by these authors was scaled to the heliocentric distance of the comet on May 13/14 by using the inverse sqare law. Considering the uncertainty in the water production rate of comet Liller there seems not to be a significant difference in the H_2O/H_2O^+ ratios of the two comets. Only somewhat less than one percent of the water released by the nucleus ends up in the ion tail as H_2O^+. Probably the main reason for this is the chemistry of water in the coma. Significant amounts of water may enter the ion tail, e.g., as OH^+ or O^+. Another possibility is that a significant amount of cometary ion flux is transported at larger lateral distances from the tail axis with high speed and a density so low that it cannot be detected by ground-based observations. McComas et al [1987] have derived from the in-situ measurements of the Giacobini-Zinner encounter that the tail current sheet, which they identify with the ion tail of Giacobini-Zinner as observed from the ground, contained only about one percent of the total ion population, not only of H_2O^+. The question of how many of the produced ions end up in the ion tail we observe from the ground, and how the fraction of ions collected in the ion tail as observed from the ground depends on the total production rate

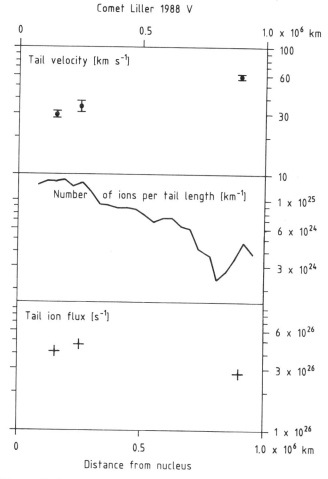

Fig. 7. Tail velocities (upper panel), H_2O^+ ion densities (centre panel), and tail ion flux (bottom panel) in comet Liller 1988 V [H. Rauer, *private communication*].

of neutrals is of considerable importance not only for the interpretation of ion abundances and fluxes in the ion tail but also for the understanding of the comet-solar wind interaction. We may speculate that large comets, of which the cometopause fills a rather large part of the neutral source cloud, i.e. the so-called

nonlinear comets [Schmidt-Voigt 1989], will collect a significant part of their ions in the observable ion tail whereas the small, linear comets like Giacobini-Zinner indeed loose most of their ions to the solar wind outside of the observable ion tail.

Ion tail kinetics

Tail kinks and associated Rayleigh-Taylor instability

Cometary ion tails respond to changing solar wind conditions. From a study of the ion tail of comet Halley as observed in 1910 Biermann [1952] discovered the close correlation between comet tail activity and what is now called solar wind high speed streams. This way he was led to postulate the existence of the solar wind. Not much progress in understanding of the effect of nonstationary solar wind phenomena on comet tail activity has been achieved since then. One of the few phenomena where, in the opinion of the writer, we may be close to an understanding concerns the physics of tail kinks [Jockers 1985].

How will a comet tail respond to a solar wind direction change? In the solar wind, because of the solar rotation, solar wind direction changes are always associated with changes of the solar wind speed and show a typical pattern. This is best documented in the case of the solar wind high speed stream interface [Gosling et al. 1978]. A simple tangential discontinuity across which the velocity component tangential to the discontinuity has a jump, is unlikely to occur in interplanetary space. Nevertheless let us assume for simplicity that such a discontinuity moves along a cometary tail. Then, as shown in Fig. 8, the comet tail will be separated into three parts. Part A is the outer tail which has not yet been affected by the discontinuity. It points into the old solar wind direction. Part C consists of plasma released after passage of the discontinuity and points into the new solar wind direction. Part B is the part of the plasma tail which was affected by the discontinuity. Note that along part B the solar wind velocity vector cannot be parallel to the tail axis. If in part B the tail is nonuniform and contains local condensations these condensations will be accelerated less by the solar wind dynamic pressure than the surrounding less dense plasma and a Rayleigh-Taylor instability will develop. An estimation of the effect using Newton's law of hypersonic flow is summarized in Table 2. It should be mentioned that the above discussion has not taken into account the solar wind aberration caused by the comet's orbital motion. It therefore applies strictly only in the coordinate system moving with the comet. Changes in the radial solar wind component of the solar wind coordinate system will in the coordinate system moving with the comet introduce changes in the solar wind aberration angle and therfore may also produce tail kinks. To illustrate the Rayleigh-Taylor instability associated with solar wind direction changes an example is shown in Fig. 9. Velocity vectors were determined by Jockers and Lüst [1973] by tracing tail condensations in comet Bennett 1970 II on exposures obtained by Beytrishvili at the Abastumani observatory. Undulations

Tail Kinks

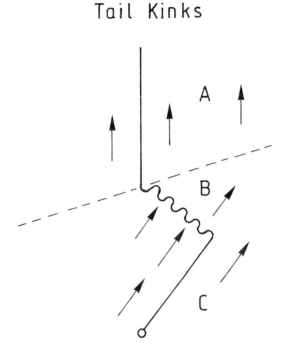

Fig. 8. Pair of kinks and associated Rayleigh-Taylor instability in a cometary tail (see text), caused by a solar wind direction change.

TABLE 2:   Raleigh–Taylor Instability caused by Solar wind cross–tail flow

Pressure balance

$M_t\, N_t\, b_t = n_p\, v_p^2 \sin^2\alpha$

M_t	= 20	molecular weight of tail ions
N_t	= 10^{11} cm^{-2}	column density
n_p	= 10 cm^{-3}	solar wind proton density
v_p	= 400 kms^{-1}	solar wind speed
α	= 10°	solar wind direction angle

gives acceleration $b_t = 240$ cm s^{-2}

In one hour tail is displaced 15 000 km into solar wind flow direction

(small arrows) are seen in the tail at places where the velocity vector is not parallel to the tail axis. This example is more complicated as the tail is split into more than three parts contrary to Fig. 8. The wavy structures appear not only in the distant tail but also in the loop close to the cometary head. A perhaps cleaner example has been published by Brandt et al [1980] and has been analyzed by Jockers [1986]. In this case the space probe Helios II was very close to the comet. The in-situ measurements confirm the flow pattern of Fig. 8 [Le Borgne 1982, Watanabe et al. 1986]. Fig. 10 shows a sequence of wide-field images of comet Kobayashi-Berger-Milon 1975 IX obtained by the author and E. Moore at the 14-inch Schmidt telescope of Joint Observatory of Cometary Research on 31 July 1975. Outward of kink (a) on the left side of the tail we see another example of such a wavy pattern. An example where we appear to look on the wavy pattern from the "front", i.e. where the observer is close to the plane of the displacement vector of the undulation, has been described by Jockers [1983a].

Tail rays

In addition to the Rayleigh-Taylor undulations outside of kink (a) Fig. 10 shows the development of a tail ray pair. Within the three hours of observations the ray pair grows to a length of several million km. This example may perhaps appear as a typical tail ray event but from his experience with many cometary photographs the author would say that such clean cases of a single tail ray pair are comparatively rare. More frequently we observe bundles of tail rays which may appear either more or less symmetrical to the tail axis or on one side of the main tail only. In comet Kohoutek 1973 XII, studied in detail by Jockers [1985], one-sided rays were most frequently observed. The appearance of a tail ray bundle on one side of the cometary main tail will always be followed by a turn of the main tail to the side of the tail rays [Jockers 1985].

A natural explanation of the folding motion of tail rays is to assume the rays embedded in a shear flow, parallel to the main tail axis, with constant velocity gradient perpendicular to the flow direction [Ness and Donn 1966]. If we denote the angle between ray and tail axis with θ, then, as shown by these authors, in a shear flow the time derivative $d\theta/dt$ should be proportional to $-\sin^2\theta$. Wurm and Mammano [1972] have measured the time evolution of this angle for a few well documented cases. The ray closure rate decreases when the tail ray approaches the axis but clearly not fast enough for the equation of Ness and Donn to be valid. E.Moore [*private communication*] has measured the tail angles in the set of observations shown in Fig. 10 and again finds clear disagreement with this rule. It follows that the ray particles must at least at times move toward the tail axis. Wurm's idea, also expressed in the Wurm and Mammano [1972] paper, is, that the tail rays are emitted in solar direction and make a turn of almost 180° into the tail. Because of their rapid initial motion the rays are not visible on the solar side of the comet. In view of our present understanding of the comet-solar wind interaction this interpretation seems unlikely. However, Wurm is generally regarded as an accurate observer, and so his descriptions should not necessarily be dismissed. Another perhaps relevant property of tail rays is that they become narrower when their turning motion decreases. This has been attributed by Wurm to reduced smearing of always narrow rays because of their reduced turning speed, but the effect could as well be real. As also noticed by Wurm and Mammano [1972], the folding motion of rays often does not converge to the main tail but to a position angle (Fig. 10 is an example) different from that of the tail axis. Also, tail ray events are frequently followed by a weakening of the main tail (arrow b in Fig. 10).

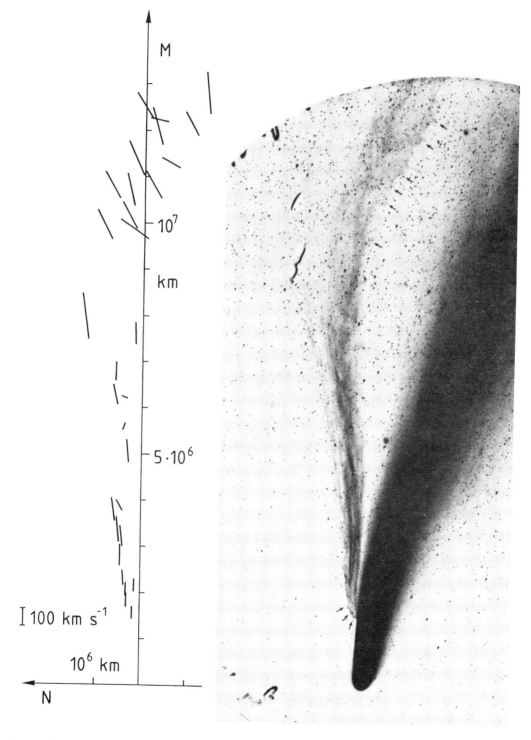

Fig. 9. Kinks in the tail of comet Bennett 1970 II. Abastumani Observatory photograph. After Jockers and Lüst [1972].

$5 \cdot 10^6 \, \text{km}$

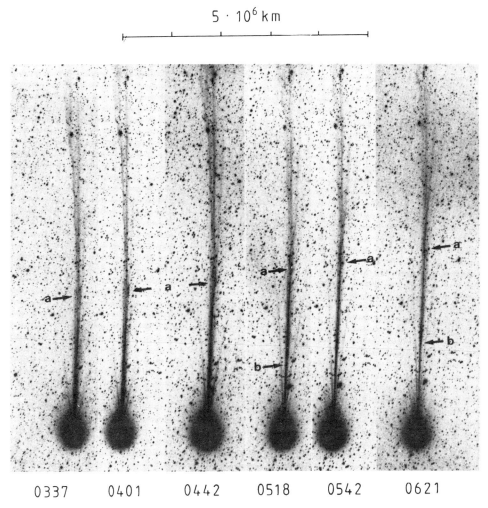

0337   0401   0442   0518   0542   0621

Fig. 10. Time sequence of exposures of comet Kobayashi-Berger-Milon 1975 IX, obtained at the Joint Observatory for Cometary Research on July 31, 1975. The sequence shows the development of a pair of tail rays. In the last frame the rays extend almost to the kink marked (a). Outward of kink (a) there is a tail ray section with Rayleigh-Taylor undulations and to the right of it a tail ray rooted at kink (a). During the evolution of the tail ray pair the main tail thinnens inward of (b).

A question frequently asked about tail rays is if they intersect the sun-comet axis on the solar or anti-solar side of the nucleus. With the limited resolution of ground-based images and the presence of the coma this question necessarily is difficult to decide. It seems that most well-developed rays join the main tail on the tailward side. This, however, needs not to be of relevance to the origin of the rays because at the time of observation the ray could already be convected past the cometary nucleus into the tail. Sometimes tail rays are seen to connect with the main tail at large distances downstream of the nucleus (e.g. see Fig. 10, arrow a, to the right of the undulation).

Ground-based images of comets only show the projection of the object on a plane perpendicular to the line of sight. It has not been possible as yet to take simultaneous pictures of a comet tail from two sufficiently well separated vantage points

(e.g. spacecraft) to yield a stereoscopic image of the comet tail and its rays. Therefore we can at present only speculate about their three-dimensional nature. Jockers [1983a] has proposed a possibility which is illustrated in Fig. 11. If the ray pairs are not co-planar with the main tail, it is easy to understand why rays sometimes appear on one side of the tail only. A reason for the deviation from co-planarity could be that, because rays form and evolve much more quickly than the main comet tail, they adjust to changes in solar wind direction more quickly than the main tail. This hypothesis also would explain the above-mentioned fact that, if tail rays appear on one side of the main tail they precede the motion of the main tail into this direction.

The tail ray phenomenon is still very poorly understood. Ness and Donn [1966] have suggested that, when an inter-

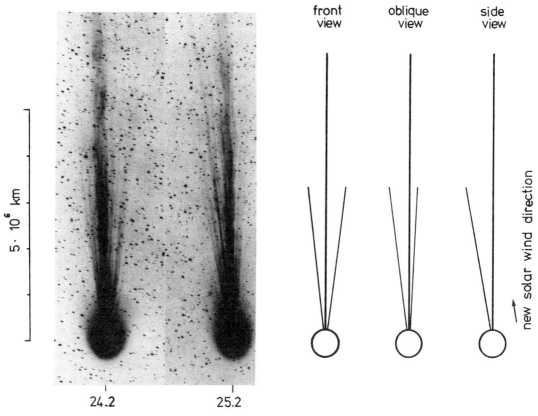

Fig. 11. Comet Ikeya 1963 I observed at Boyden Observatory at February 24 and 25 1963 (courtesy E.H. Geyer), and a suggestion concerning the three-dimensional structure of tail rays.

planetary sector boundary passes through a comet, plasma may collect at the places of zero magnetic field and a pair of tail rays may form. Unfortunately sector boundaries occur not sufficiently frequently to explain the frequent occurrence of cometary rays. Instead of a 180° turn of the interplanetary magnetic field Schmidt and Wegmann [1982] have proposed a 90° turn. Such turns occur more frequently in solar wind tangential or rotational discontinuities [Lepping and Behannon 1986]. The interchange of the meridional and equatorial magnetic planes may produce a pattern reminiscent to tail rays. Schmidt-Voigt [1989] has done three-dimensional numerical MHD modelling to explore the situation. He confirmed a finding of Fedder et al. [1984] that the 180° turn of the magnetic field produces two density enhancements similar to tail rays. These enhancements are, however not strong enough to show up in column densities, i.e. in projection on the sky, like the observed rays. In principle, strict MHD should not produce any effect with a 180° magnetic change since the MHD equations only depend on the square of the magnetic field. The produced effects are caused by magnetic diffusion and it was argued that the numerical diffusion inherent in the method tends to produce an effect more strongly than the diffusion present in nature. The 90° turn of the magnetic field produced a cloud of ions but the numerical resolution was not sufficient to show

the actual rays. Clouds are associated with the tail ray phenomenon.

Nevertheless these MHD models alone cannot tell the whole story because they cannot explain the frequently observed bundles of tail rays. It is likely that plasma instabilities are involved. The older literature has been summarized by Ip and Axford [1982]. Magnetic dips correlated with ion density enhancements were observed outbound on Vega 1 and inbound on Vega 2. They had a width of 400-1600 km, and were separated by distances of 4000 to 10000 km. These dips have been associated with the coma roots of tail rays by Yeroshenko et al. [1986]. Identification of the magnetic dips with the rays seen in Fig. 2 (this image was obtained a few hours after the Vega 1 flyby) is difficult as the Vega flybys occurred very nearly along the line of sight as seen from the earth. Russell et al. [1987, see also Russell et al., *this volume*] have suggested that these structures are mirror waves. They become unstable because the pick-up process on average enhances the temperature perpendicular to the magnetic field over the temperature parallel to the field. This certainly interesting idea has been further elaborated by Raeder [1990]. His two-dimensional model, however, tends to overestimate the growth rate of the instability and it will be necessary to study the problem in three dimensions.

Tail disconnection events

The tail ray phenomenon is intimately connected to another kind of ion tail activity, the so-called tail disconnection events. This name is due to Niedner and Brandt [1978], who revived a name coined by Barnard (see e.g. Barnard [1908a,b, 1909], Brandt [1982]). A collection of tail disconnection events has been published by Niedner [1982]. From the observational point of view a disconnection is difficult to verify because there is always a detection limit when the plasma tail becomes indistinguishable from the sky background. Because of the clear association of the disconnection events with solar wind stream interfaces (and, consequently, with the magnetic sector boundaries associated with these interfaces) the writer has proposed the name comet tail substorm instead [Jockers 1985]. In this paper [see also Jockers 1986] the following description has been given for this kind of activity.

1. A comet tail substorm starts with the development of tail rays from the coma. These rays grow almost with solar wind speed and may sometimes reach a length of several million km.

2. Frequently a cloud or kink is later observed which must have originated in the coma region of the comet at the onset of the tail substorm.

3. Upstream of this cloud the strength of the main tail is sometimes reduced (Tail disconnection).

4. It seems that the ion production rate is increased during the substorm event.

5. The tail substorm is accompanied by changes in the direction of the main tail axis.

The time scale of the event is a few hours to several days. A spectacular example is seen in Fig. 1 where, despite of the two days difference between the July 31 and August 2 images, we probably still see the same event. Another excellent example observed in comet Morehouse 1908 III has been reproduced by Brandt [1982]. As bright comets most often are observable from the same site only for a couple of hours, the phenomenology of these tail events is difficult to study. Previous writers have been aware of these events. Barnard [1908a,b, 1909] has stressed the disconnection aspect. Hoffmeister [1943], in his analysis of tail photographs of comet Whipple-Fedke-Tevzadze 1943 I and other comets, has measured the position angle of what he calls the primary tail ray, i.e. the strongest tail feature emanating from the coma. He points out that sometimes the primary ray disconnects and a new primary ray forms at another direction. He also has noticed that the overall activity of comets is on average larger when the primary tail ray deviates more strongly from the average aberrated anti-solar direction. This indicates a relation between the disconnection events and solar wind direction changes. A similar description of the tail

substorm (redirection of flow from the main tail into the rays) has been given by Wurm and Mammano [1972].

Even though the gross features of the tail disconnection or tail substorm events are known there are so far no quantitative measurements of the ion column density in the coma and the near-coma tail during their occurrence. With the recent advance in astronomical observing techniques hopefully such measurements will soon become available.

Niedner and Brandt [1978] have presented a qualitative explanation of the tail disconnection phenomenon in terms of reconnection of magnetic field lines. They view the cometary tail as a magnetotail and implicitly make the erroneous assumption that the cometary plasma will always trace out the magnetic field structure in the comet. I will not describe this model in detail but mention only that magnetic reconnection is proposed to be induced at the upstream side of the comet when it crosses sector boundaries of the interplanetary magnetic field. The number of observed tail disconnection events is, however, too high to be explainable solely by interplanetary sector boundary crossings [Jockers 1985]. The in-situ measurements of the magnetic field during the spacecraft encounters with comet Halley have not confirmed the model. The increasing complexity of the magnetic field draping found during the Vega-2, Vega-1 and Giotto flybys is not reflected in an increasingly structured ion tail shortly after the encounter periods. A "tail disconnection" associated by Niedner and Schwingenschuh [1986] with the Vega-1 field reversal seems unconvincing and was abandoned by the authors in their subsequent paper [Niedner and Schwingenschuh 1987]. These authors could, however, establish the association of the main disconnection event in comet Halley from March 10, 1986, with an interplanetary sector boundary crossing. Instead of the frontside reconnection mechanism suggested by Niedner and Brandt [1978] Russell et al. [1986] have proposed a near-tail reconnection model, similar to reconnection occurring in the geomagnetic tail during substorms. This model does not depend on magnetic sector boundary crossings.

As the disconnection events are closely related to the tail ray phenomenon, MHD models of magnetic field rotations as described in the previous section on tail rays are also of relevance here. Besides of the magnetic field, plasma effects may also influence the disconnection events. To explain the enhancement and later decay of the ion production a temporary enhancement of ionization of cometary neutrals is attractive. If the production of neutral particles remains constant, the ionization enhancement will deplete the neutral source cloud and the temporary enhancement of ion production will be followed by a temporary reduction. Ip and Mendis [1978] have proposed that the enhanced dynamical pressure in a solar wind high speed stream may trigger plasma instabilities in the comet which in turn may lead to temporary ionization by the critical velocity ionization effect. Another possibility is increased solar wind charge exchange during solar wind flux enhancements associated with the solar wind stream interfaces [Jockers 1983b, Beushausen and Jockers 1983]. Obviously the solar wind direction changes, associated with the solar wind stream interfaces, must influence the disconnections. Numerical modelling of a solar wind stream interface passing through a comet, including the associated solar wind direction changes, could clarify the situation. As this is a truely three-dimensional problem, such calculations, however, are still difficult to perform.

Acknowledgements.   I am grateful to several colleagues at the conference for fruitful discussions on the topic of this paper. E.H. Geyer has made available to me his valuable photographs of comet Humason 1962 VIII which so far have not been published elsewhere. D. Bockelée-Morvan, J. Crovisier, and E. Gérard communicated to me their unpublished OH observations of comet Liller 1988 V. H. Rauer has contributed parts of her thesis work to this overview. M.A. DiSanti and F.Scherb have sent preprints of their work prior to publication.

References

Ahnert, P., Der Komet 1942g (Whipple-Fedke), Z. Astrophys., 22, 286, 1943.

Alfvén, H., On the theory of comet tails, Tellus,9, 12, 1957.

Arpigny, C., A study of molecular and physical processes in comets, Mém. Acad. Royale Belgique, Coll. in 8°, 35, Fasc. 5, 1965.

Barnard, E.E, Comet c 1908 (Morehouse), Astrophys. J., 28, 292, 1908a.

Barnard, E.E, Photographic observations of Comet c 1908. Second paper, Astrophys. J., 28, 384, 1908b.

Barnard, E.E, Photographic observations of Comet c 1908 (Morehouse). Third paper, Astrophys. J., 29, 65, 1909.

Beushausen, R., and K. Jockers, One-dimensional, time-dependent models of the interaction of the solar wind with a comet, in Asteroids, comets, meteors I: Proceedings of a meeting held at the Astronomical Observatory of the Uppsala University, 20-22 June 1983, edited by C.-I. Lagerkvist and H. Rickman, p. 317, Uppsala Universitet-Reprocentralen HSC, Uppsala 1983.

Brandt, J.C., Observations and dynamics of plasma tails, in Comets, edited by L.L. Wilkening, p. 519, University of Arizona Press, Tucson 1982.

Brandt, J.C., N.D. Hawley, and M.B. Niedner Jr., A very rapid turning of the plasma-tail axis of comet Bradfield 1979l on 1980 February 6, Astrophys. J., 241, L51, 1980.

Biermann, L., Über den Schweif des Kometen Halley im Jahre 1910, Z. Naturf., 7a, 127, 1952.

Celnik, W.E. and Th. Schmidt-Kaler, Structure and dynamics of plasma tail condensations of comet P/Halley 1986 and inferences on the structure and activity of the cometary nucleus, Astron. Astrophys., 187, 233, 1987.

DiSanti, M.A., U. Fink and A.B.Schultz, The spatial distribution of H_2O^+ in comet P/Halley, Icarus, 86,152,1990.

Eddington, A.S., The envelopes of comet Morehouse (1908c), Mon. Not. Roy. Astron. Soc., 70, 442, 1910.

Fedder, J.A., S.H. Brecht and J.G. Lyon, MHD simulation of a comet magnetosphere, NRL memorandum report 5306, 1984.

Feldman, P.D. et al., IUE observations of comet P/Halley: evolution of the ultraviolet spectrum between September 1985 and July 1986, Astron. Astrophys., 187, 325, 1987.

Gosling, J.T., A.J. Hundhausen, S.J. Bame, and W.C. Feldman, Solar wind stream interfaces, J. Geophys. Res., 83, 1401, 1978.

Greenstein, J.L., The spectrum of comet Humason (1961e), Astrophys. J., 136, 688, 1962.

Hoffmeister, C., Physikalische Untersuchungen an Kometen. I. Die Beziehungen des primären Schweifstrahls zum Radiusvektor, Z. Astrophys., 22, 265, 1943.

Hoffmeister, C., Physikalische Untersuchungen an Kometen. II. Die Bewegung der Schweifmaterie und die Repulsivkraft der Sonne beim Kometen 1942g, Z. Astrophys., 23, 1, 1944.

Ip, W.-H., Cometary plasma tails and the inner source region, in Magnetotail physics, edited by A.T.Y. Lui, p.367, Johns Hopkins University Press, Baltimore and London, 1987.

Ip, W.-H., and W. I. Axford, Theories of physical processes in the cometary comae and ion tails, in Comets, edited by L.L. Wilkening, p.588, University of Arizona Press, Tucson 1982.

Ip, W.-H., and D. A. Mendis, The flute instability as the trigger mechanism for disruption of cometary plasma tails, Astrophys. J., 223, 671, 1978.

Jockers, K., Some possibilities concerning the three-dimensional structure of cometary ion tails, in Asteroids, comets, meteors I: Proceedings of a meeting held at the Astronomical Observatory of the Uppsala University, 20-22 June 1983, edited by C.-I. Lagerkvist and H. Rickman, p.327, Uppsala Universitet Reprocentralen HSC, Uppsala 1983a.

Jockers, K., Solar wind modulation of cometary ion production by the charge-exchange mechanism, in Cometary exploration, Proceedings of the International Conference on Cometary Exploration, Nov. 15-19 1982, Budapest, Hungary, edited by T.I. Gombosi, Vol. 1, p. 257, KFKI, Budapest 1983b.

Jockers, K., The ion tail of comet Kohoutek 1973 XII during 17 days of solar wind gusts, Astron. Astrophys. Suppl. Ser.,62, 791, 1985.

Jockers, K., Observations of cometary plasma, in Asteroids, comets, meteors II: Proceedings of a meeting held at the Astronomical Observatory of the Uppsala University, 3-6 June 1985, edited by C.-I. Lagerkvist et al., p. 411, Uppsala Universitet Reprocentralen HSC, Uppsala, 1986.

Jockers, K., and Rhea Lüst, Tail pecularities in comet Bennett caused by solar wind disturbances, Astron. Astrophys., 26, 113, 1973.

Jockers, K., E.H. Geyer, H. Rosenbauer, and A. Hänel, Observations of ions in comet P/Halley with a focal reducer, Astron. Astrophys., 187, 256, 1987.

Le Borgne, J.-F., Comet Bradfield 1979 X event on 1980 February 6: Correlation with an interplanetary solar wind disturbance, in ESO Workshop on "The need for coordinated ground-based observations of Halley's comet", Paris, France, 29-30 April 1982, edited by P. Véron et al., p. 217, European Southern Observatory, Garching bei München, 1982.

Lees, L., Interaction between the solar plasma wind and the geomagnetic cavity, AIAA Journ., 2, 1576, 1964.

Lepping, R.P. and K.W. Behannon, Magnetic field directional discontinuities: characteristics between 0.46 and 1.0 AU, J. Geophys. Res., 91, 8725, 1986.

McCarthy, P.J., M.A. Strauss, and H. Spinrad, The ionospheric extent of P/Halley from ground based observations during the 1985-1986 apparition, in 20th ESLAB symposium on the exploration of Halley's comet, Proceedings of the international symposium, Heidelberg, Germany, 27-31 October 1986, edited by B. Battric et al., Vol. 3, p.87, ESA SP-250, 1986.

McComas D.J., J.T. Gosling, and S.J. Bame, The Giacobini-Zinner magnetotail: Tail configuration and current sheet, J. Geophys. Res., 92, 1139, 1987.

Miller F.D., H_2O^+ in the tails of 13 comets, Astron. J., 85, 468, 1980.

Ness, N.F.,and B.D. Donn, Concerning a new theory of type I comet tails, in Nature et origine des comètes, treizième colloque international d'astrophysique, Liège, Belgium, 5-7 July 1965, edited by P. Swings, p. 343, Mem. Soc. Roy. Sci. Liège, 5th series, Vol. 12, 1966.

Neubauer, F.M., The ionopause transition and boundary layers at comet Halley from Giotto magnetic field observations J. Geophys. Res., 93, 7272, 1988.

Neubauer, F.M. et al., First results from the Giotto magnetometer experiment, Nature, 321, 352, 1986.

Niedner, M.B. Jr., Interplanetary gas XXVII. A catalog of disconnection events in cometary plasma tails, Astrophys. J. Suppl. Ser., 46, 141, 1981.

Niedner, M.B. Jr., and J.C. Brandt, Interplanetary gas XXIII. Plasma tail disconnection events in comets: evidence for magnetic field line reconnection at interplanetary sector boundaries? Astrophys. J., 223, 655, 1978.

Niedner, M.B. Jr., and K. Schwingenschuh, Plasma tail activity at the time of the Vega encounters, in 20th ESLAB symposium on the exploration of Halley's comet, Proceedings of the international symposium, Heidelberg, Germany, 27-31 October 1986, edited by B. Battric et al., Vol. 3, p.419, ESA SP-250, 1986.

Niedner, M.B. Jr., and K. Schwingenschuh, Plasma tail activity at the time of the Vega encounters, Astron. Astrophys., 187, 103, 1987.

Raeder, J., Ein zweidimensionales, anisotropes Multiionenmodell der Plasmawechselwirkung einer Kometenatmosphäre mit dem Sonnenwind, Mitt. Inst. Geophys. Meteor. Univ. Köln, 69, 1990.

Rauer, H. and K. Jockers, Focal reducer observations of comets Liller 1988a and P/Tempel2 1987g, in Asteroids, comets, meteors III: Proceedings of a meeting held at the Astronomical Observatory of the Uppsala University, 12-16 June 1989, edited by C.-I. Lagerkvist et al., p. 417, Uppsala Universitet Reprocentralen HSC, Uppsala 1990.

Russell, C.T., M.A. Saunders, J.L.Phillips, and J.A. Fedder, Near-tail reconnection as the cause of cometary tail disconnections, J. Geophys. Res., 91, 1417, 1986.

Russell, C.T., W. Riedler, K. Schwingenschuh, and Ye. Yeroshenko, Mirror instability in the magnetosphere of comet Halley, Geophys. Res. Letters, 14, 644, 1987.

Russell C.T., Guan Le, O.L. Vaisberg, K. Schwingenschuh, W. Riedler, Fast magnetoacoustic waves at the pile-up boundary of comet P/Halley, this volume.

Scherb, F., K. Magee-Sauer, F.L. Roesler and J. Harlander, Fabry-Perot observations of comet Halley H_2O^+, Icarus,86,172

Schmidt, H.U.and R. Wegmann, Plasma flow and magnetic fields in comets, in Comets, edited by L.L. Wilkening, p.538, University of Arizona Press, Tucson 1982.

Schmidt-Voigt, M., Time-dependent MHD simulations for cometary plasma, Astron. Astrophys., 210, 433, 1989.

Schwingenschuh, K. et al., Cometary boundaries: VEGA observations at Halley, Adv. Space Res., 6(1), 217, 1986.

Watanabe, T., T. Kakinuma, and M. Kojima, An interplanetary disturbance relevant to the tail-turning of comet Bradfield (1979l) on 1980 February 6, Proceedings Res. Inst. Atmospherics, Nagoya University 33, 19, 1986.

Wehinger, P.A., S. Wyckoff, G.H. Herbig, G.M. Herzberg and H. Lew, Identification of H_2O^+ in the tail of comet Kohoutek (1973f), Astrophys. J., 190, 143, 1974.

Wurm, K., and A. Mammano, Contributions to the kinematics of type I tails of comets, Astrophys. Sp. Sci., 18, 273, 1972.

Wyckoff, S., and J. Theobald, Molecular ions in comets, Adv. Space Res., 9(3), 157, 1989.

Yeroshenko, Ye. G. et al., Magnetic field fine structure in comet Halley's coma, in 20th ESLAB symposium on the exploration of Halley's comet, Proceedings of the international symposium, Heidelberg, Germany, 27-31 October 1986, edited by B. Battric et al., Vol. 1, p. 189, ESA SP-250, 1986.

THE 10 JANUARY 1986 DISCONNECTION EVENT IN COMET HALLEY

Malcolm B. Niedner, Jr.

Laboratory for Astronomy and Solar Physics, NASA–Goddard Space
Flight Center, Greenbelt, MD 20771

John C. Brandt and Yu Yi

Laboratory for Atmospheric and Space Physics,
University of Colorado, Boulder, CO 80309

Abstract. The disconnection event (DE) in the plasma
tail of comet Halley on January 9-12, 1986 is examined.
We measured the distances between the comet head and
the disconnected tail for a series of images for that time
period and then extrapolated to the nucleus to determine
the disconnection time, T_d = January 9.60 +/- 0.2 days.
The approximate solar-wind conditions at the time of the
DE were obtained by corotation of IMP-8 satellite data in
Earth orbit to comet Halley. At the time of the DE, comet
Halley is inferred to have been close to a magnetic sector boundary and a high-speed stream compression region,
circumstances which are consistent with models invoking
magnetic reconnection in response to either sector boundary crossings or increases in solar-wind plasma pressure
(sunward and tailward reconnection, respectively). However, a heliographic latitude separation of 22 degrees (between the comet and IMP-8), gaps in the IMP-8 data, and
other, more minor disconnection activity slightly preceding
the January 9-10 DE, render a more definitive statement
about the linkage of the DE to external conditions quite
difficult. In particular, it appears not possible at this time
to resolve the effects of magnetic changes associated with
the sector boundary and plasma pressure in the compression region. Resolution of conflicting DE theories will likely
require a statistical treatment of many events.

Introduction

Large-scale plasma structures in comets often show rapid
and major changes. The most dramatic of these is the disconnection event (or DE), where the entire plasma tail disconnects from the comet and a new tail is formed. The
appearance of DEs can be quite varied and can change
markedly with time within a given event. For some well-
evolved events the appearance can be more that of a "kink"
than of a detached structure. In the next section we will
briefly discuss the characteristics of DEs and their "definition" from an observational viewpoint.

At the very least, study of the physics of DEs requires
determination of the kinematics and the solar-wind conditions for specific events. In this paper, we present such a determination for the DE of 10 January 1986 in comet Halley,
probably the most spectacular pre-perihelion plasma-tail
event. The kinematics are well-determined for this event
due to an abundance of wide-field imaging, and a reasonable level of solar wind and interplanetary magnetic field
(IMF) monitoring existed as a result of data returned by
the IMP-8 satellite.

Review of DE Theories, Plasma-Tail Formation

The most successful and widely-accepted general theory of the cometary plasma tail is due to Alfvén (1957),
in which a plasma tail consists of swept-up interplanetary
magnetic field draped over the ionosphere of the comet,
and the molecular plasma is trapped on the field lines. Although Alfvén's model was actually developed, in part, to
explain the very high repulsive accelerations sometimes observed in "type I" tails (i.e., under disturbed conditions),
the formation of an induced magnetotail which is the heart
of the model has become the very paradigm of our concept
of a "normal" or quiet-time cometary plasma tail. This
magnetotail model was confirmed in all essential respects
by the flyby of comet P/Giacobini-Zinner by the Interplanetary Cometary Explorer (ICE) spacecraft [e.g., Smith et
al., 1986].

Cometary Plasma Processes
Geophysical Monograph 61
©1991 American Geophysical Union

Normally, according to observations, the plasma tail is attached to the head region. Sometimes, however, the plasma tail is disconnected from the head in a very spectacular way, and the comet grows a new tail from the folding of prominent tail rays. This destruction/formation sequence for plasma tails is often repeated many times for bright, well-observed comets [Niedner and Brandt, 1978], and the cycle of activity closely follows the cycle of solar-wind IMF and plasma at the comet [Niedner and Brandt, 1979]. For example, the DE model of Niedner and Brandt [1978] invokes sunward magnetic reconnection and sector boundary crossings.

Not for all images or for all events does the apperance of a DE completely resemble "disconnection", however, and it is a legitimate question to ask whether particular "events" are actually DEs as defined above. The occasional occurrence of disconnected plasma tails magnetically merging with the incipient new tails taking their place was noted by Niedner and Brandt [1980], who by analyzing a series of images of comet Kohoutek 1973XII found that the so-called "Swan cloud" discussed by Hyder et al. [1974] was actually an advanced stage of an earlier DE. The receding tail was seen to be disconnected in the earliest images, and merged with the collapsed rays/new tail in the later ones. Hyder et al. only had the later images at their disposal, in which the Swan more closely resembled a giant "kink" than a detached structure; hence its true nature was disguised to those authors. In many respects the 1986 January 10 DE in comet P/Halley (the subject of this paper) is similar to the Swan cloud in comet Kohoutek; discussion of this point is contained in later sections.

Table 1 lists the major theories of DEs, which can be broadly categorized into three main classes (the number following each theory indicates which class it belongs to):

- (1) Ion Production Effects: If the ion production rate were substantially reduced on a short timescale ($<$ hours), the imparted effect on the tail might be an interruption of ion brightness, i.e., a DE. The change in ionization rate could result from changes in solar-wind conditions or in the density of neutrals out of which the ions are created. In the latter scenario the DE is produced more by internal than external causes; the theories which take this view [e.g., Wurm and Mammano, 1972] no longer seem tenable in view of the high correlation between DEs and solar-wind/IMF structures [Niedner, 1982; Delva et al., 1991].

- (2) Pressure Effects: A large pressure increase in the solar-wind could compress the ionosphere to a very small size and either free the magnetic field lines or excite various instabilities (e.g., Rayleigh-Taylor) in

TABLE 1. Theories of Disconnection Event Mechanisms

Theory	Category
Cessation of Ion Tail Sources (Wurm and Mammano, 1972)	(1)
Interplanetary Shock Passages (Jockers and Lust, 1973)	(2)
Sector Boundary/Sunward Reconnection (Niedner and Brandt 1978)	(3)
High-Speed Stream/Flute Instability (Ip and Mendis, 1978)	(2)
High-Speed Stream/Differential Acceleration (Jockers, 1985)	(2)
High-Speed Stream/Tailward Reconnection (Ip, 1985; others)	(3)

the tail. Such instabilities could lead to "differential acceleration" effects [e.g., Jockers, 1985] and the appearance of a DE. If any of these models is correct, we would expect increases of solar-wind pressure at the time of DEs, possibly due to high-speed streams or flare-generated interplanetary shocks and pistons.

- (3) Magnetic Reconnection: There are two kinds of DE models invoking magnetic reconnection. One invokes front-side reconnection, which should occur at sector boundary crossings. But because a large percentage of sector boundaries are correlated with the leading edges of high-speed streams it is difficult to distinguish between pressure effects and frontside reconnection during magnetic sector boundary crossings. The other theory invokes tail-side reconnection, which could be triggered by a variety of events including high-speed streams and interplanetary shock passages; once again there is overlap with conditions favoring other processes (for example, sector boundaries often accompany streams).

It is our view that for the cometary plasma tail to be severed during a DE, the magnetic field lines threading the tail must somehow be freed from the cometary head. Another way of putting it is that the flow of ions must undergo a substantial alteration; one way of stopping the flow of ions to the flux tube which is the main tail (thereby creating the appearance of disconnection) is to detach the magnetic tail altogether from the ion sources which feed it in the near-nuclear ion production zone. Either of the two magnetic reconnection theories discussed above accomplish this; most of the other models do not.

Cometary and Solar-Wind Conditions

Kinematics, Disconnection Time for January 9-10 DE

During the time interval from mid-November 1985 to early-July 1986, wide-field imaging of comet Halley was carried out by the Large-Scale Phenomena Network (LSPN) of the International Halley Watch (IHW). Approximately 127 images were obtained for the interval January 9.0-12.0 UT 1986, during which a major DE, and the subject of this paper, occurred.

Figure 1 shows a sample image sequence for the event.

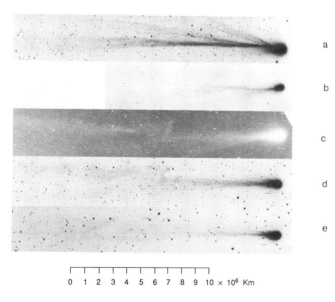

a

b

c

d

e

0 1 2 3 4 5 6 7 8 9 10 × 10⁶ Km

Fig. 1. Sample photographic sequence of wide-field imagery showing recession of disconnected plasma tail from head of comet Halley, 1986 January 10-11. Panels a,b,d,e all to same approximate scale (given below sequence).
a.) 2.1334 hours UT on 1986 January 10, IIa-O emulsion without filter, Joint Observatory for Cometary Research (JOCR).
b.) 9.3751 hours UT on January 10, IIa-O without filter, University of Kyoto Observatory.
c.) 19.3034 hours UT on January 10, IIa-O with GG-385 filter, Calar Alto Observatory.
d.) 2.5334 hours UT on January 11, IIa-O without filter, JOCR.
e.) 5.8166 hours UT on January 11, IIIa-J without filter, Mauna Kea Observatory.

Before describing the measurements made on these five and many other images showing the DE, it is necessary to describe the morphology as it evolved with time, paying special attention to the disconnection nature of the event. This description is important in view of the prior discussion about whether particular events represent true disconnections of the entire plasma tail.

Broadly speaking, Figures 1b-e show a spectacularly-kinked plasma tail, the size and amplitude of the kink growing with time (note that the image frames in Figure 1 are to the same scale, with the exception of panel c). In Fig. 1a there is no evidence of the "kink", but what is apparent is that the plasma tail is quite bright, necking down near the cometary head, and is flanked on the top by a highly prominent system of discrete tail rays. The necking-down morphology, accompanied by the development of strong tail rays, is the classic precursor activity accompanying many if not all disconnection events [Niedner and Brandt, 1979].

In fact, the tail is newly-disconnected in Fig. 1a, the point of disconnection lying quite close to the cometary head where it is difficult to see in Fig. 1a. Figure 2 is

Fig. 2. Slightly underexposed enlargement of inner tail region of Fig. 1.a., showing detached plasma tail sandwiched between pairs of strong tail rays.

an underexposed enlargement of the region of interest, and it should be obvious to the reader that the tail is in fact disconnected (even here, however, the tip of the detached tail lies close to the lower of several ray pairs). In the scenario of Niedner and Brandt [1980], the kink developed later when the receding tail encountered the tail rays of Fig. 1a during the latters' closure to define a new tail axis and a new tail. The twisted, kinked structure first seen in Fig. 1b represents an evolved phase of the initially "clean" DE of Figs. 1a and 2, and the reader is invited to make comparisons between this DE and the Swan Cloud in comet Kohoutek, which it much resembles [Niedner and Brandt, 1980].

The completeness of the photographic coverage makes it possible to track the detached tail for several days from January 10 to January 12 as it receded from the comet head. From the series in Figure 1 and many other images spanning the time interval of interest, we can determine

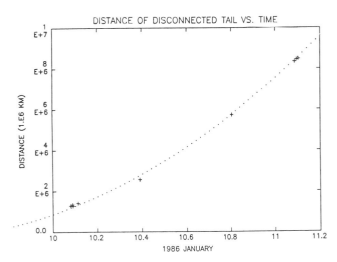

Fig. 3. Plot of measured distance between near end of disconnected tail and center of comet head (points). Curved line is predicted distance for a time of disconnection of January 9.68, an initial velocity of 21 km s^{-1}, and a (constant) acceleration of 80 cm s^{-2}.

TABLE 2. Measured Distances of January 10-13 DE

UT Date (mid-exp.)	Distance (km)	Observatory	Observer(s)
1986 January			
10.08056	1.288E6	JOCR	E. Moore/E. Marr
10.08889	1.345E6	JOCR	E. Moore/E. Marr
10.09028	1.277E6	Lowell	H. Giclas
10.11389	1.414E6	JOCR	E. Moore/E. Marr
10.39063	2.560E6	Univ. Kyoto	T. Tsujimura
10.80431	5.540E6	Calar Alto	K. Birkle
11.08889	8.010E6	JOCR	E. Moore/E. Marr
11.09722	7.999E6	JOCR	E. Moore/E. Marr
11.10556	8.073E6	JOCR	E. Moore/E. Marr
11.24236	9.556E6	Mauna Kea	A. Storrs
11.73229	1.632E7	Bulgarian Nat'l	V. Shkodrov/et al.
12.09306	2.127E7	JOCR	E. Moore/E. Marr
13.08194	4.095E7	JOCR	E. Moore/E. Marr
13.09792	4.187E7	JOCR	E. Moore/E. Marr

the kinematics assuming that the disconnected tail was oriented along the prolonged radius vector. This is probably a good approximation during the first day or two after disconnection.

Figure 3 shows the measured distance (from the nucleus) of the end of the detached tail as a function of time; the measurements and the particulars of each observation are listed in Table 2. It is immediately obvious from Figure 3 that the tail's speed of recession was not constant but increased substantially with distance from the head. We have generated several predicted distance-time curves for various choices of disconnection time (T_d), acceleration (A, assumed constant), and initial velocity (V_o), and attempted to find that curve which best fits the data. In fact the exercise is rather dependent on which measured points are included in the fitting. Ideally, because we are primarily interested in the time of disconnection, it would be advantageous to have many data points inside 1-2.E6 km and use only them in the analysis. This ideal is not realized here, however (and almost never is). We have analyzed several subsets of the data whose distances are < 1.E7 km, and find that the measurements are well-fitted by values of T_d = January 9.60 (+/- 0.2) UT, A = 75 (+/- 5) cm s^{-2}, and V_o = 15 (+/- 5) km s^{-1}. If the solar-wind/IMF feature responsible for the DE arrived, say, 0-12 hours before the actual time of disconnection, then it is on early-to-mid January 9 that we should look for causal agents in any corotated set of spacecraft data.

Disconnection Activity preceding January 9-10 DE

It is to be noted that during the two weeks preceding the spectacular January 9-10 DE under discussion here, at least two, more minor DEs were observed to take place. First, during 1985 December 30-31, the event discussed by Saito et al. [1987] occurred; and second, an event on 1986 January 7-8 occurred which has been discussed briefly by Niedner [1986]. While a detailed analysis of the latter's kinematics and disconnection time have yet to be performed, its proximity in time to the January 9-10 event is pointed out.

Solar-wind Measurements, Heliospheric Current Sheet, and Corotation to Comet Halley

Solar-wind and IMF properties were measured at the Earth by the IMP-8 satellite, and were corotated to comet Halley to provide approximate conditions local to the comet at the time of onset of the DE. In Figure 4 is shown the Sun-Earth-Comet geometry projected onto the ecliptic plane. Under the assumption that the solar plasma is drawn out into an Archimedean spiral pattern characterized by constant flow speed (w_r), the approximate time lag between arrival at the Earth and comet Halley is given by:

$$\Delta t = t_e - t_c = (l_e - l_c)/\Omega + (r_e - r_c)/w_r \qquad (1)$$

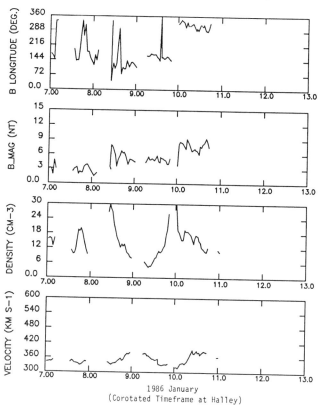

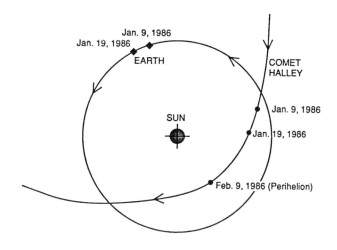

Fig. 4. Sun-Earth-Comet geometry projected onto the ecliptic plane for comet Halley for 9-19 January, 1986.

Fig. 5. IMP-8 solar-wind and IMF conditions observed at Earth and corotated to comet Halley; time axis gives corotated dates at the comet. From top to bottom the panels contain magnetic field longitude angle (135° is ideal outward sector, 315° is inward sector), magnetic field magnitude, plasma density, and plasma bulk velocity.

where l is the longitude and r the heliocentric distance with the subscripts "e" (earth) and "c" (comet). Here, the sidereal spiral pattern speed Ω=14.4 degrees day^{-1} and w_r = 400 km sec^{-1}, which is approximately the average speed of the solar wind on dates associated with the DE (as will be shown below). In corotating from comet Halley on January 9.6 UT (= T_d) to IMP-8, we calculate a time delay of 7.1 days, or arrival at IMP-8 on January 16.7 UT. As noted above, any solar-wind/IMF causal agent for the DE would presumably have arrived at the comet some 0.00-0.5 days before T_d, and hence January 16.7 UT is on the "late end" of the arrival window at IMP-8. These arrival time calculations assume, of course, that solar wind and IMF structure is rigorously organized along meridional (North-South) planes, an assumption that can produce serious error when the latitudinal separation of comet and spacecraft is large enough. In the case of Halley and IMP-8, the separation was 22.2 degrees (Halley at 17.5 deg on Jan. 9.6, IMP-8 at -4.7 deg on Jan. 16.7).

In Figure 5 IMP-8 observations of the IMF longitude angle and field magnitude, and the plasma density and flow speed, are shown for the corotated timeframe at the comet. The only one of these parameters needing explanation is the field longitude angle: it is angle between the spacecraft-sun vector and the projection of the field vector onto the heliographic plane, increasing counterclockwise when viewed from the north solar pole. At 1 AU a classical outward magnetic sector has longitude = 135°; inward sectors classically have longitude = 315°. The most important features in Figure 5 are the obvious magnetic sector boundary at the comet very late on January 9 (evidenced by a rotation of the IMF longitude angle from approximately 150 to 330 degrees, exact time masked by a

several-hour datagap), and a 4-5-fold increase in the plasma density beginning a few hours earlier but peaking at about the same time. Assuming they are oriented meridionally in the heliosphere, both of these features arrive at the comet about 0.5-0.75 days "late", i.e., that length of time after inferred onset of the DE process. The January 7-8 timeframe at the comet is characterized both by density peaks and short-term polarity reversals in the magnetic field.

Figure 6 is a Carrington Rotation plot of the coronal magnetic neutral line (heavy wavy line) obtained via extrapolation of the photospheric field to the so-called "source surface" at 2.5 solar radii (the dotted wavy line will be explained below). The '+' and '-' symbols running across the bottom and top of the figure, respectively, denote the magnetic polarity of the south and north solar poles. These coronal data were kindly supplied by J. T. Hoeksema of the Wilcox Solar Observatory. Superposed on the corona are theoretical tracks made by Halley's Comet and IMP-8, calculated on the basis that the coordinates (latitude, lon-

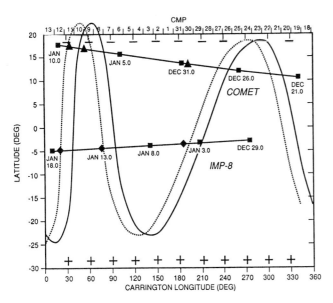

Fig. 6. Carrington map of computed coronal magnetic neutral line (heavy wavy line) for Carrington Rotation 1770. The comet and spacecraft "tracks" give the longitude and latitude of the coronal footpoint of the Archimedean spiral instantaneously linking the comet (or IMP-8) to the corona for a solar-wind speed of 400 km s^{-1}. The triangles and diamonds along the tracks of the comet and spacecraft give the zero dates of DEs and the times of observed magnetic sector boundaries, respectively. The dotted wavy line is the observed neutral line shifted later by one day to bring it into better agreement with the times of DEs and sector boundaries.

gitude) for any specified date give the instantaneous footpoint of the Archimedean spiral connecting the Sun to the comet or spacecraft if the solar-wind speed is 400 km s^{-1}. Such a (constant) value of the solar wind seems appropriate in view of the data in Figure 5. This type of plot is an excellent way of relating cometary activity to interplanetary data, the coronal neutral line serving as an important guide to the orientation of the neutral sheet when comet and spacecraft are widely-separated in heliographic latitude.

The triangles on the Halley track give the approximate zero dates of DEs (December 31, January 8, and January 9.60), whereas the diamonds on the IMP-8 track indicate interplanetary magnetic sector boundaries observed by the spacecraft. It should be said, however, that the first two sector boundaries–those of January 4 and 13 at IMP-8–are not displayed in Figure 5 due to the relatively short interval of those plots, nor were these boundaries extremely well-observed or well-timed in the data. The reasons for the latter situation were two-fold: first, as in the case of the

January 17 sector crossing (at IMP-8), datagaps are a problem; and second, the crossings consist of multiple reversals before a new uniform polarity is established. Evidently the neutral sheet geometry is quite complex in this Carrington longitude range. It is worth noting, however, that the overall polarity sequence across the sector boundaries agree with those expected on the basis of the coronal neutral line. Specifically, the crossings observed on (approximately) January 4.8, 13.0, and 17.25 were (+/-), (-/+), and (+/-), all of which agree with the polarity sequence of the nearest predicted neutral line crossing.

While the January 7-8 and December 31 tail events are not the focus of this paper, it is nonetheless apparent from the figure that the DEs and sector boundaries agree reasonably well with the predicted crossing times of the neutral sheet by the comet and IMP-8, except, perhaps, for the December 31 event [cf. Saito et al., 1987]. We do not believe, however, that the apparent lack of agreement for this one event should be viewed as necessarily detrimental to the sector boundary model: given the somewhat chaotic variation of the field polarities across IMP-8's path in Carrington longitude range 160-190, it is not obvious that the neutral line extends smoothly toward larger longitudes at higher latitudes (as the coronal data would imply). Perhaps more work on the structure of the neutral sheet in this region is needed.

Although there is good overall agreement between DEs, sector boundaries, and the neutral sheet surface in Figure 6, one could make the argument that a shift of the entire neutral line by one day (14 degrees) later brings the coronal data and interplanetary events in better agreement (dotted neutral line). Still, considering the somewhat long corotation time delay between IMP-8 and Halley, and the long extrapolation inward to the corona, the internal self-consistency in Figure 6 is impressive if sector boundary passages are in some way related to DEs. [Adoption of a constant solar-wind speed different from 400 km s^{-1} does move the DEs and interplanetary sector boundaries to different Carrington coordinates, but this is not a particularly strong effect.]

Returning to Figure 5, it is virtually impossible based on timing to resolve the sector boundary crossing from the compression region as Halley's plasma tail was disconnecting on January 9.6 UT. At that time the comet was probably immersed in both structures: the newly-changed magnetic sector and the elevated density wind.

Conclusion

The determination of solar-wind conditions at the time of cometary DEs does not lend itself to complete accuracy. In this paper we have illustrated the circumstances and uncertainties in the case of the DE of 9.60 January 1986.

The data are consistent with the DE being produced by the sector-boundary magnetic reconnection model [Niedner and Brandt, 1978] and/or the tail reconnection scenario (due to an increase in the particle density of the solar wind by a factor of 4 to 5). Note that the solar-wind speed as measured by IMP-8 for this entire interval is about 400 km^{-1} or less.

We add, with interest, that our measured distances of the rejected tail agree reasonably well with those of Tomita et al. [1987], but our interpretation of the event itself is rather different. Whereas Tomita et al. portray the event as a "kink" in the tail, we argue that the structure is a disconnected tail which has merged with the new tail forming in its place to produce one very twisted and contorted tail (refer to previous discussion).

We consider that the determination of solar-wind conditions for a large number of DEs will be necessary for a start toward understanding the basic physics of DEs. The large number is necessary because the circumstances of any specific DE can be uncertain. The situation for the 9.60 January 1986 event is relatively well determined, but at least two distinct mechanisms are possible.

Acknowledgments. We thank the many individuals and institutions who contributed the International Halley Watch imaging data which were used here, including E. P. Moore of the New Mexico Institute of Mining and Technology and the Joint Observatory for Cometary Research (JOCR). It is also a pleasure to thank J. Todd Hoeksema of the Wilcox Solar Observatory for coronal neutral line data. Finally, the comments of two anonymous referees helped considerably in improving this paper; we thank them here.

References

Alfven, H., On the theory of comet tails, *Tellus, 9,* 92-96, 1957.

Delva, M., Schwingenschuh, K., Niedner, M. B., and K. I. Gringauz, Halley remote plasma tail observations and in situ solar wind properties, *Planet. Space Sci.,* in press, 1991.

Hyder, C. L., Roosen, R. G., and J. C. Brandt, Tail structures far from the head of comet Kohoutek I., *Icarus, 23,* 601-610, 1974.

Ip, W.-H., Cometary plasma physics: large-scale interactions, in *Advances in space plasma physics,* p. 1-21, B. Buti ed., 1985.

Ip, W.-H., and D. A. Mendis, The flute instability as the trigger mechanism for disruption of cometary plasma tails, *Astrophys. J., 223,* 671-675, 1978.

Jockers, K., The ion tail of comet Kohoutek 1973XII during 17 days of solar wind gusts, *Astron. Astrophys. Suppl., 62,* 791-838, 1985.

Jockers, K., and R. Lust, Tail peculiarities in comet Bennett caused by solar wind disturbances, *Astron. Astrophys., 26,* 113-121, 1973.

Niedner, M. B., Interplanetary Gas XXVIII. A study of the three-dimensional properties of interplanetary sector boundaries using disconnection events in cometary plasma tails, *Astrophys. J. Suppl. Ser., 48,* 1-50, 1982.

Niedner, M. B., First impressions of plasma-tail activity in Halley's Comet, *IHW Newsletter, No. 9,* 2-8, 1986.

Niedner, M. B., and J. C. Brandt, Interplanetary Gas XXIII. Plasma tail disconnection events in comets: evidence for magnetic field line reconnection at interplanetary sector boundaries?, *Astrophys. J., 223,* 655-670, 1978.

Niedner, M. B., and J. C. Brandt, Interplanetary Gas XXIV. Are cometary plasma tail disconnections caused by sector boundary crossings or by encounters with high-speed streams?, *Astrophys. J., 234,* 723-732, 1979.

Niedner, M. B., and J. C. Brandt, Structures far from the head of comet Kohoutek II. A discussion of the swan cloud of January 11 and of the general morphology of cometary plasma tails, *Icarus, 42,* 257-270, 1980.

Saito, T., Yumoto, K., Hirao, K., Minami, S., Saito, K., and E. Smith, Structure and dynamics of the plasma tail of comet P/Halley I. Knot event on December 31, 1985, *Astron. Astrophys., 187,* 209-214, 1987.

Smith, E. J., Tsurutani, B. T., Slavin, J. A., Jones, D. E., Siscoe, G. L., and D. A. Mendis, International Cometary Explorer encounter with Giacobini-Zinner: magnetic field observations, *Science, 232,* 382-385, 1986.

Tomita, K., Saito, K., and S. Minami, Structure and dynamics of the plasma tail of comet P/Halley II. Kink event on January 10-11, 1986, *Astron. Astrophys., 187,* 215-219, 1987.

Wurm, K., and A. Mammano, Contributions to the kinematics of type I tails of comets, *Astrophys. Space Sci., 18,* 273-286, 1972.

Mirror Mode Waves at Comet Halley

C. T. Russell and Guan Le
Institute of Geophysics and Planetary Physics University of California,
Los Angeles, CA 90024-1567

K. Schwingenschuh and W. Riedler, Space Research Institute, Graz, Austria

Ye. Yeroshenko, IZMIRAN, Moscow Region, USSR

Abstract. High resolution VEGA magnetic field and plasma data in Halley's magnetosphere reveal out-of-phase oscillations of the type expected to be driven by the mirror mode instability. The spacecraft passes through these structures in about 20 s. The magnetic energy density drops about 6.5 x 15^9 ergs/cm^3 in a typical event. The thickness of these regions is about a water-group ion gyro diameter. While such enhancements should be invisible against the comet when viewed perpendicular to the wavefronts, they could be visible as rays when viewed tangential to the wavefronts.

Introduction

On their closest approach to Halley on March 6 and 9, 1986 the VEGA spacecraft detected numerous small-scale depressions in the magnetic field [Yeroshenko et al., 1986]. The magnetometer included a fourth sensor to measure gradients in the field [Riedler et al., 1986] but no gradients were observed in these events indicating they were large-scale phenomena occurring within the ambient medium and not on the spacecraft itself [Yeroshenko et al., 1987]. These events were first postulated to be associated with critical ionization velocity effects [Galeev et al., 1986]. However, Trotignon et al. [1989] have shown that the plasma waves that were taken to be diagnostic of the CIV process are more likely due to dust impacts. An alternate explanation of these small-scale depressions, proposed by Russell et al. [1987], was that these events were in fact the magnetic signature of the mirror mode instability. This mechanism would account for the increases in ion density observed in conjunction with the magnetic decreases on VEGA-1 [Vaisberg et al., 1989] and VEGA-2 [Russell et al., 1987].

The analyses of these waves to date have been rather brief. The properties of these waves have not been completely described, nor have all the waves been analyzed. It is the purpose of this paper both to compile what we know about these structures and to complete the analysis of these waves. We begin with the VEGA-2 data which contain the fewer events.

Observations

Only one axis of the VEGA-2 magnetometer survived closest approach, thus we can analyze the fluctuations over only the inbound portion of the trajectory. Figure 1 shows the magnetic field

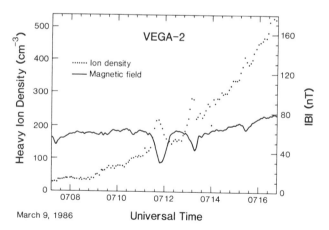

Fig. 1. The magnitude of the magnetic field and the ion density measured on the inbound leg of VEGA-2 from 0708 to 0716 UT, March 9, 1986 [Galeev et al., 1986; Gringauz et al., 1986; Riedler et al., 1986].

strength and the ion density measured by the PLASMAG-1 detector [Gringauz et al., 1986]. There are three dips in the magnetic field, centered at 0711:48, 0713:20, and 0715:31 UT. Each dip has a corresponding increase in the plasma density. The dips last about 20 s and are separated in time by

Cometary Plasma Processes
Geophysical Monograph 61
©1991 American Geophysical Union

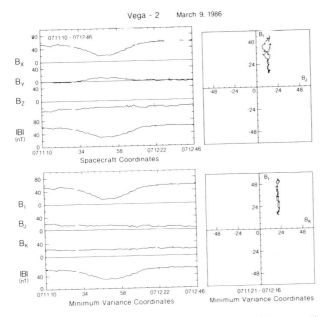

Fig. 2. High resolution measurements (10/second) of the magnetic field measured by VEGA-2 from 0711:10 to 0712:46 on March 9, 1986 at a distance of 40,000 km from the nucleus of comet Halley. The top left panel shows the data in spacecraft coordinates roughly oriented along the solar ecliptic axes. The bottom left panel shows the data rotated to the minimum variance or principal axis system. The right-hand panels show hodograms of the magnetic field as the disturbance is crossed [Russell et al., 1987].

a period greater than their duration. Figure 2 shows high resolution, 0.1 sec, measurements across the first of these fluctuations. The fluctuation is shown as a time series both in spacecraft coordinates and in principal axis coordinates, and in the form of hodograms in principal axis coordinates.

The principal axis data show that the variation occurs along a single direction in the plasma but that this direction is not parallel to the magnetic field. There is a significant non-field-aligned component of the fluctuation. The fluctuation is therefore not a simple diamagnetic depression, i.e., a non-propagating disturbance in pressure equilibrium, but rather it appears to be a slow mode wave propagating at an oblique angle to the magnetic field. We can find the direction of propagation by noting that the direction of the field perturbation must be perpendicular to the direction of propagation as should the vector cross product of the background magnetic field and the field change. Thus the triple cross product of first the background magnetic field times the field change and then times the field change again gives the direction of propagation. We defer an analysis of the direction of propagation to a later section of this paper. Table 1 lists the location of each

of these events, the background field, the field change, the duration of the event and the change in magnetic energy during the event. All vectors are given in comet-centered solar-ecliptic coordinates with x directed toward the sun, and z along the ecliptic pole. The background magnetic field is an average of the field before and after the event. The change in field is the difference in the field at the center of the depression and this background field. The duration of the event is taken to be the period over which the field depression is greater than 50% of its maximum depression. The change in magnetic energy density is the difference between the magnetic energy measured at the peak of the event and the average of the before and after energy densities.

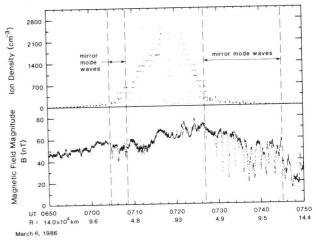

Fig. 3. (Top). Time series of heavy ion number densities measured within the hour of closest approach to comet Halley by the BD-3 plasma detector on VEGA-1. (Bottom). Simultaneously measurements of the strength of the magnetic field [Vaisberg et al., 1989].

Figure 3 shows the time series of heavy ion densities measured during the hour of closest approach on VEGA-1 together with the simultaneously measured magnetic field magnitude [Vaisberg et al., 1989]. Inbound at about 70,000 km from the nucleus are 3 fluctuations similar to those seen inbound along the VEGA-2 trajectory. Outbound there are even more of these waves. Again the events are narrow compared to their separation with few exceptions. We first examine the 3 events inbound which occurred roughly in the same location as the VEGA-2 events described above.

Table 1 lists the observed properties of the events. The background field strength is a little weaker during the VEGA-1 inbound events (50 nT versus 60 nT) because these events are observed further from the nucleus than on VEGA-2. The change in field for each of the events is weaker than the 2 larger VEGA-2 events but larger than the weakest VEGA-2 event. The duration of the events is also similar at about 20 s.

Table 1. Observed Properties of Events

Time	Background B	Change in B	Location	Duration	$\Delta(B^2/8\pi)$
0704:54 Day 65	(-30.1, 35.0, 20.3) nT	(-6.0, 4.5, 10.1)	(-1.8, 7.0, -.6)x10^4 km	15 s	4.7x10^{-9} ergs/cc
0706:19	(-23.6, 34.3, 26.7)	(-8.1, 8.9, 3.3)	(-1.6, 6.3, -.6)	26	5.1
0708:06	(-22.9, 32.5, 33.2)	(-3.7, 5.6, 8.2)	(-1.2, 5.6, -.5)	21	3.2
0726:40	(25.7, 64.0, 0.1)	(1.1, 3.5, 5.5)	(1.9, -2.5, 0.6)x10^4 km	9	2.9
0727:12	(23.0, 57.7, 19.4)	(7.7, 9.2, -1.8)	(2.0, -2.8, 0.7)	5	5.3
0729:31	(22.0, 53.2, 20.0)	(6.3, 17.6, -8.4)	(2.4, -3.8, 0.8)	12	9.0
0730:59	(24.9, 55.0, -1.0)	(24.2, 31.6, 2.7)	(2.7, -4.4, 0.9)	9	12.1
0732:22	(25.0, 55.8, 5.3)	(26.6, 33.4, 12.8)	(2.9, -5.0, 1.0)	24	13.6
0733:54	(23.8, 52.0, 21.7)	(22.3, 21.7, 16.4)	(3.2, -5.7, 1.1)	7	11.4
0734:47	(16.6, 40.4, 34.6)	(1.4, 3.9, 9.3)	(3.3, -6.1, 1.1)	5	3.7
0735:53	(12.7, 32.5, 31.6)	(9.9, 15.3, 19.9)	(3.5, -6.6, 1.2)	13	6.6
0736:30	(14.4, 37.9, 33.0)	(6.7, 7.3, 1.3)	(3.6, -6.8, 1.2)	8	2.9
0736:50	(15.4, 38.7, 31.4)	(6.4, 8.0, 13.9)	(3.7, -7.0, 1.2)	10	7.6
0737:05	(8.1, 29.3, 24.9)	(0.7, 13.3, 18.2)	(3.7, -7.1, 1.2)	16	5.2
0737:48	(14.2, 26.4, 37.4)	(2.0, 9.2, -0.8)	(3.8, -7.4, 1.3)	3	2.4
0738:27	(17.9, 41.5, 23.6)	(10.3, 16.7, 29.2)	(3.9, -7.7, 1.3)	17	9.1
0740:04	(16.6, 42.4, 23.8)	(9.2, 15.7, 12.6)	(4.2, -8.4, 1.4)	21	7.8
0741:13	(14.6, 41.0, 10.5)	(8.8, 19.3, 11.0)	(4.4, -8.9, 1.5)	20	6.3
0741:51	(9.4, 42.8, 16.9)	(7.7, 17.0, 6.3)	(4.5, -9.2, 1.5)	9	5.4
0742:32	(14.6, 47.1, 20.3)	(0.2, 5.6, 4.5)	(4.6, -9.5, 1.5)	9	3.3
0743:14	(9.7, 41.8, 22.0)	(7.1, 17.5, 16.1)	(4.8, -9.8, 1.6)	22	7.4
0711:48 Day 68	(45.2, -21.3, -15.4)	(33.0, -20.0, -0.9)	(-.8, 3.8, -.3)	28	8.3
0713:20	(47.3, -18.8, -4.6)	(17.5, -10.5, 1.5)	(-.5, 3.1, -.2)	16	7.7
0715:31	(55.6, -26.2, -1.9)	(5.1, -3.0, 3.0)	(-.1, 2.2, -.1)	9	4.2

As can be seen from Figure 3 there appear to be many more events outbound than inbound. The events occur over a greater range of radial distance and are deeper. Otherwise the fluctuations are very similar to those seen inbound. Figure 4 shows an example of the outbound fluctuations at 0733:55 UT. It is quite a smoothly varying feature, almost linearly polarized. Figure 5 shows a second example at a greater radial distance. The hodograms look similar to the earlier events except that the depressions last longer in time. This tendency is also evident in Figure 3.

The 24 events listed in Table 1 provide a good statistical sample of this phenomenon. The average event lasts 14 seconds, and during the event the magnetic energy density decreases 6.5×10^{-9} ergs/cm^3. They are found out to radial distances of 7×10^4 km when VEGA was behind the nucleus to over 10^5 km in front. There is a range of distances near the nucleus over which the events are not seen. Reference to Figure 3 shows that this is the region over which the ion density was greater than about 1000 cm^{-3}.

Figure 6 shows the locations of the events seen along the two trajectories. The top panels show projections of the two trajectories in cometary solar ecliptic coordinates. The VEGA-2 trace has only an inbound leg. The panels on the left show the view from above the solar ecliptic and on the right the view from the Sun. The bottom panels show the trajectory rotated to keep the magnetic field in the Y-direction. Symmetry has been invoked to double the coverage in the left-hand panel by assuming no right-left asymmetry and to quadruple the coverage in the right-hand panel by assuming 4-fold symmetry, east-west and north-south. From the available coverage there seems to be no preference in occurrence location except to occur within 10^5 km of the nucleus.

Derived Properties of the Events

As noted above it is possible to derive the direction of propagation of these fluctuations from the triple cross product of the background magnetic field, the field change and the field change again

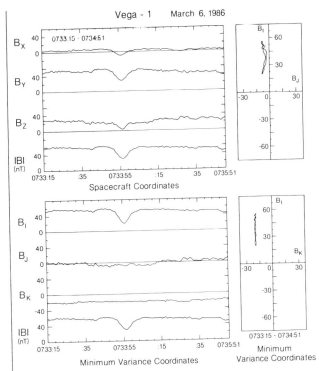

Fig. 4.   High resolution measurements of the
magnetic field from 0733:15 to 0735:51 UT on March
6, 1986 during the VEGA-1 encounter with comet
Halley.   See caption of Figure 2 for further
details.

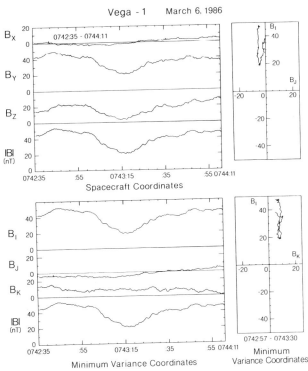

Fig. 5.   High resolution measurements of the
magnetic field from 0742:35 to 0744:11 UT on March
6, 1986 during the VEGA-1 encounter with comet
Halley.   See caption of Figure 2 for further
details.

under the assumption that these features are
magnetosonic waves. Since these fluctuations have
a strong anticorrelation between the number density
and the field strength and since the hodograms of
the waves reveal a wave-like rather than discontinuity like structure we feel this is an accurate
assumption.   Table 2 lists the direction of
propagation for those fluctuations for which we
believe the background magnetic field remains
sufficiently steady during the event and the event
was sufficiently strong (> 10 nT). This amounts to
about only 1/3 of the events.   The average
direction of propagation to the field is 72° with
a slight tendency for the direction of propagation
to be more perpendicular to the field at greater
distances from the nucleus.
    To the extent the plasma is at rest and the wave
velocity much lower than the speed of the VEGA
spacecraft relative to Halley we can calculate the
velocity of the spacecraft perpendicular to the
wavefronts and convert the event durations into
sizes. The average thickness is 800 km. Although
most of the events have thicknesses close to the
average, we note that 2 do not. This suggests that
despite our caution in analyzing steady events some
non-steady fields were included.

Plasma Properties

    As shown in Figures 1 and 3 there is an anti-
correlation between the magnetic field and the
plasma density in these events.   This anticorrelation has been treated in detail by Vaisberg
et al. [1989] using the VEGA-1 measurements of the
BD-3 plasma detector. This correlation is shown in
Figure 7 for all events for which the BD-3 detector
was operating in its sensitive mode.   Figure 7
extends beyond the end of the period for which we
believe we can accurately analyze the direction of
propagation of the events. In fact there is only
one event in Figure 7 that we believe is steady
enough and deep enough to analyze in detail with
some confidence. That is event number 2.

    However, we can use the anti-correlation between
the field strength and plasma density to derive an
ion temperature under the assumption that the sum
of the magnetic and ion pressures is constant
across the events. This is done in Figure 8 and
the results summarized in Figure 9 and Table 3.
Figure 8 also shows a model ion temperature which
demonstrates that the ion temperatures obtained are
generally as expected with the possible difference

Table 2.   Derived Properties of Events

Time	$\underline{k}$	θ_{Bk}	V_{11}	$D_{50\%}$
0704:54 Day 065	(−.37, .75, −.55)	59°	71 km/s	1060 km
0729:31	(−.25, −.34, −.91)	61°	9	110
0730:59	(−.77, 0.60, −.20)	76°	67	600
0733:54	(−.57, 0.78, −.25)	68°	76	530
0735:53	(−.42, 0.80, −.43)	77°	75	970
0741:13	(0.08, 0.47, −.88)	75°	41	810
0741:51	(−.82, 0.47, −.30)	83°	61	550
0743:14	(−.05, 0.69, −.73)	76°	73	1610
0711:48 Day 068	(0.16, 0.31, −.94)	73°	27	760

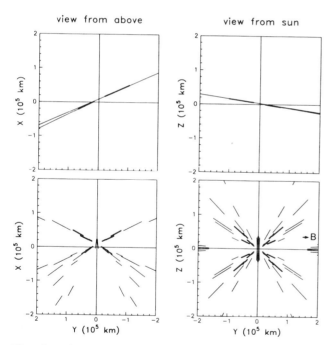

Fig. 6. The trajectory of VEGA-1 and -2 past comet Halley showing with a heavy line the location of the apparent mirror mode fluctuations. The top 2 panels show the projections as seen from above the ecliptic plane (left) and from the sun (right). The bottom 2 panels show the location in a coordinate system that keeps the magnetic field in the Y-direction. The left-hand panel assumes left-right symmetry of wave occurrence. The right-hand panel assumes left-right symmetry as well as top-bottom symmetry in occurrence.

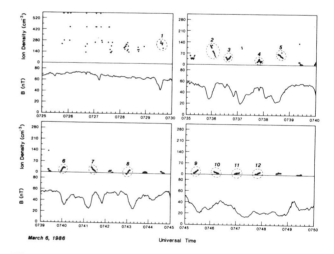

Fig. 7. Magnifications of sections of Figure 3 showing the detailed anti-correlation of the field strength and plasma density [Vaisberg et al., 1989].

that they are colder near the nucleus and warmer further from the nucleus than the model predicts.

Table 3 also displays the water group ion gyro radius for the 6 events of the 12 for which we have some confidence in our measurement of duration, ion temperature and density. The gyro radius is about 450 km. Thus these structures have a thickness of about an ion gyro diameter, if the relative velocity of the structures and the spacecraft is dominated by the spacecraft velocity relative to the nucleus. We note that the one event in which we have some confidence, the 0735:53 UT event, agrees with our statistical comparison.

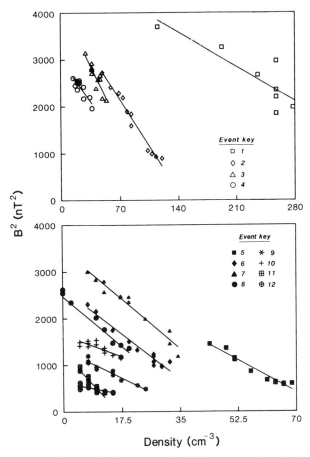

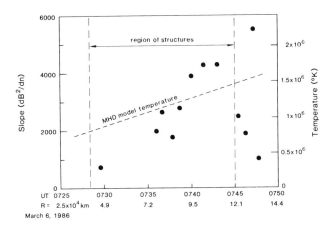

Fig. 9.   The slopes of the segments shown in Figure 8 as well as their conversion to ion temperature. Plasma temperatures from an MHD model are also shown for comparison [Vaisberg et al., 1989].

Fig. 8. Plots of the magnetic field squared versus density obtained from each of the structures numbered in Figure 7. Lines show least square fits to the data [Vaisberg et al., 1989].

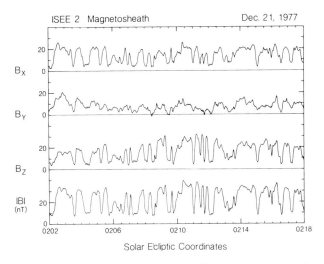

Fig. 10.   ISEE-2 magnetic field measurements in solar ecliptic coordinates in the magnetosheath just outside the magnetopause during an interval of mirror-mode like waves.

Discussion

The fluctuations discussed above seem at first unusual. They are quite different from the waves seen at larger distances and which appear to be fast magnetosonic waves (e.g., Smith et al., 1986). Rather, these waves have the characteristics of slow magnetosonic waves. Such waves are not unknown in space plasmas. Figure 10 shows a series of fluctuations observed by ISEE-2 just outside the magnetopause. These magnetosheath waves have been shown by Crooker et al. [1977] and by Moustaizis et al. [1986] to have the properties of slow mode waves perhaps driven by the mirror instability. The major difference is that in the magnetosheath the field depressions last a time equal or longer than the enhancements, whereas in the cometary magnetosphere the field depressions are relatively brief. Figure 11 shows an analysis of a typical magnetosheath slow or mirror mode fluctuation presented in the format of Figures 2, 4 and 5. This shows that the form of the fluctuation is

indeed very similar.   Figure 12 shows another example. This one is from a disturbance deep in interplanetary space [Russell, 1990]. Thus, this form of disturbance occurs under a wide range of conditions.

We have stated above that such waves have the properties of slow mode or mirror mode waves. Figure 13 illustrates the structure of such a wave [Siscoe, 1983].   The sinusoids represent the magnetic field lines. The dashed lines show wavefronts propagating at a large angle to the magnetic field. The ions occupy the weak field region, mirroring between the stronger fields to the top and bottom of the diagram. The observed thickness of these structures of about one ion gyro diameter is consistent with this picture.

Table 3.   Plasma Properties of Events

Time	M_{max}	T_i	B_{ext}	R_g^{WG}	$D_{50\%}$
0729:31	280 cm^{-3}	3×10^5 K	61	217	(110)
0735:53	120	8	47	460	970
0736:30	60	11	52	487	---
0737:48	40	7	48	421	---
0738:27	70	12	51	520	(809)
0740:04	33	16	51	600	(720)

Note: ( ) = less confidence in data.

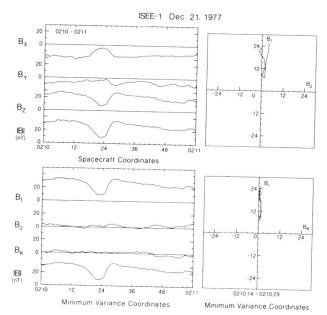

Fig. 11.   ISEE-1 magnetosheath measurements analyzed in a manner similar to that used in Figures 2, 4 and 5.

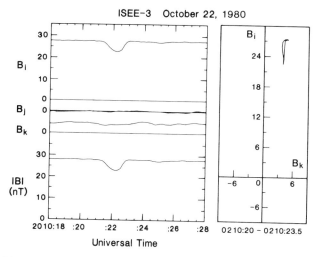

Fig. 12.   ISEE-3 magnetic field measurements of a fluctuation deep in interplanetary space which appears also to be a mirror-mode oscillation [Russell, 1989].

Under normal circumstances, however, this instability is not expected to be the most unstable. Rather the ion cyclotron instability which is also driven by a greater perpendicular than parallel temperature should have the higher growth rate. This is illustrated in Figure 14 which shows the growth rate of various plasma modes under typical plasma conditions [Barnes, 1979]. Perhaps the answer lies in the multicomponent nature of the plasma which will modify the growth rate of the ion cyclotron waves [Price et al., 1986]. It is unfortunate that we do not have measurements of the temperature anisotropies so

that we could determine if the mirror mode was in fact unstable.

An important question is whether such structures which clearly have ray like properties are in fact the cause of visible cometary rays. We note that the density enhancement in these structures drops with distance from the nucleus as the ion temperatures rise. Thus these structures will have the greatest effect closer to the nucleus. On the other hand, they disappear on both VEGA 1 and 2 close to the nucleus. If we take our largest enhancement of 280 cm^{-3} and a 220 km ion gyro radius we obtain an enhancement integrated along the line-of-sight of 10^9 ion/cm^3 if our line-of-sight is perpendicular to the plane containing the ions. We would not see this against the background of the comet which contains at least 1000 times

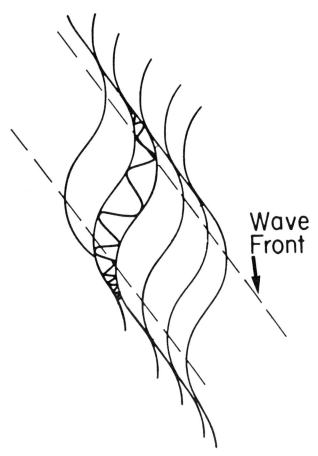

Fig. 13. The relationship between the field line oscillations (sinusoids), the wavefronts and the ion motion in a mirror-mode wave [Siscoe, 1983].

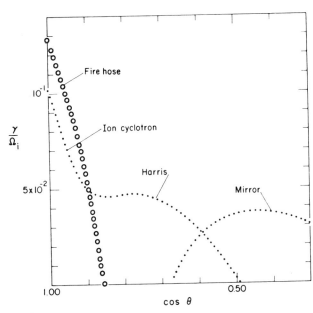

Fig. 14. The growth rate of various plasma instabilities for typical plasma conditions as a function of the direction of the propagation vector relative to the magnetic field [Barnes, 1979].

this integrated density [Russell et al., 1989]. However, if one were to look tangent to the wavefront, one might get a sufficient enhancement in visual emissions. Moreover, the enhancement would have more of the appearance of a ray under such geometry than if viewed perpendicular to the wavefront.

Conclusions

The VEGA 1 and 2 magnetometers and ion instruments detected fluctuations in the magnetic field and ion density that appear to be slow mode or mirror mode waves. These waves are propagating at a large angle to the magnetic field and have scale sizes of approximately a gyro diameter. They may be responsible for condensing ions into structures that become visible when observed edge on.

Acknowledgments. This research was supported by the National Aeronautics and Space Administration under research grant NAGW-717.

References

Barnes, A., Hydromagnetic waves and turbulence in the solar wind, in Solar System Plasma Physics, (eds. E. N. Parker, C. F. Kennel and L. J. Lanzerotti), North Holland Publ. Co., 1979.

Crooker, N. U. and G. L. Siscoe, A mechanism for pressure anisotropy and mirror instability in the dayside magnetosheaath, J. Geophys. Res., 82, 185, 1977.

Galeev, A. A., et al., Critical ionization velocity effects in the inner coma of comet Halley: Measurements by VEGA-2 Geophys. Res. Lett., 13, 845-848, 1986.

Gringauz, K. I., et al., First in-situ plasma and neutral gas measurements at comet Halley, Nature, 321, 282-285, 1986.

Moustaizis, S., et al., Magnetohydrodynamic turbulence in the earth's magnetosheath, Ann. Geophys., 4, 355-362, 1986.

Price, C. P., D. W. Swift and L. C. Lee, Numerical simulation of nonoscillatory mirror waves at the earth's magnetosheath, J. Geophys. Res., 91, 101-112, 1986.

Riedler, W., K. Schwingenschuh, Ye. G. Yeroshenko, V. A. Styashkin and C. T. Russell, Magnetic field observations in comet Halley's coma, Nature, 321, 288-289, 1986.

Russell, C. T., Interplanetary magnetic field enhancements: Evidence for solar wind dust trail interactions, Adv. Space Res., 10, (3)159-(3)162, 1990.

Russell, C. T., W. Riedler, K. Schwingenschuh and Ye. Yeroshenko, Mirror instability in the magnetosphere of comet Halley, <u>Geophys. Res. Lett.</u>, <u>14</u>, 644-647, 1987.

Russell, C. T., G. Le, J. G. Luhmann and J. A. Fedder, The visual appearance of comets under varying solar wind conditions, <u>Adv. Space Res.</u>, <u>9</u>, (3)393-(3)396, 1989.

Siscoe, G. L. Solar system magnetohydrodynamics, in <u>Solar Terrestrial Physics: Principles and Theoretical Foundations</u>, edited by R. L. Carovillano and J. M. Forbes, p. 11-100, D. Reidel Publ. Co., Boston, 1983.

Smith, E. J., et al., International Cometary Explorer encounter with Giacobini-Zinner: Magnetic field observations, <u>Science</u>, <u>232</u>, 382-385, 1986.

Trotignon, J.-G., R. Grard and M. Mogilevsky, Electric-field measurements in comet P/Halley's environment by high frequency plasma-wave analyzer, <u>Ann. Geophys.</u>, <u>7</u>, 331-340, 1989.

Vaisberg, O. L., C. T. Russell, J. G. Luhmann and K. Schwingenschuh, Small scale irregularities in comet Halley's plasma mantle: An attempt at self-consistent analysis of plasma and magnetic field data, <u>Geophys. Res. Lett.</u>, <u>16</u>, 5-8, 1989.

Yeroshenko, Ye. G., V. A. Styashkin, W. Riedler, K. Schwingenschuh and C. T. Russell, Magnetic field fine structure in comet Halley's coma, <u>Proceedings of 20th ESLAB Symposium on Exploration of Halley's Comet</u>, 189-192, ESA SP-250, European Space Agency, Paris, 1987.

SPECTRAL STRUCTURE OF ULTRALOW-FREQUENCY ELECTROMAGNETIC FIELDS NEAR COMET HALLEY

Yu. M. Mikhailov[1], O.V. Kapustina[1], G.A. Mikhailova[1], E. G. Eroshenko[1], V.A. Styazhkin[1], J.G. Trotignon[2], K. Sauer[3]

(1)IZMIRAN, Troitzk, Moscow Region 142092, USSR
(2)LPCE-CNRS, G-45045, Orleans, France
(3)IKF, Berlin, Germany

Abstract The energy spectra of electromagnetic field and plasma density in the 10^{-4} - 4×10^{-2} Hz frequency band observed on Vega 1 behind bow shock are presented. It has been shown that the spectral maxima coincide with the cyclotron frequencies of cometary molecular ions.

Introduction

The wave diagnostics in the plasma environment of comet Halley have been revealing some interesting effects, such as the occurrence of low-frequency plasma turbulence [Sagdeev et al., 1986; Galeev et al., 1986], the formation of boundaries and discontinuities [Glassmeier et al., 1989], and the occurrence of a new ion species [Tsurutani et al., 1986]. The plasma radiation in the lower ultralow frequency band (10^{-4} - 10^{-1} Hz) which comprises the cyclotron frequencies of heavy cometary ions, is of great interest at present.

The characteristic of the ultralow-frequency waves upstream from the bow shock have been described elsewhere in quite sufficient detail [Acuna et al., 1986; Fusilier et al., 1987]. For example, the measurements of plasma velocity vector V and magnetic field vector B show maximum correlation for the same components of the velocity and the magnetic field. Besides, the frequency maxima of the magnetic field spectrum closely coincide with the cyclotron frequencies of cometary molecular ions [Glassmeier et al., 1989].

At the same time, the region behind shock wave at 9×10^5 - 6.5×10^5 km from the cometary nucleus (the so called "mystery" region) has been studied in much less detail. Very high-energy electron fluxes along the magnetic field vector were detected in the region [Reme et al., 1987]. The present work is a study of just this region. Contrary to the works cited above, where magnetometer data only were used in analysis, we have used a broader set of experimental data including measurements of the electric field E_y, of plasma parameters [Grard et al., 1986] and of the magnetic field [Riedler et al., 1986]. More complete results (compared with Mikhailov et al., [1987]) of analyzing the energy spectra of the parameters N_e, E_y, B_x, B_y, B_z and B_m will be presented

below. Although the results constitute only the first stage of studying the wave structure of the plasma environment of the comet, they are nevertheless of interest in themselves.

Experimental Design

We shall describe the experiment briefly. A more detailed description may be found in Grard et al.,[1986]. The high-frequency plasma-wave analyser, APV-V, placed on board the spacecraft uses a double sphere antenna with a baseline of 11m, to detect electric fields in the bandwidth 0-300kHz [Grard et al.,1985]. The APV-V narrow band filter channels were supplemented with a constant electric field channel of $11\mu V/m$ sensitivity and 0 - 0.5 Hz bandwidth. It also measured electron density and temperature with a set of two Langmuir probes. These probes were covered with a mechanical protector against direct penetration of dust grains. The current probe, which was kept at a fixed bias of +5 V had a sensitivity of 10^{-10} A and a bandwidth of 0 - 0.5 Hz. The temperature was measured by a cylindrical electrode which was kept at a floating potential of +6V with a 32s period. When the total potential on the temperature probe reached +5V, the currents flowing to the density and temperature probes were equal thereby permitting the temperature probe to be used to infer density with 32s time resolution. In such a way, the density probe data were controlled with a temperature probe. Figure 1 shows the variations in the density probe current I (L_1) and the temperature probe current I (L_2) as Vega 1 approaches the comet. The signal from the Langmuir probe was registered by the electric field receiver and can be used for checking the electric field measurements. It is unlikely that the electric field measuring circuit was influenced by the Langmuir probe through changing the floating potential of the spacecraft. As shown below, the measured signal from the Langmuir probe as an emitter 2cm long is in good agreement with preliminary calculations

The interval from 04:10 to 04:52 UT on March 6th, 1986 was selected to use when studying in detail the spectra of the measured values of N_e, B_x, B_y, B_z and B_m, E_y. The beginning of the interval coincides with the starting time of high-speed telemetry transmission, the end corresponds to the moment of gas release by the infrared spectrometer. The total duration of the records was 2242s. The respective interval on

Cometary Plasma Processes
Geophysical Monograph 61
©1991 American Geophysical Union

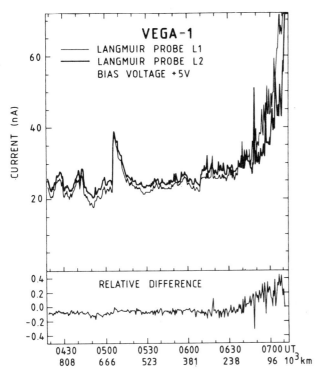

Fig.1 Density probe current $I(L_1)$ and temperature probe current $I(L_2)$ as the spacecraft approaches the comet. The relative difference between the two currents is,

$$2[I(L_1) - I(L_2)] / [I(L_1) + I(L_2)]$$

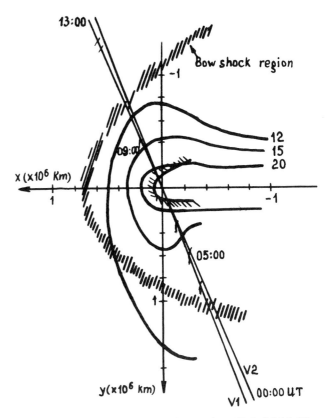

Fig 2 Vega-1 trajectory. The interval studied, 04.14:47 - 04.51:35 UT is between the 05.00 UT mark and the bow shock. The measurement base of the electric probes (thin vertical line) is parallel to the y-axis. The thick solid lines are magnetic field isointensity contours.

board VEGA 2 was shorter (smaller than 1000s) and therefore, was not analyzed in detail. The position of VEGA-1 during the selected interval on March 6th, 1986 with respect to the bow shock and to the magnetic field isointensity contours calculated in terms of the Neubauer model [Neubauer, 1986] is shown in Figure 2 which indicates also the distance to the spacecraft from the comet nucleus (900,000 to 700,000km). From the figure it is seen that during the interval, the spacecraft was immediately behind the bow shock. In this region, which was free of dust, MHD waves develop and are under the effect of the solar wind, its direction relative to the bow shock and the newly-formed cometary ions.

Initial Data and Method of Processing

To estimate and to compare the spectra of the records parameter N_e, E_y, B_x, B_y, B_z, and B_m the interval from 04:14:47 to 04:52:35 UT was chosen. Figure 3 shows the time variations of the initial records averaged over intervals of dt = 6s, so that total duration is T = N dt = 2148 s, N = 358. Figure 4 presents the amplitude spectra of these data estimated by the FFT algorithm with a frequency step dt = 8.14 x 10^{-5} Hz. Analysis of the data in Figures 3, and 4 shows that the observed processes are the sum of at least two random processes produced by different origins. One of the processes is more intensive and exhibits a pronounced maximum at f ~ 10^{-3} Hz, while the second process is less intensive and

exhibits a series of maxima at frequencies f > 0.003 Hz. The reliability of the spectra S, which were obtained by the Fourier method on a single record of a random process was verified using two more methods: namely modified periodogram method (MPM) [Welch, 1976] and the maximum entropy method (MEM) which is known to be more effective when used as a frequency meter with a high frequency resolution [Mikhailova et al., 1986]. The following parameters were selected to use the MPM: a cosine time "window"; the duration of short-term interval M = N, N/2, the number of overlaps K = 1, 3; the relative variance σ of estimating the power density spectrum (PDS) G(f) for K = 3 is 0.25. In the MEM, the autoregression order was chosen to be m = N/3; (N/3 + N/2)/2; N/2. Shown as an illustration in Figures 5, and 6 are the electric field PDS, estimated by the MPM and MEM. Similar curves have been obtained for all the records examined. Figure 6 presents strong burst at a frequency equal to 0.0315 Hz which is due to a signal emitted by the Langmuir probe. This burst PDS corresponds to a field intensity of $E_y = 150 \mu$ Vm^{-1}. This signal may be considered as a calibration signal.

The analysis of the entire set of the spectra estimated by the MPM for K = 1 and 3 and by the MME for the variable m has shown the following.

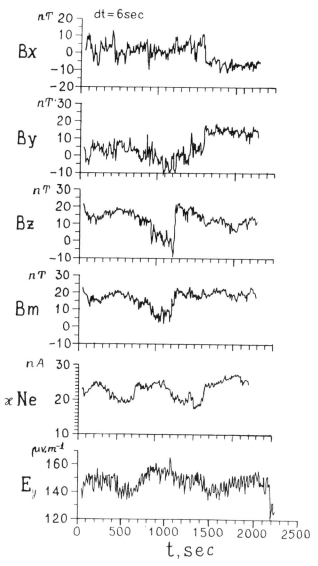

Fig.3 Variations of B_x, B_y, B_z, B_m, N_e, E_y, in the interval studied, 04:14:47 - 04:51:35 UT.

$\kappa^{-1} = 4.0$ sm $^{-3}$ /na

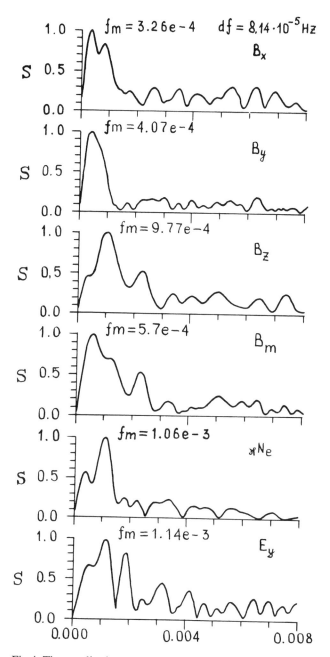

Fig.4 The amplitude spectra S(f) (in relative units), estimated by the FFT algorithm.

1. As expected, the MPM-estimated power spectra are smoother than the Fourier spectra S, but exhibit clearly (except the first intensive maximum) a series of less intensive high-frequency maxima whose PDS is an order lower than that of the lower-frequency maximum, but is more than an order of magnitude above the background;

2. The MEM is known to allow a higher frequency resolution, but the frequency estimate obtained by the method depends on numerous factors. In particular, the estimate depends on signal-to-noise ratio, on the initial phase of sinusoidal oscillations, on phase differences in the case of several sinusoidal oscillations, and on the autoregression order m. Unfortunately, the dependences have not been analyzed theoretically elsewhere, so we estimated the frequency shift using the result of calculations by varying m

from N/3 to N/2 at a sufficiently high signal-to-noise ratio (greater than 4). The results of the given approach are displayed in figure 5 showing a clear dependence of the frequency shift and of the spectral line broadening on the parameter m. Since the narrowest spectral lines occur at m = N/2, the MEM will be used below with just the given autoregression order. The limits of the frequency shift in the dependence on m lie within the frequency resolution determined at FWHM in the MPM-estimated spectrum.

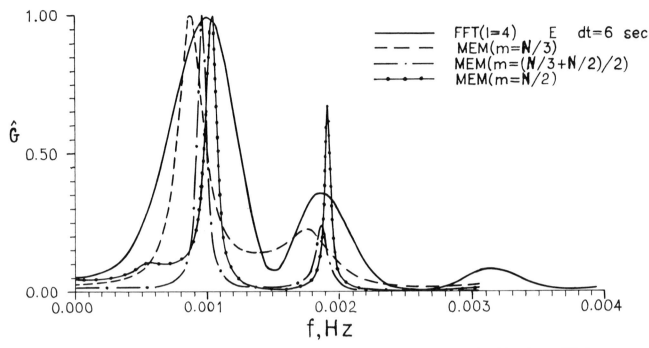

Fig.5 Power density spectrum G(f) (in relative units) of electric field E_y calculated by the modified periodogram method for M = N (the solid line) and by the maximum entropy method for M = N/3 (the dashed line), m = (N/3+N/2)/2 (the dashed-dotted line) and m = n/2 (the solid dotted line).

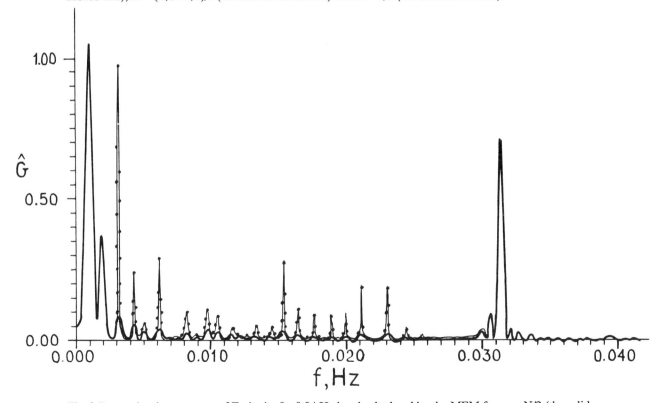

Fig.6 Power density spectrum of E_y in the 0 - 0.04 Hz band calculated by the MEM for m = N/2 (the solid dotted line) and by the MPM (the solid line).

Table 1

ion species / ion mass	κN_e		E_y		B_z		B_m	
	f, 10^{-3} Hz	G(f) nA^2 Hz^{-1}	f, 10^{-3} Hz	G(f) $(\mu$V/m$)^2$ Hz^{-1}	f, 10^{-3} Hz	G(f) nT^2 Hz^{-1}	f, 10^{-3} Hz	G(f) nT^2 Hz^{-1}
	1.10 ± 0.25	$6.65 \cdot 10^{2}$	0.98 ± 0.30	$1.86 \cdot 10^{3}$	0.98 ± 0.29	$2.27 \cdot 10^{3}$	0.65 ± 0.38	$1.20 \cdot 10^{3}$
	1.79 ± 0.15	$0.50 \cdot 10^{2}$	1.87 ± 0.25	$0.68 \cdot 10^{3}$	2.20 ± 0.36	$0.49 \cdot 10^{3}$	2.28 ± 0.28	$0.26 \cdot 10^{3}$
Fe$^+$, 56	3.33 ± 0.28	$0.45 \cdot 10^{2}$	3.17 ± 0.25	$0.15 \cdot 10^{3}$	3.50 ± 0.36	$0.10 \cdot 10^{3}$	3.42 ± 0.25	$0.21 \cdot 10^{2}$
CO$_2^+$, 44	4.23 ± 0.25	$0.11 \cdot 10^{2}$	4.22 ± 0.50	$0.8 \cdot 10^{2}$	5.05 ± 0.36	$0.14 \cdot 10^{3}$	5.13 ± 0.40	$0.52 \cdot 10^{2}$
S$^+$, 32	5.62 ± 0.25	$0.14 \cdot 10^{2}$	6.14 ± 0.50	$0.5 \cdot 10^{2}$	6.27 ± 0.35	$0.65 \cdot 10^{2}$	6.35 ± 0.20	$0.23 \cdot 10^{2}$
N$_2^+$,CN$^+$ 26	7.08 ± 0.27	$0.35 \cdot 10^{1}$	6.92 ± 0.20	$0.1 \cdot 10^{2}$	7.32 ± 0.35	$0.8 \cdot 10^{2}$	7.41 ± 0.30	$0.78 \cdot 10^{1}$
21	8.94 ± 0.40	$0.46 \cdot 10^{1}$	8.17 ± 0.50	$0.3 \cdot 10^{2}$	8.54 ± 0.35	$0.3 \cdot 10^{2}$	8.34 ± 0.30	$0.84 \cdot 10^{1}$
H$_2$O$^+$, 18	10.10 ± 0.30	$0.46 \cdot 10^{1}$	9.77 ± 0.50	$0.5 \cdot 10^{2}$	10.30 ± 0.30	$0.1 \cdot 10^{2}$	9.60 ± 0.30	$0.47 \cdot 10^{1}$
O$^+$, 16	11.70 ± 0.30	$0.67 \cdot 10^{1}$	11.60 ± 0.50	$0.2 \cdot 10^{2}$	12.00 ± 0.30	$0.2 \cdot 10^{2}$	11.60 ± 0.30	$0.71 \cdot 10^{1}$
N$^+$, 14	13.30 ± 0.30	$0.12 \cdot 10^{1}$	13.30 ± 0.50	$0.2 \cdot 10^{2}$	13.80 ± 0.30	$0.35 \cdot 10^{2}$	13.10 ± 0.30	$1.17 \cdot 10^{1}$
C$^+$, 12	16.40 ± 0.30	$0.38 \cdot 10^{1}$	15.40 ± 0.50	$0.5 \cdot 10^{2}$	14.00 ± 0.30	$0.3 \cdot 10^{2}$	15.20 ± 0.30	$2.61 \cdot 10^{1}$
10	18.00 ± 0.30	$0.21 \cdot 10^{1}$			18.90 ± 0.30		17.50 ± 0.30	$1.59 \cdot 10^{1}$
8	19.80 ± 0.40	$0.54 \cdot 10^{1}$					18.40 ± 0.30	$0.68 \cdot 10^{1}$
	20.90 ± 0.30	$0.30 \cdot 10^{1}$	21.10 ± 0.50	$0.3 \cdot 10^{2}$			20.40 ± 0.30	$0.25 \cdot 10^{1}$
			23.10 ± 0.50	$0.3 \cdot 10^{2}$	22.00 ± 0.30	$0.2 \cdot 10^{2}$	23.30 ± 0.30	$0.30 \cdot 10^{1}$
noise level	$0.40 \cdot 10^{-1}$		$0.3 \cdot 10^{1}$		$0.4 \cdot 10^{1}$		0.4	

Table 1 presents the frequencies of all reliable maxima estimated by the MMP at K = 1 with frequency resolution at FWHM and specified by the MME at m = N/2. Also presented is the estimated PDS G(f) in the absolute values of measured parameters.

Discussion of Results

The first column of Table 1 indicates the ion species whose cyclotron frequency is the same as the frequency of the maximum. The ion cyclotron frequency f_i was defined as $f_i = 1.53 \times 10^{-2} B_m/M_i$, where M_i is the mass of the ion, B_m is the modulus of the magnetic field in nT. The mean value of B_m taken for the estimates of f_i within the 04:10 - 04:47 UT interval, where the magnetic field is essentially inhomogeneous, is 12.34 nT. For ionized water molecules $f_i = 10.489 \times 10^{-3}$ Hz. Apparently, spectral lines with the frequencies of maxima in the range 10.1 - 10.6 x 10^{-3} Hz can be ascribed to emission from H$_2$O$^+$ ions. Further, using the molecular weight ratios, we can find the frequency maxima corresponding to O$^+$ ions (16 x 10^{-3} Hz), C$^+$ ions (14.8 x 10^{-3} Hz) and N$^+$ ions (13 x 10^{-3} Hz). It is of interest that the frequency 18.0 x 10^{-3} Hz corresponds to an ion of mass 10. The same ion was detected in the PLASMAG experiment [Gringauz et al., 1986] but was not identified there either.

Table 1 presents an interesting maximum at 4.22 x 10^{-3} Hz which seems to be due to CO$_2^+$ ions (M=44). As a whole, Table 1 is an attempt to construct an electromagnetic analogue of the output data of the PLASMAG instrument.

In relating the frequency of a maximum to an ion frequency, it is of importance to find out if a frequency transformation due to the Doppler effects occurs when going from the solar wind coordinate system to the spacecraft coordinate system.

Since the spacecraft velocity is much below the solar wind velocity, the transition from the solar wind coordinates to the cometocentric coordinates is of primary importance. Considering the Doppler effects, the frequency in the solar wind coordinates system f_{sw} is related to the frequency in the cometocentric coordinate system f_{sc} by,

$$f_{sw} = f_{sc} \left(1 + \frac{V_{sw} \cos \psi}{C_R}\right)$$

(1)

where V_{sw} is the solar wind velocity ; C_R is the phase velocity of the waves and ψ is the angle between the solar wind velocity vector and the wave normal vector.

Substituting V_{sw} = 510 km/s [Galeev et al., 1986] and C_R = 300 km/s, as calculated by [Mikhailov et al., 1987] using the specified data on electron number density [Grard et al., 1989] and using a tentative estimate of ψ to be 70° - 80°, we find that the correction to the frequency does not exceed 20%. This error affects the absolute value of the frequency, but does not affect the ratio of frequencies. In general it lies in the error limits of the table 1.

As to the lowest-frequency maxima, they arise probably from excitation of standing MHD waves. As shown by Motschmann et al., [1990], the heavy-ion velocity remains in practice the same at the bow shock, while the proton velocity there exhibits a jump. As a result, the bow shock having been passed, a many-component plasma with relative velocity of its ion components is formed. Such a system is a source of standing (with respect to the cometocentric coordinate system) multifluid waves. The wavelength of the standing structures is defined as:

$$\psi = \frac{<V>}{f_i}$$

(2)

where $<V>$ is the stream velocity taken to be approximately equal to the solar wind velocity V_{sw} = 510 km/s; f_i is cyclotron frequency of water group ions f = 10.4 x 10^{-3} Hz.

Substituting these values in (1) we obtain λ = 49.8 x 10^{-3} km. It nearly coincides with the wave structure shown in Figure 1. If the spacecraft is assumed to have traversed a structure with the standing MHD waves at velocity V_{sc} = 68 km/s (the relative velocity of the spacecraft and the comet), we obtain

$$f = \frac{V_s}{\lambda} = 1.38 \times 10^{-3} \text{ Hz}$$

(3)

This frequency coincides with the frequency of the first reliable maximum from Table 1.

An examination of the variation of B_z in the interval studied shows that the component B_z suffers a jump at the moment t = 1200 s (04:34:47), see Figure 3. The character of the jump corresponds to a rotational discontinuity. The spectral analysis of the intervals 04:14:47 - 04:34:47 and 04:34:47 - 04:51:35 has shown that the low-frequency component 1.3 x 10^{-3} Hz is present only before the jump. After the jump, the spectrum contains mainly the cyclotron harmonics f > 3 x 10^{-3} Hz. This may be indirect evidence for the fact that the low-frequency side of the spectrum is due to the bow shock rather than to the processes associated with the spacecraft approach to the comet.

Thus, the low-frequency component is accumulated in the region between the bow shock and the discontinuity, thereby confirming indirectly the hypothesis of the origin set forth earlier.

Conclusion

The results of a detailed spectral analysis of electromagnetic and plasma parameters at distances of 9 x 10^5 - 6.5 x 10^5 km from the nucleus of comet Halley have been presented. An attempt has been made to relate the radiation maxima to the ion cyclotron frequencies of ionized molecules. An explanation has been proposed for the ultralow-frequency maxima in the spectrum to be accounted for by excitation of standing waves in a many-component plasma with relative velocity of its components.

References

Acuna, M.N., Glassmeier, K.H., Burlaga, L.F., Neubauer, F.M., Ness, F.N.. Upstream waves of cometary origin detected by Giotto magnetic field experiment. *Exploration of Halley's Comet*. ESA SP 250, 1, 1986.

Fuselier, S.A., Anderson, K.A., Balsiger, H., Glassmeier, K.H., Goldstein, B.E., Neugebauer, M., Rosenbauer, H., Shelley, E.G.. The foreshock region upstream from comet Halley bow-shock. *Symposium on the diversity and similarity of comets.*, ESA SP-278, 77-82, 1987.

Galeev, A.A., Gribov, B.E., Gombosi, T., Gringauz, K.I., Klimov, S.I., Oberc, F., Remisov, A.P., Riedler, W., Sagdeev, R.Z., Savin, S.P., Sokolov, A. Yu., Shapiro, V.D., Shevchenko, V.I., Szego, K., Verigin, M.I., Eroshenko, E. G.. Position and structure of the comet Halley bow shock: Vega-1 and Vega-2 measurements. *Geophys Res. Lett.*, 13, 841-844, 1986.

Glassmeier K.H., Coates, A.J., Acuna, M.N., Goldstein, B.E., Johnstone, A.D., Neubauer, F.M., Reme, H.. Spectral characteristic of low-frequency Plasma Turbulence upstream of comet Halley. *J. Geophys. Res.*, 94, 37-48, 1989.

Grard, R.. The plasma and wave experiment of Vega (EPINOCHE-APV-V). *Field, particle and wave experiments on cometary missions,* (ed. K. Schwingenshuh and W. Riedler) Verlag der osterreichischen, Akademie der Wissenshaften, pp. 175-185, 1985.

Grard, R., Pedersen, A., Trotignon, J.G., Beghin, C., Mogilevsky, M., Mikhailov, Yu., Molchanov, O., Formisano, V., Observations of waves and plasma in environment of comet Halley, *Nature*, 321, 290-291, 1986.

Grard, R.M., Laakso, H., Trotignon, J.G., Pedersen, A., Mikhailov, Yu. M., Observations of plasma environment of comet Halley during the Vega flybys. *Ann. Geophys.* 7(2), 141-150, 1989.

Gringauz, K.I., Gombosi, T.I., Remisov, A.P., Apathy, I., Szemerey, I., Verigin, M.I., Denchikova, L.I., Dyachkov, A.V., Keppler, E., Klimenko, I.N., Richter, A.K., Somogyi, A.J., Szego, K., Szendro, S., Tatrallyay, M., Varga, A., Vladimirova, G.A.. First in situ plasma and neutral gas measurements at comet Halley, *Nature*, 321, 282-285, 1986.

Mikhailov, Yu. M., Mogilevsky, M.M., Eroshenko, E.G., Molchanov, O.A., Grard, R., Pedersen, A., Beghin, C., Trotignon, J.G., Formisano, V., Schwingenschuh, K.. Correlation between magnetic and electric field fluctuations during flyby near Halley's comet, *Symposium on the diversity and similarity of comets*, ESA SP-278, 109-112, 1987.

Mikhailova, G.A., Dubovoy, A.P., Kapustina, O.V., Yaroslavtsev, A.A., Application of the method of maximum entropy to estimating the spectral density of the ELF emission power in the outer ionosphere. *Magnetospheric Research, 8,* Nauka, Moscow, pp. 91-95, 1986.

Motschmann, U., Sauer, K., Ronach, T., Mackenzie, J.. Multiple-ion shocks, *Ann. Geophys.* (in press, 1990).

Neubauer, F.M.. Magnetic field model of Halley's comet, *Exploration of Halley's Comet.* ESA SP-250, 35-40, 1, 1986.

Reme, H., Sauvaud, J.A., d'Uston, C., Cros, A., Anderson, K.A., Carlson, C.W., Curtis, D.W. Lin, R.P., Korth, A., Richter, A.K., Mendis, D.A.. General features of comet P/Halley: solar wind interaction from plasma measurements, *Astron. Astrophys,* 187, 33-38, 1987.

Riedler, W., Schwingenschuh, K., Eroshenko, E.G., Styashkin, V.A., Russell, C.T.. Magnetic field observations in comet Halley's coma, *Nature* 321, 288-289, 1986.

Sagdeev, R.Z., Shapiro, V.D., Shevchenko, V.I., Szego, K.. MHD-Turbulence in the solar wind - comet interaction, *Geophys. Res. Lett.,* 13, 85-88, 1986.

Tsurutani, B.T., Smith, E.J.. Hydromagnetic waves and instabilities associated with cometary ion pick up: ICE observations, *Geophys. Res. Lett.* 13, 263-266, 1986.

Welch, P.D.. The use of the fast Fourier transform for the estimation of power spectra: a method based on time averaging over short modified periodograms, *IEEE Trans. Audio and Electroacoustic,* AU-15 N2, 70-73, 1967

ELECTRIC-FIELD DISCONTINUITIES AT COMET HALLEY DURING THE VEGA ENCOUNTERS

J.G. Trotignon[1], C. Béghin[1], R. Grard[2], A. Pedersen[2], M. Mogilevsky[3] and Y. Mikhailov[3]

[1] Laboratoire de Physique et Chimie de l'Environnement, Orléans, France.
[2] Space Science Department, Noordwijk, The Netherlands.
[3] Institute of Terrestrial Magnetism, the Ionosphere and Radio-Wave Propagation, Troitsk, USSR.

Abstract. The high-frequency plasma-wave analyser, APV-V, measured strong electric fields when the Vega spacecraft came in the inner coma of comet P/Halley. Several discontinuities were detected in the region from the bow shock crossing to the closest approach to the nucleus, i.e. at cometocentric distances lying between 10^6 km and about 8,000 km. This paper deals with the electric-field observations in the vicinity of some of these discontinuities: the bow shock, the mystery region, the cometopause and, closer to the nucleus, several sudden changes in the electric-field spectral intensity. The spectral measurements are stressed and an attempt is made to interpret the observations in terms of the plasma and dust environments in the light of recent results.

us simply recall that it is believed to be due to dust impact effects, which do not exclude that another process, like a mirror instability [Russell et al., 1987], could occur close to this event with no relationship between the two processes. The bow shock crossings and the region called "cometopause" by Gringauz et al. [1986] are not really seen on the APV-V electric-field measurements, though MHD waves are detected near the cometopause. On the contrary, the "mystery region" identified on the RPA-COPERNIC plasma experiment on board Giotto [Rème et al., 1987, 1990; d'Uston et al., 1987] is clearly seen, as also are the spiral dust-jet crossings within 140,000 km from the nucleus.

1. Introduction

This report is based on observations of comet P/Halley made by the high-frequency plasma-wave analyser, APV-V, on Vega. The APV-V experiment made use of a double sphere antenna, with a baseline of 11 m, to detect electric fields in the bandwidth 0-300 kHz. It also measured electron density and temperature with a set of two Langmuir probes. Special attention is paid to the electric-field spectrum data recorded between 1 million km and 8,000 km from the nucleus.

An overview of the major transition regions crossed by the Vega spacecraft will be found: their locations are shown in the right sketch of Figure 1. "V1" and "V2" refer to the Vega 1 and Vega 2 observations, respectively. The "V2 event" recorded by Vega 2 between 0711 UT and 0714 UT on 9 March 1986 has been studied in detail in Trotignon et al. [1989] and will not be discussed here. Let

2. E-Field Observations near the bow shock

Figure 2 shows the electric-field intensities from the 16-channel spectrum analyser during a 12-hour period around the comet Halley's bow shock crossings by Vega 1, at the top, and Vega 2, at the bottom. To improve the presentation the number of frequency channels has been multiplied artificially by four. To do that, Lagrangian polynomials of the third degree have been used to interpolate the measurements in frequency. The colour scaling shown on the right side depends on the strength of the received signals which is expressed in dB above 1 μV.m^{-1}.Hz$^{-1/2}$, so that 60 dB correspond to 1 mV.m^{-1}.Hz$^{-1/2}$.

Galeev et al. [1986] reported seeing a bow shock crossing by Vega 1 on March 6, 1986 at 0346 UT, while the bow shock was detected by Vega 2 on March 9, 1986 at about 0240 UT [Oberc et al., 1987]. Unfortunately, the electric-field spectrum data shown in Figure 2 do not allow us to detect any bow shock crossing signature. Indeed, the data are dramatically influenced by satellite subsystem interference: for example, it is the case of the Vega 2 data recorded between 0100 and 0315 UT on March 9,

Cometary Plasma Processes
Geophysical Monograph 61
©1991 American Geophysical Union

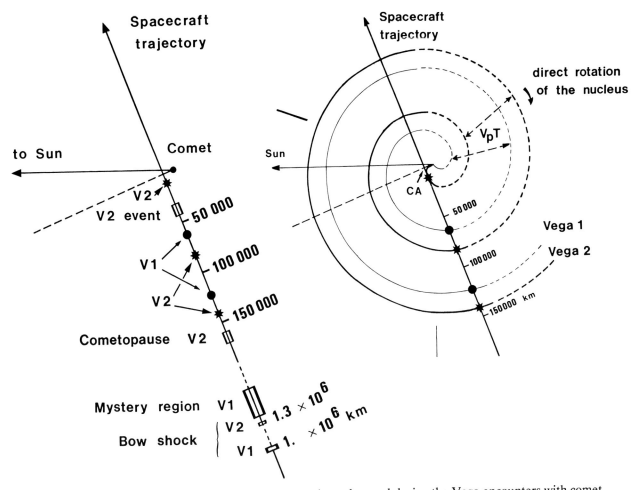

Fig. 1. The locations of the various boundary regions observed during the Vega encounters with comet Halley are shown on the left. V1 and V2 refer to the Vega 1 and Vega 2 observations. The right-hand sketch is borrowed from Trotignon et al. [1987]. It illustrates the interpretation of the large scale gradients observed in the AC electric-field intensity (full circles and stars) as crossings of spiral dust jets.

1986. The large spikes seen at 1820, 2220, 0220 UT and after 0330 UT on both spacecraft are also generated by the satellite or other instruments. Of course other strong signatures are present: sporadic bursts of activity appear at frequencies above 1 kHz, and fluctuations occur below 40 Hz, but they do not differ drastically from the ones observed in the free solar wind. A more detailed description has been given in Trotignon et al. [1989].

3.   E-Field Observations in the Mystery Region

Strong fluctuations were observed in the electron density and in the electric- and magnetic-field measurements, when Vega 1 was at a distance between 870,000 and 650,000 km from the nucleus. A spectral analysis performed in the frequency range from 0.1 mHz to 0.04 Hz by Mikhailov et al. [1987, 1990] has shown that, besides a clear peak at 1.3 mHz, several minor peaks occured at frequencies between 3 mHz and 18 mHz which coincide with the cyclotron frequencies of some of the cometary heavy ions.

A four-hour electric-field spectrogram during the Vega 1 encounter with comet Halley is shown in Figure 3. The region in which the electric-field oscillations occur extends from 0417 UT, the starting time of the high-speed telemetry transmission, to 0503 UT, where a contamination signal due to the release of nitrogen gas by the infrared spectrometer happens. Unfortunately, no such oscillations could be seen in the Vega 2 data, for nitrogen gas was released as early as 0432 UT, as shown in Figure 4.

The Vega 1 sequence is expanded further in the upper panel of Figure 5, where the electric-field intensities have

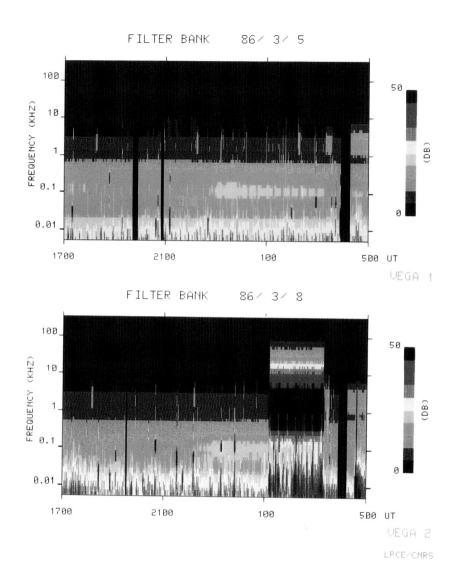

Fig. 2. Electric-field amplitudes measured in the 16 filter channels of the analyser near the bow shock crossing by Vega 1, at the top, and Vega 2, at the bottom. The signal level is expressed in dB above 1 $\mu V.m^{-1}.Hz^{-1/2}$.

been processed through a 20-point running-average filter (10-s average), and the ordinate axis covers the frequency range below 1 kHz. Another representation which bears more quantitative information is also given in Figure 6 which displays 8-s averages of the electric-field levels in the first four frequency channels on Vega 1 between 0415 and 0505 UT. Let us recall that the signals seen in these 4 elements of the filter bank correspond to the lowest 16 lines of the raster which forms the spectrogram shown in the upper panel of Figure 5, as explained in section 2. In Figure 6, the bandwidths of the filters are indicated on the left side and the respective outputs are logarithmically compressed within the same dynamic range, from -70 dB

to -40 dB: 0 dB corresponding to 1 V rms. The solid black gives the average intensity while the thin line represents the sum of the average intensity and of twice the standard deviation computed during successive 8-s intervals. For comparison we have plotted, in the lower panel of Figure 6 the 32-s averages of the current collected by the Langmuir probe, which was kept at a fixed bias of +5 V in order to detect fast variations of the electron density. The infered plasma density varies from 20 to 80 cm^{-3} [Grard et al., 1988]. As can be seen, the electric noise and the electron density exhibit well defined oscillations with a period of about 13 minutes, which corresponds quite well to the 1.3 mHz frequency cited at the beginning of this

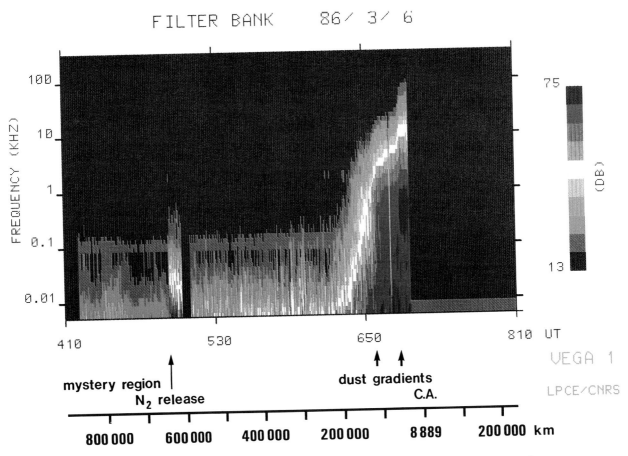

Fig. 3. Dynamic spectrogram of the Vega 1 electric-field data before the closest approach to comet Halley, C.A. The locations of events and the distance to the nucleus are indicated at the bottom. The APV-V experiment stopped working 7 minutes before C.A.

section. The electric noise extends from a few Hz to 100 Hz, i.e. below the ion plasma frequency. Moreover, the electric and plasma density oscillations look as if they would be anticorrelated rather than correlated. As, on the other hand the ion-flux oscillations observed in the TUNDE-M energetic particle instrument on Vega 1 appear to be anticorrelated with the electron plasma variations [Somogyi, private communication], it seems that the plasma wave oscillations follow the ion-flux variations. At last, it is worth noting that large-scale variations with a period of about 15 minutes also occured in the intensity of the ion fluxes measured with the PLASMAG-1 hemispherical electrostatic analysers onboard Vega 2 [Verigin et al., 1987].

This region has been called here "mystery region", for it has several things in common with the "mystery region" seen by Giotto. As a matter of fact, during the Giotto encounter of comet Halley, the RPA-COPERNIC plasma experiment identified a new region called mystery region

[Rème et al., 1987, 1990; d'Uston et al., 1987]. This region extends downstream of the bow shock from 850,000 to 550,000 km from the comet: i.e. at distances from the nucleus comparable to the ones of the region observed by APV-V onboard Vega 1, although the closest distance cannot be derived from the APV-V data due to the strong contamination induced by the nitrogen gas release. The mystery region is mainly characterized by unexpected high-energy electron fluxes (keV electrons) which are aligned with the magnetic-field direction. Moreover, the electron density and the ion fluxes detected in the ram direction appear to be anticorrelated [Rème et al., 1987], as already suggested by the Vega observations. Large-scale variations of the electron density similar to the ones observed during the Vega encounters with comet Halley are undoubtly present on the RPA-COPERNIC data.

The mystery region and its closest boundary to the nucleus the "mysterious transition" described by d'Uston et al. [1987] were not foreseen in theoretical models [see for

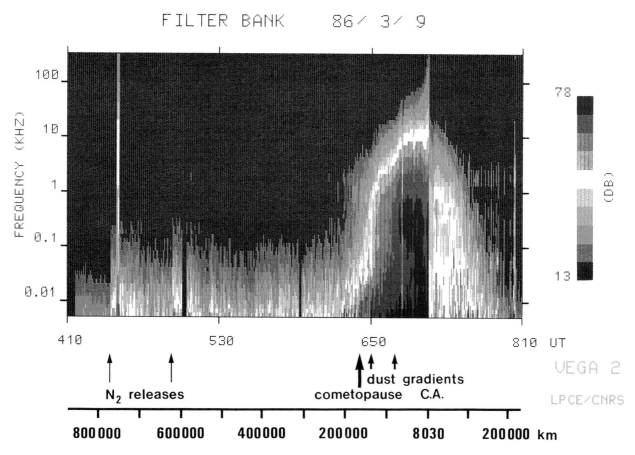

Fig. 4. Dynamic spectrogram of the Vega 2 electric-field data before the closest approach to comet Halley, C.A. A loss of sensitivity due to strong dust impacts occurs close to C.A.

example, Cravens, 1989] and they are far from being well understood at the moment. For example, the presence of energetic electrons and the mechanism which stops the decrease of the flow velocity in the mystery region have to be explained. Nevertheless, an argument in favour of the existence of the mystery region is presented in recent papers by Sauer et al. [1989, 1990]. Indeed, a multifluid plasma model, in which the thermal pressure of the picked-up cometary ions is included, predicts several plasma boundaries. The first boundary, where the flow velocity matches the ion sound velocity, coincides with the cometary shock observations. A transition from a stationary to a non-stationary flow regime takes place at this resonance point. The second boundary, called "cut-off boundary", marks the end of the non-stationary region which extends over 300,000 km from the shock. From here on begins a region, which looks like the mystery region, where MHD plasma waves occur with wavelengths of the order of several 10^4 km. They are thought to be excited at the first cut-off boundary behind the shock. The wavelength of these multifluid waves is given by

$$\lambda = (n_i + n_p) \, V_F \, / \, (n_p \, f_{ci} + n_i \, f_{cp}) \qquad (1)$$

where n_i, n_p are the ion and proton densities, f_{ci}, f_{cp} are the ion and proton cyclotron frequencies and V_F denotes the flow velocity. As the proton component dominates in the mystery region, we obtain

$$\lambda = V_F \, / \, f_{ci} \qquad (2)$$

Assuming ions of the water group, the characteristic wavelength reaches 5×10^4 km [Mikhailov et al., 1990]. Now, when the spacecraft crosses such a standing MHD wave structure at a velocity V_S, the measured frequency becomes

$$f = V_S \, / \, \lambda = f_{ci} \, V_S \, / \, V_F \qquad (3)$$

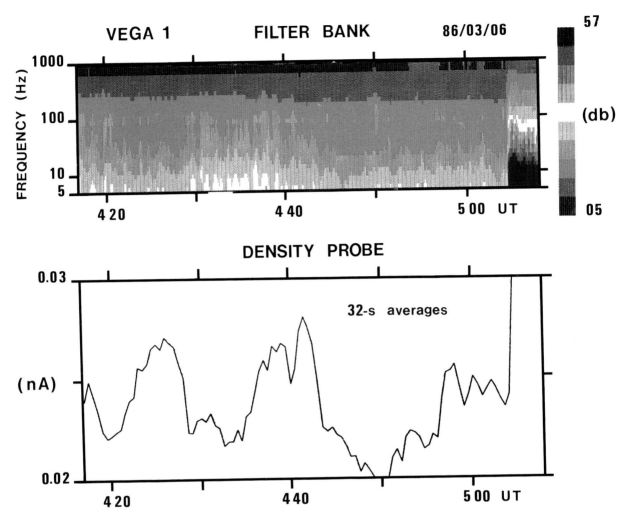

Fig. 5. Electric-field noise (at the top) and electron-density fluctuations (at the bottom) observed in the so-called "mystery region". They are believed to be associated with a standing MHD wave structure crossed by Vega 1.

which leads to f = 1.3 mHz, the frequency deduced from the spectral analysis of the APV-V data, as quoted previously. The MHD nature of the oscillations observed by the APV-V experiment and the magnetometer seems to be established, for the electric- to magnetic-field ratio is proportional to the Alfven speed [Mikhailov et al., 1987]. Moreover, the anticorrelation between the electron density and the ion fluxes exhibited on both Vega and Giotto data, is a favourable condition for MHD waves to develop.

At last, Sauer et al. [1990] have proposed a plausible explanation for the existence of the keV electrons measured with the RPA-COPERNIC plasma experiment. Indeed, electron should be accelerated slightly parallel to the magnetic field direction by lower-hybrid waves which could be excite owing to the fact that particles flow at different velocities.

4.   E-Field Observations in the Cometopause

A new solar wind-comet interaction region, called "cometopause" by Gringauz et al. [1986], was entered by the Vega 2 spacecraft on March 9, 1986 at 0643 UT. The cometopause, 10,000 km wide, begins at about 160,000 km from the comet and separates the inner region controlled by the slowly-moving cometary plasma from the outer region dominated by the shocked and mass-loaded solar wind. This transition region, characterized by a drastic change in the plasma composition, thus has been called "chemical boundary". In fact, this appellation is not well suited for physical rather than chemical processes are thought to be responsible for the boundary, as we will see later.

The location of the cometopause is indicated in Figure 4, which shows a four-hour electric-field

VEGA 1   FILTER BANK

MARCH 6, 1986

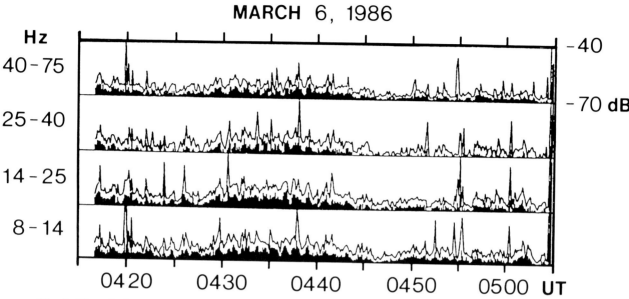

Fig. 6. Electric-field levels measured in the first four frequency channels on Vega 1 during successive 8-s intervals between 0415 and 0505 UT. The solid black gives the average intensity while the thin line shows the average intensity plus twice the standard deviation. The ordinate axis of each plot covers the dynamic range from -70 dB to -40 dB; the 0 dB level corresponds to 1 V rms.

spectrogram from the Vega 2 data, and in Figure 7, which displays a one-hour spectrogram preceding closest approach. It may be seen that the electric-field noise levels begin to increase in almost all filter channels after 0635 UT, i.e. 8 minutes before the cometopause entry. If we except a sharp increase in the intensity of the 8-14 Hz channel [Trotignon et al., 1989], no special electric-field behaviour is observed in the cometopause. A confirmation is given in Figure 8. Indeed, the cometopause was detected by the PLASMAG-1 instrument on board Vega 2 between 0643 and 0645 UT, which could hardly be noticed from the electric-field spectrum measurements shown in Figure 8. Here, the data were processed through a 20-point running-average filter in order to remove the high-frequency oscillations. 20 points correspond to a time interval of 10 s for frequencies larger than 75 Hz, i.e. for filters 4-16: the time resolution of the first four filters is twice lower.

Electric-field oscillations with a period lying from 20 s to 1 minute were reported by Trotignon et al. [1989]. They are observed from 0635 UT to the end of the cometopause and even after the cometopause. A similar wave activity was observed in the ion flux and magnetic-field data with a period of about 1 minute [Galeev et al., 1988]. These observations are in a good agreement with the 30-s period of the plasma flow oscillations predicted by a multifluid plasma model developped by Sauer et al. [1989]. Indeed, MHD waves with wavelengths of the order of 2,500 km have been

simulated. Note that, the heavy-ion and proton densities are comparable in the cometopause, so Eq. (1) becomes

$$\lambda = 2\, V_F\, /\, f_{cp} \qquad (4)$$

which leads to λ = 2,500 km.

It should be observed that the cometopause boundary must not be confused with the discontinuity crossed by Giotto at a distance of 135,000 km from the comet. The latter discontinuity has been detected in the low-energy electron measurements performed by the RPA experiment [d'Uston et al., 1987] and in the magnetic-field magnitude [Neubauer et al., 1986]. The density of 10 to 30 eV electrons decreased suddenly and the distribution function turned from isotropic to anisotropic while a sharp increase occured in the magnetic-field magnitude. This discontinuity has been called pile-up boundary, for the magnetic field began to pile up.

Up to now, the problem of the interpretation of the cometopause and the pile-up boundary is not completely solved [d'Uston et al., 1989] and even the existence of the cometopause as a real plasma boundary is controversial [Rème, 1989, private communication]. Nevertheless, using a two-fluid steady state model to predict the shocked plasma flow along the subsolar flow line, Gombosi (1987) has found a sharp ion-neutral charge exchange boundary which is in good agreement with the cometopause

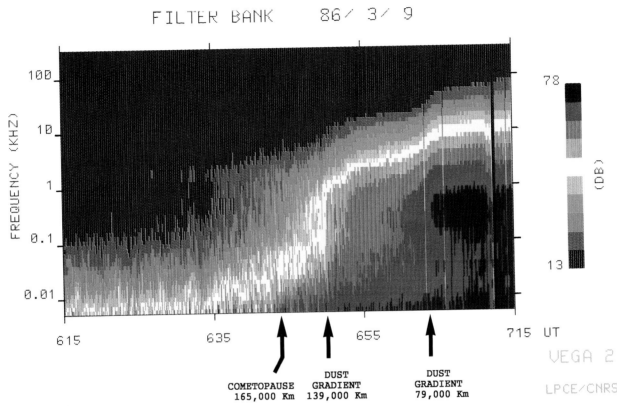

Fig. 7. Vega 2 electric-field spectrum measurements performed by the high-frequency plasma wave analyser,APV-V, for a 1-hour period preceding closest approach. Two sharp amplitude increases, called "dust gradients" are shown. The cometopause signature is questionable.

observations. Besides the deceleration flow, the most striking features are a sudden decrease of the density of the solar wind protons while their temperature remains costant and an increase of the cometary ions accompanied by a rapid cooling when the shocked flow enters the cometopause. At last, the author asserts that the cometopause is not a real plasma boundary, but rather a density crossover point where the implanted cometary ion density rapidly exceeds the depleted solar wind number density. From all of this, it follows that it is not surprising to observe no strong signatures in the wave measurements when the Vega spacecraft crossed the cometopause, as shown in Figure 8.

5.   Close-in Observations: the Dust Gradients

During the Vega spacecraft encounters with comet Halley, five electric-field spectrum discontinuities were detected within 140,000 km from the nucleus. The locations of these discontinuities are indicated in Figure 1 by full circles (for Vega 1) and stars (for Vega 2) along the spacecraft trajectories. They are seen as sharp and sudden increases of the AC electric-field amplitude. These electric-field intensity enhancements called "dust gradients" result from the passage of the spacecraft through regions of enhanced dust particle density. These regions are believed to be associated with spiral jets of radially moving dust particles. The dust gradients observed by Vega 1 are shown in Figure 3, while two of the dust gradients detected in the Vega 2 electric-field data are presented in Figures 4 and 7.

In Trotignon et al. [1987], the recurring observations of the dust gradients have been used to estimate the apparent rotation period of the cometary nucleus, T, and the radial bulk velocity of small dust particles, V_p. It appears that the nucleus rotates with a period of 54 h, which is slightly higher than the 52-h value given elsewhere. The value of V_p has been estimated to lie from 300 to 520 m.s^{-1} for dust particles with masses less than 10^{-12} g. This terminal-velocity range is remarkably close to the 380-420 m.s^{-1} range derived from the submicron dust particle fluxes recorded by the Dust Counter and Mass Analyser on board Vega [Rabinowitz, 1988].

The sketch in the right-hand side of Figure 1 shows the spiral jets, due to a discrete source, which are crossed by Vega 1 and Vega 2.

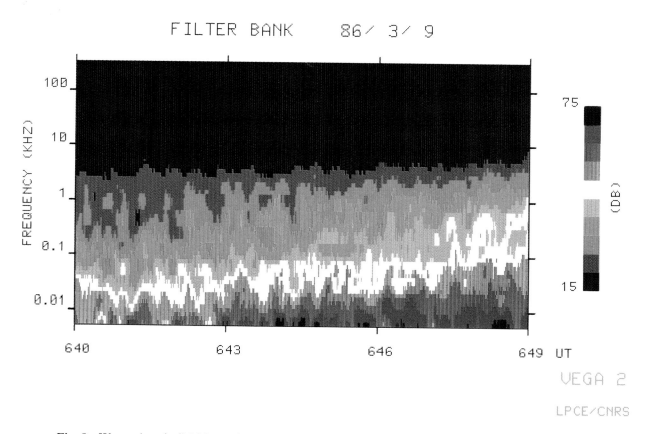

FILTER BANK   86/ 3/ 9

Fig. 8. Wave electric-field intensity measured by the filter bank during the cometopause crossing by Vega 2. This region described by Gringauz et al. [1986] extends from 160,000 km (0643 UT) to 170,000 km (0645 UT) from the comet.

The impacts of dust particles on the Vega spacecraft had also been detected by the two Langmuir probes of the APV-V experiment. They are identified by impulsive signal increases which yield an estimation of the dust particle masses [Laakso et al., 1989].

The question arises of whether the fact that dust gradients occur has consequences for the cometary plasma environment. It is well known that dust particles immersed in a plasma can affect the properties of the plasma itself, especially if the dust grains are charged. Then, the dust particles may change the dispersion properties of various, usually low-frequency, plasma waves [Goertz, 1989, and references therein]. For example, one part of the increase of the low-frequency electrostatic noise observed in the APV-V data has been attributed to the ion-acoustic mode in the presence of massive charged dust particles by de Angelis et al. [1988]. So, it is not unreasonable to claim that the presence of dust gradients in the environment of comet Halley may influence the plasma and wave characteristics, but the field of dusty plasma is still too young to say more.

6.   Conclusion

Using the electric-field spectrum measurements performed with the high-frequency plasma-wave analyser on Vega we have investigated the major transition regions of the environment of comet P/Halley. The data recorded near the bow shock crossings were too affected by spacecraft or experiment disturbances to show any significant signature. The cometopause described by Gringauz et al. [1986] is only seen as a sharp increase in the 8-14 Hz channel output signal, while MHD waves with a period lying from 20 s to 1 minute occur in a large region including the cometopause. A region characterized by strong fluctuations in the electron density and in the electric- and magnetic-field measurements with a period of 13 minutes has been crossed by Vega 1. This region observed between 870,000 and 650,000 km from the comet, looks like the mystery region identified on the RPA plasma experiment on board Giotto [Rème et al., 1987, 1990; d'Uston et al., 1987]. Finally, it must be noted that spiral dust jets were crossed within 140,000 km from the nucleus. They are seen as sudden and sharp increases in the electric-field intensity. At last, let us remark that the spectral analysis of the plasma-wave experiment data have given significant results in the mystery region and near the cometopause for the dust impacts did not affect significantly the measurements so far from the nucleus.

References

Cravens, T. E., Cometary plasma boundaries, *Adv. Space Res., 9*, (3)293, 1989.

De Angelis, U., V. Formisano, and M. Giordano, Ion plasma waves in dusty plasmas: Halley's comet, *J. Plasma Phys., 40*, 399, 1988.

Galeev, A. A., B. E. Gribov, T. Gombosi, K. I. Gringauz, S. I. Klimov, P. Oberc, A. P. Remizov, W. Riedler, R. Z. Sagdeev, S. P. Savin, A. Yu. Sokolov, V. D. Shapiro, V. I. Shevchenko, K. Szego, M. I. Verigin, and Ye. G. Yeroshenko, Position and structure of the comet Halley bow shock: Vega-1 and Vega-2 measurements, *Geophys. Res. Lett., 13*, 841, 1986.

Galeev, A. A., K. I. Gringauz, S. I. Klimov, A. P. Remizov, R. Z. Sagdeev, S. P. Savin, A. Yu. Sokolov, M. I. Verigin, K. Szego, M. Tatrallyay, R. Grard, Ye. G. Yeroshenko, M. Mogilevski, W. Riedler, and K. Schwingenschuh, Physical processes in the vicinity of the cometopause interpreted on the basis of plasma, magnetic field and plasma wave data measured on board the Vega-2 spacecraft, *J. Geophys. Res., 93*, 7527, 1988.

Goertz, C. K., Dusty plasmas in the solar system, *Rev. Geophys., 27*, 271, 1989.

Gombosi, T. I., Charge exchange avalanche at the cometopause, *Geophys. Res. Lett., 14*, 1174, 1987.

Grard, R., H. Laakso, A. Pedersen, J. G. Trotignon, and Y. Mikhailov, Observations of the plasma environment of comet Halley during the Vega flybys, *Ann. Geophysicae, 7*, 141, 1988.

Gringauz, K. I., T. I. Gombosi, M. Tatrallyay, M. I. Verigin, A. P. Remizov, A. K. Richter, I. Apathy, I. Szemerey, A. V. Dyachkov, O. V. Balakina, and A. F. Nagy, Detection of a new "chemical" boundary at comet Halley, *Geophys. Res. Lett., 13*, 613, 1986.

Laakso, H, R. Grard, A. Pedersen, and G. Schwehm, Impacts of large dust particles on the Vega spacecraft, *Adv. Space Res., 9*, (3)269, 1989.

Mikhailov, Y., O. Molchanov, R. Grard, C. Béghin, V. Formisano, K. Schwingenschuh, M. Moguilevski, E. Eroshenko, A. Pedersen, and J. G. Trotignon, Correlation between magnetic and electric field fluctuations observed during flyby near Halley's comet, *Eur. Space Agency Spec. Publ.*, ESA SP-278, 109, 1987.

Mikhailov, Yu. M., O. V. Kapustina, G. A. Mikhailova, E. G. Eroshenko, V. A. Styazhkin, J. G. Trotignon, and K. Sauer, Spectral structures of infralow-frequency electromagnetic fields near comet Halley, 1990, this issue.

Neubauer, F. M., K. H. Glassmeier, M. Pohl, J. raeder, M. H. Acuna, L. F. Burlaga, N. F. Ness, G. Musmann, F. Mariani, M. K. Wallis, E. Ungstrup, and H. U. Schmidt, First results from the Giotto magnetometer experiment at comet Halley, *Nature, 321*, 352, 1986.

Oberc, P., W. Parzydlo, P. Koperski, D. Orlowski, and S. Klimov, Some new features of plasma wave phenomena at Halley: APV-N observations, *Eur. Space Agency Spec. Publ.*, ESA SP-278, 89, 1987.

Rabinowitz, D. L., A source map for dust jets observed in the coma of comet P/Halley, *Astron. Astrophys., 200*, 225, 1988.

Rème, H., J. A. Sauvaud, C. d'Uston, A. Cros, K. A. Anderson, C. W. Carlson, D. W. Curtis, R. P. Lin, A. Korth, A. K. Richter, and D. A. Mendis, General features of comet P/Halley: solar wind interaction from plasma measurements, *Astron. Astrophys., 187*, 33, 1987.

Rème, H., C. d'Uston, C. Mazelle, J. A. Sauvaud, K. A. Anderson, C. W. Carlson, R. P. Lin, A. Korth, D. A. Mendis, A. Coates, K. H. Glassmeier, and S. A. Fuselier, Properties of the "Mystery" region and the "Mystery" transition at comet Halley, 1990, this issue.

Russell, C. T., W. Riedler, K. Schwingenschuh, and Ye. Yeroshenko, Mirror instability in the magnetosphere of comet Halley, *Geophys. Res. Lett., 14*, 644, 1987.

Sauer, K., U. Motschmann, and K. Baumgartel, Multifluid simulations of the solar wind-comet interaction, *Adv. Space Res., 9*, (3)309, 1989.

Sauer, K., U. Motschmann, and T. Roatsch, Plasma boundaries at comet Halley, *Ann. Geophysicae, 8*, 243, 1990.

Trotignon, J. G., C. Béghin, R. Grard, A. Pedersen, V. Formisano, M. Mogilevsky, and Y. Mikhailov, Dust observations of comet P/Halley by the plasma-wave analyser, *Astron. Astrophys., 187*, 83, 1987.

Trotignon, J. G., R. Grard, and M. Mogilevsky, Electric field measurements in comet P/Halley's environment by the high-frequency plasma-wave analyser, *Ann. Geophysicae, 7*, 331, 1989.

d'Uston, C., H. Rème, J. A. Sauvaud, A. Cros, K. A. Anderson, C. W. Carlson, D. W. Curtis, R. P. Lin, A. Korth, A. K. Richter, and D. A. Mendis, Description of the main boundaries seen by the Giotto electron experiment inside comet P/Halley-solar wind interaction region, *Astron. Astrophys., 187*, 137, 1987

d'Uston, C., H. Rème, J. A. Sauvaud, C. W. Carlson, K. A. Anderson, D. W. Curtis, R. P. Lin, A. Korth, and D. A. Mendis, Properties of plasma electrons in the magnetic pile-up region of comet Halley, *Ann. Geophysicae, 7*, 91, 1989.

Verigin, M. I., K. I. Gringauz, A. K. Richter, T. I. Gombosi, A. P. Remizov, K. Szego, I. Apathy, I. Szemerey, M. Tatrallyay, and L. A. Lezhen, Plasma properties from the upstream region to the cometopause of comet P/Halley: Vega observations, *Astron. Astrophys., 187*, 121, 1987.

COMETS: A LABORATORY FOR PLASMA WAVES AND INSTABILITIES

by Bruce T. Tsurutani

Jet Propulsion Laboratory, California Institute of Technology
Pasadena, California 91109

Abstract. This review discusses the various plasma waves and instabilities detected in the vicinity of comets. Particular emphasis will be placed on nonlinear wave evolution, as well as the role of nonlinear waves in wave-particle interactions. Areas that need further research will be identified to help persons who wish to contribute to the field.

Introduction

The International Cometary Explorer (ICE) mission to Comet Giacobini-Zinner and the spacecraft "armada" (Suisei, Sakigake, VEGAs 1 and 2, Giotto, and ICE) to Comet Halley have given us our first in situ measurements at comets (see Science, April 1986; Geophys. Res. Lett., March and April, 1986; Nature, May, 1986; ESA, December, 1986; Astron. Astrophys., 187, 1987). Because of the long ionization times of cometary neutrals, the ions extend to millions of kilometers from the nucleus. The interaction of these ions and electrons with the solar wind has provided us with a unique laboratory to study plasma waves and instabilities. In this article, I will first attempt to order observational results by plasma parameters and compare them with specific instabilities. Although it is not possible to control the space plasma "laboratory", the variation in the cometary ion and electron beam densities and velocities can be studied by examining regions far and near the comet nucleus. Similarly, we cannot control the initial pitch angle of the pickup ions, but by using directional discontinuities naturally provided to us by the solar wind, intervals of a variety of particle pitch angles are available. Thus by a bit of resourcefulness, a variety of plasma parameters can be made available (N_{beam}, r, etc.). Clearly, the entire available parameter space has not been fully examined. It has been partially studied, and the results obtained to date will be reviewed.

Cometary plasma wave research has also allowed us to study several new topics that are just beginning to unfold: nonlinear wave evolution, wave cascading and the development of turbulence. The second section of this review will discuss spacecraft observations relevant to these topics. I will also briefly comment on the implications of the results for wave-particle interactions: linear, nonlinear, resonant and nonresonant, and their effects on stochastic particle acceleration.

Cometary Plasma Processes
Geophysical Monograph 61
©1991 American Geophysical Union

In the third section, I will review the higher frequency ELF/VLF wave observations. Only four spacecraft were equipped with instrumentation which could make such measurements: ICE, Sakigake and the Vegas 1 and 2. Comparisons between the various measurements will be made.

I. Low Frequency Plasma Waves

Theoretical Background

The initial ion pickup pitch angle depends on the orientation of the interplanetary magnetic field (IMF) relative to the solar wind velocity. We will call this angle α. This orientation determines much of the ion distribution function and hence the specific instability that leads to wave growth and pitch angle scattering. Two extreme examples are illustrated for tutorial purposes. These are given in Figures 1 and 2.

LORENTZ FORCE ACCELERATION

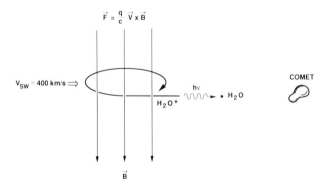

Figure 1. A schematic illustrating cometary ion pickup when the interplanetary magnetic field (IMF) is perpendicular to the solar wind velocity vector. In the plasma frame, ions form a ring distribution with velocity $\vec{V}_{sw}$ relative to the magnetic field.

Figure 1 illustrates the case where the ambient IMF is aligned orthogonal to the solar wind, $\vec{V}_{sw}$. The $\vec{V}_{sw} \times \vec{B}$ motional electric field acts on the newly created cometary ions to accelerate them into a cycloidal motion (in the inertial frame). The cometary ion will have both zero velocity and twice the solar wind velocity during its motion. It is the latter case which allows energetic particle detectors to measure the ion fluxes and distributions (Hynds et al., 1986; Ipavich et al., 1986; Somogyi et al., 1986; Mukai et al. 1986; Balsiger et al., 1986; Korth et al., 1986). With a typical solar wind velocity

ION BEAM INSTABILITY

Figure 2. An example illustrating cometary ions when the IMF is parallel to the solar wind velocity vector. There is no $\vec{V}_{sw} \times \vec{B}$ motional force. However the ions do form a beam of velocity - $\vec{V}_{sw}$, in the plasma frame.

of 400 km/sec, the peak kinetic energy of a cometary water group ion is $1/2 \ m \ (2V_{sw})^2 \approx 60$ keV.

The ion motion is more easily visualized in the solar wind (plasma) frame. The ions gyrate about the magnetic field with a velocity V_{sw}. The ions are also convected past the comet and the spacecraft with a velocity V_{sw}.

The above ion anisotropy can lead to three low frequency instabilities, the ion cyclotron instability (Sagdeev and Shafranov, 1961; Shapiro and Shevchenko, 1964; Rowlands et al., 1966; Kennel and Petschek, 1966; Thorne and Tsurutani, 1987; Brinca and Tsurutani, 1988a; Lee, 1989; Gary, 1990; Brinca, 1991), a parallel propagating nonoscillatory mode (Brinca and Tsurutani, 1987 a, b; 1989a), and a fluid mirror instability (Chandrasekhar et al., 1958; Hasegawa, 1969, 1975; Price et al., 1986; Lee et al., 1987; Price, 1989; Hasegawa and Chen, 1989; see also comments in Brinca and Tsurutani, 1989a). These various modes can be easily distinguished experimentally. Assuming the angle α is not exactly 90° and there is a parallel component of the ion velocity along the magnetic field with $V_{\parallel} > V_A$, where V_A is the Alfvén speed, the ion cyclotron instability leads to resonant left-hand waves propagating antiparallel to the ions.

The fluid mirror mode is a nonoscillatory mode. The instability condition for this is:

$$\frac{\beta_{\perp}}{\beta_{\parallel}} > 1 + 1/\beta_{\perp} \qquad (1)$$

(Crooker and Siscoe, 1977; see also formulation and discussion of a kinetic mirror mode in Wu et al., 1988). Thus this fluid instability occurs for cases of large anisotropies or in high β regions such as planetary magnetosheaths (Tsurutani et al., 1982). This is most easily recognized by the large magnetic field magnitude variations and small directional changes (Price et al., 1986; Lee et al., 1987). The latter is caused by the dominance of growth of this mode at large angles to $\vec{B}$ (60° - 80°; Price et al., 1986). An example of this mode from a computer simulation is shown in Figure 3, taken from Price et al. (1986).

It should be noted that although the above picture of the ion and electron velocity distribution is generally accurate, it is not totally correct. There are additional (small) ion velocities which must also be considered under certain situations. In most cases they are negligible. The cometary neutrals have velocities radially away from the nucleus. The velocity is ~1 km/s for the case of H_2O molecules and ~ 8 and 20 km/s for the case of hydrogen atoms. There are also thermal velocities associated with the atoms and molecules which are less than the above streaming velocities. Finally, there are small velocities associated with photoionization and charge exchange processes. These velocities are small in comparison to solar wind speeds and can be neglected for most calculations. However, because of the presence of these small velocities (and because the IMF typically has continuous directional fluctuations), it is clear that the ion pickup distribution is not a pure delta function, as has been idealized for some calculations (see Lee, 1989 for discussion).

Figure 2 illustrates the case where the ambient magnetic field is parallel to the solar wind velocity vector. In this instance, there is no $\vec{V}_{sw} \times \vec{B}$, solar wind force. On the other hand, the ions form a beam in the solar wind (plasma) frame moving at a velocity $-\vec{V}_{sw}$ relative to the ambient plasma. Two types of instabilities can result from this distribution: a right-hand resonant helical-(ring-) beam instability and a fluid, nonresonant instability.

The right-hand resonant helical beam instability can have strong wave growth, not only for a pure beam ($\alpha = 0°$), but for angles up to $\alpha \approx 70°$. This instability is fed by the free energy associated with both the gyromotion energy ($E_{\perp}$) and the energy parallel to the field ($E_{\parallel}$). This was first discussed by Wu and Davidson (1972) and Wu and Hartle (1974) for the pickup of interstellar and planetary ions. A large amount of theoretical work took place after the Ampte releases and the comet encounters. Recent results are found in Winske et al. (1985), Winske and Gary (1986), Sharma and Patel (1986), Thorne and Tsurutani (1987), Goldstein and Wong, (1987), Lee and Ip (1987), Brinca and Tsurutani (1988a), Gary and Madland (1988) and Price et al. (1988). Particularly nice reviews are found in Lee (1989), Gary (1990) and Brinca (1991). Pure beam instabilities (the case shown in Figure 2) were originally discussed in Fairfield (1969) and Barnes (1979). The conditions for cyclotron resonance is the standard relationship: $\omega = \vec{k} \cdot \vec{V} + n\Omega_i$ where ω is the wave frequency, k and $\vec{v}$ the wave k vector and particle velocity, n an integer and Ω_i the ion gyrofrequency. For this instability $\vec{k}$ and $\vec{v}$ have the same directional sense.

The nonresonant or firehose instability is a fluid instability that grows when the parallel pressure is large in comparison to the perpendicular plus magnetic pressure. The plasma is unstable when $P_{\parallel} > P_{\perp} + B^2/4\pi$. This has close analogies to the instability derived from water squirting from a hose, thus the name. Past works in this area can be found in Sagdeev and Vedenov (1958), Sentman et al. (1981), Gary et al. (1984), Winske and LeRoy (1984), Sagdeev et al. (1986), Galeev et al. (1989) and Brinca and Tsurutani (1989a). Gary et al. (1985) and Gary (1990) have shown that this wave mode has sufficient growth only for beams with very high velocities and/or densities. For proton beams, a number density of ~ 1% and a beam velocity > 10-15 V_A were shown to be necessary for wave growth. For the cometary case where water group ions are the important specie, beam densities of 0.06% are required.

These above two wave modes can be easily identified. In the right-hand resonant instability, the ion beam overtakes the waves so there is an anomalous Doppler shift in the ion frame of reference (the ion senses the waves as left-hand polarized, rotating in the same sense as the ion gyromotion about the ambient magnetic field). Since the ions form a beam in the solar wind frame, the ions and the generated waves propagate in the direction opposite to the solar wind direction, or towards

FIELD LINES AT $T\Omega_+ = 96$

$T\Omega_+ = 96$

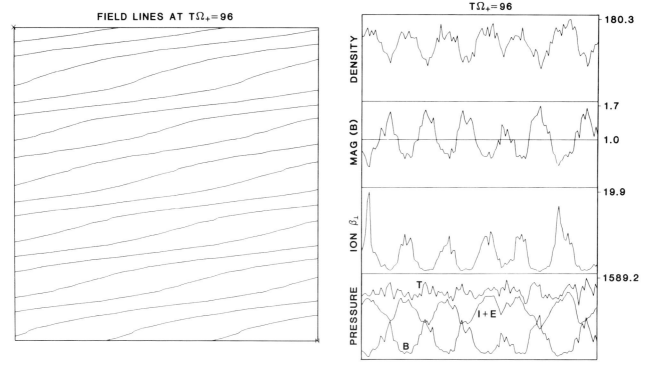

Figure 3. The density, magnetic field magnitude, perpendicular ion plasma beta, and perpendicular pressures from a 1-D simulation. The wave normal angle is 80° to $\vec{B}$. B corresponds to magnetic pressure, I + E the plasma pressure, and T the total pressure. The structures are time stationary. Note the anticorrelation between the density and field magnitude, the large variation in β, and the near-constant total pressure.

the sun. For waves propagating parallel to the ambient magnetic field, the phase velocity is substantially less than the solar wind speed (with the exception of regions very close to the comet) and the waves will be anomalously Doppler shifted in the spacecraft frame. These waves will be observed as left-hand polarized.

The nonresonant instability generates low-frequency left-hand waves propagating counter to the ion beam. To distinguish these waves from the anomalously Doppler shifted right-hand waves discussed above, we note that the waves are not necessarily generated at resonant frequencies. Because the spacecraft velocities relative to the comet were small in comparison to the solar wind speed (21 km/s in the case of ICE at Giacobini-Zinner and 65-70 km/s for the armada at Halley), one can, to first order, view the spacecraft as being in the cometary ion frame (Tsurutani and Smith, 1986a). Thus waves generated by resonant instabilities were measured at the cometary ion cyclotron frequency (Tsurutani and Smith, 1986 a,b; Saito et al., 1986; Johnstone et al., 1987 a,b; Glassmeier et al., 1987) as predicted by Wu et al. (1973). The nonresonant waves would be detected at lower frequencies. No concrete evidence for the presence of the latter waves have been obtained yet.

Observational Results
Right-hand Resonant Helical Beam Instability

Figure 4 illustrates the presence of waves at comet Halley at

the H_2O group (16-19 amu) ion cyclotron frequency. This figure is taken from Johnstone et al. (1986) and displays the solar wind proton velocity components in field-aligned coordinates. The y-axis is parallel to $\vec{B}$, the z-axis is in the $\vec{V}_{sw}$ x $\vec{B}$ direction and the x-axis forms a right-hand system. Note that the peak in power occurs at the water group cyclotron frequency, as expected from previous arguments. The velocity deviations are up to 10 km/s and the largest amplitudes are in the two transverse components. In the examples given by Glassmeier et al. (1987), the magnetic field deviations are typically $\Delta \vec{B}/B \approx 0.5$ and $\Delta |B|/B \approx 0.1$.

Figure 5 shows a similar figure for ICE at Giacobini-Zinner. The field components are in GSE coordinates. Spectra of the IMF during an "active" solar wind interval has been taken from Siscoe et al.(1968) and is superposed on the 3 panels. It is noted that the peak in the three power spectra occur near 100 seconds, the H_2O group ion cyclotron frequency for this interval. The wave fluctuation intensities are higher than during the flyby of comet Halley. Not only are the wave intensities greater, but the wave forms are highly steepened as well (shown later). The peak-to-peak $\Delta \vec{B}/B$ value was ≈ 2.0 and $\Delta |B|/B \approx 0.5$. It is thought that the smaller neutral production of comet Giacobini-Zinner, Q = 2 x 10^{28} molecules/s ($Q_{GZ} \approx Q_H/35$), led to a shock formation at an upstream distance 10 times closer to the nucleus (10^5 km, instead of 10^6 km for the case of Halley), and this closeness is related to the stronger waves at Giacobini-Zinner (V. Shapiro, personal communication, 1989).

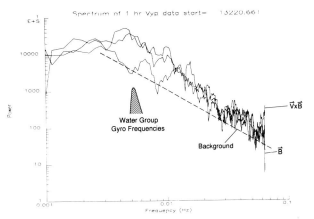

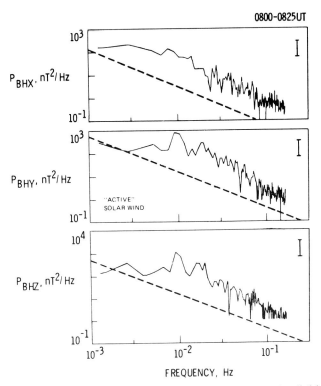

0800-0825UT

Figure 4. The presence of H_2O group (16-19 amu) ion cyclotron waves at Comet Halley. The y axis is parallel to $\vec{B}$, the z axis is in the $\vec{V}_{sw} \times \vec{B}$ direction and the x-axis forms a right-hand system. Taken from Johnstone et al., 1986.

Figure 6 illustrates the spatial extent of the waves in the cometary (Giacobini-Zinner) environment. Each point represents a 25-minute average power spectral density. The left-hand panel is for transverse fluctuations and the right-hand panel (x-axis) indicates compressional fluctuations. With a relative spacecraft-comet velocity of 21 km/s, a half day represents a distance of $\sim 9 \times 10^5$ km. Thus, from the figures it is apparent that the cometary influence extends at least one million km from the comet nucleus and perhaps further. These wave amplitudes have been modelled by Gary et al. (1988) and have been shown to fit quasilinear predictions, assuming the previously given neutral production rate, a neutral velocity of 1 km/s and an ionization rate of 10^6 sec. In comparison, the VLF electrostatic component of the plasma waves have been detected to a distance at least as far as $2-3 \times 10^6$ km and ions as far as 8×10^6 km. It is possible that the ICE instrumentation detected cometary effects as far as 28×10^6 km from Halley (Scarf et al., 1986a; Wenzel et al., 1986) but the neutral atoms/molecules would have had to have very high velocities to be able to reach such large distances. These results remain quite controversial, with Daly (1987) and Tsurutani et al. (1987a) giving arguments against this interpretation.

Using both the solar wind proton and magnetic field data, Glassmeier et al. (1989) have obtained cross spectral densities of the plasma and magnetic fields (shown in Figure 7). This figure was taken from an interval where the magnetic field was nearly aligned with the solar wind velocity and is given in a HSE coordinate system where the x-axis is towards the sun and the y-axis is opposite to the direction of planetary orbital motion. The z-axis forms a right-hand system. This particular plot shows the magnetic field z component. The top panel illustrates the presence of several peaks, with the cyclotron fundamental at 7 mHz and harmonics at 14, 21, 29 and 35 mHz. Note that the coherency is > 0.9 for all of these peaks (shown in the middle panel). Glassmeier et al. stated that the other two components were studied, and similar results were obtained. The waves at 7 mHz are linearly polarized and the higher order harmonic waves either linearly or highly elliptically polarized.

Theoretical analysis of Brinca and Tsurutani (1989b) and Goldstein et al. (1990) have shown that drifting and nondrifting loss cone types of distributions as well as ion

Figure 5. Power spectra of the (GSE) magnetic field components of Comet Giacobini-Zinner. A power law curve, representative of an active solar wind, is also shown for comparative purposes. The wave intensities are well above solar wind fluctuation levels. The waves are considerably more intense than those at Halley. From Tsurutani and Smith, 1986b.

beams with perpendicular temperature anisotropies can lead to the generation of cyclotron harmonic waves. Brinca and Tsurutani (1989b) have speculated that the lack of observations of harmonic spectra when $\alpha \approx 90°$ by Glassmeier et al. (1989) may be due to strong overlapping of individual harmonics and thus a blurring of the structures. Another possibility is a severe reduction in the growth rate of cyclotron harmonic instabilities in a nonuniform ambient magnetic field (Sharma and Patel, 1986).

Waves Detected Far from the Comet

Waves were examined at a distance up to 7×10^5 km from comet Giacobini-Zinner (~7 times the bow shock stand-off distance). A variety of polarizations have been detected, from circularly polarized waves propagating at large angles relative to the ambient field to highly elliptically polarized waves ($\lambda_1/\lambda_2 > 10$) propagating at angles less than 20° relative to $\vec{B}$. Figure 8 shows an example of a nearly circularly polarized wave propagating at an angle of 29° relative to $\vec{B}$. Even at these large distances, the wave transverse components were typically $\Delta \vec{B}/B \sim 0.5$ with a compressional component $\Delta |B|/B \sim 0.1 - 0.3$.

The wide variety of polarization and direction of propagation of the waves has not been adequately explained by theory. Linear theory predicts the growth of parallel propagating, noncompressive waves. Almost no waves with the latter properties have been detected to date.

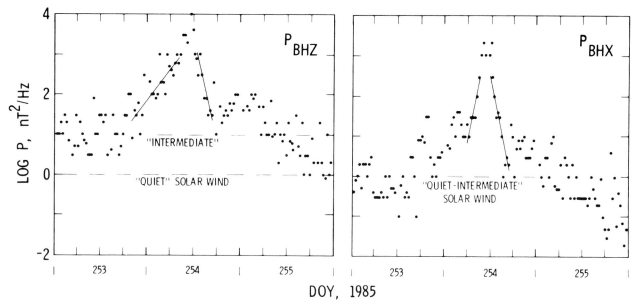

Figure 6. Twenty-five min. average power spectral densities at 10^2s, the H_2O group ion cyclotron frequency. The x-axis is aligned along the ambient field, and z is transverse to the field. Thus right-hand and left-hand panels represent compressional and transverse fluctuations, respectively. Typical interplanetary fluctuation intensities are shown for comparison. From Tsurutani and Smith, 1986b.

Waves at Comet Halley

The waves detected upstream of Comet Halley's bow shock, although smaller in amplitude than those near comet Giacobini-Zinner's bowshock, were less ordered and much more turbulent in appearance. Some examples of the waves taken from 2 to 1.75×10^6 km from the nucleus is given in figures 9 and 10 (from Johnstone et al., 1986). Figure 9 is the proton data taken from the plasma experiment (Johnstone et al., 1976) and figure 10 is the magnetic field data from the d.c. magnetometer (Neubauer et al., 1986). Both the field and plasma data are plotted in the FLD (field draped) coordinates as described previously. The wave compressional components are small (the y component is nearly constant). It is noted that at times the fluctuations appear relatively ordered and other times relatively more disordered. At the beginning and ends of this interval there are short durations of relatively correlated field and plasma data. Hodograms for the transverse components (x and z) for both the magnetic field and the plasma data (not shown) indicate that there is no dominant wave polarization. Johnstone et al. (1987), have stated that the results give "a good impression that linearly polarized waves are present".

Resonant L-mode Waves? $\alpha = 90°$, Intervals far from the Comet

Conditions when α is close to 90° have been investigated using the occurrence of interplanetary discontinuities. At these large upstream distances, the anisotropic pressure associated with the pickup cometary ions is insufficient for the growth of the mirror mode (this can be easily calculated from the expression given previously). The wave mode theoretically expected is the left-hand ion cyclotron wave (and harmonics).

An example of this solar wind condition is given in Figure 11, taken from an interval some 5.5×10^5 km upstream of Giacobini-Zinner. This Figure illustrates that both prior to and after the interval where $\alpha \approx 90°$ ($\alpha \approx 125°$: shaded interval), heavy (H_2O group) ion cyclotron waves are present. However during the interval when $\alpha \approx 90°$, there is an absence of any long period (H_2O) waves. [Since the spacecraft is almost in the ion rest frame, the waves should be observed at the local gyrofrequency without significant Doppler shift (Tsurutani and Smith, 1986a)]. Other intervals of this type have been studied, with similar results.

Figure 12 gives a plot of the logarithm of the flux of the water group ions versus the sine of the angle α. It is noted that when α is large, there are enhanced fluxes of 65-95 keV ions detected (vertical shaded regions). Such features have previously been discussed by Sanderson et al. (1986). This Figure includes the interval discussed in Figure 14. A detailed examination of the pitch angle anisotropy using the three energetic ion detectors of this instrument (Hynds et al., 1986) indicates that there is little if any, pitch angle scattering during the high α interval. More recent work on ion anisotropies by Richardson et al., (1989a) have given results which are in good agreement with the above. These conclusions are in accord with the lack of left-hand waves to provide the scattering.

The flux variation with α shown in Figure 12 is another illustration of the lack of ion isotropy during this inbound interval (it should be noted that similar α dependences were not observed at comet Halley, however). If the cometary ions were fully isotropized, there would be no α dependence at all.

The lack of left-hand waves at comet Giacobini-Zinner is presently somewhat of a mystery. Although such specific

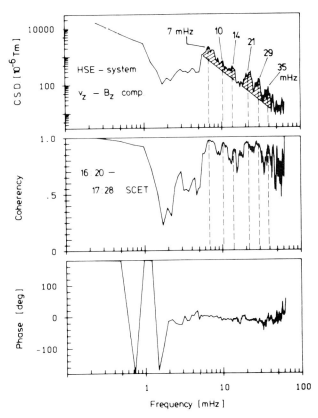

Figure 7. From top-to-bottom, the cross-spectral densities of the solar wind proton velocity and magnetic field, the coherency, and the phase relationship between velocity and $\vec{B}$. This interval analyzed occurred when the IMF $\vec{B}$ was along the solar wind velocity direction. Several significant psd peaks are: the H_2O group ion cyclotron frequency at (7 mHz) and cyclotron harmonics at 14, 21, 29 and 35 mHz. From Glassmeier et al., 1989.

relationships have not been examined at comet Halley, it is noted that Glassmeier et al. (1989) found a lack of a specific peak at the H_2O group ion cyclotron frequency when α was near 90°. The total wave power was also considerably less. An example of this will be shown later in this paper.

It is relevant to point out that the lack of left-hand waves have also been noted in the earth's magnetosheath. Price et al. (1986) have performed 1-d simulations and found a higher linear growth rate for the left-hand cyclotron mode than for the mirror mode. Yet in magnetosheath plasmas at the Earth, Jupiter and Saturn, the mirror mode is typically the most obvious wave mode in the data (Tsurutani et al., 1982). There is no accepted explanation at this time. Price et al. (1986) and Hada (1986) has suggested that the presence of minor ion species could decrease the growth of the ion cyclotron waves (but not influence the mirror mode growth rate), but the percent density necessary to accomplish this is unreasonably high (Hada, 1986).

One possible suggestion is that long convective growth times are necessary for left-hand waves to attain measurable amplitudes (Tsurutani et al. 1989a). This implies steady ambient magnetic fields, as exist in the magnetosphere during the recovery phase of magnetic storms (Erlandson et al.,

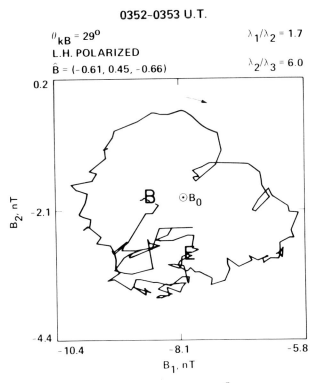

Figure 8. Hodogram for a wave ~ 5×10^5 km upstream of Comet Giacobini-Zinner.. The wave is left-hand circularly polarized in the spacecraft frame and is propagating at an angle of 29° relative to $\vec{B}$. The ambient magnetic field direction $\vec{B}$ is given for reference. In this case, $\alpha = 52°$, typical of a Parker spiral, except with a large southward component. From Tsurutani et al., 1989a.

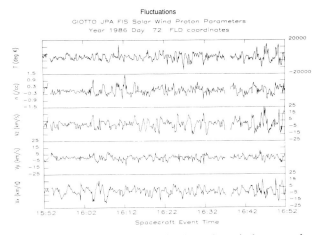

Figure 9. Example of waves in the solar wind proton data taken from 1.75 to 2.0×10^6 km from Halley (for references the Halley shock was transversed at 1.1×10^6 km). From top-to-bottom are the proton temperature, density and the velocity components. In this coordinate system, the y-axis is parallel to $\vec{B}$, $\hat{z}$ is parallel to $-V_{sw} \times \vec{B}$, and the x-axis forms a right-hand system. The wave compressional component is quite small. The fluctuations are in general relatively disordered. From Johnstone et al., 1986.

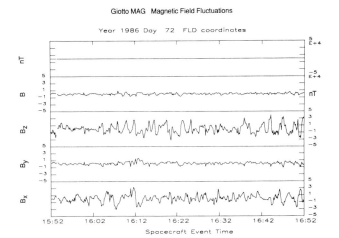

Giotto MAG   Magnetic Field Fluctuations

Year 1986 Day 72   FLD coordinates

Figure 10. The magnetic field in the same format as the proton data in Figure 9. The fluctuations are primarily transverse. From Johnstone et al., 1986.

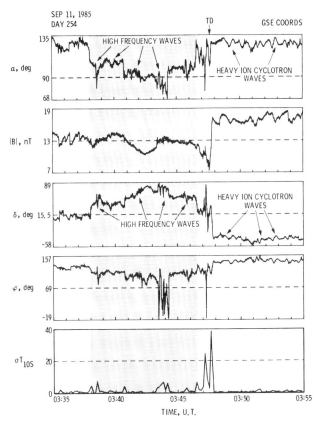

Figure 11. An interval (shaded) where α is close to 90°. There is an absence of waves with frequencies near the H_2O group ion cyclotron frequency. From Tsurutani et al., 1989b.

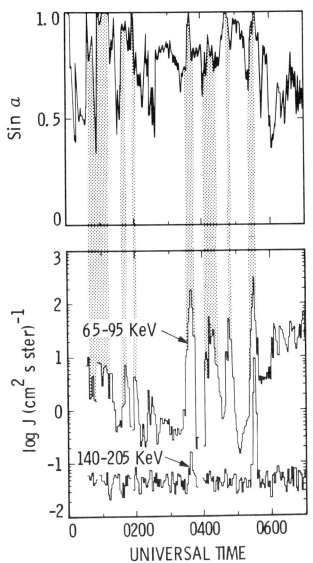

Figure 12. Sin α (the angle between $\vec{V}_{sw}$ x $\vec{B}$) and the logarithm of the energetic ion flux at comet Giacobini-Zinner. When sin α is large (shaded regions), the flux of the 65-95 keV ions is high. This is due to the pickup process discussed previously. Note that the ion flux is highly variable, indicating a general lack of pitch angle isotropy. From Tsurutani et al., 1989a.

1989). However, the fluctuating fields typically present in the magnetosheath and in a cometary environment may cause substantial wave damping, inhibiting significant growth. It is suggested that a simple computer simulation experiment first using quiet and then turbulent ambient fields could answer this important question.

The observations leave another question unanswered: how do the cometary ions become isotropized when $\alpha \approx 90°$, if they do at all? Perhaps there are ELF/VLF electrostatic waves which are present during such conditions. A search for possible modes should be undertaken and, if found, the pitch

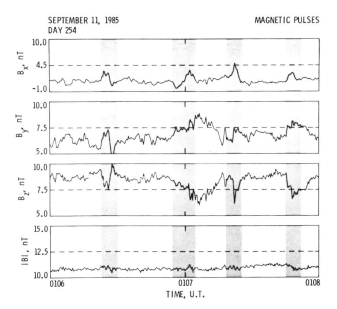

Figure 13. An example of transverse pulses in the magnetic field when $\alpha \approx 80° - 90°$ (shown in shading). The pulses have durations of 6-7s, comparable to the proton cyclotron period in the spacecraft frame. The pulses are typically linearly or highly elliptically polarized and are propagating at angles close to $\vec{B}$. The coordinate system is GSE. From Tsurutani et al., 1989b.

angle scattering rates calculated and compared to the observations.

A surprising result was the detection of short duration magnetic pulses during $\alpha \approx 90°$ intervals. This was indicated in Figure 11. High time resolution examples are given in Figure 13. Four pulses are indicated by vertical shading. The field variations have duration of 6-7s, comparable to the proton cyclotron frequency in the spacecraft frame. The pulses are essentially totally transverse oscillations (the bottom panel shows almost no variation in |B|). Analyses of a number of such events indicate that the waves typically propagate in directions parallel to $\vec{B}$, $\theta_{kB} \approx 5-15°$, and are linearly or highly elliptically polarized. The transverse amplitudes are typically 1-3nT (in a 8-12 nT field), although peak-to-peak amplitudes as large as 5-7 nT have been detected.

At the present time, the generation mechanism of these unusual wave structures is unknown. It is uncertain whether they exist only for intervals where $\alpha \approx 80° - 90°$ or whether they are present during other conditions but are masked by the presence of large amplitude H_2O group ion cyclotron waves. It is even unclear whether these structures are cometary in origin or not. The power associated with these isolated pulses is so small that standard techniques such as power spectral analyses cannot be used to determine their presence.

Lack of Proton Cyclotron Waves

Although cometary power spectral analyses have strongly indicated peaks at the H_2O group ion cyclotron frequency, both the ICE and Giotto magnetic field and plasma observations have not found any waves (other than the above) associated with the pickup of cometary protons. Gary et al. (1988) have simulated the proton pickup process and have found there is

rapid isotropization of the protons into a stable "shell" distribution. The waves saturate at small amplitudes, possibly explaining the lack of detectable waves. The heavy ions do not show similar effects because of the much larger mass of the ions.

Wu et al. (1990) have offered an alternative explanation. From their linear analyses, they have found that the instability associated with proton pickup is characterized by a broad band spectrum near the proton gyrofrequency. The authors have suggested this broad-bandedness is a possible reason for the lack of observed enhanced proton cyclotron waves.

The Nonresonant Ion Beam Instability

To date there has been no confirmation of the onset of this instability at comets, although it could be expected under certain plasma conditions. The waves generated would have $\lambda \gg \lambda_{res}$. Thus they would be hard to detect, in general. Long time intervals must be studied and this necessarily involves averaging over different plasma spatial regions, perhaps leading to ambiguous results. It should be mentioned that in the stochastic acceleration theory of Galeev et al. (1989), such long wavelength waves are necessary to acceleration ions to the observed high energies.

Drift Mirror Mode

Closer to the comet, within the magnetosheath, mirror mode waves have been detected by the VEGA magnetic field experiment. Figure 14 is from Yeroshenko et al. (1987). These observations were taken at a distance of $3-18 \times 10^4$ km from the comet, deep in the sheath region. These structures are characterized by irregular (nonsinusoidal) dips in the magnetic field.

Similar structures have also been detected at Giacobini-Zinner (Smith et al., 1987; Slavin et al., 1987) and have been observed at Halley as well (Glassmeier, private communication, 1989). However the magnetic structures observed by ICE and Giotto were nowhere nearly as striking as those found in the VEGA data set.

Russell et al., (1987) have speculated that these structures may be the cometary rays observed from ground based

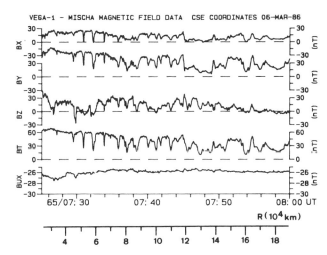

Figure 14. Magnetic "biteouts" in the Halley sheath measurements. These are identified as nonoscillatory mirror mode structures. From Yeroshenko et al., 1987.

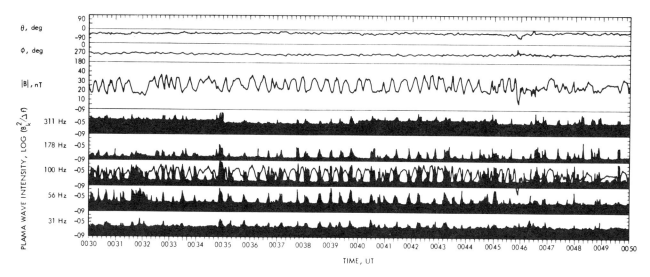

Figure 15. From top to bottom, the magnetic field in GSE coordinates and the ELF wave magnetic spectrum channels. The top three panels illustrate the presence of mirror mode waves. Characteristics are magnetic magnitude changes with little changes in direction. The magnetic magnitude plot is superposed onto the 100 Hz ELF wave channel. The electromagnetic whistler mode waves called "lion roars" occur in the dips of the magnetic field, illustrating the coupling between the MHD ion mirror instability and the electromagnetic (electron) micro-instability. From Tsurutani et al., 1982.

telescopes which extend to 10's of million kilometers downstream of cometary comas. However, little detail has been given for this supposition to date. It remains to be shown that $\sim 10^7 - 10^8$ km "wavelengths" can be generated from this mechanism, as is observed for rays. If the field-aligned wavelengths are considerably shorter, one would expect more of a "patch work" pattern rather than elongated "tubes", as are necessary for rays. More theoretical work is needed to tie this wave mode to cometary rays.

If the magnetic field is displayed in polar coordinates, the field signature becomes quite unique. An example of this is given in Figure 15, for the case of ISEE-1 in the Earth's magnetosheath. The top three panels show the magnetic field in GSE polar coordinates. Note that almost all of the variation occurs in the field magnitude and little in the angles. The bottom 5 panels illustrate the presence of ELF electromagnetic whistler mode waves. These are called "lion roars" from frequency-time variations when played through a loud speaker (Smith et al., 1971). The magnetic field magnitude has been replotted in the 100 Hz wave channel to illustrate the correlation with wave intensity to the field decreases. Thorne and Tsurutani (1981) and Tsurutani et al. (1982) have explained the growth of lion roars by the lowering of the critical energy, $Ec = B^2/8\pi N$, associated with the high β regions of the mirror mode waves.

Figure 16 shows the relationship between the magnetic and plasma pressures in mirror mode structures. They are 180° out-of-phase, such that the total pressure is essentially constant. This anticorrelation causes the plasma β to vary from values of 1-2 to 30 across these structures. Tsurutani et al. (1982) demonstrated that these structures are propagating at 60° - 80° relative to $\bar{B}$ and have scale sizes of ~ 20 proton

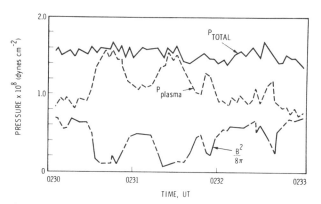

Figure 16. The relationship between the magnetic and plasma pressures in mirror mode structures. The two partial pressures are 180° out-of-phase. The total pressure is constant. Because of this feature, the low field, high β regions can have β as large as 30. This is part of what drives the electrons unstable. From Tsurutani et al., 1982.

gyroradii in the plasma frame. The scale sizes are somewhat larger than what is theoretically predicted (Hasegawa, 1975). This remains an important theoretical issue.

Figure 17 demonstrates the presence of mirror mode structures dominating the magnetosheaths of Saturn from the bow shock at ~ 1810 UT until the magnetopause at ~ 2200 UT. Again, plotting the data in spherical coordinates helps identify the wave mode. The scale of the mirror mode structures is ~ 40 r_p.

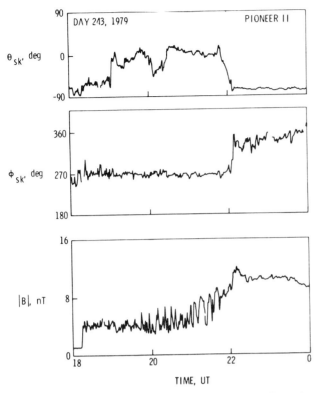

Figure 17. An example of mirror mode waves in the Saturnian magnetosheath. Pioneer 11 is at a radial distance of 18 R_S and a local time of 1155. The bow shock and magnetopause are readily apparent. From Smith et al., 1980 and Tsurutani et al., 1982.

II. Nonlinear Properties Of The Low Frequency Cometary Waves

Figure 18 gives examples of the cometary waves near the bow shock of Giacobini-Zinner. These waves have been detected from about twice the distance upstream of the bowshock to about half the distance to the nucleus. Specific features to note are: 1) the large amplitude, 100 s "sawtooth" structures in the B_z component, 2) significant magnetic field compression (bottom panel), and 3) high frequency whistlers trailing the magnetic compression (in time). As stated previously, the transverse H_2O group ion cyclotron (100 s) peak-to-peak amplitudes can be as large as $\Delta \bar{b}/B \sim 2.0$, the magnetic compression $\Delta |B|/B \approx 0.5$ or even higher, and the peak-to-peak whistler amplitudes $\Delta \bar{b}/B \approx 1.5$. Waves with the above properties were only detected at Giacobini-Zinner.

Figure 19 shows a comparison between the plasma density and magnetic field magnitude for the H_2O group ion cyclotron waves. There is a clear correlation between $|B|^2$ and density (N), establishing that this is indeed the magnetosonic mode. Thus the steepened waves in Figure 18, which have the sharp decreases on the trailing edge, are propagating towards the sun, but are convected past the spacecraft by the higher solar wind velocity. The whistler packets which are at the leading edge of the magnetosonic waves are also anomalously Doppler shifted and appear left-hand polarized in the spacecraft frame.

A simplified schematic of how a sinusoidal wave could possibly steepen is given in Figure 20. This has been adapted

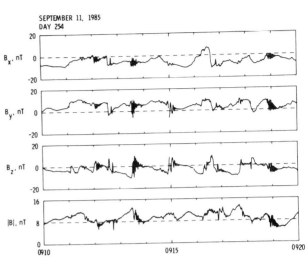

Figure 18. Cometary waves near the bow shock of comet Giacobini-Zinner. The coordinate system is GSE. The 100s oscillations are magnetosonic waves driven by the pickup of H_2O group ions. The generating mechanism is the resonant ring-beam instability, and because the spacecraft is basically in the ion rest frame, the waves appear near the ion cyclotron frequency. The high frequency 0.3 - 2.0 Hz waves are whistler mode emissions. The magnetosonic waves have a peak-to-peak amplitude as large as 2.0 and a compressional component $\Delta |B|/B \approx 0.5$. The whistler mode waves occur at the steepened, leading edge of the magnetosonic waves. From Tsurutani et al., 1987b.

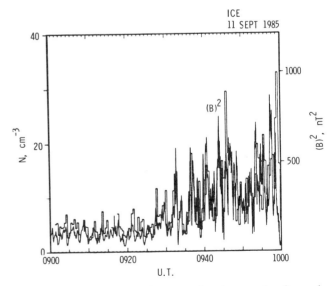

Figure 19. Relationship between the electron density and magnetic field (squared) at comet Giacobini-Zinner. In this interval, these two parameters are well correlated, indicating that the fluctuations are magnetosonic in nature. Other intervals have less correlation. From Tsurutani et al., 1987b.

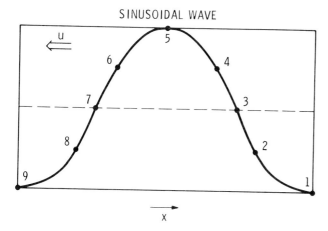

SINUSOIDAL WAVE

STEEPENED NONLINEAR WAVE

Figure 20. A schematic illustrating how a sinusoidal wave (top panel) could steepen to have a sharp phase rotation at its leading edge (bottom panel). From Tsurutani et al., 1987b. An adaptation from Cohen and Kulsrud, 1975.

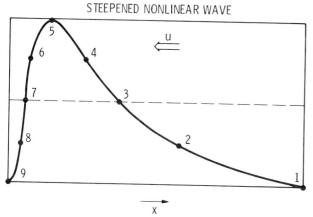

from Cohen and Kulsrud (1975). Note that the steepened nonlinear wave (bottom panel) has a 180° phase rotation at the leading edge and an elongated portion at the trailing end. The trailing end of the wave contains the remainder of the 180° phase rotation. This simple schematic is intended to give only a general description. Simulation results indicate that the distance between points 3 and 7 actually increases (N. Omidi, personal communication, 1989).

A high time resolution example of a steepened wave is given in Figures 21 and 22. Figure 21 shows the steepened wave in principal axes coordinates where the standard notation is used: B_1, B_2, B_3 correspond to the field in the maximum, intermediate and minimum variance directions. The steep "partial rotation" is observed from 1411:54 to 1412:04, indicated in the figure. The hodogram in figure 22 shows that there is a ~ 180° phase rotation associated with this phenomenon. There is a gradual monotonic change in B_1 from 1410:20 - 1411:26 UT. This change can be observed to cause a more-or-less linear variation of the wave.

Thus, the "partial rotation" is associated with a field rotation from a large magnetosonic wave distorted field to the upstream ambient field. The "partial rotation" accomplishes both field magnitude and direction changes. This field change is an

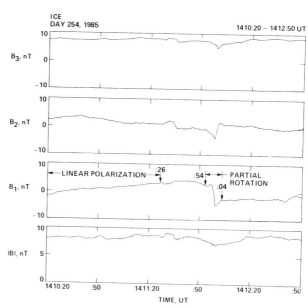

ICE
DAY 254, 1985   1410:20 – 1412:50 UT

Figure 21. An example of a steepened magnetosonic wave. The steepened wave front is located at 1411:54 to 1412:04 UT. This is a rotational discontinuity. Because it does not rotate a full 360° in phase, it has been called a "partial rotation". The trailing portion of the wave, from 1410:20 - 1411:26, also contains some phase rotation, but because this is often small, this portion of the wave appears "linear" in polarization. The next figure, Figure 22, will illustrate that all of the above features are part of the same wave. From Tsurutani et al., 1989c.

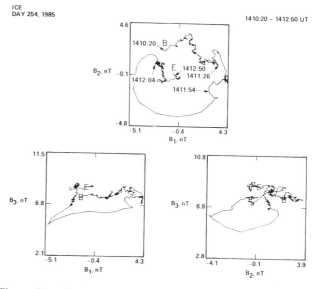

Figure 22. The hodograms for the steepened magnetosonic wave shown in Figure 21. B_1, B_2 and B_3 correspond to the field components in the maximum, intermediate and minimum variance directions. Specific times are indicated in the B_1 - B_2 plot so that the partial rotation and "linear" parts of the wave can be identified. Note that the total phase rotation is 360°. From Tsurutani et al., 1989c.

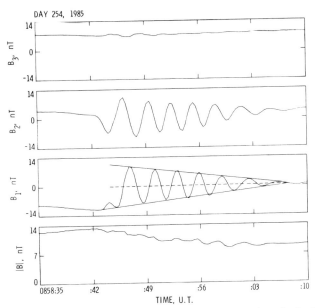

Figure 23. An example of an upstream whistler in principal axis coordinates. The whistler begins at the steepened edge of the magnetosonic wave and decreases linearly in amplitude with upstream distance. The whistler is present throughout the magnetic field magnitude gradient and presumably serves to broaden the gradient and reorient the magnetic field to the upstream ambient. From Tsurutani et al., 1989c.

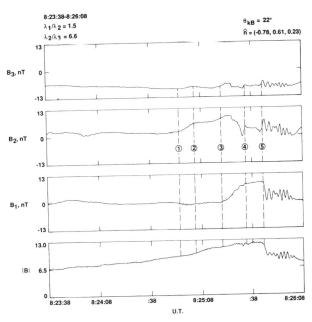

Figure 24. A more complex steepened magnetosonic wave case shown in principle axis coordinates. There are several distinct regions of wave phase changes. These are indicated in the Figure. One region contains a "back rotation" where the wave polarization reverses for a short interval. This is shown in more detail in the hodogram, Figure 25. Such back rotations are not uncommon in the data set. Note that in the trailing ~ half of the wave, from 0823:38 to 0824:55 UT, there is a gradual increase in B magnitude. However, there is essentially no change in wave phase (B_1 or B_2), indicating that this portion of the wave is almost purely compressive. From Tsurutani et al. 1990a.

important one to keep in mind. It will be contrasted to the case of upstream whistlers, which is quite different.

An example of a fully developed upstream whistler is shown in Figure 23 in Principal Axis coordinates. It begins at ~ 0858:42 UT at about the peak magnitude intensity of the compressional magnetosonic wave to which it is attached. The whistler decreases in amplitude linearly with time. The B_1 component oscillates about a value of zero. Because the whistler is propagating close to parallel to the upstream ambient magnetic field direction (θ_{kB} = 10.9°), the magnetic field magnitude decreases smoothly with time and distance. Two important points should be noted. First, it is apparent that the whistler is an integral part of the magnetosonic wave and it serves to decrease the field magnitude and alter the field direction. The partial rotation presumably has developed into a full wave packet. There is no longer an abrupt transition to the upstream ambient field, as occurs at a partial rotation. The whistler acts to gradually reorient the steepened wave field to the ambient direction and also causes a broadening of the field magnitude gradient.

It should be noted that the evolution of a steepened magnetosonic wave to a wave with a "fully developed" upstream whistler packet discussed above is only a hypothesis based on wave "snapshots". There is no evidence of this sequence from spacecraft observations. Computer simulation results of Omidi and Winske (1990) (their figures 7 and 9) indicates that this is the evolutionary sequence and show details of the whistler development.

There have been several proposal sources for the whistlers. Perhaps the most clearly demonstrated mechanism is through the generation of dispersive whistlers, as suggested by Omidi and Winske (1987; 1988; 1990) from hybrid simulation

results. They have proposed a scenario for the time sequence of whistler development. An open question that still exists is the influence of other whistler generation mechanisms that have been proposed, such as that discussed by Goldstein and Wong (1987), and Brinca and Tsurutani (1988b) and by Kaya et al. (1990). The former two mechanisms propose generation by pickup of heavy ions and protons at the distorted steepened fronts of the magnetosonic waves, and the latter mechanism involves phase trapping of heavy ions by the whistler wave train. It would be important to establish the relative wave amplitude contribution for these various mechanisms. Preliminary simulation results in (Omidi and Winske, 1990) indicate that the generation of dispersive whistlers may account for the majority of the wave amplitude.

Another interesting theoretical approach to explain the steepened magnetosonic wave plus whistlers are models based on the derivative nonlinear Schroedinger (DNLS) equation with dispersion and viscosity included (Kennel et al., 1988). Kennel et al. and recent results from Galinsky et al. (1990) have duplicated many of the observed wave features. However, present models result in two shocks per wavelength, while the observations typically indicate only one. Further effort in this area is currently ongoing.

Even though the steepened magnetosonic waves appear to be complex (above discussion), recent analyses indicate that in some cases they are even more interesting. Figure 24 and 25

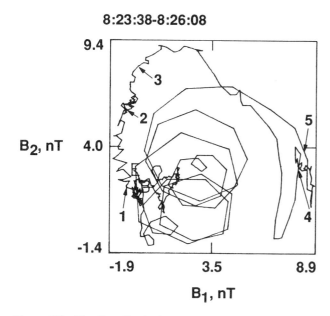

8:23:38-8:26:08

Figure 25. The B_1 - B_2 hodogram for Figure 24. Note the intervals of discrete phase changes and the "back rotation".

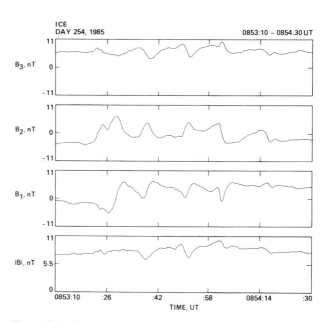

Figure 26. An example of a complex whistler packet. Not all whistlers propagate relatively parallel to $\vec{B}$. This is one such example. These whistlers are propagating at 45-51° relative to $\vec{B}$. Note the large compressional component associated with the waves.

give an example of a steepened magnetosonic wave with regions of both phase rotation and lack of rotation. In the bottom panel of Figure 24, it can be noted that there is a slow (linear) magnetic magnitude increase throughout the wave form. However, the Principal Axis components have a different time - (and distance-) evolution. The B_1 and B_2 components are constant from 0823:38 to 0824:44 UT (left-hand side of the plot to point 1), or approximately one half of the (H_2O group) wavelength. For the same interval B_3 steadily decreases indicating that this part of the wave is almost purely compressive. There is a small phase rotation in B_2 from point 1 to 2, but B_1 remains constant during this interval. B_1 has an abrupt rotation between points 2 and 3 and then remains constant until point 4, where the "partial rotation" takes place. The hodogram in Figure 25 illustrates this more graphically. There is almost no phase change from point B until point 1. From point 1 until almost point 4, there is a "back-rotation". This portion of the wave is right-hand polarized in the spacecraft frame, opposite to that of the partial rotation or the whistlers. The possibility that this shift is due to propagation effects has been examined and was ruled out. Although such "reverse rotations" have been predicted to occur in solitons (Hada et al., 1989), they are expected only within the shock structure (partial rotation plus upstream whistlers) and not well back within the body of the magnetosonic wave (Shapiro, private communication, 1989). Tsurutani et al. (1990a) offer several possible explanations, but the most likely appears to be the detection of a right-hand wave propagating away from the sun. This wave could be generated by the interaction of right-hand waves propagating toward the sun with density fluctuations associated with the steepened magnetosonic wave or whistlers, as from the decay instability (see Kojima et al., 1989).

The whistler packets are also not always as simple as the previous example shown. Figure 26 illustrates a whistler packet which is not propagating parallel to the magnetic field and therefore has a large magnetic compressional component associated with it. It has been noted that in many cases the

plane of propagation of the whistlers is not the same as that for the magnetosonic waves (Tsurutani et al., 1989c).

Highly nonlinear magnetic pulses were observed near the outbound bow shock passage at Comet Giacobini-Zinner. An example is shown in Figure 27. This particular pulse is relatively symmetric, with no indication of wave steepening, in contrast to the waves detected on the inbound pass (Figure 18). The wave peak-to-background "compression" ratio is ~ 4.0. The range for the events studied is 2.3 to 7.0. The full width of the wave is ~ 35s, comparable to the H_2O group ion gyroperiod.

The hodogram for the wave is shown in Figure 28. Although the wave is propagating at an angle of 67° relative to $\vec{B}$, it is circularly polarized. The sense of rotation is right-handed in the spacecraft frame. Because all of the magnetic pulses examined were right-handed, it has been speculated that the waves may have phase speeds comparable to or greater than the solar wind speed.

Computer simulations are needed to determine wave properties in this highly nonlinear regime. The wave polarization as a function of angle of propagation relative to $\vec{B}$ and the wave phase speed as a function of wave magnitude need to be theoretically understood.

Comparison to the Earth's Foreshock Wave

Nonlinear waves have been detected in the Earth's foreshock which in some ways are quite similar to the cometary case. The generation mechanism for the long period waves (~ 20 s in the spacecraft frame) is established as being due to solar wind ions reflected at the bow shock. The development of ions from bow shock "reflected" to "intermediate" to "diffuse" distributions proceeds from the region nearly tangent to the foreshock, to the deep foreshock, respectively (Asbridge et

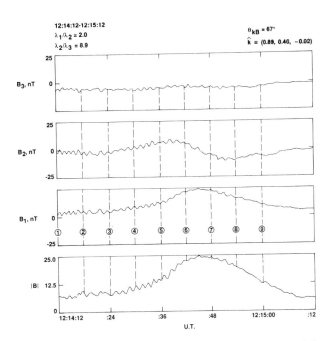

Figure 27. An example of a magnetic pulse plotted in Principal Axis coordinates. The wave is almost symmetric, showing no evidence of steepening. The peak field to background "compression" ratio is ~ 4.0. The wave is propagating at an angle 67° relative to $\vec{B}$. From Tsurutani, et al. 1990b.

al., 1968; Gosling et al., 1978; Bonifazi and Moreno, 1981 a,b; Paschmann et al., 1981).

The detailed characteristics of the waves are not well known at this time. It is believed that simple, nonsteepened waves are observed close to the shock and more complex cases (steepened with whistlers) are found further from the shock. Also, intrinsic "complex" left-hand polarized waves have been observed. The percent time this occurs and the nature of these waves is not certain at this time, but it is estimated to occur less than 10% of the time (C. T. Russell, personal communication, 1989). It is unclear whether the left-hand waves are generated by the nonresonant instability, by some other mechanism such as three wave processes as discussed previously, or associated with ion ring distributions. More effort is needed in this area.

Wave Cascading - Turbulence?

A high frequency spectrum of wave power at frequencies above the H_2O group ion cyclotron frequency is given in Figure 29, taken from Glassmeier et al. (1989). This spectrum was calculated when the angle between the IMF and V_{sw} was close to 90°. A f^{-2} line has been added to the figure. This spectral shape is significantly different than a Kolmogorov $f^{-5/3}$, or turbulence spectrum. Note that there is no apparent enhancement at the H_2O group ion cyclotron frequency, 4 mHz, in this case. This is in sharp contrast to figures 4, 5 and 7 where there is enhanced power at ~ 10^{-2} Hz. This lack of a "pump" wave is in agreement with the previous discussion of the lack of generation of (left-hand) ion cyclotron waves at large α. It is also noted that the wave power is considerably below that in Figure 7.

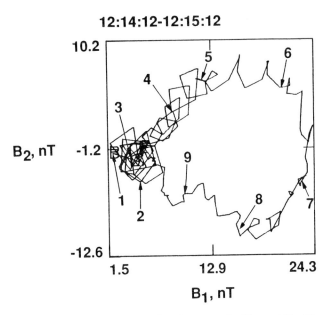

Figure 28. The hodogram for the event in Figure 27. The numbers correspond to the times given in Figure 27. The wave is circularly polarized.

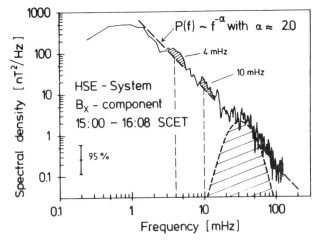

Figure 29. A power spectrum of a transverse component of the magnetic field fluctuations at Comet Halley. The coordinate system is the same as described in Figure 7. The interval analyzed occurred when $\vec{B}$ was nearly orthogonal to the solar wind velocity. The spectrum can be fit with a $f^{-2.0}$ power law, indicating that it does not fit the Kolmogorov-Obukov turbulence prediction. There is no obvious peak at the H_2O group ion cyclotron period (4 mHz), in agreement with the results in Figure 11 for Comet Giacobini-Zinner. The wave power is lower by an order of magnitude from that shown in Figure 7. From Glassmeier et al., 1989.

Although it has been speculated by many authors that this power law dependence (varying from 5/3 to 2) may be evidence for a cascade process, there are many other possible explanations which do not require such exotic (and interesting) physical processes. Some of these are: 1) although the spacecraft is nearly in the frame of the cometary ions, the spacecraft-comet relative velocity will cause small Doppler shifts in the waves. For ICE, the relative velocity was 21 km/s and for the spacecraft armada to Halley, it was 65-70 km/s; 2) Glassmeier et al. (1989), Brinca and Tsurutani (1989b) and Goldstein et al. (1990) have discussed the presence of cyclotron harmonic emissions. These modes would add power at the higher frequencies. 3) If the ions are not cold and have finite temperatures, this will shift the wave frequencies to higher values (Brinca and Tsurutani, 1988b). 4) Finally it should be stressed that these waves are highly nonlinear. In the case of Giacobini-Zinner, wave steepening and whistler precursor forming processes generate higher and lower frequency power. It is possible that some or all of the above processes could combine to give a relatively smooth power law spectrum.

Strong wave-particle interactions will dominate the H_2O group ion cyclotron wave evolution. Shapiro and Shevchenko (1988), have demonstrated that resonant wave-particle interactions and ion shell formation will lead to a wave power law spectrum with an index of ~ 2.0, similar to the observations. This may be another simple explanation for the observed spectra.

Wave-particle Interactions

The above nonlinear wave features open up a whole host of theoretical challenges concerning the interaction of these low frequency waves with the solar wind plasma or with energetic particles. Current models assume parallel propagating, noncompressive waves travelling at the Alfvén speed (Terasawa, 1989). None of these assumptions are consistent with the data.

The wave amplitudes are highly nonlinear $\Delta \bar{B}/B \approx 1.0$ - 2.0, a feature that cannot be accommodated by present theories. The waves have a significant compressive component, a feature which should lead to enhanced particle mirroring. If the waves propagate at sufficiently large angles to the ambient magnetic field, there could be gradient B drift acceleration, particularly for energetic protons which have relatively small gyroradii (see general discussion and references in Armstrong et al., 1985). The whistler mode waves and the "partial rotations" are small on the scale of the water group ion gyroradius. The interaction of energetic particles with such structures would cause a (nonresonant) breaking of the first adiabatic invariant, a topic not studied in much depth to date. The nonlinear steepening of the magnetosonic waves leads to the creation of higher frequency waves (previously discussed) and also longer wavelength emissions. The latter are particularly important for the acceleration of the high energy tail of the ion distribution. Finally, it should be mentioned that the steepened magnetosonic waves are propagating at speeds much larger than the Alfvén speed. Although, small amplitude magnetosonic waves propagate at the fast speed, the speed of the steepened waves may be 2-3 times the Alfvén speed. It is best to use the results of numerical simulations to get good values for this velocity.

According to the simulations of Winske and Gary (1986), it has been estimated that 30-50% of the particle pick-up energy go into waves. An estimate of the wave energy at Giacobini-Zinner is 5 x 10^15 joules (Tsurutani and Smith, 1986b) and for Halley 1.8 x 10^17 joules. Relevant questions are "does part of this energy go back into energizing the cometary pickup ions via energy diffusion through stochastic acceleration or via transit time damping? Are the waves damped via solar wind protons, leading to a hotter solar wind (Sharma et al., 1988)? A wave energy budget is needed to understand these potentially important wave-particle interaction effects. The Sharma et al. work is a nice start in this area.

All of the above features are relevant for wave-particle interactions, particularly for stochastic Fermi acceleration processes. Some of these features may be difficult to apply to acceleration models, but these features represent plasma reality. It will first be necessary to examine these various wave phenomena to determine how important they are, i.e., what effect they have on the overall wave-particle interaction process of interest. The most important processes should be included into the appropriate models. The recent work by Cravens (1989), represents an excellent beginning for these types of studies.

A seminal work, considering resonant and nonresonant interactions, weak turbulence, as well as strong turbulence is presented in an article in this book by Yoon and Wu (1991). In the latter process, the authors consider the direct influence of electric and magnetic fields on the ion orbits in the wave-particle interaction process. This is a general area of research that could benefit from considerably more attention.

III. ELF and VLF Plasma Waves

There were ELF (10-1500 Hz) and VLF (10^3-10^6 Hz) plasma wave detectors placed on four spacecraft flown to comets: ICE (Scarf et al., 1978), Sakigake (Oya et al., 1986) and the two VEGAs (Grard et al., 1986). Of these, the ICE electrostatic wave antennae were an order of magnitude larger, leading to approximately an order of magnitude greater sensitivity. Only ICE had the capability of measuring the electrostatic noise associated with thermal electron motion. This was used by Meyer-Vernet et al. (1986) to obtain the plasma density in the cold ion tail of Giacobini-Zinner.

In addition to electric field measurements, both ICE and Sakigake had search coil sensors which were used to measure ELF magnetic waves. Because the Sakigake coil was mounted on the spacecraft body, the spacecraft interference was unfortunately too high to make useful measurements.

Figure 30 gives the VEGA 1 wave spectrum for a variety of distances from the comet. The wave intensity increases with decreasing distance from the comet. Curve a) corresponds to a distance of 1.8 x 10^5 km and curve g), 4 x 10^4 km. The waves have a maximum in a frequency range of 200-400 Hz for all spectra.

The overview of the Giacobini-Zinner cometary waves is shown in Figure 31, taken from Scarf et al. (1986b). Both the electric and magnetic wave intensities are given. It is obvious that waves generated by cometary pickup ions and photoelectrons are present for at least 10^6 km and perhaps 2 x 10^6 km and more. Scarf et al. (1986b), stated that these emissions were the highest ever detected during the lifetime of the spacecraft, including solar flare intervals, at interplanetary shocks and within the Earth's geomagnetic tail.

A comparison of the ICE Giacobini-Zinner and the Sakigake Halley wave intensities is given in Figure 32. Although Sakigake was at greater distances, 8 x 10^6 km, the wave intensities was an order of magnitude higher.

Kennel et al. (1986) compared the plasma waves near comet Giacobini-Zinner to those at interplanetary shocks and at the Earth's bow shock. The ELF/VLF wave signatures are almost completely analogous. Figure 33 illustrate the cometary wave features at and near the inbound and outbound shocks. Upstream of the shocks are electron plasma oscillations and ion acoustic waves. There is broadband turbulence in the downstream regions. Similar features are found at almost all collisionless shocks detected in space.

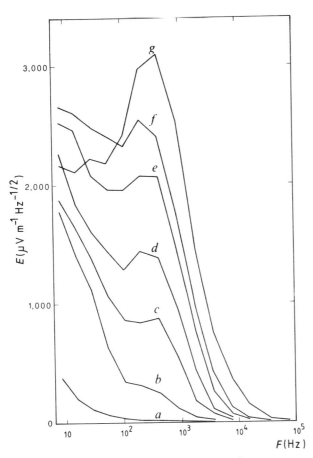

Figure 30. VEGA 1 ELF/VLF 2 min. averaged wave spectra close to Comet Halley. The various curves correspond to distances from the nucleus. Curves a) through g) correspond to 1.8×10^5, 1.4×10^5, 1.2×10^5, 1.1×10^5, 9×10^4, 6×10^4, and 4×10^4 km, respectively. A maximum in wave power is observed at 200-400 Hz. From Grard et al., 1986.

The above cometary wave signatures led Scarf et al. (1986b) to immediately identify the cometary structure as a bow shock, whereas the magnetic field and plasma signatures were far less obvious (Smith et al., 1986a and Bame et al., 1986). Later, more thorough analyses (Jones et al., 1986; Tranquille et al.,

1986; Smith et al., 1986b) have shown that the Scarf et al. (1986) initial deduction was correct.

A summary figure of some of the wave modes at comets is given in Figure 34. The top two left-hand panels illustrate the presence of electron plasma oscillations (EPO) and ion acoustic emissions shown in the previous Figure. Electromagnetic whistler modes are observed below the electron cyclotron frequency (lower right-hand panel), and at and below the range of the search coil sensors are lower hybrid resonance (LHR) waves (center, bottom panel). The LHR emissions are discussed in more detail in Coroniti et al. (1986).

Recently, Richardson et al., (1989b) found that when the interplanetary magnetic field was directed approximately parallel to the solar wind velocity, electrostatic bursts at frequencies between 300 Hz and 10.0 kHz often occurred. The parallel field orientations are caused by the distortions associated with the steepened magnetosonic waves. These relationships are shown in Figure 35. The electrostatic bursts thus have a quasiperiod of ~ 100 sec due to the modulation caused by the magnetosonic waves. The generation of the electrostatic emissions have been explained as being due to the pickup of cold cometary photoelectrons (Brinca et al., 1989; Gomberoff, 1990; Brinca and Tsurutani, 1990).

Final Comments

I have tried to give an up-to-date review and commentary on Low Frequency, ELF and VLF waves at comets, often focussing on unsolved problems. It is clear that although many of the basic wave modes and instabilities have been identified, there is still much more work to be done. Nonlinear wave phenomenon and wave-particle interactions are exciting, developing fields stimulated by cometary research. Both further observational and theoretical efforts are necessary to understand possible implications. The next few years should be exciting ones for cometary plasma studies.

Acknowledgements. Portions of this paper represents work done at the Jet Propulsion Laboratory, California Institute of Technology under contract with NASA. I would like to thank the convenors of the Chapman Conference, Alan Johnstone of the Mullard Research Laboratory, and Malcolm Niedner of the NASA Goddard Space Flight Center for organizing the meeting and making this review possible. I would also like to thank the two referees and several colleagues, who gave insightful and helpful comments to help shape this paper into its present form.

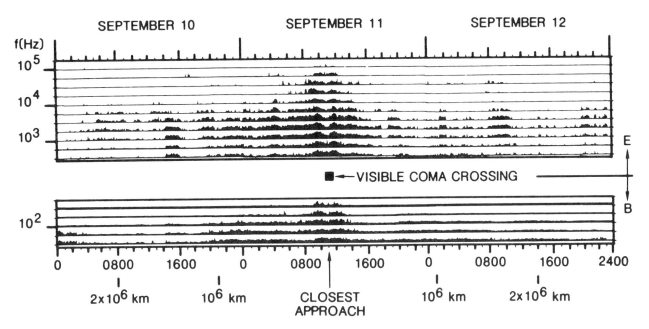

Figure 31.   An overview of the electric and magnetic components of the Giacobini-Zinner ELF/VLF waves. Peak wave intensities are located close to the visible coma crossing. Cometary waves extend to 1-2 x 10^6 km from the nucleus. These emissions were the highest ever detected during the 11 year lifetime of the ICE/ISEE-3 mission. From Scarf et al., 1986b.

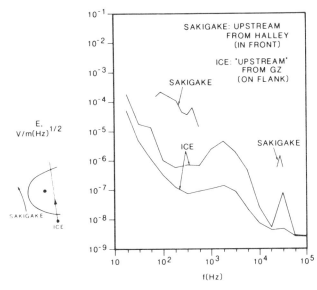

Figure 32. A comparison of the Sakigake (Halley) and ICE (Giacobini-Zinner) wave intensities. The Halley emissions were about one order of magnitude more intense. From Scarf, 1989.

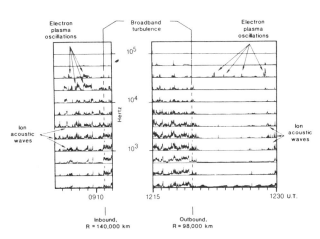

Figure 33.   Plasma waves near the inbound and outbound crossings of the Giacobini-Zinner bow shock. Upstream electron plasma oscillations and ion acoustic waves and downstream broad band turbulence are observed at both crossings. These wave modes are typical of interplanetary shocks and the Earth's bow shock. Taken from Kennel et al., 1986.

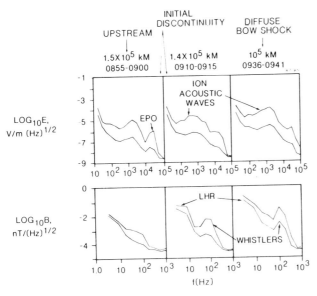

Figure 34. A summary of different ELF/VLF wave modes detected upstream, at, and downstream of the Giacobini-Zinner bow shock. From Scarf, 1989.

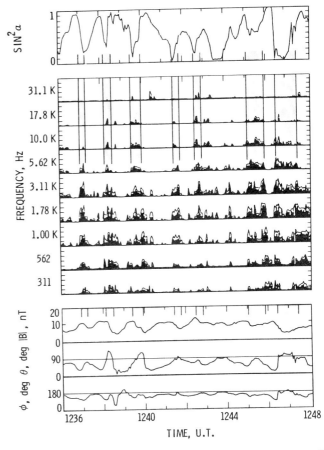

Figure 35. The relationship between bursts of ELF/VLF electrostatic bursts (middle panel) and $\sin^2\alpha$ decreases (top panel) caused by steepened magnetosonic waves (bottom panel). The correlated intervals are denoted by shading. From Richardson et al., 1989 and Brinca et al., 1989.

References

Astron. and Astrophys., Comet Halley, 187, 1987

Armstrong, T. P., M. E. Pesses, R. B. Decker, Shock Drift Acceleration, *in Coll. Shocks in the Heliosphere: Review of Current Research*, ed. B. T. Tsurutani and R. G. Stone, Am. Geophys. Un. Monograph 35, 271, 1985.

Asbridge, J. R., S. J. Bame, and I. B. Strong, Outward Flow of Protons from the Earth's Bow Shock, *J. Geophys. Res., 73,* 5777, 1968.

Balsiger, H., K. Altwegg, F. Buhler, et al., Ion Composition and Dynamics of Comet Halley, *Nature, 321,* 330, 1986.

Bame, S. J., R. C. Anderson, J. R. Asbridge, D. N. Baker, W. C. Feldman, S. A. Fuselier, J. T. Gosling, D. J. McComas, M. F. Thomsen, D. T. Young, and R. D. Zwickl, Comet Giacobini-Zinner: Plasma Description, *Science, 232,* 356, 1986.

Barnes, A., Hydromagnetic Wave and Turbulence in the Solar Wind, *Solar System Plasma Physics, 1*, edited by E. N. Parker, C. F. Kennel and L. J. Lanzerotti, 251, 1979.

Bonifazi, C. and G. Moreno, Reflected and Diffuse Ions Backscattering from the Earth's Bow Shock 1. Basic Properties, *J. Geophys., Res., 86,* 4397, 1981a.

Bonifazi, C. and G. Moreno, Reflected and Diffuse Ions Backscattering from the Earth's Bow Shock 2. Origin, *J. Geophys. Res. 86,* 4405, 1981b.

Brinca, A. L. and B. T. Tsurutani, On the Polarization, Compression, and Non-Oscillatory Behavior of Hydromagnetic Waves Associated with Pickup Ions, *Geophys. Res. Lett., 14,* 495, 1987a.

Brinca, A. L. and B. T. Tsurutani, Unusual Characteristics of Electromagnetic Waves Excited by Cometary Newborn Ions with Large Perpendicular Energies, *Astron. Astrophys., 187,* 311, 1987b.

Brinca, A. L. and B. T. Tsurutani, Survey of Low-Frequency Electromagnetic Waves Stimulated by Two Existing Newborn Ion Species, *J. Geophys. Res., 93,* 48, 1988a.

Brinca, A. L. and B. T. Tsurutani, Temperature Effects on the Pickup Process of Water-Group and Hydrogen Ions: Extension of "A Theory for Low-Frequency Waves Observed at Comet Giacobini-Zinner" by M. L. Goldstein and H. K. Wong, *J. Geophys. Res., 93,* 243, 1988b.

Brinca, A. L. and B. T. Tsurutani, Influence of Multiple Ion Species on Low-Frequency Electromagnetic Wave Instabilities, *J. Geophys. Res., 94,* 13565, 1989a.

Brinca, A. L. and B. T. Tsurutani, On the Excitation of Cyclotron Harmonic Waves by Newborn Heavy Ions, *J. Geophys. Res., 91,* 5467, 1989b.

Brinca, A. L., B. T. Tsurutani and F. L. Scarf, Local Generation of Electrostatic Bursts at Comet Giacobini-Zinner: Modulation by Steepened Magnetosonic Waves, *J. Geophys. Res., 94,* 60, 1989.

Brinca, A. L. and B. T. Tsurutani, Reply, *J. Geophys. Res., 95,* 8291, 1990.

Brinca, A. L., Cometary Linear Instabilities: From Profusion to Prospective, *Cometary Plasma Processes*, A. Johnstone, ed., the American Geophys. Un. Press, Washington, D.C., this issue, 1991.

Chandrasekhar, S., A. N. Kaufman, and K. M. Watson, The Stability of the *Pinch, Proc. Roy. Soc. A, 245,* 435, 1958.

Cohen, R. H., and R. M. Kulrud, Nonlinear Evolution of Parallel Propagating Hydromagnetic Waves, *Phys. Fluids, 17*, 2215, 1975.

Coroniti, F. V., C. F. Kennel, F. L. Scarf, E. J. Smith, B. T. Tsurutani, S. J. Bame, M. F. Thomsen, R. Hynds, K. P. Wenzel, Plasma Wave Turbulence in the Strong Coupling Region at Comet Giacobini-Zinner, *Geophys. Res. Lett., 13*, 869, 1986.

Cravens, T. E., Test Particle Calculations of Pick-up Ions in the Vicinity of Comet Giacobini-Zinner, *Planet. Space Sci., 37*, 1169, 1989.

Crooker, N. U. and G. L. Siscoe, A Mechanism for Pressure Anisotropy and Mirror Instability in the Dayside Magnetosheath, *J. Geophys. Res., 82*, 185, 1977.

Daly, P. W., Can Neutral Particles from Comet Halley Have Reached the ICE Spacecraft?, *Geophys. Res. Lett, 14*, 648, 1987.

ESA SP-250, The Exploration of Halley's Comet, ed. B. Battrick, E. J. Rolfe and R. Reinhard, ESTEC Noordwijk, Holland, 1986.

Erlandson, R. E., L. J. Zanetti and T. A. Potemra, Magnetic-Field Fluctuations from 0 to 26 Hz Observed from a Polar-Orbiting Satellite, *IEEE Trans. Plasma Science, 17*, 196, 1989.

Fairfield, D. H., Bow Shock Associated Wave Observed in the Far Upstream Interplanetary Medium, *J. Geophys. Res., 74*, 3541, 1969.

Galinsky, V. L., A. V. Khrabrov, and V. I. Shevchenko, The Evolution of High Amplitude Parallel Propagating MHD-Waves in the Vicinity of Bow Shocks, *Planet. Space Sci., 38*, 1069, 1990.

Galeev, A.A., R. Z. Sagdeev, V. D. Shapiro and V. I. Shevchenko, MHD Turbulence and Particle Acceleration in a Mass-Loaded Solar Wind, submitted *J. Geophys. Res.*, 1989.

Gary, S.P., C.W. Smith, M. A. Lee, M. L. Goldstein and D. W. Forslund, Electromagnetic Ion Beam Instabilities, *Phys. Fluids, 27*, 1852, 1984.

Gary, S. P., C. D. Madland and B. T. Tsurutani, Electromagnetic Ion Beam Instabilities II, *Phys. Fluids, 28*, 3691, 1985.

Gary, S. P. and C. D. Madland, The Electromagnetic Ion Instabilities in a Cometary Environment, *J. Geophys. Res., 93*, 235, 1988.

Gary, S. P., C. D. Madland, N. Omidi, and D. Winske, Computer Simulation of Two Pickup Ion Instabilities in a Cometary Environment, *J. Geophys. Res., 93*, 9584, 1988.

Gary, S. P., Electromagnetic Ion/Ion Instabilities and Their Consequences in Space Plasmas: A Review, *Space Science Reviews*, in press, 1990.

Geophysical Research Letters, Comets I, 13, 237, 1986

Geophysical Research Letters, Comets II, 13, 381, 1986.

Glassmeier, K. H., F. M. Neubauer, M. H. Acuna, and F. Mariani, Low-Frequency Magnetic Field Fluctuations in Comet P/Halley's Magnetosheath: Giotto Observations, *Astron. Astrophys. 187*, 65, 1987.

Glassmeier, K. H., A. J. Coates, M. H. Acuna, M. L. Goldstein, A. D. Johnstone, F. M. Neubauer, and H. Reme, Spectral Characteristics of Low-Frequency Plasma Turbulence Upstream of Comet P/Halley, *J. Geophys. Res., 94*, 37, 1989.

Goldstein, M. L. and H. K. Wong, A Theory for Low-Frequency Waves Observed at Comet Giacobini-Zinner, *J. Geophys. Res., 92*, 4695, 1987.

Goldstein, M. L., H. K. Wong and K. H. Glassmeier, Generation of Low Frequency Waves at Comet Halley, *J. Geophys. Res, 95*, 947, 1990.

Gomberoff, L., Comment on Local Generation of Electrostatic Bursts at Comet Giacobini-Zinner: Modulation by Steepened Magnetosonic Waves by A. L. Brinca et al., *J. Geophys. Res., 95*, 8287, 1990.

Gosling, J. T., J. R. Ashbridge, S. J. Bame, G. Paschmann, and N. Sckopke, Observations of Two Distinct Populations of Bow Shock Ions in the Upstream Solar Wind, *Geophys. Res. Lett. 5*, 957, 1978.

Grard, R., A. Pedersen, J.- G. Trotignon, C. Beghin, M. Magilevsky, Y. Mikhailov, O. Molchanov, and V. Formisano, Observations of Waves and Plasma in the Environment of Comet Halley, *Nature, 321*, 290, 1986.

Hada, T., personal communication, 1986.

Hada, T., C. F. Kennel and B. Buti, Stationary Nonlinear Alfvén Waves and Solitons, *J. Geophys. Res. 94*, 65, 1989.

Hasegawa, A., Drift Mirror Instability in the Magnetosphere, *Phys. Fluids, 12*, 2642, 1969.

Hasegawa, A., *Plasma Instabilities and Nonlinear Effects, Phys, and Chem. in Space, 8*, Springer-Verlag, New York, 94, 1975.

Hasegawa, A. and L. Chen, Theory of the Drift Mirror Instability in *Plasma Waves and Instabilities at Comets and in Magnetospheres, Geophys. Monogr. Series, 54*, edited by B. T. Tsurutani and H. Oya, AGU, Washington, D.C., 1989.

Hizanidis, K., P. J. Cargill and K. Papadopoulos, Lower Hybrid Waves Upstream of Comets and Their Implications for the Comet Halley "Bow Wave", *J. Geophys. Res. 93*, 9577, 1988.

Hynds, R. J., S. W. H. Cowley, T. R. Sanderson, K. P. Wenzel, and J. J. Van Rooijen, Observations of Energetic Ions from Comet Giacobini-Zinner, *Science, 232*, 361, 1986.

Ipavich, F. M., A. B. Galvin, G. Gloeckler, D. Hovestadt, B. Klecker and M. Scholer, Comet Giacobini-Zinner: In situ Observations of Energetic Heavy Ions, *Science, 232*, 366, 1986.

Johnstone, A. D., J. A. Bowles, A. J. Coates, A. J. Coker, S. J. Kellock, J. Raymont, B. Wilken, W. Studemann, W. Weiss, R. Cerulli-Irelli, V. Formisano, E. de Giori, P. Perani, M. de Bernardi, H. Borg, S. Olsen, J. D. Winningham, and D. A. Bryant, *ESA SP-1077*, 15, 1976.

Johnstone, A., K. Glassmeier, M. Acuna, H. Borg, D. Bryant, A. Coates, V. Formisano, J. Heath, F. Mariani, G. Musmann, F. Neubauer, M. Thomsen, B. Wilken, and J. Winningham, Waves in the Magnetic Field and Solar Wind Flow Outside The Bow Shock at Comet Halley, *Proc. 20th ESLAB Symp.*, ESA SP-250, Noordwijk Holland, 277, 1986.

Johnstone, A., K. Glassmeier, M. Acuna, H. Borg, D. Bryant, A. Coates, V. Formisano, J. Heath, F. Mariani, G. Musmann, F. Neubauer, M. Thomsen, B. Wilken, and J. Winningham, Waves in the Magnetic Field and Solar Wind Flow Outside the Bow Shock at Comet P/Halley, *Astron. Astrophys., 187*, 47, 1987a.

Johnstone, A. D., A. J. Coates, J. Heath, M. F. Thomsen, B. Wilken, K. Jockers, V. Formisano, E. Amata, J. D. Winningham, H. Borg, and D. A. Bryant, Alfvénic Turbulence in the Solar Wind Flowing During the Approach to Comet P/Halley, *Astron. Astrophys. 187*, 25, 1987b.

Jones, D. E., E. J. Smith, J. A. Slavin, B. T. Tsurutani, G. L. Siscoe and D. A. Mendis, The Bow Wave of Comet Giacobini-Zinner: ICE Magnetic Field Observations, *Geophys. Res. Lett., 13*, 243, 1986.

Kaya, N., H. Matsumoto and B. T. Tsurutani, Test Particle Simulation Study of Whistler Wave Packets Observed Near Comet Giacobini-Zinner, *Geophys. Res. Lett., 16*, 25, 1989.

Kennel, C. F. and H. E. Petschek, Limit on Stable Trapped Particle Fluxes, *J. Geophys. Res., 71,* 1, 1966.

Kennel, C. F., F. V. Coroniti, F. L. Scarf, B. T. Tsurutani, E. J. Smith, S. J. Bame and J. T. Gosling, Plasma Waves in the Shock Interaction Regions at Comet Giacobini-Zinner *Geophys. Res. Lett., 13,* 921, 1986.

Kennel, C. F., M. A. Malkov, R. Z. Sagdeev, V. D. Shapiro and A. V. Khrabrov, Alfvén Shock Wave Trains with a Dispersion, *Soviet Physics JETP Lett., 48,* 79, 1988.

Kojima, H., H. Matsumoto, Y. Omura, and B. T. Tsurutani, Nonlinear Evolution of High Frequency R-Mode Waves Excited by Water Group Ions near Comets: Computer Experiments, *Geophys. Res., Lett., 16,* 9, 1989.

Korth, A., A. K. Richter, A. Loidl, K. A. Anderson, C. W. Carlson, D. W. Curtis, R. P. Lin, H. Reme, J. A. Sauvaud, C. d'Uston, F. Cotin, A. Cros and D. A. Mendis, Mass Spectra of Heavy Ions Near Comet Halley, *Nature, 321,* 335, 1986.

Lee, L. C., C. S. Wu and C. P. Price, On The Generation of Magnetosheath Lion Roars, *J. Geophys. Res., 92,* 2343, 1987.

Lee, M. A., Ultra Low Frequency Waves at Comets, in *Plasma Waves and Instabilities at Comets and in Magnetospheres,* ed. B. T. Tsurutani and H. Oya, Am. Geophys. Un., Wash. D.C., *53,* 13, 1989.

Lee, M. A. and W. H. Ip, Hydromagnetic Wave Excitation by Ionized Interstellar Hydrogen and Helium in the Solar Wind, *J. Geophys. Res. 92,* 11041, 1987.

Meyer-Vernet, N., P. Couturier, S. Hoang, C. Perche, J. L. Steinberg, J. Fainberg and C. Meetre, Plasma Diagnosis from Thermal Noise and Limits on Dust Flux or Mass in Comet Giacobini-Zinner, *Science, 232,* 370, 1986.

Mukai, T., W. Miyake, T. Terasawa, M. Kitayama and K.Hirao, Plasma Observations by Suisei of Solar-Wind Interaction with Comet Halley, *Nature, 321,* 299, 1986.

Nature, Encounters with Comet Halley: The First Results, 321, 259, 1986.

Neubauer, F. M., M. H. Acuna, L. F. Burlaga, B. Franke, B. Grankow, F. Mariani, G. Musmann, N. F. Ness, H. U. Schmidt, T. Terenzi, E. Ungstrup, and M. Wallis, *ESA SP-1077,* 1, 1986.

Omidi, N. and D. Winske, A Kinetic Study of Solar Wind Mass Loading and Cometary Shocks, *J. Geophys. Res., 92,* 13409, 1987.

Omidi, N. and D. Winske, Subcritical Dispersive Shock Waves Upstream of Planetary Bow Shocks and at Comet Giacobini-Zinner, *Geophys. Res. Lett., 15,* 1303, 1988.

Omidi, N. and D. Winske, Steepening of Kinetic Magnetosonic Waves into Shocklets: Simulations and Consequences for Planetary Shocks and Comets, *J. Geophys. Res., 95,* 2281, 1990.

Oya, H., A. Morioka, W. Miyake, E. J. Smith and B. T. Tsurutani, Discovery of Cometary Kilometric Radiation and Plasma Waves at Comet Halley, *Nature, 321,* 307, 1986.

Paschmann, G., N. Sckopke, I. Papamastorakis, J. R. Ashbridge, S. J. Bame, and J. T. Gosling, Characteristics of Reflected and Diffuse Ions Upstream from the Earth's Bow Shock, *J. Geophys. Res., 86,* 4355, 1981.

Price, C. P., Mirror Waves Driven by Newborn Ion Distributions, *J. Geophys. Res., 94,* 15001, 1989.

Price, C. P., D. W. Swift and L. C. Lee, Numerical Simulations of Nonoscillatory Mirror Waves at the Earth's Magnetosheath, *J. Geophys. Res., 91,* 101, 1986.

Price, C. P., J. D. Gaffey, Jr., and J. Q. Dong, Excitation of Low-Frequency Hydromagnetic Waves by Freshly Created Ions in the Solar Wind, *J. Geophys. Res., 93,* 837, 1988.

Richardson, I. G., S. W. H. Cowley, R. J. Hynds, P. W. Daly, T. R. Sanderson and K. P. Wenzel, Properties of Energetic Water Group Ions in the Extended Pick-up Region Surrounding Comet Giacobini-Zinner, *Planet. Space Sci.,* 1989a.

Richardson, I. G., K. P. Wenzel, S. W. H. Cowley, F. L. Scarf, E. J. Smith, B. T. Tsurutani, T. R. Sanderson, and R. J. Hynds, Correlated Plasma Wave, Magnetic Field, and Energetic Ion Observations in the Ion Pickup Region of Comet Giacobini-Zinner, *J. Geophys. Res., 94,* 49, 1989b.

Rowlands, J., V. D. Shapiro and V. I. Shevchenko, Quasilinear Theory of Plasma Cyclotron Instability, *Soviet Physics, JETP, 23,* 651, 1966.

Russell, C. T., W. Riedler, K. Schwingenschuh and Y. Yeroshenko, Mirror Instability in the Magnetosphere of Comet Halley, *Geophys. Res. Lett., 14,* 644, 1987.

Sagdeev, R. Z. and A. A. Vedenov, Some Properties in a Plasma Having Anisotropic Distribution Function in Magnetic Field, in *Plasma Phys. and Prog. of Cont. Fusion,* edited by M. A. Leontovich, *3,* 278, Atomizdat, Moscow, 1958.

Sagdeev, R. Z. and V. D. Shafranov, On the Instability of a Plasma with an Anisotropic Distribution of Velocities in a Magnetic Field, *Soviet Physics JETP, 12,* 130, 1961.

Sagdeev, R. Z., V. D. Shapiro, V. I. Shevchenko, and K. Szego, MHD Turbulence in the Solar Wind-Comet Interaction Region, *Geophys. Res. Lett., 13,* 85, 1986.

Saito, T., K. Yumoto, K. Hirao, T. Nakagawa and K. Saito, Interaction Between Comet Halley and the Interplanetary Magnetic Field Observed by Sakigake, *Nature, 321,* 303, 1986.

Sanderson, T. R., K.-P. Wenzel, P. Daly, S. W. H. Cowley, R. J. Hynds, E. J. Smith, S. J. Bame, and R. D. Zwickl, The Interaction of Heavy Ions from Comet P/Giacobini-Zinner with the Solar Wind, *Geophys. Res. Lett., 13,* 411, 1986.

Scarf, F. L., R. W. Fredericks, D. A. Gurnett, and E. J. Smith, ISEE-C Plasma Wave Instrument, *IEEE Trans. Geosci. Electron., GE-16,* 225, 1978.

Scarf, F. L., F. V. Coroniti, C. F. Kennel, T. R. Sanderson, K.-P. Wenzel, R. J. Hynds, E. J. Smith, S. J. Bame and R. D. Zwickl, ICE Plasma Wave Measurements in the Ion Pickup Region of Comet Halley, *Geophys., Res. Lett., 13,* 857, 1986a.

Scarf, F. L., F. V. Coroniti, C. F. Kennel, D. A. Gurnett, W.-H. Ip, and E. J. Smith, Plasma Wave Observations at Comet Giacobini-Zinner, *Science, 232,* 377, 1986b.

Scarf, F. L., Plasma Wave Observations at Comets Giacobini-Zinner and Halley, in *Plasmas Waves and Instabilities at Comets and in Magnetospheres,* edited by B. T. Tsurutani and H. Oya, Amer. Geophys. Un., Washington, D.C., *53,* 31, 1989.

Science, The International Cometary Explorer Mission to Comet Giacobini-Zinner, 232, 353, 1986.

Sentman, D. D., J. P. Edmiston and L. A. Frank, Instabilities of Low Frequency, Parallel Propagating Electromagnetic Waves in the Earth's Foreshock Region, *J. Geophys. Res., 86,* 7487, 1981.

Shapiro, V. D. and V. I. Shevchenko, Astrophysical Plasma Turbulence, *Soviet Science Reviews,* edited by R. A. Syunyaev, Harwood Acad. Pub., London, England, *Section E,* 427, 1988.

Sharma, A. S., P. J. Cargill and K. Papadopoulos, Resonance Absorption of Alfvén Waves at Comet-Solar Wind Interaction Regions, *Geophys. Res. Lett., 15,* 740, 1988.

Sharma, O. P. and V. L. Patel, Low-Frequency Electromagnetic Waves Driven by Gyrotropic Gyrating Ion Beams, *J. Geophys. Res., 91,* 1529, 1986.

Siscoe, G. L., L. Davis, Jr., P. J. Coleman, Jr., E. J. Smith and D. E. Jones, Power Spectra and Discontinuities of the Interplanetary Magnetic Field: Mariner 4, *J. Geophys. Res., 73*, 61, 1968.

Slavin, J. A., E. J. Smith, and B. T. Tsurutani, Magnetic Field Structures in the Comet Giacobini-Zinner Ionosheath, paper presented at 19th General Assembly, IUGG, Vancouver, B. C., 1987.

Smith, C. W., M. L. Goldstein, S. P. Gary and C. T. Russell, Beam Driven Ion Cyclotron Harmonic Resonance in the Terrestrial Foreshock, *J. Geophys. Res., 90*, 1429, 1985.

Smith, C. W. and S. P. Gary, Electromagnetic Instabilities: Growth at Cyclotron Harmonic Wave Numbers, *J. Geophys. Res., 92*, 117, 1987.

Smith, E. J., A. M. A. Frandsen and R. E. Holzer, Lion Roars in the Magnetosheath (abstract), *EOS Trans. AGU, 52*, 903, 1971.

Smith, E. J., L. Davis Jr. D. E. Jones, P. J. Coleman , Jr., D. S. Colburn, P. Dyal, and C. P. Sonett, Saturn's Magnetosphere and It's Interaction with The Solar Wind, *J. Geophys. Res., 85*, 5655, 1980.

Smith, E. J., B. T. Tsurutani, J. A. Slavin, D. E. Jones, G. L. Siscoe, and D. A. Mendis, International Cometary Explorer Encounter with Giacobini-Zinner: Magnetic Field Observations, *Science, 232*, 382, 1986a.

Smith, E. J., J. A. Slavin, S. J. Bame, M. F. Thomsen, S. W. H. Cowley, I. G. Richardson, D. Hovestadt, F. M. Ipavich, K. W. Ogilvie, M. A. Coplan, T. R. Sanderson, K. P. Wenzel, F. L. Scarf, A. F. Vinas, and J. D. Scudder, Analysis of the Giacobini-Zinner Bow Wave, *Proc. 20th ESLAB Symp. Explor. Halley's Comet, ESA SP-250*, 461, 1986b.

Smith, E. J., B. T. Tsurutani, J. A. Slavin, F. L. Scarf and S. J. Bame, Waves in the Giacobini-Zinner Magnetosheath: ICE Observations, in *Proc. Chapman Conf. on Plasma Waves and Instabilities in Magsph. and at Comets.* ed. H. Oya and B. T. Tsurutani, Sohbun Insatsu, Sendai, Japan, 1987.

Somogyi, A. J., K. I. Grimgauz, K. Szego, L. Szabo, Gy. Kozma, et al., First Observations of Energetic Particle Near Comet Halley, *Nature, 321*, 285, 1986.

Terasawa, T., Particle Scattering and Acceleration in a Turbulent Plasma Around Comets, in *Plasma Waves and Instabilities at Comets* and in *Magnetospheres*, ed. by B. T. Tsurutani and H. Oya, Am. Geophys. Un. *53*, 41, 1989.

Thorne, R. M. and B. T. Tsurutani, Generation of Magnetosheath Lion Roars, *Nature, 293*, 384, 1981.

Thorne, R. M. and B. T. Tsurutani, Resonant Interactions between Cometary Ions and Low Frequency Electromagnetic Waves, *Planet. Space Sci., 35*, 1501, 1987.

Tranquille, C., I. G. Richardson, S. W. H. Cowley, T. R. Sanderson, K.-P. Wenzel, and R. J. Hynds, Energetic Ion Properties Observed Near the Periphery of the Mass-Loaded Flow Region Surrounding Comet P/Giacobini-Zinner, *Geophys. Res. Lett., 13*, 853, 1986.

Tsurutani, B. T., E. J. Smith, R. R. Anderson, K. W. Ogilvie, J. D. Scudder, D. N. Baker, and S. J. Bame, Lion Roars and Nonoscillatory Drift Mirror Waves in the Magnetosheath, *J. Geophys. Res., 87*, 6060, 1982.

Tsurutani, B. T. and E. J. Smith, Hydromagnetic Waves and Instabilities Associated with Cometary-Ion Pickup: ICE Observations, *Geophys. Res. Lett. 13*, 263, 1986a.

Tsurutani, B. T. and E. J. Smith, Strong Hydromagnetic Turbulence Associated with Comet Giacobini-Zinner, *Geophys. Res. Lett., 13*, 259, 1986b.

Tsurutani, B. T., A. L. Brinca, E. J. Smith, R. M. Thorne, F. L. Scarf, J. T. Gosling and F. M. Ipavich, MHD Waves Detected by ICE at Distance $\geq 28 \times 10^6$ km from Comet P/Halley: Cometary or Solar Wind Origin?, *Astron. Astrophys., 187*, 97, 1987a.

Tsurutani, B. T., R. M. Thorne, E. J. Smith, J. T. Gosling, and H. Matsumoto, Steepened Magnetosonic Waves at Comet Giacobini-Zinner, *J. Geophys. Res., 92*, 11074, 1987b.

Tsurutani, B. T., D. E. Page, E. J. Smith, B. E. Goldstein, A. L. Brinca, R. M. Thorne, H. Matsumoto, I. G. Richardson and T. R. Sanderson, Low-Frequency Plasma Waves and Ion Pitch Angle Scattering at Large Distances ($> 3.5 \times 10^5$ km) from Giacobini-Zinner: Interplanetary Magnetic Field α Dependence, *J. Geophys. Res., 94*, 18, 1989a.

Tsurutani, B. T., A. L. Brinca, B. Buti, E. J. Smith, R. M. Thorne and H. Matsumoto, Magnetic Pulses with Durations Near The Local Proton Cyclotron Period: Comet Giacobini-Zinner, *J. Geophys. Res., 94*, 29, 1989b.

Tsurutani, B. T., E. J. Smith, A. L. Brinca, R. M. Thorne and H. Matsumoto, Properties of Whistler Mode Wave Packets at the Leading Edge of Steepened Magnetosonic Waves: Comet Giacobini-Zinner, *Planet. Space Sci., 37*, 167, 1989c.

Tsurutani, B. T., E. J. Smith, H. Matsumoto, A. L. Brinca and B. Buti, Discrete Phase Changes within Nonlinear Steepened Magnetosonic Waves: Comet Giacobini-Zinner, *Geophys. Res. Lett.,* 1990a.

Tsurutani, B. T., E. J. Smith, H. Matsumoto, A. L. Brinca and N. Omidi, Highly Nonlinear Magnetic Pulses at Comet Giacobini-Zinner , *Geophys. Res. Lett., 17*, 757, 1990b.

Wenzel, K.-P., T. R. Sanderson, I. G. Richardson, S. W. H. Cowley, R. J. Hynds, S. J. Bame, R. D. Zwickl, E. J. Smith and B. T. Tsurutani, In-Situ Observations of Cometary Pick-Up Ions ≥ 0.2 AU Upstream of Comet Halley: ICE Observations, *Geophys. Res. Lett., 13*, 861, 1986.

Winske, D. and M. M. LeRoy, Diffuse Ions Produced by Electromagnetic Ion Beam Instabilities, *J. Geophys. Res., 89*, 2673, 1984.

Winske, D., C. S. Wu, Y. Y. Li, Z. Z. Mou, and S. Y. Gao, Coupling of Newborn Ions to the Solar Wind by Electromagnetic Instabilities and the Interaction with the Bow Shock, *J. Geophys. Res., 90*, 2713, 1985.

Winske, D. and S. P. Gary, Electromagnetic Instabilities Driven by Cool Heavy Ion Beams, *J. Geophys. Res., 91*, 6825, 1986.

Wu, C. S. and R. C. Davidson, Electromagnetic Instabilities Produced by Neutral Particle Ionization in Interplanetary Space, *J. Geophys. Res. 72*, 5399, 1972.

Wu, C. S., R. E. Hartle and K. Ogilvie, Interaction of Singly Charged Interstellar Helium Ions with the Solar Wind, *J. Geophys. Res., 78*, 306, 1973.

Wu, C. S. and R. E. Hartle, Further Remarks on Plasma Instabilities Produced by Ions Borne in the Solar Wind, *J. Geophys. Res., 79*, 283, 1984.

Wu, C. S., D. Kraus-Varban, and T. S. Huo, A Mirror Instability Associated with Newly Created Ions in a Moving Plasma, *J. Geophys. Res., 93*, 11527, 1988.

Wu, C. S., L. F. Ziebell and P. H. Yoon, On the Observation of Enhanced Hydromagnetic Waves Near Gyrofrequencies of Cometary Ions, submitted *J. Geophys. Res.*, 1990.

Yoon, P. H. and C. S. Wu, Ion Pickup by the Solar Wind via Wave-Particle Interactions in *Cometary Plasma Processes*, ed. A. Johnstone, Am. Geophys. Un. Press, Washington, D.C., this issue 1991.

Yeroshenko, Y. G., V. A. Styashkin, W. Riedler, K. Schwingenschuh, and C. T. Russell, Fine Structure of the Magnetic Field on Comet P/Halley's Comet, *Astron. Astrophys., 187*, 69, 1987.

COMETARY LINEAR INSTABILITIES: FROM PROFUSION TO PERSPECTIVE

A.L.Brinca

Centro de Electrodinâmica, Instituto Superior Técnico, P-1096 Lisboa Codex, Portugal

Abstract. This theoretical review of linear cometary instabilities attempts to provide some guidance to the nonspecialist on the interpretation of the large variety of growing wave modes fed by newborn particles. It emphasizes the underlying phenomenology, presents stability criteria of general applicability, and draws attention to a few caveats. Outlines of standard, and alternative models for the description of the intervening plasma media precede the statement of stability results, and the discussion of wave-particle interactions, resonant and nonresonant instabilities, parallel and oblique propagation, multiple-ion effects and electrostatic modes.

Introduction

The interaction of the solar wind with comets brings about a rich phenomenology in the realm of space plasma physics. Sublimation of ices from the cometary nuclei produces fluxes of particles that permeate the ambient solar wind while undergoing dissociation and ionization. Although the escape speeds are small (up to a few km/s), the large ionization time scales (of the order of 10^6 s at 1 AU) allow for the creation of huge regions (with scale sizes of millions of kilometers) affecting the flow of the solar wind. The dependence of the relevant processes taking place in these volumes on the characteristics of the interacting media determines the existence and location of boundary regions that structure the cometary environment.

The density of the neutral particles originated in the nuclei at cometocentric distances beyond the bow wave are small with respect to the solar wind density. However, the ionization of these neutrals corresponds to the abrupt injection of free energy into the solar wind (frame) and can feed a large variety of instabilities. The generated waves play an important role in the pickup of cometary ions, ensuing (mass loading) deceleration of the solar wind, and the formation of shock fronts; the associated wave-particle interactions in this collisionless magnetoplasma contribute to the heating and acceleration of plasma particles.

Recent space missions to comets Giacobini-Zinner and Halley provided the opportunity to contrast observations and theoretical anticipations. The results were mixed; however, discrepancies and unexpected behaviors are welcome to motivate and foster future research, possibly influencing the

Cometary Plasma Processes
Geophysical Monograph 61
©1991 American Geophysical Union

planning of coming cometary projects such as CRAF (Cometary Rendezvous and Asteroid Flyby), scheduled to be launched in August of 1995 (comet Kopff and asteroid Hamburga).

The characteristics and intensity of the wave activity detected in cometary environments [see review by Tsurutani, 1991] show the clear occurrence of nonlinear phenomenology. Comparison of theory with observations requires further analytical and simulation research, albeit nonlinear studies in this area have already been performed [e.g., Galeev et al., 1991]. In reviewing the linear theory of cometary instabilities, we are concerned with the first step in the interpretation of field observations. The large number and variety of parameters to be chosen for the background solar wind, ambient magnetic field, and newborn species brings about a bewildering panoply of growing modes. Our aim is to contribute to their understanding by emphasizing stability results of general applicability and a unifying description whenever possible, and to draw attention to a few (sometimes ignored) caveats. We complement recent works on similar areas [Lee, 1989 ; Gary, 1990] and thus attempt to minimize the repetitious exposition of analytical approaches and utilization of figures.

The Cometary Medium

Simple Model

The instabilities of concern occur in the interaction between the solar wind and comets, at heliocentric distances around 1 AU. A simple model of the unperturbed solar wind adopts a collisionless magnetoplasma of hydrogen, flowing in the inertial frame with a velocity $\mathbf{V_{SW}}$, at an angle α with respect to the interplanetary magnetic field (IMF) $\mathbf{B_o}$. Typical (Giacobini–Zinner, as observed by ICE) values would be $B_o=8\mathrm{nT}$, $V_{SW}=400\mathrm{km/s}$, and $\alpha=\pi/4$, but the variability of the flow and the occurrence of discontinuities warrant consideration of other values. The proton and electron populations have isotropic Maxwellian velocity distributions, a common number density ($\sim 5\,\mathrm{cm}^{-3}$), ratios of the plasma and magnetic field pressures (β_e and β_p) of the order unity, entailing an Alfvén speed $v_A \sim V_{SW}/5$.

The free energy sources lie in the cometary particles sublimated from the ices in the nuclei. Their speeds in the inertial (and comet) frame are very small when compared to V_{SW}, and have correspondingly small thermal spreads. When they become ionized (photoionization by solar radiation, charge exchange with solar wind protons, or collisional ionization by energetic electrons), the newborn charged particles are injected into the solar wind frame with a velocity close to $-\mathbf{V_{SW}}$. They describe helical trajectories about the IMF, with perpendicular (with respect to $\mathbf{B_o}$) speeds of $\sim|V_{SW}\sin\alpha|$ and parallel drifts of $\sim|V_{SW}\cos\alpha|$. A simplified, commonly adopted model of their distribution in velocity space is thus a drifting ring, with limiting cases of pure rings ($\sin\alpha=0$) or beams ($\cos\alpha=0$). Different coexisting species of newborn particles have been considered (predominantly, protons, water group ions and electrons). As already stressed, the modeled collisionless environment lies upstream of the cometary bow wave.

Alternatives

The above assumptions (viz. spatial homogeneity, unperturbed solar wind and collisionless physics) are not uniformly valid and might be reconsidered in specific problems, in particular as the cometary nucleus is approached. Alternative models for the cometary environment have been mentioned or used.

As pointed out by Lee [1989], the newborn ion velocity distributions might not be gyrotropic, i.e., instead of solely depending on the parallel, $v_\parallel$, or perpendicular, $v_\perp$, velocities, $F_s(v_\parallel,v_\perp)$, they also become a function of φ, the azimuthal angle in the perpendicular plane of the (cylindrical coordinates) velocity space. Under these circumstances, the "equilibrium" solution of the Vlasov equation cannot be simultaneously homogeneous in space and constant in time. Although not yet pursued in the context of cometary plasmas, the stability analysis for nongyrotropic distributions requires a distinct approach, as in the pioneering study of Sudan [1965].

The intrinsic spatial inhomogeneity of the source of cometary newborns (the neutral gas cloud density varies with the inverse of the square of the cometocentric distance) has been explicitly taken into consideration in analytical [e.g., Galeev et al., 1987] and simulation [e.g., Omidi and Winske, 1987] studies. Similarly, the self-consistent incorporation of the solar wind deceleration due to ion pickup mass loading in stability investigations has been done analytically [e.g., Lakhina and Verheest, 1988] and in numerical simulations [e.g., Omidi and Winske, 1987].

Consideration of the continuous injection of newborn particles in the solar wind due to ionization of the cometary neutrals requires inclusion of source terms in the kinetic equation for the distribution functions [e.g., Galeev et al., 1987], or the equivalent use of injection rates for the newborn species in computer simulations [e.g., Gary et al., 1989].

Other deviations from the simple model of the cometary environment have to do with the equilibrium velocity distribution functions of the newborn species. It was argued that, in a first approximation, the cometary charged particles would acquire in the solar wind frame a drifting-ring distribution in velocity space, sometimes modeled as $F(v_\parallel,v_\perp) = \delta(v_\parallel - v_{o\parallel})\delta(v_\perp - v_{o\perp})/(2\pi v_{o\perp})$, where $v_{o\parallel}$ and $v_{o\perp}$ stand for the initial parallel and perpendicular speeds determined by the injection velocity, $-\mathbf{V_{SW}}$, and the initial pitch angle, α. This idealization does not take into consideration the escape speeds of the cometary neutrals in the inertial frame, their thermal spreads, and changes of velocity due to ionization. These speeds are small with respect to V_{SW} but, in conjunction with IMF directional fluctuations, bring about distributions with finite spreads

in $v_\parallel$ and $v_\perp$ [Tsurutani, 1991]. The consequences of this simplification depend, however, on the problem under consideration: the idealization might be warranted. For example, as pointed out below, the stability of parallel propagation ($\mathbf{k}\|\mathbf{B_o}$) in media with separable velocity distributions, $F_s(v_\parallel,v_\perp)=F_{s\parallel}(v_\parallel)F_{s\perp}(v_\perp)$, is independent of the shape of $F_{s\perp}(v_\perp)$, being solely sensitive to the mean square perpendicular speed, $\langle v_\perp{}^2\rangle$.

Of special relevance to the cometary environment are velocity distributions of the spherical shell type. The interaction of the newborn particles with pre-existing, or comet generated, wave turbulence involves pitch angle scattering at a faster rate than energy diffusion. As a result, the initial drifting-ring distributions of the newborns evolve into slowly thickening spherical shells. Cometary observations [e.g., Neugebauer et al., 1989] and numerical simulations [e.g., Gaffey et al., 1988] have detected the formation of this type of distribution. We shall comment on their stability below.

The newborn electrons are infrequently incorporated in the analyses of low-frequency (below the proton cyclotron frequency) modes: their free energy plays a minor role in feeding those instabilities, and their much lighter mass leads to a rapid isotropization and pickup by the solar wind.

Plasma Description

Vlasov Model

Upstream of the cometary bow wave it is reasonable to assume that collisional processes can be neglected in plasma physics studies. The stability of the solar wind permeated by the newborn populations can then be analyzed with recourse to the Vlasov and Maxwell equations. The adopted equilibrium state is charge neutral, bears zero total electrical current, and the particle populations have equilibrium gyrotropic velocity distributions. Several possibilities exist of implementing this model. For example, if the drifting newborn electrons are explicitly taken into consideration, the solar wind frame is satisfactory without further ado; otherwise, assuming already picked up newborn electrons, we can impart a parallel drift to the solar wind core ions that cancels the electrical current associated with the cometary beam ions in the electron frame.

The wave dispersion equation relating the wavenumber vector, $\mathbf{k}$, and the complex oscillation frequency, $\omega=\omega_r+i\gamma$, is obtained by standard methods: linearization of the equations, Fourier (in space) and Laplace (in time) transformation of the field quantities and imposition of the existence of nontrivial solutions. The approach, for arbitrary orientations of the wavevector with respect to the uniform background magnetic field $\mathbf{B_o}$, is straightforward but laborious and yields a dispersion equation obtained from the annulment of the determinant of a third order matrix,

$$\det\{D_{ij}(\mathbf{k},\omega)\}=0 \qquad (1)$$

where the elements D_{ij} are complicated functions depending on the velocity distributions of the intervening species [e.g., Davidson, 1983]. This equation supports the linear stability analysis of the Vlasov model: in initial value problems (real $\mathbf{k}$), modes propagating as $\exp i(\mathbf{k}\cdot\mathbf{r}-\omega t)$ are unstable if the solution $\omega(\mathbf{k})$ of (1) has $\gamma>0$.

Alternatives

The kinetic Vlasov description of the plasma is appropriate to the envisaged collisionless medium, representing collective phenomena with characteristic lengths as short as the electron Debye length. Simpler, alternative descriptions can be used in their domains of validity [e.g., Cayton, 1978].

In the Vlasov-fluid (hybrid) plasma model, the ion distribution function satisfies the Vlasov equation (kinetic description), the electrons are represented as a massless, pressureless fluid, and the electromagnetic field satisfies Maxwell's equations with the displacement current neglected and charge neutrality assumed. The model can be extended to include finite electron pressure, with the electron mass still neglected. It is valid for lengths larger than both the Debye length and the electron gyroradius, and it ignores dispersion due to charge separation. It is adopted in hybrid (particle ions and fluid electrons) plasma simulation codes [e.g., Winske and Leroy, 1984].

The guiding center plasma model assumes the gyroradius and the cyclotron frequency of both electrons and ions to be zero and infinite, respectively. The double adiabatic theory, also known as the guiding center fluid model, includes anisotropic pressure in the ideal magnetohydrodynamic description. The critical length for the application of these models is much larger than the ion gyroradius. They ignore dispersion due to finite gyroradius effects and are unable to describe resonant instabilities, although containing nonresonant growing modes.

Stability

Introduction

The wave dispersion equation (1) determines the linear stability properties of the modeled medium. In particular, its solutions for real wavevector (initial value problem) define the stability of the allowed modes (an infinite number, since the equation is transcendental, albeit only a few are not heavily damped), with the growth rate γ quantifying the temporal growth or attenuation of the modes.

For each unstable wave, one would like to identify the physical mechanisms underlying the instability (free energy source feeding it, in particular), the domain of stability in the k plane (with special emphasis on the neighborhoods of maximum growth), and the dependence of its characteristics on parameter space. The linear analysis, however, is only the starting point for the understanding of the cometary wave activity and related plasma processes. The hierarchy of unstable modes based on linear growth rates does not necessarily translate to their importance in the mechanisms where they intervene. For example, a given plasma model might yield the largest growth rate for modes that saturate at very low levels and hence become irrelevant to the pickup process [Winske et al., 1985].

Although efficient numerical programs to solve the general kinetic dispersion equation for certain types of equilibrium velocity distributions (usually iso- or bi-Maxwellians and combinations thereof) are nowadays available [e.g., Rönnmark, 1982], the analytical complexity of (1) strongly recommends the adoption (initially, at least) of adequate approximations that elucidate the physical nature of the growing waves.

About two decades ago, the studies of linear stability placed great emphasis on the discrimination between absolute and convective instabilities [e.g., Bers, 1983]. Loosely speaking, one aimed at differentiating between "oscillating" (absolute instability: the perturbation grows in time to infinity everywhere) and "amplifying" (convective instability: the perturbation decays to zero for all fixed points, albeit growing to infinity at points moving with the disturbance) media. We are not aware of investigations in this area for the cometary environment, and note that it has been recently argued that the basis of the theory on absolute and convective instabilities is questionable, the relevant issue being (for space plasmas, in particular) whether the growth of unstable waves in some region will saturate by nonlinear effects, or by linear convection [Oscarsson and Rönnmark, 1986].

General Results

When embarking on linear stability studies of homogeneous, time-invariant, collisionless magnetoplasmas having (1) as their nonrelativistic dispersion equation, it is useful to be aware of a few general results that might avoid unnecessary analyses, or suggest alternative sources of energy.

The conservation of total (field plus particle kinetic) energy integrated over the region of phase space $(\mathbf{r}, \mathbf{v})$ occupied by the plasma follows from the Vlasov-Maxwell system of equations. The Vlasov equation implies the temporal invariance of the integration over phase space of smooth, differentiable functionals of the particle species distribution functions, $f_s(\mathbf{r}, \mathbf{v}, t)$. As a consequence, it can be demonstrated [Davidson, 1983] that if the equilibrium distribution functions are isotropic, $F_s(v_{\parallel}, v_{\perp}) = F_s(v^2 = v_{\parallel}^2 + v_{\perp}^2)$, and monotonic decreasing functions of v for all plasma components s, the magnetoplasma is electromagnetically stable. It has been shown that these isotropic, monotonically decreasing functions are also unable to feed growing electrostatic waves in magnetoplasmas [Tataronis, 1967].

These criteria hold for arbitrary orientations of the wavevector k with respect to the background magnetic field $\mathbf{B}_0$ (usually denoted by the angle θ). Other electromagnetic stability results require consideration of parallel propagation ($\sin \theta = 0$). Of particular interest for cometary environments, as already mentioned, is the spherical shell velocity distribution arising from the (predominantly, pitch-angle scattering by turbulence) evolution of the newborn species. They are but a special case of isotropic equilibrium distributions, and it can be demonstrated [Brinca, 1990] that electromagnetic waves propagating along the ambient magnetic field are stable when the particle populations have isotropic velocity distributions. Partially filled spherical shell distributions [Freund and Wu, 1988] and drifting complete shells [Gary and Sinha, 1989; Yoon et al., 1989] are not isotropic and can thus become unstable for parallel (and possibly oblique) propagation. Notice, however that this criterion does not hold for electrostatic waves and obliquely propagating modes. Thin spherical shells can feed growing cyclotron harmonic (Bernstein) waves [Tataronis and Crawford, 1970], and isotropic distributions can be contrived to satisfy Penrose's criterion [e.g., Davidson, 1983] for the instability of (electrostatic) parallel Langmuir waves. As to the stability of obliquely propagating waves ($\sin \theta \neq 0$), we note that parallel stability does not ensure oblique stability [e.g., Brinca and Tsurutani, 1987]. The possibility that the free energy available in non–Maxwellian, non–monotonic decreasing isotropic distributions (spherical shells, in particular) might destabilize oblique electromagnetic modes was stressed by Brinca [1990], and confirmed by Wu and Yoon [1990] who found that thin spherical shells of pickup ions can stimulate the growth of obliquely propagating, low-frequency Alfvén and magnetosonic waves.

Another useful result in the stability analysis of parallel magnetoplasma modes stems from the functional dependence of the equilibrium distributions on the perpendicular and parallel velocities. When the distributions are separable, $F_s(v_{||},v_\perp)=f_{s||}(v_{||})f_{s\perp}(v_\perp)$, the dispersion equation for parallel propagation (and, hence, the stability of parallel modes) is insensitive to the shape of the perpendicular velocity part of the distribution, $f_{s\perp}(v_\perp)$. It can readily be shown [e.g., Davidson, 1983] that only the mean square value of the perpendicular velocity, $\langle v_\perp^2 \rangle_s$, occurs in the dispersion equation. Several frequently used models of velocity distributions are separable (e.g., bi-Maxwellians, drift rings, some loss cone types) and the stability analysis can thus be conducted as if their perpendicular component were an ideal delta function ring distribution, with the inherent simplification in the treatment.

The electrostatic dispersion equation for parallel propagation in a magnetoplasma does not depend on the background magnetic field: the force on the particles originated in the wave electric field is aligned with $\mathbf{B_o}$. Electrostatic modes physically similar to those occurring in unmagnetized plasmas can thus be envisaged in cometary environments [e.g., Brinca et al., 1989]. Their stability might be inferred from the Penrose criterion which provides a necessary and sufficient condition for wave growth when a given functional of the reduced equilibrium velocity distribution (projection along k of a weighted composite of the distributions of the intervening populations) is positive [e.g., Davidson, 1983]. Basically, the familiar necessary "positive slope" in the distribution, $\partial F/\partial v > 0$, accompanying Landau growth must be associated with a sufficiently large depression ("valley") in F.

Resonant Instabilities

Many criteria have been devised to try and categorize linear instabilities in magnetoplasmas [e.g., Cap, 1976]. In cometary studies it is common to find the broad distinction between resonant and nonresonant modes, to adopt designations for the instabilities that provide information on some of their characteristics (e.g., free energy source, polarization, direction of propagation) or carry over from other plasma contexts, to differentiate between parallel and oblique propagation, and the electrostatic or electromagnetic nature of the waves. Concrete situations might, however, invalidate the applicability of rigid classification schemes, as discussed below.

Particle Resonances

General. Charged particles in magnetoplasmas exhibit unperturbed helical motions about the background uniform magnetic field $\mathbf{B_o}$, with cyclotron frequencies Ω_s and parallel velocities $\mathbf{v_{pa}}$. Resonance with a plane wave of frequency ω_r and wavevector k occurs when: i) the parallel velocity of the particle coincides with the parallel phase velocity of the wave, $\omega_r/k_{||}$, so that it senses a stationary parallel wave electric field (Landau

resonance with electrostatic or oblique electromagnetic waves); ii) the particle feels a Doppler shifted wave frequency, $\omega'=\omega_r-\mathbf{k}.\mathbf{v_{pa}}$, equal to its cyclotron frequency, Ω_s, with the fields rotating about $\mathbf{B_o}$ in the same sense as the particle (fundamental cyclotron resonance); iii) the Doppler shifted wave frequency, ω', is a multiple of the particle cyclotron frequency (higher order cyclotron resonances at oblique propagation). These resonances comply with the equation

$$\omega_r-\mathbf{k}.\mathbf{v_{pa}}\pm n\Omega_s=0 \;;n=0,1,2,... \qquad (2)$$

where the Landau, fundamental cyclotron, and higher order cyclotron resonances are associated, respectively, with $n=0, n=1,$ and $n>1$; the $+$ or $-$ signs depend on the type of charge and polarization of the wave, as discussed below for parallel propagation.

Clearly, resonance conditions describe a strong interaction between the particle and the wave, and net energy flows from one to the other. The overall contribution of the resonant population might yield wave growth or damping. Growth implies that the concurrent particle diffusion in velocity space leads to a decrease in the global particle kinetic energy. Gendrin [1981] discusses the conditions that bring about wave generation for parallel propagation.

When the wave mode is predominantly fed by a limited group of particles through this interaction mechanism, the instability is termed resonant. Typically, either the resonant species has an average parallel velocity $\mathbf{v_{pa}}$ approximately satisfying (2) and a thermal speed $v_{ts} \ll v_{pa}$, or the parallel resonant velocity defined by (2) lies within a thermal speed of the peak of the distribution function, ensuring that a significant number of particles have a strong wave-particle interaction [e.g., Gary, 1990]. Nonresonant instabilities are fluid like, arise from a generalized contribution of the particles that can be characterized by macroscopic parameters (e.g., pressure), and depend on moments of the velocity distributions, not detailed features of their shapes. It should be stressed, however, that the distinction between these two types of instabilities might fade away. Resonant and nonresonant aspects of the wave-particle interaction affect each other. In particular, nonresonant particles of the resonant species, or wholly nonresonant populations might drastically modify the characteristics of resonant instabilities by influencing the real dispersion (ω_r,k) of the wave modes: the resulting alteration in the parallel resonant velocity defined in (2) brings into resonance different particles and thus possibly different contributions to wave growth. Also, the distinction between resonant and nonresonant particles is lost when the growth rate is sufficiently large [Kennel and Scarf, 1968].

Consequences. A useful consequence of Eq. (2) in observations of cometary waves generated by resonant instabilities, as first pointed out by Tsurutani and Smith [1986], is that the measured spacecraft frequency spectrum is approximately centered at (multiples of) the cyclotron frequency of the resonant species. This result requires, nevertheless, some qualifications.

It assumes that the cometary particles are almost at rest in the spacecraft frame; however, Giotto passed the nucleus of Halley with a relative speed around 70 km/s. As mentioned below, even neglecting this relative motion, the spacecraft frequency of resonant modes can strongly deviate from the particles cyclotron frequency. A simple Doppler shift relates the wave frequency in the spacecraft frame (ω_{sc}) and the solar wind frame (ω_r), $\omega_{sc}=\omega_r+\mathbf{k}.\mathbf{V_{SW}}$ (the speed of the spacecraft is very small in the inertial frame when compared to V_{SW}). Each unstable oblique $\mathbf{k}$ excites a distribution of modes in the solar wind frame with a common frequency ω_r and wavevectors lying on a conical surface with semiaperture θ, apex and axis on, and along $\mathbf{B_o}=B_o\hat{z}$. Taking without loss of generality the solar wind velocity to lie in the $x-z$ plane, at an angle α to $\mathbf{B_o}$, and using ϕ to denote an azimuthal coordinate defining the angular position of the projection on the $x-y$ plane of the tip of one of the continuously distributed unstable wavevectors, we find that the spacecraft frequency excited by a specific $\mathbf{k}$ is given by $\omega_{sc}=\omega_r+kV_{SW}(\cos\alpha\cos\theta+\sin\alpha\sin\theta\cos\phi)$. Variation of ϕ in its 2π domain yields the range of spacecraft frequencies generated by one unstable wave vector cone associated with one wave solar wind frequency. Only for parallel propagation $(\sin\theta=0)$, or parallel injection of newborns into the solar wind $(\sin\alpha=0)$, does there exist a one-to-one correspondence between wave frequency in the spacecraft and solar wind frames. In general, for each unstable frequency ω_r in the solar wind, there corresponds a spacecraft frame frequency spectrum centered at $\langle\omega_{sc}\rangle=\omega_r+kV_{SW}\cos\alpha\cos\theta$, with bandwidth $\triangle\omega_{sc}=2kV_{SW}|\sin\theta\sin\alpha|$. When the instability is fed by harmonic cyclotron resonance with a group of particles of cyclotron frequency Ω_s and average parallel velocity $\mathbf{v_{pa}}=v_{pa}\hat{z}$ in the solar wind frame, Eq. (2) is approximately satisfied for a given $n\geq1$. In the cometary environment, and assuming that the speed of the ejected cometary neutrals that originate the newborn charged particles is very small in the inertial frame when compared to the solar wind speed, the initial velocity of newborn ions is $-\mathbf{V_{SW}}$, so that $v_{pa}=-V_{SW}\cos\alpha$. We therefore conclude that cometary harmonic cyclotron resonant waves have spacecraft spectra centered near (multiples of) the cyclotron frequency of the feeding species, $\langle\omega_{sc}\rangle=n\Omega_s$. It should be stressed, however, that the spacecraft frequency of resonant modes can significantly differ from Ω_s, when $\omega_r\gg\Omega_s$, if the thermal spread of the resonant particles is taken into consideration [Brinca and Tsurutani, 1988a]; for example, resonant water group ions can excite modes with spacecraft frequencies of the order of the proton cyclotron frequency.

Parallel Electromagnetic Propagation

General. The free energy feeding the instabilities in the modeled cometary environment lies in the newborn particles. In resonant wave growth, a restricted and identifiable group of particles strongly interacts with the wave. For parallel electromagnetic propagation, only the fundamental cyclotron resonances are possible, and the kinetic dispersion equation becomes drastically simplified. It is thus not surprising to find the vast majority of the treatments providing detailed descriptions of the phenomenology underlying linear instabilities centered on the parallel resonant modes. They frequently assume that the contribution of the resonant particles to the real wave dispersion can be ignored and that the growth rate can be determined from a functional of the velocity distribution evaluated at the parallel cyclotron resonant velocity. Mathematically, the parallel kinetic dispersion equation contains a quadrature in the parallel velocity $v_\parallel$ which is a (Hilbert transform) Cauchy integral with the ("resonant") denominator $(\omega-kv_\parallel\pm\Omega_s)$; in the limit of small growth rate $(\gamma\rightarrow0)$, use is made

of the Plemlj relation [e.g., Kennel and Scarf, 1968] to perform the $v_\parallel$-quadrature, and derive the approximate expression for the growth rate that explicates the dominant role of the resonant particles [e.g., Lee, 1989].

Of course, occurrence of resonance is not tantamount to wave growth. The intense wave-particle interaction might as well lead to damping. Gendrin [1981] determines the nature of the interaction by considering the relative position of three curves in the $(v_\parallel,v_\perp)$ plane: isodensity, constant energy and diffusion. Wave growth implies an overall concurrent decrease in the particle kinetic energy.

Brillouin diagrams. An appealing method of visualizing potential domains of wave-particle interactions, and possibly associated resonant instabilities, utilizes Brillouin diagrams to represent the real wave dispersion (ω_r,k) and the fundamental cyclotron resonance condition (straight lines crossing the frequency axis at Ω_s and with slope determined by the parallel velocity, $\omega_r-kv_\parallel=\Omega_s$). The classical depiction of parallel wave dispersion in the first quadrant $(\omega_r>0,k>0)$ enables the identification of the regions of potentially strong coupling between the particle beams and the background modes. However, this representation rapidly becomes awkward because one needs to keep track of the wave polarization (right- or left-hand circular), the sign of the particle charge, and the direcion of the beam parallel drift, in order to locate the intersections of the wave dispersion curves with the resonance lines which correspond to real resonances.

An alternative depiction [Gendrin, 1965] uses the whole Brillouin plane (positive and negative ω_r and k) and permits the separation into different quadrants of the four basic wave types. We shall be concerned with left-hand modes propagating along (LF, first quadrant) or against (LB, second quadrant) $\mathbf{B_o}$, and right-hand modes along (RF, third quadrant) or against (RB, fourth quadrant) $\mathbf{B_o}$. The particle beams are always assumed to have parallel drifts along $\mathbf{B_o}$, so that "F" ("B") waves are co(counter)streaming; the resonance condition lines should thus have positive slopes and cross the real frequency (vertical) axis at $\omega_r=\Omega_s>0$ (positive-charge beams) or $\omega_r=-\Omega_s<0$ (negative-charge beams).

The exact real wave dispersion is obtained from the solution of the kinetic dispersion equation. Qualitative anticipation of the resonant domains, nevertheless, requires utilization of approximate dispersion curves in the Brillouin diagrams. When the free energy lies in dilute (with respect to the background magnetoplasma) populations, as in cometary situations, an acceptable compromise might use real dispersion curves derived from simpler (cold or magnetohydrodynamic, MHD) descriptions for the background medium. It should be emphasized, however, that the coupling between the modes of the dilute populations and of the background plasma may bring about significant modifications in the unperturbed dispersion, possibly invalidating the expectations based on the Brillouin diagrams. In Figure 1 (2), we sketch Brillouin planes with the MHD, low-frequency, parallel electromagnetic dispersion of a hydrogen magnetoplasma without (with) a minority heavy ion species, and resonance lines for the intervening particle types (their slopes represent the, assumed positive, parallel drift velocity of the associated beam). The right-hand wave is the magnetosonic mode that, at higher frequencies, goes into the whistler branch and becomes damped when approaching the Ω_e resonance. The Alfvén/ion cyclotron left-hand wave resonates at Ω_p. The addition of a heavy ion (e.g., oxygen) population to a hydrogen magnetoplasma modifies the dis-

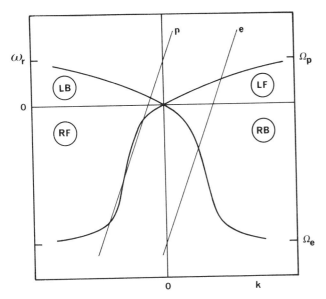

Fig. 1. Sketch in the Brillouin plane of the real dispersion of parallel electromagnetic modes in a hydrogen magnetoplasma, and (straight lines) resonance conditions for protons (p), and electrons (e) with an arbitrary common parallel velocity. The quadrants correspond to (LF) left-hand circularly polarized waves costreaming with respect to the particles, (LB) counterstreaming left-hand waves, (RF) costreaming right-hand waves, and (RB) counterstreaming right-hand waves. The intersections of the straight (cyclotron resonance) lines with the dispersion curves locate domains of strong wave-particle interaction. The particle masses (and, hence, their cyclotron frequencies, Ω_s) are ficticeous to facilitate the depiction.

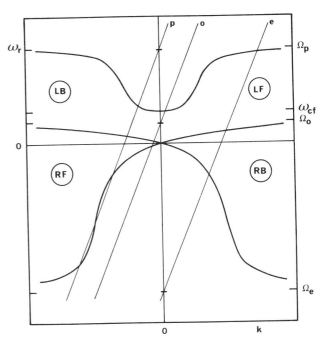

Fig. 2. Same as Figure 1, for a hydrogen magnetoplasma with a heavy (oxygen) ion species and an additional resonance condition for the oxygen ion (o)particles.

persion of this left-hand mode: the original wave below Ω_p splits into two branches, the lower resonating at Ω_o and the upper evolving from a cutoff frequency $\omega_{cf} > \Omega_o$ to a resonance at $\Omega_p > \omega_{cf}$ (this splitting is erased by high temperatures of the heavy ion species, removing its resonance).

Consequences. Observation of these figures suggests potential domains of strong coupling and reveals the impossibility of certain wave-particle interactions. As expected, positively (negatively) charged beams are unable to resonate with counterstreaming right-(left-) hand modes (particles counterstreaming with respect to a wave sense a higher, Doppler upshifted frequency but the same polarization). Costreaming right-hand modes may resonate up to three times with positively charged beams (an "anomalous" interaction because the beam parallel velocity must be larger than the wave phase velocity, involving an inversion of the wave polarization sensed by the particles). Proton beams resonate once with (each branch of) counterstreaming left-hand Alfvén waves (the Doppler upshifted wave frequency approximates Ω_p and the sense of the polarization coincides with the proton cyclotron motion), but do not interact with costreaming left-hand waves (there are no left-hand modes whose frequencies above Ω_p could be Doppler downshifted to resonate with the protons). Heavy ion (oxygen) beams resonate once with the counterstreaming lower branch of the left-hand Alfvén wave and can resonate up to three times with the

costreaming upper branch of the left-hand Alfvén wave (whose frequency can be appropriately Doppler downshifted to Ω_o).

The negatively charged newborn electrons are not usually taken into consideration in the discussion of the excitation of low-frequency cometary waves; it is assumed that their light masses bring about a rapid isotropization and pickup by the solar wind. Photoelectrons with typical energies of 1 eV (in the cometary frame) have speeds of the order of the solar wind speed and random velocity orientation; their free energy is much smaller than in cold drift-ring distributions. Energetic electrons play important roles, however, in other space plasma environments (e.g., the magnetospheric counterstreaming whistler-electron interaction).

Applications. Application of these considerations to comets should bear in mind that the slope of the resonance lines associated with "cold" newborn beams is determined by $V_{SW} \cos \alpha$. Modifications in the relative orientation of the solar wind velocity and magnetic field (caused by solar wind discontinuities or the influence of large cometary-wave magnetic fields) change this slope. Large parallel velocity spreads in the newborn particles ("hot" beams) are associated with correspondingly large slope ranges that broaden the resonance. (In particular, a significant number of beam particles may counterstream with respect to the average beam drift and thus resonate with wave modes not acessible to the "average" beam particle: we shall mention examples of this behavior below.) It is clear from the Brillouin diagrams that certain interactions require the beam parallel velocities to surpass given thresholds or lie within given ranges. The nature of the excited instabilities, therefore, varies with α (which not only defines the parallel drift but also determines the available perpendicular energy). We reiterate, however, that (i) the exact real dispersion of the perturbed medium may significantly deviate from the simplified description adopted in the Brillouin diagrams, and (ii) occurrence of resonance does not necessarily imply instability (the global characteristics of the velocity distributions and of the interaction must be taken into consideration).

Cometary resonant right-hand instabilities originated in the interactions depicted in the RF quadrant of the Brillouin diagrams (ion beams faster than the costreaming magnetosonic/whistler wave) usually exhibit the largest growth rates and α-ranges of existence [e.g., Thorne and Tsurutani, 1987 ; Brinca and Tsurutani, 1988b ; Gary and Madland, 1988]. As α approaches $\pi/2$, the parallel drifts eventually fall below the instability threshold and resonant right-hand wave growth is quenched (large parallel beam temperatures smear this evolution by providing sizeable number of particles above threshold for large α). Decreases in α provide parallel drifts satisfying the instability threshold but diminish the perpendicular energy of the newborn particles: the maximum wave growth for these modes is attained at intermediate values of α. Sufficiently small α and dense beams bring about the nonresonant firehose instability discussed below. Examples of resonant RF instabilities excited by hidrogen or oxygen cometary newborn ions can be found, for instance, in the aforementioned studies and in Goldstein and Wong [1987].

The LB quadrant of the diagrams shows that positively charged beams resonate with counterstreaming left-hand modes. The interaction does not, however, always translate into instability. Hot costreaming beams are able to stimulate left-hand growing waves fed by resonant counterstreaming beam ions [e.g., Gary, 1990]; in its own frame the beam may be isotropic. Cold ion beams can strongly modify the dispersion of the unperturbed background medium, giving rise to unstable modes whose real dispersion smoothly evolves from the LB to the RF quadrant: the transition takes place through nonoscillatory, $\omega_r=0$, purely growing structures, $\gamma>0$ [Brinca and Tsurutani, 1988b]. Hence, the left-hand counterstreaming waves thus excited do not display an autonomous identity. Increases in α tend to favor the left-hand instabilities fed by anisotropy, enhancing the perpendicular energy of the cometary ion beams. As $\alpha=\pi/2$ is reached, the distinction between co- and counter-streaming modes should become meaningless, with only left-hand instabilities fed by the ion anisotropy. The picture is, however, a bit more complicated at $\alpha=\pi/2$. For zero drift and parallel temperature, the wave becomes of the negative energy type, and a reactive instability [Melrose, 1986] arises for $\omega_r \lesssim \Omega_s$ above ($k>k_m$) a given threshold wavenumber k_m [Hasegawa, 1975]. As the parallel thermal spread increases, the unstable wavenumber range becomes finite, the corresponding real frequencies might fall well below Ω_s, and the instability evolves from reactive to kinetic [Brinca et al., 1990] : the resonant counterstreaming ions (although the average drift is zero) are feeding the left-hand wave growth.

Interactions in the LF quadrant involving positively charged particles only occur when the medium contains heavy ion populations. We are not aware of cometary studies identifying left-hand costreaming instabilities stimulated by these resonances. However, the LF interactions have been invoked in other (magnetospheric) contexts to explain the acceleration of heavy ions [Gendrin and Roux, 1980] : these resonances would, therefore,

correspond to wave damping in the envisaged situation. The modification introduced in the dispersion of the left-hand Alfvén waves by the heavy ion population contributes, however, to the onset of a nonresonant costreaming left-hand instability fed by the perpendicular energy of cometary ions injected into the solar wind with large initial pitch angles [Brinca and Tsurutani, 1987], as discussed below.

Nonresonant Instabilities

Linear cometary studies have disclosed the occurrence of the two basic, firehose and mirror, magnetohydrodynamic (fluid, nonresonant) instabilities driven by pressure anisotropies.

The firehose instability requires the effective parallel plasma pressure to exceed the perpendicular pressure by at least the value of the magnetic tension. Its essential phenomenology can be understood by picturing a magnetic flux tube with a perturbation bend [Siscoe, 1983]. The magnetic tension tends to straighten the tube, and a similar effect arises from the perpendicular pressure (motion) of the particles: the centrifugal force exerted by this perpendicular cyclotron motion acts against the tension on the outside (where the particule density decreases because of the bend) and with the tension on the inside (where the particle density increases), so that the overall effect of the perpendicular pressure is to reinforce the staightening (stabilizing) action of the magnetic tension. Destabilization comes from the centrifugal force of the parallel motion of the particles which tends to enhance the bend: instability arises when this centrifugal force due to parallel particle pressure overcomes magnetic tension and perpendicular particle pressure.

In the cometary environment, the destabilizing influence is associated with the parallel motion and, therefore, small initial injection (in the solar wind frame) pitch angles α favor the onset of the firehose instability. Theoretical cometary analyses have indeed obtained this instability [Sagdeev et al., 1986; Brinca and Tsurutani, 1988b] for small α and (possibly too) large relative densities of the newborn species; its threshold is reduced with an increase of the newborn mass, but the growth rate is relatively insensitive to the species type [Winske and Gary, 1986] and maximizes for parallel propagation.

The mirror instability relates to the other regime of particle pressure anisotropy: $p_\perp > p_\parallel$. It arises from a compressional wave that creates local (magnetic) mirror fields which trap the plasma. When the anisotropy is sufficiently large, the diamagnetic repulsion of the trapped plasma tends to exclude the magnetic field. This accelerates the flow of the plasma into the thus deepened well of the local mirror (via the $-\nabla \mu B$ force, where μ is the particle magnetic moment), and therefore the perturbation grows [Hasegawa, 1969]. The plasma density of the anisotropic species will have an out-of-phase variation with respect to the magnetic flux density, and the wavevector of the mirror mode must be oblique ($\sin \theta \neq 0,1$): linear parallel propagation does not yield compressional waves and strict perpendicular propagation would preclude the occurrence of the necessary field nodes for plasma trapping. In the cometary environment, the necessary pressure anisotropy would be provided by the newborn particles when injected into the solar wind with large initial pitch angles α. Drifting ring newborn ion distributions have been found to drive mirror waves [Price, 1989] and an instability that resembles the hydromagnetic mirror wave [Wu et al.,

1988a].

Both the firehose and mirror instabilities coexist with other linear instabilities that, under cometary conditions, usually display lower thresholds
and/or higher growth rates (the right-hand polarized resonant electromagnetic ion beam instability and the left-hand ion cyclotron instability, in
the domains $\alpha \to 0$ and $\alpha \to \pi/2$). It has been suggested that the presence
of other ionic species can have a strong damping effect on ion cyclotron
waves, thus allowing for the enhancement of the role played by the mirror
mode [Price, 1989].

Newborn heavy ions with large initial pitch angles can also feed other
growing waves without becoming cyclotron resonant [Brinca and Tsurutani, 1987]. This nonresonant instability, fed by the perpendicular heavy
ion energy, has somewhat unusual characteristics at oblique propagation
(maximum growth, almost linear polarization and high compression ratios at small obliquities), and its basic characteristics can be derived from
an anisotropic fluid description [B. D. Scott, 1988; private communica-
tion]; the influence of multiple ions on its complex dispersion [Brinca and
Tsurutani, 1989a] also confirms the nonresonant-like behavior.

Multiple Ion Effects

The cometary environment is a multi-ion medium, with predominance
of water group and hydrogen ions. Apart from the free energy carried by
the newborn particles, and its decisive role in feeding growing modes, the
existence of dilute heavier ion populations in hydrogen magnetoplasmas
(as assumed in the simple solar wind model) modifies the real wave dispersion of the medium. As already mentioned in the discussion of parallel
electromagnetic propagation, the original Alfvén/ion cyclotron left-hand
mode in cold plasmas with a resonance at Ω_p splits into separate branches
ordered by the ion cyclotron frequencies: the lowest frequency branch
starts with Alfvén dispersion and goes into resonance at the lowest ion
cyclotron frequency, Ω_{sm}; the next branch evolves from a cutoff frequency above Ω_{sm} into the next higher ion cyclotron frequency, and the
pattern repeats itself until the highest resonance at Ω_p is reached [Young
et al., 1981]. Although this fine structure might be erased by high ion
temperatures (cyclotron resonances are removed), its occurrence at lower
temperatures opens the possibility for mode coupling at oblique propagation with polarization conversion [Young et al., 1981], and the excitation of
new instabilities, as studied by Brinca and Tsurutani [1987] for a hydrogen
magnetoplasma permeated by newborn oxygen ion beams.

Analysis of the stability of low-frequency electromagnetic modes generated by coexisting newborn ion species shows that the effect of multiple
ions on wave growth, for given background magnetoplasma conditions
and relative densities, depends not only on their masses but also on the
physical nature of the wave modes. Whereas each one of the coexisting
ion beams tends to stimulate resonant instabilities without undue influence
from the other species if the ion masses are disparate, newborn ions of similar masses can strongly catalyze growth of fluid like, nonresonant modes,
but only bring about weak growth enhancements in cyclotron resonant
instabilities.

This behavior has a simple physical interpretation. When conditions favor cyclotron wave growth by a given ion species, the other populations,

because of their different masses and hence distinct cyclotron frequencies,
cannot be in exact cyclotron resonance with the same wave (ω,k): their
perturbed average perpendicular velocity vectors are not stationary with
respect to the wave electric field. As a result, if their masses are very
different (e.g., hydrogen and water group ions), each newborn species
resonates with its own wave, bringing about distinct and independent cyclotron instabilities whose growth rates and real dispersions are not affected
by the presence of the nonresonant ions; the global complex dispersion diagram is approximately obtained by the superposition of the individual
cyclotron resonant complex dispersion diagrams generated by each newborn ion species. When the ion masses are similar, the coexistence of
the various newborn species yields only one unstable resonant mode, as
if stimulated by a single species with given effective mass and density;
the global resonant growth arises from the particle contributions of the
coexisting species but, since they cannot interact simultaneously with the
wave under ideal resonance conditions, the end result caused by this partial
phase mixing is a growth rate much smaller than the one encountered in
the same cyclotron instability of a single species with a number density
equal to the sum of the densities of the coexisting ions of similar masses.

In contrast to the above kinetic phenomenology, the fluid, nonresonant
instabilities do not depend on delicate relative orientations (between wave
fields and particle velocities) brought about by satisfying the cyclotron
resonant conditions. The instability criteria can be expressed in terms of
macroscopic parameters that depend on the partial contributions of the
various species. The global fluid behavior obtained for coexisting newborn species of similar masses can be approximately reproduced with a
single ion species whose number density equals the total densities of the
coexisting populations.

This behavior has been illustrated for the case of newborn ions with
disparate [Brinca and Tsurutani, 1988b; Gary and Madland, 1988] and
similar [Brinca and Tsurutani, 1989a] masses.

Instabilities at Oblique Propagation

The instability classification schemes are not always precise. Not only
growing waves occurring at parallel propagation maintain their unstable
behavior within a finite range of obliquities, but modes owing their existence to oblique propagation are discussed elsewhere (e.g., the nonresonant
mirror mode and the electrostatic cyclotron Bernstein waves). Here, we
stress some peculiar features of oblique stability in the cometary context
that are not addressed in other sections of this review.

The imbalance between the number of theoretical studies for parallel
and oblique stability is frequently justified on the grounds that most magnetoplasma models yield maximum growth for parallel propagation. The
argument is not thoroughly convincing: apart from the purely oblique instabilities missed in parallel analyses, modes with larger growth rates at
parallel propagation may play more important roles in the ensuing nonlinear evolution when propagating obliquely because of their inherent oblique
properties (e.g., compression and shock formation). The fact remains that
oblique analysis is much more complex: it deals with a complicated dispersion equation and explores a two-dimensional domain of the wavevector
k. In general, the solutions of the oblique kinetic dispersion equation (1)

can only be obtained numerically, thus obfuscating the physical interpretation of the instabilities. Warranted simplifications clarify the nature of the modes, but confirmation of their domain of validity still requires the exact solution of (1). Available analytical solutions of the kinetic oblique dispersion equation [e.g., Thorne and Summers, 1986] are useful but they apply only to situations of weak growth stimulated by particle resonances.

Study of the oblique characteristics of parallel electromagnetic instabilities in cometary environmemts would conclude, in general, that maximum growth obtains indeed for parallel propagation [Brinca and Tsurutani, 1989b; Gary, 1990]. Exceptions have been reported for growing modes stimulated by newborn oxygen [Brinca and Tsurutani, 1987] and hydrogen [Wu et al., 1988b] ions. As already suggested, however, a survey of the relevant properties of oblique instability cannot be confined to this approach: several cases of potential cometary interest would be completely missed.

Purely oblique instabilities include the nonresonant mirror mode (invoked by Russell [1988] in the interpretation of visible structures found in conjunction with comet Halley), the predominantly electrostatic lower-hybrid mode (whose fluctuating field, according to Buti and Lakhina [1987], is responsible for the stochastic acceleration of cometary ions), and the electrostatic Bernstein cyclotron modes.

Stimulation of wave growth by ions in higher-order cyclotron resonance with oblique modes is also relevant to comets. As shown by Brinca and Tsurutani [1989c] and Goldstein et al. [1990], these wave-particle interactions may feed oblique instabilities with mean spacecraft frequencies centered near multiples of the resonant ion cyclotron frequency. This harmonic structure in the excited spectrum may help explain similar harmonic structures observed by Giotto upstream of comet Halley [Glaßmeier et al., 1989].

A comprehensive survey of oblique stability under cometary conditions implies the systematic exploration of the wavevector plane $(k_{\parallel}, k_{\perp})$ when solving the dispersion equation. To each real valued k (initial value problem), one associates the corresponding real frequency and growth rate, and other wave parameters of interest for the problem under investigation (e.g., polarization, magnetic and mass density compression ratios). As the wavevector plane is surveyed, with emphasis on the neighborhoods of growth rate maxima, these wave parameters define, for each mode, a set of sheets (surfaces) better visualized in 3-D, or contour plots over the k plane.

A first step in this direction was taken by Brinca and Tsurutani [1989b] who studied the oblique behavior of low-frequency electromagnetic waves excited by newborn cometary oxygen and hydrogen ions for large values of the initial injection pitch angle ($\alpha > \pi/4$), characterizing the unstable modes with contour plots of the growth rate, real frequency, polarization, and wave magnetic field compression ratio ($|B_{\parallel}/B|$, where **B** is the wave magnetic field). The results unveil a complicated dispersion topology, with the tracking of the solutions of the oblique kinetic dispersion equation occasionally hindered by steep gradients, or splittings and intersections found in the wave parameter sheets. Although, in general, the unstable cometary modes at parallel propagation reduce their growth rates as the obliquity angle θ increases (some exceptions pointed out above), exploration of the

wave characteristics on the k plane discloses the frequent occurrence of islets of oblique growth which, as a rule, do not connect to the parallel growing modes.

In contrast to the parallel propagation situation where electromagnetic wave–particle interactions always imply fundamental cyclotron resonances, $n=1$ in Eq. (2), we can find all types of resonances, ($n=0,1,2,...$; Landau, fundamental and higher–order cyclotron) in the interactions involving oblique modes. We would expect the real dispersion of oblique resonant instabilities to approximately satisfy Eq. (2) near maximum growth, albeit bearing in mind that finite parallel temperatures of the newborn particles might smear the signatures of the underlying resonances [Brinca and Tsurutani, 1988a]. The oblique survey results of Brinca and Tsurutani [1989b] include nonresonant growing modes (e.g., the mirror wave) and hint at the association of several growth islets with Landau, fundamental and higher–order cyclotron interactions, thus also suggesting the existence of oblique resonant instabilities.

On Electrostatic Instabilities

The pioneering work of C. S. Wu and collaborators on the electromagnetic [Wu and Davidson, 1972] and electrostatic [Hartle and Wu, 1973] instabilities fed by newborn ions, and further extensions and clarifications [Wu and Hartle, 1974], albeit performed for planetary and interstellar ions in the solar wind, is readily adaptable to cometary ions.

The free energy available in the newborn particles can feed growing electrostatic waves, although the theoretical work in this area is much sparser than in the electromagnetic case. Relative motions of particle species (cometary newborn particles and solar wind populations) are typical sources of electrostatic instabilities. The newborn ion parallel drift has been invoked to stimulate a slow ion acoustic mode, whereas the perpendicular motion, in the approximation of unmagnetized ions, can be thought of as a perpendicular beam capable of Landau feeding (lower hybrid) electrostatic waves of appropriate perpendicular phase velocity [Lakhina, 1987]. The parallel drifts of newborn photoelectrons and protons can generate, respectively, an ion acoustic mode and ion/ion acoustic waves [Brinca et al., 1989]. The free energy in the perpendicular motion of the newborn particles might excite electrostatic Bernstein cyclotron waves both initially, when modeled by ring-like distributions [Sagdeev et al., 1987], and possibly, by analogy with similar analyses in different contexts [Tataronis and Crawford, 1970], after evolving into thin spherical shells by pitch angle scattering.

Away from parallel propagation, the linear wave dispersion precludes the exact distinction between electrostatic and electromagnetic modes: in general, they exhibit mixed behavior [Wu et al., 1988b]. Also, conditions occur when the simpler electrostatic approach needs to be revised; for example, in the case of the lower hybrid instabilities, sufficiently energetic ion rings warrant incorporation of electromagnetic effects in the wave dispersion [Akimoto et al., 1985].

Summary

The richness of the interaction between the solar wind and comets provides ample ground for the development of theoretical stability analyses.

As evidenced in this review, even with the adoption of simplified models for the plasma media, a large number of linear instabilities may be excited. Consideration of the mechanisms involved in the transfer of energy between the cometary particles and the wave fields contributes to the clarification of the physical nature of the instabilities and suggests flexible schemes for their classification. Assessment of their relevance to the cometary phenomenology involves investigations beyond the linear stage, and requires comprehension of the associated nonlinear evolution and saturation mechanisms [e.g., Galeev et al., 1991].

Because the freedom of choice and imagination of the theoretician surpass the data constraints confronting the experimentalist, it is not surprising to encounter anticipated instabilities not (yet ?) validated by observations [e.g., Tsurutani, 1991]. We find no harm here: theoreticians and experimentalists should keep challenging each other for the benefit of space plasma physics.

Acknowledgments. This work was supported by a research contract from INIC (Instituto Nacional de Investigação Científica) and by NATO research grant 669/84.

References

Akimoto, K., K. Papadopoulos, and D. Winske, Lower–hybrid instabilities driven by an ion velocity ring, *J. Plasma Phys., 34*, 445, 1985.

Bers, A., Space-time evolution of plasma instabilities–absolute and convective, in *Handbook of Plasma Physics (Vol. 1)*, ed. by M. N. Rosenbluth and R. Z. Sagdeev, North–Holland, New York, 1983.

Brinca, A. L., On the electromagnetic stability of isotropic populations, *J. Geophys. Res., 95*, 221, 1990.

Brinca, A. L., and B. T. Tsurutani, Unusual characteristics of electromagnetic waves excited by cometary newborn ions with large perpendicular energies, *Astron. Astrophys., 187*, 311, 1987.

Brinca, A. L., and B. T. Tsurutani, Temperature effects in the pickup process of water group and hydrogen ions: Extensions of "A theory for low–frequency waves observed at comet Giacobini–Zinner" by M. L. Goldstein and H. K. Wong, *J. Geophys. Res., 93*, 243, 1988a.

Brinca, A. L., and B. T. Tsurutani, Survey of low-frequency electromagnetic waves stimulated by two coexisting newborn ion species, *J. Geophys. Res., 93*, 48, 1988b.

Brinca, A. L., and B. T. Tsurutani, Influence of multiple ion species on low–frequency electromagnetic wave instabilities, *J. Geophys. Res., 94*, 13565, 1989a.

Brinca, A. L., and B. T. Tsurutani, The oblique behavior of low-frequency electromagnetic waves excited by newborn cometary ions, *J. Geophys. Res., 94*, 3, 1989b.

Brinca, A. L., B. T. Tsurutani, On the excitation of cyclotron harmonic waves by newborn heavy ions, *J. Geophys. Res., 94*, 5467, 1989c.

Brinca, A. L., B. T. Tsurutani, and F. L. Scarf, Local generation of electrostatic bursts at comet Giacobini–Zinner: Modulation by steepened magnetosonic waves, *J. Geophys. Res., 94*, 60, 1989.

Brinca, A. L., N. Sckopke, and G. Paschmann, Wave excitation downstream of the low–β, quasi-perpendicular bow shock, *J. Geophys. Res., 95*, 6331, 1990.

Buti, B., and G. S. Lakhina, Stochastic acceleration of cometary ions by lower hybrid waves, *Geophys. Res. Lett., 14*, 107, 1987.

Cap, F. F., *Handbook on Plasma Instabilities (Vol. 1)*, ch. 10, Academic Press, New York, 1976.

Cayton, T. E., Energy anisotropy instabilities in high–beta plasmas: A comparison of various kinetic and fluid descriptions, *Phys. Fluids, 21*, 1790, 1978.

Davidson, R. C., Kinetic waves and instabilities in a uniform plasma, in *Handbook of Plasma Physics (Vol. 1)*, ed. by M. N. Rosenbluth and R. Z. Sagdeev, North–Holland, New York, 1983.

Freund, H. P., and C. S. Wu, Stability of a spherical shell distribution of pickup ions, *J. Geophys. Res., 93*, 14277, 1988.

Gaffey, J., Jr., D. Winske, and C. S. Wu, Time scales for formation and spreading of velocity shells of pickup ions in the solar wind, *J. Geophys. Res., 93*, 5470, 1988.

Galeev, A., A. Polyudov, R. Sagdeev, K. Szegö, V. Shapiro, and V. Shevchenko, MHD turbulence caused by a comet in the solar wind, *Sov. Phys. JETP, 65*, 1178, 1987.

Galeev, A., R. Sagdeev, V. Shapiro, V. Shevchenko, and K. Szego, Quasi-linear theory of the ion cyclotron instability and its application to the cometary plasma, this volume, 1991.

Gary, S. P., Electromagnetic ion/ion instabilities and their consequences in space plasmas: A review, *Space Sci. Rev.*, in press, 1990.

Gary, S. P., and C. D. Madland, Electromagnetic ion instabilities in a cometary environment, *J. Geophys. Res., 93*, 235, 1988.

Gary, S. P., and R. Sinha, Electromagnetic waves and instabilities from cometary ion velocity shell distributions, *J. Geophys. Res., 94*, 9131, 1989.

Gary, S. P., K. Akimoto, and D. Winske, Computer simulations of cometary ion/ion instabilities and wave growth, *J. Geophys. Res., 94*, 3513, 1989.

Gendrin, R., Classification des interactions possibles de gyrorésonance entre un plasma et un faisceau de particules, dans le cas d'une propagation longitudinale, *Ann. Geophys. , 21*, 414, 1965.

Gendrin, R., General relationships between wave amplification and particle diffusion in a magnetoplasma, *Rev. Geophys., 19*, 171, 1981.

Gendrin, R., A. Roux, Energization of helium ions by proton–induced hydromagnetic waves, *J. Geophys. Res., 85*, 4577, 1980.

Glaßmeier, K. H., A. J. Coates, M. H. Acuña, M. L. Goldstein, A. D. Johnstone, F. M. Neubauer, and H. Rème, Spectral characteristics of low–frequency plasma turbulence upstream of comet P/Halley, *J. Geophys. Res., 94*, 37, 1989.

Goldstein, M. L., and H. K. Wong, A theory for low–frequency waves observed at comet Giacobini–Zinner, *J. Geophys. Res., 92*, 4695, 1987.

Goldstein, M. L., H. K. Wong, and K. H. Glaßmeier, Generation of low–frequency waves at comet Halley, *J. Geophys. Res., 95*, 947, 1990.

Hartle, R. E., and C. S. Wu, Effects of electrostatic instabilities on planetary and interstellar ions in the solar wind, *J. Geophys. Res., 78*, 5802, 1973.

Hasegawa, A., Drift mirror instability in the magnetosphere, *Phys. Fluids*, *12*, 2642, 1969.

Hasegawa, A., *Plasma Instabilities and Nonlinear Effects*, p.79, Springer-Verlag, Berlin, 1975.

Kennel, C. F., and F. L. Scarf, Thermal anisotropies and electromagnetic instabilities in the solar wind, *J. Geophys. Res.*, *73*, 6149, 1968.

Lakhina, G. S., and F. Verheest, Alfvén wave instabilities and ring current during solar wind–comet interactions, *Astrophys. Space Sci.*, *143*, 329, 1988.

Lee, M. A., Ultra-low frequency waves at comets, in *Plasma Waves and Instabilities at Comets and in Magnetospheres*, ed. by B. T. Tsurutani and H. Oya, Am. Geophys. Union, Washington D.C., 1989.

Melrose, D. B., *Instabilities in Space and Laboratory Plasmas*, ch.3, Cambridge Un. Press, Cambridge, 1986.

Neugebauer, M., A. Lazarus, H. Balsinger, S. Fuselier, and H. Rosenbauer, The velocity distribution of cometary protons picked up by the solar wind, *J. Geophys. Res.*, *94*, 5277, 1989.

Omidi, N., and D. Winske, A kinetic study of solar wind mass loading and cometary bow shocks, *J. Geophys. Res.*, *92*, 13409, 1987.

Oscarsson, T. E., and K. G. Rönnmark, Comments on the theory of absolute and convective instabilities, *Geophys. Res. Lett.*, *13*, 1384, 1986.

Price, C. P., Mirror waves driven by newborn ion distributions, *J. Geophys. Res.*, *94*, 15001, 1989.

Rönmark, K., WHAMP-Waves in homogeneous, anisotropic multicomponent plasmas, *Rep.179*, Kiruna Geophys. Inst., Kiruna, Sweden, 1982.

Russell, C. T., The interaction of the solar wind with comet Halley: Upwind and downwind, *Q. Jl. R. astr. Soc.*, *29*, 157, 1988.

Sagdeev, R. Z., V. D. Shapiro, V. I. Shevchenko, and K. Szegö, MHD turbulence in the solar wind–comet interaction region, *Geophys. Res. Lett.*, *13*, 85, 1986.

Sagdeev, R. Z., V. D. Shapiro, V. I. Shevkenko, and K. Szegö, The effect of mass loading outside cometary bow shock for the plasma and wave measurements in the coming cometary missions, *J. Geophys. Res.*, *92*, 1131, 1987.

Siscoe, G. L., Solar system magnetohydrodynamics, in *Solar-Terrestrial Physics*, ed. by R. L. Carovillano and J. M. Forbes, Reidel Publishing, Dordrecht, 1983.

Sudan, R. N., Growing waves in a nongyrotropic plasma, *Phys. Fluids*, *8*, 1915, 1965.

Tataronis, J. A., Cyclotron harmonic wave propagation and instabilities, *Ph.D. Thesis*, Stanford University, California, December, 1967.

Tataronis, T. A., F. W. Crawford, Cyclotron harmonic wave propagation and instabilities, I, Perpendicular propagation, *J. Plasma Phys.*, *4*, 231, 1970.

Thorne, R. M., and D. Summers, Analytical solutions to the general problem of oblique wave growth and damping, *Phys. Fluids*, *29*, 4091, 1986.

Thorne, R. M., and B. T. Tsurutani, On the resonant interaction between cometary ions and low frequency electromagnetic waves, *Planet. Space Sci.*, *35*, 1501, 1987.

Tsurutani, B. T., Comets: A laboratory for plasma waves and instabilities, this volume, 1991.

Tsurutani, B. T., and E. J. Smith, Hydromagnetic waves and instability associated with cometary ion pickup: ICE observations, *Geophys. Res. Lett.*, *13*, 263, 1986.

Winske, D., and M. M. Leroy, Hybrid simulation techniques applied to the earth's bow shock, in *Computer Simulation of Space Plasmas*, ed. by H. Matsumoto and T. Sato, D. Reidel, Hingham, 1984.

Winske, D., and S. P. Gary, Electromagnetic instabilities driven by cool heavy ion beams, *J. Geophys. Res.*, *91*, 6825, 1986.

Winske, D., C. S. Wu, Y. Y. Li, Z. Z. Mou, and S. Y. Guo, Coupling of newborn ions to the solar wind by electromagnetic instabilities and their interaction with the bow shock, *J. Geophys. Res.*, *90*, 2713, 1985.

Wu, C. S., and R. C. Davidson, Electromagnetic instabilities produced by neutral particle ionization in interplanetary space, *J. Geophys. Res.*, *77*, 5399, 1972.

Wu, C. S., and R. E. Hartle, Further remarks on plasma instabilities produced by ions born in the solar wind, *J. Geophys. Res.*, *79*, 283, 1974.

Wu, C. S., and P. H. Yoon, Kinetic hydromagnetic instabilities due to a spherical shell distribution of pickup ions, *J. Geophys. Res.*, *95*, 10273, 1990.

Wu, C. S., D. Krauss–Varban, and T. S. Huo, A mirror instability associated with newly created ions in a moving plasma, *J. Geophys. Res.*, *93*, 11527, 1988a.

Wu, C. S., X. T. He, and C. P. Price, Excitation of whistlers and waves with mixed polarization by newborn cometary ions, *J. Geophys. Res.*, *93*, 3949, 1988b.

Yoon, P. H., M. E. Mandt, and C. S. Wu, Evolution of unstable shell distribution of pickup cometary ions, *Geophys. Res. Lett.*, *16*, 1473, 1989.

Young, D. T., S. Perrault, A. Roux, C. de Villedary, R. Gendrin, A. Korth, G. Kremser, and D. Jones, Wave–particle interactions near Ω_{He^+} observed on Geos 1 and 2, 1. Propagation of ion cyclotron waves in He^+-rich plasma, *J. Geophys. Res.*, *86*, 6755, 1981.

QUASILINEAR THEORY OF THE ION CYCLOTRON INSTABILITY AND ITS APPLICATION TO THE COMETARY PLASMA

A.A. Galeev[1], R. Z. Sagdeev[3], V. D. Shapiro[1], V. I. Shevchenko[1], K. Szego[2]

(1)Space Research Institute, Profsoyuznaya 84/32, 117810 Moscow GSP-7, USSR

(2)Central Research Institute for Physics, Budapest, Hungary.

(3) University of Maryland, Maryland 20742, USA.

Abstract. The quasilinear theory of Alfven turbulence in the mass-loaded solar wind is reviewed and compared with observations obtained in the upstream region of comet Halley's bow shock. The theory explains the formation of the shell distribution of the cometary ions from the initial ring distribution as the result of the pitch angle diffusion caused by Alfven waves. In situ measurements of wave intensity, frequency spectrum and polarisation of Alfven waves also agree well with the predictions of the ion cyclotron instability, driven by cometary ions. The turbulence in the vicinity of the comet Giacobini-Zinner seems to be somewhat different - while being more intense it is essentially more regular in its nature. The main reason for this difference is due to the fact that in the latter case, with smaller gas production rates, turbulence develops significantly closer to the comet.

The possibility of the excitation of oblique magnetosonic waves accompanied by density fluctuations is analysed. The mechanism for such excitation is wave refraction on the transverse inhomogeneities in the solar wind. It is shown that in the vicinity of the bow shock the intensity of magnetosonic waves grows rapidly.

The acceleration of the cometary ions in the mass-loaded solar wind by MHD turbulence (stochastic Fermi acceleration) is also analysed in the paper. It is shown that the softest part of the spectrum of the accelerated ions could be attributed to the acceleration by oblique magnetosonic waves while the main acceleration is due to the long wavelength firehose magnetic field perturbations. The ion energy spectra obtained by solving the energy diffusion equation are compared with observations.

1. Introduction

It was Bell [1978] who first suggested that in the region upstream from supernova shocks there exists intensive Alfvenic turbulence excited by the beam of cosmic ray particles accelerated at the shock front and expanding into the interstellar plasma. Pitch angle diffusion isotropises the ion

Cometary Plasma Processes
Geophysical Monograph 61

beam distribution as the result of its interaction with Alfven waves. The high energetic particles staying in the vicinity of the shock makes possible Fermi acceleration at the shock front. The result of this acceleration is a power-like spectrum of cosmic ray particles, in fairly good agreement with observations [Krymski, 1977; Axford et al., 1977; Blanford and Ostriker, 1978]. Similar MHD (Magnetohydrodynamic) turbulence has been observed in the region upstream from the Earth's bow-shock [Hoppe et al., 1981; Sentman et al., 1981]; in that case the source of free energy for the excitation of the Alfven waves is provided by the reflected protons forming an ion beam in the solar wind plasma.

In recent cometary missions MHD turbulence was also observed in the upstream regions of the bow-shock of comets Giacobini-Zinner [Tsurutani and Smith, 1986] and P/Halley [Johnstone et al., 1986; Yumoto et al., 1986]. In that case the Alfven waves were excited by heavy cometary ions implanted into the solar wind by photoionisation of neutral gas evaporated from the cometary surface. The typical expansion velocity of the gas, v_g, is about 1 km/s, the time of photoionisation, τ, is 10^6 s. Accordingly, the heavy ions are born in the solar wind at a typical distance of the order of several million kilometers from the nucleus i.e. far upstream in the solar wind. Newly born ions travel in the crossed electric and magnetic fields of the solar wind in cycloidal trajectories in a plane perpendicular to the magnetic field (Figure 1). This motion is the superposition of the drift with velocity $u_\perp$ equal to the perpendicular solar wind velocity cE/H, and of the Larmor rotation with velocity $u_\perp$. The ion velocity along the magnetic field, $u_\parallel$ remains equal to gas velocity or is extremely small in comparison with the solar wind velocity (Figure 1). This means that in the solar wind frame the newly born ions form a beam expanding along the magnetic field lines with the velocity $-u_\parallel$, and gyrating about these lines with the velocity $u_\perp$.

This problem is similar in many respects to the cosmic ray shocks. A high energy component appears in the upstream plasma; in the cometary case this component is formed by the heavy ions of cometary origin created by photoionisation. Their interaction with the solar wind plasma causes the excitation of Alfven waves, the pressure of the heavy ions decelerates the mass-loaded solar wind in the

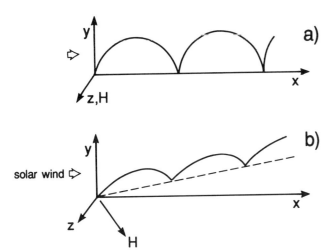

Fig 1. Cometary ion pick-up in the cases of normal (a) and oblique (b) solar wind flow towards the magnetic field. α is the angle between solar wind flow and magnetic field.

upstream region of the cometary bow-shock, in a full analogy with the diffusive precursor of the cosmic ray shock. The nonlinear theory of MHD turbulence in the upstream region of cometary bow-shocks [Sagdeev et al, 1986; Galeev et al, 1987; Galeev et al., 1988] is based on the quasilinear approximation, and, consequently in many respects is similar to the quasilinear theory of Alfven wave turbulences in the vicinity of cosmic ray shocks [Lee,1982; Lee and Ip,1987]. That theory predicts the isotropisation of the cometary ion distribution via their interaction with Alfven waves and the formation of the shell distribution from the initial ring one. This result was first obtained theoretically [Sagdeev et al., 1986] and was confirmed later by in situ observations [Neugeubauer et al, 1989; Coates et al, 1989] . Other results of the quasilinear theory of MHD turbulences, such as wave spectra, intensity and polarization of the waves are also in qualitative agreement with observations.

In sections 2 and 3 the quasilinear theory of Alfven turbulences in the mass-loaded solar wind is reviewed [Sagdeev et al., 1986; Galeev et al., 1988], and compared with the observations obtained in the upstream region of comet Halley's bow-shock. The turbulence in the vicinity of the bow-shock of comet Giacobini-Zinner seems to be somewhat different: in spite of the smaller gas production rate ($Q = 10^{30}$ mol/s for Halley and $Q = 5\times10^{28}$ mol/s for Giacobini-Zinner) the intensity of waves is significantly higher for Giacobini-Zinner. A simple explanation for this is based on the fact that the source term driving the instability is proportional to Q/r^2 (r is the distance from the cometary nucleus). Whereas the typical space scales in the mass-loaded solar wind including the position of the bow-shock are proportional to Q (equations 3.3. and 3.3a below), the source term for the wave excitation is approximately proportional to Q^{-1} (at least when the distance to the bow-shock is larger than ion Larmor radius, i.e. $Q > 10^{28}$ mol/s). In the case of Giacobini-Zinner less distance is available to develop turbulences which yields a less developed pitch angle diffusion of resonant ions. MHD waves detected upstream of Giacobini-Zinner's bow-shock, although larger in amplitude,

seem to be more regular and not as turbulent as in the case of Halley's comet.

In section 4 the excitation mechanism of oblique magnetosonic waves, accompanied by density fluctuations, is investigated. The mechanism is wave refraction on transverse inhomogeneity in the vicinity of the cometary bow-shock. Usually a strong Cherenkov damping of the oblique magnetosonic waves prevents their excitation. The analysis presented in section 4 is based on the numerical solution of the quasilinear equations; it is shown that 'plateau' formation in the distribution of protons that are in Cherenkov resonance with the waves switches off the damping and allows the excitation of oblique magnetosonic waves in the vicinity of the cometary bow-shock.

Finally, in section 5, the acceleration of the cometary ions to energies exceeding significantly the injection energy $m_i u^2/2$ (in the solar wind frame) is discussed. The most likely mechanism is second order (stochastic) Fermi acceleration by MHD waves; but as is well known, for such acceleration the relative motion of magnetic field humps created by MHD waves is necessary, which is equivalent to dispersion of the wave phase velocity. In this paper two different options are analysed; one is the acceleration by oblique magnetosonic waves, the other is acceleration by oppositely moving magnetic field perturbations of Alfvenic type created by fire-hose instability. It is shown that acceleration by short wavelength ($k > \omega_{Hi}/u$) magnetosonic waves produces a distribution function which decreases rapidly with energy and it seems that only the soft part of the observed ion tail could be connected with this acceleration. The main acceleration mechanism is due to the long wave length ($k < \omega_{Hi}/u$) fire-hose magnetic field perturbations. The ion energy spectra obtained by solving the energy diffusion equation are compared to the observations.

2. Quasilinear theory of MHD turbulence in the mass-loaded solar wind

As was already mentioned in the introduction, in the upstream region of the cometary bow-shock the solar wind is loaded by heavy ions of cometary origin, composed mainly of water ions. In the solar wind frame they form a beam of particles moving along the magnetic field and simultaneously taking part in cyclotron rotation. Their distribution function,

$$f_i \sim \delta(v_\perp - u_\perp)\, \delta(v_{||} + u_{||})$$

(2.1)

corresponds to a ring in velocity space with $\theta \approx \alpha$ (Figure 2), α is the angle between the magnetic field and the solar wind flow. Such a distribution function is unstable and, due to the ion cyclotron instability, Alfven waves are excited, their maximum growth rate corresponding to parallel propagation. The particles in the beam scatter in pitch angle due to the interaction with the Alfven waves. When they diffuse towards smaller pitch angles the source of free energy is the perpendicular motion of the beam particles (Figure 2); the waves in that pitch angle interval (marked as I in Figure 2) are excited by normal Doppler cyclotron resonance,

$$\omega - kv \cos\theta = \omega_{Hi}$$

(2.2)

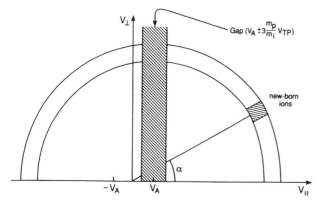

Fig 2. Evolution of the velocity space of cometary ions in the solar wind frame. $v_\parallel$ is the component of velocity antiparallel

to the magnetic field; $\theta = \arctan v_\perp/v_\parallel$ is the pitch angle. Also shown are the diffusion lines being the circles with the centre at the point $v_\parallel = -v_A$ for $\theta < \alpha$ and $v_\parallel = v_A$ for $\theta > \alpha$.

where $\omega = kv_A$ is the wave frequency and ω_{Hi} is the heavy ion gyrofrequency. Their growth rate, given by Wu and Davidson [1972], and Galeev et al.[1987] for the case of a cold ion beam, is sufficiently large,

$$\gamma = \frac{\sqrt{3}}{2^{5/3}} \omega_{Hi} \left[\varphi(x) \tan^2\alpha \, \frac{|u_\parallel|}{v_A} \right]^{1/3}$$

(2.3)

where $\varphi(x) = n_i m_i / n_0 m_p$ is the relative change of the mass density due to mass-loading of the solar wind, $n_i(x)$ is the cometary ion density, n_0 the proton density in the solar wind, m_p and m_i the proton and heavy ion masses respectively, and the x axis is directed along sun-comet line towards the comet. Since $v \sim u \gg v_A$, it follows from the resonance condition (2.2) that the waves excited for $\theta < \alpha$ have $k < 0$, i.e. for $\alpha < \pi/2$ they propagate towards the comet. If $\theta > \alpha$ (marked as II in Figure 2) the source of free energy is the parallel motion of the beam particles and the waves are excited due to anomalous Doppler cyclotron resonance,

$$\omega - kv \cos\theta = -\omega_{Hi}$$

(2.4a)

It follows from the resonance condition that for $\theta > \alpha$ the excited waves with $k > 0$ propagate sunward. For a cold ion beam and $\tan^2\alpha > \gamma/\omega_{Hi}$ the growth rate is still given by equation (2.3). In the opposite case, for quasiparallel propagation in the solar wind when $\tan^2\alpha < \gamma/\omega_{Hi}$ the growth rate is equal to,

$$\gamma = \omega_{Hi} \left[\varphi(x) \frac{|u_\parallel|}{2v_A} \right]^{1/2}$$

(2.4b)

In both cases the growth rate for a cold beam is of the order of ω_{Hi} and the resonant ion cyclotron instability accompanied by pitch angle diffusion of cometary ions develops quickly.

One important peculiarity of this diffusion is due to the fact that near $\theta \approx \pi/2$ the wave number increases infinitely, i.e. $k \approx \omega_{Hi}/(v_\parallel - v_A) = \infty$ when $v_\parallel = v_A$. In that case the cyclotron absorption of Alfven waves by the solar wind protons becomes important. Due to this there is a gap in the velocity diffusion coefficient at $\theta \approx \pi/2$, as is well known from the theory of cosmic ray shock acceleration. Such cyclotron wave-damping happens if $k \geq 1/3 \, \omega_{Hp}/v_{Tp}$, and it follows from the resonance condition (2.4a) that in the cometary case the gap is narrow,

$$|v_\parallel - v_A| < 3 \, (m_p/m_i) v_{Tp}$$

(2.5)

If $v_\parallel$ is sufficiently small, i.e.,

$$\frac{v_\parallel^2}{u^2} \leq \frac{\delta|H|}{H_0} \cong \sum_k \frac{|H_k|^2}{2H_0^2}$$

(2.6)

the magnetic mirroring effect caused by the modulation of the magnetic field intensity with MHD waves [Sagdeev et al., 1986] can reverse the direction of the particles motion.

In the cometary case, as follows from the comparison of equation (2.5) and (2.6), even a small magnetic field modulation $\delta H/H_0 < 10^{-2}$ helps the particles to overcome the diffusion gap (2.5) and to appear in the region $v_\parallel < v_A$. In that region - marked as III in Figure 2 - the free energy source is again the perpendicular motion of the resonant particles. The waves are excited by normal Doppler cyclotron resonance (2.2) and $k > 0$. Hence the possibility of wave excitation and pitch angle diffusion exists in the whole interval $0 < \theta < \pi$ and as a result of pitch angle diffusion the initial ring distribution of cometary ions is transformed into a shell distribution which is almost isotropic in the solar wind frame. Such isotropisation is the meaning of the so-called 'collective pick-up' of cometary ions by the solar wind. The existence of such a 'collective pick-up' mechanism and the formation a shell distribution of cometary ions was predicted theoretically [Sagdeev et al., 1986], and later confirmed by in-situ observations.

The dynamics of the beam in phase space as measured by Giotto near Halley's comet is illustrated in Figure 3 taken from Coates et al., [1989]. The initial ring distribution of the cometary ions that exists in the solar wind at a large distance from the comet (Figure 3a) is transformed into a shell distribution closer in with $v \approx u$ (Figure 3b). The subsequent broadening of the shell that has been observed near the cometary bow-shock (Figure 3c) is the result of the energy diffusion of the resonant ions and is discussed in section 5. The distribution function of the resonant ions could be derived by solving the following kinetic equation,

$$(u - v_\parallel \cos\alpha) \frac{\partial f}{\partial x} - \frac{v_\perp}{2} \frac{du}{dx} \frac{\partial f}{\partial v_\perp} -$$

$$\frac{\pi e^2}{2m_i^2} \frac{\partial}{\partial v_\parallel} \left[\sum_{k,\pm} |H_k|^2 \frac{v^2}{c^2} \delta(kv_\parallel - \omega \pm \omega_{Hi}) \frac{\partial f}{\partial v_\parallel} \right] =$$

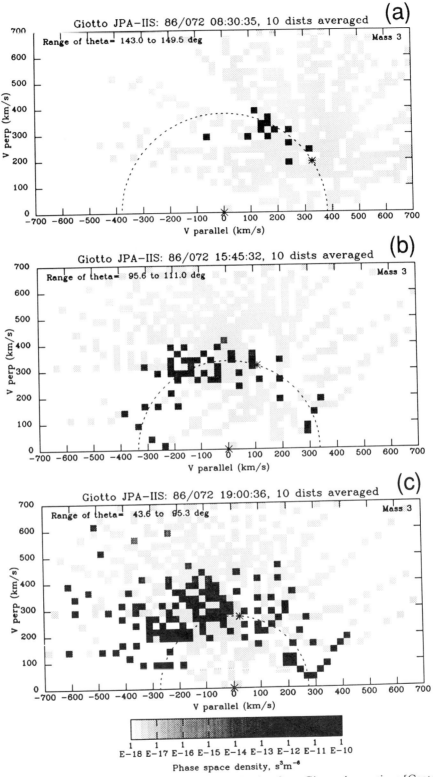

Fig.3. Evolution of the heavy ion (water group) distribution from Giotto observations [Coates et al., 1989]. 3a - initial distribution; 3b - pitch angle diffusion and formation of the shell at r $\simeq$ 2.0 x 10⁶ km; 3c.- energy and pitch diffusion at r $\cong$ 0.5 x 10⁶ km.

$$\left[\frac{Q\,e^{-r/v_g\tau}}{4\pi r^2 v_g\tau}\right]\delta(v_\perp^2 - u_\perp^2)\,\delta(v_{||} + u_{||})$$

(2.7)

The first term in the kinetic equation is the usual convective term; the second one is responsible for the adiabatic heating of cometary ions in the inhomogeneous magnetic field of the solar wind flow. The last term on the l.h.s. of equation (2.7) is the quasilinear stoss-term; the derivatives over $v_{||}$ in this term are to be calculated along the diffusion lines,

$$v_\perp^2 + (v_{||} - \omega/k)^2 = \text{const.}$$

(2.8)

Diffusion along these lines corresponds to the energy conservation of the particles in the wave frame. The reason is very simple; for Alfven waves the electric field is of an inductive nature, so it vanishes in the wave reference frame and the particle energy in this frame is conserved. The r.h.s. of equation (2.7) is the source term, describing the photoproduction of new ions with a distribution function (2.1); τ is the typical time for photoionization, Q is the gas production rate, and the exponential factor reflects the decrease of gas flow due to photoionisation.

For the waves under consideration $\omega/k \sim v_A \ll v_{||} \sim u$, hence the diffusion lines (2.8) are close to the lines of constant particle energy and the diffusion actually occurs over the pitch angle θ. More accurately, the diffusion lines are concentric circles in the velocity space with the center shifted over the value $\omega/k = -v_A$ for $\theta < \alpha$ and over the value $\omega/k = v_A$ for $\theta > \alpha$. When the particles diffuse along such lines their energy changes very little, only of the order of v_A/u. A necessary condition for energy diffusion is to have a pitch angle interval in which resonance with waves having different ω/k values is possible, hence having diffusion lines which intersect. This situation is discussed in section 5. If such an intersection is absent only pitch angle diffusion is possible, the characteristic time τ_s for pitch angle diffusion is small, and can be obtained from equation (2.7),

$$\tau_s^{-1} \approx \left(\frac{\omega_{Hi}^2}{\Delta k\, v_{||}}\right)\sum_k \frac{|H_k|^2}{H_0^2} \approx \omega_{Hi}\left[\frac{\delta H}{H_0}\right]^2$$

(2.9)

where δH is the amplitude of magnetic fluctuations in the Alfven waves, and Δk is the spectral width of those waves. The effectiveness of the particle's isotropisation by the waves can be characterised by comparing this time to the average time τ_c spent by the photoions in the solar wind. The latter is calculated in Coates et al., [1989],

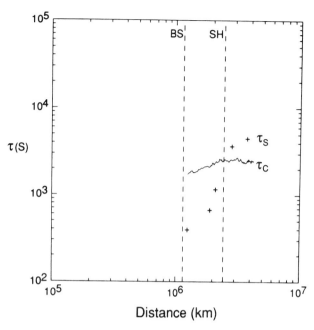

Fig.4. Characteristic times for pick-up ions in the solar wind, τ_c [equation (2.10)] is the average time spent by photoion in the solar wind, τ_s [equation (2.9)] is the typical time of pitch angle diffusion.

$$\tau_c = \frac{\int \frac{e^{-r/v_g\tau}}{r^2(l)}\cdot l\, dl}{v_g \int \frac{e^{-r/v_g\tau}}{r^2(l)}\, dl}$$

(2.10)

In this formula the integration goes along the path of the incoming pick-up ions. Equation (2.10) takes into account the change of the ion photoproduction along the trajectory in the cometary gas flow. The dependence of τ_c and τ_s on the distance from the comet is shown in Figure 4 obtained from Giotto measurements near Halley [Coates et al, 1989]. There is one difference compared with Coates et al, [1989]. Their value of $\tau_s = (4\omega_{Hi}^{-1})(H_0/\delta H)$ was taken from numerical simulations [Price and Wu, 1987], but that expression is not appropriate for quasilinear diffusion. In Figure 4 the vertical line corresponds to the distance $r \approx 3 \times 10^6$ km where τ_c and τ_s are equal. For smaller distances the pitch angle diffusion is faster and in the solar wind frame the isotropisation actually takes place.

In a recently published paper [Coates et al., 1990] the velocity distribution of heavy ions was investigated at comet Halley by evaluating the mean absolute deviation of the pitch angle from its mean value. Basically the same conclusions were obtained:

1) A fairly big increase was observed in pitch angle width and also the formation of the shell distribution at distances 2.5 - 2.6 $\times 10^6$ km was seen.

2) The typical time for energy diffusion is much larger than for pitch angle diffusion and only in the vicinity of the cometary bow-shock is the energy increase significant.

Hence the observational results for pitch angle diffusion are in good agreement with the predictions of quasilinear theory.

It is a widespread statement in the literature that the observed pitch angle diffusion develops much slower than predicted theoretically [Coates et al., 1990]. This is mainly due to the comparison of observations with the results of numerical simulations. However, in the latter, large amplitude regular waves are considered which causes rapid isotropisation in 1-2 gyroperiods. The stochastic pitch angle diffusion, as described by the quasilinear equations, develops more slowly, on a time scale τ_s of about 10 - 20 gyroperiods for Halley's comet (Figure 4). As a result of isotropisation, the parallel velocity of heavy ions in the solar wind frame decreases from the value $u\cos\alpha$ to,

$$<v_{||}> = \frac{2\pi}{n} \int f(v_\perp^2 + v_{||}^2 \pm \frac{2\omega v_{||}}{k}) v_{||} v_\perp dv_\perp dv_{||}$$

$$= -\frac{3\omega}{2k} \int_0^\alpha \cos^2\theta \ \sin\theta \ d\theta + \frac{3\omega}{2k} \int_\alpha^\pi \cos^2\theta \ \sin\theta \ d\theta$$

$$= v_A \cos^3\alpha$$

(2.11)

This result is in good agreement with observations [Coates et al., 1990]. If we neglect small terms of the order of ω/ku, only the continuous creation of new ions leads to a deviation from the isotropic distribution of cometary ions. The velocity distribution function of heavy ions could be given in the form

$$f = f_0(u(x), v) + f_1(u(x), v, \theta)$$

(2.12)

where the anisotropic part of the distribution function is $f_1 \sim f_0 \tau_s/\tau_c$. Sufficiently close to the comet where $\tau_s/\tau_c << 1$ the anisotropic part of the distribution function is also small. In this case the following relations can be derived from equation (2.7) for the anisotropic part,

$$\frac{\partial f_1}{\partial \theta} = -\frac{1}{\omega_{Hi}^2} \left[\frac{H_0^2}{\sum_\pm |H_k^\pm|^2} \right] \{ 2v^2 \sin\theta |\cos\theta| \cos\alpha \frac{\partial f_0}{\partial x} +$$

$$\frac{2}{3} \cos^2\theta \sin\theta \frac{du}{dx} v^2 \frac{\partial f_0}{\partial v} +$$

$$\frac{Q e^{-r/v_g\tau}}{2\pi^2 r^2 v_g \tau u} \frac{|\cos\theta|}{\sin\theta} \delta(v - u) \left(\sigma(\theta-\alpha) + \frac{(\cos\theta - 1)}{2} \right) \}$$

(2.13)

where $-kv = \pm\omega_{Hi}/\cos\theta$, and $\sigma(x)$ is the step-function. For the isotropic part of the distribution function we have the following equation,

$$u \frac{\partial f_0}{\partial x} + <v \cos\theta \ \frac{\partial f_1}{\partial x}> - \frac{1}{3} \frac{du}{dx} v \frac{\partial f_0}{\partial v}$$

$$= \frac{Q e^{-r/v_g\tau}}{16\pi^2 r^2 v_g^2 \tau u^2} \delta(v - u)$$

(2.14)

3. The motion of the mass/loaded solar wind; the excitation of Alfven waves, their wave spectra and polarisation.

The second term in equation (2.14) describes the particle diffusion caused by Alfvenic turbulence. This can be represented in the form $-\partial/\partial x(D_{xx}\partial f_0/\partial x)$, where the diffusion coefficient, in order of magnitude, is equal to $D_{xx} \sim (v^2/\omega_{Hi})(H_0^2/\sum|H_k|^2)$.

In the upstream region such diffusion is important only on the characteristic dimension of the diffusive precursor of the shock which is $L_D \sim (u/\omega_{Hi}) (H_0^2/\sum|H_k|^2)$. Far upstream from the shock or at distances $v_g\tau >> L_D$ where the mass-loading of the solar wind begins, the diffusion is negligible. There the motion of the mass-loaded wind can be obtained from the solution of the hydrodynamic equations,

$$\frac{\partial(\rho u)}{\partial x} = \frac{Q m_i e^{-r/v_g\tau}}{4\pi r^2 v_g \tau}$$

$$\frac{\partial}{\partial x} (\rho u^2 + P) = 0$$

(3.1)

The main contribution to the pressure comes from the cometary ions, and, in that case, $P = 1/3 \int m_i v^2 f_0 d^3v$. This system of equations must be solved together with the kinetic equation (2.14) omitting the diffusion term on the l.h.s. The solution was obtained by Galeev et al.[1987]: the distribution function of the cometary ions in the interval $u < v < u_\infty(u_\infty/u)^{1/3}$ has the form,

$$f_0(x,u) = \frac{15}{64\pi} \frac{\rho_\infty u_\infty^2}{mu^{5/4}v^{15/4}} \left[1 - \frac{2}{5} \frac{u_\infty}{u^{1/4}v^{3/4}} \right]$$

(3.2)

where u_∞ is the unperturbed solar wind velocity and $f_0=0$ outside.

The finite width of the velocity distribution function is due to the presence of cometary ions convected from the upstream solar wind region, and the factor $(u_\infty/u)^{1/3}$ reflects the increase of ion velocity due to adiabatic heating in an inhomogeneous magnetic field of the mass-loaded solar wind flow. The motion of the solar wind is governed by the following equations for the mass density ρ and for the average velocity u,

$$\frac{\rho u}{\rho_\infty u_\infty} = \frac{5}{4}\left[\frac{u_\infty}{u}\right] - \frac{1}{4}\left[\frac{u_\infty}{u}\right]^2$$

$$\rho u = \rho_\infty u_\infty \left[1 + \frac{r_0}{r}\varphi\left(\frac{r}{v_g\tau}\right)\right]$$

(3.3)

where the following notation is used,

$$r_0 = \frac{Qm_i}{4\pi\rho_\infty u_\infty v_g\tau}$$

$$\varphi(y) = e^{-y} - y\int_y^\infty \frac{e^{-\xi}}{\xi}d\xi$$

(3.3a)

The growth of mass flux ρu the solar wind is accompanied by the deceleration of the wind only for $u/u_\infty > 2/5$. Since the sonic velocity in a mass-loaded wind is equal to,

$$c_s = \sqrt{\frac{\gamma P}{\rho}} = \frac{u}{\sqrt{3}}\sqrt{\frac{(u_\infty - u)}{(u - 0.2u_\infty)}}, \qquad \gamma = \frac{5}{3}$$

the singularity at $u = 0.4\, u_\infty$ corresponds to a Mach number $M = u/c_s = 1$. This singularity appears since a smooth transition from supersonic to subsonic flow in the decelerated solar wind is impossible. Hence in the supersonic part of the flow there must be a shock. It follows from numerical simulations [Schmidt and Wegman, 1982] that the shock is located at M=2 ($\rho u = 4/3\ \rho_\infty u_\infty$) and the stand-off distance at the subsolar point is $r = 1.9\ r_0 \sim 3.3 \times 10^5$ km for Halley's Comet. The shape of the shock is paraboloid, with the ratio of the sunward and the perpendicular axes 1:2. The results of the calculations for the shock positions both for the case of the isotropised shell distribution of cometary ions (I) and for the case of ring distribution (II) are shown in Figure 5. taken from Galeev [1986]. The positions of the shock intersections during various encounters with Halley's Comet is also indicated.

Due to the isotropisation of implanted ions the shock position is shifted closer to the cometary nucleus and this position is in better agreement with the observations. In the upstream region of the bow-shock the presence of a small anisotropic part of the resonant ion distribution is attributed to the continuous creation of new ions. According to equation (2.13) this leads to the excitation of Alfven waves with a growth rate given by the equation,

$$\gamma_k^\pm = -\frac{2\pi^2 e^2}{m_i \omega_k}\frac{k}{|k|}\frac{\omega_{Hp}^2}{\omega_{pp}^2}\int_0^\infty dv_\perp\, v_\perp^2 \frac{\partial f_1}{\partial\theta}\Big|_z$$

(3.4)

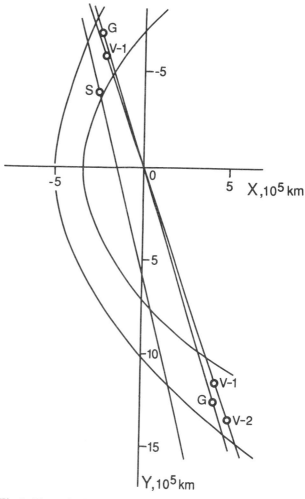

Fig.5. The calculated positions of the cometary bow-shock and its crossings by various spacecraft; V-1: Vega 1, V-II: Vega 2, G: Giotto, S: Suisei.

where $|_z$ indicates that the derivative should be taken at $-v_\parallel = \pm\omega_{Hi}/k$. With the help of equations (3.2) and (2.13), sufficiently far upstream in the solar wind where the shell distribution is quite narrow, the anisotropic part of the distribution function can be expressed in the following form which is proportional to the gas production rate,

$$\frac{\partial f_1}{\partial\theta} = -\frac{1}{\omega_{Hi}^2}\left[\frac{H_0^2\delta(v-u)}{\sum_\pm |H_k^\pm|^2}\right]\frac{Q\, e^{-r/v_g\tau}}{4\pi^2 r^2 v_g\tau u}$$

$$\{0.5\cos\alpha\ |\cos\theta|\sin\theta + \frac{2|\cos\theta|}{\sin\theta}[\sigma(\theta-\alpha)+(\cos\theta-1)/2]\}$$

(3.5)

where the k value should be taken at $k = \pm\omega_{Hi}/v\cos\theta$. The growth rate of the Alfven waves corresponding to the anisotropy distribution function (3.5) above is equal to,

$$\gamma_k = A(\alpha,k) \frac{H_0^2}{k^2|H_k|^2} \frac{Q\, e^{-r/v_g\tau}}{16r^2 v_g n_0} \frac{m_i}{m_p} \frac{\omega_{Hi}}{v_A\tau}$$

(3.6)

where for the circular-polarised waves with k < 0, excited by normal Doppler cyclotron resonance,

$$A(\alpha,k) = (1 - \cos\alpha)(1 - \lambda^2) + (1 - \lambda)^2 ; \quad \lambda = \frac{\omega_{Hi}}{ku_\infty}$$

(3.7a)

For these waves k (or λ) is in the range $1 > |\lambda| > \cos\alpha$. For right hand polarised waves with k > 0 excited by anomalous Doppler resonance,

$$A(\alpha,k) = (1 - \cos\alpha)(1 - \lambda^2) + 2(1 + \lambda)$$

(3.7b)

For these waves k (or λ) is in the range $\cos\alpha > \lambda > 0$, and finally for left hand polarised waves with k > 0 excited by normal Doppler resonance,

$$A(\alpha,k) = \cos\alpha (1 - \lambda^2) + 2(1 - \lambda)$$

(3.7c)

where for these waves k (or λ) is in the range $1 > \lambda > 0$.

In the short wavelength region ($k \gg \omega_{Hi}/u_\infty$) the spectrum, governed by ion-cyclotron instability, behaves as $|H_k|^2 \sim k^{-2}$, in a fairly good agreement with the observations near comet Halley (Figure 6).

In Figure 6 the frequency spectrum of the MHD waves observed near the bow shock of Halley's comet is presented. The frequency of these waves as measured by the spacecraft is modified by Doppler shift and is equal to,

$$f \sim |ku_{||} + |k|v_A|/2\pi \quad \text{or} \quad f \sim f_{Hi} |\cos\alpha \pm \frac{v_A}{u}|$$

In this formula the upper sign stands for the case when the waves propagate in the solar wind direction (k < 0), the lower one for the opposite case (k > 0). The maximum of the wave spectral density is at the cyclotron frequency of cometary ions $f_{Hi} \sim 10^{-2}$ Hz (for water ions), as well as the observed behaviour $|H_f|^2 \sim f^{-\alpha}$ with $\alpha \sim 2$ agrees well with the above results of quasilinear approach. Although the main part of the observed power behaviour lies in the region $f > f_{Hi}$; the magnetic field and plasma data both at Giacobini-Zinner and Halley's comet indicated the presence of very low frequency MHD oscillations at frequencies far below the ion cyclotron frequency f_{Hi} (Figure 6).

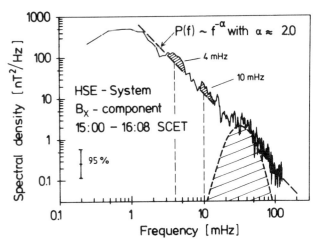

Fig.6. The frequency spectrum of MHD waves as measured by the Giotto spacecraft for a quasiperpendicular magnetic field orientation [Glassmeier et al., 1989]. In a broad frequency range starting from the water ion gyrofrequency f_{Hi} = 4 mHz there is a good agreement between the measured spectrum $|H_f|^2 \sim f^{-2}$ and those predicted on the basis of quasilinear theory of the ion cyclotron instability. Some minor spectral enhancement in the hatched regions from the author's point of view could indicate the possibility of some additional excitation mechanism.

In this part of the spectrum the waves are of long wavelength ($k \ll \omega_{Hi}/u$) while the previously described resonant ion cyclotron instability could lead only to the excitation of sufficiently short wavelength ($k > \omega_{Hi}/u$) oscillations. There are two possible mechanisms to excite the long wavelength oscillations. One is nonlinear wave transportation in k space due to the induced scattering of waves on solar wind protons. A detailed discussion of this nonlinear process can be found in Galeev et al., [1987]. It is shown there that such nonlinear interaction could not be responsible for the excitation of the above mentioned wave power behaviour, at least for the parameters of Halley's Comet. Another possible mechanism is the presence of firehose instability excited due to anisotropic cometary ion pressure in the mass-loaded solar wind. To develop firehose instability, the following condition should be satisfied [Sagdeev and Vedenov, 1958],

$$P_{||} - P_\perp = H_o^2/4\pi$$

(3.8)

where $P_{||}$ and $P_\perp$ are the parallel and transverse components of the solar wind gas-kinetic pressure, respectively. Hence the firehose instability exists only sufficiently close to the comet where the gas-kinetic pressure at least exceeds the magnetic pressure (a typical distance is $r < 2 - 2.5 \times 10^6$ km). The firehose instability is always preceded by the resonant ion cyclotron instability that has no threshold to develop. Nevertheless, since the isotropisation of cometary ions due to this latter instability is not complete, the remaining anisotropy is sufficient to excite the firehose instability, but only in the case of quasiparallel propagation $\alpha < 45°$ when the parallel pressure component exceeds the perpendicular one. For example, measurements carried out during the Suisei encounter with Halley's comet [Terasawa et al., 1986] confirm that $P_{\parallel}/P_{\perp} \sim 1.5$ which is enough to excite the firehose instability if $P_{\parallel} > 3H_o^2/4\pi$. The lowest wavelength for the firehose instability is just the ion Larmor radius u/ω_{Hi} [Sagdeev and Vedenov, 1958].

In our opinion the theoretical assumption that there are two different instability mechanisms (the ion cyclotron and firehose) responsible for short wavelength ($k > \omega_{Hi}/u$) and long wavelength ($k < \omega_{Hi}/u$) MHD oscillations is a realistic one. However, there is a big difference between these two instabilities. While the resonant cyclotron instability is responsible for the dominant part of the power behaviour of the MHD waves and experimentally has been investigated in detail, there is no unambiguous experimental confirmation of the onset of the firehose instability, in spite of the widespread opinion that this instability is possible upstream from the cometary bow-shock. This is probably due to the fact that long time intervals must be studied to investigate the very low frequency waves connected with the firehose instability. During such long time intervals the ambient solar wind flow could change thereby complicating the whole picture.

Since the dominant wave energy of the MHD fluctuations corresponds to short wavelength ($k > \omega_{Hi}/u$) oscillations excited by the ion-cyclotron instability, the increase of the total wave energy approaching the comet can be obtained using the cyclotron instability growth rate (3.7) only. Integrating the quasilinear equation for the wave spectral density $u\,\partial/\partial x\,|H_k|^2 = 2\gamma_k|^2 = 2\gamma_k|H_k|^2$ over $k \geq \omega_{Hi}/u$, it is possible to obtain the following equation for the growth of wave energy in MHD fluctuations,

$$u\frac{\partial}{\partial x}\sum_k |H_k|^2 = \frac{Q\,e^{-r/v_g\tau}}{8\pi r^2 v_g \tau}\,m_i v_A\,u\,P(\cos\alpha)$$

$$(3.9)$$

where $P(x) = 1 + 2x^2 - x^4/3$.

It is important that the growth rate in the quasilinear approximation (equation 3.7) is inversely proportional to $|H_k|^2$ and the variation of the wave intensity as a function of the distance is not exponential but polynominal. With the help of equation (3.1) the r.h.s. of equation (3.9) can be expressed by the deceleration rate of the solar wind. Far upstream where the deceleration is small i.e. $u_\infty - u \ll u_\infty$, the intensity of magnetic field fluctuations is proportional to the deceleration,

$$\sum_{k>0}\frac{|H_k|^2}{4\pi} \approx \frac{3}{16}\rho_\infty v_A\,(u_\infty - u)\,P(\cos\alpha)$$

$$(3.10)$$

In Figure 7 the theoretical results are compared with the observations of Giotto; for those deceleration values which are consistent with the above theoretical assumptions the agreement is reasonably good. Some dispersion in the

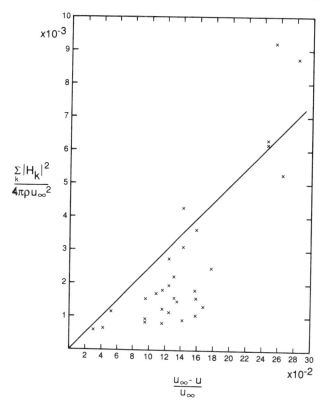

Fig.7. The normalized energy of the magnetic fluctuations.

$$\sum_k |H_k|^2 / 4\pi\rho u_\infty^2$$

Its value is shown as a function of the solar wind deceleration u/u_∞. The data marked by dots are from the Giotto measurements, the continuous line shows the theoretical results as obtained from equation (3.9) for the parameters of the Giotto flyby. The agreement (within the accuracy 40-50%) seems to be reasonably good for the idealized theoretical model used in our calculations. Some spread in the experimental results is probably due to the change of α in the solar wind flow.

observational data is probably due to variations in α. For ion cyclotron waves it is also possible to estimate the ratio of the amplitudes for left hand and right polarized waves,

$$\varepsilon^2 = \frac{\sum_k |H_k^+|^2}{\sum_k |H_k^-|^2}$$

If the wave is linearly polarized $\varepsilon=1$, while $\varepsilon=0$ or ∞ corresponds to circular polarization; ε is a function of $\cos\alpha$ only,

$$\varepsilon^2 = \frac{(1 - \cos\alpha + \cos^2\alpha - (1/6)\cos^4\alpha)}{(\cos\alpha + \cos^2\alpha - (1/6)\cos^4\alpha)}$$

$$(3.11)$$

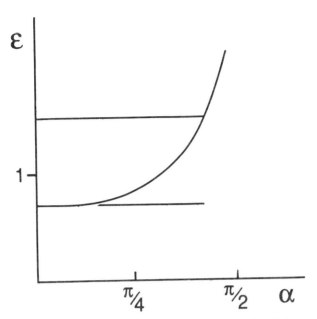

Fig.8. The ellipticity factor ε changes from 0.7 to 1.5.

The plot of $\varepsilon(\alpha)$ is shown in Figure 8. For $0 < \alpha < 80°$, ε changes from 2/3 to 3/2 which corresponds to elliptical polarization, and only as $\alpha \to \pi/2$ does $\varepsilon \to \infty$ corresponding to left hand circular polarization.

Finally, with the help of the growth-rate equations (3.6) and (3.7), calculated in the quasilinear approximation, it is possible to obtain the ratio I between the wave energy for waves with $k > 0$ (propagating sunward) and those with $k < 0$ (propagating towards the comet). This ratio is also only a function of $\cos\alpha$,

$$I = \frac{[1 + (8/3)\cos\alpha + 2\cos^2\alpha - (1/3)\cos^4\alpha]}{[1 - (8/3)\cos\alpha + 2\cos^2\alpha - (1/3)\cos^4\alpha]}$$

(3.12)

The plot of this function is shown in Figure 9. For $\alpha \ne \pi/2$, I $\gg 1$ which means that the excited waves are propagating mainly in sunward direction, which also agrees well with in-situ observations [Tsurutani and Smith, 1986].

4. Excitation of oblique magnetosonic waves in the upstream region of the cometary bow-shock.

As was stated in the introduction, far upstream from the bow-shock the MHD turbulence is composed of almost parallel propagating Alfven waves, with the angle of propagation to the ambient magnetic field not exceeding 10-15° [Johnstone et al., 1986]; such Alfven turbulence is incompressible. The observations made during the encounter with comet Giacobini-Zinner show that closer to the bow-shock the character of the turbulence changes and there are strong correlations between magnetic field and density fluctuations [Tsurutani et al., 1987]. The existence of such correlations is illustrated in Figure 10. The most probable

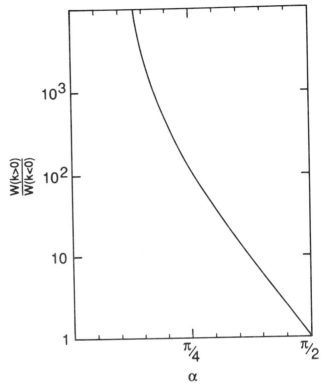

Fig.9. The ratio of the wave intensities with $k > 0$ (sunward direction) and $k < 0$ (comet direction). Except for a quasiperpendicular field orientation ($\alpha > 80°$) the waves are excited predominantly in sunward direction.

mechanism for this is proposed in Kennel [1986]; i.e. wave refraction is generated on transverse inhomogeneities in the foreshock region due to which a large portion of oblique magnetosonic waves is excited. Far upstream from the bow-shock the waves are propagating parallel to the ambient magnetic field. However, they can refract while being convected by the solar wind flow in the direction of the curved shock front. From the usual equation of wave refraction $dk/dt = -\nabla\omega$ it follows that the perpendicular component of wave vector relative to the ambient magnetic field is,

$$k_\perp \sim k \frac{v_A}{u} \frac{x}{L_\perp}$$

(4.1)

where $L_\perp$ is the characteristic space scale of plasma inhomogeneity perpendicular to the magnetic field. The main obstacle to the excitation of oblique magnetosonic waves is the strong Landau damping of these waves. Since in the solar wind cometary interaction the value of β (β is the ratio of gas kinetic pressure to the magnetic pressure) is of the order of unity, the Alfven velocity is comparable with the thermal velocity of the solar wind protons and that leads to strong Landau damping of oblique magnetosonic waves with the damping rate,

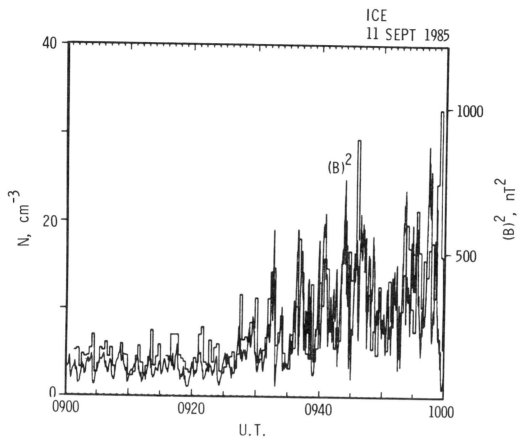

Fig.10. Magnetic field and density fluctuations near the Giacobini-Zinner bow-shock according to Tsurutani et al., [1987].

$$\gamma_L = \frac{\sqrt{\pi}}{2}\,\omega\,\frac{k_\perp^2}{k^2}\,\sqrt{\beta}\,\exp\,(-1/\beta)$$

$$(4.2)$$

As was shown in [Kotelnikov et al., 1990] only the formation of a quasilinear 'plateau' in the resonant velocity range $\omega/k_\parallel = v_\parallel$ can prevent such damping. The dispersion relation of oblique magnetosonic waves has the form,

$$\omega = k_\parallel v_A \left(1 + \lambda(\beta)\,\frac{k_\perp^2}{k_\parallel^2} \right)$$

$$(4.3)$$

where the function $\lambda(\beta)$, when solar wind protons have a Maxwellian distribution, is equal to,

$$\lambda = -\frac{i\beta}{2}\,\sqrt{\pi}\,s\,\exp\,(-s^2)\,\phi(is)$$

$$(4.3a)$$

where $s = \omega/(k_\parallel\sqrt{2T_p/m_p}) \sim 1/\sqrt{\beta}$ and $\phi(x) = 2/\sqrt{\pi}\int_o^x \exp(-t^2)dt$ is the probability integral. The change of λ with β is illustrated in Figure 11. Since the resonance velocity interval is quite narrow $\Delta v_\parallel \sim v_A k_\perp^2/k_\parallel^2$, the quasilinear 'plateau' forms easily and 'switches off' the Landau damping. In order to investigate this process we solved the quasilinear system of equations taking into account cyclotron excitation of magnetosonic and Alfven waves, wave refraction, Landau damping of magnetosonic waves, and quasilinear diffusion of particles in resonance with these waves. The system of equations obtained has the following form,

$$u\frac{\partial}{\partial x}|H_k^m|^2 - \frac{\partial\omega}{\partial z}\frac{\partial}{\partial k_\perp}|H_k^m|^2 =$$

$$\frac{S\,|H_k^m|^2}{\int (\,|H_k^m|^2 + |H_k^A|^2\,)\,dk_\perp}$$

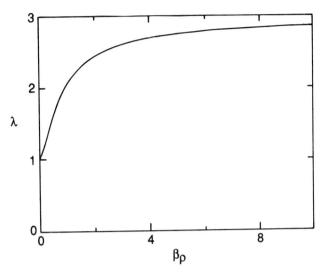

Fig.11. The behaviour of λ in the dispersion law (4.3) as function of β.

$$+ \pi \frac{\omega_{Hp}^2}{\omega_k} \frac{|H_k^m|^2}{n_0} \frac{k_{||}}{|k_{||}|} \int J_1^2 (k_\perp v_\perp / \omega_{Hp}) v_\perp^2 \frac{\partial f_p}{\partial v_{||}} d^2 v_\perp$$

$$(4.4)$$

$$u \frac{\partial}{\partial x} |H_k^A|^2 - \frac{\partial \omega}{\partial z} \frac{\partial}{\partial k_\perp} |H_k^A|^2 =$$

$$\frac{S |H_k^A|^2}{\int (|H_k^m|^2 + |H_k^A|^2) \, dk_\perp}$$

$$(4.5)$$

and,

$$u \frac{\partial f_p}{\partial x} = \frac{\pi e^2}{m_p^2} \frac{\partial}{\partial v_{||}}$$

$$\left[\int J_1^2 (\frac{k_\perp v_\perp}{\omega_{H_i}}) \frac{v_\perp^2}{c^2} |H_k^m|^2 \delta(\omega - k_{||} v_{||}) \frac{\partial f_p}{\partial v_{||}} \, dk_\perp dk_{||} \right]$$

$$(4.6)$$

In these equations for simplicity it is assumed that the transverse inhomogeneity is one dimensional being directed along the z-axes and the excited waves propagate away from the comet. Under these assumptions the driving term in the equations for Alfven and magnetosonic waves can be simplified significantly,

$$S(x,k_{||}) = \frac{Q \, e^{-r(x)/v_g \tau}}{8 r^2 v_g \tau n_0} H_0^2 \frac{\omega_{Hp}}{k_{||}^2 v_A}$$

$$\left[\cos\alpha (1 - \frac{\omega_{Hi}^2}{k_{||}^2 u^2}) + 2\sigma(\frac{k_{||} u}{\omega_{Hi}} - \frac{1}{\cos\alpha}) \right]$$

$$(4.7)$$

for $k_{||} \geq \omega_{Hi}/u$, and

$$S = 0 \qquad \text{for } k_{||} < \omega_{Hi}/u.$$

In order to solve this set of equations numerically, further simplification of the initial set of equations (4.4 - 4.6) is required. First, we neglect the Landau damping when evaluating the total wave energy,

$$\sum (x,k_{||}) = \int dk_\perp (|H_k^m|^2 + |H_k^A|^2)$$

The reason for that is that the main contribution comes from parallel waves as well as from the Alfven waves that do not experience Landau damping. It follows from equations. (4.4) and (4.5) that,

$$\sum (x,k_{||}) = \sum (0,k_{||}) + \int_0^x dx' \, S(x',k_{||})/u$$

$$(4.8)$$

It is also assumed that the rate of wave refraction $\delta = -1/u \, \partial\omega/\partial z$ is constant. Using κ ($\kappa = k_\perp - x\delta$) and x as independent variables one can integrate equation (4.4) for $|H_k^m|^2$ and the following result can be obtained,

$$|H_k^m|^2 = \frac{|H_k^m(x=0)|^2}{\sum (x=0)} \int_0^x dx' \, S(x',k_{||})/u$$

$$\exp \left[\int_0^x dx' \, \gamma_L (\kappa + x\delta, k_{||}, x')/u \right]$$

$$(4.9)$$

where we have used the notation,

$$\gamma_L (k_\perp, k_{||}, x) = \frac{\pi \omega_{Hp}^2}{2\omega_k n_0} \int J_1^2 (\frac{k_\perp v_\perp}{\omega_{Hp}}) v_\perp^2 \frac{\partial f_p}{\partial v_{||}} d^2 v_\perp$$

which is the Landau damping rate of magnetosonic waves. At $x = 0$ or far upstream in the solar wind it is reasonable to assume that equipartition of energy between Alfven and magnetosonic waves is valid, and that the initial distribution of magnetosonic waves over transverse wave number $k_\perp$ is Gaussian with some spread k_0. Hence in equation (4.9) we may write,

$$\frac{|H_k^m (x=0)|^2}{\sum (x=0)} = \frac{1}{2k_0} \sqrt{\pi} \, \exp(-k_\perp^2 / k_0^2)$$

When solving the diffusion equation (4.6) we separate the variables and represent the proton distribution function in the following approximate form,

$$f_p(x,v_{||},v_\perp) = \frac{n_0 m_p}{2\pi T_p} \exp\left[-\frac{m_p v_\perp^2}{2T_p}\right] F_p(x,v_{||})$$

F_p being defined later. First equation (4.6) is integrated over $v_\perp$. Then, using in this equation the magnetosonic wave power $|H_k^m|^2$ from equation (4.9) and $\xi = 2/(\lambda v_A)(v_{||}-v_A)$ as the independent variable, it is possible to rewrite equation (4.6) in the following form,

$$\frac{\partial\varphi}{\partial\zeta} = \frac{A}{\rho}\,\psi(\rho,\eta)\,\frac{\partial}{\partial\xi}\left\{\int_\varepsilon^\infty \frac{d\kappa_{||}}{\kappa_{||}^2 \sqrt{\xi}}\sum_\pm R(\kappa_{||}^2\,\xi)\right.$$

$$\left[\cos\alpha(1-\frac{\varepsilon^2}{\kappa_{||}^2}) + 2\sigma(\frac{\varepsilon}{\cos\alpha}-\kappa_{||})\right]\}$$

$$\exp\left\{B\int_0^\zeta d\zeta'\, R[\,(\kappa_{||}\sqrt{\xi}\pm v\,(\zeta-\zeta'))^2]\,\frac{\partial\varphi}{\partial\xi'}\right\}\frac{\partial\varphi}{\partial\xi}$$

$$(4.10)$$

where $\varepsilon = (v_{Tp}/u)\,(m_p/m_i)$ and $\xi' = (\sqrt{\xi}\pm v(\zeta'-\zeta)/\kappa_{||})^2$. The distribution function substituted into this equation was,

$$F_p(x,v_{||}) = \sqrt{\frac{m_p}{2\pi T_p}}\,\varphi(\zeta,\xi)$$

and the following notation was used,

$$R(\mu) = \mu\, e^{-\mu}\,[\,I_0(\mu) - I_1(\mu)]\;;\quad \mu = \xi\frac{k_{||}^2 v_{Tp}^2}{\omega_{Hp}^2}$$

$$A = \frac{\sqrt{\pi}}{3}\frac{\beta^2}{\lambda^3}\frac{Q}{v_g n_{0\infty}\tau}\frac{\omega_{Hi}}{u^2}\frac{m_i}{m_p}\frac{\omega_{Hp}}{k_0 v_{Tp}}$$

$$B = \sqrt{\frac{\pi}{2}}\frac{\beta}{\lambda}\frac{u}{v_{Tp}}\frac{m_i}{m_p}$$

$$\eta = \frac{u^2}{v_A\omega_{Hi}}\frac{1}{v_g\tau}$$

$$v = \delta\,(u^2/\omega_{Hi}\omega_{Hp})\sqrt{\frac{\beta}{2}}$$

and the dimensionless units for length and wave number were,

$$\zeta = \frac{x}{(u^2/v_A\omega_{Hi})}\,,\quad \kappa_{||} = k_{||}v_{Tp}\omega_{Hp}\,,\quad \rho = \frac{r}{(u^2/v_A\omega_{Hi})}$$

Equation (4.10) has been solved numerically by Kotelnikov et al., [1990]. A solution was obtained in the interval $0 < \xi < 2$ with the initial condition corresponding to a Maxwellian distribution,

$$\varphi(0,\xi) = \exp\,(-1/2\beta\,[1+\xi\lambda/2]^2,$$

$$(4.11a)$$

and with the following boundary conditions,

$$\partial\varphi/\partial\xi(\zeta,0) = 0;\qquad \varphi(\zeta,\xi_{max}) = \varphi\,(0,\xi_{max})$$

$$(4.11b)$$

Since the diffusion coefficient in equation (4.10) is not equal to zero at $\xi = 0$, the first of these boundary conditions corresponds to the absence of the diffusive flux through the boundary $\xi = 0$, and the second boundary condition corresponds to the fact that there is no diffusion of the particles at large $\xi \Rightarrow \xi_{max}$.

The results of the numerical solution are shown in Figures 12 - 14. In Figure 12 the formation of the 'plateau' in ξ on the distribution function is shown for $\beta = 0.3$. It can be seen that as the comet is approached the 'plateau' is present for a wide range of $\xi < 0.5$. The formation of the 'plateau' causes the excitation of oblique magnetosonic waves; in Figure 13 the square of the angular width of the wave spectrum is exhibited as function of β, far upstream in the solar wind and near the bow-shock. The square of angular width is defined as,

$$\frac{<k_\perp^2>}{<k_{||}^2>} = \frac{\int dk_\perp dk_{||}k_\perp^2\,|H_k^m|^2}{\int dk_\perp dk_{||}k_{||}^2\,|H_k^m|^2}$$

The parameters used in this calculations are those obtained during the encounter with comet Giacobini-Zinner: $Q = 5 \times 10^{28}$ mol/s, $v_g = 1$ km/s, $\tau = 10^6$ s, $H = 8$ nT, $n_o = 5$cm^{-3}, $u = 500$ km/s. For the angular width of the initial spectrum $\kappa_0 = 10^{-4}$ was used.

The excitation of oblique magnetosonic waves is accompanied by the appearance of density fluctuations. They are growing monotonically towards the comet, reaching for $\beta = 0.3$ the relative value $\sqrt{(\delta n^2/n_o^2)} \sim 0.3 - 0.6$ near the bow-shock, c.f. Figure 14.

5. Acceleration of cometary ions

Ion detectors in the vicinity of Giacobini-Zinner's bow-shock registered a high energy ion component of cometary origin with energies up to 400 keV, significantly exceeding the injection energy $2m_i u^2$ or ~ 90 keV for the parameters of

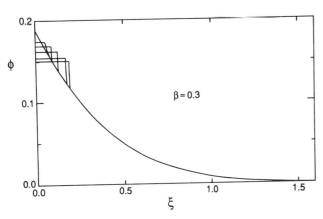

Fig.12. Evolution of the resonant proton distribution function for $\beta = 0.3$. The plots of the distribution function for $\zeta = 0$, 20, 40, 60, 80, 90 are also shown.

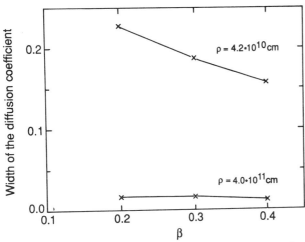

Fig.13. Angular width of the magnetosonic wave spectra for $\beta = 0.2; 0.3; 0.4$.

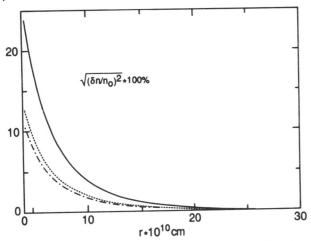

Fig.14. Mean square value of density fluctuations as function of the distance from the comet for $\beta = 0.2; 0.3; 0.4$.

the encounter [Ipavich et al., 1986]. Similarly, a high energy ion component was observed by VEGA and Giotto spacecraft near Halley's comet [Somogyi et al., 1986; McKenna-Lawlor et al., 1986]. This high energy ion component forms an ion tail of the shell distribution equation (3.2). As was proposed first in [Ip and Axford, 1986; Sagdeev et al., 1986], the most probable explanation for the appearance of this tail is second order Fermi (stochastic) acceleration by MHD waves in the upstream region of the bow-shock. That is the principal difference between this and the cosmic ray acceleration by supernova shocks, where the main acceleration of cosmic ray particles is due to first order Fermi acceleration at the shock front; stochastic Fermi acceleration by MHD waves could only be important in generating cosmic rays by injecting particles from hot interstellar plasma [Galeev et al., 1986]. The most important reason for this difference is that in the cometary case the energetic particles appear in the upstream solar wind at a typical distance defined by photoionisation $L \sim v_g \tau$, contrary to the cosmic ray case where the energetic particles appear in the upstream region of the shock at significantly smaller diffusive distance L_D. It is well known that for stochastic Fermi acceleration the relative motion of magnetic field humps created by MHD waves is required. That is equivalent to the dispersion of phase velocity of MHD waves. In that case in the quasilinear Stoss-term of the equation (2.7) there exists a range of resonant pitch angles where resonance with waves of different phase velocities is possible. This means that the different diffusion lines defined by equation (2.8) intersect and in addition to the pitch angle diffusion the velocity (energy) diffusion is also possible. This latter is described by the following equation,

$$u \frac{\partial f_0}{\partial x} = \frac{1}{v^2} \frac{\partial}{\partial v} \left(D_{vv} v^2 \frac{\partial f_0}{\partial v} \right)$$

(5.1)

where the velocity diffusion coefficient could be easily obtained from equation (2.7) by averaging over fast pitch angle diffusion. This yields [Galeev et al., 1987],

$$D_{vv} = \frac{\omega_{Hi}^2}{4H_0^2 v} \int_0^\pi d\theta \frac{\sin^3 \theta}{|\cos\theta|}$$

$$\sum_{\alpha,\beta} \frac{|H_k^\alpha|^2 |H_k^\beta|^2 \left((\omega/k_{||})^\alpha - (\omega/k_{||})^\beta \right)^2}{\sum_\alpha |H_k^\alpha|^2}$$

$$\pm \omega_{Hi} = -k_{||} v \cos\theta$$

(5.2)

Here the summation is to be carried out over all branches of the MHD waves having different phase velocities $(\omega/k_{||})^\alpha \neq (\omega/k_{||})^\beta$.

There are two different possibilities to excite these branches. One is connected with the excitation of oblique magnetosonic waves as discussed in the previous section. These waves propagate at different angles $(k_\perp/k_{||})$ to the magnetic field and - in accordance with the dispersion law equation (4.3) - phase velocity dispersion takes place, as required. Another possibility is the acceleration by large wavelength ($k \leq \omega_{Hi}/u$) oscillations when the magnetic field fluctuations are excited by the fire-hose instability. Since this instability is of a nonresonant nature, the waves propagating in opposite directions have different phase velocities $\pm v_A$. In recent publications [Isenberg, 1987; Gombosi et al., 1989] it was also proposed to use ambient magnetic field fluctuation propagating towards the comet $|H_k^-|^2$ for the acceleration mechanism. However, as it follows from equation (5.2), in the $|H_k^-|^2 \ll |H_k^+|^2$ case the energy diffusion rate is defined by the smallest wave power which means that the energy diffusion in that case would be as slow as in the unperturbed solar wind.

Let us consider first the acceleration by magnetosonic waves. In this case the main difficulty is due to the fact that these waves are excited by cyclotron instability and are of short wavelength $k > \omega_{Hi}/u$. Hence the resonant acceleration ($-kv\cos\theta = \pm\omega_{Hi}$) of fast ($v \gg u$) particles is possible only in the narrow pitch angle interval $\cos\theta < u/v \ll 1$ that corresponds to a diffusion coefficient decreasing with v,

$$D = \frac{\omega_{Hi}^2 v_A^2 \lambda^2}{4H_0^2 v} \int_{\omega_{Hi}/u}^{\infty} \frac{dk_{||}}{k_{||}} dk_\perp dk_\perp'$$

$$\frac{|H_k|^2 |H_k|^2 [(k_\perp^2/k^2) - (k_\perp'^2/k'^2)]^2}{\int dk_\perp |H_k|^2}$$

$$(5.3)$$

where λ was defined by equation (4.3a). It follows from equation (5.1) that the resulting distribution of high energetic ions is decreasing rapidly with v [Galeev et al., 1987],

$$f(x,v) \sim \eta(x)\exp(-\eta(x)v^3) \qquad (5.4)$$

where,

$$\eta^{-1} = \frac{9\lambda^2}{4} \int_0^x dx' \frac{\omega_{Hi}^2 v_A^2}{H_0^2 u} \int_{\omega_{Hi}/u}^{\infty} \frac{dk_{||}}{k_{||}} dk_\perp dk_\perp'$$

$$\frac{|H_k|^2 |H_k|^2 [(k_\perp^2/k^2) - (k_\perp'^2/k'^2)]^2}{\int dk_\perp |H_k|^2}$$

The measurements of the velocity distribution function of these high energy ions near Giacobini-Zinner's bow-shock [Ipavich et al., 1986; Richardson et al., 1986] are shown in Figure 15. The results are exhibited in the solar wind reference frame where - in accordance with the previous analysis - the distribution function is fairly isotropic. In Figure 15 the theoretical distribution (5.4) is also shown for the flyby parameters of comet Giacobini-Zinner listed before (line I). In the vicinity of the cometary shock $r \sim 0.1\ v\tau \sim 10^{10}$ cm, the distribution function (5.4) can be rewritten in the form,

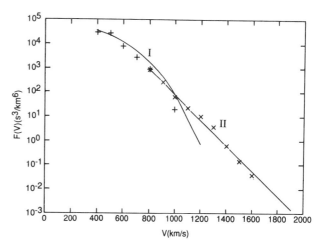

Fig.15. Energetic particle spectrum in the vicinity of the cometary bow-shock. Line I is the self-consistent distribution function equation (5.4) for the particles accelerated by magnetosonic turbulence and calculated for the parameters of the ICE flyby with comet Giacobini-Zinner. Line II is the theoretical distribution function equation (5.6) for the particles accelerated by long wavelength ($k < \omega_{Hi}/u$) turbulence. The calculations correspond to the case when only 20% of the MHD wave energy is in these long wavelength oscillations. Data marked by dots are from the ICE observations, + show data from ULEICA detector [Ipavich et al., 1986], x are data from EPAS detector [Richardson et al., 1986].

$$f_0 \sim 8.5 \cdot 10^{-26} \exp(-0.81\ v^3/u_\infty^3)$$

This self-consistent distribution function fits the observations quite well in the soft part of the energy spectrum (5×10^7 cm/s $< v < 10^8$ cm/s). In the more energetic part of the spectrum the acceleration is due to long wavelength magnetic fluctuations $k \ll \omega_{Hi}/u$ and is probably connected with the fire-hose instability. As was mentioned before, the main difficulty here is that such slow magnetic field fluctuations, with a time duration of the order of ten minutes or longer, are influenced by the global variations of the solar wind, which complicates significantly the interpretation of the measurements and the theoretical calculation of this part of the spectrum.

The low frequency part of the wave spectrum of MHD fluctuations ($f < f_{Hi}$) as measured by the ICE spacecraft near comet Giacobini-Zinner [Tsurutani and Smith, 1986] and by the Sakigake spacecraft near Halley's comet [Yumoto et al., 1986] are quite flat. However, the measurements in this part of the spectrum are somewhat ambiguous because of the long time intervals in question. A more frequently sampled measurement for this very low frequency part was done during the VEGA flyby and it gave a power like spectrum $|H_k|^2 \sim k^{-\alpha}$ with $\alpha = 1.5 - 2$ as is shown in Figure 16 [Gribov et al., 1987].

The spectrum of the long wavelength magnetic field fluctuations used in the calculation of the energy spectrum is approximated by the following expression,

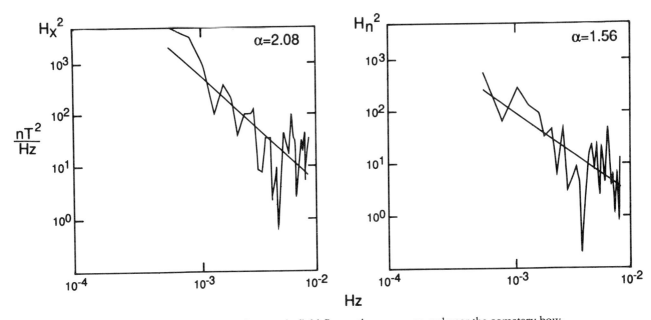

Fig.16. The spectral density of magnetic field fluctuations as measured near the cometary bow-shock during the VEGA-1 flyby. H_x is the component of magnetic field along the sun-comet line, H_n is the component in the perpendicular plane. Note that near the bow shock the heavy ion gyrofrequency $f_{Hi} \cong 1.5 \times 10^{-2}$ Hz, so these figures show the long wavelength ($k < \omega_{Hi}/u$) fluctuations which could be attributed to fire-hose instability and which are responsible for the stochastic Fermi acceleration of heavy ions.

$$|H_k|^2 = C(x) \frac{\omega_{Hi}}{k^2 u}, \qquad k \le \frac{\omega_{Hi}}{u} \tag{5.5}$$

In spite of the fact that these oscillations have similar functional dependence on k as for $k > \omega_{Hi}/u$, these two parts of the spectrum do not match. The long wavelength part goes considerably below the other, since only a small part of the wave energy is contained in it. We will assume that waves propagating in both directions with phase velocities $\pm v_A$ have the same low frequency spectral density. Solving the diffusion equation (5.1) with this assumption it is possible to obtain the following distribution function of the accelerated ions,

$$f(v) \sim w^{-3}(x) \exp(-v/w(x))$$

$$w(x) = w(0) + 0.5 \int_0^x dx \, \omega_{Hi} C(x) \frac{v_A^2}{H_0^2 u^2} \tag{5.6}$$

This distribution function is much flatter than the one described by equation (5.4). The significantly slower decrease of the distribution function with v is due to the fact that for the long wavelength ($k < \omega_{Hi}/u$) MHD fluctuations having wave spectrum described by equation (5.5) the diffusion coefficient is increasing with v,

$$D = C \frac{\omega_{Hi}}{2H_0^2} \frac{v_A^2}{u} v$$

The distribution function (5.6) is shown in Figure15 (line marked as II). The distribution function was calculated for the parameters of the Giacobini-Zinner's flyby using the realistic estimate,

$$\int dx \frac{C}{H_0^2} \cong 2.7 \times 10^{10} cm \tag{5.7}$$

In this case the distribution function has the form,

$$f(v) \sim \exp(-v/82 \text{ km/s}) \tag{5.7a}$$

It can be seen from Figure 15 that the most rigid part of the observed energy spectrum 1000 km/s $< v <$ 2000 km/s can be attributed to the distribution function calculated above.

For the Giacobini-Zinner's bow shock the amplitude of the magnetic field fluctuations is $|\Delta H/H_0| \sim 2$ [Tsurutani, 1991]. If we assume that the wave energy is changing with distance as $1/r$ in the interval $r_0 < r < r_1$, and $r_0 = 10^{10}$ cm is the position of the bow-shock, $r_1 \cong v_g \tau \cong 2 \times 10^{11}$ cm, it is possible to write the following relation,

$$C(x) \cong 4\alpha \frac{r_0}{r} H_0^2, \qquad r_0 < r < r_1$$

where α is the part of the wave energy contained in long wavelength oscillations. In that case near the bow shock,

$$\int dx \, \frac{C}{H_0^2} = 4\alpha \, r_0 \ln(\frac{r_1}{r_0})$$

It is easy to see by comparing with equation (5.7) that the second order Fermi acceleration by long wavelength oscillations is quite effective. It may yield the observed fluxes of energetic particles if a small part ($\alpha \cong 20\%$) of the total wave energy is contained in these oscillations.

Other possible mechanisms of particle acceleration are; adiabatic heating by inhomogeneous magnetic field in the decelerated solar wind flow and various mechanisms for particle acceleration at the shock front. With respect to the adiabatic heating it is necessary to note that - as it follows from equation (3.2) - due to this mechanism the particle energy in the solar wind frame could only increase by the factor $(u_\infty/u)^{2/3}$ which is only (20-25)% of the injection energy for Giacobini-Zinner's case. Accordingly, this acceleration mechanism is not sufficient to explain the observed ion acceleration up to 400 keV.

As far as shock acceleration is concerned, different acceleration mechanisms are possible at the shock front, such as first order Fermi acceleration and v x B acceleration. However, the accelerated particles cannot escape from the shock front into the upstream region at distances exceeding the diffusion length $L_D \sim (u/\omega_{Hi}) \, (H_0^2/\sum|H_k|^2) \sim 10^{10}$ cm for the parameters of the ICE flyby. This means that the shock acceleration could not be responsible for the high energy component in the upstream region. In general, in spite of some uncertainty connected with the absence of good experimental data for the low frequency part of the spectrum of the magnetic fluctuations, it seems reasonable to assume from our analysis that the most important mechanism for the acceleration of cometary ions in the upstream region is the stochastic Fermi acceleration by MHD waves.

The situation in the cometosheath region is quite different. Particles can be accelerated at the shock front and then convected by the solar wind into the sheath region. In that region the lower hybrid waves are excited in a very effective way; and these waves can accelerate ions, but these questions are outside the scope of the present paper.

Acknowledgments The authors are indebted to Dr. B Gribov who calculated the frequency spectra for magnetic field fluctuations from the VEGA measurements. They also would like to express their gratitude to the referees for their useful criticism and assistance.

References

Achterberg A.,On the propagation of relativistic particles in high beta plasma, *Astron. Astrophys.*, 98, 161-172, 1981.

Axford W. I., E. Leer, G. Scadron, *Proc. 15th Int.Cosmic Ray Conf.*, 2, 132-137, 1977.

Bell A. R., *Mon. Not. R. astr. Soc.*,183, 142-156, and 443-455, 1978.

Blanford R. D., J.P. Ostriker, *Ap.J.*, 221, L29-32, 1978.

Coates A. J., A. D. Johnstone, B. Wilken, K Jockers, K.-H. Glassmeier, Velocity-space diffusion of pick-up ions of water group at comet Halley, *J. Geophys. Res.* 94, 9983-9993, 1989.

Coates A. J., B. Wilken, A. D. Johnstone, K. Jockers, K-H. Glassmeier, D.E. Huddleston, Bulk properties and velocity distribution of water group ions; Giotto measurements, *J. Geophys. Res.,* 90, 10249-10260, 1990.

Galeev A. A., Theory and observation of solar wind/cometary plasma interaction processes, *Exploration of Halley's Comet*, ESA SP-250, 1, pp 3-18, 1986.

Galeev A. A., R. Z. Sagdeev, V. D. Shapiro, Injection mechanism for cosmic ray acceleration. *Plasma Astrophysics*, ESA SP-251, 297-305, 1986.

Galeev A. A., R. Z. Sagdeev, V. D. Shapiro, V. I. Shevchenko, K. Szego, MHD turbulence in a mass-loaded solar wind and the cometary bow-shock similarities and the differences with the cosmic ray shock, in *Collisionless shocks, Balatonfured*, pp 237-261, 1987.

Galeev A. A., A. N. Polyudov, R. Z. Sagdeev, V. D. Shapiro, V. I. Shevchenko, K. Szego, MHD turbulence in the solar wind flow interacting with comets, *JETP*, 192, 2090-2105, 1987.

Galeev A. A., R. Z. Sagdeev, V. D. Shapiro, V. I. Shevchenko, K. Szego, MHD turbulence and particle acceleration in a mass-loaded solar wind, *Adv. Space Sci.*, 9, 331-336, 1988.

Gribov B. E. et al, Stochastic Fermi acceleration of ions in the pre-shock region of comet P/Halley, *Astron. Astrophys.* 187, 293-296, 1987.

Glassmeier K. H., A. J. Coates, M. H. Acuna, M. L. Goldstein, A. D. Johnstone, F. M. Neubauer, H. Reme, Spectral characteristics of low frequency plasma turbulence upstream of comet P/Halley, *J. Geophys, Res.,* 94, 37-48, 1989.

Gombosi T., K. Lorencz, J. R. Jokipii, Combined first and second order Fermi acceleration in cometary environments, *J. Geophys. Res.*, 94, 15011-15024, 1989.

Hoppe M. M., C. T. Russell, L. A. Frank, T. E. Eastman, E. W. Greenstadt, Upstream hydromagnetic waves and their association with backstreaming ion population, ISEE 1 and 2 observations, *J. Geophys. Res.,*86, 4471-4492, 1981.

Ip, W.-H., Axford, W.I..The acceleration of particles in the vicinity of comets, *Planet. Space Sci.*, 34, 1061-1065, 1986.

Ipavich F. M., A. B. Galvin, G. Gloecker et al., Comet Giacobini-Zinner in situ observations of energetic heavy ions, *Science*, 232, 366, 1986.

Isenberg, P. A., Energy diffusion of pick up ions upstream of comets, *J. Geophys. Res*, 92, 8795-8799, 1987.

Johnstone A. D., K. H. Glassmeier, M. Acuna, H. Borg, D. Bryant, A. J. Coates, V. Formisano, J. W. Heath, S. Mariani, G. Musmann, F. M. Neubauer, M. Thomsen, B. Wilken, J. Winningham, Waves in the magnetic field and solar wind flow outside the bow shock at comet Halley, *Exploration of Halley's Comet*, ESA SP-250, 1, 277-281, 1986.

Kennel, C. F., Quasiparallel Shocks, *Adv. Space Res*, 6, 5, 1986.

Kotelnikov A. D., A. N. Polyudov, M. A. Malkov, R. Z. Sagdeev, V. D. Shapiro, High amplitude magnetosonic waves in the upstream region of the cometary bow shock, *Astron. Astrophys., in press,* 1990.

Krymski G. F., *Sov. Physics Doklady,* 23, 327-328, 1977.

Lee M. A., Coupled hydromagnetic wave excitation and ion acceleration upstream of the Earth's bow shock, *J. Geophys. Res.* 87, 5063-5080, 1982.

Lee M. A., W. -H. Ip, Hydromagnetic wave excitation by ionised interstellar hydrogen and helium in the solar wind. *J. Geophys. Res.,* 92, 11041-11052, 1987.

McKenna-Lawlor S. E., E. Kirsch, D. O'Sullivan et al, Energetic ions in the environment of comet Halley, *Nature,* 321, 347-348, 1986.

Neugeubauer M., A. J. Lazarus, H. Balsiger, S. A. Fuselier, F. M. Neubauer, H. Rosenbauer, The velocity distribution of cometary protons picked-up by the solar wind, *J. Geophys. Res.,* 94, 5227-5240, 1989.

Price C. P., C. S. Wu, The influence of strong hydromagnetic turbulences on newborn cometary ions, *Geophys. Res. Lett.* 14, 856-859, 1987.

Richardson I. C., S. W. H. Cowley, H. J. Hynds, T. R. Sanderson, K. -P Wenzel, P. W. Daly, Three dimensional energetic bulk flows at comet P/Giacobini-Zinner, *Geophys. Res. Lett.,* 13, 415-418, 1986.

Sagdeev R.Z., A. A. Vedenov, Some properties of a plasma having an anisotropic distribution function in a magnetic field, *Plasma Physics and Controlled Fusion 3,* (ed.by Leontovich M. A.) Atomizdat, Moscow, pp. 278-290, 1958.

Sagdeev R. Z., J. Blamont, A. A. Galeev, V. D. Shapiro, V. I. Shevchenko, K. Szego, VEGA spacecraft encounters with comet Halley, *Nature,* 321, 259-261, 1986.

Sagdeev R. Z., V. D. Shapiro, V. I. Shevchenko, K. Szego, MHD turbulence in the solar wind-comet interaction region, *Geophys. Res. Lett.,* 13, 85-88, 1986.

Schmidt H., R. Wegmann, Plasma flow and magnetic fields in the comets, in *Comets,* ed. by Wilkening L. L., Univ. of Arizona, Tucson, Arizona, 538-560, 1982.

Sentman D. D., J. P. Edmiston, L. A. Frank, Instabilities of low frequency parallel propagating electromagnetic waves in the Earth foreshock region, *J. Geophys. Res.* 86, 7487-7497, 1981.

Somogyi A. J., K. I. Gringauz, K. Szego et al., First observations of energetic particles near comet Halley, *Nature,* 321, 285-287, 1986.

Terasawa J., T. Mukai, W. Mijake et al., Detection of cometary pick-up ions up to 10 km from comet Halley, Suisei Observations, *Geophys. Res.Lett.,* 13, 837-840, 1986.

Tsurutani B. T., E. S. Smith, Hydromagnetic waves and instabilities associated with cometary ion pick-up. ICE observations, *Geophys. Res. Lett.,* 13, 263-266, 1986.

Tsurutani B., R. M. Thorne, E. J. Smith, J. T. Gosling, H. Matsumoto, Steepened magnetosonic waves at comet Giacobini-Zinner, *J. Geophys. Res.* 92, 11074-11082, 1987.

Tsurutani B., Comets: a laboratory for plasma waves and instabilities, *this issue,* 1991.

Wu C. S., R. C. Davidson, Instabilities produced by neutral particle ionisation in interplanetary space, *J. Geophys. Res.,* 72, 5399-5406, 1972.

Yumoto K., T. Saito, K. Nakagawa, Long period HM waves associated with cometary C^+ (or H_2O^+) ions: Sakigake observations, *Exploration of Halley's Comet,* ESA SP-250, 249-253, 1986.

ION PICKUP BY THE SOLAR WIND VIA WAVE-PARTICLE INTERACTIONS

Peter H. Yoon and C. S. Wu

Institute for Physical Science and Technology
University of Maryland, College Park, Maryland 20742

Abstract. Ion pickup by the solar wind is a topic of current interest. It has received much attention and has been discussed extensively in recent years, particularly after the cometary flybys. In situ observations and theoretical studies have enabled us to establish a fairly good understanding of the physics of ion pickup process. In a weakly turbulent solar wind, hydromagnetic waves that are either intrinsic or excited by the newborn ions can lead to pitch angle diffusion which may be approximately described by the quasi-linear theory with or without incorporating the resonance broadening mechanism due to weak turbulence. According to the weak turbulence theory, it is found that in general, the pickup ions do not form a complete spherical shell distribution in the time-asymptotic state, unless the wave field is sufficiently strong. In some cases, the ions can only possess a partial or incomplete shell. On the other hand, when the solar wind is highly turbulent, direct interaction between the wave fields and the newborn ions can result in rapid pitch angle scattering which may lead to swift formation of a spherical shell distribution on the time scale of an ion gyro-period.

1. Introduction

One of the fundamental processes of great significance in heliospherical physics or cometary plasma physics is the pickup of newborn protons or heavier ions (e.g., helium or other species) by the solar wind. Neutral particles associated with a cometary exosphere can subsequently become ionized by solar ultraviolet radiation or other processes. After ionization, these newborn ions are anticipated to interacted with, and eventually be assimilated into the rapidly moving solar wind. Conceptually, one may conjecture that the entire process is comprised of three basic stages which may be described below:

Creation of the newborn ions
↓
Pickup of the newborn ions
by the solar wind
↓
Thermalization of the pickup ions
↓
Complete assimilation of the
newborn ions into the solar wind

Cometary Plasma Processes
Geophysical Monograph 61
©1991 American Geophysical Union

The first-stage evolution, or the pickup process, has been studied and discussed extensively in recent years. The research efforts have been mainly stimulated by the AMPTE program and cometary observations. To date, although the complete assimilation process is still far from comprehended, theoretical understanding of the basic physical picture of the pickup process has been established. The purpose of this article is to present a brief review of the topic. Since in the present discussion emphasis is placed on the theories relevant to ion pickup processes, citations of the literature concerning observational results, which are reviewed in a separate article by Coates [1991], are by no means complete.

The structure of the paper is as follows. In section 2, we first make a few historical remarks and review briefly the early publications in the literature. Section 3 introduces the definitions as well as the concepts of pitch angle scattering and pitch angle diffusion processes. Then, in section 4, pitch angle diffusion based on the quasi-linear theory developed in plasma physics is discussed for a weakly turbulent plasma. Section 5 is devoted to the discussion of pitch angle scattering in a strongly turbulent plasma. Finally, summary and conclusions are presented in section 6.

2. Historical Remarks and Early Research Efforts

The question of how newly ionized particles are picked up by the solar wind first arose many years ago when scientists were interested in the helium ions created in the interplanetary space that were of interstellar or planetary origins [Feldman et al.,1971; Hozer and Axford, 1971; Wu et al., 1973; Hartle et al., 1973; Fahr, 1974; Hozer, 1977; Thomas, 1978]. It was proposed that the newborn ions can give rise to excitations of a variety of wave modes [Wu and Davidson, 1972; Hartle and Wu, 1973; Wu and Hartle; 1974] which in turn, can lead to the pickup of newborn ions [Wu et al., 1973]. However, most of these discussions were restricted to either linear analyses or hypotheses. The conjecture that collective wave-particle interactions could result in the pickup was not concretely demonstrated until the work by Winske et al. [1984]. Clearly, the research efforts on ion pickup in the early Seventies did not attract much attention for many years until the AMPTE program was launched in the early Eighties.

The AMPTE spacecraft carried out a series of lithium and barium gas release experiments both inside and outside of the Earth's magnetosphere. Although these active experiments only yielded limited information as far as the study of ion pickup is concerned, the spacecraft did make important observations of interstellar He$^+$ in the solar wind [Mobius et al., 1985]. In support of the program, a number of theoretical discussions using test particle models [Decker et al., 1983, 1984; Brinca, 1984] and hybrid-code simulations [Winske et al., 1984, 1985] were carried out. These preliminary studies were very valuable and have enabled us to gain important physical insights into the pickup process. One of the most significant conclusions is that low frequency hydromagnetic waves can play a very important role in the pickup process [Decker et al., 1984; Brinca, 1984]. If the intrinsic turbulence level in the solar wind is very low, the newborn ions can excite Alfvén waves, as first predicted by Wu and Davidson [1972] and Wu and Hartle [1974]. Then, the instability and its ensuing turbulence can lead to pitch angle diffusion and scattering so that the newborn ions become picked up eventually [Winske et al., 1984, 1985].

In situ observations made during the flybys at comets Giacobini-Zinner and Halley which occurred in 1985 are much more interesting from the viewpoint of plasma physics and the research of wave-particle interactions. The wave measurements made by ICE, Giotto, and Vega as well as the Japanese spacecraft during the flybys of Halley and Giacobini-Zinner immediately attracted considerable attention. It was found that the observed hydromagnetic turbulence associated with the newborn ions [Tsurutani and Smith, 1986a, b; Smith et al., 1986; Neubauer et al., 1986; Terasawa et al., 1986; Saito et al., 1986; Yumoto et al., 1986; Tsurutani et al., 1987] are indeed very consistent with those predicted by the early theories [Wu and Davidson, 1972; Wu and Hartle, 1974]. In short, in situ observations prompted many researchers to conclude that instabilities attributed to newborn ions are responsible for the observed turbulence. The wave observations also stimulated many subsequent theoretical investigations [e.g. Brinca and Tsurutani, 1987, 1988, 1989; Goldstein and Wong, 1987; Price et al., 1988; Thorne and Tsurutani, 1987; Wu et al., 1988a, b]. Recent research efforts in this area are reviewed by Brinca [1991] and therefore will not be discussed in detail in this discussion.

According to ICE measurements [Gloeckler et al., 1986; Hynds et al., 1986; Richardson et al., 1986; Sanderson et al., 1986; von Rosenvinge et al., 1986] cometary ions can be picked up very rapidly, particularly in regions where the turbulence levels are high. This finding spurred the theoretical studies by Wu et al. [1986] and Price and Wu [1987]. The hybrid-code numerical simulations and test-particle calculations discussed in these articles show that in the presence of strong magnetic fluctuations rapid pickup of newborn ions can occur, a conclusion also consistent with the results of in situ observations at comets Giacobini-Zinner and Halley.

Another significant and interesting result derived from theoretical studies is that pickup ions sometimes possess a hollow shell distribution function in velocity space [Wu et al., 1986; Price and Wu, 1987; Gary et al., 1986]. This result is also in agreement with in situ measurements which is discussed by Coates [1991] in this monograph. The finding that pickup ions can form shell distribution functions has further stimulated a series of investigations concerning the stability of such a distribution [Freund and Wu, 1988; Gary and Sinha, 1989; Yoon et al., 1989; Brinca, 1990; Wu and Yoon, 1990; Yoon, 1990] as well as other relevant theoretical problems [Ziebell et al., 1990; Yoon and Ziebell, 1990].

However, careful examinations of observational results acquired with Giotto spacecraft which are reported by Neugebauer et al. [1989, 1990], for example, reveal that in the far upstream regions, the pickup ion distributions often possess incomplete spherical shell distributions. Such features seem to deserve theoretical attention, and have subsequently motivated us at the University of Maryland group to carry out a series of studies. In the subsequent sections, this issue will be addressed.

3. Pitch Angle Diffusion and Pitch Angle Scattering

In general, the formation of an ion shell distribution function may be attributed to the wave-particle interaction which may be described by the pitch angle diffusion process for a weakly turbulent plasma, and by the pitch angle scattering process when the turbulence level is sufficiently high. In the following, these two processes are defined and explained in order. It should be remarked at the outset that in the present discussion the term plasma "turbulence" is only used to describe enhanced hydromagnetic fluctuations.

3.1. Pitch Angle Diffusion in a Weakly Turbulent Field

In a weakly turbulent plasma in which the wave energy density is much lower than the kinetic energy density of the ions, we may assume that the effect of wave fields on the orbit of each ion is negligible to the lowest order. However, in this case hydromagnetic fluctuations can still result in a diffusion process in velocity space. Such a diffusion process may be described by the quasi-linear theory (or the weak turbulence theory) in plasma physics. In what follows, we assume that the electrons are unimportant in the pickup process and that the hydromagnetic waves are propagating predominantly in the direction parallel and/or anti-parallel to the ambient magnetic field. The assumption of parallel propagation is justifiable from the viewpoint of linear stability theory in which, it is established that for a typical newborn ion distribution (e.g., a ring-beam distribution) the most unstable waves occur for the parallel propagation. We are primarily interested in the interaction between the newborn ions and the hydromagnetic wave field.

The quasi-linear diffusion equation for each ion species involved in the pickup process (i.e., the newborn ions, the background protons, etc.) can be written as

$$\frac{\partial}{\partial t} f_j(v_\perp, v_{||}, t) = \operatorname{Re} i \frac{2\pi e_j^2}{m_j^2} \sum_{+,-} \int_{-\infty}^{\infty} dk$$

$$\times \frac{1}{v_\perp} \left[\left(1 - \frac{k v_{||}}{\omega} \right) \frac{\partial}{\partial v_\perp} + \frac{k v_\perp}{\omega} \frac{\partial}{\partial v_{||}} \right] \frac{\varepsilon_k(t)}{\omega \pm \Omega_j - k v_{||}} \tag{1}$$

$$\times v_\perp \left[\left(1 - \frac{k v_{||}}{\omega} \right) \frac{\partial}{\partial v_\perp} + \frac{k v_\perp}{\omega} \frac{\partial}{\partial v_{||}} \right] f_j(v_\perp, v_{||}, t)$$

where $\omega = \omega_k + i\gamma_k$ is generally complex, $\gamma_k > 0$ corresponding to instability. Here, f_j denotes the distribution function of species j, and f_j is normalized to unity ($\int d^3\mathbf{v}\, f_j = 1$), e_j and m_j are the electric charge and mass for each species, respectively, and the quantity $\Omega_j = e_j B_0/m_j c$ is the gyro-frequency associated with each species, B_0 being the strength of the ambient magnetic field. Moreover, the quantity $\varepsilon_k(t)$ denotes the spectral wave energy density defined by $\varepsilon_k(t) = \langle \delta E_k \delta E_k^* \rangle / 8\pi$, where δE_k is the spectral component of the fluctuating electric field.

The spectral wave energy $\varepsilon_k(t)$ satisfies the wave kinetic equation given by

$$\frac{\partial \varepsilon_k(t)}{\partial t} = 2\gamma_k \varepsilon_k(t) \tag{2}$$

The complex frequency $\omega = \omega_k + i\gamma_k$ at a given instant is determined from the dispersion equation

$$\frac{c^2 k^2}{\omega^2} = \pm \sum_j \frac{n_j}{n_0} \frac{\omega_{pj}^2}{\omega \Omega_j} + \sum_j \frac{n_j}{n_0} \frac{\omega_{pj}^2}{2\omega} \int d^3\mathbf{v}\, \frac{v_\perp}{\omega \pm \Omega_j - k v_\parallel}$$
$$\times \left[\left(1 - \frac{k v_\parallel}{\omega}\right) \frac{\partial}{\partial v_\perp} + \frac{k v_\perp}{\omega} \frac{\partial}{\partial v_\parallel} \right] f_j(v_\perp, v_\parallel, t) \tag{3}$$

where we have neglected the displacement current and have assumed a cold electron population. Moreover, we have assumed $\omega \ll \Omega_e$, where $\Omega_e = e B_0/m_e c$ is the electron gyro-frequency, and have used the quasi-neutrality condition. In (3) $\omega_{pj} = (4\pi n_0 e_j^2/m_j)^{1/2}$ is the plasma frequency associated with each species, n_0 and n_j being the total plasma density and the ambient density associated with each species, respectively. The set of self-consistent equations (1) - (3) describes the quasi-linear wave-particle interaction between the self-consistently generated turbulence and/or the existing weakly turbulent wave fields and the newborn ions as well as the background solar wind ions. In the following discussion, we pay attention mainly to the time evolution of the newborn ions and neglect the thermal effects of the background solar wind protons on the dispersion relation. Furthermore, for definiteness, we consider that the newborn ions are protons, although the analysis can be generalized to any other kind of ions.

It is instructive to rewrite the set of equations (1) - (3) in terms of the following normalized variables:

$$\mathbf{u} = \mathbf{v}/v_A$$
$$\tau = t/t_0 \tag{4}$$
$$t_0^{-1} = \frac{2\pi^2 e^2 \varepsilon_0}{m_i^2 v_A^2 \Omega_i} = \frac{\pi \Omega_i}{4} \left(\frac{\delta B}{B_0}\right)^2$$

where $\varepsilon_0 = \sum_\pm \int dk\, \varepsilon_k(0)$ defines the initial level of the pre-existing turbulence, and δB denotes the strength of the initial fluctuating magnetic field. In the definition of t_0^{-1}, we have made use of the relation $\varepsilon_0 = \langle \delta E^2 \rangle / 8\pi = (v_A/c)^2 \langle \delta B^2 \rangle / 8\pi$, where v_A is the Alfvén speed defined by $v_A = B_0/(4\pi n_0 e^2)^{1/2} = c\Omega_i/\omega_{pi}$. In other words, in writing $\delta E = (v_A/c)\, \delta B$, we have assumed that the waves satisfy the dispersion relation $\omega_k = k v_A$. Let us further introduce the following notation for the normalized newborn ion distribution function and the normalized wave density

$$f_i(u_\perp, u_\parallel, \tau) \equiv v_A^3 f_i(v_\perp, v_\parallel, t)$$
$$N_k(t) = \varepsilon_k(t)/\varepsilon_0 \tag{5}$$

such that $\int d^3\mathbf{u}\, f_i(\mathbf{u}, \tau) = 1$ and $\sum_\pm \int dk\, N_k(0) = 1$.

In terms of these new quantities we find that the quasi-linear equations governing the time-evolution of the newborn ion distribution function as well as the self-consistent wave field can be expressed as

$$\frac{\partial f_i(u_\perp, u_\parallel, \tau)}{\partial \tau} = \operatorname{Re} \frac{i}{\pi} \sum_{+,-} \int_{-\infty}^{\infty} dk$$
$$\times \frac{1}{u_\perp}\left[\frac{\partial}{\partial u_\perp} - \frac{k v_A}{\omega}\left(u_\parallel \frac{\partial}{\partial u_\perp} - u_\perp \frac{\partial}{\partial u_\parallel}\right)\right] \frac{\Omega_i N_k(\tau)}{\omega \pm \Omega_i - k v_A u_\parallel}$$
$$\times u_\perp \left[\frac{\partial}{\partial u_\perp} - \frac{k v_A}{\omega}\left(u_\parallel \frac{\partial}{\partial u_\perp} - u_\perp \frac{\partial}{\partial u_\parallel}\right)\right] f_i(u_\perp, u_\parallel, \tau) \tag{6}$$
$$\frac{\partial N_k(\tau)}{\partial \tau} = 2\gamma_k t_0\, N_k(\tau)$$

where the complex frequency $\omega = \omega_k + i\gamma_k$ is determined from the dispersion equation which, as a result of the assumption of cold background protons and a single component of newborn ions, is now written as:

$$\frac{k^2 v_A^2}{\Omega_i^2} = \frac{\omega^2}{\Omega_i(\Omega_i \pm \omega)} + \frac{n_i}{n_0}\frac{\omega}{2} \int d^3\mathbf{u}\, \frac{u_\perp}{\omega \pm \Omega_i - k v_A u_\parallel}$$
$$\times \left[\frac{\partial}{\partial u_\perp} - \frac{k v_A}{\omega}\left(u_\parallel \frac{\partial}{\partial u_\perp} - u_\perp \frac{\partial}{\partial u_\parallel}\right)\right] f_i(u_\perp, u_\parallel, \tau) \tag{7}$$

One immediate consequence of the above reformulation is that one may establish the criterion based on which the relative importance of self-consistently generated field versus the extraneous wave fields may be determined. This can be achieved by considering the wave kinetic equation in (6). Specifically, if the newborn ion density is sufficiently small such that $n_i \ll n_0$, then the growth rate γ_k, which is essentially determined by the second term on the right-hand side of the dispersion equation (7), is expected to be small. Moreover, if the level of pre-existing turbulence is sufficiently large (i.e., if t_0 is not too large, since t_0 is proportional to $B_0^2/\delta B^2$), then one may satisfy the condition:

$$\gamma_k t_0 \ll 1 \tag{8}$$

If the inequality (8) is satisfied, then one may ignore the self-consistent effect by taking a constant spectral wave energy density in time $\partial N_k(\tau)/\partial \tau = 0$. Also, we may neglect the second term on the right-hand side of (7), and evaluate the dispersion relation by keeping the background contribution only; $k^2 v_A^2 = \omega_k^2/(1 \pm \omega_k/\Omega_i)$, while setting $\gamma_k = 0$.

On the other hand, if the newborn ion density is sufficiently large so that the growth rate is no longer small, and/or if the background fluctuation level is sufficiently low so that the opposite inequality:

$$\gamma_k t_0 \gg 1 \qquad (9)$$

is satisfied, then the self-consistent approach must be used, and thus, one has to solve the full set of coupled equations (6) and (7). In the cometary environment, however, even in the far upstream of the cometary bow shock where the turbulence level is low, the self-consistent effect can sometimes be ignored, since the density of the newborn ions are very low and the level of intrinsic turbulence is moderate. Thus one may assume the inequality (8) to hold. This greatly simplifies many theoretical aspects of the analysis.

For the time being, therefore, let us consider first the case in which the inequality (8) holds. In this case, the turbulent wave fields which may result from intrinsic solar wind turbulence, or turbulence generated by newborn ions in other regions, which subsequently are convected to the region of interest, are assumed to be sufficiently large in amplitude, and are thus assumed to be in a steady state. Of particular interest to us is the low-frequency hydromagnetic turbulence which satisfy the dispersion relation

$$\omega_k = k v_A, \qquad \text{parallel propagation} \qquad (10)$$

$$\omega_k = -k v_A, \qquad \text{anti-parallel propagation}$$

for both the (+) and the (−) modes. The turbulent wave fields may possess both components corresponding to right-hand and left-hand circular polarizations, respectively, and in general, the wave field may consist of waves propagating parallel and anti-parallel to the ambient magnetic field, respectively. For each polarization, (+) and (−), we denote the time-independent wave fields associated with turbulence propagating parallel to the ambient field as N'_k, and those associated with turbulence propagating in the anti-parallel direction as N''_k, so that the total normalized wave energy density can be written as follows:

$$N_k = (1/2)\,(N'_k + N''_k) \qquad (11)$$

In the above, the factor (1/2) is introduced so that the relation $\Sigma_\pm \int dk\, \varepsilon_k = \varepsilon_0 \Sigma_\pm \int dk\, N_k = \varepsilon_0$ may hold.

Using (10) and (11) one may recast the quasi-linear diffusion equation given in (6) as follows

$$\frac{\partial f_i}{\partial t} = \frac{1}{u_\perp}\left[\frac{\partial}{\partial u_\perp} - \left(u_\parallel \frac{\partial}{\partial u_\perp} - u_\perp \frac{\partial}{\partial u_\parallel}\right)\right]$$

$$\times D' u_\perp \left[\frac{\partial}{\partial u_\perp} - \left(u_\parallel \frac{\partial}{\partial u_\perp} - u_\perp \frac{\partial}{\partial u_\parallel}\right)\right] f_i$$

$$+ \frac{1}{u_\perp}\left[\frac{\partial}{\partial u_\perp} + \left(u_\parallel \frac{\partial}{\partial u_\perp} - u_\perp \frac{\partial}{\partial u_\parallel}\right)\right]$$

$$\times D'' u_\perp \left[\frac{\partial}{\partial u_\perp} + \left(u_\parallel \frac{\partial}{\partial u_\perp} - u_\perp \frac{\partial}{\partial u_\parallel}\right)\right] f_i$$

$$(12)$$

where

$$D' = \sum_{+,-}\int_0^\infty dk\, N'_k\, \delta\left[\frac{k v_A}{\Omega_i}(1 - u_\parallel) \pm 1\right]$$

$$(13)$$

$$D'' = \sum_{+,-}\int_0^\infty dk\, N''_k\, \delta\left[\frac{k v_A}{\Omega_i}(1 + u_\parallel) \pm 1\right]$$

In the above, we have rendered the relevant k-integrals to be restricted to the positive domain only by making use of the symmetry relations $\omega_k = -\omega_{-k}$ and $N_k = N_{-k}$.

Equation (12) may actually describe pitch angle diffusion of newborn ions in two velocity frames, one co-moving with the waves propagating in the parallel direction, and other moving in an opposite direction. To demonstrate this point, let us rewrite (12) by introducing the following velocity coordinates in each reference frame moving with the waves:

$$u' = [u_\perp^2 + (u_\parallel - 1)^2]^{1/2}, \qquad \mu' = (u_\parallel - 1)/u'$$

$$u'' = [u_\perp^2 + (u_\parallel + 1)^2]^{1/2}, \qquad \mu'' = (u_\parallel + 1)/u''$$

$$(14)$$

The physical meaning of the velocity coordinates (14) is graphically depicted in Figure 1. Of particular interest to us are the quantities μ' and μ''. As shown in the figure, μ' represents the cosine of the pitch angle in the reference frame moving

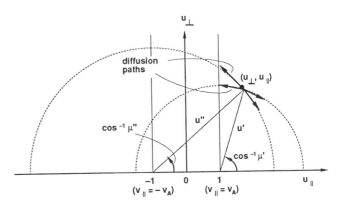

Fig. 1. Pitch angle diffusion paths and the wave frame coordinates (u', μ') and (u'', μ'') when the waves have counter-propagating components. In each wave frame, the diffusion paths are defined as a constant energy surface in the wave frame.

with the waves in the positive-k direction, with speed $v_\parallel = v_A$ (or $u_\parallel = 1$). Similarly, μ'' represents the cosine of the pitch angle in the velocity frame moving in the negative-k direction with speed $v_\parallel = -v_A$ (or $u_\parallel = -1$). As a result of the transformation (14), we find that the diffusion equation (12) can be recast into the following form

$$\frac{\partial f_i}{\partial \tau} = \frac{\partial}{\partial \mu'}\left[D'\left(1-\mu'^2\right)\frac{\partial f_i}{\partial \mu'}\right] + \frac{\partial}{\partial \mu''}\left[D''\left(1-\mu''^2\right)\frac{\partial f_i}{\partial \mu''}\right]$$

$$D' = \sum_{+,-}\int_0^\infty dk\, N'_k\,\delta\left(-\frac{kv_A}{\Omega_i}u'\mu' \pm 1\right) \qquad (15)$$

$$D'' = \sum_{+,-}\int_0^\infty dk\, N''_k\,\delta\left(\frac{kv_A}{\Omega_i}u''\mu'' \pm 1\right)$$

Each term on the right-hand side of the diffusion equation in (15) only involves derivatives with respect to the pitch angle variable μ' or μ''. Thus it is clear that each term describes the pitch angle diffusion process about the velocity centered at $u_\parallel = 1$ and $u_\parallel = -1$, respectively, as shown in Figure 1. Physically, (15) describes the ions undergoing diffusion along the paths shown in Figure 1 with arrows. Since, at a given time, the ions can follow either of the two diffusion paths each centered at $u_\parallel = 1$ and -1, respectively, it can be seen that the ions will also undergo a stochastic acceleration (or deceleration).

The resonance conditions (determined by equating the arguments of the delta functions in (15) to zero) reveal that for the waves propagating in the positive-k direction (henceforth designated as the forward direction), the (+) mode component may resonate with the ions possessing $u_\parallel > 1$, while the (−) mode component resonate with the ions possessing $u_\parallel < 1$. On the other hand, for the waves propagating in the negative-k direction (designated as the backward direction), the ions with $u_\parallel > -1$ interact with the (−) mode component, while the ions possessing $u_\parallel < -1$ resonate with the (+) mode component. This is shown clearly in Figure 2.

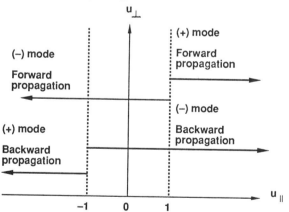

Fig. 2. Velocity space domains associated with the wave-particle resonance corresponding to four components of the wave.

One final remark which may naturally lead us to the next subsection is the following. As is well known, in quasi-linear theory it is assumed that, to lowest order, particle orbits are unaffected by the wave field. The pitch angle diffusion process described so far is the result of collective interaction between the particles and the wave. In this description, an ensemble of particles redistributes itself in such a way as to produce a net

diffusion. However, the motion of an individual particle is described by an unperturbed orbit (that is, a combination of simple gyro-motion about the ambient magnetic field, and a drift motion along the field). On the other hand, if the level of fluctuating wave field is sufficiently strong, the particles may directly interact with the waves such that pitch angle scattering can take place. In this case, particles no longer follow the unperturbed orbits. That is, the orbit of a particle may significantly deviate from the simple gyro-motion plus the drift along the ambient field. Thus, in principle, the two processes (i.e, diffusion and scattering) can be distinguished if one could follow the individual orbit of each ion. In a general situation both processes may be at work simultaneously. However, when the level of fluctuating fields is low, then one may use the diffusion approximation. On the other hand, when the turbulence level is high, pitch angle scattering is expected to prevail and the diffusion approximation becomes unjustifiable.

3.2. Pitch Angle Scattering by Large Amplitude Waves

Theoretically it is exceedingly difficult to deal with the case of strong turbulence in the context of plasma kinetic theory. Even when the turbulence level is moderately strong, the particle diffusion equation (1) has to be modified to incorporate the effect of nonlinear wave-wave and wave-particle interactions. Typically, such a treatment should result in terms that are of order $\varepsilon_k(t)^2$ or higher, on the right-hand side of (1). Moreover, the wave kinetic equation (2) becomes a nonlinear equation in $\varepsilon_k(t)$, which should describe a variety of nonlinear wave processes. One of the most noteworthy developments in the strong plasma turbulence theory is the so-called renormalized plasma kinetic theory developed by Kadomtsev [1965], Dupree [1966], and Weinstock [1969]. These theories are successfully applied to a variety of physical applications. Notwithstanding these developments, to date, a satisfactory and rigorous but practically useful theory of strong plasma turbulence does not exist.

To illustrate the essential difference between the pitch angle scattering and pitch angle diffusion, we consider a simplest case in which a single test ion is under the action of an externally generated Alfvén wave with a fixed frequency and wavenumber satisfying the dispersion relation $\omega_k = kv_A$. We assume that the wave propagates along the ambient magnetic field $\mathbf{B}_0 = B_0\,\hat{\mathbf{z}}$. Thus, an ion should experience the following electric and magnetic fields:

$$\mathbf{E} = \delta\mathbf{E}(t,\mathbf{r})$$
$$\mathbf{B} = \mathbf{B}_0 + \delta\mathbf{B}(t,\mathbf{r}) \qquad (16)$$

where $\delta\mathbf{E}$ and $\delta\mathbf{B}$ denote the fields associated with the wave. Since we are considering an Alfvén wave, we have the relation

$$\delta\mathbf{E} = (v_A/c)\,\hat{\mathbf{z}} \times \delta\mathbf{B} \qquad (17)$$

The equation of motion of a test ion in the combined ambient and Alfvén wave fields is

$$\frac{d\mathbf{v}(t)}{dt} = \frac{e}{m_i}\left\{\delta\mathbf{E}(t,\mathbf{r}(t)) + \frac{\mathbf{v}(t)}{c}\times\left[\mathbf{B}_0 + \delta\mathbf{B}(t,\mathbf{r}(t))\right]\right\} \qquad (18)$$

where e and m_i are the ion charge and mass. The ion velocity $\mathbf{v}(t)$ and the displacement $\mathbf{r}(t)$ at each instant of time are related by

$$d\mathbf{r}(t)/dt = \mathbf{v}(t) \qquad (19)$$

The two equations (18) and (19) should be discussed with given initial conditions.

We now consider the case where the wave is circularly polarized so that

$$\delta\mathbf{B}(t,\mathbf{r}) = \delta\mathbf{B}^{\pm}(t,\mathbf{r}) = \delta B \,(\hat{\mathbf{x}}\cos\phi \pm \hat{\mathbf{y}}\sin\phi) \qquad (20)$$

where $(\pm)$ correspond to the right- and left-hand circular polarizations, respectively, and the phase ϕ is defined by

$$\phi = \varphi_0 - \mathbf{k}\cdot\mathbf{r} + \omega t = \varphi_0 + k\,(v_A t - z) \qquad (21)$$

Since φ_0 is a constant, without loss of generality, we henceforth set $\varphi_0 = 0$.

Making use of the above equations, one may show that for each ion there is a constant of motion

$$\frac{1}{2}m_i\left[\mathbf{v}(t) - v_A\hat{\mathbf{z}}\right]^2 = \frac{1}{2}m_i\left[\mathbf{v}(0) - v_A\hat{\mathbf{z}}\right]^2 \qquad (22)$$

According to this equation, the ions move on a surface defined by $v_\perp^2(t) + [v_\parallel(t) - v_A]^2 = $ const. That is, in the reference frame co-moving with the wave, the total energy of the ions is conserved.

In order to understand the motion of a test ion in a combined fields of constant magnetic field and a large amplitude Alfvén wave, we solve the equation of motion (18) and (19) numerically. For convenience, we introduce a set of normalized variables

$$\mathbf{r}' = (\Omega_0/v_A)\,\mathbf{r}$$
$$t' = \Omega_0\,t$$
$$\mathbf{v}' = \mathbf{v}/v_A \qquad (23)$$
$$k' = kv_A/\Omega_0$$
$$\varepsilon = \delta B/B_0$$

where $\Omega_0 = eB_0/m_ic$ is the gyro-frequency associated with the ambient magnetic field. Note that $\mathbf{r}'$ may also be written as $\mathbf{r}' = (\omega_{pi}/c)\mathbf{r}$, where $\omega_{pi} = (4\pi n_0 e^2/m_i)^{1/2}$ is the ion plasma frequency. The dimensionless parameter t' is a measure of the time in units of inverse ion gyro-frequency Ω_0^{-1}. Then, in terms of these new variables, the relevant components of the equation of motion are given by

$$dz'/dt' = v'_\parallel$$
$$dv_x'/dt' = \mp\,\varepsilon\,(v_\parallel' - 1)\sin k'(t' - z') + v_y'$$
$$dv_y'/dt' = \varepsilon\,(v_\parallel' - 1)\cos k'(t' - z') - v_x' \qquad (24)$$
$$dv_\parallel'/dt' = \varepsilon\,[\pm\,v_x'\sin k'(t' - z') - v_y'\cos k'(t' - z')]$$

These equations are subject to initial conditions. If a group of test ions is uniformly distributed in configuration space, then in general, the orbit of ions at different locations within a unit wavelength of the wave will not be identical. Thus, in order to discuss all possible ion trajectories, one has to consider an ensemble of ions uniformly distributed in z'-direction within one wavelength (i.e., $0 \leq z'(0) \leq 2\pi/k'$). However, for the sake of illustration, let us consider the case in which a test ion is launched at a point $z'(0) = 0$ with a velocity having finite components perpendicular and parallel to the ambient field. Then the initial conditions for (24) can be written as follows

$$z'(0) = 0$$
$$v_x'(0) = V_0 \cos\psi \sin\alpha$$
$$v_y'(0) = V_0 \sin\psi \sin\alpha \qquad (25)$$
$$v_\parallel'(0) = V_0 \cos\alpha$$

Notice that as a result of normalizations introduced in (23), V_0 is the initial speed of the test ion measured in units of v_A. In the numerical analysis we choose $V_0 = 5$. In (25), α is the polar angle in velocity space, which is the initial pitch angle of the test ion, and ψ is the initial azimuthal angle in velocity space. Without loss of generality we set $\psi = 0$, since the set of equations (24) does not depend on x' and y' explicitly.

In obtaining the numerical solution of (24), we assume that $k' = kv_A/\Omega_0 = 0.1$, and we restrict ourselves to a finite time interval $0 < t' = \Omega_0 t < 20$. In Figure 3, the motion of a test ion is shown in three-dimensional representation. In the figure, the range of velocity components is defined as follows: $-5 \leq v_x'$, v_y', $v_\parallel' \leq 5$. The plot is generated for both the left-hand (upper panels) and the right-hand (lower panels) circular polarizations, and for each polarization, three choices of initial pitch angle corresponding to $\alpha = 0°$, $\alpha = 45°$, and $\alpha = 90°$ are shown. For this figure, we have assumed $\delta B = B_0$ (or, $\varepsilon = 1$). The initial positions for the ion are: for $\alpha = 0°$, $(v_x', v_y', v_\parallel') = (0, 0, 5)$; for $\alpha = 45°$, $(5\sin45°, 0, 5\cos45°)$; and for $\alpha = 90°$, $(5, 0, 0)$. The ambient magnetic field vector is pointing in the $v_\parallel'$-direction, whereas the magnetic field component associated with the wave is pointing in a direction orthogonal to the ambient magnetic field and rotating in v_x'-v_y' plane. From the figure one can see that the motion of test ion is basically made up of two gyro-motions: one gyro-motion with angular frequency Ω_0 in v_x'-v_y' plane due to the ambient magnetic field, and another with frequency $\delta\Omega = e\delta B/m_ic$ due to the wave field. Notice that for the case of right-hand mode, because the field vector is rotating in the opposite direction with respect to the ion gyro-motion, the orbit exhibits a complicated behavior.

To understand the complex three-dimensional motion in further detail, in Figure 4, we show the projection of the orbit into the v_x'-v_y' pane. Let us first pay our attention to the case of $\alpha = 0°$. In the unperturbed orbit approximation such as is implied in the quasi-linear theory, the ion is supposed to remain in its initial position indefintely. However, in the present case, because of the strong wave field, the ion motion exhibits a complex behavior as shown in the figure. For $\alpha = 45°$ and $90°$, the unperturbed orbit in the projected plane should be simple circles.

Here, we reiterate that the set of examples shown above is but a sample of many possible trajectories. If we launched the test ion at a different location along $z'(0)$, or if we used a

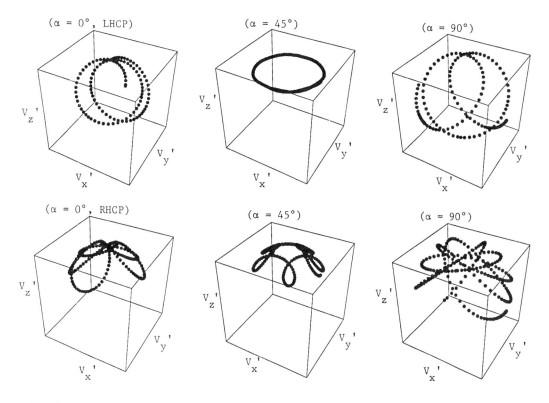

Fig. 3. Three-dimensional representation of the ion orbit for three choices of initial pitch angle; $\alpha = 0°, 45°,$ and 90°. The upper panels correspond to the left-hand circular polarization (LHCP), while the bottom panels are for the case of right-hand circular polarization (RHCP). The velocity ranges are $-5 \leq v_x', v_y', v_z' \leq 5$ in each panel.

different value of k', the result would be quite different from that shown in the above two figures. Nevertheless, we choose to present these figures for the purpose of illustration.

Needless to say, as one can clearly see from these figures, when the wave amplitude is suffiently high, the ion motion considerably deviates from the unperturbed orbit. This deviation can be understood as the basis of the pitch angle scattering mechanism. When a distribution of ions is present with each ion undergoing a motion influenced by both the ambient field as well as the wave field, the resulting evolution process of the ion distribution can be described by both pitch angle diffusion and scattering. However, as stated earlier, when the wave level is suffiently high, the diffusion approximation fails, since the diffusion approximation implies the unperturbed orbit approximation, which is clearly violated for the case of strong wave fields as evidenced by the simple example given above.

Finally, in passing, we remark that in order to model a realistic cometary environment, one may construct the turbulent magnetic and electric wave fields in the following manner

$$\delta\mathbf{B}(t,z) = \sum_{+,-} \sum_{k \geq 0} \delta B_k^{\pm} [\, \hat{\mathbf{x}} \cos \phi_k^{\pm} \pm \hat{\mathbf{y}} \sin \phi_k^{\pm}]$$
$$+ \sum_{+,-} \sum_{k \geq 0} \delta B_k'^{\pm} [\, \hat{\mathbf{x}} \cos \phi_k'^{\pm} \pm \hat{\mathbf{y}} \sin \phi_k'^{\pm}] \tag{26}$$

$$\delta\mathbf{E}(t,z) = \frac{v_A}{c} \sum_{+,-} \sum_{k \geq 0} \delta B_k^{\pm} [\, \pm \hat{\mathbf{x}} \sin \phi_k^{\pm} - \hat{\mathbf{y}} \cos \phi_k^{\pm}]$$
$$- \frac{v_A}{c} \sum_{+,-} \sum_{k \geq 0} \delta B_k'^{\pm} [\, \pm \hat{\mathbf{x}} \sin \phi_k'^{\pm} - \hat{\mathbf{y}} \cos \phi_k'^{\pm}] \tag{27}$$

where the quantities without the prime are those corresponding to the waves propagating in the parallel direction, while the primed quantities are those that correspond to the waves propagating in the opposite direction. In the above, the random phase for each wave mode is given by

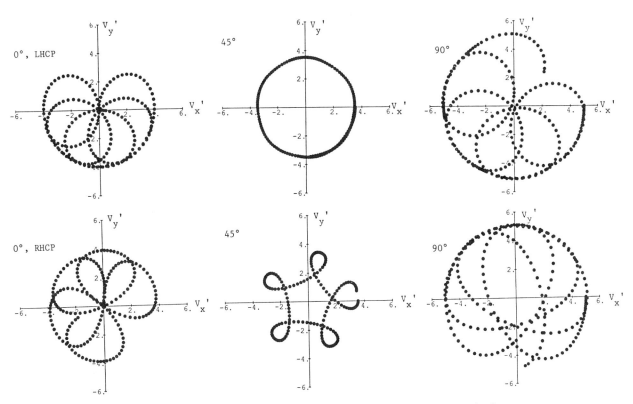

Fig. 4. Projections of the ion orbits shown in Figure 3 unto the plane v_x'-v_y'.

$$\phi_k^{\pm}(t,z) = \varphi_k^{\pm,\text{ random}} + k(v_A t - z) \qquad (28)$$

$$\phi_k'^{\pm}(t,z) = \varphi_k'^{\pm,\text{ random}} + k(v_A t + z)$$

In (26) and (27) suitable models for spectral components δB_k and $\delta B_k'$ can be constructed. For example, in the context of the present section, we may use the following model (see (5) and (11))

$$\langle \delta B \rangle^2 = \sum_{+,-} \sum_{k \geq 0} \left[\left(\delta B_k^{\pm} \right)^2 + \left(\delta B_k'^{\pm} \right)^2 \right]$$

$$= \delta B^2 \sum_{+,-} \sum_{k \geq 0} \left(N_k^{\pm} + N_k'^{\pm} \right) \qquad (29)$$

Here, the spectral function is normalized as follows: $\sum_{\pm} \sum_{k \geq 0}$ $(N_k + N_k') = 1$, where the superscript ($\pm$) is omitted. With the turbulence modelled by (26) - (29) the particle equations (18) and (19) can be used in the test particle simulations as was done by Price and Wu [1987], Cravens [1989], and Terasawa [1989].

The preceding discussion aims to clarify the physical concepts concerning the two processes. It is important to point out that in general both aspects of pitch angle diffusion and pitch angle scattering can exist simultaneously. Nevertheless, in a weakly turbulent plasma, pitch angle diffusion approximation is expected to be valid, and the resultant quasi-linear mechanism plays a dominant role. Whereas in a strongly turbulent plasma, diffusion approximation breaks down and the pickup process is mainly determined by the pitch angle scattering mechanism.

4. Quasi-Linear Theory for Weak Turbulence Regime

In the far upstream regions of a comet, the level of intrinsic turbulence is usually moderate or low. Thus, in these regions, quasi-linear theory of pitch angle diffusion is anticipated to be valid. In the literature, discussions along this line have been given by a number of authors [Sagdeev et al., 1986a,b; Isenberg, 1986, 1987a,b; Lee and Ip, 1987; Cravens, 1989; Gaffey and Wu, 1989; Ziebell et al., 1990; Yoon and Ziebell, 1990; Ziebell and Yoon, 1990; Yoon et al., 1990]. A discussion of the ion pickup process in moderately turbulent plasmas as well as conclusions obtained in earlier investigations was given by Gaffey et al. [1988]. In the following, we emphasize those issues which have been addressed more recently.

Gaffey and Wu [1989] made the first attempt to study the time-asymptotic distribution of the newborn ions when these ions are created continuously by ionization. The problem is studied with several simplifying approximations. One of them is that the quasi-linear diffusion coefficient which appears in the kinetic equation is assumed to be independent of pitch angle. In fact, such an approximation is equivalent to the model that the spectral density of the fluctuating field is assumed to have a spectral index $\alpha = 1$, with an infinite spectral range [Ziebell et al., 1990].

Quasi-linear pitch angle diffusion is a process mainly due to resonant wave-particle interactions. As it will be discussed shortly, as the pitch angle approaches $\pi/2$, or μ approaches 0, the resonance condition cannot be satisfied [Galeev, 1986]. Thus, under general conditions, mathematically $\mu = 0$ is a barrier where pitch angle diffusion is forbidden. This difficulty is first addressed by Galeev [1986] in the context of the cometary physics. Similar discussion can also be found in the works of Cravens [1989] and Ziebell et al. [1990]. These authors suggest that the sharp resonance condition, which appears in the kinetic equation, may be spurious in real situation and, as a result, diffusion of ions across the barrier $\mu = 0$ can occur. For example, a numerical model is suggested in the theoretical discussion by Cravens [1989] who showed that diffusion flux at $\mu = 0$ can be finite. On the other hand, Ziebell et al. [1990] considers the effect of resonance broadening due to weak turbulence.

The most recent discussion of Giotto measurements by Neugebauer et al. [1989] shows that in the upstream regions of the comet Halley, pickup ions consistently possess partial shell distributions, rather than complete spherical shell distributions. The observed partial shell distributions seem to indicate a time-asymptotic characteristics of the ions, rather than a transient phenomenon. In an attempt to address this issue, based on the quasi-linear theory as well as the hybrid-code simulation, Ziebell et al. [1990] found that the formation of partial shell distributions can indeed be formed in a quasi-time asymptotic state. Inspite of these results, we agree that, in order to settle this issue conclusively, further investigations may be called for.

4.1. Pitch Angle Diffusion in a Steady-State Turbulence

In the following, let us consider the case in which hydromagnetic waves predominantly propagates in one direction, say, $k > 0$. Thus we assume that the dispersion relation $\omega_k = k v_A$ is valid. We are particularly interested in the case when the turbulent spectrum of the waves is time-independent. This means that N_k = constant in time (i.e., the inequality (8) holds). Since we are considering waves propagating in the direction $k > 0$ only, going back to (15), we may drop terms that involve μ''. That is, we may choose to work in the wave frame denoted by $(u', \mu') = (u, \mu)$ (Henceforth, we drop the primes in u and μ). Notice also that, according to the diffusion equation (15), in this case particles undergo only pitch angle diffusion in the wave frame. To proceed with the discussion, let us introduce the normalized wavenumber

$$\kappa = k v_A / \Omega_i \qquad (30)$$

Then one may represent the spectral wave energy distribution in terms of the new variable (30) as follows: $N_k dk = N_\kappa d\kappa$.

Generally, the waves may be composed of two components; right-hand circular component (designated with the + sign), and the left-hand circular component (designated with the − sign). When both components are present, the overall polarization will be that of an elliptical polarization. However, for simplicity, we shall consider two limiting cases only; one is the linearly polarized waves, with both (+) and (−) modes present with equal wave amplitudes, and the other case is when only one type of circular polarization exists, say the (+) mode component. Let us assume that the turbulence is

distributed in the wavenumber space according to a power law. Then we may model the wave spectra as follows: for the linearly polarized turbulence spectrum we may write

$$N_\kappa^{(+)} = N_\kappa^{(-)} = N \kappa^{-\alpha} \qquad \kappa_{\min} < \kappa < \kappa_{\max}$$

$$N = \frac{(\alpha - 1) \kappa_{\max}^{\alpha - 1} \kappa_{\min}^{\alpha - 1}}{2(\kappa_{\max}^{\alpha - 1} - \kappa_{\min}^{\alpha - 1})} \qquad (31)$$

while for the circularly polarized waves, we may have

$$N_\kappa^{(+)} = N \kappa^{-\alpha} \qquad \kappa_{\min} < \kappa < \kappa_{\max}$$

$$N_\kappa^{(-)} = 0 \qquad (32)$$

$$N = \frac{(\alpha - 1) \kappa_{\max}^{\alpha - 1} \kappa_{\min}^{\alpha - 1}}{(\kappa_{\max}^{\alpha - 1} - \kappa_{\min}^{\alpha - 1})}$$

In (31) and (32) $\kappa_{\min}$ is introduced so that the spectral wave energy density can be normalized to unity, $\sum_\pm \int d\kappa N_\kappa = 1$, and $\kappa_{\max}$ is introduced because the dispersion relation $\omega_k = k v_A$ is valid only for ω_k sufficiently less than Ω_i. In order to determine these quantities we proceed as follows: we first note from the wave-particle resonance condition in the wave frame $k v_A / \Omega_i = \pm 1/u\mu$ (see the argument of the delta function which appears in the definition of the diffusion coefficient in (15)), that in order to ensure wave-particle interaction for all particles involved, one needs to choose the minimum value of the wavenumber such that $\kappa_{\min} < 1/u$. Secondly, one may readily see that for k greater than $k_{\max} = \Omega_i / v_A$, the dispersion relation $\omega_k = k v_A$ cannot be applied. From these considerations, we make the following choices for $\kappa_{\min}$ and $\kappa_{\max}$

$$\kappa_{\min} = 1/u = v_A/v \qquad \kappa_{\max} = 1 \qquad (33)$$

where we have used the definition $u = v/v_A$.

With the spectra given by (31) and (32), one may reduce the quasi-linear diffusion equation (15) to an equation that describes only the pitch angle diffusion in the velocity frame moving with the waves. The result is the following

$$\frac{\partial f_i}{\partial \tau} = \frac{\partial}{\partial \mu}\left[D(1 - \mu^2) \frac{\partial f_i}{\partial \mu} \right]$$

$$\qquad (34)$$

$$D = N u^{\alpha - 1} |\mu|^{\alpha - 1} \Theta(\mu^2 - u^{-2}) \quad \text{linear polarization}$$

$$D = N u^{\alpha - 1} |\mu|^{\alpha - 1} \Theta(\mu - u^{-1}) \quad \text{circular polarization}$$

where we have dropped the primes (compare (34) with (15)). Notice that as a result of the cutoff at $\kappa = \kappa_{\max}$ there is a resonance gap in the μ-space defined by $\mu^2 < 1/u^2$, for the case of linearly polarized waves. It is interesting to note that the effect of cutoff at the low end of spectrum, namely $\kappa = \kappa_{\min}$ enters only in the normalization constant N. In other words, $\kappa_{\min}$ does not affect the wave-particle interaction. For the case of circular polarization, the waves cannot interact with particles possessing $\mu < -1/u$, since the wave-particle resonance condition prohibits particles with negative μ to interact with the waves with positive κ. Moreover, the cutoff at κ_{max} further prohibits resonance for the ions with $-1/u < \mu < 1/u$.

In the diffusion equation (34) one may note that the velocity variable u appears only as a parameter. Therefore, we may introduce a normalized "temporal" variable

$$T \equiv N\, u^{\alpha-1}\, \tau \qquad (35)$$

and rewrite the diffusion equation (34) as

$$\frac{\partial f_i}{\partial T} = \frac{\partial}{\partial \mu}\left[D\,(1-\mu^2)\,\frac{\partial f_i}{\partial \mu}\right]$$

$$D = |\mu|^{\alpha-1}\,\Theta(\mu^2 - u^{-2}) \quad \text{(LP)} \qquad (36)$$

$$D = |\mu|^{\alpha-1}\Theta(\mu - u^{-1}) \quad \text{(CP)}$$

where (LP) and (CP) stand for the linear and the circular polarizations, respectively. The solution of this equation is shown in Figure 5 which is taken from the paper by Ziebell et al. [1990].

In Figure 5, the plot of the distribution $f_i(\mu,T)$ is shown versus μ, for time intervals that correspond to $T = 0, 0.1, 0.2, 0.5,$ and 10. In the figure, the initial position of the ion distribution in μ-space is located at $\mu_0 = 0.5$, with characteristic spread $\Delta\mu = 0.05$. The initial distribution is given by a Gaussian distribution; $f_i(\mu,0) = \text{const}\ \exp\,[-\,(\mu -$

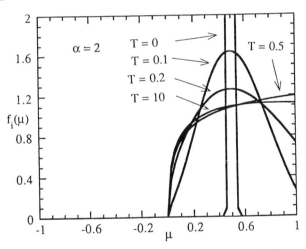

Fig. 5. Pitch angle distribution $f_i(\mu)$ versus μ for several different time intervals. The spectral index α is set equal to 2, and the initial distribution is located at $\mu = 0.5$ with the spread $\Delta\mu = 0.05$. Time-asymptotic distribution is a half-shell distribution in velocity space.

$\mu_0)^2/\Delta\mu^2]$. The spectral index α is set equal to 2. Because of the narrow profile of the initial distribution, there are virtually no ions in the negative μ-space initially. In the figure, we have assumed that the characteristic velocity u is sufficiently large so that the resonance gap effectively becomes a point $\mu = 0$ (this is equivalent to taking $\kappa_{max} \to \infty$). The results shown in the figure can be applied to both the linear and circular polarizations. That is, the result for both polarization is identical. Observe from the figure that a characteristic half-shell distribution is formed by the time the system has evolved to $T = 0.5$. Further evolution beyond $T = 0.5$ shows very little

change, and one may conclude that at $T = 10$, the system has reached a virtual time-asymptotic state.

According to Figure 5, the time-asymptotic solutions of (36) represents basically a half-shell velocity distribution in the wave frame. The physical reason for the formation of half-shell is that the resonance condition prohibits the ions to diffuse into the negative μ-space, regardless of the wave polarization.

Actually, however, the motion of an individual ion may be affected by the fluctuating wave fields (that is, they are pitch angle scattered – see section 3), although in the standard quasi-linear theory, we assume the ion motions to be unaffected by the wave fields. The effect of waves on the ion motion may be relatively unimportant in those regions in velocity space where the resonant wave-particle interaction can take place. However, in the region where the resonant interaction is not allowed (i.e., in the resonance gap), we expect the effect of waves on the ion motion to play an important role. Such an effect can broaden the condition for wave-particle resonance, hence it is known as the resonance broadening effect. In the quasi-linear theory, one may include the resonance broadening mechanism [Dupree, 1966; Weinstock, 1969; Davidson, 1972] by replacing the delta function inside the resonant diffusion coefficient by a more general "resonance function". That is, in the standard quasi-linear theory, the typical diffusion coefficient is defined by

$$D = \frac{2\pi^2 e^2}{m_i^2}\int dk\,\varepsilon_k\,\delta(\omega_k \pm \Omega_i - k v_\parallel) \qquad (37)$$

In the resonance-broadened (modified, or renormalized) quasi-linear theory, one may write the diffusion coefficient as

$$D = \frac{2\pi e^2}{m_i^2}\,\text{Re}\int dk\,\varepsilon_k \int_0^\infty d\tau$$

$$\times \exp\left[i\,(\omega_k \pm \Omega_i - k v_\parallel)\tau - (k^2 D/3)\,\tau^3\right] \qquad (38)$$

The above equation is now an integral equation for D, and analytical solution of this equation is generally not tractable. However, for the present purpose, since we are interested in the weakly turbulent situation, we may assume $D \ll 1$, and obtain an approximate solution.

In the paper by Ziebell et al. [1990] an approximate solution of (38) is carried out, and the quasi-linear diffusion equation is modified to incorporate the effect of waves on the ion motions, and it was found that the modified pitch angle diffusion equation can be effectively written in terms of the following diffusion coefficients, which now retains the effect of the resonance broadening

$$D = |\mu|\,\Theta(\mu^2 - u^{-2}) + \sigma \quad \text{(LP)}$$

$$D = \mu\,\Theta(\mu - u^{-1}) + \sigma \quad \text{(CP)} \qquad (39)$$

$$\sigma = \frac{1}{2\sqrt{3}\,\pi u}\left(\frac{\delta B}{B_0}\right)$$

In the above, the term σ represents the resonance broadening effect. We note that δB, which is the strength of the fluctuating field, explicitly enters in the definition of σ. Since δB is

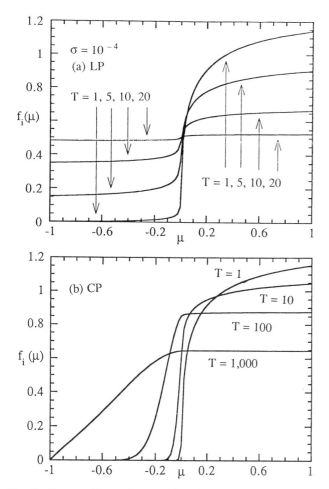

formation ($T = 0.5$; see Fig. 5) and the complete shell formation ($T = 20$; this figure). On the other hand, for the circularly polarized waves (b), the only mechanism that allows the ions to diffuse into the negative μ-space is the resonance broadening mechanism, which is much less effective than the resonant wave-particle interaction. Thus, the characteristic partial shell structure remains for very long time, up to $T = 100$ as shown in the figure. Even for $T = 1000$, most of the ions are concentrated in the positive μ-space.

Evidently, the degree of broadening of wave-particle resonance is determined by the level of the fluctuating field $\delta B/B_0$: the higher the level of waves, the more prominent the effect of resonance broadening. Of course, when the turbulence level is exceedingly strong and pitch angle scattering prevails over diffusion, even the modified quasi-linear theory becomes invalid.

We see that the modified quasi-linear theory incorporates some aspects of the pitch angle scattering by allowing the waves to modify the wave-particle resonance condition. In this sense, the modified quasi-linear theory contains the effects of both pitch angle diffusion and weak pitch angle scattering.

4.2. Pitch Angle and Velocity Diffusion in a Steady-State Turbulence

We now discuss the case of quasi-stationary state turbulence with both directions of propagation, parallel and anti-parallel to the ambient magnetic field. In this case, as a result of pitch angle diffusion in the two reference frames each moving at the speed $v_{\parallel} = v_A$ and $v_{\parallel} = -v_A$, respectively, as discussed in section 3 (see also, Figure 1), we expect a net diffusion in velocity space as well as in pitch angle space. In this case, as the particles pitch angle diffuse in both wave frames, they will suffer a random kick which results in a net diffusion in velocity space. The result can be viewed as a net acceleration of ions by the waves. The initial ring-beam distribution of newborn ions, when they undergo a net acceleration as well as pitch angle diffusion, will result in a broadened shell-like distribution in the rest frame of ions (as compared to the thin-shell distribution centered at $v_{\parallel} = v_A$, as discussed in the previous subsection). Consequently, this type of process is known as the second order Fermi acceleration process in the literature [Sagdeev et al. 1986a, b; Ip and Axford, 1986; Gribov et al., 1986; Gombosi, 1987; Isenberg, 1987a, b; Gaffey et al., 1988; Coates et al., 1989; Terasawa, 1989].

At this point we note that although equation (15) provides a convenient interpretation of the mechanism of pitch angle diffusion in both wave frames, for practical purposes it is not very useful, since we have introduced two sets of variables (v', μ') and (v'', μ'') from one set of variables ($v_{\perp}, v_{\parallel}$). Hence, we go back to (12) and, for convenience, rewrite this equation in spherical coordinates of the ion rest (or, solar wind) frame (u, μ), where $u = v/v_A$. The resulting equation is given by [Yoon and Ziebell, 1990; Ziebell and Yoon, 1990]

$$\partial f_i / \partial \tau = \tag{40}$$
$$\frac{1}{u}\left[\frac{\partial}{\partial u} + \frac{\partial}{\partial \mu}\left(1 - \frac{\mu}{u}\right)\right] D\,(1 - \mu^2)\,u\left[\frac{\partial}{\partial u} + \left(1 - \frac{\mu}{u}\right)\frac{\partial}{\partial \mu}\right]f_i$$
$$+ \frac{1}{u}\left[\frac{\partial}{\partial u} - \frac{\partial}{\partial \mu}\left(1 + \frac{\mu}{u}\right)\right] D'\,(1 - \mu^2)\,u\left[\frac{\partial}{\partial u} - \left(1 + \frac{\mu}{u}\right)\frac{\partial}{\partial \mu}\right]f_i$$

Fig. 6. Same as Figure 6 except that the resonance broadening effect is included. The resonance broadening parameter σ is set equal to 0.0001. Case (a) corresponds to the linear polarization, and (b) corresponds to the circular polarization. It can be seen that for the case of linear polarization, the effect of resonance broadening produces a complete shell distribution in the quasi-time asymptotic state, whereas in the case of circular polarization, only a partial shell emerges in the long-time limit.

assumed to be a small quantity, we expect that σ is a small quantity.

Figure 6 presents numerical results taken from the article by Ziebell et al. [1990]. In this figure, as in Figure 5, $\alpha = 2$, and the initial distribution is a Gaussian centered at $\mu = 0.5$ with spread $\Delta\mu = 0.05$. The parameter σ is taken to be 10^{-4}. It is seen that due to resonance broadening, ions are able to diffuse across the gap $\mu = 0$. However, the result depends upon the polarization of the waves. The case of linearly polarized waves is shown in the upper panel (a), whereas the case of circularly polarized waves is displayed in the lower panel (b). For the case of linear polarization (a), the plot of the distribution function is shown for $T = 1, 5, 10$, and 20. Notice that for $T = 20$, an almost complete spherical shell distribution is formed. The physical reason is that once the ions cross the resonance gap via resonance broadening mechanism, they may resonate with the left-hand mode very effectively. One should note, however, the difference between the time scales of half-shell

where

$$D = \kappa_r \left[N^{(+)}_{\kappa_r} \Theta(u\mu - 2) + N^{(-)}_{\kappa_r} \Theta(-u\mu) \right]$$

$$D' = \kappa_r' \left[N'^{(+)}_{\kappa_r'} \Theta(2 - u\mu) + N'^{(-)}_{\kappa_r'} \Theta(u\mu) \right] \qquad (41)$$

In the above, κ_r and κ_r' denote the resonant wavenumbers for the positive and negative propagating waves, respectively;

$$\kappa_r = 1/|1 - u\mu|, \qquad \kappa_r' = 1/|1 + u\mu| \qquad (42)$$

Here, quantities without the primes denote those associated with the waves propagating in $k > 0$ direction, while quantities with ($'$) denote those corresponding to waves propagating in the opposite direction, $k < 0$. The superscripts (+) and (−) are used to denote the polarization.

We note that the above equation can be also written as

$$\frac{\partial f_i}{\partial \tau} = \frac{1}{u^2} \frac{\partial}{\partial u} \left[u^2 \left(D_{uu} \frac{\partial f_i}{\partial u} - \frac{D_{u\mu}}{u} \frac{\partial f_i}{\partial \mu} \right) \right]$$

$$- \frac{1}{u} \frac{\partial}{\partial \mu} \left(D_{\mu u} \frac{\partial f_i}{\partial u} - \frac{D_{\mu\mu}}{u} \frac{\partial f_i}{\partial \mu} \right) \qquad (43)$$

where

$$D_{uu} = (1 - \mu^2)(D + D')$$

$$D_{u\mu} = D_{\mu u} = (1 - \mu^2) \left[D(\mu - u) + D'(\mu + u) \right] \qquad (44)$$

$$D_{\mu\mu} = (1 - \mu^2) \left[D(\mu - u)^2 + D'(\mu + u)^2 \right]$$

In the above, D and D' are defined in (41). The two representations (40) and (43) are identical. The expression (43) is useful in the interpretation of the physical meaning of each term on the right-hand side. Particularly, the first term represents the energy (or the velocity) diffusion, while the last term represents the pitch angle diffusion in the rest frame of the ions. The middle two terms represent the cross diffusion terms.

In equations (40) and (43), we have already imposed the cutoffs $\kappa_{min} = 1/u$ and $\kappa_{max} = 1$ in the wave spectra. As a result of the cutoff at $\kappa = \kappa_{max}$, unlike Figure 1, the point $\mu = 0$ becomes a resonance gap (this is evident from the four step functions which appears in the definition of the diffusion coefficients). That is, the particles cannot cross the point $\mu = 0$, if one relied solely on the resonant wave-particle interaction. However, on the basis of the same physical argument as given in the previous subsection, we expect that the gap at $\mu = 0$ is spurious which can be overcome by the resonance broadening mechanism. In order to incorporate the effect of waves on the ion motions, which is expected to broaden the wave-particle resonance condition, we add the term

$$\sigma = \frac{1}{2\sqrt{3}\pi u} \left(\frac{\delta B}{B_0} \right) \qquad (45)$$

on the right-hand sides of the diffusion coefficients defined in (41) or (44).

At this point, we note that in the study of newborn ion acceleration by counter-propagating hydromagnetic waves (i.e., the second order Fermi process), it is often assumed that the pitch angle diffusion process is much faster than the energy

diffusion process, so that the initial ring-beam distribution quickly evolves into a shell-like distribution. Hence, an isotropic shell distribution is often used as an initial condition, instead of a ring-beam distribution. Then, it is customary to assume that the isotropy will remain throughout the rest of the diffusion process, so that an average over the variable μ is carried out in the diffusion equation (43). The resulting equation represents a pure velocity diffusion which involves only u. However, it should be pointed out that such an approximation scheme might tend to significantly exaggerate the actual acceleration rate. The reason is that, by taking the average of (43) over μ, the resonance gap at $\mu = 0$ is artificially removed. In actual situation, because of the resonance gap at $\mu = 0$, the spreading of the distribution in u-space will be considerably hindered, as it will be shown below.

Figure 7, which is taken from Ziebell and Yoon [1990], shows the time evolution of an initial cold ring-beam distribution due to pitch angle and velocity diffusion under the influence of intrinsic turbulence which is considered to be quasi-stationary. In the figure, the spectral wave energy density is assumed to be a power law of the type given in (31). In particular, all four components $N_\kappa^{(+)}$, $N_\kappa^{(-)}$, $N'_\kappa^{(+)}$, and $N'_\kappa^{(-)}$ are assumed to have same amplitudes and same spectral index α which is set equal to 2. Therefore, the turbulence under consideration is comprised of counter-propagating linearly polarized hydromagnetic waves. The time variable is normalized with respect to the normalization constant associated with waves, $N = \{ \sum_\kappa (N_\kappa^{(+)} + N_\kappa^{(-)}, + N'_\kappa^{(+)}, + N'_\kappa^{(-)}) \}^{-1}$; $T = N\tau$. The parameter σ is taken as in Figure 6, namely, $\sigma = 10^{-4}$. The initial distribution is assumed to be a Gaussian in both u and μ, centered at $u = 5$ and $\mu = 0.5$, with the spreads $\Delta u = \Delta \mu = 0.05$. The numerical procedure as well as detailed discussions are explained more fully in the works by Ziebell and Yoon [1990] and Yoon and Ziebell [1990]. Schematically, the initial cold ring-beam distribution forms a characteristic half-shell distribution as shown in panels (a), (b), and (c), in a time scale roughly corresponding to the time scale for half-shell formation as discussed in Figures 5 and 6. Then, a more or less complete shell distribution is formed, as shown in panels (d), (e), and (f), in a much slower time scale which is comparable to the time scale for complete shell formation in the case of Figure 6(a). Of particular interest is the panel (d) and (e). As a result of the resonance gap at $\mu = 0$ (or, equivalently, $u_\| = v_\|/v_A = 0$), the complete shell distribution possesses an appreciable degree of anisotropy at $\mu = 0$ which subsequently smears out at later times (see panel f). The point is that the resonance gap hinders both pitch angle and velocity diffusion process, and if one does not consider this carefully, one may tend to obtain results which may significantly exaggerate actual diffusion rates.

4.3. Self-Consistent Pitch Angle Diffusion

Let us now consider the case when the inequality (9) holds. In this case, a self-consistent approach is necessary and one has to solve the coupled set of equations (6) and (7). For the sake of simplicity, let us assume that only low-frequency hydromagnetic waves are excited. In the preceding case the wave fields are assumed to be given and independent of time, whereas now the waves are self-consistently generated by the newborn ions. Of course, in general, depending on the initial ion distribution, the excited wave modes may or may not

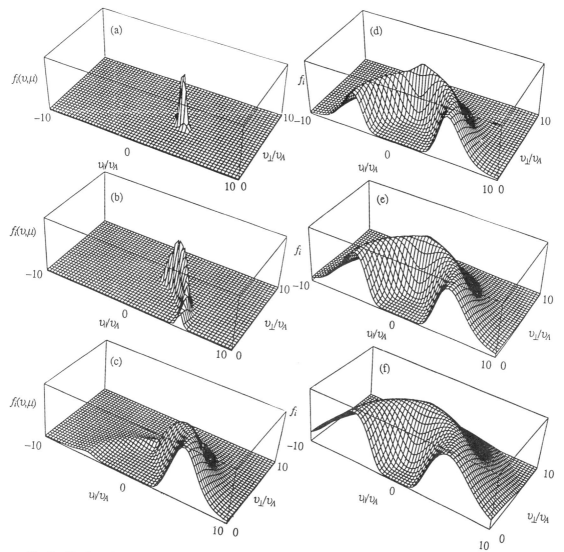

Fig. 7. Newborn ion distribution $f_i(u,\mu,t)$ versus $u_\perp = v_\perp/v_A$ and $u_\parallel = u_\parallel = v_\parallel/v_A$ in three-dimensional representation. The characteristic time scale for half-shell formation (panels (a) - (c)) is comparable to the time scale for half-shell formation in Figure 6, whereas the time scale associated with the complete shell formation and the ensuing velocity diffusion (panels (d) - (f)) is much larger (for details see Ziebell and Yoon [1990]).

satisfy the hydromagnetic dispersion relation. It is known from the linear theory that for a quasi-parallel pickup process, that is, when the characteristic velocity component associated with the newborn ions along the ambient magnetic field (i.e., beam velocity) is greater than the component perpendicular to the ambient field (i.e., ring velocity), the excited wave modes are predominantly in the hydromagnetic regime, and that the excited waves propagate primarily along the direction parallel to the beam velocity. Thus, in the following analysis we restrict ourselves to the case when the beam velocity exceeds the ring velocity:

$$\langle v_\parallel \rangle \equiv \int d^3\mathbf{v}\; v_\parallel f_i(v_\perp,v_\parallel,0) > \langle v_\perp \rangle$$

$$\equiv \int d^3\mathbf{v}\; v_\perp f_i(v_\perp,v_\parallel,0)$$

(46)

When (46) is satisfied, we expect that the wave excitation will occur in the low-frequency hydromagnetic regime, and that the excited waves travel in the direction parallel to the beam velocity, say $k > 0$. Moreover, to further simplify the matter, we assume that the real part of the dispersion relation is mainly determined by the background plasma so that it is time-independent. Then we may write

$$\omega_k = k v_A, \quad (0 < k < k_{max}) \qquad (47)$$

where $k_{max} = \Omega_i/v_A$ was introduced before.

The self-consistent quasi-linear diffusion equation in this case can be expressed in the wave frame [Yoon et al., 1990] as follows

$$\frac{\partial f_i(u,\mu,\tau)}{\partial \tau} = \frac{\partial}{\partial \mu}\left[D(u,\mu,\tau)\,(1-\mu^2)\,\frac{\partial f_i(u,\mu,\tau)}{\partial \mu} \right] \qquad (48)$$

$$D(u,\mu,\tau) = \text{Re}\,\frac{i}{\pi}\sum_{+,-}\int_0^1 d\kappa \frac{N_\kappa(\tau)}{\pm 1 - \kappa u\mu + i\Gamma_\kappa(\tau)}$$

Here, we have made use of the dispersion relation $\omega_k = k v_A$ and the definition (30) to write $\kappa = k v_A/\Omega_i$. Furthermore, the normalized growth rate

$$\Gamma_\kappa(\tau) = \gamma_k(\tau)/\Omega_i \qquad (49)$$

$$= \frac{n_i}{n_0}\frac{\pi^2}{2\kappa}\int_{1/\kappa}^\infty du\,u^2\left[(1-\mu^2)\frac{\partial f_i(u,\mu,\tau)}{\partial \mu}\right]_{\mu = \pm 1/\kappa u}$$

is used in (48). In (48), if we let $\Gamma_\kappa = 0$, then we recover (34). In writing the second line of (49), we have assumed that the growth rate is determined from the newborn ion term in (7), and that the instability is of the kinetic type.

The wave kinetic equation (6) can be formally solved and it can be written as

$$N_\kappa(\tau) = N_\kappa(0)\exp\left[2\Omega_i t_0\int_0^\tau d\tau'\,\Gamma_\kappa(\tau')\right] \qquad (50)$$

The diffusion coefficient can be written as a combination of two parts: resonant and non-resonant diffusion coefficients. The resonant diffusion coefficient can be written as

$$D^r = \kappa_r\sum_{+,-} N_{\kappa_r}(\tau)\,\Theta(\pm\mu - 1/u) \qquad (51)$$

where

$$\kappa_r = 1/u|\mu| \qquad (52)$$

is the resonant wavenumber in the wave frame, while the non-resonant diffusion coefficient can be approximated by the following expression

$$D^{nr} = \frac{1}{\pi}\Theta(u^{-2} - \mu^2)\sum_{+,-}\int_0^1 d\kappa\,\Gamma_\kappa(\tau)\,N_\kappa(\tau) \qquad (53)$$

We first discuss the resonant wave-particle interaction. The resonant interaction can take place in the range of pitch angles that corresponds to $1/u < \mu < 1$ for the (+) mode, and $-1 < \mu < -1/u$ for the (−) mode, as can be seen from the two conditions $\Theta(\pm\mu - 1/u)$ in (51). Therefore the region

corresponding to $-1/u < \mu < 1/u$ constitutes the resonance gap in which the resonance cannot occur. Let us assume that initially the distribution peaks at $u = u_0 \gg 1$ and $\mu = \mu_0 > 0$. Then, according to (49), for the range $0 < \mu < \mu_0$, the distribution has a positive slope, $\partial f_i/\partial\mu > 0$, and for $\mu_0 < \mu < 1$, $\partial f_i/\partial\mu < 0$. For a given value of μ, the resonance occurs for a wavenumber given by $\kappa = 1/u\mu$ (for the (+) mode). This means that for $0 < \mu < \mu_0$, for which $\partial f_i/\partial\mu > 0$, the waves will grow in the range of wavenumbers corresponding to $1/u\mu_0 < \kappa < \infty$. But since we must impose the cutoff at large κ, namely $\kappa_{max} = 1$, the effective range of wave growth would correspond to $1/u\mu_0 < \kappa < 1$. Using a similar argument, one may conclude that initial waves in the range $1/u < \kappa < 1/u\mu_0$ will begin to damp. To summarize, the wave-particle resonance occurs in the range $1/u < \mu < 1$. This region is subdivided into two regions; the region $1/u < \mu < \mu_0$ contributes to the wave growth in the wavenumber space $1/u\mu_0 < k < 1$, and the region $\mu_0 < \mu < 1$ contributes to the wave damping in the range $1/u < \kappa < 1/u\mu_0$ (see Figure 8).

Subsequent quasi-linear evolution of the resonant portion of the ion distribution corresponding to the region $1/u < \mu < 1$ is described by (48). It is expected to be very similar to the case of well-known electrostatic "bump-in-tail" instability [Davidson, 1972], only in this case, we are working in the μ-space. That is, since the region $1/u < \mu < \mu_0$ corresponds to the growing portion of the wave spectrum, the waves will enhance the diffusion process and result in the formation of the "plateau" distribution in μ-space. Therefore, in the time-asymptotic limit we expect the following

$$D^r(u,\mu,\tau) \to \text{finite}, \qquad \partial f_i(u,\mu,\tau)/\partial\mu \to 0 \qquad (54)$$

$$\text{for } 1/u < \mu < \mu_0, \quad \text{as } \tau \to \infty$$

The plateau formation in the resonant portion of the μ-space results in the decrease of the growth rate until it reaches zero at $\tau \to \infty$. At this point, the system would reach an asymptotic state, and further evolution will not take place. Careful consideration of (49) reveals that if the initial distribution is concentrated in $\mu > 0$, only the (+) mode undergoes amplification, while the (−) mode remains virtually unchanged, until a significant slope of the distribution is attained in the negative μ-space at a much later time when enough particles have crossed the resonance gap. However, it is expected that the (−) mode will be relatively unimportant when compared to the (+) mode, since by the time sufficient amount of ions have diffused into the negative μ-space so as to cause the (−) waves to grow, the (+) mode would have reached a high saturated level.

For the resonant μ-space that corresponds to $\mu_0 < \mu < 1$, it is expected that the initial shape of the distribution will not show noticeable change, since the only appreciable diffusion may occur at very early time as a result of the initial wave level, which subsequently will damp out rapidly.

In the resonance gap $-1/u < \mu < 1/u$, the resonant diffusion cannot take place, but the non-resonant diffusion becomes very important. Particularly, the non-resonant diffusion provides the channel through which the ions may cross the resonance gap and diffuse into the negative μ-space. The non-resonant diffusion coefficient (53) may be rewritten as

$$D^{nr} = \frac{1}{2}\Theta(1/u^2 - \mu^2)\frac{\partial}{\partial\tau}\int d\kappa\sum_{+,-} N_\kappa(\tau) \qquad (55)$$

where $\sum \int d\kappa\, N_\kappa(\tau)$ is the total wave energy. Equation (55) shows that the non-resonant diffusion depends on the time rate of increase in the total wave energy. It is expected that the rate of non-resonant diffusion assumes maximum value initially, since the slope of the distribution in the resonant μ-region is the largest initially, implying that the initial growth rate is the largest. However, if the initial distribution is concentrated near $\mu \sim \mu_0 >> 0$, then although the initial non-resonant diffusion rate may be large, few actual particles may cross the gap and appear in the negative μ-region, since initially there are virtually no ions in the diffusion gap. Subsequent time evolution tends to decrease the slope of the distribution in the resonant region, thereby reducing the non-resonant diffusion rate. Thus, in the time-asymptotic state, the non-resonant diffusion rate will approach zero and the distribution will remain "frozen":

$$D^{nr}(u,\mu,\tau) \to 0, \quad \partial f_i(u,\mu,\tau)/\partial\tau \to 0 \tag{56}$$

$$\text{for } -1/u < \mu < 1/u, \quad \text{as } \tau \to \infty$$

It is important to point out that in the time-asymptotic state, a substantial slope in the distribution may remain in the non-resonant μ-space (see Figure 8). However, this feature is not expected to contribute to the wave growth since the slope is out of the resonance region. The overall feature will be that of a partial plateau distribution as schematically shown in Figure 8.

Figure 8 summarizes the theoretical scenario presented in this section. The figure shows the initial distribution in μ-space as well as the spectrum of unstable (+) mode. The initial distribution has a positive slope for $1/u < \mu < \mu_0$, and a negative slope for $\mu_0 < \mu < 1$. The resonance gap $-1/u < \mu < 1/u$ is also shown. The figure also depicts amplification of the waves in the range $1/u\mu_0 < \kappa < 1$, and the damping in the region $\kappa < 1/u\mu_0$, and it schematically shows the plateau formation in the range $1/u < \mu < \mu_0$ for large time.

One final remark is in order. The final distribution depicted in Figure 8 may be stable with respect to the hydromagnetic waves with the growth rate given by (49), since the slope of the distribution is outside the range of wave-particle resonance. However, this type of distribution may excite other waves, most notably the ion cyclotron waves. The ion cyclotron waves might be excited in both directions, parallel and anti-parallel to the ambient magnetic field. Such waves not only will cause further pitch angle diffusion, but may also cause velocity diffusion (or acceleration). Thus, one should practice caution when interpreting the final distribution depicted in Figure 8 as the "asymptotic" state. It is asymptotic only within the context of the discussion presented here.

5. Wave-Particle Interaction in a Highly Turbulent Plasma

In the presence of strong magnetic fluctuations, the nature of wave-particle interactions is very different from that discussed in section 4. Effects of the electric and magnetic fields associated with the waves on the motion of the ions become significant. It is therefore apparent that in this case pitch angle scattering may prevail over diffusion and the quasi-linear theory is no longer appropriate.

From the data acquired with ICE and other spacecraft, we have learned that in regions very close to a comet where turbulence level is usually high, ion pickup seems to occur in a very rapid manner, at least much faster than what we understood from the early theoretical research [Winske et al., 1984, 1985]. Motivated by the ICE observations reported by Gloeckler et al. [1986], we launched a series of theoretical studies.

Ion pickup process in a strongly turbulent solar wind was first addressed by Wu et al. [1986]. These authors investigated the problem by means of hybrid-code simulations. Using a code which was previously used by Winske et al. [1984, 1985], they first initialize the system with a 0.5 percent of water group ions which possess a ring-beam distribution and let it evolve for a time interval. By the time the earlier water group ions are picked up and a moderately strong level of magnetic field fluctuations has been attained, a new population of fresh ions with the same initial distribution is introduced into the system. This approach facilitates the study of the pickup process in the presence of strong turbulence. The major conclusion of the study is that moderately strong turbulence can expedite the pickup process. The pickup time scale may be of the order of several gyro-periods of the newborn ions.

However, there are two difficulties in the above-mentioned simulation study. First, the simulations can only generate moderately strong turbulence and, second, a ring-beam distribution for the newborn ions is used for initial conditions

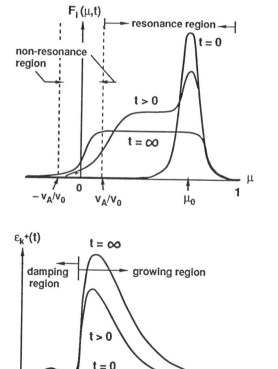

Fig. 8. Schematic representations depicting the self-consistent pitch angle diffusion of the ion distribution (upper panel), and the evolution of the self-consistently generated waves (lower panel). In the time-asymptotic limit, a partial plateau (i.e., a partial shell) distribution is formed.

(because the simulation code requires a gyrotropic distribution of the ions). The second point is troublesome, because, strictly speaking, such an initial distribution is not really consistent with the conclusion. The reason is that when the level of turbulence is strong or moderately strong and the pickup time is of the order of ion gyro-period, as shown in Wu et al. [1986], the ring-beam distribution may never form. The distribution of the newborn ions may evolve directly from an oblique beam which occurs initially in a real situation to a spherical shell at a later time. To resolve these difficulties, Price and Wu [1987] took a "test particle calculation" approach. That is, in this investigation, the motion of the newborn ions is computed in prescribed magnetic fluctuations but the fields are not affected by the newborn ions. The magnetic fields are generated in such a way that they fit closely to those observed at Giacobini-Zinner by ICE. The justification of the test-particle approach is that when the magnetic field fluctuations have an energy density larger than the kinetic energy density of the newborn ions, the influence of the turbulence on the motion of the ions prevail. Using this approach, one can study the strongly turbulent case without any difficulty and, moreover, the initial ion distribution can be arbitrary.

Price and Wu [1987] found that in a strongly turbulent case, an initial ion beam can evolve into a spherical shell distribution within one gyro-period of the newborn ion. Thus for water group ions the pickup time is on the order of 100 seconds. This result is consistent with that estimated from observations. The result obtained by Price and Wu [1987] also confirms that the major conclusions obtained from the test particle calculation are consistent with that discussed in Wu et al. [1986] which, as mentioned in the preceding paragraph, uses hybrid-code simulations but with an initial distribution that is not very realistic. The stricking feature is that, nevertheless, the two results are in agreement with each other.

6. Summary and Conclusions

Our current understanding of the ion pickup by the solar wind may be summarized briefly in the following. In general, the pickup process may be attributed to both pitch angle diffusion and pitch angle scattering. The former is a collective process mainly due to resonant wave-particle interactions while the effects of the wave fields on individual particles are unimportant and the latter is dictated by direct interaction between the wave fields and the particles.

Intuitively it is anticipated that pitch angle diffusion is comparatively more important than pitch angle scattering in a moderately or weakly turbulent plasma. The process may be described by the quasi linear theory in plasma physics. However, to depict the real situation, it is very important to include the effect of resonance broadening which can be particularly important for particles with large pitch angles, since for these particles the sharp resonance condition required in the usual quasi-linear theory is not satisfied. It is also possible that for ions with large pitch angles, pitch angle scattering is more important than diffusion, even when the turbulence level is low. This is an interesting theoretical issue which should be investigated in the future.

Intuitively, we expect that in a moderately turbulent solar wind both pitch angle diffusion and scattering can take place. When the turbulence level becomes progressively higher, the process of pitch angle scattering plays a more prevailing role. In this case the resonance condition which is important in the quasi-linear theory is no longer a serious constraint. Consequently pickup ions can form fairly complete shell distributions.

In a highly turbulent case, direct influence of the electric and magnetic fields on the ion orbits determines the pickup process. In other words, pitch angle scattering should prevail over diffusion. Both test particle calculations and hybrid-code simulations [Price and Wu, 1987; Wu et al. 1986] show that in the presence of strong magnetic fluctuations ($\delta B \sim B_0$), the typical time scale of ion pickup in general can be of the order of an ion gyro-period. Thus, for a newborn ion, by the time they complete one gyration, the ion can be picked up by the solar wind. Such a conclusion is almost independent of the initial pitch angle of the newborn ions.

In the literature, there are discussions which may cover both weakly and strongly turbulent cases. For example, the articles by Gary and his colleagues [Gary and Madland, 1988a; Gary et al., 1986, 1988b, 1989] consider continuous injection of newborn ions in their numerical simulations which generalizes the discussion by Wu et al. [1986].

To summarize we tend to believe that when pitch angle diffusion is dominant, partial shell distributions with pitch angle widths around 90° are consistent with quasi-linear theory (even with resonance broadening effect included); whereas when pitch angle scattering prevails, pickup ions should form more complete spherical shell distributions. These conclusions are consistent with observational results in the literature.

In space physics, many important phenomena are attributed to wave-particle interactions. Among them, the ion pickup process is perhaps most efficient and dramatic. Many years ago, skeptics doubted whether in the absence of Coulomb collisions, newborn ions can ever be picked up by the rapidly moving solar wind. Now both theoretical predictions and observational verifications show that not only is ion pickup possible, but also it can happen on a very short time scale.

Acknowledgments. The present work was supported in part by the National Aeronautics and Space Administration and in part by the National Science Foundation under grant ATM 9014356. Numerical computations were carried out at the San Diego Supercomputer Center. We want to thank many of our colleagues for fruitful and stimulating discussions. We are particularly grateful to Dr. L. F. Ziebell for his enthusiastic collaborations during his recent visit to the University of Maryland.

References

Brinca, A. L., On the coupling of test ions to magnetoplasma flows through turbulence, *J. Geophys. Res*, *89*, 115, 1984.

Brinca, A. L. and B. T. Tsurutani, Unusual characteristics of electromagnetic waves excited by cometary newborn ions with large perpendicular energies, *Astron. Astrophys.*, *187*, 311, 1987.

Brinca, A. L. and B. T. Tsurutani, Survey of low frequency electromagnetic waves stimulated by two coexisting newborn ion species, *J. Geophys. Res.*, *93*, 48, 1988.

Brinca, A. L. and B. T. Tsurutani, On the excitation of cyclotron harmonic waves by newborn heavy ions, *J. Geophys. Res.*, *94*, 5467, 1989.

Brinca, A. L., On the electromagnetic stability of isotropic population, *J. Geophys. Res.*, *95*, 221, 1990.

Brinca, A. L., Cometary linear instabilities: from profusion to perspective, in this monograph, 1991.

Coates, A. J., A. D. Johnstone, B. Wilken, K. Jockers, and K. H. Glassmeier, Velocity space diffusion of pickup ions from the water group at comet Halley, *J. Geophys. Res., 94,* 9983, 1989.

Coates, A. J., Observations of the velocity distribution of pickup ions, in this monograph, 1991.

Cravens, T. E., Test particle calculations of pickup ions in the vicinity of comet Giacobini-Zinner, *Planet. Space Sci., 37,* 1169, 1989.

Davidson, R. C., *Methods in Nonlinear Plasma Theory,* Academic Press, New York, 1972.

Decker, R. B., A. T. Y. Lui, and S. M. Krimigis, Modeling of interactions of artificially released lithium with the Earth's bow shock, *Geophys. Res. Lett., 10,* 525, 1983.

Decker, R. B., L. Vlahos, and A. T. Y. Lui, Predictions of lithium interaction with the Earth's bow shock in the presence of wave activities, *J. Geophys. Res., 89,* 7331, 1984.

Dupree, T. H., A perturbation theory for strong plasma turbulence, *Phys. Fluids, 9,* 1773, 1966.

Fahr, H. J., The extraterrestrial UV-background and the nearby interstellar medium, *Space Sci. Rev., 15,* 483, 1974.

Feldman, W. C., J. J. Lang, and F. Sherb, Interstellar helium in interplanetary space, paper presented at Solar Wind Conference, Univ. of California, and NASA Ames Center, Pacific Grove, Calif., march 21-26, 1971.

Freund, H. P. and C. S. Wu, Stability of a spherical shell distribution of pickup ions, *J. Geophys. Res., 93,* 14277, 1988.

Gaffey, J. D., D. Winske, and C. S. Wu, Time scales for formation and spreading of velocity shells of pickup ions in the solar wind, *J. Geophys. Res., 93,* 5470, 1988.

Gaffey, J. D. and C. S. Wu, Distribution functions of continuously created newborn and pickup ions in outer cometary exospheres, *J. Geophys. Res., 94,* 8685, 1989.

Galeev, A. A., Theory and observations of solar wind/cometary plasma interaction processes, in Proceedings of 20th ESLAB Symposium on the Exploration of Halley's Comet, edited by B. Battrick, E. J. Rolfe, and R. Reinhard, *Eur. Space Agency Spec. Publ., SP-250* (1), 3, 1986.

Gary, S. P., S. Hinata, C. D. Madland, and D. Winske, The development of shell-like distribution from newborn cometary ions, *Geophys. Res. Lett., 13,* 1364, 1986.

Gary, S. P. and C. D. Madland, Electromagnetic ion instabilities in a cometary environment, *J. Geophys. Res., 93,* 235, 1988a.

Gary, S. P., C. D. Madland, N. Omidi, and D. Winske, Computer simulation of two pickup ion instabilities in a cometary environment, *J. Geophys. Res., 93,* 9584, 1988b.

Gary, S. P. and R. Sinha, Electromagnetic waves and instabilities from cometary ion velocity shell distributions, *J. Geophys. Res., 94,* 9131, 1989.

Gary, S. P., K. Akimoto, and D. Winske, Computer simulations of cometary ion/ion instabilities and wave growth, *J. Geophys. Res., 94,* 3513, 1989.

Gloeckler, G., D. Hovestadt, F. M. Ipavich, M. Scholar, B. Klecker, and A. V. Galvin, Cometary pickup ions observed near Giacobini-Zinner, *Geophys. Res. Lett., 13,* 251, 1986.

Goldstein, M. L. and H. K. Wong, A theory for low frequency waves observed at comet Giacobini-Zinner, *J. Geophys. Res., 92,* 4695, 1987.

Gombosi, T. I., Preshock region acceleration of implanted cometary H^+ and O^+, *J. Geophys. Res., 93,* 35, 1987.

Gribov, G. E., K. Kecskemety, R. Z. Sagdeev, V. D. Shapiro, V. I. Shevchenko, A. J. Somogyi, K. Szego, G. Erdos, E. G. Eroshenko, K. I. Gringauz, E. Keppler, R. G. Marsden, A. P. Remizov, A. K. Richter, W. Riedler, K. Schwingenschuh, and K. P. Wenzel, Stochastic Fermi acceleration of ions in the pre-shock region of comet Halley, Proceedings of 20th ESLAB Symposium on the Exploration of Halley's comet, *Eur. Space Agency, Spec. Publ., 250,* 1, 271, 1986.

Hartle, R. E. and C. S. Wu, Effects of electrostatic instabilities on planetary and interstellar ions in the solar wind, *J. Geophys. Res., 78,* 5802, 1973.

Hartle, R. E., K. W. Ogilvie, and C. S. Wu, Neutral and ion exospheres in the solar wind with applications to Mercury, *Planet. Space Sci., 21,* 2181, 1973.

Hozer, T. E., and W. I. Axford, Interaction between interstellar helium and the solar wind, *J. Geophys. Res., 76,* 6965, 1971.

Hozer, T. E., Neutral hydrogen in interplanetary space, *Rev. Geophys., 15,* 467, 1977.

Hynds, R. J., S. W. F. Cowley, T. R. Sanderson, J. J. van Rooijan, and R. P. Wenzel, Observations of energetic ions from comet Giacobini-Zinner, *Science, 232,* 361, 1986.

Isenberg, P. A., Interaction of the solar wind with interstellar neutral hydrogen: Three fluid model, *J. Geophys. Res., 91,* 9965, 1986.

Isenberg, P. A., Evolution of interstellar pickup ions in the solar wind, *J. Geophys. Res., 92,* 1067, 1987a.

Isenberg, P. A., Energy diffusion of pickup ions upstream of comets, *J. Geophys. Res., 92,* 8795, 1987b.

Ip, W. H. and W. I. Axford, The acceleration of particles in the vicinity of comets, *Planet. Space Sci., 34,* 106, 1986.

Kadomtsev, B. B., *Plasma Turbulence,* Academic Press, New York, 1965.

Lee, M. A. and W. H. Ip, Hydromagnetic wave excitation by ionized interstellar hydrogen and helium in the solar wind, *J. Geophys. Res., 92,* 11,041, 1987.

Mobius, E., et al., Direct observation of He^+ ions of interstellar origin in the solar wind, *Nature, 318,* 426, 1985.

Neubauer, F. M., K. H. Glassmeier, M. Pohl, J. Raeder, M. H. Acuna, L. F. Burlaga, N. F. Ness, G. Musmann, F. Mariani, M. K. Wallis, E. Ungstrup, and H. U. Schmidt, First results from the Giotto magnetometer experiment at comet Halley, *Nature, 321,* 352, 1986.

Neugebauer, M., A. J. Lazarus, H. Balsiger, S. A. Fuselier, F. M. Neubauer, and H. Rosenbauer, The velocity distribution of cometary protons picked up by the solar wind, *J. Geophys. Res., 94,* 5277, 1989.

Neugebauer, M., A. J. Coates, and F. M. Neubauer, Comparison of pickup-up protons and water group ions upstream of comet Halley's bow shock, *J. Geophys. Res.,* in press, 1990.

Price, C. P., and C. S. Wu, The influence of strong hydromagnetic turbulence on newborn cometary ions, *Geophys. Res. Lett., 14,* 856, 1987.

Price, C. P., J. D. Gaffey, and J. Q. Dong, Excitation of low frequency electromagnetic waves by freshly created ions in the solar wind, *J. Geophys. Res., 93,* 837, 1988.

Richardson, I. G., S. W. H. Cowley, R. J. Hynds, T. R. Sanderson, K. P. Wenzel, and P. W. Daly, Three dimensional energetic ion bulk flows at comet P/Giacobini-Zinner, *Geophys. Res. Lett., 13,* 415, 1986.

Sagdeev, R. Z., V. D. Shapiro, and V. I. Shevchenko, MHD turbulence in the solar wind-comet interaction region, *Geophys. Res. Lett., 13,* 85, 1986a.

Sagdeev, R. Z., J. Belmont, A. A. Galeev, V. I. Moroz, V. D. Shapiro, V. I. Shevchenko, and K. Szegö, Vega spacecraft encounters with comet Halley, *Nature, 321,* 259, 1986b.

Saito, T., K. Yumoto, K. Hirao, T. Nakagawa, and K. Saito, Interaction between comet Halley and the interplanetary magnetic field observed by Sakigake, *Nature, 321,* 303, 1986.

Sanderson, T.R., K. P. Wenzel, P. Daly, S. W. H. Cowley, R. J. Hynds, E. J. Smith, S. J. Bame, and R. D. Zwickl, The interaction of heavy ions from comet P/Giacobini-Zinner with the solar wind, *Geophys. Res. Lett., 13,* 411, 1986.

Smith, E. J., et al., International Cometary Explorer encounter with Giacobini-Zinner, *Science, 232,* 382, 1986.

Terasawa, T., T. Mukai, W. Miyake, M. Kitayama, and K. Hirao, Detection of cometary pickup ions up to 107 km from comet Halley: Suisei observation, *Geophys. Res. Lett., 13,* 837, 1986.

Terasawa, T., Particle scattering and acceleration in a turbulent plasma around comets, in *Plasma Waves and Instabilities at Comets and in Magnetospheres*, eds. B. T. Tsurutani and H. Oya, Geophysical Monograph 53, p. 41, AGU, Washington DC, 1989.

Thomas, G. E., The interstellar wind and its influence on the interplanetary environment, *Annu. Rev. Earth Planet. Sci., 6*, 173, 1978.

Thorne, R. M., and B. T. Tsurutani, On the resonant interaction between cometary ions and low-frequency electromagnetic waves, *Planet. Space Sci., 35*, 1501, 1987.

Tsurutani, B. T., and E. J. Smith, Strong hydromagnetic turbulence associated with comet Giacobini-Zinner, *Geophys. Res. Lett., 13*, 259, 1986a.

Tsurutani, B. T., and E. J. Smith, Hydromagnetic waves and instabilities associated with cometary ion pickup: ICE observations, *Geophys. Res. Lett., 13*, 263, 1986b.

Tsurutani, B. T., A. L. Brinca, E. J. Smith, R. M. Thorne, F. L. Scarf, J. T. Gosling, and F. M. Ipavich, MHD waves detected by ICE at distances > 28×10^6 km from comet Halley: cometary or solar wind origin?, *Astron. Astrophys., 187*, 97, 1987.

von Rosenvinge, T. T., J. C. Brandt, and R. W. Farquhar, The International Cometary Explorer to comet Giacobini-Zinner, *Science, 232*, 353, 1986.

Weinstock, J., Formulation of statistical theory of strong plasma turbulence, *Phys. Fluids, 12*, 1045, 1969.

Winske, D., C. S. Wu, Y. Y. Li, and G. C. Zhou, Collective capture of released lithium ions in the solar wind, *J. Geophys. Res., 89*, 7327, 1984.

Winske, D., C. S. Wu, Y. Y. Li, Z. Z. Mou, and S. Y. Guo, Coupling of newborn ions to the solar wind by electromagnetic instabilities and their interactions with the bow shock, *J. Geophys. Res., 90*, 2713, 1985.

Wu, C. S. and R. C. Davidson, Electromagnetic instabilities produced by neutral particle ionization in interplanetary space, *J. Geophys. Res., 77*, 5399, 1972.

Wu, C. S., R. E. Hartle, and K. Ogilvie, Interaction of singly charged interstellar helium ions with the solar wind, *J. Geophys. Res., 78*, 306, 1973.

Wu, C. S. and R. E. Hartle, Further remarks on plasma instabilities produced by ions born in the solar wind, *J. Geophys. Res., 79*, 283, 1974.

Wu, C. S., D. Winske, and J. D. Gaffey, Rapid pickup of cometary ions due to strong magnetic turbulence, *Geophys. Res. Lett., 13*, 865, 1986.

Wu, C. S., D. Krauss-Varban, and T. S. Huo, A mirror instability associated with newly created ions in moving plasma, *J. Geophys. Res., 93*, 11527, 1988a.

Wu, C. S., X. T. He, and C. P. Price, Excitation of Whistlers and waves with mixed polarization by newborn cometary ions, *J. Geophys. Res., 93*, 3949, 1988b.

Wu, C. S. and P. H. Yoon, Kinetic hydromagnetic instabilities due to a spherical shell distribution of pickup ions, *J. Geophys. Res., 95*, 10,273, 1990.

Yoon, P. H., M. E. Mandt, and C. S. Wu, Evolution of an unstable shell distribution of pickup cometary ions, *Geophys. Res. Lett., 16*, 1473, 1989.

Yoon, P. H., Kinetic instability associated with spherical shell distribution of cometary pickup ions, *Geophys. Res. Lett., 17*, 1033, 1990.

Yoon, P. H., and L. F. Ziebell, Development of pitch-angle anisotropy and velocity diffusion of pickup ion shell distribution by solar wind turbulence, *J. Geophys. Res.*, in press, 1990.

Yoon, P. H., L. F. Ziebell, and C. S. Wu, Self-consistent pitch angle diffusion of newborn ions, *J. Geophys. Res.*, submitted, 1990.

Yumoto, K., T. Saito, and T. Nakagawa, Hydromagnetic waves near O^+ (or H_2O^+) ion cyclotron frequency observed by Sakigake at the closest approach to comet Halley, *Geophys. Res. Lett., 13*, 825, 1986.

Ziebell, L. F. and P. H. Yoon, Pitch angle and velocity diffusions of newborn ions by turbulence in the solar wind, *J. Geophys. Res.*, accepted for publication, 1990.

Ziebell, L. F., P. H. Yoon, C. S. Wu, and D. Winske, Pitch angle diffusion of newborn ions by intrinsic turbulence in the solar wind, *J. Geophys. Res.*, in press, 1990.

THE SPECTRUM AND ENERGY DENSITY OF SOLAR WIND TURBULENCE
OF COMETARY ORIGIN

A.D. Johnstone, D.E. Huddleston, A.J. Coates

University College London, Department of Physics and Astronomy,
Mullard Space Science Laboratory, Holmbury St. Mary, Dorking, Surrey RH5 6NT.

Abstract. The spectrum and energy density of solar wind turbulence of cometary origin can be related directly to the development of a bispherical shell distribution from the initial ring distribution of newly created ions. Although quasilinear theory has been very successful in obtaining such relationships we show how the power-law spectrum with an exponent of approximately minus 2 can be derived more simply from energy conservation and the resonant wave interaction characteristics. Observations of both the spectrum and the wave energy density agree well with these theoretical relationships. However the mechanisms by which the wave energy is dissipated are significant and have yet be evaluated.

1. Introduction

When cometary particles are injected into the solar wind they acquire the solar wind velocity component perpendicular to the magnetic field quasistatically within one gyroperiod. The wave-particle interactions by which they pick up the parallel component is a much slower process and is also of much greater scientific interest. Its starting point is the ring beam the new ions form in the velocity space as a result of the perpendicular pickup (figure 1). The instability of this distribution leads to the generation of Alfvenic turbulence in the solar wind flow, simultaneously scattering the cometary ions in pitch angle until they are almost isotropic in the frame of reference moving with the solar wind. The bulk velocity of this quasi-isotropic distribution is then close to the solar wind's parallel velocity. Our purpose here is to review the theory and observations of the spectrum and energy density of this turbulence and its relation to the particle distributions. As part of this review we have collected together a number of the simple relationships involved in resonant wave-particle interactions for easy reference. We have found that authors often approximate these relationships by ignoring the Alfven speed in comparison to the parallel velocity of the ions. Many of the results we use disappear if this approximation is made so we quote them in full.

Cometary Plasma Processes
Geophysical Monograph 61
●1991 American Geophysical Union

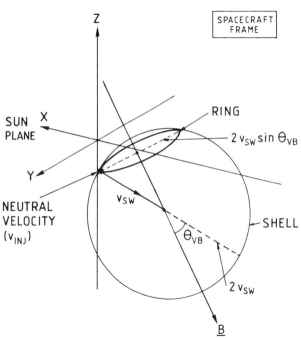

Fig. 1. Ion pickup velocity space geometry in the spacecraft frame, for an arbitrary orientation of the magnetic field $\mathbf{B}$. Neutral particles are injected at the point indicated by the star ($v_{inj} \sim (0,0.15,68.4)$ km s^{-1} in this frame) and their initial trajectory is in a ring around $\mathbf{B}$. Later they scatter on to a shell centred on the solar wind speed to a first approximation.

2. Observations of turbulence

The turbulence has been observed in the magnetic field, and in the velocity of the solar wind protons [e.g., Tsurutani and Smith, 1986a,b; Gosling et al., 1986; Yumoto et al., 1986; Glassmeier et al.,1987; Johnstone et al., 1987a,b]. As an example, figure 2 shows an overview of turbulence along the Giotto trajectory from 4.5×10^6 km up to the bow shock at 1.15×10^6 km from the nucleus of Comet Halley. In

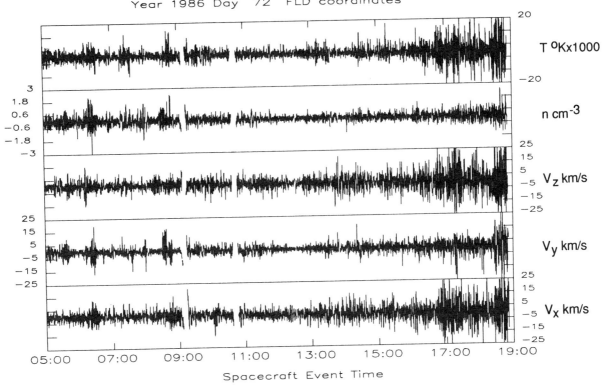

WAVES 3mHz<f<50mHz

GIOTTO JPA FIS Solar Wind Proton Parameters

Year 1986 Day 72 FLD coordinates

Spacecraft Event Time

Fig. 2. The turbulence in the solar wind proton parameters over the period from 0500 SET to 1900 SET on March 13th 1986 when the spacecraft was travelling from 4.5 million km to 1.2 million km from the nucleus. The FLD coordinate system has the y axis magnetic field aligned, the z axis in the -v x **B** direction and the x axis forming a right handed set. The mean value of the parameters has been subtracted. The frequency range indicated is set at the low frequency end by the digital filter and at the high end by the sample rate of the instrument.

producing this plot the mean value of the parameters has been subtracted by passing the data through a digital filter whose low frequency cut-off is set at 3 mHz, just below the gyrofrequency of the cometary ions and therefore just below the minimum frequency at which they generate the waves (see section 5). The mean values, including also magnetic field values, were then used to generate the FLD coordinate system used in this figure which has the y axis magnetic field aligned, the z axis in the -v x **B** direction and the x axis forming a right handed set. The frequency range indicated in the figure is set at the low frequency end by the digital filter and at the high end by the sample rate of the instrument.

During the early part of this period, up to say 1300 SET(Spacecraft Event Time), the turbulence does not appear to be related to the comet but to be the pre-existing level of turbulence in the solar wind. Over the last third the level increases, though not monotically, as the flux of cometary ions mass-loading the solar wind becomes significant. Note that the components perpendicular to the mean field (x,z) are largest though the field-aligned component (y) is not insignificant. The same comment applies to the magnetic field components which are not shown here but which are included

in figure 3. Note also that there is a significant amount of energy in density and temperature fluctuations in the solar wind protons. The waves are therefore not simple shear Alfven waves propagating parallel to the magnetic field though, for convenience in the analysis which follows, we shall assume that they are.

Figure 3 shows the frequency spectrum of both velocity and magnetic field waves in the spacecraft frame for the same data as figure 2. The power spectral density (PSD) at an average distance of 1.87×10^6 km from the nucleus (1552 to 1652 SET), where the turbulence is dominated by the input from cometary ions, is compared with the background spectrum measured in the solar wind some hours before. Note that the fluctuations have been passed through a high-pass filter which reduces the PSD at frequencies below the filter cut-off at 3 mHz. This was done to facilitate the transformation to a field-aligned frame. The increase in PSD generated by the cometary ions at their gyrofrequency is obvious, but it is also quite clear that the PSD is increased at higher frequencies as well. Glassmeier et al [1989] analysed the same magnetic data from the same general period though without the high-pass filtering. They characterised the

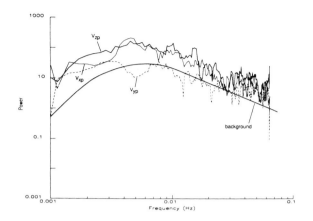

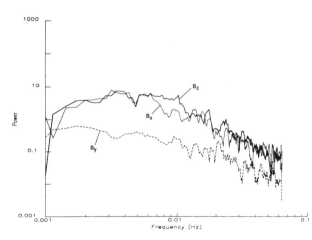

Fig. 3. The frequency spectrum of the proton velocity and magnetic waves for one hour of 1.87 million km from the nucleus. Both components have been filtered in the same way as in figure 2. The power spectral density is expressed in units of wave amplitude squared per unit frequency range. This is proportional to energy density per unit frequency range but the constant of proportionality is different for magnetic and velocity waves. The background for the velocity components was measured four hours before when there was no cometary activity. The power in the components parallel to the magnetic field (y axis) is much smaller than in the other two.

spectrum above the ion gyrofrequency as a power law $f^{-\gamma}$ with an exponent γ of the order of 2. A similar analysis by [Tsurutani and Smith, 1986 a,b] at Comet Giacobini-Zinner found a smaller exponent ~ 1.7.

The effect of the turbulence on the pitch angle distributions of the cometary ions is shown by figure 4 [Coates et al., 1990], which gives the width of the pitch angle distributions of the cometary ions along the Giotto trajectory. The distributions start to increase in width as the cometary component of the turbulence grows but do not become isotropic until the spacecraft is just outside the bow shock. This demonstrates that, unlike the assumption often made, the distributions do not isotropise instantly, or even within a few gyroperiods, in the upstream region.

3. Resonant wave particle interactions

We will pursue here an explanation of the turbulence and the pitch angle scattering in terms of resonant interactions between parallel-propagating Alfven waves in the solar wind and injected cometary ions. We begin by collecting together the basic relationships for such interactions. The resonance conditions are given by,

$$\omega - kv_{\parallel} = -\Omega_i \qquad \omega - kv_{\parallel} = +\Omega_i \qquad (1)$$

where the two signs correspond to two different polarisations of the waves; the negative sign in the left-hand equation is a right-hand polarised wave in which the electric vector rotates in the direction of the cyclotron motion of electrons. In the resonance, the ions are travelling in the same direction as the waves but moving faster and therefore seeing the opposite polarisation, i.e. the anomalous Doppler effect. The opposite situation is covered by the right hand equation.

In the analysis which follows we do not restrict ourselves to ions at the initial parallel velocity of the ring beam but are concerned with ions at all parallel velocities as they diffuse around the spherical shell. Ions do not just give up their free energy to the waves at the parallel velocity of the ring beam as is sometimes implied. This leads to the following four possible resonant interactions (figure 5);

i) waves travelling upstream $k > 0$; $v_{\parallel} > v_A$; RH polarised; anomalous Doppler.

ii) waves travelling upstream $k > 0$; $v_{\parallel} < v_A$; LH polarised; normal Doppler.

iii) waves travelling downstream $k < 0$; $v_{\parallel} > -v_A$; LH polarised; normal Doppler.

iv) waves travelling downstream $k < 0$; $v_{\parallel} < -v_A$; RH polarised; anomalous Doppler.

Equations (1) link frequencies and velocities measured in the plasma frame, i.e. the one moving with the solar wind. The resonant frequency in this frame ω_{sw} is given by,

$$\omega_{sw} = \frac{\Omega_i v_A}{\left| v_{\parallel} - v_{ph} \right|} \qquad (2)$$

where $v_{ph} = \pm v_A$ depending on the direction of propagation of the waves. The parallel velocity can take any value in the range $-v_S < v_{\parallel} < v_S$ where v_S is the solar wind speed.

The frequency of the same waves when observed Doppler-shifted in the spacecraft frame is given by,

$$\omega_{sc} = \Omega_i \frac{(v_{sc} - v_{ph})}{(v_{\parallel} - v_{ph})} \qquad (3)$$

Since the spacecraft velocity v_{sc} in the solar wind frame is approximately equal to the original parallel velocity, $v_{\parallel 0}$, of the ring distribution, the waves generated by the ions in the *ring* distribution are observed by the spacecraft to be at the ion gyrofrequency [Tsurutani and Smith, 1986a]. However as the ions diffuse around the shell towards smaller parallel velocities, equation (3) shows that the apparent frequency ω_{sc} of the resonant interaction increases.

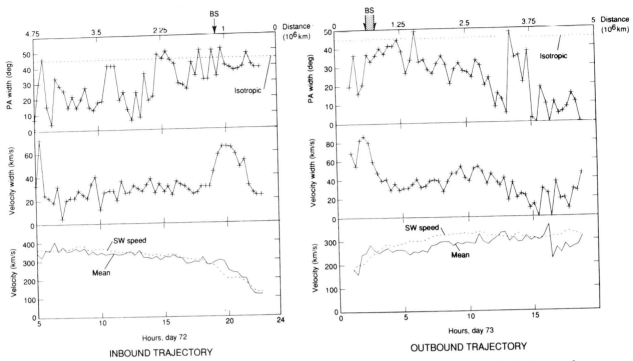

Fig. 4. The widths of the pitch angle (top panel) and velocity (middle panel) distributions and mean speed of the distributions (bottom panel), based on the mean absolute deviation, for the inbound and outbound legs of the trajectory of Giotto.

Since the interactions are evaluated in terms of an integral over the k spectrum, a more important value is the wavenumber, which is given by,

$$k = \frac{\Omega_i}{\left| v_{\parallel} - v_{ph} \right|}$$

(4)

Finally the relation between the frequency in the spacecraft frame and the wave number is,

$$\omega_{sc} = k \left(v_{sc} - v_{ph} \right)$$

(5)

Equations (2) and (4) show that as $v_{\parallel}$ approaches v_{ph}, the wave number k and the frequency ω_{sw} become very large. In fact as k increases, the waves begin to resonate with the thermal protons in the solar wind and the waves are damped. An effective upper limit for the wavenumber at which the interactions can take place is given by [Galeev et al., 1986],

$$k \leq \frac{\Omega_p}{3 v_{tp}}$$

(6)

where v_{tp} is the thermal velocity of the solar wind protons parallel to the magnetic field. The factor 3 is a qualitative value based on the argument that since only 0.1% of the distribution has a greater thermal velocity than $3v_{tp}$ there will be a negligible effect on the wave distribution at smaller k values. Combining this relation with equation (4) gives,

$$\left| v_{\parallel} - v_{ph} \right| > \frac{3 \Omega_i}{\Omega_p} v_{tp} \approx \frac{v_{tp}}{6}$$

(7)

This constraint also ensures that the waves involved in the interaction have frequencies in the solar wind frame less than the proton gyrofrequency so the dispersion relation for Alfven waves is still appropriate. The condition expressed by equation (7) leads to a gap in velocity space where there are no resonant waves. The width of the gap is inversely proportional to the mass of the ions, being a factor of 18 or so narrower for cometary ions than for protons. It has a total width $v_{tp}/3$, and is centred on $v_{\parallel} = +v_A$ for upstream waves and $v_{\parallel} = -v_A$ for downstream waves (figure 5).

Since $v_{tp} \leq 30$km/s and $v_A \approx 50$km/s the gap is relatively small for oxygen ions, as we have tried to illustrate in figure 5, and the up- and down-stream gaps do not overlap. Therefore ions may diffuse across the "upstream wave gap" by interacting with downstream propagating waves, so there is no gap in diffusion. If the injected ions were protons, the gap would be much wider than the Alfven speed and would include $v_{\parallel} = 0$. The up- and downstream gaps would then overlap considerably and there would then be a gap in diffusion. This gap is particularly important for the pitch angle diffusion of cosmic ray protons and has been referred to in the literature [e.g. Achterberg, 1981] as "the gap at $v_{\parallel} = 0$ which cannot be crossed". Since the gap *is* crossed, various nonlinear effects have been invoked to account for it. Two

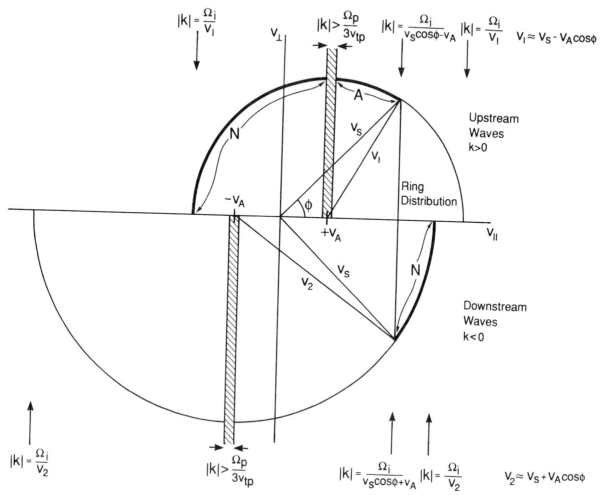

$|k| = \dfrac{\Omega_i}{V_I}$

$|k| > \dfrac{\Omega_p}{3v_{tp}}$

$|k| = \dfrac{\Omega_i}{v_s \cos\phi - v_A}$ $|k| = \dfrac{\Omega_i}{V_I}$ $V_I \approx V_S - V_A \cos\phi$

$|k| = \dfrac{\Omega_i}{V_2}$

$|k| > \dfrac{\Omega_p}{3v_{tp}}$

$|k| = \dfrac{\Omega_i}{v_s \cos\phi + v_A}$ $|k| = \dfrac{\Omega_i}{V_2}$ $V_2 \approx V_S + V_A \cos\phi$

Fig. 5. The diagram illustrates resonant wave particle interactions in the velocity space of the particles in a frame moving with the solar wind and aligned with the magnetic field. The upper half of the diagram shows the interaction between a ring distribution and waves travelling upstream and the lower half the same for waves travelling downstream. The thicker portions of the curve indicate the sections of the diffusion characteristic which are unstable. The letter A marks the anomalous Doppler resonance and N the normal Doppler resonance.

nonlinear mechanisms have been suggested by which the gap might be bridged; mirroring and resonance broadening [Fisk et al., 1974; Goldstein et al., 1975]. In the first, long wavelength fluctuations in the field strength associated with the waves reflect particles with large pitch angles. The condition for reflection is,

$$\cos^2\alpha \geq \dfrac{\Delta B}{B}$$

(8)

where α is the pitch angle. Such a reflection reverses the component of the parallel velocity and moves the particle to the opposite side of the gap. Since the mechanism is based on the invariance of the magnetic moment of the particle the wavelength should be much greater than the gyroradius of the particle. In the case of observations of the cometary ions, the frequency in the spacecraft frame should be less than the gyrofrequency of the cometary ions. Both conditions are easily met in the cometary environment. In the second mechanism the fluctuating magnetic fields perturb the particle orbit and broaden the resonance condition statistically so that waves at a given k can interact with particles in a finite range of parallel velocities. So although neither mechanism is required in the cometary situation to enable particles to jump the gap, both are probably operating. Their influence on the situation has yet to be determined. The problem has been examined recently in more detail in the cosmic ray context by Smith et al. [1990].

In a resonant interaction between an electromagnetic wave and a particle the following relationship is obeyed [Kennel and Petschek, 1966; Lyons and Williams, 1984]

$$v_\perp^2 + (v_\parallel - v_{ph})^2 = \text{constant}$$

(9)

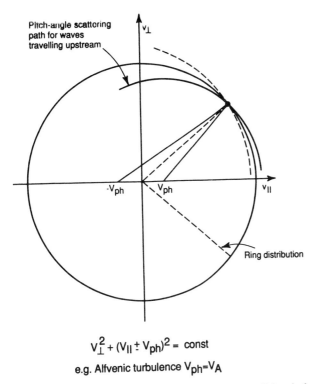

$$v_\perp^2 + (v_\parallel \pm v_{ph})^2 = \text{const}$$

e.g. Alfvenic turbulence $v_{ph}=v_A$

Fig. 6. A schematic diagram showing the possible pitch angle scattering paths of injected ions from their initial ring distribution. See text for further explanation.

where v_{ph} is the phase velocity of the waves. Equation (9) gives rise to a family of curves in velocity space along which particles diffuse under the influence of resonant wave-particle interactions. When the waves are parallel-propagating Alfven waves, v_{ph} is constant (the waves are dispersionless) and equal to $\pm v_A$, the sign depending on the direction of propagation. For the ring distribution this leads to two alternative, circular diffusion paths that the particles may take through velocity space centred on $\pm v_A$ respectively. These are shown in figure 6, drawn in a reference frame comoving with the solar wind.

Ions are created in a ring distribution at R and may then diffuse along the paths RA, RB, RC and RD. On the circle ARC the waves involved in the interaction are propagating downstream ($v_{ph}=-v_A$), while on the circle BRD they are propagating upstream ($v_{ph}=v_A$). At any position in velocity space the particles may diffuse along any one of the four paths equivalent to RA, RB, RC, and RD shown in figure 6 for the ring distribution. Since the interaction is a random occurrence it depends only on the relative intensity of the waves moving upstream or downstream at the appropriate wave number for resonance. However as the particles move along two of the paths, RD and RC, they give energy to the waves while on the other two they absorb energy from the waves. Hence the distributions are unstable to the generation of waves along paths RC and RD and the intensity of the waves in resonance will grow.

4. Free energy from the ion distribution

Suppose initially that only upstream propagating waves are present so that the ring distribution eventually diffuses uniformly over the spherical shell BRD, i.e. ignore the stability considerations mentioned in the paragraph above. The difference in energy between this asymptotic distribution and the initial ring is easily calculated.

- Ring

Kinetic energy	$1/2\,\rho_i v_S^2 \cos^2\phi$
Thermal energy	$1/2\,\rho_i v_S^2 \sin^2\phi$

- Shell

Kinetic energy	$1/2\,\rho_i v_A^2$
Thermal energy	$1/2\,\rho_i(v_S^2\sin^2\phi + (v_S\cos\phi - v_A)^2)$

$$\Delta E = E_{ring} - E_{shell} = \rho_i[v_A v_S \cos\phi - v_A^2] \qquad (10)$$

where v_S is the solar wind speed and ϕ is the angle between the magnetic field and the flow direction. In making this calculation we have assumed that the density of the ions is too small to have had a significant effect on the solar wind speed through mass-loading. Note that the asymptotic distribution is diffused around a spherical shell centred on $v_\parallel = v_A$ in the solar wind frame and not on $v_\parallel = 0$ and therefore its final bulk velocity in this frame is v_A. In other words, however much pitch angle diffusion takes place, the ions as a whole will never acquire all the parallel velocity of the solar wind. We can also calculate the change in momentum of the distribution as follows,

$$P_{ring} - P_{shell} = \rho_i\,(v\cos\phi - v_A) = \frac{\Delta E}{v_A} \qquad (11)$$

The quantity ΔE we will call the free energy F, since it is the energy released by the particle distribution and, initially at least, increases the energy density of the waves.

Bearing in mind the stability considerations of the previous section and remembering that the background distribution in the solar wind contains waves propagating in both directions [e.g.Roberts et al., 1987] a more likely asymptotic distribution for the particles will be a uniform distribution over the two partial spheres RD and RC in figure 6 [Galeev and Sagdeev 1988]. Coates et al., [1990] have calculated the bulk velocity of this bispherical asymptotic distribution in the solar wind frame and compared the theoretical values with measured ones They found that, as the Giotto spacecraft approached the comet Halley bow shock, the measured velocity converged on the bispherical value and that the bispherical value is a better fit to the data than the simple spherical shell value of v_A. This result therefore provides some encouragement that the asymptotic distribution is the bispherical one. The free energy calculation then becomes more complex and includes energy given to the distribution of waves propagating downstream by particles diffused over RC. These calculations have been performed exactly by Huddleston [1990] but are too long to be included here so we give the simplified approximate expressions to illustrate the results.

$$F_+ = (1/4) \, \rho_i \, v_A \, (v - v_A) \, (1 + \cos \phi \,)^2$$
(12)

$$F_- = (1/4) \, \rho_i \, v_A \, (v - v_A) \, (1 - \cos \phi)^2$$
(13)

where the subscripts +, -, indicate up-, and down-stream propagating waves respectively. Note that as $\phi \to 0$, i.e. when the magnetic field is parallel to the flow, $F_+ \to F$, and $F_- \to 0$, in accordance with the result given above in equation (10). As $\phi \to \pi/2$, then $F_+ = F_-$ as would be expected. The maximum free energy is available when $\phi = 0$ but it does not go to zero as $\phi \to \pi/2$.

5. Quasi-linear theory

The change in the distribution function of the ions can be related to the development of the wave spectrum by quasilinear theory. Using the formulation of Lee and Ip [1987, see also Lee, 1989], the equations are

$$\frac{\partial F_i}{\partial t} = \frac{q_i^2}{2m_i^2 v_\perp} \, \mathrm{Re} \, i \int_D dk \sum_{j=\pm} \frac{\omega_j^2}{k^2 c^2} G^j \left(\frac{v_\perp I_j(k) G^j F_i}{(\omega_j - k v_\parallel + \Omega_j)} \right)$$

$$\frac{\partial I_j}{\partial t} = 2 \gamma_j I_j$$
(14)

The linear growth rate γ_j is given by the dispersion relation,

$$\gamma_j = \frac{2 \pi^3 v_A^2 q^2}{m_i^2 c^2} \int dv_\parallel dv_\perp v_\perp^2 \, \delta(\omega - k v_\parallel + \Omega_i) \, GF_i$$
(15)

where the operator G is,

$$G = \frac{\partial}{\partial v_\perp} + \frac{k}{\omega} \left(v_\perp \frac{\partial}{\partial v_\parallel} - v_\parallel \frac{\partial}{\partial v_\perp} \right)$$

The subscript j denotes either upstream or downstream waves. This formulation does not include the effects of mirroring at large pitch angles [Goldstein et al .,1975].

Lee and Ip [1987] have applied these equations to the calculation of the wave spectrum resulting from the injection of interstellar helium and hydrogen into the solar wind. We take their simplest case when they assumed that only upstream waves were present so that they could transform to a frame moving with the waves. In this frame the initial ring distribution has the form,

$$f(\mu,0) = \frac{n_i}{2 \pi v_0^2} \delta(\mu - \mu_0) \, \delta(v - v_0)$$
(16)

and the asymptotic distribution is a uniform distribution over a spherical shell,

$$f(\mu,\infty) = \frac{n_i}{4 \pi v_0^2} \delta(v - v_0)$$
(17)

These expressions are identical to the distributions given in equation (10) taking into account that this frame is moving with the waves. They then obtain, for the time asymptotic spectrum of the waves,

$$I(k,\infty) - I(k,0) = \frac{2 \pi n_i m_i v_A |\Omega_i|}{k^2} \left\{ \frac{\Omega_i}{k v_0} - \frac{\left(\dfrac{\Omega_i}{k v_0} - \mu_0 \right)}{\left| \dfrac{\Omega_i}{k v_0} - \mu_0 \right|} \right\}$$
(18)

Integrating over k the total amount of energy E_{mwaves} in the magnetic fluctuations of the waves is,

$$E_{mwaves} = \frac{1}{2} \, n_i m_i v_A v_0 \mu_0$$
(19)

The total wave energy density is double this quantity because there is an equal amount of kinetic energy in the particles. Recalling that in the solar wind frame $v_0 \mu_0 = v_S \cos\phi - v_A$, then from (19),

$$E_{waves} = n_i m_i v_A \, (v_S \cos\phi - v_A)$$

- and when we compare again with equation (10) we see that,

$$E_{waves} = E_{ring} - E_{shell} = F$$

The analysis of Lee and Ip [1987] assumed that all the energy released by the particles as they diffused went into the waves and that the waves continued to grow without loss. Overall, energy has been conserved as the quasilinear formulation [Galeev and Sagdeev, 1988; Swanson, 1989] ensures it must. This result, obtained through a application of the quasilinear formulation, also confirms that the energy calculated in equation (10) is in fact the free energy in the particle distribution in the sense that it is the energy going into the waves.

The spectrum of the waves can be calculated without using quasilinear theory from the following simple argument based on the free energy calculation in section 4. This time we assume that the asymptotic distribution is the bispherical one and that both propagation directions are involved. As a particle diffuses in pitch angle through a range $-\delta v_\parallel$ at a parallel velocity $v_\parallel$ it gives energy to the waves at wavenumber k, determined from the resonance condition. In the asymptotic state all particles with a smaller $v_\parallel$ will have had to diffuse through this value of $v_\parallel$ and will have given this amount of energy to the waves.

Considering only upstream waves to begin with, then, in the frame moving with the waves the wave-particle interaction characteristic gives, as in equation (9),

$$v_{\perp w}^2 + v_{\parallel w}^2 = v_1^2$$
(20)

where,

$$v_1^2 = v^2 \sin^2\phi + (v\cos\phi - v_A)^2 \qquad (21)$$

and the energy in the solar wind frame is,

$$E_{sw} = \frac{1}{2} m_i \left(v_{\perp w}^2 + \left(v_{\|w} + v_A \right)^2 \right) \qquad (22)$$

From these equations we obtain,

$$\frac{\partial E_{sw}}{\partial v_{\|w}} = \frac{\partial E_{sw}}{\partial v_{\|sw}} = m_i v_A \qquad (23)$$

which gives the rate of change of particle energy along the characteristic. A similar expression applies for downstream waves except that the sign on the RHS is negative. From equation (4) we obtain,

$$\frac{\partial k}{\partial v_{\|w}} = -\frac{k^2}{\Omega_i} \qquad (24)$$

We have to consider three cases (figure 5);

a) $v_A < v_{\|sw} < v\cos\phi \;\rightarrow k > 0$; RH polarised.

$$P_a(k) = \frac{n_i m_i v_A \Omega_i}{1 + \cos\phi} \left(1 - 2\frac{v_A}{v_S}\cos\phi \right) \left(1 - \frac{k_-}{k} \right) \frac{dk}{2k^2}$$

$$k_{min} = \frac{\Omega_i}{v_S \cos\phi - v_A} \qquad (25)$$

b) $-v_1 + v_A < v_{\|sw} < v_A \;\rightarrow k > 0$; LH polarised.

$$P_b(k) = \frac{n_i m_i v_A \Omega_i}{1 + \cos\phi} \left(1 - 2\frac{v_A}{v_S}\cos\phi \right) \left(1 - \frac{k_+}{k} \right) \frac{dk}{2k^2}$$

$$k_{min} = k_+ = \frac{\Omega_i}{v_S - v_A\cos\phi} \qquad (26)$$

For both case a) and case b) the maximum k value is given by equation (6),

$$k_{max} = \frac{\Omega_p}{3v_{tp}} \qquad (27)$$

c) $v\cos\phi < v_{\|sw} < v \;\rightarrow k < 0$; LH polarised.

$$P_c(k) = \frac{n_i m_i v_A \Omega_i}{1 + \cos\phi} \left(1 + 2\frac{v_A}{v_S}\cos\phi \right) \left(1 - \frac{k_-}{k} \right) \frac{dk}{2k^2}$$

$$k_{min} = k_- = \frac{\Omega_i}{v_S + v_A\cos\phi}$$

$$k_{max} = \frac{\Omega_i}{v_S\cos\phi + v_A} \qquad (28)$$

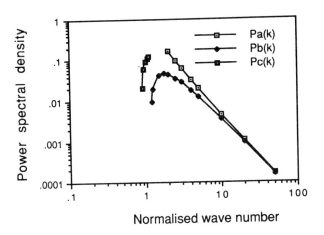

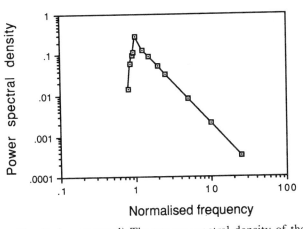

Fig. 7. (upper panel) The power spectral density of the waves in k space for the three expressions $P_a(k)$, $P_b(k)$, $P_c(k)$ plotted as a function of |k|, for $\cos\phi = 0.7$, $v_S/v_A = 5$. The power is given in units of $(n_i m_i v_A v_S/2)$ per unit wave number. The wave number is normalised to units of Ω_i/v_S. (lower panel) The three expressions have been transformed from wave number to frequency in the spacecraft frame and added together to give a single spectrum as observed by a spacecraft. Since the frequency is normalised to the gyrofrequency of the cometary ions it is obvious that the maximum power spectral density is seen at the cometary ion gyrofrequency. Fitting a power law to the spectrum above the ion gyrofrequency gives an exponent of -2.04, in good agreement with the observations.

These expressions are sketched in figure 7 for $\cos\phi = 0.7$, and $v_S/v_A = 5$. The upper panel shows the three segments of the spectrum separately though all in the positive range for the k vector even though the term $P_c(k)$ concerns waves travelling downstream. In the lower panel the three segments have been transformed into the spacecraft frame separately and added together to produce the frequency spectrum as it would be recorded by a spacecraft. The result is close to the observations. All these calculations, based on resonant wave particle interactions, give the same form for the k spectrum of the waves, i.e. that $P(k) \sim k^{-2}$ which transformed into the frequency spectrum in the spacecraft frame with the aid of equation (5) gives $P(\omega_{sc}) \sim \omega_{sc}^{-2}$. The peak in the spectrum is close to the gyrofrequency of the cometary ions and the spectrum cuts off at lower frequencies. This is close to what has been observed at comet Halley. The ω_{sc}^{-2} dependence is the natural outcome of pitch angle diffusion by resonant interactions and it is not necessary to invoke the possibility of spectral cascading through nonlinear effects to account for the wave energy at frequencies above the cometary ion gyrofrequency, as has been suggested by some authors [Glassmeier et al., 1989; and others cited therein]. The polarisation of the waves is a complex function of frequency in the plasma frame which is difficult to measure in the spacecraft frame. The observations do not yet reveal a clear pattern which could confirm or contradict this analysis.

So far the calculations described assume that the free energy released is held in the wave distribution without loss. A more complete approach to the problem, also using quasilinear theory, has been taken by Sagdeev and coworkers [Sagdeev et al., 1986; Galeev et al., 1986; Shapiro and Shevchenko, 1988]. They suppose that the ion distribution evolves to an asymptotic state where there is just sufficient anisotropy to maintain wave growth as new particles are added at the ring position. Their approach may be summarised as follows. The development of the ion distribution function is given by Boltzmann's equation with a source term for the implanted cometary ions,

$$\frac{DF_i}{Dt} = \frac{\partial F_i}{\partial t} + \frac{Q}{4\pi r^2 L} \exp(-\frac{r}{L}) \, \delta(\mathbf{v} - \mathbf{v}_0)$$

(29)

and the collisional term $\partial F_i/\partial t$ is given by the quasilinear diffusion equation (14). They then assume that the distribution function has the following form,

$$F_i = F_{i0}(|v|) + f_1(\mu) \qquad f_1 \ll F_{i0}$$

(30)

consisting of an isotropic part F_{i0} and a much smaller anisotropic term, f_1. Substituting (30) in (29) and averaging over pitch angle enables one to obtain an equation involving F_{i0} alone and a separate equation for $\partial f_1/\partial \theta$ in terms of F_{i0}. The equation for F_{i0} includes the effects of massloading on the plasma flow etc, and can be solved in the normal way. Then this solution can be substituted into the equation for $\partial f_1/\partial \theta$. With $\partial f_1/\partial \theta$ it is possible to obtain the wave growth rate γ from equation (15) and then the wave intensity from equation (14). From these two equations f_1 and the wave spectrum can be calculated [Galeev et al., 1987; Shapiro and Shevchenko, 1988]. They include the spectrum of downstream waves generated by the interaction with particles of pitch angle $\theta < \phi$; of upstream waves interacting with particles $\phi < \theta < \pi/2$ by the anomalous Doppler effect; and of upstream waves interacting with particles $\theta > \pi/2$ through the normal Doppler effect [see e.g. figure 3 Galeev et al., 1987].

They investigated the possibility that the wave intensity might be limited by the loss rate due to non-linear wave particle interactions of the form [Sagdeev et al., 1986; Galeev et al., 1987],

$$\omega_1 \pm \omega_2 = (\mathbf{k}_1 \pm \mathbf{k}_2) \cdot \mathbf{v}_p$$

(31)

where $\mathbf{v}_p$ is the velocity of a solar wind proton. The negative sign corresponds to the scattering of a wave quantum ω_1 into a quantum ω_2 of the same polarisation with the energy and momentum being absorbed by resonant thermal protons of the solar wind. Achterberg[1981] obtained the following expression for the loss rate with the appropriate approximations if,

$$k \ll \frac{2v_A \Omega_p}{v_{tp}^2}$$

(32)

then,

$$\gamma_{nl} = -\frac{11}{8} \sqrt{\frac{\pi}{8}} \, |k| \, v_{tp} \, \text{sgn}(\omega_1 - \omega_2) \int_0^\infty dk' \, \frac{|B_{k'}^2|}{B_0^2}$$

(33)

where v_{tp} is the thermal velocity of the protons and B_0 is the mean magnetic field in the solar wind. Note that γ_{nl} is negative for $\omega_1 > \omega_2$ which corresponds to a loss of wave energy to the particles. Then from,

$$\gamma + \gamma_{nl} = 0$$

(34)

it is possible to find the wave energy density in terms of the solar wind and cometary ion characteristics [Galeev et al., 1986, 1987],

$$\frac{\delta B^2}{B_0^2} = \frac{v_s^2}{\Omega_p v_{tp}} \left\{ \frac{Q}{4\pi r^2 L n_s \Omega_i} \right\}^{\frac{1}{2}}$$

(35)

where Q is the production rate of cometary molecules, L is the scale length for their photoionisation, n_s is the upstream density of solar wind protons, and r is the distance from the nucleus.

Expression (35) gives the order of magnitude of the wave amplitudes if they were limited by non-linear effects. It was found that although the effects of the non-linear wave-particle interaction (31) were too weak to limit the amplitude they did modify the wave spectrum in a not-insignificant way. Galeev et al. [1987] calculated the wave energy density by a numerical integration of their equations along the streamline including the modifying effects of the nonlinear processes for a comparison with measurements made from the Vega spacecraft at comet Halley (section 8). A plot of the wave spectrum, equivalent to the one obtained by our simplified approach has already been given by Galeev et al. [1987] based on these calculations using quasilinear theory.

6. A continuity equation for the wave intensity

The wave intensity has been calculated with the aid of quasilinear theory and a number of different assumptions; first that all the free energy released by the particles is present in the waves [e.g.Lee and Ip, 1987]: secondly that the energy in the waves is just that which balances the energy released by ions created locally with the nonlinear loss rate, equation(35), and finally that the quasilinear spectrum is modified by nonlinear processes. We write here a simplified form of continuity equation for the wave intensity U in order to illustrate the issues involved as follows,

$$ u \frac{\partial U}{\partial s} = \eta \, E_F \frac{\partial n_i}{\partial t} + u n_i E_F \frac{\partial \eta}{\partial s} - \frac{U}{\tau_D} $$
(36)

- where η is the fraction of the ion's free energy which has been released, and τ_D is the loss rate expressed as a decay time constant. This equation allows for the situation actually observed in that the pitch angle diffusion takes a significant time comparable with the flow time through the comet, i.e. $\eta \neq 1$, [Coates et al., 1989] and for the development of the wave intensity along the flow line, but it does not take into account the propagation of wave energy relative to the plasma flow velocity. This equation gives us the cases already described in the previous section if we make the appropriate assumptions. If we put $\eta = 1$ everywhere for instant pitch angle diffusion, ignore losses by putting $\tau_D = \infty$, and integrate along the flow line then we get, equivalent to the simple quasilinear approach,

$$ U = n_i E_F $$
(37)

which gives the wave energy proportional to the ion density.

If we balance losses by local input, again with $\eta = 1$, and ignore development along the flow line then we get,

$$ U = \tau_D E_F \frac{\partial n_i}{\partial t} $$
(38)

the same as the second case in which the quasilinear growth is balanced by nonlinear losses. This gives the wave energy proportional to the ion injection rate.

7. Results from computer simulations

The problem has also been addressed by computer simulation. The model of Gary et al., [1986, 1988, 1989] injects ions at a constant rate into an initially quiet solar wind. They distinguish two types of development of the wave intensity which they call linear and exponential temporal growth. The former occurs at relatively weak free energy injection rates and leads to a characteristic equation of the form,

$$ \frac{d}{dt}\left(\frac{\delta B^2}{2\mu_0} \right) = \sigma_i \frac{dn_i}{dt} \frac{m_i v_{\|0}^2}{2} $$
(39)

Integrating this equation with respect to time from an upstream location where the wave intensity is at the solar wind background level, leads to equation (37) with wave intensity proportional to ion density, and the coefficient σ_i obtained empirically from the simulation. We can derive the coefficient theoretically from equations (10) and (37) by ignoring v_A^2 in

equation (10) and putting $\phi = 0$,

$$ \sigma_i = 2 \left(\frac{v_A}{v_{\|0}} \right) $$
(40)

Putting the simulation parameters of Gary et al., [1988] into this equation gives a value of 0.2 compared with the much smaller empirical value of 0.03 they obtained.

The characteristic equation for exponential growth is,

$$ \frac{\delta B^2}{B_0^2} = 10 \left(\frac{m_i}{m_p} \right) \left(\frac{v_{\|0}}{v_A} \right)^2 \frac{1}{n_p \Omega_i} \frac{\partial n_i}{\partial t} $$
(41)

This is the same form as equation (38) with the wave intensities proportional to ion injection rate although there is no scientific relation between the coefficients of the injection rate $(\partial n_i/\partial t)$ in equations (38) and (41). Exponential growth is to be expected at early stages of growth, i.e. in the linear regime. [Gary et al., 1988] state that in these circumstances the wave intensity grows to a peak value, declines somewhat and then undergoes a transition to linear temporal growth. The exponential regime is characterised by phase bunching and particle trapping while the linear regime is associated with pitch angle scattering and random phase distributions. At the comet, phase bunching and related particle trapping is unlikely because the solar wind is turbulent and the phases randomised before cometary ions are injected. The quasi-coherent effects of the simulation are therefore unlikely to develop. It would probably be more fruitful to compare the results of the simulations at very late times with the observations.

8. Comparison with observations

There have been several comparisons between the measured wave intensity and these simple theoretical expressions. Johnstone et al., [1987a], using measurements from Giotto, estimated the free energy in a simplistic way as the kinetic energy of the particle associated with motion parallel to the magnetic field i.e.,

$$ F = \frac{1}{2} n_i m_i v_s^2 \cos^2\phi $$
(42)

They effectively compared the wave energy density with the injected ion density seeking a relation of the form of equation (37) and found that the wave energy density was up to, but usually less than, a straight line corrresponding to a fraction K of the free energy given by equation (42) with $0.16 < K < 0.26$ (figure 8). Comparing (42) with the proper calculation of the free energy given by equation (10) we see that the theoretical value of should be $K \sim \sigma_1 \sim 2(v_A/v_s \cos\phi)$ as in (40). This value of K, according to measurements of v_A, v_s, and ϕ, was ~ 0.45, so that only about half the available free energy was found in the wave distribution. Note that the measured $K \sim 0.2$ still exceeds the value of the corresponding coefficient $\sigma_1 \sim 0.03$ obtained in the simulation of Gary et al., [1988]. Gary et al., [1988] compared equation (41) with observations of the wave energy density at comet Giacobini-Zinner measured by Tsurutani and Smith [1986a,b] and the measurements at comet Halley reported by Johnstone et al.,

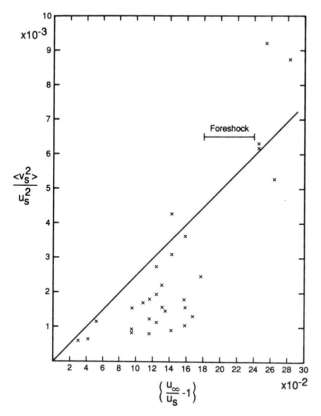

Fig. 8. A plot of the wave energy density measured by the normalised velocity variance against the free energy density which is proportional to the normalised solar wind velocity decrease. The gap in the distribution of points is due to the decrease in mean velocity on passing through what has been called the foreshock at 1825 SET.

[1987b]. While they agreed reasonably well with the measurements at comet Giacobini-Zinner they underestimated the observations at comet Halley by an order of magnitude. This corresponds to the difference in the value of the coefficient σ_1 already mentioned. Le et al., [1989], using an empirical correction for non-zero values of the angle ϕ,

$$\frac{U(\phi)}{U(0)} = \cos^8 \phi$$

(43)

compared equation (41) with measurements by the ICE spacecraft at comet Giacobini-Zinner and by the Vega spacecraft at comet Halley. There is good agreement for the former but not for the latter. Gary et al., [1988] claim that the comparisons show that the wave intensity is proportional to $(\partial n_i / \partial t)$ the rate of injection and equations (38) or (41), rather than n_i and equation (37). However with these empirical relationships it is difficult to see how such a distinction can be made; the two possibilities have not been tested in a comparable way and the data are not well enough fitted by either relationship. Equation (37) must give an upper limit to the wave energy density because the most that can be there is the total available in the free energy. Other effects such as

slow pitch angle diffusion and nonlinear wave energy losses will decrease it. The observations are consistent with this argument.

A more accurate comparison should be based on a proper comparison with all the implications of equation (36). Galeev [1988; Galeev et al., 1987; Shapiro and Shevchenko, 1988] compared the wave intensity measured by the magnetometer on Vega 2 with the wave intensity calculated as described in section 5. The density of cometary ions was obtained from the deceleration of the solar wind by calculating the mass-loading. This includes a theoretical calculation of the the anisotropy in the cometary ion distribution but did not allow for varying magnetic field directions. The calculated value exceeded the measured wave energy density throughout but not by a large factor.

We have attempted a more accurate experimental comparison in the following way. The free energy has been calculated with the use of equations (12) and (13) from the bispherical distribution and includes the effects of changing magnetic field directions. An empirical model of the massloading effects based on solar wind deceleration [Huddleston et al., 1990] was used to obtain the density of implanted cometary ions. The model was checked against direct measurements. The amount of energy, E_R, that has been released from the distribution can be estimated from the change in momentum of the particle distribution (equation (11)) as follows,

$$E_R = \left\{ \frac{v_S \cos \phi - v_i}{v_S \cos \phi - v_{BS}} \right\} F$$

(44)

where v_{BS} is the velocity of the asymptotic bispherical

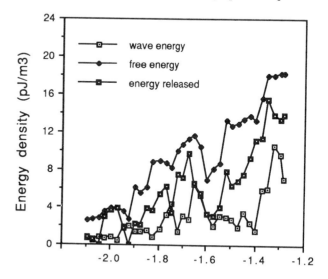

Distance from nucleus (million km)

Fig. 9. A comparison of the free energy available from the cometary ion ring distribution as it pitch-angle diffuses to an asymptotic bispherical distribution, the energy actually released at a point along the trajectory of Giotto according to measurements of the ion distribution and the measured wave energy density.

distribution [Coates et al., 1990], and v_i is the measured velocity of the implanted ion distribution. This comparison of v_i and v_{BS} effectively measures the progress of the cometary ion distribution towards isotropy.

Then F, E_R and the total measured wave power have been compared in figure 9 as a function of time along the trajectory of Giotto. As would be expected from equation (36) there is less energy currently in the wave distribution than has been released from the particles which is in turn less than would be available from the ring distribution if it had been completely diffused in pitch angle. This indicates that there has been some loss of wave energy, perhaps by the mechanism described above, i.e., induced scattering of the Alfven waves. Further study is required to verify this possibility.

Acknowledgements

ADJ thanks the Space Research Institute of the Academy of Sciences of the USSR, Moscow and in particular Prof A.A.Galeev and O.L. Vaisberg for their hospitality during the preparation of this paper. It is a pleasure to acknowledge the value of discussions on the techniques of quasilinear theory with Dr. V.D.Shapiro of the Space Research Institute.

References

Achterberg, A. : On the propagation of relativistic particles in a high beta plasma, *Astron. Astrophys.*, 98, 161-172, 1981.

Coates, A.J., Johnstone, A.D., Wilken, B., Jockers, K., Glassmeier, K.-H.: Velocity space diffusion of pickup-ions from the water group at comet Halley, *J. Geophys. Res.*, 94, 9983-9993, 1989.

Coates, A.J., Wilken, B., Johnstone, A.D., Jockers, K., Glassmeier, K.-H., Huddleston, D.E.: Bulk properties and velocity distributions of water group ions at comet Halley: Giotto measurements, *J.Geophys. Res.*95, 10249-10260, 1990.

Fisk, L.A., Goldstein, M.L., Klimas, A.J., Sandri,G. : The Fokker-Planck coefficient for pitch-angle scattering of cosmic rays, *Ap. J.*,190, 417 -428, 1974.

Galeev, A.A.: Theory and observations of solar wind/cometary plasma interaction processes, *Exploration of Halley's Comet*, ESA SP-250, pp 3 - 17, 1988.

Galeev, A.A., Polyudov, A.N., Sagdeev, R.Z., Szego, K., Shapiro, V.D., Shevchenko, V.I.: MHD turbulence caused by a comet in the solar wind, *Sov. Phys. JETP.*, 65, 1178, 1987.

Galeev, A.A., Sagdeev, R.Z.: Alfven waves in a space plasma and their role in the solar wind interaction with comets, *Astrophys. Space Sci.*, 144, 427-438, 1988.

Galeev, A.A., Sagdeev, R.Z., Shapiro, V.D., Shevchenko, V.I.: Mass loading and MHD turbulence in the solar wind/comet interaction region. *Plasma Astrophysics*, ESA SP-251, pp 307 - 316, 1986.

Gary, S.P., Akimoto, K., Winske, D.: Computer simulation of cometary-pickup-ions/ion instabilities and wave growth, *J. Geophys. Res.*, 94, 3515 - 3525, 1989.

Gary, S.P. Hinata, S., Madland, C.D., Winske, D.: The development of shell-like distributions from newborn cometary ions, *Geophys. Res. Lett.*, 13, 1367-1367, 1986.

Gary, S.P., Madland, C.D., Omidi, N., Winske, D. : Computer simulation of two-pickup-ion instabilities in a cometary environment, *J. Geophys. Res.*, 93, 9584-9596, 1988.

Glassmeier, K-H., Coates, A.J., Acuna, M.H., Goldstein, M.L., Johnstone, A.D., Neubauer, F.M., Reme, H.: Spectral characteristics of low frequency plasma turbulence upstream of comet P/Halley, *J. Geophys. Res.*, 94, 37-48, 1989.

Glassmeier, K.H., Neubauer, F.M., Acuna, M.H., Mariani, F.: Low frequency fluctuations in comet Halley's magnetosheath: Giotto observations, *Astron. Astrophys.*, 187, 65-68, 1987.

Goldstein, M.L., Klimas, A.J., Sandri, G. : Mirroring in the Fokker-Planck coefficient for cosmic ray pitch-angle scattering in homogeneous magnetic turbulence, *Ap. J.*, 195, 787 -799, 1975.

Gosling, J.T. Asbridge, J.R., Bame, S.J., Thomsen, M.F., Zwickl, R.D. : Large amplitude, low frequency, plasma fluctuations at comet Giacobini-Zinner, *Geophys. Res. Letts.*, 13, 267-270, 1986.

Huddleston, D.E., *The interaction between a comet and the solar wind*, Ph.D. thesis, University of London, August 1990

Johnstone, A.D., Coates, A.J., Heath, J., Thomsen, M.F., Wilken, B., Jockers, K., Formisano, V., Amata, E., Winningham, J.D., Borg, H., Bryant, D.A.: Alfvenic turbulence in the solar wind flow during the approach to comet P/Halley, *Astron. Astrophys*, 187, 25-32, 1987a.

Johnstone, A., Glassmeier, K., Acuna, M., Borg, H., Bryant, D., Coates, A., Formisano, V., Heath, J., Mariani, F., Musmann, G., Neubauer, F., Thomsen, M., Wilken, B., Winningham, J.: Waves in the magnetic field and solar wind flow outside the bow shock at comet P/Halley. *Astron. Astrophys.*, 187, 47-54, 1987b.

Kennel, C.F., Petschek, H.E.: Limit on stably trapped particle fluxes, *J. Geophys. Res.*, 71, 1-28, 1966.

Le, G., Russell, C.T., Gary, S.P., Smith, E.J., Riedler, W., Schwingenshuh, K.: ULF waves at comets Halley and Giacobini-Zinner: comparison with simulations, *J. Geophys. Res.*, 94, 11889 - 11895, 1989

Lee, M.A. : Ultra-low frequency waves at comets, *Plasma waves and instabilities at comets and in magnetospheres*, American Geophysical Union, Washington, pp 13-29, 1989.

Lee, M.A., Ip, W.-H.: Hydromagnetic wave excitation by ionized interstellar hydrogen and helium in the solar wind, *J. Geophys. Res.*, 92, 11,041-11,052, 1987.

Lyons, L.R., Williams, D.J.: *Quantitative aspects of magnetospheric physics*, 231pp, D.Reidel Publishing Co., Dordrecht, Holland, 1984.

Roberts, D.A., Goldstein, M.L., Klein, L.W., Matthaeus, W.H.: Origin and evolution of fluctuations in the solar wind: Helios observations and Helios-Voyager comparisons, *J. Geophys. Res.*, 92, 12023-12035, 1987.

Sagdeev, R.Z., Shapiro, V.D., Shevchenko, V.I., Szego, K.: MHD turbulence in the solar wind-comet interaction region. *Geophys. Res. Lett.*, 13, 85-88, 1986.

Shapiro, V.D., Shevchenko, V.I.: Astrophysical plasma turbulence. I, *Sov. Sci. Rev.E. Astrophys. Space Phys.*, 6, 425-543, 1988.

Smith, C.W., Bieber, J.W., Matthaeus, W.H., Cosmic ray pitch angle scattering in isotropic turbulence II: sensitive dependence on the dissipation range spectrum, *Ap. J.,* in press, 1990.

Swanson, D.G.: *Plasma Waves*, Academic Press Inc., San Diego, 1989.

Tsurutani, B.T. Smith, E.J.: Strong hydromagnetic turbulence associated with comet Giacobini-Zinner, *Geophys. Res. Lett.,* 13, 259-262, 1986a.

Tsurutani, B.T., Smith, E.J.: Hydromagnetic waves and instabilities associated with cometary-ion pick-up: ICE observation, *Geophys. Res. Lett.,* 13, 263-266, 1986b.

Yumoto, K., Saito, T., Nakagawa, T.: Hydromagnetic waves near 0^+(or H_2O^+) ion cyclotron frequency observed by Sakigake at the closest approach to comet Halley, *Geophys. Res. Lett.,* 13, 825-828, 1986.

THE MAGNETIC FIELD TURBULENCE AT COMET HALLEY OBSERVED BY VEGA 1 AND 2

G. Le and C. T. Russell

Department of Earth and Space Sciences and Institute of Geophysics and Planetary Physics

University of California at Los Angeles

K. Schwingenschuh and W. Riedler

Space Science Institute, Graz, Austria

Abstract. The magnetic field observed by the VEGA spacecraft seems to be disturbed by the comet over a wide frequency band, and over an extended range surrounding the comet. The level of magnetic fluctuations seems higher than that of the solar wind at least as low as - 0.3 mHz and as far as - 10 Mkm. The observed fluctuation level near water group ion cyclotron frequency does not show a clear dependence on the distance from the comet when the spacecraft is far from the comet, but in the cometary magnetosheath the fluctuation level decreases with the decreasing distance from the comet near both proton and water group ion cyclotron frequency.

Introduction

The interaction of the solar wind with a comet is characterized by enhanced magnetic field fluctuations near the cyclotron frequency of the newborn cometary ions, and by the extended region of the fluctuations surrounding the comet. The basic physical process which causes this fluctuation is the pick-up of the newborn cometary ions by the electro-magnetic fields of the solar wind flow. During the encounter with the comet Giacobini-Zinner, the ICE spacecraft observed magnetic fluctuations with characteristic frequency corresponding to the water group ion cyclotron frequency [Smith et al., 1986; Gosling et al., 1986; Tsurutani and Smith, 1986a, b; Tsurutani et al., 1987], and the discrete wave packets associated with the leading edge of the steepened low-frequency waves [Smith et al., 1986; Tsurutani et al., 1987; Le et al., 1989a]. The radial variation of the fluctuation level is an increase with decreasing distance from comet G-Z for both low-frequency waves near the proton and water group ion cyclotron frequency band and discrete wave packets [Le et al., 1989a, b]. The observations at comet Halley also show strong MHD turbulence surrounding the comet [Riedler et al., 1986,; Glassmeier et al., 1987, 1989; Saito et al., 1986; Yumoto et al., 1986]. The fluctuations corresponding to the MHD mirror mode instability have been observed within the induced magnetosphere of comet Halley [Russell et al., 1987], but the similar waves were not seen at comet G-Z. No evidence of steepened low-frequency waves and discrete wave packets have been reported at comet Halley. In this paper, we will present results of an examination of the VEGA 1 and 2 magnetic measurements at comet Halley during their flights by Halley on March 6 and March 9, 1986. The highest time resolution data (0.1 s) are available from three hours before until one hour after the closest approach. Occasional periods of 6 sec resolution data, which are averaged on board the spacecraft, are also available within two days from the closest approach.

Observations

In this study, the fluctuation level of the magnetic field is represented by the normalized wave amplitude or the normalized wave energy. The wave amplitude and wave energy within a specific frequency band are derived from the power spectra, and contain both transverse and compressional components. Then we normalized them by the average magnetic field strength and energy, respectively.

We have looked at the magnetic fluctuations at low frequencies and large encounter distances. Figure 1 shows 40 hours of magnetic field data observed simultaneously by VEGA 1 and 2. The time resolution of the data is 2.5 minutes. During this interval, VEGA 1 is traveling along the outbound trajectory from 5.9 Mkm to 17.3 Mkm from the comet nucleus, and VEGA 2 along the inbound trajectory from 14.2 Mkm to 3.1 Mkm. The magnetic fluctuations at both spacecraft seems more turbulent than those in the undisturbed solar wind. The power spectra from VEGA 1 and 2 during this interval (Figure 2) are very similar in the frequency range 0.1 to 3 mHz. Comparing to undisturbed solar wind, Figure 2 shows that the fluctuation level is even higher than that of active solar wind discussed by Siscoe et al. [1968] down to frequencies at least as low as 0.3 mHz. Figure 3 presents the variation of the amplitude of the fluctuation in the frequency range 0.1 to 3 mHz as a function of distance from the comet, where the solid lines are VEGA 1 data and dotted lines VEGA 2 data. The fluctuation level during the outbound pass is higher than that of inbound pass, on average. This may be caused by the facts that the inbound and outbound trajectories are not symmetric, and that the outbound trajectory is more upstream from the comet than the inbound trajectory. The upstream region may be more unstable because the solar wind and cometary ion distributions are more narrow here. Downstream the anisotropies should be weaker and the waves, which grow upstream, damped by the warm plasma.

From the 6 sec resolution data available we have calculated the wave energy near the water group ion cyclotron frequency ($.75 f_{H_2O^+}$ to $1.25 f_{H_2O^+}$) at large distances from the comet as shown in Figure 4. We have used the local average magnetic field strength to calculate the ion cyclotron frequency. The energy of fluctuations near the water

Cometary Plasma Processes
Geophysical Monograph 61

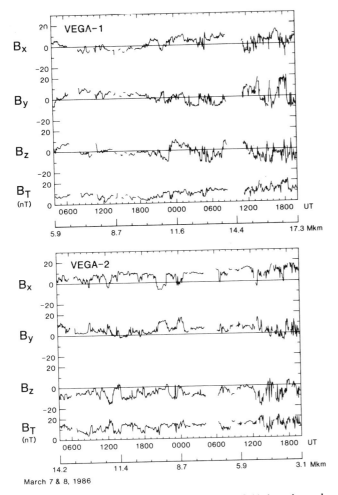

March 7 & 8, 1986

Fig. 1. Low time resolution (2.5 min) magnetic field data observed simultaneously by VEGA 1 and 2. The coordinate system is cometary solar ecliptic with x directed toward the sun.

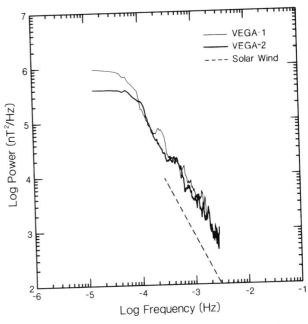

Fig. 2. The power spectra of the magnetic field in the time interval corresponding to Figure 1. The lighter trace is the total power of three magnetic field components measured by VEGA 1, and the heavier one by VEGA 2.

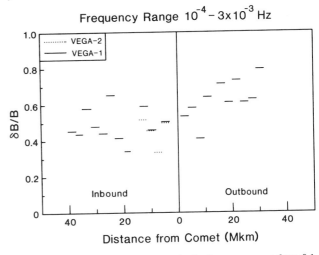

Fig. 3. The amplitude of fluctuations in the frequency range from 0.1 to 3 mHz vs. distances from the comet.

group ion cyclotron frequency is quite variable. We can not see any clear dependence of the wave energy on the distance from the comet. But we notice that the fluctuation level seems higher for smaller background magnetic fields. Figure 5 shows the wave amplitudes vs. background magnetic field magnitudes. We can see that both absolute wave amplitudes (upper panel) and normalized amplitudes (lower panel) are anticorrelated with the magnitude of the background magnetic field. Because of the lack of higher resolution data, the wave amplitudes near proton cyclotron frequency are not available at large distance from the comet.

Three hours before until one hour after the closest approach, the magnetometer was switched to the direct transmission mode with 10 vectors per sec [Riedler et al., 1986]. From these highest resolution data, we can study the fluctuations near the water group ion and proton cyclotron frequency very close to comet Halley, i.e., in the cometary magnetosheath. The magnetic field data within the magnetosheath show that the level of fluctuation appears more and more quiet when we get closer to the comet. This is true for both proton and water group ion cyclotron frequency band. Figure 6 shows the wave energy near the proton cyclotron frequency ($.75f_{H^+}$ to $1.25f_{H^+}$) vs. distance from the comet. The wave energy drops when the distance becomes smaller. The wave energy

near the water group ion cyclotron frequency has a similar variation, as shown in Figure 7.

Discussion and Summary

The magnetic field observed by VEGA 1 and 2 seems to be disturbed over a wide frequency band and an extended range surrounding the comet by the cometary new born ions. The fluctuation level is even higher than that of active solar wind [Siscoe et al., 1968] at least as low as 0.3 mHz, and as far as 17 Mkm from the nucleus (cf. Figure 2). We don't see any evidence of steepened low frequency waves and the discrete wave

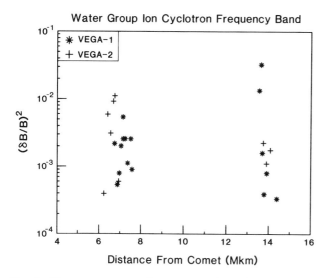

Fig. 4. The wave power near the water group ion cyclotron frequency at large distances from the comet.

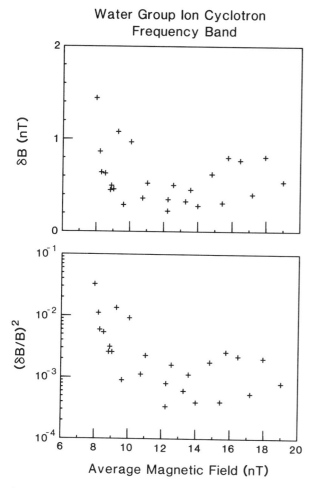

Fig. 5. The wave amplitude (upper panel) and power (lower panel) vs. background average magnetic field at large distances from the comet.

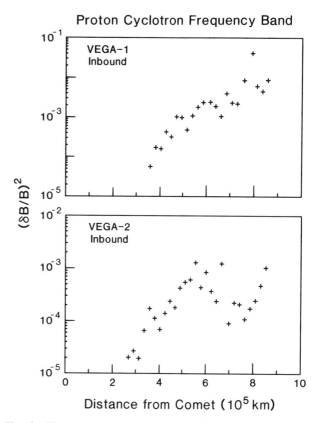

Fig. 6. The wave power near the proton cyclotron cyclotron frequency as a function of distance from the comet in the cometary magnetosheath.

packets, as observed upstream from the comet Giacobini-Zinner. The Giotto observation at comet Halley also shows the lack of steepened waves [Glassmeier et al, 1987]. This may be due to the lack of high resolution magnetic field data in the upstream region, too. The fluctuation level near the water group ion cyclotron frequency does not show the increase with decreasing distance from the comet, expected from quasilinear theory [Sagdeev et al., 1986] and computer simulations [Gary et al., 1988, 1989]. This is in contrast to the results of Galeev [1986], who found a clear increase of magnetic field energy density with decreasing distance from the comet, although the same data set has been used. We are unable to determine how Galeev's results were derived. The values reported herein are correct as can be verified by visual inspection of the data.

The fluctuation level in the cometary magnetosheath and magnetosphere decreases with decreasing distance from the comet at both the proton and water group ion cyclotron frequencies, while the magnitude of the magnetic field increases to a peak of 70–80 nT due to the field pile-up. The reason for the decrease in wave power may be that the free energy of the instabilities becomes smaller closer to the comet. When cometary ions become the dominant species in the plasma, the solar wind ions will provide the free energy to the instabilities. As the solar wind becomes heavily mass loaded with cometary ions, it moves more slowly as it approaches the nucleus. As a result, the fluctuation level will decrease as we get closer to the comet. The situation is different at large distance from the comet, where the solar wind ion is dominant in the plasma, and the cometary pick-up ions will provide the free energy. In this case, an increase of the pick-up ion density will enhance the fluctuation level.

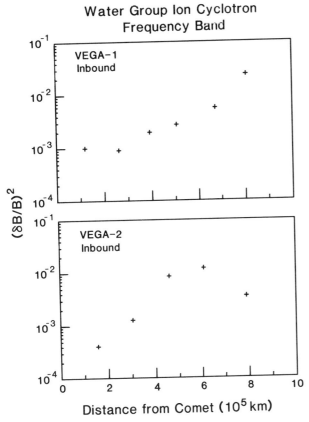

Fig. 7. The wave power near the water group ion cyclotron frequency as a function of distance from the comet in the cometary magnetosheath.

Acknowledgments. The work at UCLA was supported by the National Aeronautics and Space Administration under research grant NAGW-717.

References

Galeev, A. A., Theory and observations of solar wind/cometary plasma interaction processes, Porc. 20th ESLAB Symposium on the Exploration of Halley's Comet, ESA SP-250, pp.3–17, Heidelberg, 1986.

Gary, S. P., K. Akimoto, and D. Winske, Computer simulations of cometary-ion/ion instabilities and wave growth, J. Geophys. Res., 94, 3515–3525, 1989.

Gary, S. P., C. D. Madland, N. Omidi, and D. Winske, Computer simulations of two-pickup-ion instabilities in a cometary environment, J. Geophys. Res., 93, 9584, 1988.

Glassmeier, K.-H., A. J. Coates, M. H. Acuna, M. L. Goldstein, A. D. Johnstone, F. M. Neubauer, and H. Reme, Spectral characteristics of low-frequency plasma turbulence upstream of comet P/Halley, J. Geophys. Res., 94, 37–48, 1989.

Glassmeier, K. H., F. M. Neubauer, M. H. Acuna, and F. Mariani, Low-frequency magnetic field fluctuations in comet P/Halley's magnetosheath: Giotto observations, Astron. Astrophys., 187, 65–68, 1987.

Gosling, J. T., J. R. Asbridge, S. J. Bame, M. F. Thomsen, and R. D. Zwickl, Large amplitude, low frequency, plasma fluctuations at comet Giacobini-Zinner, Geophys. Res. Lett., 13, 267, 1986.

Le, G., C. T. Russell, and E. J. Smith, Discrete wave packets upstream from the Earth and comets, J. Geophys. Res., 94, 3755–3760, 1989a.

Le, G., C. T. Russell, S. P. Gary, E. J. Smith, W. Rielder, and K. Schwingenschuh, ULF waves at comets Halley and Giacobini-Zinner: Comparison with simulations, J. Geophys. Res., 94, 11,889–11,995, 1989b.

Rielder, W., K. Schwingenschuh, Y. G. Yeroshenko, V. A. Styashkin, and C. T. Russell, Magnetic field observations in comet Halley's coma, Nature, 321, 288–289, 1986.

Russell, C. T., W. Riedler, K. Schwingenschuh, and Y. Yeroshenko, Mirror mode instability in the magnetosphere of comet Halley, Geophys. Res. Lett., 14, 644–647, 1987.

Sagdeev, R. Z., V. D. Shapiro, V. I. Shevchenko, and K. Szego, MHD turbulence in the solar wind-comet interaction region, Geophys. Res. Lett., 13, 85–88, 1986.

Saito, S., K. Yumoto, K. Hirao, T. Nakagawa, and K. Saito, Interaction between comet and the interplanetary magnetic field observed by Sakigake, Nature, 321, 303, 1986.

Siscoe, G. L., L. D. Davis, Jr., E. J. Smith, and D. E. Jones, Power spectra and discontinuities of the interplanetary magnetic field: Mariner 4, J. Geophys. Res., 73, 61, 1986.

Smith, E. J., B. T. Tsurutani, J. A. Slavin, D. E. Jones, G. L. Siscoe, and D. A. Mendis, International Cometary Explorer encounter with Giacobini-Zinner: Magnetic field observations, Sciences, 232, 383–385, 1986.

Tsurutani, B. T., and E. J. Smith, Strong hydromagnetic turbulence associated with comet Giacobini-Zinner, Geophys. Res. Lett., 13, 259, 1986a.

Tsurutani, B. T., and E. J. Smith, Hydromagnetic waves and instabilities associated with cometary-ion pick-up: ICE observations, Geophys. Res. Lett., 13, 263, 1986b.

Tsurutani, B. T., R. M. Thorne, E. J. Smith, J. T. Gosling, and H. Matsumoto, Steepened magnetosonic waves at comet Giacobini-Zinner, J. Geophys. Res., 92, 11,074–11,082, 1987.

Tsurutani, B. T., E. J. Smith, A. L Brinca, and R. M. Thorne, Properties of whistler mode wave packets at the leading edge of steepened magnetosonic waves: Comet Giacobini-Zinner, Planet. Space Sci., 37, 167–182, 1989.

Yumoto, K., T. Saito, and T. Nakagawa, Hydromagnetic waves near O$^+$ (or H$_2$O$^+$) ion cyclotron frequency observed by Sakigake at the closest approach to comet Halley, Geophys. Res. Lett., 13, 825–828, 1986.

ACCELERATION MECHANISMS FOR COMETARY IONS

T. Terasawa

Department of Geophysics, Kyoto University

Abstract. In situ observation showed that efficient particle acceleration occurs in the region around comets. Emphasis has been given on the second-order Fermi acceleration process, by which cometary ions can be efficiently accelerated provided that the hydromagnetic turbulence around comets consists of waves of different phase velocities. In the light of accumulated information on the cometary environment, we shall re-evaluate several acceleration models.

Introduction

After sublimation, cometary molecules are eventually ionized and start to move under the effect of electromagnetic fields of the solar wind. In general geometrical condition, these cometary ions form a velocity space ring. Such distribution is unstable to the excitation of electromagnetic waves, and as a reaction of the wave excitation, ions themselves are pitch-angle scattered to form a velocity space shell. Various observations at comet Halley identified this shell structure of cometary ions in the velocity space, either as one/two dimensional cross sections of the shell, or as fully three dimensional images of the shell [Mukai et al., 1986a,b,c, 1987; Gringauz et al., 1986; Terasawa et al., 1986b,c; Neugebauer et al., 1986, 1987, 1989; Verigin et al., 1987; Coates et al., 1989].

Along with this pitch angle scattering, cometary ions are accelerated up to the maximum pickup energy, $E_{max} = 2m_i V_{sw}^2$, where m_i is the ion mass and V_{sw} the solar wind velocity. Observations at comets Halley and Giacobini-Zinner, however, showed that there appear more energetic ions, whose energies are in excess of the above maximum energy, E_{max} ($\sim$ several tens of keV). It is now believed that some kind of stochastic acceleration process is

Cometary Plasma Processes
Geophysical Monograph 61
©1991 American Geophysical Union

working in solar wind-comet interaction regions [Gloeckler et al., 1986; Hynds et al., 1986; McKenna-Lawlor et al., 1986, 1989; Somogyi et al., 1986; Johnstone et al., 1986; Richardson et al., 1986, 1987, 1988; Daly et al., 1986; Kecskeméty et al., 1989; Coates et al., 1989; Bavassano Cattaneo et al., 1990]. The purpose of this article is to give a brief review of the current understanding of the acceleration mechanisms.

Pitch Angle and Energy Diffusion Process: A Quick Look

The details of the interaction process between newborn ions and hydromagnetic waves have been theoretically studied in depth [Wu and Davidson, 1972; Wu and Hartle, 1974; Ip and Axford, 1982, 1986; Winske et al., 1985; Amata and Formisano, 1985; Omidi and Winske, 1986, 1987; Sagdeev et al., 1986, 1987; Wu, Winske and Gaffey, 1986; Gary et al., 1986, 1988; Gribov et al., 1986, 1987; Isenberg, 1987a,b; Price and Wu, 1987; Goldstein, Roberts, and Mattaeus, 1987; Thorne and Tsurutani, 1987; Goldstein and Wong, 1987; Tsurutani et al., 1987; Ip, 1988; Gary and Madland, 1988; Gombosi, 1988; Brinca and Tsurutani, 1988a, 1988b; Gaffey et al., 1988; Kojima et al., 1989; Barbosa, 1989; Gary, Akimoto, and Winske, 1989; Kaya, Matsumoto, and Tsurutani, 1989; Cravens, 1989; Lee, 1989; Terasawa, 1989; Gombosi et al., 1989; Yoon et al., 1989]. The interested readers are referred to the original references. In this section I shall summarize the essence of the mechanism of stochastic acceleration.

The efficient wave-particle interaction occurs, when the cyclotron resonance condition is satisfied. For positive ions, this resonance is possible with the right-hand polarized waves propagating in the same direction as the ions, or with the left-hand polarized waves propagating in the opposite direction. One important fact is that in the waves propagating unidirectionally parallel (or antiparallel) to the magnetic field direction, only the pitch-angle scattering occurs. This is because that the particle energy is conserved in the frame comoving with the waves. The energy diffusion becomes possible if there exist waves of different phase velocities [e.g. see, Skilling, 1975; Barbosa, 1979].

For the cometary environment, turbulence consisting of waves propagating sunward and anti-sunward along the average magnetic field has been considered. For the origin of these waves, there are two different possibilities: One is to consider the sunward propagating waves excited by cometary ions (cometary waves for abbreviation, CW^+, where + designates sunward propagation) and the anti-sunward propagating waves intrinsic to the solar wind (SWW^-, where − designates anti-sunward propagation). Another possibility is to consider that the cometary waves include both components (CW^+ and CW^-). Giotto observations suggest that the first case is more likely at least in the solar wind upstream of the cometary bow shock [Glassmeier et al., 1989]. The second case might be important in the cometosheath [Glassmeier et al., 1987].

Let an ion start from the $0°$ pitch angle and interact with waves propagating in both directions. Consider a case where the ion first makes a $90°$ change along a circle centered on $(-V_A, 0)$ and then comes back along another circle centered on $(+V_A, 0)$. (Here V_A is the Alfvén velocity). After this excursion, the ion changes its velocity by an amount of $2V_A$ (Figure 1). Similarly, a velocity change ΔV of $\sim V_A$ occurs along with $\sim$one radian excursion of the pitch angle, for which it takes the pitch angle diffusion time,

$$\tau_D \sim \Omega_i^{-1} \left(\frac{\delta B_{\text{eff}}}{B_0} \right)^{-2} \qquad (1)$$

where Ω_i is the ion cyclotron frequency, B_0 is the magnitude of the average magnetic field. The effective amplitude of the turbulence, δB_{eff} is defined as, $\delta B_{\text{eff}}^2 \equiv |k_{\text{res}}|P(k_{\text{res}})$ (k_{res} is the wavenumber for the cyclotron resonance between the ion and waves). The power spectrum density $P(k)$ is normalized as,

$$\int_{-\infty}^{+\infty} P(k)dk = B_{\text{w}}^2 \qquad (2)$$

where B_{w}^2 is the total variance of the turbulent magnetic field.

With τ_D, the velocity space diffusion coefficient D_{VV} is estimated as,

$$D_{VV} = \frac{\Delta V^2}{\tau_D} \sim \frac{V_A^2}{\tau_D} \qquad (3)$$

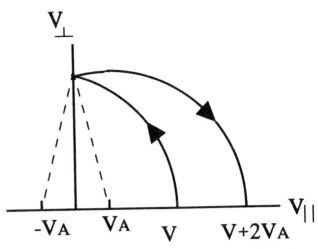

Figure 1. Schematic illustration of pitch angle and energy diffusion process. $V_\parallel$ and $V_\perp$ are velocity components parallel and perpendicular to the magnetic field direction, respectively.

Therefore, the energy diffusion time, τ_E, which is needed to accelerate ions up to $m_i V^2/2$, is estimated as,

$$\tau_E \sim \frac{V^2}{D_{VV}} \sim \tau_D \left(\frac{V}{V_A} \right)^2 \qquad (4)$$

It can be seen that this acceleration process, namely the second-order Fermi process, is a slow process, which takes time by a factor $(V/V_A)^2$ longer than the the pitch angle diffusion process. A further remark is that (4) holds when the waves propagating away and toward the sun have the same amplitudes ($\delta B_{\text{eff}}^+ = \delta B_{\text{eff}}^-$). If they differ, we should use instead,

$$\tau_E \sim \tau_D \left(\frac{V}{V_A} \right)^2 \cdot \frac{((\delta B_{\text{eff}}^+)^2 + (\delta B_{\text{eff}}^-)^2)^2}{4(\delta B_{\text{eff}}^+)^2 \cdot (\delta B_{\text{eff}}^-)^2} \qquad (4')$$

When $(\delta B_{\text{eff}}^-)^2 << (\delta B_{\text{eff}}^+)^2$, (4') becomes,

$$\tau_E \sim \tau_D \left(\frac{V}{V_A} \right)^2 \cdot \frac{(\delta B_{\text{eff}}^+)^2}{4(\delta B_{\text{eff}}^-)^2} \qquad (4'')$$

Spatial Variation of Scattering Efficiency

As expected, the wave amplitude increased as the spacecraft approached the cometary nucleus. For quantitative discussion, we compare the pitch-angle diffusion time τ_D with the pickup time τ_P, or the "resident" time of cometary ions in the solar wind. τ_P is defined as,

$$\tau_P \equiv \frac{\int \frac{ldl\exp(-r/L)}{r^2}}{V_{\text{SW}} \int \frac{dl\exp(-r/L)}{r^2}} \qquad (5)$$

where r is the distance from the cometary nucleus, and L the ionization scale length ($\sim 1\ \text{Gm} = 10^6 \text{km}$) [Coates et al., 1989]. The integration in (5), $\int dl$, is taken along the stream line of the solar wind.

If $\tau_D > \tau_P$, the effect of scattering should be weak, so that we expect to observe an intermediate status from the pickup ring to the shell (namely, a partially filled shell). If, on the other hand, $\tau_D < \tau_P$, the effect of scattering should be strong and we expect that the shell is filled completely. In Figure 2, we plot these time scales against

Pickup Time τ_P
vs.
Pitch Angle Diffusion Time τ_D

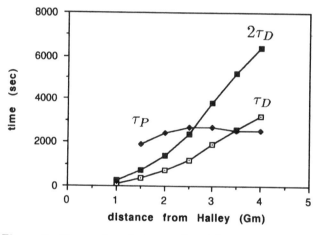

Figure 2. Comparison between the pickup time τ_P and the pitch angle diffusion time τ_D (see text).

the distance from the Halley's nucleus r: The pickup time τ_P is based on the Giotto solar wind observation [Coates et al., 1989, Fig. 7], and the pitch angle diffusion time τ_D (for water-group ions) is calculated theoretically [Gombosi et al., 1989, Fig. 6]. Considering uncertainties in the theoretical estimation, we have plotted $2\tau_D$ also. As seen in the figure, we expect that $(1\text{-}2)\tau_D < \tau_P$ (strong scattering) in the region $r \lesssim 2.5\text{Gm}$, and that $(1\text{-}2)\tau_D > \tau_P$ (week scattering) in the region $r \gtrsim 3.5\text{Gm}$. This cross over distance between τ_D and τ_P ($r \sim 2.5\text{-}3.5\text{Gm}$) is close to the distance where the transition from the partially filled shell to the completely filled shell was observed for water-group ions ($r \sim 2.5\text{Gm}$, [Coates et al., 1989]). Therefore,

the observed change in the pitch-angle distribution is consistent with the theoretical expectation. For cometary protons the anisotropic-isotropic transition was observed at $\sim 1.2\text{Gm}$ from the Halley's nucleus [Neugebauer et al., 1989].

A similar conclusion has been obtained from the ICE observation of comet Giacobini/Zinner [Richardson et al., 1988]: They observed anisotropic-isotropic transition at (0.4-0.8)Gm from the nucleus of comet Giacobini-Zinner. This distance is consistent with the theoretical expectation for this comet.

Models of Cometary Ion Acceleration

In Table 1, stochastic acceleration models which have been proposed to explain the origin of energetic cometary ions are summarized. Primary emphasis has been placed on the second-order Fermi process, in which the energy diffusion occurs as a byproduct of the pitch-angle diffusion. Another process considered so far is the first-order Fermi process, or the diffusive shock acceleration process, which becomes possible through the combination of the pitch-angle diffusion and the convergence of the macroscopic plasma flow at the cometary bow shock. Buti and Lakhina [1987] proposed a different model of a stochastic acceleration process by intense lower hybrid waves. They claimed that O^+ ions could be accelerated up to ~ 32 MeV by intense hybrid wave turbulence. On the other hand, Hizanidis, Cargill, and Papadopoulos [1988] showed that these waves relate to the ion heating process (namely, the energization of bulk ion population) rather than the ion acceleration process (namely, the energization of selected ions). To determine which is a more likely process, heating or acceleration, more work should be done.

For the stochastic acceleration process by hydromagnetic turbulence, we have estimated the acceleration efficiency in section 2. Since the energy diffusion time τ_E is a factor $(V/V_A)^2$ longer than the pitch-angle diffusion time τ_D (eq. (4) or (4')), the acceleration effect is expected to be quite weak in the far upstream region where $\tau_E \gg \tau_D > \tau_P$. In some of the early acceleration models, the acceleration efficiency is assumed high in the far upstream region. This assumption is not consistent with the current view. In the following we shall concentrate on two of the latest models of the acceleration process of cometary ions by Gombosi et al. [1989] and by Barbosa [1989] (see also Isenberg, 1990).

Model by Gombosi et al.

Gombosi et al. [1989] solved the convection-diffusion equation with acceleration terms,

TABLE 1. Stochastic Acceleration Models of Cometary Ions

author(s)		Model	scattering efficiency	wave ingredient/amplitude
Amata & Formisano	(1985)	1st Fermi	moderate	CW ($\sim 30\gamma^2$/Hz)
Ip & Axford	(1986)	2nd Fermi	strong $(\lambda/\rho_g = 1 - 5)^1$	CW$^{+/-}$
Ip	(1988)	2nd Fermi	moderate $(\lambda/\rho_g = 5 - 30)^1$	CW$^{+/-}$
Gribov et al.	(1987)	2nd Fermi	moderate	CW$^{+/-}$ ($\sim 100\gamma^2$/Hz)
Isenberg	(1987b)	2nd Fermi	weak	CW$^+$ ($\sim 500\gamma^2$/Hz) +SWW$^-$ ($\sim 10\gamma^2$/Hz)
Gombosi	(1988)	2nd Fermi	moderate	CW$^{+/-}$ ($\sim 100\gamma^2$/Hz)
Barbosa	(1989)	2nd Fermi	strong (in the cometosheath)	CW$^{+/-}$ ($B_w/B_0 \sim 0.5$)
Gombosi et al.	(1989)	1st+2nd Fermi	weak 2nd Fermi + strong 1st Fermi	CW$^+$+SWW$^-$
Buti & Lakhina	(1987)	electrostatic		lower hybrid waves (~ 15mV/m)

^{1}Here scattering efficiency is represented in terms of the ratio between the mean free path (λ) and the gyroradius of ions (ρ_g).

$$\frac{\partial F}{\partial t} + V_{\text{sw}}\frac{\partial F}{\partial x} = Q(x, V) + \frac{1}{V^2}\frac{\partial}{\partial V}\left(V^2 D_{VV}\frac{\partial F}{\partial V}\right)$$

$$+ \frac{\partial}{\partial x}\left(\kappa_{xx}\frac{\partial F}{\partial x}\right) + \frac{1}{3}\frac{dV_{\text{sw}}}{dx}\frac{\partial F}{\partial \ln V} \qquad (6)$$

where $F = F(t, x, V)$ is the phase space distribution function of cometary ions, $Q(x, V)$ the source function, and κ_{xx} the spatial diffusion coefficient. The second term of the RHS describes the effect of energy diffusion. The last term of the RHS represents the effect of the adiabatic compression. The combination of the spatial diffusion and the adiabatic compression gives the effect of the first-oder Fermi acceleration, which works where the solar wind flow has convergence ($dV_{\text{sw}}/dx < 0$).

In (6), κ_{xx} and D_{VV} depend on the amplitude and spectrum of the hydromagnetic waves. Gombosi et al. assumed that the cometary waves propagate unidirecitionally toward the sun. To have counterstreaming of waves, they included the waves intrinsic to the solar wind propagating away from the sun. The amplitude and spectrum of the cometary waves are estimated on the basis of empirical formula obtained in extensive numerical simulations by Gary et al. [1988].

In the case where only the adiabatic compression is included, Gombosi et al. obtained a quite soft spectrum (Figure 3a). If they combine the adiabatic compression and the energy diffusion, they got a moderate acceleration effect (Figure 3b). However, the resultant spectrum is still too soft to explain the observed spectrum. In the third case, they combine the adiabatic compression and the spatial diffusion, both of which are essential ingredients of the first order Fermi acceleration process. The re-

sultant energy spectrum (Figure 3c) is much harder than the previous case (Fig. 3b), especially in the far upstream region ($r \sim 4$ Gm). Finally, Gombosi et al. included all the effects (Figure 3d). From the change Fig. 3c → Fig. 3d, we see that while the energy diffusion alone is not quite effective, it is effective in boosting up the energy spectrum, which has been cocked by the first order Fermi process.

Gombosi et al. showed that their results are consistent with the energy spectrum of energetic ions (several tens to ~ 300 keV) observed by the Giotto and Vega spacecraft (Figure 4). Note that the cometary shock itself has a rather low Mach number (~ 2) due to the preshock deceleration of the mass-loaded solar wind [see e.g. Galeev, Cravens, and Gombosi, 1985]. However, ions are accelerated not only by a small velocity jump across the shock front but also by the velocity difference over the whole foreshock region, since the mean free path of cometary ions is comparable to the scale length of the foreshock region.

Model by Barbosa/Isenberg

A possible alternative mechanism is the second-order Fermi acceleration process in the cometosheath. Barbosa [1989] proposed that this process works preferentially in the cometosheath, where the strongest hydromagnetic turbulence was observed. What he solved is the steady state equation,

$$\frac{\partial F}{\partial t} = 0 = Q(V) + \frac{1}{V^2}\frac{\partial}{\partial V}\left(V^2 D_{VV}\frac{\partial F}{\partial V}\right) - \frac{F}{\tau(V)} \qquad (7)$$

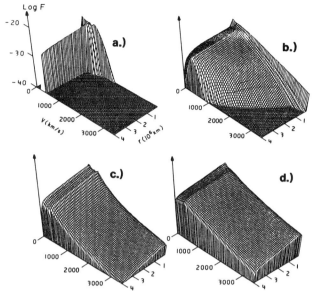

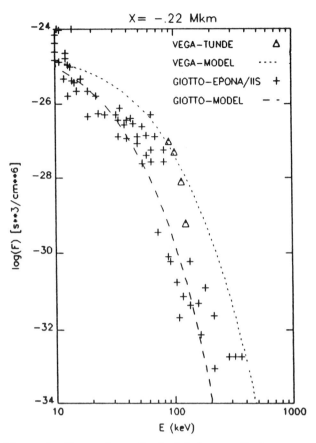

Figure 3. Steady phase space distribution functions $F(r, V)$ of cometary ions obtained with various acceleration processes (reproduced from Gombosi et al. [1989]). The phase space distribution function is given in units of s^3/cm^6. Parameters needed for the model calculation are adjusted to what were observed at comet Halley. Individual panels show solutions with (a) adiabatic compression; (b) adiabatic compression and energy diffusion; (c) adiabatic compression and spatial diffusion; (d) adiabatic compression, energy and spatial diffusions.

The energy spectrum is determined through the balance between the energy diffusion (the second term on the RHS of (7)) and the loss from the system (the third term). In the estimation of the energy diffusion coefficient D_{VV} he assumed that the large amplitude turbulence consists of waves propagating in both sunward and anti-sunward directions. Diffusive escape of ions from the cometosheath (size $2L$) determines the velocity-dependent lifetime $\tau(V)$. In the model, the parameters were adjusted to what were observed at the G/Z encounter (For example, L is set at 10^5 km, V_A at 140 km/s, $B_w/B_0 \sim 0.5$, etc.)

Isenberg [1990] recently re-evaluated the acceleration mechanism proposed by Barbosa, correcting the functional form of the assumed turbulence. In Figure 5, the ICE observation near the comet Giacobini-Zinner (solid squares) is compared with the Barbosa's result (a dashed curve) and the result corrected by Isenberg (a solid curve). Isenberg concluded that the stochastic acceleration process in the cometosheath alone cannot explain the mostly energetic ions observed near the comet Giacobini-Zinner.

Figure 4. Comparison of calculated energetic ion spectra with observations made by the Giotto and VEGA spacecraft at comet Halley. (reproduced from Gombosi et al. [1989]). The data points were taken from McKenna-Lawlor et al. [1989] and Kecskeméty et al. [1989].

Discussion and Summary

We have seen that two different ideas on the acceleration of cometary ions co-exist. One idea is to accelerate ions mainly in the upstream and bow shock regions, and the other is to accelerate ions in the cometosheath. For the first idea, the primary importance is placed on the first-order Fermi process working in the deceleration region of the solar wind [Gombosi et al., 1989]. The second-order Fermi process is contributing to form the energy spectrum, but its contribution is of minor importance. This is because that the efficiency of the latter process is determined by the weak turbulence intrinsic to the solar wind, not by the strong cometary wave turbulence (cf. eq. (4″)). This assumption on the wave characteristic is consistent with the upstream observation [Glassmeier et al., 1989].

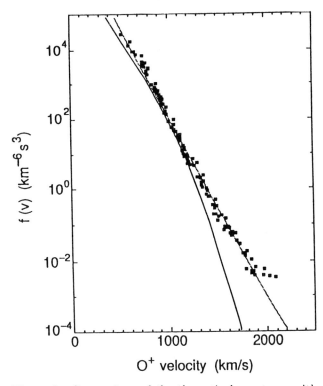

Figure 5. Comparison of the theoretical spectrum with the energetic ion data (assuming O$^+$) of Richardson et al. [1987] obtained during the inbound pass of ICE through the cometosheath of Giacobini-Zinner (reproduced from Isenberg [1990]).

It is noted that the observations do not exclude the existence of counterstreaming cometary waves in the cometosheath. If we assume that the turbulence there consists of counterstreaming waves of equal magnitude, the second-order Fermi process works much more efficiently than it does in the Gombosi's model. For the comet Giacobini-Zinner, however, the amplitude of the turbulence is found not enough to explain the observed energetic ions in terms of the second-order Fermi acceleration process [Isenberg, 1990]. For the case of comet Halley, on the other hand, it remains to be investigated how the second-order acceleration process contributed to the production of energetic ions in the cometosheath.

The crucial point, anyway, is to determine the characteristic of the hydromagnetic turbulence in the cometosheath. Theoretically, we have several possibilities for counterstreaming waves:

(1) Cometary waves excited in the upstream region eventually are convected into the bow shock. At the shock front, the wave mode conversion occurs, so that counter-streaming wave components arise from the upstream (unidirctional) waves [e.g., Westhal and McKenzie, 1969].

(2) In the nonlinear evolution of large amplitude hydromagnetic waves, such as the nonlinear cascade process [e.g. Goldstein et al., 1987] or the inverse cascade process [e.g. Terasawa et al., 1986], the counterstreaming wave components are excited.

(3) If the angle θ between the solar wind flow and the magnetic field is close to 90°, the pickup ring distribution of cometary ions, which is formed right after the ionization, becomes unstable not only to the excitation of the right-hand resonant waves but also to the excitation of the left-hand resonant waves [Gary and Madland, 1988], so that counterstreaming waves can appear. Although the saturation level of these waves for the case of $\theta \sim 90°$ is much lower than that for $\theta \sim 0°$ [Gary et al., 1989], the actual magnitude of these waves can be large due to the large production rate of cometary ions in the cometosheath.

Adding to the first- or second-Fermi processes, we can think of the following possibilities: Since compressional components exist in the observed hydromagnetic turbulence in the cometosheath, the transit-time damping process becomes possible [Glassmeier et al., 1987]. If the amplitude of compressional waves exceeds $\sim 10\%$ of B_0, ions could be accelerated also through the cyclotron subharmonic resonance [Terasawa and Nambu, 1989]. To assess these possibilities we need further quantitative treatments.

After the in situ observation of the comet-solar wind interaction region, we now realize that this is a "gold mine" of various plasma processes. While the pitch-angle diffusion process is proved to be well described by the quasi-linear theory, much remains to be done to get the full understanding of the acceleration process of cometary ions.

References

Acuña, M. H., K. H. Glassmeier, L. F. Burlaga, F. M. Neubauer, and N. F. Ness, Upstream waves of cometary origin detected by the Giotto magnetic field experiment, *Proc. 20th ESLAB Symposium on the exploration of Halley's Comet, ESA SP-250*, *3*, 447-449, 1986.

Amata, E., and V. Formisano, Energization of positive ions in the cometary foreshock region, *Planet. Space Sci.*, *33*, 1243-1250, 1985.

Barbosa, D. A., Stochastic acceleration of solar flare protons, *Astrophys. J.*, *233*, 383-394, 1979.

Barbosa, D. A., Stochastic acceleration of cometary pickup ions: The classic leaky box model, *Astrophys. J.*, *341*, 493-496, 1989.

Bavassano Cattaneo, M. B., V. Formisano, E. Amata, R. Giovi, and P. Torrente, Acceleration of pickup ions in the vicinity of comet Halley, *this issue*, 1990.

Brinca, A. L., and B. T. Tsurutani, Survey of LF electromagnetic waves stimulated by two coexisting newborn ion species, *J. Geophys. Res.*, *93*, 48-58, 1988a.

Brinca, A. L., and B. T. Tsurutani, Temperature effects on the pickup process of water-group and hydrogen ions: extensions of "A theory for low-frequency waves observed at comet Giacobini-Zinner", by M. L. Goldstein and H. K. Wong, *J. Geophys. Res.*, *93*, 243-246, 1988b.

Buti, B., and G. S. Lakhina, Stochastic acceleration of cometary ions by lower hybrid waves, *Geophys. Res. Lett.*, *14*, 107-110, 1987.

Coates, A. J., A. D. Johnstone, B. Wilken, K. Jockers, and K.-H. Glassmeier, Velocity space diffusion of pickup ions from the water group at comet Halley, *J. Geophys. Res.*, *94*, 9983-9993, 1989.

Cravens, T. E., Test particle calculations of pick-up ions in the vicinity of comet Giacobini-Zinner, *Planet. Space Sci.*, *37*, 1169-1184, 1989.

Daly, P. W., E. Kirsch, S. McKenna-Lawlor, D. O'Sullivan, A. Thomspon, T. R. Sanderson, K.-P. Wenzel, Comparison of energetic ion measurements at comets Giacobini-Zinner and Halley, *Proc. 20th ESLAB Symposium on the exploration of Halley's Comet, ESA SP-250, 3*, 179-183, 1986.

Gaffey, J. D., D. Winske, and C. S. Wu, Time-scales for the formation and spreading of velocity shells of pickup ions in the solar wind, *J. Geophys. Res.*, *93*, 5470-5486, 1988.

Galeev, A. A., T. E. Cravens, and T. I. Gombosi, Solar wind stagnation near comets, *Astrophys. J.*, *289*, 807-819, 1985.

Gary, S. P., S. Hinata, C. D. Madland, and D. Winske, The development of shell-like distributions from newborn cometary ions, *Geophys. Res. Lett.*, *13*, 1364-1367, 1986.

Gary, S. P., and C. D. Madland, Electromagnetic ion instabilities in a cometary environment, *J. Geophys. Res.*, *93*, 235-241, 1988.

Gary, S. P., C. D. Madland, N. Omidi, and D. Winske, Computer simulations of two-ion pickup instabilities in a cometary environment, *J. Geophys. Res.*, *93*, 9584-9596, 1988.

Gary, S. P., K. Akimoto, and D. Winske, Computer simulations of cometary-ion/ion instabilities and wave growth, *J. Geophys. Res.*, *94*, 3513-3525, 1989.

Glassmeier, K.-H., F. M. Neubauer, M. H. Acuña, and F. Mariani, Low-frequency magnetic field fluctuations in comet P/Halley's magnetosheath: Giotto observations, *Astron. Astrophys.*, *187*, 65-68, 1987.

Glassmeier, K.-H., A. J. Coates, M. H. Acuña, M. L. Goldstein, A. D. Johnstone, F. M. Neubauer, and H. Rème, Spectral characteristics of low-frequency plasma turbulence upstream of comet P/Halley, *J. Geophys. Res.*, *94*, 37-48, 1989.

Gloeckler, G. D., D. Hovestadt, F. M. Ipavich, M. Scholer, B. Klecker, and A. B. Galvin, Cometary pick-up ions observed near Giacobini-Zinner, *Geophys. Res. Lett.*, *13*, 251-254, 1986.

Goldstein, M. L., and H. K. Wong, A theory for low-frequency waves observed at comet Giacobini-Zinner, *J. Geophys. Res.*, *92*, 4695-4700, 1987.

Goldstein, M. L., D. A. Roberts, and W. H. Matthaeus, Numerical simulation of the generation of turbulence from cometary ion pick-up, *Geophys. Res. Lett.*, *14*, 860-863, 1987.

Gombosi, T. I., Preshock region acceleration of implanted cometary H^+ and O^+, *J. Geophys. Res.*, *93*, 35-47, 1988.

Gombosi, T. I., K. Lorencz, and J. R. Jokipii, Combined first- and second-order Fermi acceleration in cometary environment, *J. Geophys. Res.*, *94*, 15,011-15,023, 1989.

Gribov, B. E., K. Kecskeméty, R. Z. Sagdeev, V. D. Shapiro, V. I. Shevchenko, A. J. Somogi, K. Szegö, G. Erdös, E. G. Eroshenko, K. I. Gringauz, E. Keppler, R. G. Marsden, A. P. Remizov, A. K. Richter, W. Riedler, K. Schwingenschuh, and K. P. Wenzel, Stochastic Fermi acceleration of ions in the pre-shock region of comet Halley, *Proc. 20th ESLAB Symposium on the exploration of Halley's Comet, ESA SP-250, 1*, 271-275, 1986.

Gribov, B. E., K. Kecskeméty, R. Z. Sagdeev, V. D. Shapiro, V. I. Shevchenko, A. J. Somogi, K. Szegö, G. Erdös, E. G. Eroshenko, K. I. Gringauz, E. Keppler, R. G. Marsden, A. P. Remizov, A. K. Richter, W. Riedler, K. Schwingenschuh, and K. P. Wenzel, Stochastic Fermi acceleration of ions in the pre-shock region of comet Halley, *Astron. Astrophys.*, *187*, 293-296, 1987.

Gringauz, K. I., T. I. Gombosi, A. P. Remizov, I. Apàthy, I. Szemerey, M. I. Verigin, L. I. Denchikova, A. V. Dyachkov, E. Keppler, I. N. Klimenko, A. K. Richter, A. J. Somogyi, K.Szegö, M. Tátrallyay, A. Varga, and G. A. Vladimirova, First *in situ* plasma and neutral gas measurements at comet Halley, *Nature*, *321*, 282-285, 1986.

Hizanidis, K., P. J. Cargill, and K. Papadopoulos, Lower hybrid waves upstream of comets and their implications for the comet Halley "Bow wave", *J. Geophys. Res.*, *93*, 9577-9583, 1988.

Hynds, R. J., S. W. H. Cowley, T. R. Sanderson, K.-P. Wenzel, and J. J. Van Rooijen, Observations of energetic ions from comet Giacobini-Zinner, *Science*, *232*, 361-365, 1986.

Ip, W.-H., Cometary ion acceleration processes, *Computer Phys. Commun.*, *49*1-7, 1988.

Ip, W.-H., and W. I. Axford, Theories of physical processes in the cometary comae and ion tails, in *Comets*, ed. L. L. Wilkening, The Univ. Arizona Press, Tuscon, pp. 588-634, 1982.

Ip, W.-H., and W. I. Axford, The acceleration of particles in the vicinity of comets, *Planet. Space Sci.*, *34*, 1061-1065, 1986.

Isenberg, P. A., The evolution of interstellar pickup ions in the solar wind, *J. Geophys. Res.*, *92*, 1067-1073, 1987a.

Isenberg, P. A., Energy of pickup ions upstream of comets, *J. Geophys. Res.*, *92*, 8795-8799, 1987b.

Isenberg, P. A., Comment on "stochastic acceleration of cometary pickup ions: The classical leaky box model" by D. D. Barbosa, submitted to *Astrophys. J.*, 1990.

Johnstone, A., A. Coates, S. Kellock, B. Wilken, K. Jockers, H. Rosenbauer, W. Studemann, W. Weiss, V. Formisano, E. Amata, R. Cerulli-Irelli, M. Dobrowolny, R. Terenzi, A. Egidi, H. Borg, B. Hultquist, J. Winningham, C. Gurgiolo, D. Bryant, T. Edwards, W. Feldman, M. Thomsen, M. K. Wallis, L. Biermann, H. Schmidt, R. Lust, G. Haerendel, and G. Paschmann, Ion flow at comet Halley, *Nature*, *321*, 344-347, 1986.

Kaya, N., H. Matsumoto, and B. T. Tsurutani, Test particle simulation study of whistler wave packets observed near comet Giacobini-Zinner, *Geophys. Res. Lett.*, *16*, 25-28, 1989.

Kecskeméty, K., T. E. Cravens, V. V. Afonin, G. Erdös, E. G. Eroshenko, Lu Gan, T. I. Gombosi, K. I. Gringauz, E. Keppler, I. N. Klimenko, R. Marsden, A. F. Nagy, A. P. Remizov, A. K. Richter, W. Riedler, K. Schwingenschuh, A. J. Somogyi, K. Szegö, M. Tátrallyay, A. Varga, M. I. Verigin, and K.-P. Wenzel, Pickup ions in the unshocked solar wind at comet Halley, *J. Geophys. Res.*, *94*, 185-196, 1989.

Kojima, H., H. Matsumoto, Y. Omura, B. T. Tsurutani, Nonlinear evolution of high frequency R-mode waves excited by water group ions near comets: Computer experiments, *Geophys. Res. Lett.*, *16*, 9-12, 1989.

Lee, M. A., ULF waves at comets, ed. B. T. Tsurutani and H. Oya, *Geophysical Monograph*, *53*, American Geophysical Union, Washington, D. C., pp. 13-29, 1989.

McKenna-Lawlor, S., E. Kirsch, D. O'Sullivan, A. Thompson, and K.-P. Wenzel, Energetic ions in the environment of comet Halley, *Nature*, *321*, 347-349, 1986.

McKenna-Lawlor, S., S. P. Daly, E. Kirsch, B. Wilken, D. O'Sullivan, A. Thompson, K. Kecskeméty, A. Somogyi, and A. Coates, In situ energetic particle observations at comet Halley recorded by instrumentation aboard the Giotto and Vega 1 missions, *Ann. Geophys.*, *7*, 121, 1989.

Mukai, T., W. Miyake, T. Terasawa, M. Kitayama, and K. Hirao, Plasma observation by Suisei of solar-wind interaction with comet Halley, *Nature*, *321*, 299-303, 1986a.

Mukai, T., M. Kitayama, T. Terasawa, W. Miyake, and K. Hirao, Plasma characteristics around comet Halley observed by Suisei, *Proc. 20th ESLAB Symposium on the exploration of Halley's Comet, ESA SP-250*, *1*, 71-75, 1986c.

Neugebauer, M., A. J. Lazarus, K. Altwegg, H. Balsiger, B. E. Goldstein, R. Goldstein, F. M. Neubauer, H. Rosenbauer, R. Schwenn, E. G. Shelley, and E. Ungstrup, The pick-up of cometary protons by the solar wind, *Proc. 20th ESLAB Symposium on the exploration of Halley's Comet, ESA SP-250*, *1*, 19-23, 1986.

Neugebauer, M., A. J. Lazarus, K. Altwegg, H. Balsiger, B. E. Goldstein, R. Goldstein, F. M. Neubauer, H. Rosenbauer, R. Schwenn, E. G. Shelley, and E. Ungstrup, The pick-up of cometary protons by the solar wind, *Astron. Astrophys.*, *187*, 21-24, 1987.

Neugebauer, M., A. J. Lazarus, H. Balsiger, S. A. Fuselier, F. M. Neubauer, and H. Rosenbauer, The velocity distributions of cometary protons picked up by the solar wind, *J. Geophys. Res.*, *94*, 5227-5239, 1989.

Omidi, N., and D. Winske, Simulation of the solar wind interaction with the outer regions of the coma, *Geophys. Res. Lett.*, *13*, 397-400, 1986.

Omidi, N., and D. Winske, A kinetic study of solar wind mass loading and cometary bow shock, *J. Geophys. Res.*, *92*, 13,409-13,426, 1987.

Price, C. P., and C. S. Wu, The influence of strong hydromagnetic turbulence on newborn cometary ions, *Geophys. Res. Lett.*, *14*, 856-859, 1987.

Richardson, I. G., S. W. H. Cowley, V. Moore, K. Staines, R. J. Hynds, T. R. Sanderson, K.-P. Wenzel, and P. W. Daly, Spectra and bulk parameters of energetic heavy ions in the vicinity of comet P/Giacobini-Zinner, *Proc. 20th ESLAB Symposium on the exploration of Halley's Comet, ESA SP-250*, *3*, 441-445, 1986.

Richardson, I. G., S. W. H. Cowley, R. J. Hynds, C. Tranquille, T. R. Sanderson, K.-P. Wenzel, and P. W. Daly, Observation of energetic water-group ions at comet Giacobini-Zinner: Implications for ion acceleration processes, *Planet. Space Sci.*, *35*, 1323-1345, 1987

Richardson, I. G., S. W. H. Cowley, R. J. Hynds, P. W. Daly, T. R. Sanderson, and K.-P. Wenzel, Properties of energetic water-group ions in the extended pick-up region surrounding comet Giacobini-Zinner, *Planet. Space Sci.*, *36*, 1429-1450, 1988

Sagdeev, R. Z., V. D. Shapiro, V. I. Shevchenko, and K. Szegö, MHD turbulence in the solar wind-comet interaction region, *Geophys. Res. Lett.*, *13*, 85-88, 1986.

Sagdeev, R. Z., V. D. Shapiro, V. I. Shevchenko, and K. Szegö, The effect of mass loading outside cometary bow shock for the plasma and wave measurements in the coming cometary mission, *J. Geophys. Res.*, *92*, 1131-1137, 1987.

Skilling, J., Cosmic ray streaming-1, Effect of Alfven waves on particles, *Mon. Not. R. astr. Soc.*, *172*, 557-566, 1975.

Somogyi, A. J., K. I. Gringauz, K. Szegö, L. Szabó, Gy. Kozma, A. P. Remizov, J. Erö Jr, I. N. Klimenko, I. T.-Szücs, M. I. Verigin, J. Windberg, T. E. Cravens, A. Dyachkov, G. Erdös, M. Faragó, T. I. Gombosi, K. Kecskeméty, E. Keppler, T. Kovács Jr, A. Kondor, Y. I. Logachev, L. Lohonyai, R. Marsden, R. Redl, A. K. Richter, V. G. Stolpovskii, J. Szabó, I. Szentpétery, A. Szepesváry, M. Tátrallyay, A. Varga, G. A. Vladimirova, K. P. Wenzel, and A. Zarándy, First observation of energetic particles near comet Halley, *Nature*, *321*, 285-288, 1986.

Terasawa, T., M. Hoshino, J.-I. Sakai, and T. Hada, Decay instability of finite-amplitude circularly polarized Alfven Waves: A numerical simulation of stimulated Brillouin Scattering, *J. Geophys. Res.*, *91*, 4171-4187, 1986a.

Terasawa, T., T. Mukai, W. Miyake, M. Kitayama, and K. Hirao, Detection of cometary pickup ions up to 10^7 km from comet Halley: Suisei observation, *Geophys. Res. Lett.*, *13*, 837-840, 1986b.

Terasawa, T., S. Takahashi, T. Mukai, M. Kitayama, W. Miyake, and K. Hirao, Ion-pickup/mass-loading process around Halley observed by Suisei, *Proc. 20th ESLAB Symposium on the exploration of Halley's Comet, ESA SP-250*, *1*, 281-283, 1986c.

Terasawa, T., Particle scattering and acceleration in a turbulent plasma around comets, ed. B. T. Tsurutani and H. Oya, *Geophysical Monograph*, *53*, American Geophysical Union, Washington, D. C., pp. 41-49, 1989.

Terasawa, T., and M. Nambu, Ion heating and acceleration by magnetosonic waves via cyclotron subharmonic resonace, *Geophys. Res. Lett.*, *16*, 357-360, 1989.

Thorne, R. M., and B. T. Tsurutani, Resonant interactions between cometary ions and low frequency electromagnetic waves, *Planet. Space Sci.*, *35*, 1501, 1987.

Tsurutani, B. T., R. M. Thorne, E. J. Smith, J. T. Gosling, and H. Matsumoto, Steepened magnetosonic waves at comet Giacobini-Zinner, *J. Geophys. Res.*, *92*, 11,074-11,082, 1987.

Verigin, M. I., K. I. Gringauz, A. K. Richter, T. I. Gombosi, A. P. Remizov, K. Szegö, I. Apáthy, I. Szemerey, M. Tátrallyay, and L. A. Lezhen, Plasma properties from the upstream region to the cometopause of comet P/Halley: Vega observations, *Astron. Astrophys.*, *187*, 121-124, 1987.

Westhal, K. O., and J. F. McKenzie, Interaction of magnetoacoustic and entropy waves with normal magnetohydrodynamic shock waves, *Phys. Fluids*, *12*, 1228-1236, 1969.

Winske, D., C. S. Wu, Y. Y. Li, Z. Z. Mou, and S. Y. Gao, Coupling of newborn ions to the solar wind by electromagnetic instabilities and their interaction with the bow shock, *J. Geophys. Res.*, *90*, 2713-2726, 1985.

Wu, C. S. and R. C. Davidson, Electromagnetic instabilities produced by neutral particle ionization in the interplanetary space, *J. Geophys. Res.*, *77*, 5399-5406, 1972.

Wu, C. S. and R. E. Hartle, Further remarks on plasma instabilities produced by ions born in the solar wind, *J. Geophys. Res.*, *79*, 283-285, 1974.

Wu, C. S., D. Winske, and J. D. Gaffey, Rapid pickup of cometary ions due to strong magnetic turbulence, *Geophys. Res. Lett.*, *13*, 865-868, 1986.

Yoon, P. H., M. E. Mandt, and C. S. Wu, Evolution of an unstable shell distribution of pickup cometary ions, *Geophys. Res. Lett.*, *16*, 1473-1476, 1989.

Toshio Terasawa, Department of Geophysics, Kyoto University, Sakyo-ku, Kyoto 606, Japan.

THE SECOND ORDER FERMI ACCELERATION OF PICK-UP IONS

Peter Duffy

School of Cosmic Physics, Dublin Institute for Advanced Studies
5 Merrion Square, Dublin 2, Ireland

Abstract. A theoretical model for the second order Fermi acceleration of photo-ionised, pick-up ions is presented. This acceleration is caused by the scattering off magnetic turbulence in the solar wind upstream of the comet's bow shock. Exact solutions for the energetic particle spectrum are found and compared with measurements made near comet Halley.

Introduction

Particles with energies up to 500 keV were detected during the encounters with comet Halley [Somogyi et al.,1986; and McKenna-Lawlor et al.,1986]. The mechanisms responsible for the acceleration from pick-up to these observed energies have been reviewed in the past [Ip and Axford,1986; Teresawa,1991]. This paper addresses the contribution of second order Fermi acceleration to the overall process with an analytic, time dependent model.

Second order Fermi acceleration occurs when two or more wave branches are present in a plasma. Under these conditions there is no reference frame in which the turbulence is magnetostatic. Consequently net energy gains and losses are possible. For the specific case of forward and reverse Alfven waves it can be shown [Skilling, 1975] that this process is described by momentum space diffusion with coefficient,

$$D = \langle \frac{(\Delta p)^2}{2\Delta t} \rangle = 4 \frac{v_A^2}{v^2} p^2 \langle \frac{1-\mu^2}{2} \frac{v_+ v_-}{v_+ + v_-} \rangle + O(\frac{v_A^3}{v^3})$$

(1)

where $v_\pm$ can be interpreted as collision frequencies with the forward and reverse waves relative to the ambient magnetic field. In the upstream region of a comet's bow shock, and in what follows, waves travelling toward the sun are generated by the pick-up ions while those in the opposite direction are the ambient solar wind waves.

Acceleration with an Extended Source

In the model below [Duffy,1989] molecules originate from the comet's nucleus, are photoionised and then picked-up by the local interplanetary fields on a scale length L. In the solar wind frame this process produces a monoenergetic spectrum with about 1 keV per a.m.u.. These ions are scattered and accelerated by the MHD turbulence described above. If the background flow is taken to be uniform (i.e. adiabatic compression is ignored) and the solution is derived in the solar wind frame, the source term becomes (in one dimension),

$$Q(x,p,t) = \frac{Q_0 \delta(p - p_0)}{4\pi p_0^2 L} e^{(x+Ut)/L}$$

(2)

This describes injection at momentum p_0 extended over the upstream region $(x < 0)$ with a scale length L. There is no r^{-2} decay in this term as the present treatment is one dimensional. However, it is possible to derive solutions for more complicated source terms as will be shown later. It is also necessary to determine the momentum dependence of the collision frequencies to insert in equation (1). If it is assumed that the wave spectrum obeys a power law in wavenumber,

$$I_\pm = I_\pm^0 k^{-a}$$

and if only the first resonant interaction is considered, $k = m\Omega_0/p$, then it can be shown that equation (1) becomes,

$$D = \frac{4\pi\mu_0 v_A^2 q^2 m}{3} \frac{I_+ I_-}{I_+ + I_-} p^{a-1} = D_0 p^{a-1}$$

(3)

With spatial diffusion ignored (which amounts to omitting first order Fermi acceleration at the bow shock) the particle distribution function then evolves according to,

$$\frac{\partial f}{\partial t} = \frac{1}{p^2} \frac{\partial}{\partial p} [D_0 p^{a+1} \frac{\partial f}{\partial p}] + \frac{Q_0 \delta(p - p_0)}{4\pi p_0^2 L} e^{(x+Ut)/L}$$

(4)

If the injection begins at t = 0 this becomes an initial value problem and equation (4) may be solved by a Laplace transform with respect to time [Duffy 1989]. This can be carried out for a range of values of a. For the relevant case of a = 1 the differential number density becomes,

Cometary Plasma Processes
Geophysical Monograph 61
©1991 American Geophysical Union

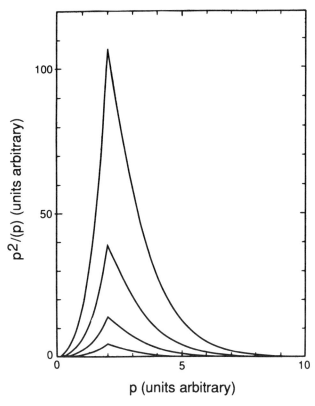

Fig.1 The evolution of the differential number density spectrum. The spectrum consists of both freshly injected particles (peak at p=2) and those which have been accelerated.

$$p^2 \, f(x,p,t) =$$

$$\frac{Q_0 e^{(x+Ut)/L}}{16\pi L^2 D_0 h^{3/2}} \left[\frac{p}{p_0} \right] \left[4e^{-hp} \sinh(hp_0) + H(p,t) \right]$$

$$(5)$$

where,

$$h = \sqrt{\frac{U}{LD_0}}$$

and,

$$H(p,t) = \frac{4}{\pi} \sqrt{\frac{U}{L}} \int_0^{\infty} \frac{e^{-(z^2 + U/L)t}}{z^2 + U/L}$$

$$\left[\cos(p + p_0) \frac{z}{\sqrt{D_0}} - \cos(p - p_0) \frac{z}{\sqrt{D_0}} \right] dz$$

which can be expressed as a combination of error functions. The evolution of the differential number density is plotted out in Figure 1 with arbitrary units. At any given time the spectrum consists of both freshly injected particles (peak at p=2) plus those particles which have been accelerated. This

solution differs from that of Ip and Axford [1986] who do not include the photo-ionisation length in injection and only consider a constant rate of change of kinetic energy. Isenberg [1987] includes an exponential term in the source function but derives a steady state solution.

Returning to the question of the source term it is clear that the differential operator in equation (4) has no dependence on L. Consequently it is possible, by a formal trick [Duffy, 1989], to derive solutions for new source functions in terms of equation (5).

Comparison with Observations

The following values are taken from Ip and Axford [1986], Gribov et al. [1987] and Isenberg [1987]. The solar wind speed, Alfven speed and photo-ionisation length are 4 x 10^5 ms^{-1}, 50 x 10^3 ms^{-1} and 10^9 m respectively. Gas is produced at a rate of 1.2 x 10^{30} mol^{-1} while the power spectrum density is 10^3(nT)^2 at 10^{-3} Hz. An assumption must now be made about ion species as the EPA experiment on board GIOTTO could not discriminate between molecules of different type. It is a useful first approximation to assume that all the molecules measured were of the water group (H$_2$0$^+$). This gives an initial pick-up energy of 18keV.

The only remaining parameter that needs to be inserted in equation (5), for comparison with measurements at a particular point, is time.   From equation (1) and equation (3)

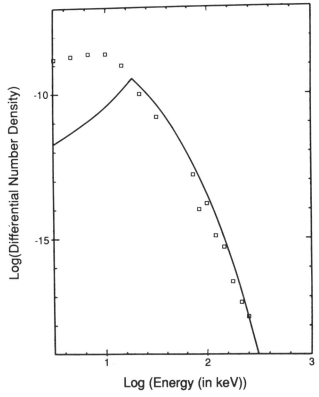

Fig.2 Logarithmic plot of differential number density vs energy (in keV) for x= 0.0 m. Data (open circles) are taken from Giotto EPONA experiment in the interval 19:25:39 to 19:45:55 UT on March 13th, 1986.

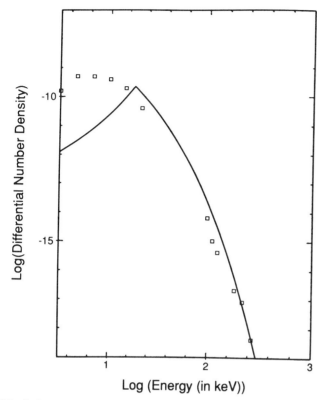

Fig.3 Logarithmic plot of differential number density vs energy (in keV) for x= -4 x 10⁶ m. Data (open circles) are taken from Giotto EPONA experiment in the interval 18:00:27 to 18:20:39 UT on March 13th, 1986.

Conclusions

The analysis and results above have shown that a one dimensional, time dependent treatment of the second order Fermi acceleration of ions injected mono-energetically over an extended region of space goes at least some way towards explaining the acceleration from pick-up to higher energies. In spite of the disagreement below injection it would seem reasonable therefore to conclude that this acceleration mechanism plays an important, if not (at least in the upstream region) decisive, role in producing energetic particles in a cometary region.

References

Duffy P., The acceleration of cometary ions by Alfven waves, *J. Plasma Phys.* 42, 13-25, 1989.

Gribov, E.E., Kecskemety, K., Sagdeev, R.Z., Shapiro, V.D., Shevchenko, V.L., Somogyi, A.J., Szego, K., Erdos, E.G., Gringauz, K.I., Keppler, E., Marsden, R.G., Remizov, A.P., Richter, A.K., Riedler, W., Schwingenschuh, K., Wenzel, K.P.. Stochastic Fermi acceleration of ions in the preshock region of Comet P/Halley, *Astron. Astrophys.* 187, 293-296, 1987.

Ip, W.-H., Axford, W.I..The acceleration of particles in the vicinity of comets, *Planet. Space Sci.*, 34, 1061-1065, 1986.

Isenberg, P.A.. Energy diffusion of pickup ions upstream of comets. *J. Geophys. Res.*, 92, 8795-8799, 1987.

McKenna-Lawlor S. E., E. Kirsch, D. O'Sullivan et al, Energetic ions in the environment of comet Halley, *Nature,* 321, 347, 1986.

Somogyi, A.J., Gringauz, K.I., Szego, K., Szabo, L., Kosma, D., Remizov, A.P., Ero, J., Klimenko, I.N., Szucs, I.-T., Verigin, M.I., Windberg, J., Cravens, T.E., Dyachkov, A., Eroda, G., Farago, M., Gombosi, T.I., Kecskemety, K., Keppler, E., Kovacs, T., Logachev, Yu.I., Lohonyai., L., Marsden, R., Redl, R., Richter, A.K., Stolpovskii, V.G., Szabo, L., Szentpetery, I., Szepesvary, A., Tatrallyay, M., Varga, A., Vladimirova, G.A.. First observations of energetic particles near comet Halley. *Nature,* 321, 285-287, 1986.

Skilling J., Cosmic ray streaming-I effect of Alfven waves on particles, *Mon.Not.R.astr.Soc.*, 172, 557-566, 1975.

Teresawa, T., Acceleration mechanisms for cometary ions, *these proceedings*, 1991.

it can be seen that the mean square change in momentum in a given time is $2D_o p^{a-1} \Delta t$. Inserting the above values, the acceleration timescale $(p/\dot{p})$ then becomes $t_a \sim 10^5 s$. In the plots in Figures 2 and 3, $t = 2.5 t_a$. The data are taken from McKenna-Lawlor et al. [1987].

It is readily apparent from the above plots that the second order Fermi mechanism can provide enough acceleration to explain the spectrum above injection after reasonable timescales. There is, however, a problem below the pick-up energy which may be due to either the exclusion of all other ion species or a failure of the theory at these low speeds.

ION PICKUP, SCATTERING, AND STOCHASTIC ACCELERATION
IN THE COMETARY ENVIRONMENT OF P/GIACOBINI-ZINNER

D.D. Barbosa

Institute of Geophysics and Planetary Physics, University of California,
Los Angeles, California 90024

Abstract. 'This paper gives a brief review of
observations and theory related to the scattering
and acceleration of cometary pickup ions with
emphasis on comet P/Giacobini-Zinner. A comparison of the regions upstream and downstream of the
bow shock is made to assess the relative merits of
each as a site for stochastic acceleration of ions
above the pickup energy through interaction with
low-frequency magnetohydrodynamic waves. In the
far upstream region the data are most consistent
with a model where pickup ions generate a low
level of MHD waves but remain relatively scatter-
free following trajectories transverse to the
interplanetary magnetic field at an oblique angle
to the solar wind velocity. In the downstream
region the intense level of magnetic fluctuations
gives rise to a rapid isotropization of the ions
and a second-order stochastic (Fermi) acceleration.
The properties of the MHD power spectrum are
related to the energetic ion spectrum in the
framework of a leaky box model where the bulk of
the acceleration occurs downstream of the shock
throughout the cometosheath. Very good agreement
of the observations with theory is evident for
both P/Giacobini-Zinner and P/Halley.

Introduction

The recent spacecraft flybys of comets
Giacobini-Zinner and Halley have provided much
valuable information on the plasma and magnetic
field environment of the objects and have confirmed many early theoretical notions regarding
the nature of the cometary interaction with the
solar wind (Ip and Axford, 1982; Mendis and Houpis,
1982). While a lot of new and interesting results
have been obtained providing fresh insights into
cometary physics, at the same time some results
have brought to light several lingering questions
concerning the nature of the ion pickup process.
One major issue dealt with whether newly
created ions would experience a strong interaction
with the solar wind through collective plasma wave

processes. Prior to the comet encounters, it was
generally believed that pickup ions would be
rapidly scattered in pitch angle and in energy
resulting in a thermalization of the fast ions
defined as a coalescence in energy of the pickup
and thermal ion populations (Ip and Axford, 1982).
The in situ data indicated however that no substantial energy degradation was occurring and that
the identity of the pickup ions was preserved --
in fact, the data suggested on the contrary that a
further energization of the ions was required.
Still, evidence was presented that an isotropization of the pickup ions in the plasma rest frame
was occurring in a region $\stackrel{<}{\sim} 10^5$ km around the comet
(the so-called mass loading region) to bring the
pickup and thermal ions to a common bulk velocity
(Gloeckler et al., 1986; Richardson et al., 1986).

Several years have passed since the encounters
took place during which time the data analysis has
progressed significantly and theory has matured.
It is appropriate now to review critically how
things stand and where they are or should be going.
The emphasis in this paper is on cometary pickup
ions with regard to whether they follow scatter-
free trajectories or are scattered by small scale
irregularities in the solar wind magnetic field.
What seems evident from the Giacobini-Zinner
results is that there are spatial regions where
each situation prevails. Close to the comet
($\stackrel{<}{\sim} 10^5$ km) rapid pitch angle scattering is likely
occurring where the magnetic fluctuations are
large (Tsurutani and Smith, 1986), whereas farther
away the magnetic turbulence is weaker and the
effects of it may be expected to be diminished.
In the far region it is convenient to study the
limit of no particle scattering and purely transverse pickup by the motional electric field, a
situation discussed earlier by Wallis and Johnstone (1982). In this case when the interplanetary magnetic field (IMF) is not at right angles
to the solar wind velocity, i.e., $\theta_{BV} \neq 90°$, the
pickup ions drift in the $\vec{E} \times \vec{B}$ direction producing
a hot cometary pickup ion tail which is aberrated
from the direction of the solar wind velocity.
In the near region the scattering is intense
enough that the pickup ions are rendered isotropic
in the solar wind frame and obliged to move with a

Cometary Plasma Processes
Geophysical Monograph 61

bulk velocity equal to the solar wind velocity. A by-product of the pitch angle scattering by MHD waves is that some energy diffusion to second order in the ratio of the Alfvén speed to particle speed occurs. Thus, the MHD wave-particle scattering theory also accounts naturally for the presence of super-Alfvénic ions at energies far in excess of the pickup energy as well as the isotropic pitch angle distributions that are observed in the near-comet region. The dividing line between these two distinct spatial regions is the comet's bow shock, and thus the analysis conveniently separates into a discussion of pickup ion motion and dynamics upstream and downstream of the shock. In what follows we shall compare the important observational features of the two regions with a view towards demonstrating that the primary acceleration region is located downstream of the shock. We then examine in more detail the aspect of ion pickup upstream of the shock in the scatter-free approximation. The downstream region is then discussed in terms of a leaky box model of rapid pitch angle scattering, spatial diffusion, and second-order stochastic Fermi acceleration with a comparison of the theory to Giacobini-Zinner observations.

Before proceeding we should note the work of Richardson et al. (1988) whose analysis of Giacobini-Zinner ion data led to the conclusion that a strong scattering regime existed downstream of the shock and a weak scattering regime far upstream of the shock. However, these authors also found an adjoining region immediately upstream of the shock extending from about 1×10^5 km to 4×10^5 km in radial distance where an intermediate scattering regime was present. Based on several properties of the low-energy ion fluxes in the anti-sunward direction, the pickup ions were inferred to be "quasi-isotropic" in the solar wind frame in this preshock region. They would thus argue that a sharp distinction between the upstream scatter-free region and the downstream strong scattering region cannot be made as we have suggested in this paper. Allowing for the existence of such an intermediate regime, it is a matter of degree as to how much an effect the waves have on ions in this preshock region. A gradual increase in the amount of scattering is certainly possible but Richardson et al. (1988), without full pitch angle converage of the ions, were unable to quantify the term "quasi-isotropic". We note that the Coates et al. (1989, 1990) study of comet Halley ions also showed some broadening of the pitch angle distribution in the preshock region, but the ions were rendered "near-isotropic" only in a small region of extent $\Delta r \sim 0.5 \times 10^6$ km just ahead of the inbound shock at $r \simeq 1.2 \times 10^6$ km (although a dramatic broadening of the distribution in energy only occurred immediately behind the shock). Thus, some "renormalization" of our definitions may be appropriate in light of these data: in order to accommodate the presence of such an intermediate region, the

"upstream" region referred to throughout this paper should strictly be considered the "far upstream" region where the scatter-free limit is most likely to be an appropriate description of ion motion.

Upstream Versus Downstream Ion Acceleration

A major result to come out of the in situ measurements by the International Cometary Explorer (ICE) satellite at comet Giacobini-Zinner was the detection of pickup ions (Ipavich et al., 1986; Hynds et al., 1986) and the waves excited by them in the solar wind extending to distances greater than 10^6 km from the comet (Tsurutani and Smith, 1986; Scarf et al., 1986). This observation prompted theoretical models of second order Fermi acceleration by MHD waves far upstream of the shock (Ip and Axford, 1986; Isenberg, 1987; Gombosi, 1988) which were based on the hypothesis that rapid pitch angle scattering of particles (a necessary requirement for the second order Fermi mechanism to work) was occurring there. Barbosa (1986) challenged this assumption and argued that the upstream ion data were consistent instead with the pickup ions following relatively scatter-free trajectories transverse to the magnetic field in which case any acceleration from MHD waves would be insignificant. Downstream of the shock, however, the conditions were right for stochastic acceleration and the entire suite of cometary observations supported the premise that the cometosheath was the primary acceleration site. Here we review the arguments for postshock scattering and acceleration of pickup ions.

The following is a summary of the main observational and theoretical differences between the upstream and downstream regions insofar as ion acceleration is concerned.

1. IMF Dependence - upstream of the bow shock the flux of ions displays a sensitive dependence on the angle between the interplanetary magnetic field and the solar wind velocity θ_{BV} whereas downstream it does not (Hynds et al., 1986). The pickup energy E_p certainly depends on $\sin^2 \theta_{BV}$ but any flux variation due to E_p should also be manifest downstream as well. Thus, the very sensitive dependence on θ_{BV} found by Hynds et al. is more likely the result of a highly anisotropic distribution function being measured by an instrument with a restricted field of view. As verified by later measurements (e.g., Coates et al., 1989) the distribution is indeed anisotropic upstream signifying that not much scattering by MHD waves has occurred there and consequently very little non-adiabatic acceleration as well.

2. Phase space density gradient - measurements of ions at energies $E \gg E_p$ show a large spatial gradient at the bow shock with the most intense fluxes behind the bow shock throught the cometosheath (Richardson et al., 1987; Coates et al.,

1989). The variation of the phase space density demonstrates that the source of these high-energy particles, with speeds much greater than the solar wind speed, is downstream rather than upstream of the shock.

3. Theoretical prerequisites - downstream of the shock the level of magnetic fluctuations is very high ($\delta B/B \to 1$) whereas upstream the intensity is much lower (Tsurutani and Smith, 1986). Moreover, the low-frequency MHD waves have a distinct bi-directional anisotropy such that in the solar wind frame the upstream waves are propagating primarily away from the comet in the sunward rather than the anti-sunward direction. In terms of wave intensity components parallel and antiparallel to the magnetic field I_+, this condition of $I_+ \gg I_-$ is consistent with a beam of super-Alfvénic pickup ions generating fast mode MHD waves in the direction of beam motion (Barbosa and Eviatar, 1986) with relatively little quasilinear scattering of the ions and little redistribution of wave energy in the opposite direction. With regard to stochastic acceleration the conditions are ill-suited since the effective energy diffusion coefficient behaves as (Isenberg, 1987):

$$D_{TT} \propto I_{eff} = \frac{4I_+ I_-}{I_+ + I_-} \simeq 4I_- \qquad (1)$$

Thus, the acceleration rate depends on I_- which is generally interpreted as the ambient undisturbed fluctuation level of the solar wind (Tsurutani and Smith, 1986). Not only is this level rather weak but it makes for an unsettling theory of cometary ion acceleration when the acceleration rate depends on the ambient solar wind conditions instead of cometary parameters.

Downstream of the shock the conditions are such that stochastic acceleration theory is matched to the observations in a natural fashion as one might expect from a priori reasoning (Barbosa, 1979). In the sub-magnetosonic downstream region waves can propagate up to the shock and undergo reflection so that the condition $I_+ \simeq I_-$ is more or less obtained throughout the cometosheath. The high level of magnetic fluctuation results in a small mean free path for pitch angle scattering so that the ion energization rate which can be written approximately as (Barbosa et al., 1984)

$$\frac{dE}{dt} = \frac{\frac{1}{2}MV_A^2}{\tau_{mfp}} \qquad (2)$$

is sufficiently large to accelerate ions to the required energies all within the cometosheath which at Giacobini-Zinner encompasses a region with a scale size $L \sim 10^5$ km. Thus, the primary argument for the need of upstream acceleration, that one requires a lot of room

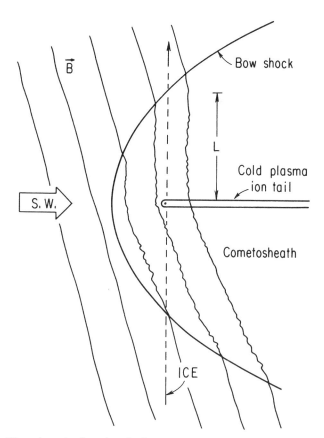

Fig. 1. A sketch of the cometary environment at P/Giacobini-Zinner where intense MHD turbulence in the cometosheath is present downstream of the bow shock over a characteristic length $L \simeq 10^5$ km. The ICE trajectory is also shown, the spacecraft penetrating the Type 1 ion tail and passing 7800 km behind the comet at closest approach (from Barbosa, 1989).

in order to accelerate ions effectively (T. Gombosi, private communication, 1989), is found to be true in a very general sense but wanting in the particular case of cometary acceleration at Giacobini-Zinner (and also comet Halley).

The Upstream Region

Figure 1 shows a sketch of the magnetic field environment of comet Giacobini-Zinner with regard to conditions around the bow shock. Upstream the interplanetary magnetic field (IMF) is generally oriented at an oblique angle to the solar wind direction. If in this case pickup ions are essentially scatter-free, they will move on $\vec{E} \times \vec{B}$ trajectories transverse to the magnetic field forming a hot ion tail that is aberrated from the solar wind direction.

The pickup ion tail is distinctly different from the more familiar Type 1 ion tail (Wurm, 1968; Brandt, 1982) which is readily observable in the emission spectra of the molecular ions CO^+ and H_2O^+ (Wyckoff, 1982). In fact it is unlikely that the pickup ion tail has ever been observed through telescopes since the ions are hot ($T_O^+ \sim$ 16 keV) and diffuse having a geometrical scale length $\sim 10^6$ km corresponding to the ionization length at 1 AU. On the other hand cold plasma Type 1 ion tails are formed within the contact surface resulting in a much smaller breadth (10^4 km) and are generally aligned with the solar wind direction.

One consequence of an aberrated pickup ion tail is that the calculation of pickup ion densities will be affected and certain inbound/outbound asymmetries in ion flux will become apparent for a spacecraft trajectory like that of ICE illustrated in Figure 1. The density of pickup ions is found by evaluating the so-called accumulation integral,

$$n_i = \frac{Q}{4\pi\ell V_{sw}\sin\theta_{BV}} \int_{x'}^{\infty} \frac{dx'}{r^2} e^{-r/\ell} \qquad (3)$$

where $r = (x'^2 + y'^2 + z'^2)^{\frac{1}{2}}$, Q is the cometary neutral source strength, and $\ell = V\tau$ is the ionization length for neutrals with outflow speed V and ionization lifetime τ. The coordinate system denoted by primed variables is based on a transformation of the usual cometocentric-sun-ecliptic coordinates (Slavin et al., 1986) to a system aligned with the pickup ion tail. Here x' points away from the comet along the pickup ion tail axis towards the inner heliosphere. Previously, authors have tended to ignore the IMF angle in calculations of pickup ion density for simplicity but its effect is quite substantial.

Figure 2 shows theoretical results for the flux of pickup ions (Barbosa, 1986)

$$J = \eta n_i V_{sw}\sin\theta_{BV} \qquad (4)$$

superimposed on the 65-95 keV ion data of Hynds et al. (1986). The factor η represents an efficiency per unit solid angle for measuring the pickup ions; that is, only a fraction of the ions (per steradian) given by the expression (3) for the pickup ion density are above the 65 keV instrument threshold and are detected. There are three curves corresponding to $\theta_{BV} = 45°$ (bottom most), 30°, and 15° (topmost). The neutral source strength $Q = 10^{28}s^{-1}$, the efficiency $\eta = (4\pi)^{-1}$, and the ionization length was set at $\ell = 10^6$ km.

Sanderson et al. (1986) have pointed out that the pickup ion theory provides a reasonably good fit to the data albeit with an arbitrary normalization. The present calculation amplifies on this in several ways.

First, with the above parameters, the theoretical curves match the data in absolute flux values without needing any drastic renormalization. The

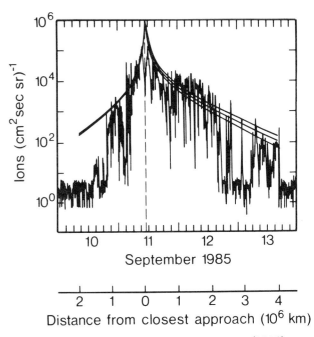

Fig. 2. 65-95 keV ion fluxes of Hynds (1986) are displayed along with theoretical curves for the pickup ion flux in the case $\theta_{BV} = 45°$, 30°, and 15° (from Barbosa, 1986).

efficiency factor η has a large uncertainty to it, but nonetheless the deduced value is plausible given the expected energy range of the new pickup ions (Ipavich et al., 1986; Sanderson et al., 1986) and the nature of the distribution of ions in solid angle (e.g., Möbius et al., 1985). Alternatively, one may accept the fit in Figure 2 as being good and, adopting the consensus value of $Q \simeq 2 \times 10^{28}s^{-1}$, then infer that $\eta \simeq (8\pi)^{-1}$.

Second, the inbound/outbound asymmetry is reproduced fairly well within the context of the model without having to invoke additional assumptions. The values preferred here for the field angles $\theta_{BV} = 45°$ and $\psi = 45°$ (which correspond to an azimuth $\phi = 145°$ and latitude $\delta = -30°$) are compatible with the measured upstream magnetic field (Smith et al., 1986; Tsurutani and Smith, 1986) and other independent estimates of the tail orientation angle ψ (Slavin et al., 1986; Daly et al., 1986). These values are obtained by visual inspection of the quality of the fit in Figure 2 and also by the measure of the inbound/outbound asymmetry ratio. Note that at equal distances away from the point of closest approach, the inbound/outbound fluxes can differ by as much as an order of magnitude.

In addition to the disparity in flux levels on the inbound and outbound legs, there is another asymmetry that is very apparent in the ion data. Whereas an exponential behavior for the spatial dependence extending to all distances is predicted, instead a sharp onset occurs at $r \sim 1 \times 10^6$ km inbound and a sharp dropout occurs at 4×10^6 km outbound. Sanderson et al. (1986) have assumed a geometric model that is symmetric, and in order to account for any inbound/outbound differences have suggested that a temporal change in the solar wind velocity might be responsible.

It is highly possible that the solar wind figures prominently in this matter. An increase in the solar wind speed modifies the pickup ion energy and can influence the duration of the ion measurement by elevating ions above the 65 keV energy threshold of the instrument (assuming that the IMF angle remains constant and does not offset the change in energy). However, the threshold effect principally specifies whether ions will be detected or not but it is uncertain how the magnitude of the flux will be affected. According to equations (3) and (4), V_{sw} drops out of the combined expression and the flux is not directly dependent on solar wind speed. There may be an indirect dependence entering through the factor η having to do with the spectral shape of the ion distribution in connection with the energy-dependent threshold (Richardson et al., 1988); quantitative models for such an effect are needed.

It is also noted that the magnetic field data display measurable levels of turbulence as far out as 10^6 km (Tsurutani and Smith, 1986). The plasma wave data show signs of enhanced activity even farther out (4×10^6 km) suggesting that pickup ions are being produced essentially everywhere (Scarf et al., 1986) but that their detection (either direct or indirect) is a matter of instrument sensitivity to exponentially decreasing levels of pickup ion fluxes and their consequent effects.

Much closer in (e.g., $r < 10^5$ km), significant fluxes occur in the higher energy channels which are attributed to pickup ions which have been further accelerated. Hynds et al. (1986) propose that these ions have been strongly scattered by the magnetic turbulence, which is more pronounced at smaller radial distances, and are isotropic in the solar wind frame. If additional energization (e.g., stochastic acceleration) accompanies the pitch angle scattering, the ions would be observed at progressively higher energies closer to the comet (Gloeckler et al., 1986) with progressively smaller signatures of inbound/outbound asymmetry. This scenario is in keeping with the full suite of measurements made by the ICE spacecraft.

The Downstream Region

It is evident from Figure 2 that upstream of the bow shock pickup ions are observed to large distances $\gtrsim 10^6$ km corresponding to the ionization length of the neutral gas coma. The pickup ions display a rapid variability which has been attributed to temporal variations of the IMF direction (Sanderson et al., 1986). Note that the downstream region in this plot corresponds to only $\pm 1\frac{1}{2}$ hr about the time of closest approach; in

this interval when ICE is in the cometosheath the ion fluxes have a much smoother temporal profile consistent with their achieving a state of pitch angle isotropy whereas upstream the ion distributions are highly anisotropic.

The comprehensive study of Richardson et al. (1987) has detailed many of the important properties of the cometary pickup ion environment and illuminated the physical processes taking place. They have shown that the pickup ions generally have an energy spectrum that behaves as an exponential in ion speed $f(\vec{v}) \propto \exp(-v/v_0)$ and that the characteristic speed of the exponential v_0 increases sharply at the bow shock to a value $v_0 \simeq 80$ km/s. The hardening of the spectrum was further shown to be associated with a large increase in the level of magnetic fluctuations observed throughout the cometosheath. A comparison of the data with theoretical predictions of Ip and Axford (1986) showed a great discrepancy in the general behavior of the theoretical spectrum thus calling for improved models to be developed.

In view of the Richardson et al. (1987) results, Barbosa (1989) concluded that the ICE observations were in large part consistent with a previously developed model (Barbosa, 1979) where the bulk of the stochastic ion acceleration occurred downstream of the bow shock in the region of the most intense magnetic turbulence and energy dissipation. Several other important ingredients of the theory were also shown to be present. The ions were isotropic downstream of the shock, a prerequisite for efficient acceleration by a second-order Fermi mechanism. The spectrum was unambiguously an exponential in ion speed as predicted earlier; inasmuch as theoretical predictions vary a great deal with functional dependences like $\exp[-(E/E_0)^q]$ and q ranging from 0.25-1 (Forman et al., 1986), the Richardson et al result is rather significant. And finally, the magnetic turbulence had the right spectral properties; according to theory an exponential in ion speed required the magnetic noise to have a spectral index $\zeta = 2$. The latter condition having been satisfied, an important check on the internal consistency of the theory was met.

Figure 3 shows a plot of the power spectrum of magnetic noise obtained by ICE just after the inbound shock crossing (Tsurutani and Smith, 1986). The dashed curve is superimposed on the spectrum corresponding to the transverse y-component of noise (middle panel) to indicate a spectral density that behaves as $P(f) \propto f^{-2}$. Tsurutani and Smith (1986) noted that the slope was suggestive of a Kolmogorov-type spectrum with a spectral index $\zeta = 5/3$. However, the comparison with the f^{-2} curve shows that the data are equally or better fit by a curve having a spectral index $\zeta = 2$. In any case, the data are surely inconsistent with $\zeta = 1$ which value leads to the specific prediction that the ion spectrum behaves as $\exp(-E/E_0)$ (Barbosa, 1989). The exponential in

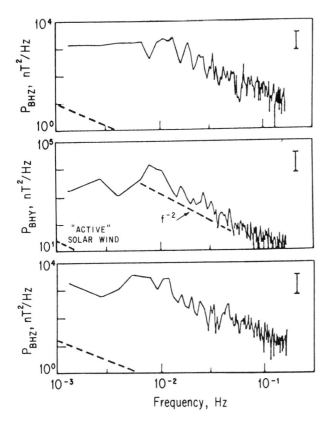

Fig. 3. MHD wave power spectral density along (BHX) and orthogonal to (BHY and BHZ) the average field direction obtained downstream of the bow shock of Giacobini-Zinner on the inbound leg of ICE (from Tsurutani and Smith, 1986).

context of solar flares. The steady-state solution to the combined energy and spatial diffusion equaiton including a pickup ion source [equation 1 of Barbosa (1989)] is given as,

$$N(E) = \frac{S_o}{qDE^{-\zeta/4}} \; E^p \; I_\nu \left[(E_</E_B)^q \right] \; K_\nu \left[(E_>/E_B)^q \right]$$

(5)

in terms of variables and parameters defined in that paper. In the case that $\zeta = 2$ and for energies greater than the pickup energy $\vec{E} > E'$, the phase space distribution function $f(\vec{v}) = MN(E)/4\pi v$ behaves as

$$f(\vec{v}) = AK_1(v/v_o)/v$$

(6)

Here K_1 is a modified Bessel funciton, A is a normalization constant and v_o is a characteristic speed parameterizing the distribution. Specifically,

$$v_o = \frac{3\pi}{8} \; L \; f_o \; \frac{\delta B^2}{B^2}$$

(7)

and

$$A = \frac{3LS_o I_1(v'/v_o)}{2\pi^2 v' v_o V_A}$$

(8)

where δB is the root-mean-square variation in the transverse magnetic field component and f_o is the frequency where the slope of the power spectrum begins to turn over. A distinguishing feature of the model is that the energy dependence of the energy diffusion coefficient is related to the spatial diffusion loss time since the same wave scatterers are involved resulting in the particular form for v_o which depends only on the properties of the MHD wave power spectrum and is species-independent having the same value for all mass-to-charge ratios. The pickup ion source strength S_o is written as (Galeev, Cravens, and Gombosi, 1985),

$$S_o = \frac{Q}{4\pi r^2 \ell} \; e^{-r/\ell}$$

(9)

in terms of the neutral source strength Q as per Section III. With reference to Figure 3 we take $f_o = 5 \times 10^{-3}$ Hz, $\delta B/B = 0.5$, and $L = 10^5$ km which leads to a predicted value of $v_o \simeq 150$ km/s.

In Figure 4 we show a plot of ICE ion data taken from Figure 4b of Richardson et al. (1987).

ion energy has been shown to provide inferior fits to the ICE data in general (Richardson et al., 1987).

The "leaky box" model proposed by Barbosa (1989) assumes that the ions are rapidly scattered in pitch angle by MHD waves downstream of the shock, isotropized, and energized by a small amount to second order in the expansion parameter $V_A/v \ll 1$ (Barbosa, 1979). In this picture the mean free path for pitch angle scattering λ is small compared to the macroscopic scale length over which the scattering takes place, here $\lambda < L \simeq 10^5$ km. The ions are thus rendered near-isotropic in pitch angle distribution and ion transport is governed by the equations of spatial diffusion along the magnetic field with MHD waves acting as scattering centers (Jokipii, 1971). Stochastic acceleration by the self-same scattering waves occurs in higher order because of the finite electric field associated with propagating MHD waves and can be construed kinematically as a second-order Fermi acceleration in overall effect (Barbosa et al., 1984).

Relevant details of the theory are presented in Barbosa (1979, 1989) and Forman et al. (1986) have given an excellent review of the theory in the

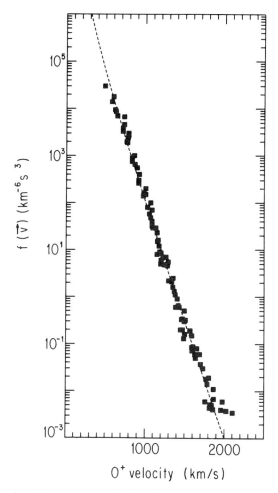

Fig. 4. Comparison of the theoretical spectrum with the energetic ion data (assuming O^+) of Richardson et al. (1987) obtained during the inbound pass of ICE through the cometosheath (from Barbosa, 1989).

The dashed curve corresponds to equation (5) with the adopted values $A = 10^{10} km^{-5} s^2$ and $v_o = 95$ km/s. The difference between the value employed here for $v_o = 95$ km/s and the best fit characteristic speed of the exponential 83 km/s found by Richardson et al. (1987) stems from the different functional forms used, i.e., modified Bessel function as opposed to simple exponential. It is clear that the Bessel function gives an excellent fit to the data over a wide range of ion speeds (note the slight curvature of the Bessel function). At the low end of the spectrum some caution is needed where the speed of the ion $v \simeq V_{SW}$ and plasma convection, which has not been included in the theory, can modify the results (see, e.g., Ip and Axford, 1982). At the high end the apparent upturn of the data may not be real possibly being influenced by counting statistics (M. Neugebauer, private communication, 1989). The factor of 2

difference in the empirical versus theoretically predicted value of v_o may simply be the result of the questionable assumption of equal intensities of Alfvén waves $I_+ = I_-$ along the magnetic field. A slight departure from this assumption would reconcile the theory with observations by reducing the acceleration efficiency a bit (Skilling, 1975). If we now ask what theory predicts for the normalization factor A, then using $Q = 2 \times 10^{28} s^{-1}$, $r = 8 \times 10^4 km$, $\ell = 10^6$ km. $V = 1$ km/s, $v' = V_{SW} = 400$ km/s, $V_A = 140$ km/s, and the empirical value for $v_o = 95$ km/s, we find from equation (8) that $A = 8 \times 10^9 km^{-5} s^2$ in excellent agreement with the empirical value $10^{10} km^{-5} s^2$. There is thus an agreeable self-consistency in the model and its predictions.

One final note regards the justification for using quasi-linear theory in the limit of large field fluctuations $\delta B/B \rightarrow 1$. In this limit the pitch angle diffusion equation derived from quasi-linear theory may only provide a qualitative measure of the particle motion in phase space rather than a precise quantitative description. The MHD waves are still predominately magnetic fluctuations so that pitch angle scattering still dominates over energy changes. However, each wave scattering is likely to produce a large deflection in pitch angle $\Delta\theta/\theta \sim 1$ and a Monte Carlo approach to calculating the change in pitch angles is to be preferred over the quasi-linear method which assumes small deflections in pitch angles $\Delta\theta/\theta \ll 1$. Nevertheless, the quasi-linear result is probably adequate to give a rough order-of-magnitude estimate of the pitch angle scattering time scale. For the present problem dealing with stochastic acceleration, the precise rate of pitch angle scattering is not important as it is only assumed to be fast enough that the distribution function is rendered near-isotropic. In this regime energy changes resulting from interactions with waves in the limit $\delta B/B \rightarrow 1$ are still small (i.e., $\Delta E/E \ll 1$) having a more fundamental dependence on the smallness of the ratio $V_A/c \ll 1$ so that the energy diffusion equation (1) of Barbosa (1989) properly describes the stochastic acceleration of ions.

Conclusion

Stochastic ion acceleration by MHD waves downstream of the bow shock can account for the ICE observations of superthermal ions with energies above the pickup energy. The so-called leaky box model can explain many features of the data in a quantitative fashion and there is an appealing internal consistency where the properties of the MHD waves are reconciled with the properties of the ions. That is, the pitch angle distribution, the spectral behavior, and the spectral slope of the ions are in good agreement with predictions based on the properties of the MHD wave power spectrum.

These results are not restricted to Giacobini-Zinner but are applicable to comet Halley also.

In the upstream region Neugebauer et al. (1989a) showed that several features of the pickup proton data could be successfully modeled in terms of scatter-free trajectories transverse to an obliquely oriented IMF in a manner similar to the calculations of Barbosa (1986) (cf., Section III). The downstream region also afforded the most favorable site for stochastic acceleration. The magnetic field fluctuations increase dramatically at the bow shock and are the most intense in the cometosheath (Neubauer et al., 1986) while the power spectrum of low-frequency waves has a spectral index of $\zeta = 2$ (Glassmeier et al., 1987). Correspondingly, the pitch angle distributions of pickup protons (Neugebauer et al., 1989b) and water-group ions (Coates et al., 1989) are highly anisotropic upstream of the shock and abruptly become near-isotropic in the cometosheath contemporaneous with substantial acceleration to energies in excess of the pickup energy. Moreover, the energy spectrum also behaves as an exponential in ion speed as evident in Figure 5 of Coates et al. (1989). Thus, the two comets are very similar in their properties insofar as the aspect of stochastic acceleration of pickup ions is concerned.

In a broad sense, the model of ion acceleration downstream of the shock is in good harmony with the large suite of spacecraft observations and is reinforced by the various aspects of the data. On the contrary, models of ion acceleration upstream of the shock in many ways conflict with the observations and have to sidestep a number of important issues. The question of the large pitch angle anisotropy in connection with the theory of second-order Fermi acceleration is a major stumbling block for upstream models in a general theoretical sense. From an observer's viewpoint upstream theories are hard pressed to explain why the spectrum is significantly harder in the cometosheath and more intense than upstream (Richardson et al., 1987).

It is abundantly clear that the comet encounters have provided a solid scientific basis for and a fresh new perspective of models dealing with ion acceleration in the vicinity of shocks. We look to find further application of these results in other, more remote astrophysical settings.

Note added in proof. P.A. Isenberg (private communication, 1990) has recently noted an apparent inconsistency in the value for f_o employed in (7). The frequency where the wave spectrum in Figure 3 turns over is approximately 5 mHz which is the value chosen for discussion of the acceleration efficiency. However, the frame of reference in which the energy diffusion equation is cast is the solar wind (SW) frame as opposed to the spacecraft (SC) frame in which the power spectrum of Figure 3 was obtained. In the solar wind frame the maximum velocity to which an O^+ ion can be accelerated is $v_{max} = V_A(1 + f_{cO^+}/f_o^{SW})$ which indicates that f_o^{SW} is the more

appropriate quantity to consider in the theory rather than $f_o^{SC} = 5$ mHz.

The mean magnetic field $<B> = 20$ nT (Tsurutani and Smith, 1986) so that $f_{cO^+} \simeq 20$ mHz. If we now assume that $f_o^{SW} = 1.5$ mHz, then the maximum attainable ion speed $v_{max} \simeq 2000$ km/s which is compatible with the ion data of Richardson et al. (1987) and motivates this choice for f_o^{SW}. The transformation to the spacecraft frame involves $f_o^{SC} = f_o^{SW}(1 + V_{SW}/V_A) \simeq 6$ mHz which compares favorably with the data in Figure 3. Other computed quantities need to be revised correspondingly; i.e., if we take $\delta B/B = 0.8$ as quoted by Tsurutani and Smith (1986) and allow for a slightly larger $L = 1.3 \times 10^5$ km, then λ/L and v_o are unchanged from the values of the previous discussion. Future work in this area will benefit from a complete, self-consistent theory of the MHD wave spectrum, both upstream and downstream, when it becomes available.

Acknowledgments. Thanks are due P.A. Isenberg and the referees for insightful comments and criticisms. Parts of this work were supported by the National Science Foundation under grant NSF ATM 86-06857. Support for the preparation of this review paper was provided by NASA grant NAGW-1558.

References

Barbosa, D.D., Stochastic acceleration of solar flare protons, Astrophys. J., 233, 383, 1979.
Barbosa, D.D., Cometary pickup ion tail and its relation to the interplanetary magnetic field: P/Giacobini-Zinner, UCLA PPG Report No. 996, Univ. of Calif., 1986.
Barbosa, D.D., Stochastic acceleration of cometary pickup ions: the classic leaky box model, Astrophys. J., 341, 493, 1989.
Barbosa, D.D. and A. Eviatar, Planetary fast neutral emission and effects on the solar wind: a cometary exosphere analog, Astrophys. J., 310, 927, 1986.
Barbosa, D.D., A. Eviatar, and G.L. Siscoe, On the acceleration of energetic ions in Jupiter's magnetosphere, J. Geophys. Res., 89, 3789, 1984.
Brandt, J.C., Observations and dynamics of plasma tails, in Comets, edited by L.L. Wilkening, pp. 519-537, Univ. of Arizona Press, Tucson, 1982.
Coates, A.J., A.D. Johnstone, B. Wilken, K. Jockers, and K.-H Glassmeier, Velocity space diffusion of pickup ions from the water group at comet Halley, J. Geophys. Res., 94, 9983, 1989.
Coates, A.J., B. Wilken, A.D. Johnstone, K. Jockers, K.-H Glassmeier, and D.E. Huddleston, Bulk properties and velocity distributions of water group ions at comet Halley: Giotto measurements, J. Geophys. Res., 95, 10249, 1990.
Daly, P.W., T.R. Sanderson, K.-P. Wenzel, S.W.H. Cowley, R.J. Hynds, and E.J. Smith, Gyroradius effects of the energetic ions in the tail lobes of comet P/Giacobini-Zinner, Geophys. Res. Lett., 13, 419-4-2, 1986.

Forman, M.A., R. Ramaty, and E.G. Zweibel, The acceleration and propagation of solar flare energetic particles, in Physics of the Sun Vol. II., edited by P.A. Sturrock, pp. 249-289, D. Reidel, Dordrecht, Holland, 1986.

Galeev, A.A., T.E. Cravens, and T.I. Gombosi, Solar wind stagnation near comets, Astrophys. J., 289, 807, 1985.

Glassmeier, K.-H., F.M. Neubauer, M.H. Acuña, and F. Mariani, Low-frequency magnetic fluctuations in comet P/Halley's magnetosheath: Giotto observations, Astron. Astrophys., 187, 65, 1987.

Gloeckler, G., D. Hovestadt, F.M. Ipavich, M. Scholer, B. Kleckler, and A.B. Galvin, Cometary pickup ions observed near Giacobini-Zinner, Geophys. Res. Lett., 13, 251-254, 1986.

Gombosi, T.I., Preshock region acceleration of implanted cometary H^+ and O^+, J. Geophys. Res., 93, 35, 1988.

Hynds, R.J., S.W.H. Cowley, T.R. Sanderson, K.-P Wenzel, and J.J. Van Rooijen, Observations of energetic ions from comet Giacobini-Zinner, Science, 232, 361-365, 1986.

Ip, W.-H. and W.I. Axford, Theories of the physical processes in the cometary comae and ion tails, in Comets, L.L. Wilkening, ed., pp. 538-634, University of Arizona Press, Tucson, 1982.

Ip., W.-H. and W.I. Axford, The acceleration of particles in the vicinity of comets, Planet. Space Sci., 34, 1061, 1986.

Ipavich, F.M., A.B. Galvin, G. Gloeckler, D. Hovestadt, B. Klecker, and M. Scholer, Comet Giacobini-Zinner: in situ observations of energetic heavy ions, Science, 232, 366-369, 1986.

Isenberg, P.A., Energy diffusion of pickup ions upstream of comets, J. Geophys. Res., 92, 8795, 1987.

Jokipii, J.R., Propagation of cosmic rays in the solar wind, Rev. Geophys. Space Phys., 9, 27, 1971.

Mendis, D.A. and H.L.F. Houpis, The cometary atmosphere and its interaction with the solar wind, Rev. Geophys. Space Phys., 20, 885-928, 1982.

Möbius, E., D. Hovestadt, B. Klecker, M. Scholer, G. Gloeckler, and F.M. Ipavich, Direct observation of He^+ pick-up ions of interstellar origin in the solar wind, Nature, 318, 426-429, 1985.

Neubauer, F.M., K.H. Glassmeier, M. Pohl, J. Raeder, M.H. Acuña, L.F. Burlaga, N.F. Ness, G. Muswann, F. Mariani, M.K. Wallis, E. Ungstrup, and H.V. Schmidt, First results from the Giotto Magnetometer experiment at comet Halley, Nature, 321, 352, 1986.

Neugebauer, M., B.E. Goldstein, H. Balsiger, F.M. Neubauer, R. Schwenn, and E.G. Shelley, The density of cometary protons upstream of comet Halley's bow shock, J. Geophys. Res., 94, 1261, 1989a.

Neugebauer, M., A.J. Lazarus, H. Balsiger, S.A. Fuselier, F.M. Neubauer, and H. Rosenbauer, The velocity distribution of cometary protons picked up by the solar wind, J. Geophys. Res., 94, 5227, 1989b.

Richardson, I. G., S.W.H. Cowley, R.J. Hynds, T.R. Sanderson, K.-P. Wenzel, and P.H. Daly, Three dimensional energetic ion bulk flows at comet P/Giacobini-Zinner, Geophys. Res. Lett., 13, 415-418, 1986.

Richardson, I.G., S.W.H. Cowley, R.J. Hynds, C. Tranquille, T.R. Sanderson, K.-P. Wenzel, and P.W. Daly, Observations of energetic water-group ions at comet Giacobini-Zinner: implications for ion acceleration processes, Planet. Space Sci., 35, 1323, 1987.

Richardson, I.G., S.W.H. Cowley, R.J. Hynds, P.W. Daly, T.R. Sanderson, and K.-P. Wenzel, Properties of energetic water-group ions in the extended pick-up region surrounding comet Giacobini-Zinner, Planet. Space Sci., 36, 1429, 1988.

Sanderson, T.R., K.-P. Wenzel, P. Daly, S.W.H. Cowley, R.J. Hynds, E.J. Smith, S.J. Bame, and R.D. Zwickl, The interaction of heavy ions from comet P/Giacobini-Zinner with the solar wind, Geophys. Res. Lett., 13, 411-414, 1986.

Scarf, F.L., F.V. Coroniti, C.F. Kennel, D.A. Gurnett, W.-H. Ip, and E.J. Smith, Plasma wave observations at comet Giacobini-Zinner, Science, 232, 377-381, 1986.

Skilling, J., Cosmic ray streaming I: effect of Alfvén waves on particles, Mm. Na. R. Astron. Soc., 172, 557, 1975.

Slavin, J.A., E.J. Smith, B.T. Tsurutani, G.L. Siscoe, D.E. Jones, and D.A. Mendis, Giacobini-Zinner magnetotail: ICE magnetic field observations, Geophys. Res. Lett., 13, 283-286, 1986.

Smith, E.J., B.T. Tsurutani, J.P. Slavin, D.E. Jones, G.L. Siscoe, D.A. Mendis, International Cometary Explorer encounter with Giacobini-Zinner: magnetic field observations, Science, 232, 382-385, 1986.

Tsurutani, B.T. and E.J. Smith, Strong hydromagnetic turbulence associated with comet Giacobini-Zinner, Geophys. Res. Lett., 13, 259-262, 1986.

Wallis, M.K. and A.D. Johnstone, Implanted ions and the draped cometary field, in Cometary Exploration I, ed. by T. Gombosi, pp. 307-311, European Physical Society, Budapest, 1982.

Wurm, K., Structure and kinematics of cometary Type I tails, Icarus, 8, 287-300, 1968.

Wyckoff, S., Overview of comet observations, in Comets, edited by L.L. Wilkening, pp. 3-55, Univ. of Arizona Press, Tucson, 1982.

OBSERVATIONS OF THE VELOCITY DISTRIBUTION OF PICKUP IONS

A. J. Coates

Mullard Space Science Laboratory, University College London, Holmbury St Mary, Dorking RH5 6NT, UK

Abstract. Ion pick-up is well-established as the process whereby comets interact with the solar wind. Although this mechanism was postulated as such over 30 years ago it was not until the 1985-6 cometary armada reached their goals that direct measurements of the process in the cometary environment could be performed. Here we review the measurements of pickup ion distributions from the various spacecraft, concentrating on the observed pitch angle and energy scattering in the turbulent solar wind upstream of comets. Comparison with simulation and with theory is attempted by way of the inferred timescales for pitch angle and energy diffusion. Measurements of the bulk velocity of the picked-up ions are compared with the guiding centre approximation far upstream of the comet and with the wave phase velocity nearer in.

Introduction

Although Biermann [1951] originally proposed the existence of a continuously flowing solar wind from observations of comet tails, Alfvén [1957] provided the first picture of the solar wind-comet interaction involving the draping of magnetic field lines in the cometary coma. This draping picture is close to today's understanding of the interaction and is now supported by experimental evidence. Here we are concerned with the process which causes that draping: the deceleration of the magnetized solar wind via the ion pickup process.

The picture of the solar wind-comet interaction which has become classical begins with the neutral atoms which drift away from the cometary nucleus when it is heated by the Sun. The neutrals may ionise either due to the solar radiation or by charge exchange with the incoming solar wind. Immediately on ionisation the particles are "picked up" in the solar wind electromagnetic field; first they are accelerated by the $v \times B$ motional electric field of the solar wind and subsequently they also gyrate around the magnetic field. In real space the resultant orbit is a cycloid but in velocity space it is a ring around the magnetic field vector.

In giving these ions energy the solar wind loses energy; here we are concerned with the mechanism by which this interaction occurs. The ring distribution is highly unstable to various wave modes and the presence of the ring distribution thus causes additional waves to those already present in the background solar wind, ultimately causing turbulence. These wave-particle interactions cause scattering of the particle distributions in pitch angle, such that the distribution has less energy in the solar wind frame. The process proceeds in two stages: first, the diffusion in pitch angle to produce a shell distribution and second, diffusion in energy to produce a thick shell. After long enough the pickup ions would evolve from a thick shell to eventually

Cometary Plasma Processes
Geophysical Monograph 61
●1991 American Geophysical Union

form a Maxwellian distribution moving with and collectively coupled to the solar wind; the shell formation can be seen as an early stage of assimilation into the flow.

The cometary encounters by spacecraft provided the opportunity to observe the pickup process in-situ. Although pickup ions had been observed previously in other contexts (ionization of interstellar neutrals [e.g., Mobius et al., 1988] and in the artificial AMPTE releases [e.g., special section in Journal of Geophysical Research vol. 91 on lithium releases, January 1986, special section in Nature vol. 320 on barium artificial comet releases, April 1986]), the cometary encounters have provided the first measurements in the presence of an extended and relatively intense source of pickup ions. This has allowed detailed studies of the evolution of the particle distribution, together with measurements of the turbulence caused by this evolution. Two main species of pickup ion are important in the outer parts of cometary comae (outside the contact surface) which we consider here, namely protons (H^+) and water group ions (e.g., O^+, OH^+) [Balsiger et al., 1986].

Although only predicted in one paper which appeared immediately before the cometary encounters [Amata and Formisano, 1985], another process which the spacecraft results have shown to be particularly important at comets is acceleration of the implanted cometary ions [e.g., Hynds et al., 1986, Somogyi et al., 1986, McKenna-Lawlor et al., 1986]. These observations of accelerated ions led to a surge of theoretical interest in the phenomenon and a coherent explanation is now emerging based on velocity diffusion (second order Fermi acceleration) supplemented by adiabatic and first-order Fermi effects downstream and just upstream of the bow shock [e.g., Ip and Axford, 1986, Isenberg, 1987, Ip, 1988, Gombosi et al., 1989].

In this paper we concentrate on in-situ observations of pickup ions in the outer coma of comets, concentrating mainly on the peak of the pickup population rather than the accelerated ions. We start with a brief review of the pickup process and an introduction to some of the frames of reference to be used later. Then we recall what the various spacecraft have told us about pitch-angle and energy scattering of pickup protons and water group ions, with some inferences on the timescales for these processes. For water group ions we consider the bulk speed of pickup ions at different positions in the cometary coma. We conclude with a discussion and ideas for future work.

Scattering of Pickup Ions

The cometary situation is illustrated in Figure 1, which shows the drift of neutral atoms and dissociation products away from the cometary nucleus in real space. They are ionised by solar radiation or by charge exchange with the incoming solar wind, the typical scale length for ionization is $\sim 10^6$ km. Consider a particle which is ionised at the point P. The new ion is immediately accelerated along the solar wind motional electric field direction, $E = -v \times B$. It also gyrates around the

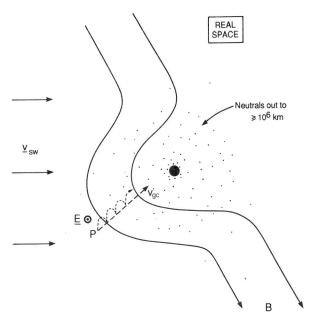

Fig. 1. Schematic view of the pickup process at a comet in real space. A particle drifting from the comet is ionised at point P and subsequently follows a cycloidal trajectory with guiding centre speed $v_{gc}=(\mathbf{E}\times\mathbf{B})/B^2$, unless scattering occurs.

magnetic field, and its resultant path is a cycloid in real space with a guiding centre drift velocity $\mathbf{v}_{gc}=(\mathbf{E}\times\mathbf{B})/B^2$.

In velocity space, using the frame of reference of a cometary spacecraft, the situation is illustrated in Figure 2(a) [from Coates et al., 1989]. The injection point $\mathbf{v}_{inj}$, marked by a star, is given by ($\mathbf{v}_{sc}$-$\mathbf{v}_n$), where $\mathbf{v}_{sc}$ is the comet-spacecraft relative velocity and $\mathbf{v}_n$ is the initial particle drift velocity away from the comet ($v_n \sim 1$km/s, e.g., Lämmerzahl et al. [1987]). For the arbitrary magnetic field direction shown (line $\mathbf{B}$) and the solar wind velocity in the comet frame $\mathbf{v}_{sw}$, the ring forms in the plane perpendicular to $\mathbf{B}$ (gyration around the field), and has a radius $v_{sw}\sin\theta_{vB}$, where θ_{vB} (sometimes referred to as α) is the angle between $\mathbf{v}_{sw}$ and $\mathbf{B}$. Thus in the ring distribution the particles immediately attain the perpendicular flow velocity relative to the solar wind. The generation of turbulence leads, as a first approximation, to pitch-angle scattering in the solar wind frame around a spherical shell of radius v_{sw} centred on the solar wind speed. This is represented by the circle in Figure 2(a) and we will refer to it as a "simple shell". From Figure 2(a) we can see that the maximum velocity of pickup ions in a ring distribution in the comet frame of reference is $2v_{sw}\sin\theta_{vB}$ and for a shell distribution it is $2v_{sw}$. We note that in the spacecraft frame the location of $\mathbf{v}_{inj}$ must be taken into account when calculating the maximum speed but this speed still clearly depends on $\sin\theta_{vB}$ for the ring and is independant of the θ_{vB} for the shell.

We now move to the solar wind frame and orient the z-axis along the magnetic field, the y direction along $\mathbf{E}$ and x along $\mathbf{E}\times\mathbf{B}$, the velocity space picture becomes that shown in Figure 2(b) [from Coates et al., 1990]. In this coordinate system it is more straightforward to calculate the injection point speed ($v_{inj\parallel}=\mathbf{v}_{sw}\cdot\mathbf{B}/B$, $v_{inj\perp}=\mathbf{E}\times\mathbf{B}/B^2$), and the bulk velocities for uniformly filled ring [= $(0,0,v_{inj\parallel})$] and simple shell [= $(0,0,0)$] distributions.

We now examine the validity of the simple shell approximation using the conservation of energy [Coates et al., 1990]. The scattering of the particles from the ring proceeds, as we have pointed out, by

the production of waves and subsequent wave-particle interactions. At the comet the instabilities produce Alfvén waves moving predominantly upstream along the field [e.g., Tsurutani, 1990, Johnstone et al., 1990], although a small fraction of the cometary waves and pre existing solar wind turbulence moves downstream. Due to conservation of energy between the waves and the particles, the particle velocities may diffuse along paths which lie on spherical shells in velocity space centred on the field-aligned speed of $\pm v_A$, see Figure 2(c). The shells, which intersect the plane of Figure 2(c) in circles, are given by the equation

$$v_\perp + (v_\parallel \pm v_A)^2 = const \qquad (1)$$

It has been suggested [Galeev and Sagdeev, 1987] that the most likely paths for the ions are the lower energy paths B and D in Figure 2(c), so that energy is given to the waves. This has implications for the wave modes which are excited by the unstable ring. We call the path BD the "bispherical shell". The bulk velocity of the uniformly filled bispherical shell is shown as $(0,0,v_{bulk\parallel})$. The value of $v_{bulk\parallel}$ may be calculated [Coates et al., 1990] and it is found to depend on θ_{vB} and to range between $\pm v_A$.

In addition to the pitch angle diffusion discussed above, we note that particle energisation (and deceleration) is possible by velocity diffusion involving scattering centres where the particle velocities diffuse along paths of different radii centred on $\pm v_A$ along the field. This process can be illustrated using an extension of Figure 2(c) covering higher and lower energy branches than those shown here [Terasawa and Scholer 1989; see Figure 1 of Terasawa, 1990]. Velocity diffusion acting alone would cause a shell distribution to become broader in velocity, causing fluxes of particles well above (and also below) the pickup energy. While the density of the higher energy particles is some orders of magnitude lower than the original pickup ions their presence was one of the major observational surprises of the cometary encounters. Other energisation processes may occur near to the bow shock, for example adiabatic acceleration and first order Fermi acceleration, and these are reviewed in this volume [Terasawa 1990], as well as separately elsewhere [Coates, 1990].

Having summarised the current view of the pickup geometries, some of which has evolved as the results have been analysed, we now discuss the experimental observations of pickup ions at comets.

Observations

The observations are reported for protons and for water group ions in separate sections, to give the current view of what is known about pickup ions of each species. Only results which clearly illustrate specific points are included, which unfortunately means that many interesting results must be omitted.

Protons

Detectors on the Vega, Suisei and Giotto spacecraft were all able to detect pickup protons. An example from the Suisei experiment, a 2-dimensional plasma analyser, is shown in Figure 3 [from Terasawa et al, 1986]. Since there was no magnetometer on Suisei the magnetic field direction was inferred from the shape of the solar wind proton distribution, leading to the inferred position of the pickup shell indicated by dashed circles - the smaller radius for protons and the larger radius for water group. The example shown is for approximately 3×10^6km, but Suisei detected protons at distances less than about 10-15×10^6km from Halley and water group at less than about 4×10^6km [Mukai et al., 1986, Terasawa et al., 1986]. It is possible to recognise the pickup proton signature but this is made difficult by the solar wind peak; this difficulty is present with all the pickup proton observations and the solar wind must be carefully subtracted for quantitative work. From these plots and similar ones obtained

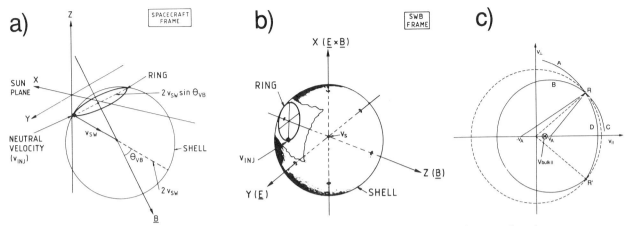

Fig. 2. Three schematic views of the pickup process in velocity space: (a) in the spacecraft frame, newborn ions appear at the point shown by the star (b) solar wind magnetic field-aligned (SWB) frame showing the ring and simple shell distributions, (c) SWB frame showing the bispherical shell, details in text (from Coates et al. [1989, 1990]).

closer to the comet where the solar wind was slower and the water group ions could be fully measured (inside the shock in this case) the experimenters concluded that they observe partially filled shells, indicating that pitch-angle scattering had occurred, and the shell filling was assumed to be of order 50% [Mukai et al., 1986].

The most detailed analysis of pickup protons has been performed on data from the IMS HERS sensor, which was able to distinguish pickup protons from the solar wind on the Giotto inbound pass and had the advantage of 3-dimensional coverage (but with some unmeasured gaps). Figure 4 [from Neugebauer et al., 1989] shows the 1-dimensional pitch angle and velocity distributions of cometary protons in the SWB frame at various distances from Halley. These were obtained from the three-dimensional measurements, by summing the distribution function in the solar wind frame over all pitch angles for the velocity distributions (plots on the right) and over all velocities for the pitch angle distributions (plots on the left). From the pitch angle plots the general impression is that the distribution becomes somewhat more isotropic as the shock is approached (e.g., set k compared to the rest), although the use of an algorithm for the mean absolute deviation [Neugebauer et al., 1990] to determine the pitch angle widths of the cometary protons showed that the pitch angle width of the protons was not a strong function of distance upstream of the bow shock.

Figure 5 shows some SWB-frame $v_\perp$-$v_\parallel$ plots from Neugebauer et al. [1989] from the sample times b,g and k shown in the 1-d plots. These intervals are chosen here since they reveal some of the unexpected features mentioned in the original paper. Interval b illustrates the format of the plots and is a relatively "expected" sample. The solar wind appears, of course, at the origin. A simple shell distribution would lie along the dashed circle while the injection point, i.e. the location of the cold ring, is marked by the circled star. The dotted circle is centred on the upstream propagating Alfvén speed (cf Figure 2(c)). The colour-enhanced phase space density distribution for interval b shows some spreading around the shell direction along a path between the dashed and the dotted circles. The distribution from interval g is unexpected in that the distribution is quite anisotropic and is enhanced near to 180 degrees pitch angle. This is difficult to understand theoretically since a relatively even rate of spreading to larger and smaller pitch angles appears more likely from simulations [c.f. Wu et al., 1972, 1986; Gary et al., 1986, 1988] However it has recently been shown that this anisotropy disappears at larger distances

7 MARCH 2351–8 MARCH 0251

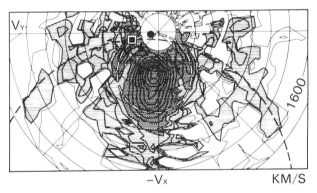

Fig. 3. Suisei measurement of cometary protons and water group ions at (3.4-2.7) x 10^6 km from comet Halley (from Terasawa et al. [1986]).

from the comet (M. Neugebauer et al., paper presented at Chapman conference on Cometary Plasma Processes, 1989). The distribution in interval k was measured just outside the bow shock, and the unexpected result here was that the peak of the pickup ion distribution was outside the expected location based on the local solar wind speed. This is further illustrated in Figure 6, which shows the increase of the pickup proton bulk speed (open circles) immediately in front of the bow shock. Various suggestions for this increase were put forward, namely adiabatic acceleration, the presence of older pickup ions convected from upstream to the observation point or Fermi acceleration effects. The authors preferred the adiabatic explanation but a combination of the three processes is likely to contribute to the effect.

Water group ions

The first observations of what were justifiably assumed to be cometary pickup water group ions were made by the ICE spacecraft [e.g., Hynds et al, 1986, Ipavich et al., 1986]. The most detailed ICE analysis has been performed on the EPAS data (Hynds et al., op cit, and later papers). However the threshold of EPAS detector for water

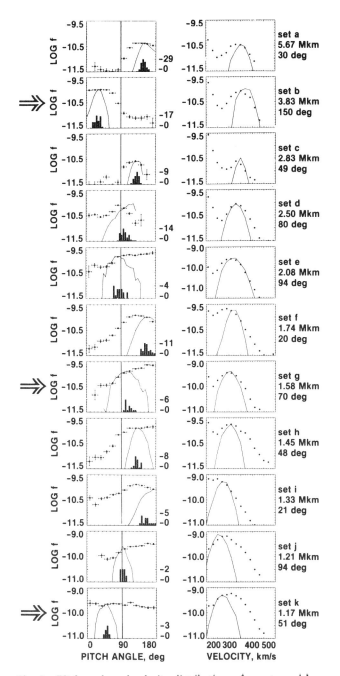

Fig. 4. Pitch angle and velocity distributions of cometary pickup protons at various distances from comet Halley (adapted from Neugebauer et al. [1989]).

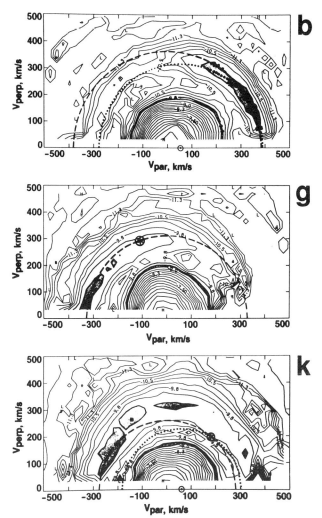

Fig. 5. $v_\perp - v_\parallel$ pickup proton distributions for the intervals b,g and k shown in Figure 4 (adapted from Neugebauer et al. [1989]).

group ions was 65 keV, which was greater than the pickup energy for most of the ICE encounter with Giacobini-Zinner. The majority of the particles which were measured were therefore the products of at least some further acceleration beyond the pickup energy. Nevertheless, some of the first analyses successfully correlated the observed flux to $\sin^2\theta_{vB}$ (proportional to the maximum energy of a pickup ring as discussed above), in the region well upstream of the shock [Sanderson et al., 1986]. This implies that the particles displayed some proper-

ties of a ring-like distribution even above the pickup energy. This comparison is shown in Figure 7, where from the top the panels show (1) the observed and predicted (using the integral of the production rate along a streamline - i.e. assuming rapid pitch angle scattering) flux, (2) the observed flux normalised to the prediction, (3) the maximum pickup energy for a pickup ring ($4E_{sw}\sin^2\alpha$), (4) the solar wind speed inferred from electron measurements, (5) and $\sin^2\alpha$. The authors claimed a good correlation between panels (2) and (3) showing that the angle controls the fluxes, i.e. something akin to a ring-like distribution is observed, but also pointed out the need for additional acceleration mechanisms due to the high threshold of the detector.

In a later paper Richardson et al. [1988] calculated the ion flow direction from a spherical harmonic analysis of the observed fluxes within the energy range of the EPAS detector and found that at distances greater than 4×10^5 km from the nucleus, the ion flow direction was intermediate between $\mathbf{E}\times\mathbf{B}$ (the guiding centre drift direction, see Figure 1) and the solar wind direction. This indicated partial but incomplete isotropisation of the particles at those distances. The data are shown in Figure 8 [from Richardson et al., 1988], where the first panel shows observed flux and the second panel shows the slope

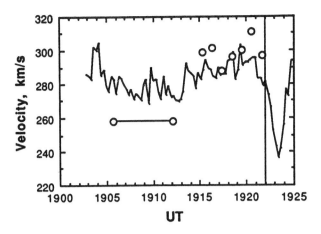

Fig. 6. Cometocentric solar wind speed (solid trace) and pickup proton shell radius (open circles) near the bow shock structure which starts at 1922 SCET (adapted from Neugebauer et al. [1989]).

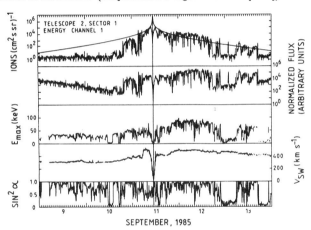

Fig. 7. Comparison of observed and predicted pickup ion fluxes measured by ICE at comet Giacobini-Zinner (see text) (from Sanderson et al. [1986]).

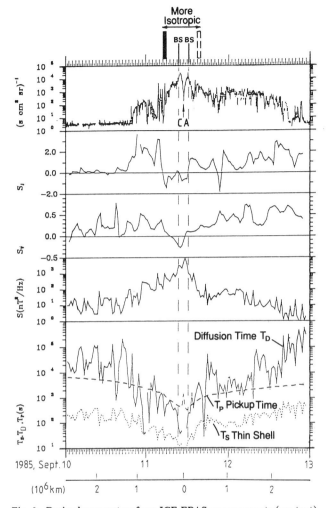

Fig. 8. Derived parameters from ICE-EPAS measurements (see text) (adapted from Richardson et al. [1988]).

of a least squares fit between log(flux) and $\sin^2\alpha$, which indicates that field-related flux modulations are stronger in the outer pickup region. The third panel shows the slope of a least squares fit between the ion streaming direction and the $\mathbf{E}\times\mathbf{B}$ direction, indicating that the streaming direction is intermediate between the solar wind and $\mathbf{E}\times\mathbf{B}$ directions above 4×10^5 km. The fourth panel shows wave power densities. Also shown in Figure 8 are various measures of the timescale for ring and shell formation, as inferred from the data. The pickup time τ_R (=T_p in the diagram) was estimated from the ratio of the predicted ion density to the ion production rate N_i/P, using the equation

$$\tau_R = \frac{R^2}{v_{sw}\exp(-R/L)} \int \frac{dl\exp(-r/L)}{r^2} \qquad (2)$$

where r is the radial distance from the comet, R is the position of the spacecraft, L the scale length for neutral particle ionization and the integral is performed over distance l along the pickup ion streamline from the observation point to infinity upstream. This equation multiplied by v_{sw} gives the pickup region size only if the upstream ionization rate is everywhere the same as the local ionization rate, clearly an inaccurate assumption. τ_R is compared to the time for

thin partial (the amount of pitch-angle scattering being proportional to $\Delta B/B$) shell formation T_s and the time for quasi-linear diffusion in pitch angle and energy T_D, both estimated from Gaffey et al. [1988]. T_D exceeded T_p by at least a factor 5 at more than 1.5×10^6 km, indicating weak scattering in that region. Inside this distance $T_D \sim T_p$, becoming less than this nearer to closest approach. The authors concluded an agreement between the flow directions and expectations from the timescales.

The Vega plasma detector was able to see the water group ions only inside of the bow shock and did not measure the 3-dimensional distribution [Gringauz et al., 1986], so that the results were not suitable for studies of the pickup process in the outer coma. The Vega energetic particle detector measured the high energy tail of the distribution but in this case found that the fluxes showed no correlation with α [Kekskeméty et al., 1989], probably due to the fact that the detector looked perpendicular to the sun-comet axis, so that the ions which were detected would have been scattered in both pitch angle and energy, and this processing would mask the α control.

On Suisei the plasma instrument was an E/q analyser which, as mentioned above, could only see the whole of the pickup water group

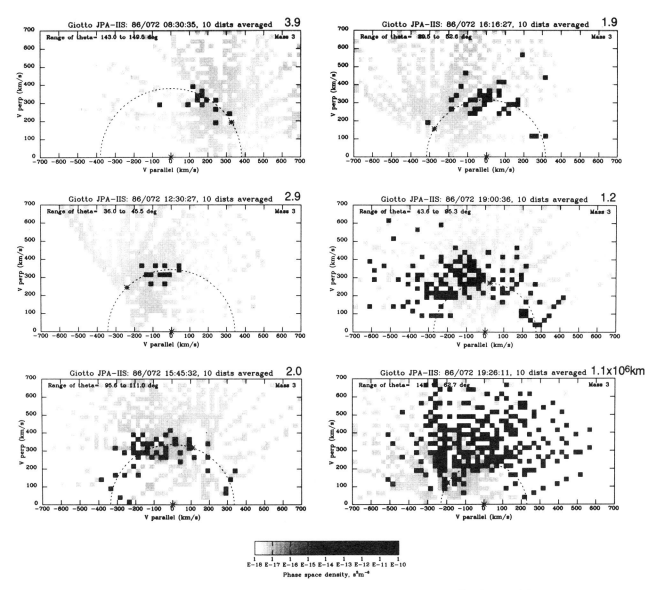

Fig. 9. $v_\perp - v_\parallel$ pickup water group ion distributions at various distances from comet Halley (corrected from Coates et al. [1989]).

population when it the solar wind speed was low enough, downstream of the shock. In the upstream region a 50% shell filling factor was assumed. Again because the full distribution was not measured in the outer coma the results are not suitable for quantitative studies of the pickup process in that region.

Turning to the Giotto results, the JPA IIS instrument covered the velocity range which included the peak of the pickup ion distribution using five electrostatic analyser-time of flight systems which provided mass discrimination over the full 3d sphere. It was able to follow the distribution as it evolved outside 10^5 km from Halley, i.e. from first detection at a few million km to approximately the magnetic pile-up boundary at $\sim10^5$ km (well inside the bow shock at $\sim 10^6$ km). We therefore concentrate on the water group ion results from this sensor. Figure 9 shows a sequence of SWB-frame $v_\perp$-$v_\parallel$ plots (corrected versions, originally from Coates et al. [1989]). In these plots mea-

sured zeros are indicated as the majority of the grey squares in the first frame and the measured phase space densities are grey scaled according to the bar at the bottom of the plot. The sequence shows the evolution of the distribution while approaching the comet, from 3.9×10^6 km, where the distribution is relatively narrow in pitch angle and in energy, through 2.9×10^6 km (similar but a different field orientation so that the distribution is shifted). At 2.0×10^6 km the distribution has scattered in pitch angle but not significantly in energy; the distribution is roughly shell-like. By 1.2×10^6 km, just outside the bow shock, the broadening in energy has started, and is even more pronounced at 1.1×10^6 km, just inside the shock.

The inbound pitch angle and energy distributions are examined further in Figure 10, which shows pitch angle distributions (left panel) and energy distributions (right panel) against time. Closest approach is above the top of the plot in each panel. At early times the pitch

a)

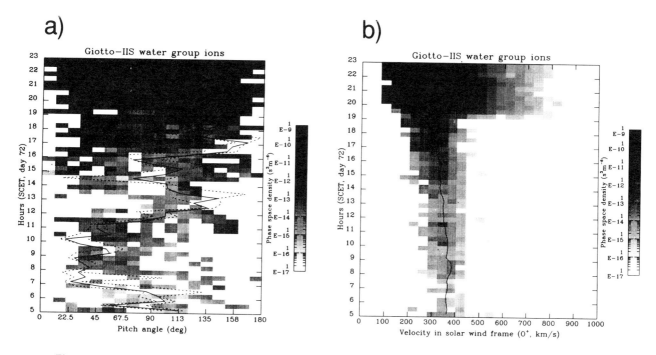

b)

Fig. 10. Water group ion phase space density grey-scale distributions on the Giotto inbound pass from $(4.75-0.25) \times 10^6$ km. Closest approach is above the top of each plot. (a) Pitch angle (abscissa) versus time (ordinate), the pitch angle of a ring distribution is superimposed. (b) Velocity (abscissa) versus time (ordinate), the expected pick-up location is superimposed. (from Coates et al. [1990]).

angle distribution is controlled by α, the expected dependance is over-laid in the left panel. Some pitch angle scattering occurs after about 1430 SCET but the distribution is not nearly isotropic until the bow shock region, and even downstream of this some α control of the ions is still evident. From Figure 10 it is therefore clear that pitch angle scattering occurs well outside the bow shock (i.e., before 1922 SCET) but that, in contrast, energy scattering does not become pronounced until the bow shock region. Thus we can conclude that pitch angle scattering is more rapid than energy scattering, which confirms theoretical ideas and simulations [e.g., Wu et al., 1972, 1986; Gary et al., 1986, 1988]. A similar conclusion can be drawn from the experimental results on the Giotto outbound pass [Coates et al., 1990].

The mean widths of the pitch angle and velocity distributions were also calculated from these data and are shown in Figure 11. The top panel shows the mean absolute deviation of the pitch angle distribution [see Coates et al., 1990], which shows a relatively sharp increase of the width at approximately 2.5×10^6 km, but the distribution never becomes permanently and completely isotropic. The second panel is the width of the velocity distribution; this is relatively constant up to the bow shock region but then shows a dramatic increase. This parameter is a sensitive indicator of the velocity diffusion process and shows that velocity diffusion is approximately constant with distance in the upstream region. Now the quasi-linear velocity diffusion co-efficient D_{vv} depends on the upstream (I_+) and downstream (I_-) propagating wave intensities [Isenberg, 1987] as follows:

$$D_{vv} \propto \frac{I_+ I_-}{I_+ + I_-} \qquad (3)$$

For $I_- \ll I_+$ this reduces [Isenberg, 1987] to

$$D_{vv} \propto I_- \qquad (4)$$

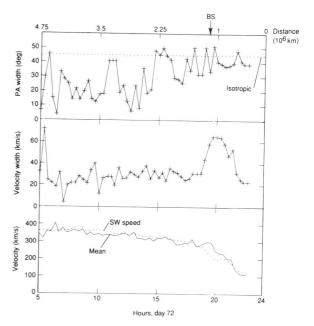

Fig. 11. Widths of the pitch angle (top panel) and log velocity (bottom panel) distributions calculated from the mean absolute deviation for the data in figure 10. (from Coates et al. [1990]).

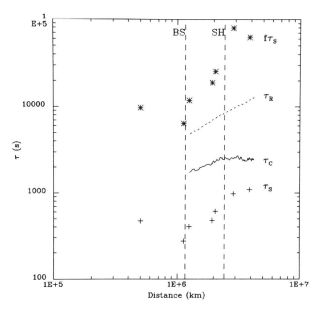

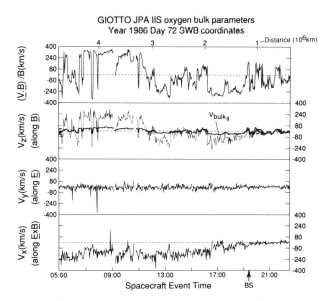

Fig. 12. Pickup times: τ_R is the transition time across the pickup region assuming constant production rate [Richardson et al., 1988]), τ_C is the average time ions spend in solar wind flow [Coates et al., 1989]), τ_S is the pitch angle scattering time [Gaffey et al., 1988], and f $(= (v_{sw}/v_A)^2)$ times τ_S is the energy scattering time [Lyons and Williams, 1984]). The vertical lines indicate the maximum distance where the observations show a filled shell (SH) and the position of the bow shock (BS). (from Coates et al. [1989]).

We have already seen that in the upstream region I_- is predominantly solar wind waves. Thus the constancy of the velocity width in this region is a good indicator that quasi-linear theory describes this region quite well. This is also qualitatively consistent with the relatively low levels of turbulence seen outside the Halley bow shock.

The scattering times are examined in Figure 12 from [Coates et al., 1989]. Here another measure of the time spent in the flow (or "pickup time") was calculated:

$$\tau_C = \frac{\int \frac{ldl \exp(-r/L)}{r^2}}{v_{sw} \int \frac{dl \exp(-r/L)}{r^2}} \qquad (5)$$

This is more accurate than τ_R (equation (2)) used by Richardson et al. [1988] as τ_C allows for a profile of ionization rates along the streamline. τ_C was compared to the shell formation time τ_S from the results of Gaffey et al. [1988], which is 4 times the cometary ion gyroperiod in the wave field

$$\tau_S = \frac{4}{\Omega_i} \left[\frac{\Delta B}{B_o} \right]^{-1} \qquad (6)$$

The energy scattering time was taken to be a factor f $(=v_{sw}/v_A)^2$ times τ_S [Lyons and Williams 1984]. τ_C is always greater that τ_S, which would imply that filled shells should always be observed. A reason for this could be the use of the locally measured turbulence level at all points, rather than an average over the pickup region which would be more appropriate and could increase τ_S by the necessary factor of ~3. If τ_S were increased by this factor we would conclude that pitch angle scattering was occurring in approximately the right

Fig. 13. The lower three panels show the measured water group velocity components in the SWB frame for the inbound leg. The top panel shows the ring prediction calculated from solar wind velocity and magnetic field data (from Coates et al. [1990]).

place although the energy scattering time fτ_S would then be too large.

The bulk velocity of the pickup ions has recently been presented by Coates et al. [1990]. The velocity transformed into the SWB frame is shown in Figure 13. In this Figure the lower 3 panels show the components of the bulk speed, the z component is parallel to the magnetic field in the solar wind frame. From the discussion in the section on scattering of pickup ions, we expect the z component to be $v_{inj,\parallel}$ (ring prediction), 0 (simple shell) or $v_{bulk,\parallel}$ (bispherical). The y and x components should be zero (the deviation of v_x from this is probably due to noise, as discussed by Coates et al. [1990]). For comparison with the v_z component, the top panel shows the ring prediction, while the value of $v_{bulk,\parallel}$ is overlaid on the v_z trace (second panel). Inspection of Figure 13 shows that earlier than ~ 1500 SCET v_z correlates well with the ring prediction although at a reduced level, whereas particularly after the bow shock the correlation of v_z with $v_{bulk,\parallel}$ is quite striking. This would indicate that in the region well upstream of the bow shock the distribution retains ring-like features; the lower magnitude of v_z seems to be a feature of the pitch-angle scattering process as free energy is lost from the particle distribution to the turbulence created in the solar wind. The water group ion distribution becomes more isotropic (but still not completely isotropic) inside of approximately 2.5 x 10^6 km, while downstream of the shock the distribution apparently becomes bispherical to a good approximation.

Another way of displaying the 3-dimensional pickup ion data is shown in Figure 14, which illustrates how the distribution was collapsed into 3 planes in the SWB frame. In each view the injection point is shown as a star and the ring appears as the smaller dashed circle or straight line, depending on the view, with the edge of the simple shell shown by the larger circle in each case. Figure 15 shows some data displayed in this way. The top panel shows the phase space density averaged over 1280 seconds at ~ 2.9 x 10^6 km; the non-zero values are scattered around the expected ring location although they are biassed towards zero parallel speed, this is also evident from the bulk parameter calculations as noted above. The bottom panel shows data from 0.4 x 10^6 km, well inside the shock, where the distribution is much more isotropic. In the lower right $(v_z - v_x)$ subpanel there

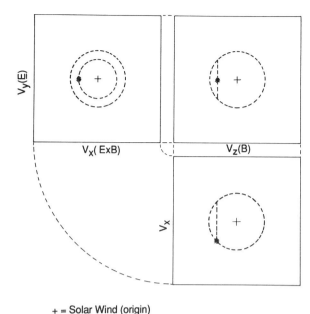

+ = Solar Wind (origin)

Fig. 14. Format for three dimensional plots: measured distributions are collapsed onto the three planes in the SWB frame as shown.

is a clear remnant of what must be a shell distribution from further upstream [e.g., Thomsen et al., 1987].

Discussion and Conclusions

The data discussed above allow us to draw a number of conclusions. The overall picture of ion pickup at comets has been amply confirmed, and the scattering has been experimentally shown to be faster in pitch angle than in energy, in line with expectations from theory and simulation. Cometary protons were observed further from the comet than were water group ions also as expected, and water group ions dominate the mass-loading process as their mass density is higher at all points from their first detection [Neugebauer et al., 1990; Coates et al., 1990]. The scattering of cometary protons in pitch angle is faster than for water group ions for which the pitch angle width is a stronger function of distance, but neither species scatters much in energy outside the bow shock [Neugebauer et al., 1990].

The details of the interaction, in particular the fact that ring-like features were observed in the water group distributions by both ICE and Giotto at distances well outside the bow shock, show that the usual assumption of isotropy of pickup ions is only valid for a region inside of a fairly sharp shell-formation "boundary" upstream of the shock (by about 4 x 10[5] km for G-Z and 1.5 x 10[6] km for Halley). However it is not yet clear whether this "boundary" is a real feature or whether it is a function of the magnetic field geometry in the extended pickup region, although there are some indications that the time that ions spend in the pickup region is of the same order as predicted in shell-formation theories when they reach the appropriate distance [Richardson et al., 1988, Coates et al., 1989]. Downstream of the shock, the bispherical shell picture provides a better model than the simple shell picture in the Giotto measurements of water group pickup ions [Coates et al., 1990]. This is an elegant result as it illustrates conservation of energy between the particles and the waves. However it is somewhat surprising that the bispherical bulk speed provides such a good fit to the observations since we know that at least some particles are accelerated by velocity diffusion and possibly first-order

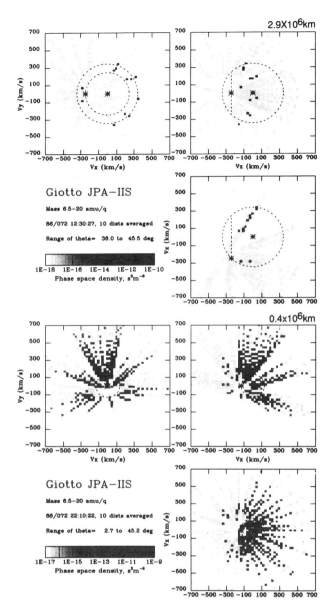

Fig. 15. Three dimensional plots of water group pickup ions in the SWB frame, (a) at 2.9 x 10[6] km and (b) at 0.4 x 10[6] km.

Fermi acceleration in this region, and also that another population of pickup ions is still present which originated upstream [Thomsen et al., 1987].

Work on the Giotto data is continuing vigorously on comparisons between proton and water group ion scattering [Neugebauer et al., 1990], and on comparison of both pickup populations to models [Gombosi et al., 1990]. Comparisons of the pickup populations with energetic particle data are just starting. There is also room for more detailed comparisons of the observed scattering rates to perhaps a simulation within a global cometary model. None of these results are presented here. We are therefore entering a new stage of analysis of the data we have and before the next Chapman conference on this topic there are sure to be many more answers from Giotto but also questions for GEM and CRAF, on the measured velocity distribution of pickup ions.

Acknowledgments. I am grateful for the support of the Royal Society, UK.

References

Alfvén, H., On the theory of comet tails, *Tellus, 9,* 92–96, 1957

Amata E., and V. Formisano, Energisation of positive ions in the cometary foreshock region, *Planet. Space Sci., 33,* 1243–1250, 1985.

Balsiger, H., et al., Ion composition and dynamics at comet Halley, *Nature, 321,* 330–334, 1986.

Biermann L., Kometenschweife und solare korpuscularstrahlung, *Z. Astrophys., 29,* 274–286, 1951

Coates, A. J., A. D. Johnstone, B. Wilken, K. Jockers, K. H. Glassmeier, Velocity space diffusion of pickup ions from the water group at comet Halley, *J. Geophys. Res., 94,* 9983–9993, 1989; correction *J. Geophys. Res., 95,* 4343, 1990.

Coates, A. J., B. Wilken, A. D. Johnstone, K. Jockers, K. H. Glassmeier, and D. E. Huddleston, Bulk properties and velocity distributions of water group ions at comet Halley: Giotto measurements, *J. Geophys. Res., 95,* 10249–10260, 1990.

Coates, A. J., Cometary plasma energisation, *Ann. Geophys.,* in press, 1990.

Gaffey, J. D., Jr., D. Winske, and C. S. Wu, Time scales for formation and spreading of velocity shells of pickup ions in the solar wind, *J. Geophys. Res., 93,* 5470–5486, 1988.

Galeev, A. A., and R. Z. Sagdeev, Alfvén waves in a space plasma and its role in the solar wind interaction with comets, *Astrophys. Space Sci., 144,* 427–438, 1988.

Gary, S. P., S. Hinata, C. D. Madlund, and D. Winske, The development of shell-like distributions from newborn cometary ions, *Geophys. Res. Lett., 13,* 1364–1367, 1986.

Gary, S. P., C. D. Madlund, N. Omidi, and D. Winske, Computer simulations of two-pickup-ion instabilities in a cometary environment, *J. Geophys. Res., 93,* 9584–9596, 1988.

Gombosi, T. I., K. Lorencz, and J. R. Jokipii, Combined first- and second-order Fermi acceleration in cometary environments, *J.Geophys.Res., 94,* 15011–15023, 1989.

Gombosi, T. I., Neugebauer, M., Johnstone, A. D., Coates, A. J. and Huddleston, D. E., Comparison of observed and calculated implanted ion distributions outside comet Halley's bow shock, *J.Geophys.Res.,* in press, 1990

Gringauz, K. I., et al., First in situ plasma and neutral gas measurements at comet Halley, *Nature, 321,* 282–285, 1986.

Hynds, R. J., S. W. H. Cowley, T. R. Sanderson, K. P. Wenzel and J. J. van Rooijen, Observations of energetic ions from comet Giacobini-Zinner, *Science, 232,* 361–365, 1986.

Ip, W. H. and Axford, W. I., The acceleration of particles in the vicinity of comets, *Planet. Space Sci., 34,* 1061–1065, 1986.

Ip, W. H., Cometary ion acceleration processes, *Computer Phys. Comm., 49,* 1–7, 1988.

Ipavich, F. M., A. B. Galvin, G. Gloeckler, D. Hovestadt, B. Klecker, and M. Scholer, Comet Giacobini-Zinner: In situ observations of energetic heavy ions, *Science, 232,* 366–369, 1986.

Isenberg, P. A., Energy diffusion of pickup ions upstream of comets, *J. Geophys. Res., 92,* 8795–8799, 1987b.

Johnstone A. D., D. E. Huddleston and A. J. Coates, The spectrum and energy density of solar wind turbulence of cometary origin, this volume, 1990

Kecskeméty, K., et al., Pickup ions in the unshocked solar wind at comet Halley, *J.Geophys.Res., 94,* 185–196, 1989.

Lämmerzahl P., et al., Expansion velocity and temperatures of gas and ions measured in the coma of comet P/Halley, *Astron. Astrophys., 187,* 169–173, 1987

Lyons, L. R., and D. J. Williams, *Quantitative Aspects of Magnetospheric Physics,* p.141, 231 pp., D. Reidel, Hingham, Mass., 1984.

McKenna-Lawlor, S., E. Kirsch, D. O'Sullivan, A. Thompson and K. P. Wenzel, Energetic ions in the environment of comet Halley, *Nature, 321,* 347–349, 1986.

Möbius, E., B. Klecker, D. Hovestadt, and M. Scholer, Interaction of interstellar pickup ions with the solar wind, *Astrophys. Space Sci., 144,* 487–505, 1988.

Mukai, T., W. Miyake, T. Terasawa, M. Kitayama, and K. Hirao, Plasma observation by Suisei of solar-wind interaction with comet Halley, *Nature, 321,* 299–303, 1986.

Neugebauer, M., A. J. Lazarus, H. Balsiger, S. A. Fuselier, F. M. Neubauer and H. Rosenbauer, The velocity distributions of cometary protons picked up by the solar wind, *J. Geophys. Res., 94,* 5227–5239, 1989.

Neugebauer, M., A. J. Coates and F. M. Neubauer, Comparison of picked-up protons and water group ions upstream of comet Halley's bow shock, *J.Geophys.Res.,* in press, 1990.

Richardson I. G., S. W. H. Cowley, R. J. Hynds, P. W. Daly, T. R. Sanderson, and K. P. Wenzel, Properties of energetic water-group ions in the extended pickup region surrounding comet Giacobini-Zinner, *Planet. Space Sci., 36,* 1429–1450, 1988.

Sanderson T. R., K. P. Wenzel, P. Daly, S. W. H. Cowley, R. J. Hynds, E. J. Smith, S. J. Bame and R. D. Zwickl, The interaction of heavy ions from comet P/Giacobini-Zinner with the solar wind, *Geophys.Res.Lett., 13,* 411–414, 1986.

Somogyi, A. J., et al., First observations of energetic particles near comet Halley, *Nature, 321,* 285–288, 1986.

Terasawa, T., T. Mukai, W. Mikaye, M. Kitayama, and K. Hirao, Detection of cometary pickup ions up to 10^7 km from comet Halley: Suisei observation, *Geophys. Res. Lett., 13,* 837–840, 1986.

Terasawa, T. and M. Scholer, The heliosphere as an astrophysical laboratory for particle acceleration, *Science, 244* 1050–1057, 1989.

Terasawa, T., Acceleration mechanisms for cometary ions, this volume, 1990.

Thomsen, M. F., et al., In-situ observations of a bi-modal ion distribution in the outer coma of comet Halley, *Astron. Astrophys., 187,* 141–148, 1987.

Tsurutani, B. T., Comets: a laboratory for plasma waves and instabilities, this volume, 1990.

Wu, C. S., and R. C. Davidson, Electromagnetic instabilities produced by neutral-particle ionization in interplanetary space, *J. Geophys. Res., 77,* 5399–5406, 1972.

Wu, C. S., D. Winske, and J. D. Gaffey, Jr., Rapid pickup of cometary ions due to strong magnetic turbulence, *Geophys. Res. Lett., 13,* 865–868, 1986.

OBSERVATIONS OF PICKUP IONS ACCELERATED IN THE VICINITY OF COMET HALLEY

M.B. Bavassano Cattaneo, V. Formisano, E. Amata, R. Giovi, P. Torrente

Istituto di Fisica dello Spazio Interplanetario
Consiglio Nazionale delle Ricerche, Frascati, Italy

Abstract. The distribution functions of protons and water group ions measured by the Giotto Implanted Ion Spectrometer have been examined systematically over several million kilometers upstream of comet Halley, in order to obtain informations on the acceleration of cometary ions above the pick up energy. The ability of two spectral shapes (namely a power law and an exponential in energy) to represent the data has been investigated. Between cometocentric distances 2.5×10^6 and 2.5×10^5 km there are indications that, for both species, the low energy part of the distribution functions is better represented by a power law, suggesting that first order Fermi acceleration is the dominant process, although the possibility of another mechanism operating simultaneously cannot be excluded.

Introduction

One of the most interesting findings of the cometary fly-bys is the observation of particles with energies larger than the pickup energy [see e.g. McKenna-Lawlor et al., 1989]. This fact suggests that other acceleration mechanisms, besides the pickup process, are effective in the vicinity of comets. The peculiar characteristics of the cometary environment — great extension of the mass loading region, large amplitude turbulence, plasma deceleration due to momentum transfer—allow in principle several processes. Among the possible candidates are adiabatic compression and first and second order Fermi mechanisms. Adiabatic compression, occurring in the region of plasma deceleration, is not able, alone, to produce the observed energies [Ip and Axford, 1986; Gombosi et al., 1989]. Second order Fermi mechanism is a stochastic process of energy diffusion, leading both to a deceleration and to an acceleration of particles. Comets are surrounded by a several million kilometers wide mass loading region where a high level of wave activity is present [Tsurutani and Smith, 1986a and 1986b], and energy diffusion is believed to be a possible process [Ip and Axford, 1986]. The last process, first order Fermi acceleration, is due to spatial diffusion of ions between converging scattering centres. In a cometary environment this process had first been suggested, long before the comet encounter, by Formisano and Amata [1984] and Amata and Formisano [1985]. Different spectral shapes are expected, depending on which acceleration process is operating. So, when first order Fermi acceleration is operative, and when ion losses are inhibited (as is the case at comets, due to the great extent of the mass loading region) the ion distribution function is a power law. On the contrary, in the case of second order Fermi acceleration, the ion distribution in the solar wind frame is expected to fall off exponentially at high energies [Gribov et al,1986; Ip and Axford,1986].

Recently a number of studies have been devoted to the problem of ion acceleration in the vicinity of comets. Richardson et al. [1987], studying the form of the energetic ion spectra with an approach similar to the one presented here, find that the energetic particle distribution functions near comet Giacobini -Zinner are fitted either by a power law or by an exponential in velocity. Moreover a comparison of their spectra with those predicted by Ip and Axford [1986] for second order Fermi acceleration shows that the observed spectra are much softer than the theoretical ones suggesting a mean free path larger than the one used in the model. More sophisticated theoretical models [Isenberg, 1987; Gombosi, 1988] include continuous injection of pick up ions and determine the evolution of the plasma rest frame distribution function under the effect of convection, adiabatic compression and velocity diffusion. More recently, however, it has been suggested that both first and second order Fermi acceleration play a role in the vicinity of comets [Ip, 1988; Gombosi et al., 1989a and b]. In particular Gombosi et al. [1989 a

Cometary Plasma Processes
Geophysical Monograph 61

and b] study the effect, in the preshock region, of combined adiabatic, first and second order Fermi acceleration. According to this scenario, far from the comet, further than any appreciable plasma deceleration, velocity diffusion would accelerate particles, thus creating a seed population which first order Fermi mechanism could accelerate more efficiently. The resulting preshock oxygen spectra would be exponential in energy below 100 keV, and power law above 100 keV. Calculations carried out along Giotto's trajectory predict for water group ions an e-folding temperature between 10 and 20 keV in the low energy part of the spectrum, and a velocity spectral index around -30 for the high energy tail. On the contrary, if only first order Fermi acceleration is present together with adiabatic acceleration, the velocity spectra would be power law with spectral index below -20.

In order to allow the comparison between theoretical models and observations it is important to characterize the shape of the observed energy distributions above the pick up energy, and for this purpose the plasma rest frame distribution functions of protons and of water group ions, measured by the Giotto plasma experiment, have been separately fitted to a power law and to an exponential in energy, systematically over a distance of several million kilometers upstream of the comet. The results of this analysis will be discussed below.

Instrumentation and data treatment

As Giotto approached Halley comet, cometary ions (water group ions and protons) have been detected by the JPA plasma experiment as far out as 4.5 million km (cometocentric distance) and at around the same time a steady deceleration of the plasma was observed from an initial speed of ~ 400 km s^{-1} to ~ 300 km s^{-1} just before the bow shock. The magnetic field, on the other hand, showed enhanced wave activity at a cometocentric distance $\sim 11 \times 10^6$ km when the relative total variance in the $10^{-2} \div 8. \times 10^{-4}$ Hz frequency range increased by a factor $2 \div 3$. (The Giotto magnetic field data have been kindly provided by F. Neubauer and the magnetometer has been described by Neubauer et al., 1987).

In this study cometary ions at cometocentric distances between 2.5×10^6 to 2.5×10^5 km will be considered, namely upstream and downstream of the bow shock (located at a cometocentric distance of ~ 1.2 million km). At larger distances particle fluxes are too low to allow any quantitative consideration.

The Giotto JPA experiment consists of two separate analysers: one 270° electrostatic analyser (FIS) and one time of flight analyser (IIS) capable of separating masses in five mass bins. In order to determine the shape of the distribution function it is crucial to separate masses cor-

rectly: for this reason we have used data from the IIS analyser, which collects, in two separate bins, protons and particles with masses in the $6.5 \div 20$ amu range. For details on the experiment, see Johnstone et al, 1986. Here we shall only recall that IIS consists of five polar sensors, separated by gaps in polar angle, which provide, as the spacecraft rotates, a complete angular coverage in the azimuthal direction. Energy varies between 70 eV and 85 keV in 32 energy channels.

Throughout this study we have averaged the distribution function over ~ 8.5 minutes in order to smear out temporal fluctuations. The distribution functions have been transformed to the solar wind frame, using the solar wind velocity obtained by the FIS sensor upstream of the bow shock and by the combination of FIS and polar sector 5 of IIS [as described by Formisano et al., 1990, and by Amata et al., 1990] downstream of the shock.

Data obtained by one polar sector alone (i.e. over one conical surface in velocity space) have been used, in order to have a homogeneous data set. The choice of the polar sector depends on which sector has the greatest coverage in velocity space in the solar wind frame, and this in turn depends on the position of the local solar wind velocity with respect to the sensor's field of view. We have used polar sector 3 (with field of view containing the spacecraft/sun direction) upstream of the shock and polar sector 5 (with field of view pointing towards the spacecraft/comet direction) downstream. The next step was to consider the one dimensional velocity distributions in the solar wind frame. Only in the case of perfect isotropy would the data points in these spectra match smoothly on one curve, otherwise a great spread in the observational points would be expected. Observations of both protons and water group ions show that this latter fact occurs in most of the upstream region. This is consistent with the observations of Coates et al. [1989] and of Richardson et al. [1988] who noticed that upstream of the shock the implanted ion shell is complete, but not isotropic, and becomes isotropic just outside the bow shock. In the preshock region we therefore considered the velocity spectra in one radial sector, namely in a $\sim 67.5°$ wide sector in the antisunward direction. On the contrary inside the bow shock the isotropy of the distribution allowed to examine the whole three dimensional distribution.

Observations

a) Protons

Protons have been considered in the energy range 400 $\div$ 3800 eV upstream of the bow shock and in the range 130 $\div$ 3800 eV downstream of the shock. Due to gaps

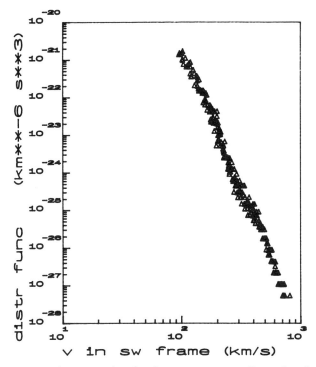

Fig. 1. An example of a downstream one dimensional velocity distribution in the plasma rest frame prior to subtraction of the solar wind contribution. This spectrum starts at 2143 SCET, day 72, 1986. Only the tail of the main population is detected, together with the pickup protons.

in the polar angle coverage, the peak of the solar wind population is missed upstream of the shock (as it lies in the gap between two adjacent polar sectors), and appears only well downstream as soon as the solar wind velocity rotates close to the field of view of polar sector 5. However the high energy tail of the main population, together with the pickup protons, is constantly clearly detected. This is evident in figure 1 where a proton one dimensional velocity distribution in the solar wind frame is shown. This downstream distribution function starts at 21 43 SCET (Spacecraft Event Time): as we mentioned, the peak of the main population is not detected, but its tail (between 90 and 200 km s^{-1}) is visible at low energies and the change in slope around 350 km s^{-1} is due to pickup protons. It is interesting to note the isotropy of this spectrum, which can be deduced noting that all of the 238 data points (which correspond to measurements taken in different directions in the spacecraft frame) lie on a smooth curve in the one-dimensional velocity spectrum.

After transformation of the distribution function to the solar wind frame, it is necessary to separate the pickup protons from the main population. This is achieved by

performing a maxwellian fit on the low energy part of the distribution (consisting only of solar wind protons), extrapolating it to higher energies and subtracting it from the high energy tail (which contains implanted protons). Finally, now that the high energy tail of the distribution function consists only of pickup protons, the determination of its shape can be performed. Each pickup distribution function has been fitted separately to a power law and to an exponential in energy, and the corresponding linear correlation coefficient (i.e. the linear correlation coefficients in the log f vs. log v scale, corresponding to a power law, and in the log f vs. energy scale, corresponding to an exponential in energy) have been determined. One important point should be kept in mind: as we are interested in the contribution of the acceleration process to the shape of the distribution function, the fitting has been performed for energies above the energy of the ions picked up upstream and convected to the observation point (or above the pickup energy whenever it did not differ greatly from the energy of convected pickup ions).

The time history of the two linear correlation coefficients over a nine hour period, i.e. over 2.2 million km upstream of the comet is shown in the lower panel of figure 2. The full line refers to the power law and the dotted line

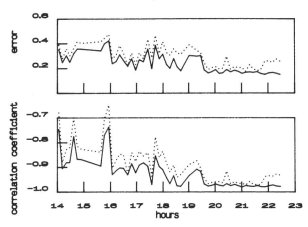

Fig. 2. Lower panel: time history of the linear correlation coefficient for a power law (full line) and an exponential in energy (dotted line) for pickup protons. The plot ranges between cometocentric distances 2.5×10^6 km and 2.5×10^5 km. Upper panel: time history of the errors of the two fits (the power law is the full line, the exponential in energy is the dotted line).

to the exponential in energy. Earlier than 14 SCET (corresponding to a cometocentric distance of 2.5×10^6 km) the fits are not meaningful (due to low fluxes) and the results have not been shown. Inspection of figure 2 shows that in the upstream region (i.e. before $\sim$ 19 SCET) both correlation coefficients lie between -0.8 and -0.9 and

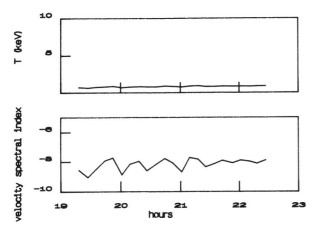

Fig. 3. Velocity spectral index (lower panel) and e-folding temperature (upper panel) obtained by fitting the pickup proton distribution function with a power law and with an exponential in energy, respectively.

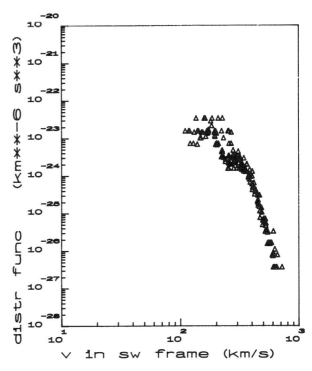

Fig. 4. A downstream distribution function of water group ions starting at 2126 SCET. The two peaks around ~ 180 km s^{-1} and ~ 300 km s^{-1} are due to locally picked up ions and to upstream picked up ions respectively.

have large fluctuations, due to considerable spread in the data points. However, in spite of the large fluctuations, starting from the distance of $\sim 2.5 \times 10^6$ Km, the full line is always closer to -1. Downstream of the shock both curves are quite close to -1 and close to one another, indicating that both fits are satisfactory, although the power law line lies also here closer to -1 and is stable and $\sim$ -0.97. The striking point is that, although the difference between the two correlation coefficients is sometimes not too large, however the correlation coefficient of the power law is systematically closer to -1, over the entire period (i.e. over a distance greater than 2 million km). This enables us to say with sufficient confidence, that during the whole period the data are better fitted by a power law. In the upper panel the error on each fit (full and dotted lines for the power law and the exponential in energy respectively) has been shown. It appears that upstream of the shock both fits have a larger error than downstream (due to smaller fluxes and consequently fewer data points used in the fits), but the power law always has a smaller error than the exponential in energy.

It is interesting to look at the velocity spectral index. We considered its time history only after the shock crossing. The results are shown in the lower panel of figure 3. The spectral index fluctuates around -8, and does not vary appreciably over the considered distance of 1 million km. For the sake of completeness we show in the upper panel of the figure the e-folding temperature of the exponential fit. It is almost constant and is around ~ 800 eV. On the other hand, in the few isotropic spectra before the shock, the fittings (not shown in the figure) give values between $\sim$ -8 and -9 for the velocity spectral index, and ~ 700 eV for the e-folding temperature.

b) Water group ions

Water group ions have been studied in the energy range $3.8 \div 55$ keV, but downstream of the bow shock the lower limit has been gradually decreased down to 1.5 keV. Similarly to what was observed for protons, in the preshock region the solar wind frame distribution function is not isotropic and only one radial section of it has been considered. On the contrary, downstream of the bow shock the complete distribution function can be studied. Soon after the bow shock traversal, as already noted by Thomsen et al. [1987] and by Amata et al. [1989], the distribution function shows two peaks: the higher energy population (peaked at a velocity of ~ 300 km s^{-1}) consists of ions picked up upstream of the shock and convected downstream, the lower energy peak (partially missed in the furthest spectra due to incomplete angular coverage) is due to locally picked up ions. An example of water group ions distribution function in the plasma frame downstream of the shock is shown in figure 4. Here the two peaks appear: one at $\sim$180 km s^{-1} (which is the local bulk speed) and the other at ~ 300 km s^{-1} (which is close to the upstream bulk speed).

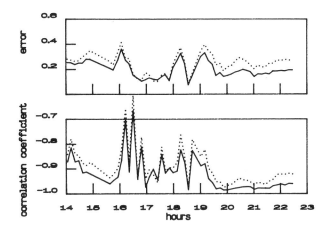

Fig. 5. The same as figure 2 for water group ions.

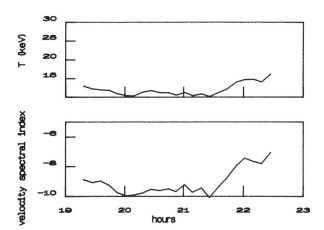

Fig. 6. The same as figure 3 for water group ions.

This bimodal distribution poses the question whether the low energy peak can influence the form of the high energy peak and whether therefore the contribution of the low energy population should be subtracted. In order to check this we assumed that the two peaks had the same shape, namely a power law (the reason for the choice of this shape will be given later in this section), extrapolated the low energy peak to higher energies and subtracted it from the high energy population. We found that the resulting high energy part of the spectra did not differ significantly from the original one, and therefore we considered this subtraction negligible.

In the determination of the shape of the distribution function, as for protons, we considered only the part of the distribution function above the upstream pickup energy. The time history of the linear correlation coefficient for the two functional shapes is shown in the lower panel of Figure 5. Similarly to the protons results, the curves in the preshock region show large fluctuations due to the spread of the data points. However, the power law (full line) has, upstream aswell as downstream, a linear correlation coefficient closer to -1. As for protons, the important point to note from figure 5 is that systematically, over the entire distance of 2.2 million km, an advantage for the power law with respect to the exponential in energy is observed. This again enables us to say that, also for water group ions, the power law gives a better fit. In the upper panel of the figure the error of the fit for the power law (full line) and for the exponential in energy (dotted line) are shown,and the former is almost always smaller than the latter.

We next determined the velocity spectral index. Upstream of the shock it was determined only for the few isotropic spectra just outside the bow shock (the spectral index is meaningful only for isotropic spectra) and was found to lie between -10 and -7. After the shock the velocity spectral index is shown in the lower panel

of figure 6. Until 2130 SCET it is almost constant, between -9 and -10, implying that these spectra are slightly softer than the corresponding proton spectra. One interesting feature is observed around 2130 SCET when the spectral index starts gradually to increase and becomes ~ -7.5 after 22 SCET. This change of the spectral index corresponds to a change in shape and to a hardening of the spectra, and was not observed in the protons data. The reason for this hardening is still unclear and needs further study. Although the exponential was seen to fit the data less accurately than the power law, we show in the upper panel of figure 6, the corresponding e-folding temperature. In the upstream region (not shown in the figure) it lies between 8 and 11 keV, and then downstream it becomes ~ 10 keV.

We may try to compare our observations with Gombosi et al.'s [1989a] and [1989b] model. (Note that the comparison can be performed only for preshock oxygen spectra, as the model extends only in the upstream region). The power law shape was predicted by the model in the case of first order Fermi acceleration alone, but the observed spectra are much harder than the model's. The reason for this difference could either be due to the fact that the parameters used in the model are not appropriate for Giotto's encounter (for instance the smaller extent of the foreshock region in the model limits the time the ions spend in the acceleration region), or could indicate that although first order Fermi acceleration is the dominant mechanism, other processes are effective. On the other hand in the case of combined first and second order Fermi mechanism the model predicts an exponential shape at low energies and a power law only above 100 keV. As we already mentioned the data, at least in the low energy range considered, do not show any change in shape. Unfortunately the model does not extend downstream of the shock, so that no comparison of the observations with the model can be done in that region.

And also no masses different from oxygen are included in the model, so that protons,for which we have independent observations, cannot be compared with the model's predictions.

Conclusions

We have systematically examined the cometary protons and water group ions over an extended region of several million kilometers around comet Halley in order to find the functional form which best approximates the rest frame distribution functions. This has been obtained by fitting them with a power law and with an exponential in energy. The fittings were done between ~ 600 and ~ 3800 eV for protons, and between ~ 10 and ~ 55 keV for water group ions. In spite of the limited energy range considered, the results of our analysis show that

i) at cometocentric distances between 2.5×10^6 km and 2.5×10^5 km, there are indications that at low energies both species upstream and downstream of the shock are better fitted by a power law;

ii) for protons, immediately up− and downstream of the shock, the corresponding velocity spectral index lies between -8 and -9;

iii) for water group ions the velocity spectral index ranges between -7 and -10;

iiii) the comparison of the observations with a theoretical model [Gombosi et al., 1989a and 1989b] shows that the upstream observed spectra are harder than those predicted by the model in the case of first order Fermi acceleration alone, although they have the same functional shape.

Acknowledgements.   We would like to thank F.M. Neubauer for kindly providing the magnetic field one minute averages.

This research has been supported by the Agenzia Spaziale Italiana.

References

Amata, E. and V. Formisano, Energization of positive ions in the cometary foreshock region, *Planet. Space Sci.* 33, 1243, 1985.

Amata, E., V. Formisano, P. Torrente, M.B. Bavassano Cattaneo, A.D. Johnstone and B. Wilken, Experimental plasma parameters at comet Halley, *Adv. Space Res.*,9, 313, 1989.

Amata, E., V. Formisano, P. Torrente, R. Giovi, The plasma parameters during the inbound and outbound legs of the Giotto trajectory, this issue.

Coates, A.J.,A.D. Johnstone, B. Wilken, K. Jockers and K.-H. Glaßmeier, Velocity space diffusion and pickup ions from the water group at comet Halley, *J. Geophys. Res.*, 94, 9983, 1989.

Formisano, V. and E. Amata, Energetic positive ions in the cometary foreshock region, *Adv. Space Res.* 4, 253, 1984.

Formisano, V., E. Amata, M.B. Bavassano Cattaneo, P. Torrente, A. Coates, A. Johnstone, B. Wilken, K. Jockers, M. Thomsen, D. Winningham,, H. Borg, Plasma flow inside comet P/Halley submitted to *Astron. Astrophys.*, 1990.

Gombosi, T.I., Preshock region acceleration of implanted cometary H^+ and O^+, *J. Geophys. Res.*, 93, 35, 1988.

Gombosi, T.I., K. Lorencz and J.R. Jokipii, Combined first and second order Fermi acceleration at comets, *Adv. Space Res.*,9, 337, 1989a.

Gombosi, T.I., K. Lorencz and J.R. Jokipii, Combined first and second-order Fermi acceleration in cometary environments, *J. Geophys. Res.* 94, 15011, 1989b.

Gribov, B. E., K. Kecskeméty, R. Z. Sagdeev, V. D. Shapiro, V. I. Shevchenko, A. J. Somogyi, K. Szegö, G. Erdös, E. G. Eroshenko, K. I. Gringauz, E. Keppler, R. Marsden, A. P. Remizov, A. K. Richter, W. Riedler, K. Schwingenschuh, and K.P. Wenzel, Stochastic Fermi acceleration of ions in the preshock region of comet Halley, Proc. 20th ESLAB Symposium on the Exploration of Halley's comet (eds. B. Battrick, E.J. Rolfe and R. Reinhard), *EUR. Space Agency Spec. Publ.,ESA SP-250*, Vol.1, p. 271, 1986.

Ip, W.-H., Cometary ion acceleration processes, *Computer Physics Communications*, 49, 1, 1988.

Ip, W.-H. and W.I. Axford, The acceleration of particles in the vicinity of comets, *Planet. Space Sci.*, 34,1061, 1986.

Isenberg, P.A., Energy diffusion of pickup ions upstream of comets, *J. Geophys. Res.* 92, 8795, 1987.

Johnstone, A.D., J.A. Bowles, A.J. Coates, A.J. Coker, S.J. Kellock, J. Raymont, B. Wilken, W. Stüdemann, W. Weiss, R. Cerulli Irelli, V. Formisano, F. de Giorgi, P. Perani, M. de Bernardi, H. Borg, S. Olsen, J.D. Winningham, D.A. Bryant, The Giotto three-dimensional positive ion analyser, *Eur. Space Agency Spec. Publ.,ESA SP-1077*, 1986.

McKenna-Lawlor, S., P. Daly, E. Kirsch, B. Wilken, D. O'Sullivan, A. Thompson, K. Kecskeméty, A. Somogyi, and A. Coates, In situ energetic particle observations at comet Halley recorded by instrumentation aboard the Giotto and Vega 1 missions, *Ann. Geophys.*, 7, 121, 1989.

Neubauer, F.M., M.H. Acuna, L.F. Burlaga, B. Franke, B. Gramkow, F. Mariani, G. Musmann, N.F. Ness, H.U. Schmidt, R. Terenzi, E. Ungstrup and M. Wallis, The Giotto Magnetometer experiment, *J. Phys. E: Sci. Instrum.* 20, 714, 1987.

Richardson, I. G., S. W. H. Cowley, R. Hynds, C. Tranquille, T. R. Sanderson, K.-P. Wenzel and P. W. Daly, Observations of energetic water group ions at comet Giacobini- Zinner: implications for ion acceleration processes, *Planet. Space Sci.*, 35, 1323, 1987.

Richardson, I.G., S.W.H. Cowley, R.J. Hynds, P.W. Daly, T.R. Sanderson and K.-P. Wenzel, Properties of energetic water group ions in the extended pickup region surrounding comet Giacobini-Zinner, *Planet. Space Sci.*, 36, 1429, 1988.

Thomsen, M.F., W.C. Feldman, B. Wilken, K. Jockers, W. Stüdemann, A.D. Johnstone, A. Coates, V. Formisano, E. Amata, J.D. Winningham, H. Borg, D. Bryant and M.K. Wallis, In-situ observations of a bi-modal ion distribution in the outer coma of comet P/Halley, *Astron. Astrophys.* 187, 141, 1987.

Tsurutani, B.T.and E.J. Smith, Strong hydromagnetic turbulence associated with comet Giacobini-Zinner, *Geophys. Res. Lett.* 13, 259, 1986a.

Tsurutani, B.T. and E.J. Smith, Hydromagnetic waves and instabilities associated with cometary ion pickup: ICE observations, *Geophys. Res. Lett.* 13, 263, 1986b.

ENERGETIC WATER-GROUP IONS AT COMET GIACOBINI-ZINNER: AN OVERVIEW OF OBSERVATIONS BY THE EPAS INSTRUMENT

S.W.H. Cowley[1], A. Balogh[1], R.J. Hynds[1], K. Staines[1], T.S. Yates[1], P.W. Daly[2], I.G. Richardson[3], T.R. Sanderson[4], C. Tranquille[4], K.-P. Wenzel[4]

Abstract. An overview is presented of the observations of energetic (45-480 keV) cometary water group ions made by the EPAS instrument on the ICE spacecraft during the fly-by of comet Giacobini-Zinner in September 1985. The discussion covers ion observations in the large-scale ion pick-up region surrounding the comet on $\sim 10^6$ km spatial scales, observations of the cometary shock at $\sim 10^5$ km and the region of slowed mass-loaded flows which occur downstream, and the finite gyroradius effects which occur near closest approach.

(1) Introduction

The first in situ observations of the interaction between a comet and the solar wind plasma were made in September 1985, when the International Cometary Explorer (ICE) spacecraft flew by comet Giacobini-Zinner, passing centrally through the tail of the comet within 7800 km of the nucleus on the antisunward side [von Rosenvinge et al., 1986]. The ICE spacecraft was not specifically instrumented for cometary studies, having originally been designed to study plasma particles and electromagnetic fields in the solar wind upstream from the Earth's magnetosphere as part of the ESA/NASA International Sun-Earth Explorer programme. Consequently, no in situ observations of the neutral atmosphere of Giacobini-Zinner were made during the encounter, and only rudimentary measurements of the cometary dust were possible. However, detailed observations were obtained of

the plasmas and fields in the vicinity of the comet. In particular, the spacecraft instrumentation which was originally intended to study solar energetic particles provided detailed observations of heavy cometary ions during the encounter, since these ions are in general accelerated to energies of tens to hundreds of keV in their interaction with the solar wind [Hynds et al., 1986a; Ipavich et al., 1986]. This is significant because the heavy cometary ions are the principal agents which mediate the interaction between the comet and the ambient solar wind plasma, leading, ultimately, to induced magnetotail and plasma tail formation [see e.g. the theoretical reviews by Ip and Axford (1982), Mendis and Houpis (1982), Schmidt and Wegmann (1982), and Johnstone (1985)].

The purpose of this paper is to provide an overview of some recent results on the properties of heavy cometary ions in the vicinity of Giacobini-Zinner, derived from measurements made by the Energetic Particle Anisotropy Spectrometer (EPAS) experiment on the ICE spacecraft. A review of the initial EPAS findings may be found in Hynds et al. [1986b], while Cowley [1987] has provided an overall summary of ICE field and plasma data. In the next section we begin by giving a brief outline of the theoretical considerations which are required to understand and interpret the ion data. The third section then provides details of the ICE instrumentation which is relevant to our study. Subsequent sections then review the principal results derived from the ion data in the various regions surrounding the comet, setting them within their plasma physical context derived from other ICE observations.

(2) Theoretical Overview

The cometary ions discussed in this review originate from molecules (mainly water) which sublime from the comet nucleus and expand outwards, unrestrained by the comet's gravity. The molecules are photodissociated into atomic components on time scales of $\sim 10^5$ s, and ionized, by either solar UV, charge exchange with solar wind protons, or solar wind electron impact, on time scales $\sim 10^6$ s. Combining these times with a molecule/heavy atom expansion speed of ~ 1 km s^{-1} shows that there will exist an extensive region of

[1]Blackett Laboratory, Imperial College, London SW7 2BZ, UK
[2]Max-Planck-Institut fur Aeronomie, D-3411 Katlenburg-Lindau, FR Germany
[3]NASA/Goddard SFC, Greenbelt, MD 20771, USA
[4]European Space Agency, ESTEC, NL-2200 AG Noordwijk, The Netherlands

Cometary Plasma Processes
Geophysical Monograph 61
©1991 American Geophysical Union

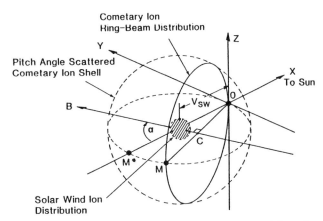

Fig. 1. Sketch showing the form of the cometary pick-up ion distribution function in velocity space and its relationship to the distribution of solar wind ions. Point O is the origin of velocity space in the rest frame of the cometary neutral atoms, and thus the point at which cometary ions are created. To a good approximation it is also the rest frame of the ICE spacecraft. The X-axis is antiparallel to V_{SW}, while the X-Y plane contains the magnetic field vector $\mathbf{B}$ lying at angle α to V_{SW}. The solar wind ion distribution function is indicated by the hatched sphere lying on the X axis, while the ring-beam distribution of newly created pick-up ions is indicated by the solid circle centred on C, passing through O and M. Efficient pitch-angle scattering of the ions would result in the formation of a "shell" distribution indicated by the dashed lines, representing a sphere of radius V_{SW} centred on the solar wind ion distribution. [After Richardson et al., 1988]

heavy ion production surrounding the comet's nucleus, several million kilometers in extent. We will show below in Section (3) that at Giacobini-Zinner cometary ions were observed to distances of $\sim 4 \times 10^{6}$ km.

The heavy ions are created essentially at rest in the comet's frame (also in the spacecraft's frame, due to the low encounter speed, 21 km s^{-1}, between ICE and Giacobini-Zinner), so that they subsequently move on cycloidal trajectories in the crossed electric and magnetic fields of the solar wind. Thus the ions immediately (on gyro time scales) acquire the component of the solar wind velocity transverse to the interplanetary magnetic field (i.e. the ExB drift), but not the parallel component. In addition, they also acquire a gyration speed around the field lines which is equal to the ExB drift speed. The "pick-up" ion distribution in velocity space is thus expected to be of a "ring-beam" form, as shown in Figure (1). In this diagram O is the origin of velocity space in the neutral atom (comet) rest frame, and hence the point from which ions start after their creation. The X axis is antiparallel to the solar wind velocity vector V_{SW} and points (approximately) towards the Sun, while the XY plane contains the interplanetary magnetic field vector $\mathbf{B}$, which lies at angle α to V_{SW}. The

solar wind ion distribution is indicated schematically by the hatched sphere lying on the X axis at speed V_{SW}, while the ring-beam distribution of cometary pick-up ions is indicated by the solid circle centred on C. In the rest frame of the solar wind the ions gyrate about the field lines with speed $V_{SW} \sin \alpha$, while drifting "sunward" along the field lines with speed $V_{SW} \cos \alpha$. The ion speed in this frame is constant and equal to V_{SW}, so that the ion energy is also constant and equal to AE_{SW}, where A is the ion mass in amu, and E_{SW} is the mass of a proton moving at the solar wind speed. In the neutral atom rest frame (and spacecraft rest frame to a good approximation), however, the ions ExB drift perpendicular to the field with speed $V_{SW} \sin \alpha$, while gyrating about the field lines with the same speed. Consequently in this frame the ion energy varies periodically during the cycloidal motion between zero (point O), and a maximum (point M) given by

$$E_{MAX} = 2m_i(V_{SW} \sin \alpha)^2 = 4A\, E_{SW} \sin^2\alpha \qquad (1)$$

Thus, for example, when the field is orthogonal to the flow ($\sin^2\alpha = 1$), E_{MAX} for water group ions (taking A=16 for oxygen) lies in the range 30-85 keV as V_{SW} varies between 300 and 500 km s^{-1} (as it did during the fly-by of Giacobini-Zinner). The ion energy in the solar wind frame varies between 7.5 and 21 keV as V_{SW} varies over the same range. It is notable that these latter energies are much higher than the thermal energies of the solar wind protons and electrons in the solar wind frame, which are instead of order ~ 10 eV. Consequently, the pressure of the heavy ions will become comparable with or larger than the ambient solar wind thermal pressure when the heavy ion number density exceeds a factor $\sim 10^{-3}$ of the solar wind number density.

The ring-beam distributions of pick-up ions are expected to be unstable to the generation of MHD waves, which will initially scatter the particles in pitch angle in the solar wind frame, giving rise to a spherical shell in velocity space of radius V_{SW}, centred on V_{SW}, as shown in Figure (1) [Wu et al., 1986; Gary et al., 1986; Gaffey et al., 1988]. In this case the bulk velocity of the pick-up ions is just V_{SW}, while the maximum pick-up ion energy in the spacecraft frame corresponds to point M*, and is given by

$$E_{MAX}^* = 4A\, E_{SW} \qquad (2)$$

Subsequently, the ions will also be scattered in energy, filling in the region inside the shell in velocity space, at least partially, and forming a high-energy tail outside [Ip and Axford, 1986; Isenberg, 1987]. It can therefore be seen that the ion fluxes which are observed by a fixed energy detector (such as EPAS), observing ions of energy E, will depend significantly on the degree of scattering which the ions have undergone. If pitch angle scattering is unimportant, then ions will be observed at energy E only for a sufficiently large angle between the solar wind and interplanetary magnetic field vectors (assuming, of course, that E is less than E* MAX). The fluxes observed by the detector will then be modulated by the rapidly-varying direction of the magnetic field, in addition to being modulated by the speed of the solar wind. The ions will also have a net drift in the **ExB** direction, rather than along V_{SW}. Alternatively, if the ions are pitch-angle scattered into near isotropy, then the mean ion flow should remain consistently along the direction of V_{SW}, and the fluxes will depend only on V_{SW}, and not upon the field direction. In this case we would also expect the ions to be scattered in energy to values above and below that of the spherical "shell". In Section (3) below we will show that a transition from ring-beam to quasi-isotropic pick-up ion distribution functions (with weaker residual anisotropies) took place at cometocentric distances of $\sim 4 \times 10^5$ km, in association with increasing levels of MHD wave power as the comet was approached [Tsurutani and Smith, 1986a,b; Gosling et al., 1986]. This distance corresponds to a point which is well upstream from the cometary shock (located at $\sim 10^5$ km along the spacecraft track) and the region of slowed, mass-loaded flow, to which subject we now turn.

As an element of the solar wind flows through the comet's coma it accumulates an increasing population of pick-up ions, which will cause the flow to slow. Ultimately, the flow on the comet-Sun line will be brought to rest at the contact surface (ionopause), where the force of the oncoming field and flow is balanced by outflowing cometary material. For comets with molecule production rates of $\sim$ few x 10^{28} s^{-1} (like Giacobini-Zinner) this surface is expected to lie at distances of only ~ 500 km from the nucleus [Schmidt and Wegmann, 1982]. Under usual conditions, however, a smooth transition from supersonic to subsonic flow cannot occur. Instead, a shock will form in the supersonic flow at some point upstream from the nucleus [Biermann et al., 1967]. Theory and numerical analysis indicate that on the comet-Sun line the shock should be located at a point where the accreted mass density of the cometary ions reaches ~ 30 % of the solar wind mass density (corresponding to ~ 2 % by number density of water group ions), resulting in ~ 20 % reductions in the flow speed just upstream [Wallis, 1973; Schmidt and Wegmann, 1980; Galeev et al., 1985]. More importantly, however, the above discussion then indicates that at these concentrations the thermal pressure of the cometary ions will exceed that of the solar wind by about an order of magnitude, reducing the Mach number (M) of the flow to $M \approx 2$, compared with values of $M \approx 510$ in the undisturbed flow. Consequently, the shock will be weak, with a compression ratio of ~ 2 in the hydrodynamic limit. On the basis of these results the shock at Giacobini-Zinner is expected to lie at cometocentric distances of $\sim 4 \times 10^4$ km along the comet-Sun line [Mendis et al., 1986; Fuselier et al., 1986]. Numerical computations then show that away from this line the shock distance should flare to larger values, reaching distances of ~ 2.4 times the subsolar distance on a line orthogonal to the comet-Sun line passing through the comet, corresponding, approximately, to the ICE

trajectory relative to Giacobini-Zinner [Schmidt and Wegmann, 1980; 1982]. It is therefore anticipated on this basis that the cometary shock should lie at a cometocentric distance of $\sim 10^5$ km along the ICE trajectory, an expectation which will be shown to be justified in Section (5) below. Downstream from the shock the flow speed should then decline continuously as the comet is approached.

The shear in the flow which occurs across the region where the plasma is mass-loaded and slowed leads to distortions of the frozen-in magnetic field, which becomes draped over the obstacle as originally suggested by Alfven [1957]. However, analysis of the ICE thermal plasma data obtained at closest approch [Bame et al., 1986; Meyer-Vernet et al., 1986] has indicated that a narrow, sharply-bounded region of very high mass loading occurred close to the nucleus, where the ion production rates were ~ 20 times adjacent values [McComas et al., 1987]. The mass addition on the streamlines passing through this region caused the flow to slow to ~ 10 km s^{-1} near to the comet, with a flow shear in the region outside which was at least an order of magnitude larger than the general shears across the mass-loaded region. This region of high flow shear then resulted in the formation of a narrow, well-defined magnetic tail in which the field strength increased, in the inner part of the lobes, to ~ 60 nT, comparable with the stagnation pressure of the solar wind flow [Siscoe et al., 1986; Smith et al., 1986a; Slavin et al., 1986a]. The "plasma sheet" between the lobes contained high densities (peaking at 670 cm^{-3}) of cold (electron temperature ~ 1 eV) cometary plasma, corresponding to the streamtubes which map back to the "prime" pick-up region, and corresponding to the visible "ion tail" of the comet [Slavin et al., 1986b]. The width of this layer was found to be $\sim 1.5 \times 10^3$ km, comparable with the expected diameter of the ionopause, suggesting that the "prime" region might correspond to a layer where the flux tubes had diffused inside the ionopause in the upstream flow [McComas et al., 1987]. The radius of the tail lobes themselves (and region of strong flow shear) was found to be $\sim 5 \times 10^3$ km, embedded within a wider region of radius $\sim 2 \times 10^4$ km where the plasma number density was about an order of magnitude less than that in the plasma sheet (varying as the inverse square of the cometocentric distance), but was still dominated by cold cometary material. In Section (6) we will discuss the energetic ion observations in the vicinity of the plasma and magnetic tails, where large reductions in the ion flux were observed, together with the effects produced by the resulting density gradients.

In summary, it can be seen that discussion of the plasma interaction between the comet and the solar wind can be divided into three main regions, namely the "pick-up region", the "mass-loaded region", and the "cold ion tail", which will form the subjects of sections (4) to (6) respectively. The "pick-up region" is the outermost region, extending from the limit of cometary ion detectability, $\pm 4 \times 10^6$ km, inwards to the cometary shock at $\pm 10^5$ km. With an encounter speed of 21 km s^{-1}, these distances correspond to time periods of ± 2 days and ± 1.5 hours about closest approach, respectively. The characteristic feature of this region is the presence of cometary ions streaming in the solar wind, the latter being essentially unaffected by their presence, other than due to the presence of waves which are generated by the unstable pick-up particle velocity distributions. This region is further subdivided into "outer" and "inner" zones according to the degree of pitch angle scattering which the ions have undergone. In the "outer" zone the ring-beam anisotropy is a dominant factor determining the properties of the ion fluxes, while in the "inner" zone (within about $\pm 4 \times 10^5$ km, or ± 5 hours about closest approach) the ions are sufficiently scattered that the anisotropy is either absent, or greatly reduced.

The "mass-loaded region", bounded between the cometary shock and cold ion tail (± 1.5 hours to ± 15 min about closest approach), is then the region where the solar wind flow becomes dominated by the accreted mass of cometary ions, and becomes significantly slowed by their presence. At Giacobini-Zinner this region was also characterized by very high fluctuation levels in the field and thermal plasma, and energetic ion distributions which were highly isotropized. Finally, the "cold ion tail" is taken to consist of the region within ± 15 min of closest approach where the cold cometary particles dominate the number density of the plasma. Within this region lies the induced cometary magnetotail (traversed in a total interval of ~ 8 minutes), and the central cold plasma sheet which separates the lobes as described above (interval of ~ 1 min). In the following sections we will describe the ion oberservation in each of these regions. First, however, in Section (3) we will briefly describe the ICE instrumentation which is relevant to our study.

(3) Instrumentation

The EPAS experiment on the ICE spacecraft measures the energetic ion distribution in three dimensions in velocity space [Balogh et al.. 1978; van Rooijen et al., 1979]. It consists of three identical semiconductor particle telescopes, each with a conical field-of-view of half angle 16° and a geometrical factor of 0.05 cm^2 sr. These telescopes, designated T1-T3, are inclined at 30°, 60°, and 135° to the spacecraft spin axis, which is directed northward, perpendicular to the ecliptic (the Z axis, see the left side of Figure (2)). As the spacecraft spins, the particle counts are binned into eight equi-angular (45°) azimuthal sectors S1-S8 (see the right side of Figure (2)). The viewing direction of S1 is centred on the azimuth of the Sun (X axis), S2 on 45° to the east of the Sun, and so on, giving flux measurements in a total of 24 independent directions. For the five primary energy channels (E1-E5) a 24-point angular distribution is obtained

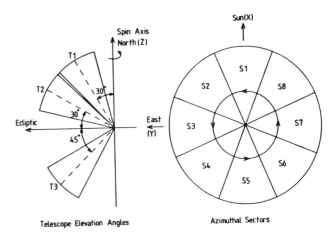

Fig. (2): Sketch showing the geometry of the EPAS viewing directions. The left hand figure shows the fields of view of the three EPAS telescopes, T1-T3, relative to the spacecraft spin axis. The half angle of the viewing cones is 16°. The right hand figure shows the viewing directions of the azimuthal sectors S1-S8, looking down on the ecliptic plane from the north. [After Richardson et al., 1988]

every 32 s. The energy ranges of these channels depend upon ion mass. EPAS has no direct mass-discriminating capability, but analyses of ion bulk flows and spectra demonstrate that at Giacobini-Zinner the energetic ions are predominantly of the water group [Richardson, et al., 1986, 1987a]. Similar conclusions have been derived using other ICE instruments which have mass-resolving capability [Ipavich et al., 1986; Ogilvie et al., 1986]. For these ions, energy channels E1 to E5 cover the ranges 65-95, 95-140, 140-205, 205-310, and 310-480 keV [Hynds et al., 1986a]. In addition, data are also available from an additional lower energy channel (E0) which usually monitors instrument noise levels, but which recorded fluxes of 45-65 keV water group ions well above noise levels during the encounter. A 24-point distribution of E0 data is available every 256 s.

Data from two other ICE experiments are also used in this study i.e. thermal electron measurements made by the Los Alamos plasma experiment [Bame et al., 1978], and magnetic field data obtained from the JPL vector helium magnetometer [Frandsen et al., 1978]. The electron experiment provides 3 s (spin period) snapshots of the electron distribution in the energy range 10 eV to 1 keV every 24 s, from which the solar wind bulk parameters have been derived [Bame et al., 1986]. However, during the Giacobini-Zinner encounter the instrument was operated in a two-dimensional mode, such that it was only possible to derive the ecliptic plane components of the solar wind velocity vector. Magnetic field data were obtained with a resolution of 3 vectors per second during the encounter [Smith et al., 1986a].

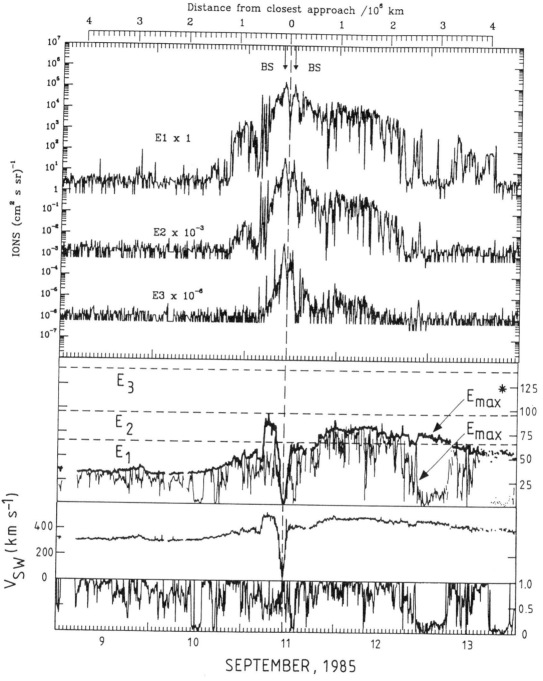

Fig. 3. Overview of energetic ion, thermal plasma and magnetic field data for the 5-day interval starting 0000 UT 9 September 1985, centred near closest approach (vertical dashed line). The top panel shows the ion flux in the first three primary EPAS energy channels E1 to E3 in T2, S1. The second panel shows E_{MAX} (thin line) and E^{*}_{MAX} (thick line), derived from equations (1) and (2), compared with the EPAS energy channels (bounded by the horizontal dashed lines). The third and fourth panels then show the data used to derive the pick-up energies, namely the solar wind speed V_{SW}, and $\sin^{2}\alpha$, where α is the angle between the magnetic field and solar wind velocity. The arrows at the top of the plot show the positions of the bow shock crossings. [Adapted from Sanderson et al., 1986a,b]

(4) Cometary Ions in the Pick-up Region

In Figure (3) we give an overview of the energetic ion observations in the large-scale pick-up region surrounding Giacobini-Zinner, adapted from Sanderson et al. [1986a,b]. This shows data obtained over a five day interval starting at 0000 UT on 9 September 1985, centred near closest approach. During this interval the comet passed nearly directly from north to south across the spacecraft at a relative speed of 21 km s^{-1}. Thus at the start of the interval the comet was located at a distance of 4.5×10^6 km to the north of the spacecraft, subsequently receding to a similar distance to the south at its end. Closest approach from the nucleus (7800 km) occurred at 1102 UT on 11 September (vertical dashed line). The two arrows at the top of the figure mark the shock crossings at the outer boundary of the mass-loaded region, thus serving to emphasize the enormous extent of the large-scale pick-up region relative to the latter.

The top panel of Figure (3) shows 256 sec averages of the ion fluxes in the first three primary EPAS energy channels in telescope T2 sector S1. This is the telescope and sector which views closest to the solar direction (see Figure (2)), and which generally recorded the highest fluxes throughout the encounter (i.e. the ions stream strongly in a generally antisunward direction; the detailed streaming direction and its relationship with the solar wind velocity and the **ExB** drift will be discussed further below). Beneath this we show the maximum energy of unscattered water group (A=16) pick-up ions, E_{MAX}, derived from equation (1) (thin line), togther with the maximum energy of a pitch-angle scattered "shell" population, E^*_{MAX}, obtained from equation (2) (thick line). It should be noted that $E_{MAX} \leq E^*_{MAX}$, with equality occurring when the magnetic field is orthogonal to the flow ($\sin^2\alpha = 1$). For purposes of comparison, the horizontal dashed lines in this panel indicate the limits of the primary EPAS energy channels. The lower two panels then show the data from which the E_{MAX}, E^*_{MAX} values were derived, namely the solar wind speed V_{SW} obtained from the thermal electron data, and $\sin^2\alpha$, obtained from the magnetometer data.

The ion measurements in Figure (3) demonstrate four important characteristics of the ion fluxes in the large-scale pick-up region. First, the results confirm that the comet acts as a source of energetic ions in the solar wind over spatial scales of several million kilometers. Cometary ion fluxes were observed starting at ~ 1900 UT on 9 September at a distance of ~ 3.0×10^6 km inbound, and continue, albeit somewhat intermittently, until ~ 1700 UT on 13 September at a distance of ~ 4.1×10^6 km outbound. The fluxes are generally largest closest to the comet and decline with increasing distance, thus reflecting the changing density of the neutral coma of the comet, and hence of the ion source strength, on spatial scales of ~ 10^6 km.

Second, although the density of the coma and the ion source strength are expected to be smoothly varying with distance from the comet [see e.g. Schmidt and Wegmann (1982), and Sanderson et al. (1986a,b)], it can be seen that the observed ion fluxes show large temporal modulations on time scales of minutes and hours. Comparison of the flux profiles with the data in the lower panels of the figure shows that the rapid flux modulations are associated with rapid variations in the direction of the magnetic field relative to the flow (rather then e.g. to rapid variations in V_{SW}), with the largest fluxes being observed when the field is close to orthogonal to the flow ($\sin^2\alpha = 1$), such that E_{MAX} is large, and the smallest fluxes when it is parallel ($\sin^2\alpha = 0$). The flux modulations tend to be larger further from the comet, with an amplitude of two to three orders of magnitude, reducing to one order of magnitude or less just upstream from the shocks. In accordance with the theoretical discussion in Section (2), these observations show that throughout most of the large-scale pick-up region the cometary ion distribution function is strongly anisotropic in the solar wind frame, and retains a ring-beam form. The transition to more scattered quasi-isotropic distributions in the inner part of the pick-up region will be discussed further below.

Although these observations show that the cometary ions are not effectively pitch-angle scattered to near isotropy in the solar wind frame throughout most of the pick-up region, the third property shows that significant scattering does indeed take place, in that ions are consistently observed at energies above the local pick-up energy E_{MAX}, and indeed above E^*_{MAX}. It can be seen, for example, that both E_{MAX} and E^*_{MAX} are less than the lower threshold of energy channel E1 (65-95 keV) throughout most of the inbound region, despite which modest fluxes were intermittently observed in E1 on both 9 and 10 September, and also in E2 (95-140 keV) closer to the comet. Similarly, during the last quarter of 11 September and the first half of 12 September on the outbound pass, E_{MAX} and E^*_{MAX} lay generally within the range of channel E1, yet ions were also observed in channels E2 and E3 (140-205 keV). Examination of the energy spectra shows that these ions form a steeply falling high-energy tail at energies above the local pick-up energy [Sanderson et al., 1986a,b]. It may be noted that the fluxes of the scattered ions at energies above E^*_{MAX} are also modulated by $\sin^2\alpha$, showing that despite the energy diffusion which must have taken place, the ions remain anisotropic in pitch angle in the sense of the ring-beam form. This result is rather curious, since the time scale for energy diffusion is expected to be considerably longer than that for pitch angle diffusion, though, of course, small numbers of ions will be accelerated to high energies even at early times. Thus whether these observations require modification of existing theory remains to be determined.

The steeply falling nature of the ion distribution function at energies above the local pick-up energy also provides a simple explananation of the fourth property which is evident in Figure

(3), namely that there is a significant difference in the overall flux levels between the inbound and outbound pick-up regions, with larger fluxes being observed outbound than inbound. In principle, an asymmetry of this nature could result from **ExB** streaming of the pick-up ions produced from the coma source in an average inclined interplanetary field (for the ICE encounter geometry we would require $B_X B_Z$ positive on average to give greater fluxes outbound than inbound). Although modest effects of this nature may indeed be present in the data, it seems clear that this cannot be the principal cause of the asymmetry observed in Figure (3). If, for example, we examine points of approximately equal distance from the comet on the inbound and outbound pass, each of which occur under $\sin^2\alpha \approx 1$ conditions such that even unscattered pick-up ions stream anti-sunward with the solar wind and hence have equivalent paths with respect to the ion source region, then we still find that the inbound fluxes are less than the outbound fluxes by between one to three orders of magnitude. A simple explanation may be found, however, in terms of the steeply falling nature of the spectrum above the local pick-up energy, on noting that the solar wind speed (and hence the pick-up energy) is consistently lower during the inbound pass (prior to ~ 0600 UT on 11 September) than during the outbound pass, as previously pointed out by Sanderson et al. [1986b]. Let us consider, for example, how the cometary ion distribution function varies with the solar wind speed at a certain point in space, and for a fixed direction of the magnetic field relative to the flow. Suppose, as a simple approximation, we assume that the shape of the ion distribution in the solar wind frame does not change as the solar wind speed varies, but that the distribution simply scales with the solar wind speed in the rest frame of the comet V_{SW}, i.e. that normalized f is a function only of v/V_{SW}, where $\mathbf{v}$ is the ion velocity in the solar wind frame. Suppose also that when the solar wind speed is V_{SW} the distribution function along the antisolar direction above the local pick-up energy in the solar wind frame can be represented by a falling exponential in ion speed i.e. that f is proportional to $\exp(-v/V_0)$, as indicated by the results presented by Richardson et al. [1987a,b]. It is then simple to show that at a given speed V in the comet (spacecraft) rest frame, the ratio of distribution function values at solar wind speed V'_{SW} and at V_{SW} is given by

$$\frac{f'}{f} = \left(\frac{V_{SW}}{V'_{SW}}\right)^4 \exp-\left(\left(\frac{V_{SW}}{V'_{SW}}-1\right)\frac{V}{V_0}\right) \qquad (3)$$

where we have scaled the absolute value of f such that the total ion flux remains constant, as should be the case. If we take $V = 900$ km s^{-1} corresponding to EPAS channel E1, $V_0 = 45$ km s^{-1} giving a steeply falling tail (see below), and take $V_{SW} = 450$ km s^{-1}, and $V'_{SW} = 350$ km s^{-1}, we then find $f'/f = 0.9 \times 10^{-2}$. With this simple system we therefore find

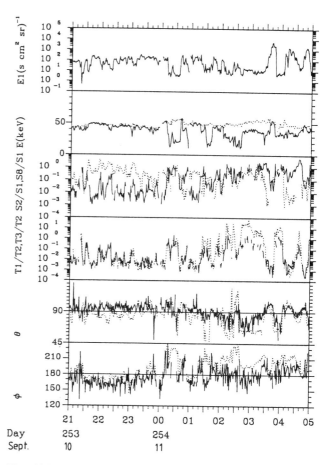

Fig. 4(a). Data for 2100 UT on 10 September to 0500 UT on 11 September, showing 64 s values of the direction-averaged E1 (65-95 keV) ion intensity (top panel); E_{MAX} (solid line) and E^*_{MAX} (dotted line) (second panel); the ratios of the ion fluxes in the telescope and sector viewing closest to the solar direction (T2 S1) and the four adjacent sectors, as described in the text (third and fourth panels); the polar and azimuthal angles of ion streaming (solid lines) and **ExB** drift direction (dotted lines) in CSE coordinates (fifth and sixth panels).

that the flux varies by two orders of magnitude as the solar wind speed changes from 350 km s^{-1} to 450 km s^{-1}, thus illustrating how the sharply falling nature of the spectrum can produce large changes in the cometary ion flux at a given energy above the pick-up energy, for relatively modest variations in the solar wind speed. Further investigation of the observed variations is warranted to substantiate this picture.

Another question which is raised by these observations concerns the direction of ion streaming in the solar wind, since unscattered ions will stream in the direction of **ExB** (e.g. at 45° to the comet-Sun line for a garden-hose field), while an isotropized population will, of course, stream directly with the solar wind in the antisolar direction. This question has been addressed by Richardson et al. [1988, 1989], with results illustrated in Figure (4). Figure (4a) shows data from the inbound outer pick-up region from 2100 UT on 10 September to 0500 UT on 11 September, corresponding to cometocentric distances from 1.0×10^6 to 4.5×10^5 km, while Figure (4b) shows similar data from a contiguous interval, 0500 to 1700 UT on 11 September, which spans both traversals of the inner pick-up region and mass-loaded region ($\pm 4.5 \times 10^5$ km about closest approach). In the latter figure the arrows show the positions of the bow shock crossings (BS), as well as the point of closest approach (CA). The top panel of each figure shows the direction-averaged E1 fluxes at 64 s resolution, while the second panel shows E_{MAX} (solid line) and E^*_{MAX} (dotted line), as in Figure (3). Large modulations of the ion flux which are (imperfectly) correlated with E_{MAX} are evident prior to ~ 0600 UT on 11 September, as can be seen in Figure (4a) and on the extreme left-hand side of Figure (4b), after which the ion fluxes vary much more smoothly prior to closest approach, both upstream and downstream from the shock, and exhibit little relation to E_{MAX}. The boundary between the "outer" and "inner" inbound pick-up regions is

1985 Day 253, 2102:20 to Day 254, 0459:58

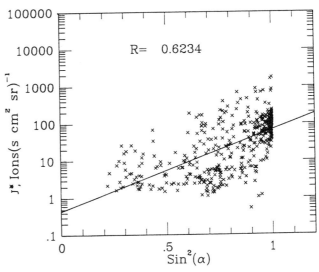

Fig. 4(c). Scatter plot of 64 s averages of the E1 ion intensity, normalized to the pick-up ion density at a cometocentric distance of 10^6 km, versus $\sin^2\alpha$, for the same interval as in (a).

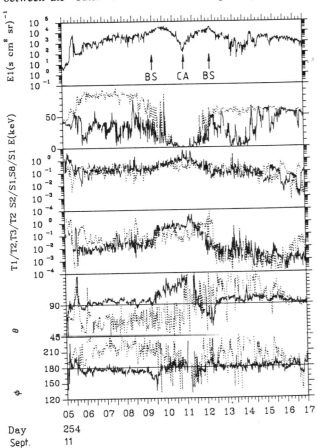

Fig. 4(b). As for (a), except for the interval 0500 to 1700 UT on 11 September. Closest approach (CA) to the comet, and the bow shock (BS) crossings are indicated by arrows.

1985 Day 254, 0600:46 to 0839:36

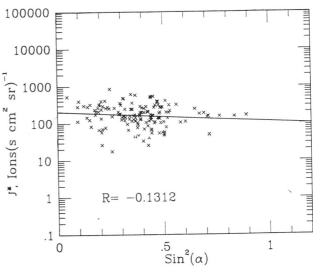

Fig. 4(d). As for (c), except for the interval 0600 to 0840 UT on 11 September.

thus taken to occur at 0600 UT, at a cometocentric distance of ~ 4×10^5 km, well upstream of the shock. Outbound, the situation is less clear-cut, with weak modulations in flux being present in the region immediately upstream from the outbound shock, which increase more gradually in amplitude with increasing distance from the comet. The lower four panels in each figure then give information about the direction of ion streaming. Panels three and four show the ratio of the E1 fluxes observed in the four adjacent viewing directions to that observed in T2 S1, where the third panel shows S2/S1 (solid line) and S8/S1 (dotted line) for T2, and panel four shows T1/T2 (solid) and T3/T2 (dotted) for S1. Panels five and six then show the direction of ion flow determined from a spherical harmonic fit to the directional intensities (solid), together with the **ExB** drift direction (dotted). Panel five shows the polar angle of the flow, $\theta = 0^o$ being northward, while panel six shows the flow azimuth, with $\varphi = 0^o$ sunward, and $\varphi = 90^o$ to the east of the Sun. It can be seen that during the inbound pass rapid variations in the direction of ion streaming are observed prior to 0600 UT, which respond at least partially to the direction of **ExB**, while after this time the ions stream consistently in an antisolar direction (upstream from the shock) uninfluenced by **ExB**. The inference therefore is that prior to 0600 UT on the inbound pass the ion population is only partially scattered, while after this time the ions are essentially isotropized. This change took place in coincidence with an interval in which the solar wind speed rapidly increased, such that E^*_{MAX} lay just below the E1 band prior to the change, but lay within it afterwards. On the outbound pass, weak responses to the direction of the **ExB** drift are present in the region immediately upstream from the shock, indicating that weak vestiges of the ring-beam anisotropy remain present in this region, these features slowly increasing in amplitude with increasing cometocentric distance. The behaviour of the ion flows in the mass-loaded region downstream from the shocks will be discussed in the next section.

The correlations noted visually above are substantiated in Figures (4c) to (4f). Figures (4c) and (4d) show scatter plots of normalized 64 s direction-averaged E1 fluxes, J^*, versus $sin^2\alpha$ for the intervals 2100 UT on 10 September to 0500 UT on 11 September, corresponding to the inbound outer pick-up region shown in Figure (4a), and 0600-0840 UT on 11 September, corresponding to the inbound inner pick-up region contained in Figure (4b). To remove the effect of varying cometocentric distance we have normalized the observed flux values to the equivalent flux at a distance of 10^6 km along the inbound track, by multiplying the measured flux by the ratio of the theoretically expected number densities. An (imperfect) positive correlation is evident in the former interval, but is wholly absent in the latter. In Figures (4e) and (4f) (which correspond to the same time intervals as Figures (4c) and (4d), respectively) we show the results of further

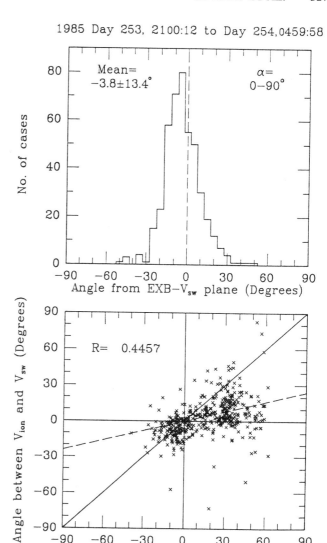

Fig. 4(e). The top panel shows a histogram of the angle between the ion streaming direction and the plane defined by the solar wind velocity and **ExB** vectors. The bottom panel shows a plot of the angle between V_{SW} and the projection of ion streaming onto this plane (φ_i), versus the angle between V_{SW} and **ExB** (φ_D). The angles are taken positive anticlockwise from V_{SW} when looking from north of the plane. The interval corresponds to that shown in (a) and (c).

analysis of the ion streaming direction. The upper panels show histograms of the direction of ion streaming relative to the plane defined by the solar wind velocity and **ExB** vectors, positive values indicating flows directed northward of this plane. It can be seen than on average the ion streaming lies close to this plane during both intervals. The lower panels then compare the angle in this plane between the solar wind velocity and the projection of the ion streaming direction (φ_i,

1985 Day 254, 0600:46 to 0839:36

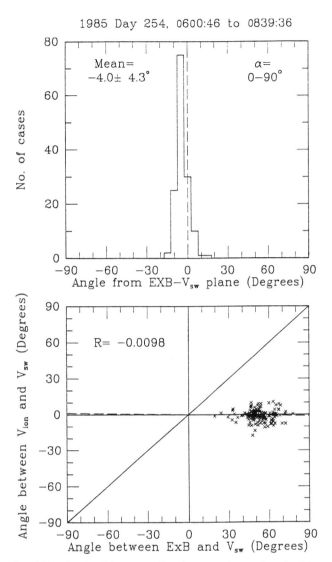

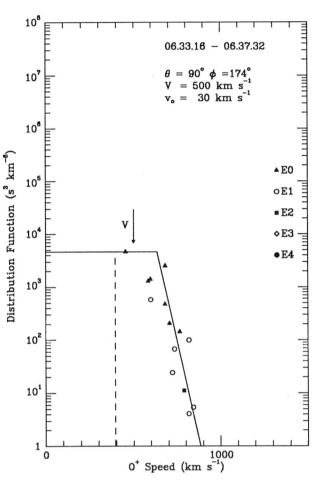

Fig. 4(f). As for (e), except for the same interval as in (d). [After Richardson et al., 1988]

Fig. 5. The water group ion distribution function f is plotted versus ion speed in the bulk flow frame for three representative 256 s periods in the inbound inner pick-up region. Details of the velocity vector used to transform the data from the spacecraft frame are shown in the top right hand corner of each panel. Data from the different energy channels are shown using different symbols, as indicated by the key. The vertical arrow marked "V" indicates the speed of the local "shell" ions, equal to the speed of the transformation velocity vector. The vertical dashed line shows the lowest solar wind frame speed sampled by EPAS during each period. The fitted straight lines are used to represent the data for purposes of estimating bulk parameters. [After Staines et al., 1990]

vertical axis), and the solar wind velocity and the **ExB** direction (φ_D, horizontal axis). The angle is taken as positive anticlockwise from **V**$_{SW}$ when looking from north of the plane. If the ions purely **ExB** drift then the points will lie along the solid line at 45° to the horizontal, while if they stream with the solar wind they will lie along the horizontal axis. In Figure (4e) the majority of points lie in the wedge formed by these lines, with a best fit line indicating $\varphi_i \approx 0.24 \varphi_D$. Thus the partially scattered population in the outer pick-up region has, on average, a direction of ion streaming which is intermediate between the solar wind and **ExB** directions. In the inner pick-up region shown in Figure (4f), however, the ions stream consistently in the solar wind direction, with no response to **ExB**.

In Figure (5) we show three sample ion distribution functions, transformed to the solar wind rest frame, taken from the quasi-isotropic inbound inner pick-up region [Staines et al., 1990]. The 256 s intervals corresponding to each of these distributions are indicated in the top right hand corner of each plot, together with details of the velocity vector (V, θ, φ) used to transform the data (determined principally from the thermal electron measurements, see Staines et al. [1990] for

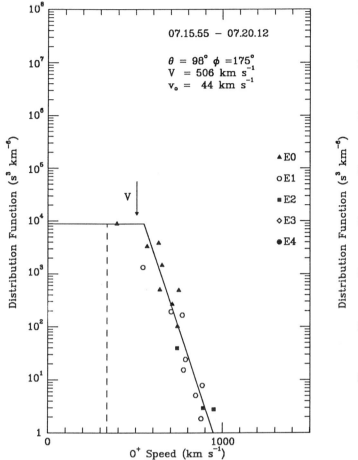

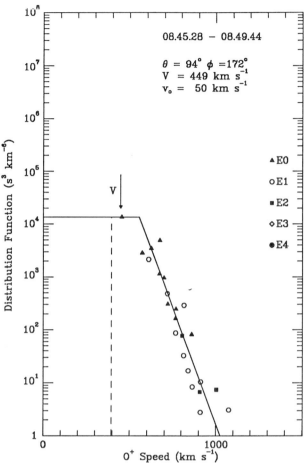

details). Values from the various EPAS energy channels are shown by different symbols, and data from channel E0 (45-65 keV) are included. The vertical arrow labelled "V" is drawn at a rest frame speed equal to the transformation speed, which hence indicates the local pick-up speed of the ions. From the discussion given in Section (2) we would then anticipate that the ion distribution function should peak near this point, being partially filled in at speeds below, and with a high-energy tail at speeds above. It can be seen that the vast majority of the EPAS data correspond to points lying on the high-energy tail above the local pick-up speed, and that in this regime the data can be well represented by an exponential in ion speed as mentioned above, as indicated by the inclined straight line which has been fitted to the data (the characteristic speed V_0 is also given in the top right hand corner). (We note that this empirically-determined spectral form has little pedigree from a theoretical viewpoint [see Richardson et al. (1987a) and the discussion therein]; Barbosa [1989] has claimed good agreement with a "leaky box" diffusion model, but this has recently been questioned by Isenberg [unpublished manuscript, 1990].) However, it can also be seen in Figure (5) that the point which lies at the

lowest speed (corresponding to T2, S1, E0) lies at or below the local pick-up speed, and that this point does indeed depart in value from the fit to the high-energy tail, indicating the anticipated turn-over in the spectrum.

In order to derive bulk parameters from these ion data, the observed portion of the distribution must first be extrapolated to lower energies. In the absence of any direct information about the form of the distribution in this region, the simplest assumption is to take f to be constant through the peak value lying near V, as shown by the horizontal lines in Figure (5). The bulk parameters of the cometary plasma can then be estimated by integrating under the straight lines shown. A trend of increasing number and energy density as the comet is approached is already apparent from the increasing values of f with time in the figure. We note, however, that the Giotto results at comet Halley indicate that the low-energy portion of the spectrum may not be "filled in" to the degree implied by the "constant f" approximation, especially in the region upstream from the shock [Coates et al., 1990]. The bulk parameters derived on the basis of this approximation would then represent over-estimates. An alternative extreme approximation would be to take f to be zero at all energies

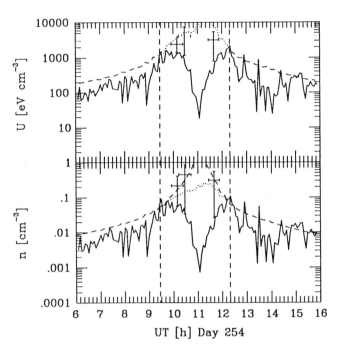

Fig. 6. Comparison of theoretical and experimental water group ion bulk parameters for the 10-hour interval about closest approach to Giacobini-Zinner. The thermal energy density is shown in the upper panel, and the number density in the lower panel. Theoretical estimates of the total bulk parameters are shown by the dashed lines, while the dotted lines downstream from the shock (vertical dashed lines) represent an estimate of the partial densities of ions picked up upstream from the shock. EPAS values are shown by the solid lines, where only the partial densities are shown in the mass-loaded region. Bulk parameters determined from ULECA measurements in the mass-loaded region are shown as individual data points. [After Staines et al., 1990].

below the lowest energy sampled by EPAS, such that the resulting bulk parameter values would then in general represent under-estimates. However, Staines et al. [1990] have shown that the values derived using these two extreme hypotheses typically differ only by a factor ~ 2 or less, such that the results are still highly meaningful within the estimated uncertainties. Figure (6) shows the results of integrations using the "constant f" extrapolation for the inner pick-up and mass-loaded region (0600-1600 UT), where the upper panel shows the water group ion thermal energy density, and the lower panel the number density [Staines et al., 1990]. Downstream from the shocks (vertical dashed lines) the ion data have been analysed on the assumption that only the energetic ions produced in the fast flow region upstream from the shock, and which are subsequently convected downstream, contribute over the observed energy range. Thus in this region the computed parameters represent only partial values and exclude the contribution of ions created more

locally downstream from the shock. (The individual data points shown in this region represent estimates of the total values based on measurements from the ULECA instrument on ICE, determined by Gloeckler et al. [1986].) The plausibility of this division results from the significant reduction in plasma flow speed which occurs across the cometary shock and in the mass-loaded region, such that the pick-up ion energy also rapidly decreases in this region, leading to the expectation of a two-component cometary ion plasma. This explanation was originally invoked by Thomsen et al. [1987] to explain the two branches in the water group pick-up ion spectrum observed in the Giotto plasma measurements at comet Halley, where detailed observations of the low-energy components were also made.

The dashed lines in Figure (6) represent the results of theoretical estimates of the water group ion total bulk parameters, while the dotted lines represent estimates of the partial values downstream from the shock of ions picked up upstream [Staines et al., 1990]. These estimates were derived from a simple model in which the cometary ions are produced by ionization of a conventional spherically symmetric neutral coma (characterized by a water molecule production rate $Q = 2.5 \times 10^{28}$ s^{-1}, as determined from remote spacecraft and ground-based measurements, and a length scale for ionization $L = 10^6$ km), followed by immediate pick-up by an antisunward-directed solar wind flowing at the observed speed [see Staines et al. [1990] for details]. The energy density calculation also assumes that each ion retains its initial pick-up energy in the solar wind frame, though there is a correction for adiabatic compression in the mass-loaded region. The experimental results in the inner pick-up region show generally good overall agreement with the theoretical estimates, though the theoretical values are generally higher by a factor of about two, indicating that the quotient Q/L given by the above numbers may be too large by this factor. However, the difference corresponds approximately to the estimated uncertainties in the bulk parameter determinations, as mentioned briefly above.

Closer inspection of Figure (6) shows, however, that the number density and energy density of the cometary ions rapidly increase over a ~ 30 minute interval just upstream from both of the shock crossings, such that the derived values reach and exceed the theoretical values at the shocks themselves. It should be noted that it is these rapid increases which finally result in the energy density of the cometary ions exceeding that of the solar wind plasma just ahead of the shock, thus also resulting in a significant reduction in the upstream flow Mach number [Staines et al., 1990]. At the largest distance from the comet shown in Figure (6) $(3.8 \times 10^5$ km) the water group ion thermal energy density is less than the solar wind thermal energy density by factors of 2 to 4, and the magnetosonic Mach number is ~ 5. Just upstream from the shock, however, the water group ion energy density has increased to exceed the solar wind plasma energy density by a factor ~ 4, and the magnetosonic Mach number is correspondingly reduced to ~ 2, in line with the theoretical expectations outlined in Section (2). The underlying effects which result in the rapid increase in bulk parameters will be discussed further in the next section.

With regard to Figure (6), we finally note that in the mass-loaded region the bulk parameters remain relatively constant for an interval of 30-40 minutes downstream from the shocks, before decreasing to low values near closest approach. This behaviour is fundamentally at variance with the theoretical estimates, even with those based solely on ion production upstream from the shock and subsequent transport downstream. These results therefore indicate that some energetic ion removal process must be operative, as will be discussed further in Section (6).

(5) The Cometary Shock and Outer Mass-Loaded Region

In the last section we concluded our survey of ion observations in the pick-up region with a discussion of the rapid changes in ion bulk parameters which occur just upstream from the shock transitions. In this section we will begin by considering these changes in more detail, together with the detailed structure of the shock as observed in the ion data.

Figure (7) shows details of the cometary ion bulk velocity vector and spectral characteristics over an eight hour interval centred near closest approach (0700-1500 UT 11 September 1985), adapted from Richardson et al. [1987a]. The cometary ion bulk velocities were determined by assuming that the distribution is isotropic in some frame of reference, and then successively transforming the data until the distribution function values from all the EPAS telescopes, sectors and energies lie on a single continuous curve. Operationally, the procedure was carried out by fitting the transformed data to an exponential speed distribution (which, as previously indicated, is found empirically to give an excellent fit), and iterating the transformation speed until the scatter of the data about the best fit line reaches a minimum [see Richardson et al. (1987a) for full details]. The top three panels of the figure show the cometary ion bulk speed and direction so determined from 256 s averages of the data (including channel E0) where angles theta and phi are defined as in Figure (4a,b). The fourth panel shows the variation of the ion distribution function with time at four rest frame energies which span the usual EPAS energy range (thus removing the effect on the observations at a fixed spacecraft frame energy which are due to varying bulk speed), while the bottom panel shows the variation of the characteristic speed V_0 of the exponential fit. The vertical dashed lines indicate the centres of the shock structures (to be discussed further below), and are drawn at 0928 UT inbound and 1220 UT outbound (as in Figures (3), (4a) and (6)).

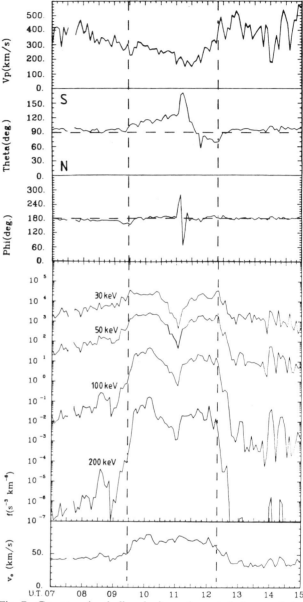

Fig. 7. Cometary ion bulk velocity and spectral characteristics for the interval 0700 to 1500 UT on the day of encounter. The vertical dashed lines show the "centres" of the shock transitions at 0928 UT inbound, and 1220 UT outbound. The upper three panels show the ion flow speed and direction, the fourth panel the value of the ion distribution function at four rest frame energies spanning the usual EPAS range, and the lower panel the characteristic speed V_0 of the ion distribution (such that f is proportional to $\exp(-v/V_0)$, where v is the ion speed in the bulk rest frame). [Adapted from Richardson et al., 1987a]

The results show that the region of the shock is associated with four principal features. First, the distribution function curves, and the variation of V_0, show that the shocks are associated with intervals of rapid spectral hardening, which begin about 30 minutes upstream from the shock crossings themselves (corresponding to a distance of 3.5×10^4 km along the spacecraft track), at ~ 0900 UT inbound and (ending at) ~ 1250 UT outbound. It is this effect which leads to the rapid changes in upstream bulk parameters noted in the previous section. The hardening continues across the region of the shock (being particularly intense inbound), and into the mass-loaded region downstream for a further ~ 40 minutes. Overall, V_0 increases from a typical value of ~ 40 km s^{-1} in the inner pick-up region, to ~ 75 km s^{-1} in the mass-loaded region well downstream from the shock. The high-energy tail observed in the pick-up region well upstream from the shock, and the hardening which occurs downstream, presumably result from ion heating (second order Fermi) in the wave fields present in these regions, the downstream turbulence being particularly intense. However, the spectral hardening which occurs in the region immediately upstream from the shock may be due either to spatial diffusion of the heated ions out of the mass-loaded region, or, possibly, to a first order Fermi process operating between the upstream waves and the turbulent fields behind the shock, as suggested by Richardson et al. [1987a], and as previously discussed theoretically by Amata and Formisano [1985].

The second shock feature seen in Figure (7) is the deflection of the angle of the ion flow away from the comet tail axis (see also Figure (4b)). Since the trajectory of the comet was principally from north to south across the spacecraft, the observed deflection occurs pricipally in the polar angle of the flow (θ), corresponding to a deflection southward in the inbound outer mass-loaded region, and northward in the outbound outer mass-loaded region, with an amplitude of ~ 15°. (The flow angles near closest approach are related to gradient anisotropy effects to be discussed in the next section.) These deflections of the flow direction take place over a 4-5 minute interval on the *upstream* side of the dashed line shown, between 0924 and 0928 UT inbound, and 1220 and 1225 UT outbound (corresponding to distances of ~ 5.5×10^3 km along the spacecraft track), and represent the first indication of the shock structure proper [Tranquille et al. 1986a,b]. It may also be noted that the flow azimuth is also deflected in the region upstream from the shock crossings, in a manner not reflected in the electron observations [Staines et al., 1990]. This may result from a gradient anisotropy effect at the shock surface.

The third shock effect indicated in Figure (7) concerns the slowing of the flow in the region immediately downstream from the directional deflections. Over the adjacent 6 minute intervals (i.e. between 0928 and 0934 UT inbound, and 1214 and 1220 UT outbound, corresponding to distances of 7.5×10^3 km along the spacecraft track), the ion distributions were observed to broaden substantially in elevation and azimuth angle, indicative of a rather sudden reduction in the flow speed, combined with continued hardening of the

spectrum. The change in the flow speed is somewhat obscured in the upper panel of Figure (7) due to the fact that the ion bulk speeds derived for the region upstream from the shock are rather scattered and unreliable, presumably due to residual anisotropies in the ion distribution function. The electron data, however, indicate a speed of 500 km s^{-1} in the inbound region, falling gradually after 0840 UT to reach ~ 400 km s^{-1} at the shock (0928 UT). Outbound, the electron speeds are steady at just over 400 km s^{-1} in the region upstream from the shock (1220 UT). However, the ion speeds downstream from the shock, ~ 300 km s^{-1}, are more reliably determined than those upstream, due to the strong scattering of the distribution to isotropy by the field turbulence. At the same time, the electron values become much more scattered in the downstream region (again due to the turbulence), though agreeing on average with the ion determinations in the ~ 40-50 minute intervals downstream. From the combined data, therefore, we infer that a reduction in flow speed of ~ 100 km s^{-1} took place across the above intervals.

These results are summarized in the expanded views of the shock transitions shown in Figures (8a) and (8b) for the inbound and outbound regions respectively [Tranquille et al., 1986a]. In each figure the upper panels show the ion bulk speed and direction determined from full resolution 32 s EPAS data (thus neglecting channel E0), together with the characteristic speed V_0 in the fourth panel, and the magnetic field magnitude (3 s averages) in the lower panel [Smith et al., 1986a]. The variation of V_0 shows the overall hardening of the spectrum with decreasing distance from the comet, which occurred over nearly the whole of the ~ 1 hour intervals shown, straddling the sharper shock transitions themselves. As indicated above, the latter transitions last for an interval of ~ 10 minutes (1.3×10^4 km along the spacecraft track), and consist of two adjacent features as indicated by the vertical dashed lines, namely the outer deflection of the polar angle of the flow (0924-0928 UT inbound, 1220-1225 UT outbound), and the subsequent slowing of the flow downstream (0928-0934 UT inbound, 1214-1220 UT outbound).

The lower panels of Figure (8) also show that the slowing of the flow is accompanied by a sudden onset of strong compressional magnetic turbulence. In addition, the electron data (not shown) also indicate the occurrence of compression and heating of the thermal plasma [Bame et al., 1986]. Smith et al. [1986b] have examined the conditions upstream and downstream from these transitions, and have concluded that they are consistent with weak Mach number perpendicular shocks. The transitions also certainly occur at the expected spatial positions relative to the comet (see the discussion in Section (2)), where the cometary ion number density reaches of order ~ 1% of the solar wind number density (0.1 cm^{-3}, see Figure (6), compared with ~ 10 cm^{-3}).

The fourth and final effect which the shock transitions have on the cometary ions, already alluded to above, is the rapid

ICE EPAS 11 SEPTEMBER 1985

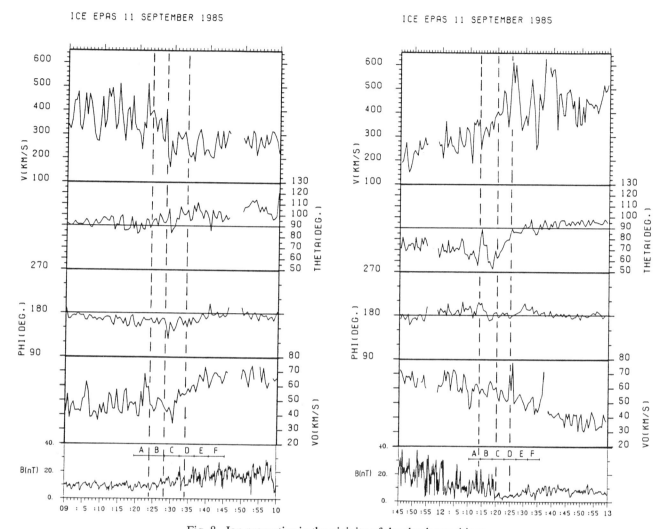

Fig. 8. Ion properties in the vicinity of the shock transitions,
(a) inbound and (b) outbound. The top three panels show the
ion bulk speed and direction determined from 32 s EPAS data,
the fourth panel the characteristic spectral speed V_0, while the
bottom panel shows the 3 s averaged magnetic field strength
(the letters in this panel refer to the spectra shown in Figure
(9)). The vertical dashed lines indicate the intervals during
which ion flow is deflected in polar angle (outer intervals),
and subsequently slowed (inner intervals). [Adapted from
Tranquille et al., 1986a]

isotropization which takes place in the turbulent fields in the
downstream region. This aspect is illustrated in Figures (9a)
and (9b) [from Richardson et al., 1987a], each of which show
a sequence of six contiguous 256 s rest frame spectra which
span the inbound and outbound transitions respectively. The
corresponding intervals are indicated in Figure (8) by the
letter-coded bars in the magnetic field panel. In the inbound
sequence, distribution A is typical of the rather scattered

spectra obtained in the quasi-isotropic inner pick-up region
upstream from the shock. Distribution B is similar, and
represents the interval during which the angular deflection
takes place (see the transformation velocity information in the
top right hand corner of each panel), while distributions C and
D correspond to the initial slowing of the flow. Distributions
E and F are then characteristic of the smooth, isotropic, rest
frame spectra observed in the outer mass-loaded region.

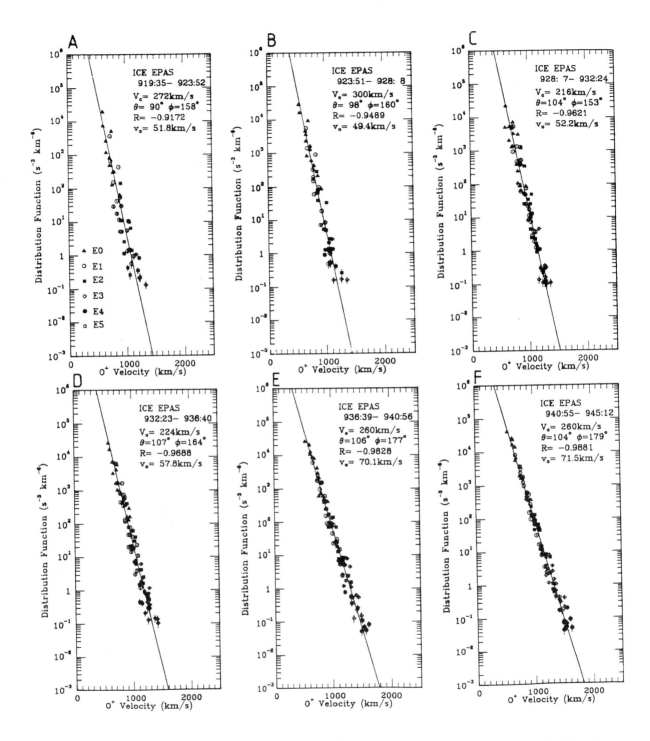

Fig. 9(a). Six consecutive 256 s averaged E0-E5 rest frame ion spectra are shown, spanning the inbound shock transition. The data given in the top right hand corner of each plot indicate the time interval corresponding to the spectrum, the transformation speed (V,θ,φ) determined iteratively from the ion data themselves, the correlation coefficient R between f and the rest frame ion speed v representing a normalized quantitative measure of the smoothness of the spectrum, and the characteristic speed V_0 of the best fit line (also shown). [After Richardson et al., 1987a]

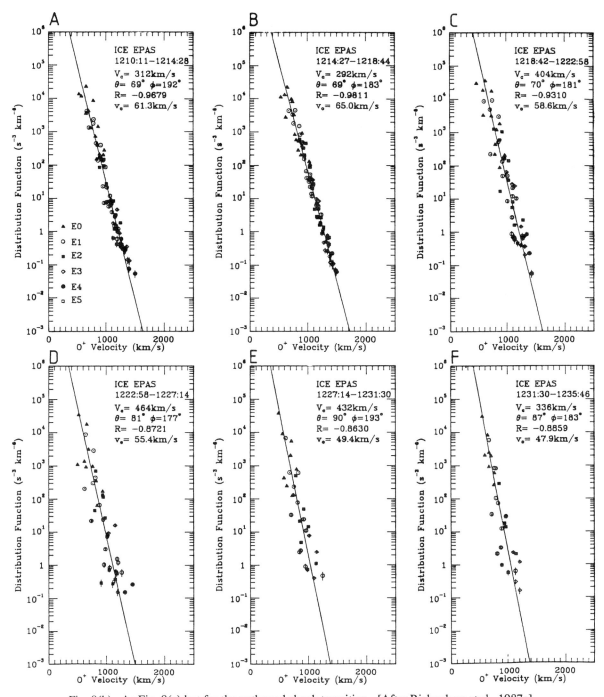

Fig. 9(b). As Fig. 9(a) but for the outbound shock transition. [After Richardson et al., 1987a]

Similarly, for the outbound data, working from upstream to downstream, distributions F and E are characteristic of the upstream inner pick-up region, the angular deflection of the flow occurs in D and C, and the slowing of the flow mainly in B. A strong "smoothing" of the rest-frame spectrum is then evident in A and B, indicating a closely isotropic population, relative to C to F. In both inbound and outbound cases it is therefore clear that the transition from the rather scattered, quasi-isotropic spectra of the inner pick-up region to the smooth, isotropic spectra of the outer mass-loaded region takes place rather abruptly as the flow slows, starting at the shock, and occurs in association with the onset of large-amplitude compressional field turbulence. Finally, we should mention that due to the slowing of the flow in the mass-loaded region, the range of rest frame cometary ion speeds encompassed by the EPAS measurements moves to somewhat higher values, and does not generally encompass the pick-up speed even of ions created upstream from the shock which are subsequently convected downstream [Staines et al., 1990]. Consequently, all the EPAS data in this region usually correspond to points on the high-energy tail of the distribution, and no turn-over is generally evident at the lowest speeds, as was the case in the pick-up region data such as those shown here in Figure (5).

(6) The Inner Mass-Loaded Region and Cold Ion Tail

From the results already presented in Figures (6) and (7) it can be seen that in the outer part of the mass-loaded region, 40-50 minutes downstream from the shocks, the bulk flow speeds gradually decline from ~ 300 to ~ 200 km s^{-1} as the comet is approached, in approximate agreement with the (averaged) electron speeds, while the ion rest frame distribution function undergoes only modest changes, the spectrum slowly hardening in the turbulent downstream flow. Correspondingly, the ion bulk parameters change only relatively slowly in this region, and are in approximate agreement with the theoretical estimate based on the assumption that the ions observed were created upstream from the shock and were subsequently convected downstream. However, in the central region, within about ± 40 minutes of closest approach (i.e. between ~ 1020 and ~ 1140 UT, corresponding to ± 5x10^4 km along the spacecraft track), the ion properties clearly depart significantly from simple theoretical expectation. Specifically, the value of the ion distribution function at a fixed rest frame energy (fourth panel of Figure (7)) exhibits a deep minimum of about two orders of magnitude in this region, minimizing near closest approach, such that there is a corresponding decline in the ion bulk parameters (Figure (6)). We note that the region where this decline occurs is much wider than that occupied by the cold ion tail and induced magnetotail. As noted in Section (2), the cold ion tail, where the plasma density and temperature is dominated by cold cometary plasma, and where the electron bulk velocity generally lies below the 30 km s^{-1} lower limit of the electron measurements, was traversed between ~ 1050 and ~ 1115 UT (± 1.6x10^4 km along the spacecraft track), while the induced magnetotail was traversed from ~ 1100 to ~ 1108 UT (~ 5x10^3 km radius), and the plasma sheet separating the lobes between 1102 and 1103 UT (~ 1.5x10^3 km thickness).

The decline of the energetic cometary ion population in the inner mass-loaded region is further substantiated in Figure (10), which compares two 256 s rest frame ion spectra, one from the outer inbound mass-loaded region

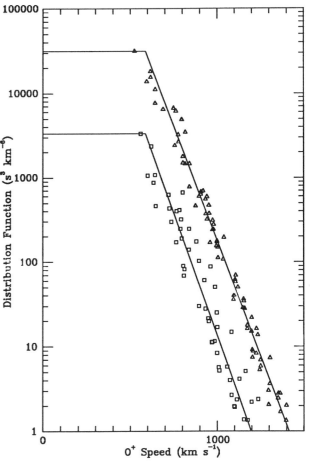

Fig. 10. Comparison of 256 s rest frame ion distribution functions from the outer mass-loaded region (triangles), and the inner mass-loaded region just outside the cold ion tail (squares). The intervals are 1002:14-1006:30 and 1044:53-1049:09 UT respectively. The straight lines show fits to the data used for estimating the ion bulk parameters, assuming the ions correspond only to those picked up upstream from the shock. [After Staines et al., 1990]

(1002:14-1006:30 UT, triangles), the other from the region just outside the inbound cold ion tail (1044:53-1049:09 UT, squares). It can be seen that the spectral form does not differ significantly between the two cases, but that the value of f at a given ion speed is less in the inner region than in the outer by about one order of magnitude. The origin of this decline remains to be determined in detail, but the most likely cause is energetic ion outflow along the compressed, draped field lines of the inner mass-loaded region due to the action of the magnetic mirror force, as discussed e.g. by Kimmel et al. [1987] and Luhmann et al., [1988].

Be this as it may, the existence of the strong flux gradient (with an overall exponentiation scale length of $l \approx 10^4$ km) has two further important effects on the ion properties in the inner

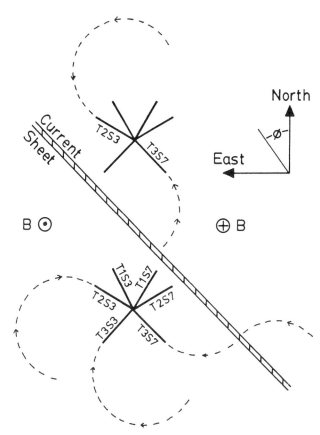

Fig. 11. Schematic diagram illustrating the gyroradius effects observed near to closest approach, looking along the comet tail towards the Sun. The circled symbols indicate the direction of the draped magnetic field, which are separated by the inclined current sheet. The EPAS look directions in this plane are indicated by the "stick" diagrams, together with the schematic trajectories of the ions observed in these directions (dashed lines), for spacecraft locations both to the north and the south of the current sheet. [After Daly et al., 1986]

region. The first is that the derived ion bulk speed shown in Figure (7) does not decline with the electron speed to small values near to closest approach, but instead exhibits a shallow plateau, with a minimum speed of ~ 150 km s^{-1}. The second effect is that the ion flow has a strong component directed consistently from north to south throughout closest approach, which distorts the expected antisymmetric behaviour of the polar angle of the ion flow resulting from plasma deflection away from the comet tail axis.

The basic explanation for these effects is shown in Figure (11), taken from Daly et al. [1986]. This shows a view of the system looking along the comet tail towards the Sun. Due to the direction of the upstream magnetic field, the draped fields in the vicinity of the comet are directed into the page in the upper right hand part of the diagram, and out of

the page in the lower left, and the current sheet separating them lies from top left to bottom right, as indicated by the magnetic observations [Slavin et al., 1986a]. In agreement with the above discussion, the ion fluxes may be taken to be weakest closest to the current sheet, and to increase with increasing distance on either side. The "stick" diagrams then indicate the look directions of the EPAS telescopes and sectors which lie in this plane, together with the trajectories of the ions which are detected in these directions (dashed lines). If we first consider the situation in the inbound region, when the spacecraft was located to the south of the current sheet, then we can see that the guiding centres of the ions detected e.g. in T1, S3 or in T2, S3 will be displaced furthest away from the current sheet where the ion flux is highest, while those detected e.g. in T3, S7 will have guiding centres displaced closest to the current sheet where the fluxes are weakest. Consequently, the gradient will give rise to a net ion flux at the spacecraft directed from northwest to southeast, additional to any net fluxes which are due to field-aligned flows or ExB drift. As the spacecraft crosses the current sheet the magnetic field changes direction, and hence so does the sense of ion motion. The consequence is that the guiding centres of the ions detected e.g. in T1, S3 or in T2, S3 are again displaced furthest from the current sheet where the ion flux is largest (but now to the north of the sheet rather than to the south), while the guiding centres of the ions detected e.g. in T3, S7 are again displaced closest to the sheet where the fluxes are weakest. Consequently there will again exist a gradient-associated net flux directed from northwest to southeast. This then explains the consistent southward-directed flows noted above. The argument is further substantiated in Figure (12), where we compare the fluxes observed in T2, S3 (solid line) with those in T3, S7 (dotted line) (both for energy channel E1), for the interval 0900-1300 UT on the day of encounter. The vertical dashed lines mark the shocks (BS), the boundary of the cold ion tail (IT), and the boundary of the induced magnetotail (MT). The fluxes in both channels increase sharply near to the shock transitions due to the hardening of the spectrum and the slowing of the flow, but whereas the T3, S7 fluxes (with guiding centres displaced towards the current sheet on both sides) show a significant minimum about closest approach, the effect in T2, S3 (where the guiding centres are displaced away on both sides) is much weaker. Simple analysis shows that the effective ion bulk speed associated with the gradient anisotropy is

$$V_G \approx V_0 \left(\frac{R_g}{l} \right) \qquad (4)$$

where, as before, V_0 is the characteristic speed of the exponential spectrum, R_g is a typical ion gyroradius, and l is the exponential spatial scale length. Using $V_0 = 75$ km s^{-1}, $R_g = 1.5 \times 10^4$ km, and $l = 10^4$ km we then find

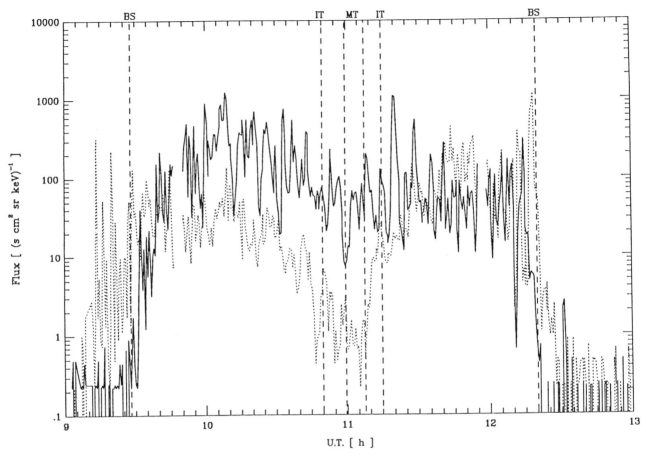

Fig. 12. Comparison of 32 s E1 fluxes in T2, S3 (solid line) and in T3, S7 (dotted line), for the interval 0900-1300 UT on 11 September 1985. The vertical dashed lines indicate the location of the shocks (BS), the boundary of the cold ion tail (IT), and the boundary of the induced magnetotail (MT).

$V_G \approx 100$ km s^{-1}, which is at least of the correct order to explain the minimum flow speeds derived in Figure (7). However, a detailed quantitative analysis of this problem has yet to be carried out.

Acknowledgements. The EPAS is a joint project of the Blackett Laboratory, Imperial College, the Space Science Department of ESA, ESTEC, and the Space Research Laboratory, Utrecht. KS acknowledges the support of a BNSC/SERC Research Assistantship, and TSY the support of a SERC postgraduate studentship.

References

Alfven, H., On the theory of comet tails, *Tellus, 9,* 92, 1957.

Amata, E., and V. Formisano, Energisation of positive ions in the cometary foreshock region, *Planet. Space Sci., 33,* 1243, 1985.

Balogh, A., G. van Dijen, J. Genechten, J. Henrion, R. Hynds, T. Sanderson, G. Stevens, and K.P. Wenzel, The low energy proton experiment on ISEEC, *IEEE Trans. Geosci. Electron., GE16 (3),* 176, 1978.

Bame, S.J., J.R. Asbridge, H.E. Felthauser, J.P. Clore, H.L. Hawk, and J. Chavez, ISEE-C solar wind plasma experiment, *IEEE Trans. Geosci. Electron., GE16 (3),* 160, 1978.

Bame, S.J., R.C. Anderson, J.R. Asbridge, D.N. Baker, W.C. Feldman, S.A. Fuselier, J.T. Gosling, D.J. McComas, M.F. Thomsen, D.T. Young, and R.D. Zwickl, Comet Giacobini-Zinner: a plasma description, *Science, 232,* 356, 1986.

Barbosa, D.D., Stochastic acceleration of cometary pickup ions: the classic leaky box model, *Ap. J., 341,* 493, 1989.

Biermann, L., B. Brosowski, and H.U. Schmidt, The interaction of the solar wind with a comet, *Solar Phys., 1,* 254, 1967.

Coates, A.J., B. Wilken, A.D. Johnstone, K. Jockers, K.-H. Glassmeier, and D.E. Huddleston, Bulk properties and velocity distributions of water group ions at comet Halley: Giotto measurements, *J. Geophys. Res.*, *95*, 10249, 1990.

Cowley, S.W.H., ICE observations of comet Giacobini-Zinner, *Phil. Trans. R. Soc. Lond. A*, *323*, 405, 1987.

Daly, P.W., T.R. Sanderson, K.P. Wenzel, S.W.H. Cowley, R.J. Hynds, and E.J. Smith, Gyroradius effects on the energetic ions in the tail lobes of comet P/Giacobini-Zinner, *Geophys. Res. Lett.*, *13*, 419, 1986.

Fuselier, S.A., W.C. Feldman, S.J. Bame, E.J. Smith, and F.L. Scarf, Heat flux observations and the location of the transition region boundary of Giacobini-Zinner, *Geophys. Res. Lett.*, *13*, 247, 1986.

Frandsen, A.M.A., B.V. Connor, J. van Amersfoort, and E.J. Smith, The ISEE-C vector helium magnetometer, *IEEE Trans. Geosci. Electron., GE16 (3)*, 195, 1978.

Gaffey, J.D., Jr., D. Winske, and C.S. Wu, Time scales for formation and spreading of velocity shells of pick-up ions in the solar wind, *J. Geophys. Res.*, *93*, 5470, 1988.

Galeev, A.A., T.E. Cravens, and T.I. Gombosi, Solar wind stagnation near comets, *Ap. J.*, *289*, 807, 1985.

Gary, S.P., S. Hinata, C.D. Madland, and D. Winske, The development of shell-like distributions from newborn comet ions, *Geophys. Res. Lett.*, *13*, 1364, 1986.

Gloeckler, G., D. Hovestadt, F.M. Ipavich, M. Scholer, B. Klecker, and A.B. Galvin, Cometary pick-up ions observed near Giacobini-Zinner, *Geophys. Res. Lett.*, *13*, 251, 1986.

Gosling, J.T., J.R. Asbridge, S.J. Bame, M.F. Thomsen, and R.D. Zwickl, Large amplitude, low frequency plasma fluctuations at comet Giacobini-Zinner, *Geophys. Res. Lett.*, *13*, 267, 1986.

Hynds, R.J., S.W.H. Cowley, T.R. Sanderson, K.P. Wenzel, and J.J. van Rooijen, Observations of energetic ions from comet Giacobini-Zinner, *Science*, *232*, 361, 1986a.

Hynds, R.J., S.W.H. Cowley, I.G. Richardson, T.R. Sanderson, C. Tranquille, K.P. Wenzel, and P.W. Daly, Energetic ion obervations during comet Giacobini-Zinner encounter, *Adv. Space Res.*, *5*, No. 12, 17, 1986b.

Ip, W.-H., and W.I. Axford, Theories of physical processes in cometary comae and ion tails, in *Comets,* edited by L.L. Wilkening, p. 588, University of Arizona Press, Tuscon, USA, 1982.

Ip, W.-H., and W.I. Axford, The acceleration of particles in the vicinity of comets, *Planet. Space Sci.*, *34*, 1061, 1986.

Ipavich, F.M., A.B. Galvin, G. Gloeckler, D. Hovestadt, B. Klecker, and M. Scholer, Comet Giacobini-Zinner: in situ observations of energetic heavy ions, Science, 232, 366, 1986.

Isenberg, P.A., Energy diffusion of pick-up ions upstream of comets, *J. Geophys. Res., 92, 8795, 1987.*

Johnstone, A.D., Comets, in Solar System Magnetic Fields, edited by E.R. Priest, p. 257, D. Reidel Publ. Co., Dordrecht, The Netherlands, 1985.

Kimmel, C.D., J.G. Luhmann, J.L. Phillips, and J.A. Fedder, Characteristics of cometary picked up ions in a global model of Giacobini-Zinner, *J. Geophys. Res.*, *92*, 8536, 1987.

Luhmann, J.G., J.A. Fedder, and D. Winske, A test particle model of pick-up ions at comet Halley, *J. Geophys. Res.*, *93*, 7532, 1988.

McComas, D.J., J.T. Gosling, S.J. Bame, J.A. Slavin, E.J. Smith, and J.L. Steinberg, *J. Geophys. Res.*, *92*, 1139, 1987.

Mendis, D.A., and H.L.F. Houpis, The cometary atmosphere and its interaction with the solar wind, *Rev. Geophys. Space Phys.*, *20*, 885, 1982.

Mendis, D.A., E.J. Smith, B.T. Tsurutani, J.A. Slavin, D.E. Jones, and G.L. Siscoe, Comet-solar wind interaction: dynamical length scales and models, *Geophys. Res. Lett.*, *13*, 239, 1986.

Meyer-Vernet, N., P. Couturier, S. Hoang, C. Perche, J.L. Steinberg, J. Fainberg, and C. Meetre, Plasma diagnosis from thermal noise and limits on dust flux or mass in comet Giacobini-Zinner, *Science, 232*, 370, 1986.

Ogilvie, K., M.A. Coplan, P. Boschler, and J. Geiss, Ion composition results during the International Cometary Explorer encounter with Giacobini-Zinner, *Science, 232*, 374, 1986.

Richardson, I.G., S.W.H. Cowley, R.J. Hynds, T.R. Sanderson, K.P. Wenzel, and P.W. Daly, Three-dimensional energetic ion bulk flows at comet P/Giacobini-Zinner, Geophys. Res. Lett., 13, 415, 1986.

Richardson, I.G., S.W.H. Cowley, R.J. Hynds, C. Tranquille, T.R. Sanderson, K.P. Wenzel, and P.W. Daly, Observations of energetic water group ions at comet Giacobini-Zinner: implications for ion acceleration processes, *Planet. Space Sci., 35,* 1323, 1987a.

Richardson, I.G., S.W.H. Cowley, V. Moore, K. Staines, R.J. Hynds, T.R. Sanderson, K.P. Wenzel, and P.W. Daly, Energy spectra of energetic ions in the vicinity of comet P/Giacobini-Zinner, *Astron. Astrophys.*, *187*, 276, 1987b.

Richardson, I.G., S.W.H. Cowley, R.J. Hynds, P.W. Daly, T.R. Sanderson, and K.P. Wenzel, Properties of energetic water-group ions in the extended pick-up region surrounding comet Giacobini-Zinner, *Planet. Space Sci., 36*, 1429, 1988.

Richardson, I.G., S.W.H. Cowley, R.J. Hynds, P.W. Daly, T.R. Sanderson, and K.P. Wenzel, Energetic cometary ion flows in the pick-up region of comet Giacobini-Zinner, *Adv. Space Res., 9*, (3)381, 1989.

von Rosenvinge, T.T., J.C. Brandt, and R.W. Farquhar, The International Cometary Explorer mission to comet Giacobini-Zinner, *Science, 232*, 353, 1986.

van Rooijen, J.J., G.D. van Dijen, H.T. Lafleur, and P. Lowes, A low energy proton spectrometer fordirectional distribution measurements in space, *Space Sci. Inst., 4,* 373, 1979.

Sanderson T.R., K.P. Wenzel, P.W. Daly, S.W.H. Cowley, R.J. Hynds, E.J. Smith, S.J. Bame, and R.D. Zwickl, The interaction of heavy ions from Giacobini-Zinner with the solar wind, *Geophys. Res. Lett., 13,* 411, 1986a.

Sanderson T.R., K.P. Wenzel, P.W. Daly, S.W.H. Cowley, R.J. Hynds, E.J. Smith, S.J. Bame, and R.D. Zwickl, Observations of the interaction of heavy ions from comet P/Giacobini-Zinner with the solar wind, *Adv. Space Res., 6,* No. 1, 209, 1986b.

Schmidt, H.U., and R. Wegmann, MHD calculations for cometary plasmas, *Computer Phys. Commun., 19,* 309, 1980.

Schmidt, H.U., and R. Wegmann, Plasma flow and magnetic fields in comets, in *Comets,* edited by L.L. Wilkening, p. 538, University of Arizona Press, Tucson, USA, 1982.

Siscoe, G.L., J.A. Slavin, E.J. Smith, B.T. Tsurutani, D.E. Jones, and D.A. Mendis, Statics and dynamics of Giacobini- Zinner magnetotail, *Geophys. Res. Lett., 13,* 287, 1986.

Slavin, J.A., E.J. Smith, B.T. Tsurutani, G. L. Siscoe, D.E. Jones, and D.A. Mendis, Giacobini-Zinner magnetotail: ICE magnetic field observations, *Geophys. Res. Lett., 13,* 283, 1986a.

Slavin, J.A., B.A. Goldberg, E.J. Smith, D.J. McComas, S.J. Bame, M.A. Strauss, and H. Spinrad, The structure of a cometary Type I tail: ground-based and ICE observations of P/Giacobini-Zinner, *Geophys. Res. Lett., 13,* 1085, 1986b.

Smith, E.J., B.T. Tsurutani, J.A. Slavin, D.E. Jones, G. L. Siscoe, and D.A. Mendis, International Cometary Explorer Encounter with Giacobini-Zinner: magnetic field observations, *Science, 232,* 382, 1986a.

Smith, E.J., J.A. Slavin, S.J. Bame, M.F. Thomsen, S.W.H. Cowley, I.G. Richardson, D. Hovestadt, F. M. Ipavich, K.W.

Ogilvie, M.A. Coplan, T.R. Sanderson, K.P. Wenzel, F.L. Scarf, A.F. Vinas, and J.D. Scudder, Analysis of the Giacobini-Zinner bow wave, Proc. 20th ESLAB Symposium on the Exploration of Halley's Comet, ESA SP-250, p. 461, 1986b.

Staines, K., A. Balogh, S.W.H. Cowley, R.J. Hynds, T.S. Yates, I.G. Richardson, T.R. Sanderson, K.P. Wenzel, D.J. McComas, and B.T. Tsurutani, Cometary water-group ions in the region surrounding comet Giacobini-Zinner: Distribution functions and bulk parameter estimates, *Planet. Space Sci.,* in press, 1990.

Thomsen, M.F., W.C. Feldman, B. Wilken, K. Jockers, W. Studemann, A.D. Johnstone, A. Coates, V. Formisano, E. Amata, J.D. Winningham, H. Borg, D.A. Bryant, and M.K. Wallis, In-situ observations of a bi-modal ion distribution function in the outer coma of comet P/Halley. *Astron. Astrophys., 187,* 141, 1987.

Tranquille, C., I.G. Richardson, S.W.H. Cowley, T.R. Sanderson, K.P. Wenzel, and R.J. Hynds, Energetic ion properties observed near the periphery of the mass-loaded region surrounding comet P/Giacobini-Zinner, *Geophys. Res. Lett., 13,* 853, 1986a.

Tranquille, C., I.G. Richardson, S.W.H. Cowley, T.R. Sanderson, K.P. Wenzel, and R.J. Hynds, Energetic ion observations of a cometary bow shock-like structure, *Adv. Space Res., 6,* No.1, 235, 1986b.

Tsurutani, B.T., and E.J. Smith, Strong hydromagnetic turbulence associated with comet Giacobini-Zinner, *Geophys. Res. Lett., 13,* 259, 1986a.

Tsurutani, B.T., and E.J. Smith, Hydromagnetic waves and instabilities associated with cometary ion pick-up: ICE observations, *Geophys. Res. Lett., 13,* 263, 1986b.

Wallis, M., Weakly-shocked flows of the solar wind plasma through atmospheres of comets and planets, *Planet. Space Sci., 21,* 1647, 1973.

Wu, C.S., D. Winske, and J.D. Gaffey, Jr., Rapid pick-up of cometary ions due to strong magnetic turbulence, *Geophys. Res. Lett., 13,* 865, 1986.

MODEL CALCULATIONS OF OXYGEN ION FLUXES FROM THE DISSOCIATION OF H_2O, CO, AND CO_2 AT GIGAMETER DISTANCES FROM COMET HALLEY

P. W. Daly

Max-Planck-Institut für Aeronomie, W-3411 Katlenburg-Lindau, Germany

Abstract. In a previous work, a method was developed to calculate densities of neutral particles emitted from comet Halley taking into account the Kepler trajectories that they describe. In that work, densities of the products of the dissociation of H_2O were found out to distances of 24 Gm from the nucleus. Additionally ion fluxes were determined by integrating along solar wind streamlines to the point of "observation". In the present work, densities and fluxes of oxygen atoms and ions are found from H_2O, CO, and CO_2. It is discovered that the last two dominate over the first beyond about 3 Gm from the nucleus mainly because of their larger ionization scale lengths. This causes the O density to be more circularly symmetric than predicted by the earlier study. The effect of changing the path of integration has been looked at and it is found that the ion fluxes are very sensitive to such variations. The usual practice of integrating along solar wind streamlines in the presence of strong pitch angle scattering is shown to be incorrect: one must do the calculation for many pitch angles and average the results. In the absence of pitch angle scattering the proper integration path is determined from the local solar wind and magnetic field conditions.

Introduction

With the discovery of substantial fluxes of cometary ions at large distances from comets Giocobini-Zinner [Hynds et al., 1986; Ipavich et al., 1986] and Halley [Terasawa et al., 1986; Somogyi et al., 1986; McKenna-Lawlor et al., 1986; Balsiger et al., 1986], efforts have been made to calculate these fluxes from outgassing models of the cometary nucleus. Initially researchers have used the Haser model [e.g. Festou, 1981] in which the source neutral atoms and molecules move with a uniform velocity relative to the nucleus before being ionized at some fixed ionization scale length. Such applications have for example been made by Terasawa et al. [1986], Ipavich et al. [1986], Gloeckler et al. [1986], Sanderson et al. [1986], and Neugebauer et al. [1989].

In a previous paper [Daly, 1989] (to be called Paper I) this author presented an alternative model in which the neutral particles move along Kepler trajectories around the Sun. Their motion relative to the comet nucleus deviates from that of a straight line with constant speed. The densities of neutral products from the dissociation of H_2O, the most prolific parent molecule, were calculated out to distances of 24 Gm (24×10^6 km) from the nucleus of comet Halley during the time of the Giotto encounter. The model shows that the O atoms, with an assumed ejection speed of 1 km s^{-1}, are distributed not in spheres but in elongated ovoids; they also exhibit cut-offs between 4 and 9 Gm, depending on direction. The H atoms (8 and 20 km s^{-1})

Cometary Plasma Processes
Geophysical Monograph 61

are further subject to radiation pressure which has the effect of limiting the slower component on the sunward side to about 7 Gm. The results for the faster component hardly deviate at all from the Haser model predictions.

Paper I also calculated ion fluxes of the three "species" by integrating the ion production rate along the solar wind streamline. The production rate is the neutral density divided by the ionization time constant. This is the same procedure that has been used by previous authors to find the ion fluxes but with the difference that the neutral densities are found from the Kepler, not the Haser model.

In this work, two extensions to Paper I are to be made. First, oxygen fluxes are calculated not only for the O from H_2O, but also from CO and CO_2. Although these constituents are produced at rates more than a factor of 10 less than that for H_2O, the higher ejection speed for these O atoms means that beyond a certain distance they dominate over the O from H_2O. This is mainly due to their longer ionization scale length. Secondly, integration paths other than the solar wind streamline are taken. Ions move with the $\mathbf{E} \times \mathbf{B}$ drift velocity and with the component of velocity parallel to the magnetic field $\mathbf{B}$. This means that different pitch angles have different guiding center velocities. The effect of this on the ion fluxes is investigated.

The Model

The model used to calculate the densities of neutral particles emitted from comet Halley is described fully in Paper I [Daly, 1989]. Here I will present a brief description of its main features.

Coordinates

The coordinate system used for the calculations is that of the ecliptic and vernal equinox of 1950.0. This is an inertial coordinate system in which the unit of distance is the astronomical unit AU, and the unit of time is the ephemeris day. Time is measured in days from 1986 Jan. 0.0 ET (Ephemeris Time). During 1986 Ephemeris Time was 55.183 s ahead of Coordinated Universal Time (UTC). For presentation of the data, the Halley-Solar-Ecliptic (HSE) system is employed, in which the origin is at the Halley nucleus, the $+x$ axis points towards the Sun, and the north ecliptic pole lies in the $+zx$ plane. This means that the y axis is parallel to the ecliptic plane. The convenient unit of distance for our purposes is the gigameter (Gm) equivalent to one million km. The closest approach of Giotto to the Halley nucleus occurred on day 73.0027, or March 14, 1986, UT 0003.

Neutral Densities

Figure 1 illustrates the main features of the model. Neutral particles are emitted uniformly at a fixed speed with a total rate of Q

Density and Flux Calculation

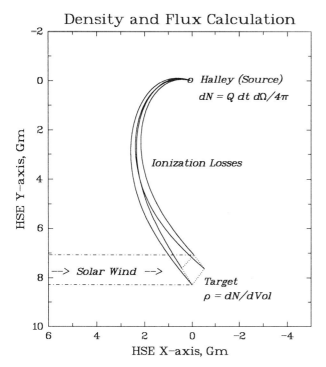

Fig. 1. Illustration of how neutral densities and ion fluxes are calculated. Particles are emitted uniformly in all directions from the comet at a rate $Q/4\pi$ s^{-1} sr^{-1}. An element of solid angle and time $d\Omega\, dt$ maps to a volume element dVol at some other point, the target, by means of Kepler trajectories. In flight the particles are subject to losses by ionization. The density at the target is the number of surviving emitted particles divided by the volume element. This entire calculation can be done analytically. The ion flux at the target is found by integrating the neutral density times ionization rate along a solar wind streamline sunward of the target. This calculation can only be done numerically.

particles s^{-1}, a rate that varies with the inverse square of the distance from the Sun R_s. It is also set to zero at times earlier that day -70.0 (October 22, 1985) when the comet was 2 AU from the Sun. This simulates the observed "turn-on" of the emissions and also limits the search for solutions as described below. The emission rate is quoted as its value at 1 AU.

Ideally one should have Q as a function of emission speed as a continuous distribution. However, to carry out such a computation with a large number of emission speeds would increase the calculation time prohibitively. Since the purpose of this work is not to do a detailed determination of the densities but only to discover their overall behavior, I therefore take the speed distribution to be a sum of several delta functions, one per parent neutral.

An emitted particle traverses a path from the source (the comet nucleus) to some target point in space along a Kepler trajectory. The effect of radiation pressure can be included by correcting the effective gravitational constant since both gravity and radiation fall off as $1/R_s^2$. In Paper I it is shown that radiation pressure is significant only for H atoms and can be neglected for the heavier species.

During the flight from source to target the particle is subject to losses by photo-ionization and charge exchange with solar protons. The time constant for these losses also goes as $1/R_s^2$ and is therefore also quoted as the value at 1 AU. It turns out that because of the

nature of Kepler trajectories, the total loss factor including the spatial dependence of the time constant depends only on the orbit parameters and the angular displacement from source to target: the loss factor can be found analytically.

The number of particles arriving in a volume element dVol at the target is the number that were emitted from the nucleus in the solid angle element $d\Omega$ during the time element dt, reduced by the ionization loss factor. A major part of the mathematics goes into the calculation of the relation between dVol and $d\Omega\, dt$, which can be found analytically. With this, we have the density at the target, at least for one trajectory joining the comet at some earlier time to the target at the desired time. The total density is the sum of those for all such solutions.

Another major part of the calculation is the determination of the source times when a trajectory joins the (moving) nucleus with the (fixed) target for the given emission speed. Even if only a single emission speed is allowed, there can still be multiple solutions to this problem, as illustrated in Paper I. Although it is not possible to find these emission times analytically, the emission speed for any given source time can be found. Thus the emission time can be stepped to find the solutions roughly and a Newton-Ralphson iteration method can determine them to any desired accuracy. In this way all the emission times contributing to the density at the target are found.

Ion Densities

Once the density of the neutral particles ρ_{neu} has been established, and the ionization time constant τ is known, the ion density ρ_{ion} can be calculated from the continuity equation with source terms:

$$\frac{\partial \rho_{\text{ion}}}{\partial t} + \nabla\cdot(\rho_{\text{ion}}\mathbf{v}) = \frac{\rho_{\text{neu}}}{\tau}. \tag{1}$$

The general solution to equation 1 is

$$\rho_{\text{ion}}(\mathbf{r}, t) = \frac{1}{\Omega(\mathbf{r}, t)} \int^t dt'\, \rho_{\text{neu}}(\mathbf{r}', t')\, \Omega(\mathbf{r}', t')/\tau \tag{2}$$

$$\Omega(\mathbf{r}, t) = \exp \int^t dt'\, \nabla\cdot\mathbf{v} \tag{3}$$

where the integrals are carried out along the path of the particles moving with velocity $\mathbf{v}$. The factor Ω is a correction for the local volume element due to the divergence of $\mathbf{v}$. In the case of a constant, radially outwards velocity, $\nabla\cdot\mathbf{v} = 2/R_s$ and $\Omega = R_s^2$.

The uncertainty in the lower limits of the integrals in equations 2 and 3 means that ρ_{ion} is determined only to within an additive constant, and Ω to within a multiplicative constant that cancels out in equation 2. We assume ρ_{ion} is zero at those earlier locations where the integrand is insignificant and start the integration there. (In fact, one integrates from t backwards in time until the integral becomes stable.)

As presented above, the story is still not perfectly correct, for I have assumed all the ions move with the same velocity $\mathbf{v}$. In fact there will be a distribution in phase space and the equations should all be extended to include 6 dimensional positions and their time derivatives. The factor Ω can be dropped because the phase space volume is constant (Liouville's Theorem). This ambitious integration scheme is beyond the scope of the present work, but it will have to be done at some time to calculate velocity space ion distributions.

In this work, I limit myself to the 3 dimensional configuration space and integrate over the path of the average ion velocity, that is, the guiding center velocity. The gyromotion of the ions is neglected. How this average velocity is related to the solar wind speed and ion pitch angle is described later in the section on integration paths.

TABLE 1. Chemistry for O Production

Parent (total lifetime)		Products			Branching Ratio	$Q(O)$ (relative)
H_2O 8.3×10^4 s	$\rightarrow$	H 20 km s^{-1}	+	OH ~1 km s^{-1}	0.80	
OH 2×10^5 s	$\rightarrow$	H 8 km s^{-1}	+	O ~1 km s^{-1}		1.000
CO 1.5×10^6 s	$\rightarrow$	C 4.9 km s^{-1}	+	O 3.6 km s^{-1}	0.43	0.086
CO_2 5.0×10^5 s	$\rightarrow$	CO 3.3 km s^{-1}	+	O 5.7 km s^{-1}	0.46	0.016
	$\rightarrow$	CO 2.2 km s^{-1}	+	O 3.9 km s^{-1}	0.14	0.005

Using $Q(CO)/Q(H_2O) = 0.20$ and $Q(CO_2)/Q(H_2O) = 0.035$

TABLE 2. Input Values to the Oxygen Model

Quantity	from H_2O	from CO	from CO_2
O Production Rate*	3.5×10^{29} s^{-1}	3.5×10^{28} s^{-1}	5.5×10^{27} s^{-1}
Production Turn-On[†]	−70.0	−70.0	−70.0
Ejection Speed	1 km s^{-1}	3.6 km s^{-1}	5.7 km s^{-1}
Ionization Rate*	10^{-6} s^{-1}	10^{-6} s^{-1}	10^{-6} s^{-1}
Radiation Pressure β	0	0	0

*At 1 AU;   [†]Days from 1986 Jan. 0.0 ET

Neutral Oxygen Densities from Halley

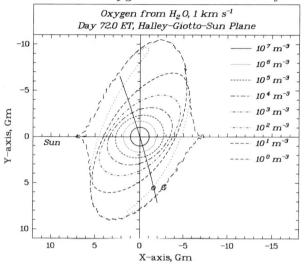

Fig. 2. A map of the density of neutral O atoms from the dissociation of H_2O, ejection speed = 1 km s^{-1}, in the plane containing the Sun, comet Halley, and Giotto, one day before the encounter. The Giotto trajectory is drawn as a solid line, and the position of Giotto is indicated with a small circle labeled "G". The interval between isodensity contours is a factor of 10. The calculation has been carried out to a distance of 12 Gm. The word "Sun" marks the direction, not the position of the sun.

The ion fluxes are found simply by multiplying ρ_{ion} by v. It turns out that the fluxes are not very sensitive to the solar wind speed selected since ρ_{ion} is roughly inversely proportional to v. If it were not for the weak time dependence of ρ_{neu}, this statement would be exactly true. Thus for the modeling of ion fluxes it is not necessary to have an accurate number for the solar wind speed.

Oxygen Densities and Fluxes

The Oxygen Model

The water group densities and fluxes presented in Paper I represent the contributions that come from the most abundant parent species, H_2O. However, there are other sources of O atoms, such as CO and CO_2, that, although having considerably lower production rates, do lead to larger numbers of O atoms at greater distances.

The chemistry of O production from H_2O, CO, and CO_2 is listed in Table 1, in which the lifetimes, branching ratios, and resultant speeds are taken mainly from Huebner [1985]. The speed of the fast H atom comes from Crovisier [1989] and the data for the OH dissociation from Keller [1976]. The production rate of H_2O has been measured in situ [Krankowsky et al., 1986] and from ground observations [Feldman et al., 1987] to be about 5.5×10^{29} molecules s^{-1} at the time of the Giotto encounter ($R_s = 0.9$ AU). Converting to 1 AU and allowing only 80% of the H_2O to create O atoms, the production rate is 3.5×10^{29} s^{-1} for O from H_2O. The density of CO is less than 20% of that of H_2O [Balsiger et al., 1987; Eberhardt et al., 1987], leading to a relative O production of 8.6%. CO_2 is produced at 3.5% of the H_2O rate [Krankowsky et al., 1986] so that the two O atoms have relative production rates of 1.6% and 0.5%. Since this second O has a speed similar to that of the O from CO, it will be included with that component.

The actual numbers used in the Kepler trajectory calculations for the three components of oxygen are given in Table 2. The production rates have been taken as convenient numbers so that their relative values differ slightly from those in Table 1.

One problem with applying the Kepler trajectory model to the O products of CO and CO_2 is the large dissociation length. The model assumes that the neutral O atoms are created at the nucleus, whereas in fact they are produced within a sphere around it. With an expansion speed of about 1 km s^{-1} for all parent molecules, this sphere has a mean radius of 1.5 Gm and 0.5 Gm for CO and CO_2, respectively. Since these sizes are of the same order of magnitude as the scale of the calculations, this effect cannot be negligible; however, the model has no way of handling it at this point.

Oxygen Density Distribution

To visualize the spatial distribution of the O atoms from different parents, I have generated contour plots of the densities within the plane containing the comet, Giotto, and the Sun. Figure 2 shows the distribution for O from H_2O, calculated out to distances of 12 Gm. The time of this "snapshot" is 24 hours before the encounter with Giotto, the orbit and position of which are also indicated. It is clear that the iso-density contours are elliptical, not circular, and that there is a cut-off where the contours for densities 10^1 and 10^0 m^{-3} merge. The sharpness of this cut-off is a result of the delta function for the emission speed distribution; a spread-out distribution would yield an extended cut-off. This figure is discussed in more detail in Paper I.

A similar density map for the O atoms from CO is presented in Figure 3. Here the deviation of the contours from circles is still present if not so drastic as in Figure 2. The contours for the two highest densities, 10^7 and 10^6 m^{-3}, are closer to the nucleus for CO than for H_2O, but all others are further out. That is, at about 3 Gm, the O from CO begins to dominate over that from H_2O.

The density map for O from CO_2 is not shown. However, its contour lines are nearly perfect circles and it also begins to dominate over the O from H_2O at about 3 Gm.

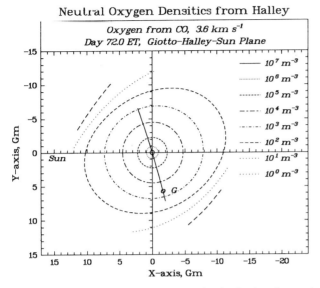

Fig. 3. A map similar to that in Figure 2 for the density of neutral O atoms from the dissociation of CO.

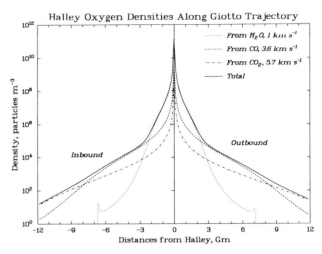

Fig. 4. Plots of the densities of O atoms from the dissociation of H_2O, CO, and CO_2, against distance of Giotto from the Halley nucleus. The solid curve is the sum of the three components of O.

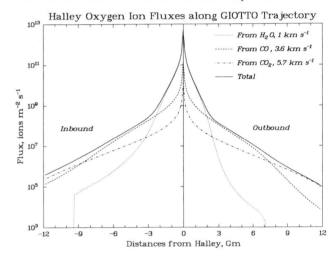

Fig. 5. Plots of the fluxes of O^+ ions from the dissociation of H_2O, CO, and CO_2, against distance of Giotto from the Halley nucleus. The solid curve is the sum of the three components of O^+.

The reason that the O with the lower production rates has a higher density at larger distances is due to its shorter flight time and correspondingly less ionization. For example, O from H_2O requires 66.9 days to reach the the location of Giotto on day 72.0, which is 12.33 ionization time constants. The O from CO, on the other hand, requires 19.9 days or 3.10 time constants. Without ionization effects, the ratio of O from CO to that from H_2O is 0.036; with ionization, the ratio becomes 360. The O from CO_2 has a flight time of 12.2 days, or 1.64 time constants, and its corresponding ratios to H_2O are .0027 and 119.

Along the Giotto Trajectory

The next step is to illustrate the O densities and O^+ fluxes along the Giotto trajectory. For the O densities, this is simply a cut through Figure 2 along the solid line representing the flight path. These densities are plotted as a function of distance from the nucleus for each of the three parents in Figure 4. Here one sees clearly how the O atoms from CO and CO_2 become more abundant than those from H_2O at about 3 Gm. The exact location of this cross-over depends on the direction of the cut, as is obvious from the elliptical contours in Figure 2.

The O^+ fluxes along the Giotto trajectory are found according to equation 2 using an integration path radially outward from the Sun and a solar wind speed of 400 km s^{-1}. This is a typical value, and as mentioned earlier, the fluxes are only weakly dependent on the solar wind speed. The results for O^+ from the three parents are plotted in Figure 5.

The O^+ from H_2O is asymmetric between inbound and outbound, a result that can be understood with the help of Figure 2. The flux at any point is virtually a column integral through the neutral densities on the sunward side. Inbound, the maximum densities are sunward of Giotto, whereas outbound they are on the anti-sunward side. For the O^+ from the other parents, there is less asymmetry in the neutral density distribution so these fluxes show much the same profile inbound and outbound. The result is that the asymmetry in the H_2O vanishes almost entirely in the total flux curve.

Other Integration Paths

As explained earlier in the model section, the proper calculation of the ion fluxes involves an integral through 6-dimensional phase space but I use as an approximation an integral in 3-dimensional configuration space. I do this by taking the velocity of the ion's guiding center as the average over gyrophase for a given pitch angle. The ion fluxes in the previous section were found using the solar wind streamline as the integration path. Now I wish to demonstrate what happens when different paths are selected.

Pick-up Ions

At the moment of creation, ions possess the velocity that the dying neutral particle had, which is negligible compared to the solar wind speed. In the frame moving with the solar wind, the ion has instan-

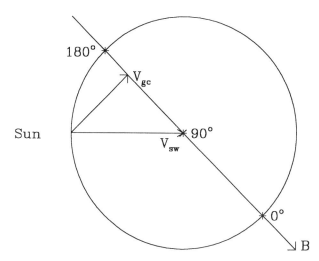

Fig. 6. Diagram in velocity space illustrating the relationship between pitch angle and guiding center velocity. The Sun is to the left and the solar wind velocity points to the right. In the solar wind frame (origin at the tip of the $\mathbf{v}_{sw}$ vector) all pick-up ions have the same speed v_{sw} and gyrate on the surface of a sphere, the intersection of which with the plane of the diagram is shown as a circle. The average velocity over the gyromotion is the guiding center velocity, located on the line of the magnetic field, $\mathbf{B}$. This velocity $\mathbf{v}_{gc}$ is drawn for the ion with the injection pitch angle. Also indicated are the guiding center velocities for pitch angles $0°$, $90°$, and $180°$.

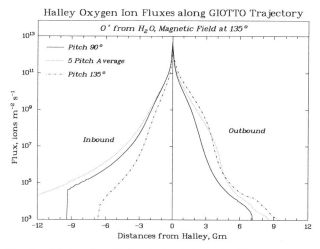

Fig. 7. O^+ flux from H_2O calculated along the Giotto trajectory for several different integration paths. The magnetic field has HSE coordinates $\theta_B = 90°$, $\phi_B = 135°$, as illustrated in Figure 6. The three curves are for pitch angle $\beta = 90°$ (the radially outward solar wind), an average over 5 pitch angles, and for the pitch angle at ion creation, $\beta = 135°$.

taneously the negative solar wind velocity, $-\mathbf{v}_{sw}$. In this frame there is no electric field so the ion gyrates about the magnetic field $\mathbf{B}$ and has a constant component parallel to $\mathbf{B}$: the average velocity over the gyromotion is this parallel component. This is the velocity $\mathbf{v}_{gc}$ in Figure 6.

As pitch angle scattering takes place, the total velocity is unchanged in the solar wind frame. The ion traces out a circle in velocity space lying on the surface of a sphere centered on the solar wind velocity. The intersection of this sphere with the plane containing $\mathbf{B}$ and $\mathbf{v}_{sw}$ is the large circle in Figure 6. For pitch angle β the guiding center velocity in the spacecraft frame is

$$\mathbf{v}_{gc} = \mathbf{v}_{sw} + v_{sw} \cos\beta \, (\mathbf{B}/B). \qquad (4)$$

At creation the ion has pitch angle $\beta = \pi - \alpha$, where α is the angle between $\mathbf{v}_{sw}$ and $\mathbf{B}$. From equation 4 it can be seen that the ion's guiding center velocity is identical with that of the solar wind only for $\beta = 90°$, and this can occur for the newly injected ions only when $\alpha = 90°$.

Spacecraft observations at comets Giacobini-Zinner [Richardson et al., 1988] and Halley [Coates et al., 1989] indicate that pick-up ions at large distances from the nucleus maintain their injection pitch angle, but that at smaller distances they become uniformly distributed in pitch angle. Coates et al. [1989] see this change from a ring to a shell distribution near 2 Gm from the Halley nucleus on the inbound leg of the Giotto flyby.

The presence of a uniform, or shell, distribution of pitch angles is often used to justify $\mathbf{v}_{sw}$ as the velocity in the ion integral of equation 2. It is true that this is the average velocity of the whole pick-up ion population, but in fact each pitch angle has a different guiding center velocity and thus a different integration path. To investigate this effect, I compare the average flux over several pitch angles to that for $90°$ and to that for the injection pitch angle.

Method of Integration

Before proceding, some comments should be made on the derivation of the integration path from the ion velocity and on the method of integration. The integral in equation 2 is actually a time integral, but the spatial coordinates are functions of time according to the ion velocity. For a radial velocity, there is no difficulty in finding $\mathbf{r}(t)$ in the inertial frame. For a generalized velocity, something more must be assumed since a constant velocity in a solar centered system is not realistic.

The ion velocity is specified in the HSE coordinate system. One possibility would be to require that the HSE components of velocity are constant, so that $\mathbf{v}$ rotates with location. This, however, gives the ecliptic a special role in the model that should, if anything, be played by the solar equator. Instead, I demand that $\mathbf{v} \cdot (\mathbf{r}/r)$ and $\mathbf{r} \times (\mathbf{v} \times \mathbf{r})/r^2$ both be constant in the inertial frame. The particle path can be found explicitly as a function of time and $\nabla \cdot \mathbf{v} = 2/r$, so this is a convenient generalization of the constant radial velocity. (Note that r is the distance from the Sun, the same as R_s used previously.)

The integration itself is carried out as follows: starting at the point of observation, I integrate in steps of 1 Gm using a Gaussian method along the path from which the ions come, and continue until the next contribution is less than 0.001 times the total. Each step is a separate Gaussian integral, so it would be more accurate to integrate over the whole interval with the same number of total points. However, it is not known beforehand where the integrand becomes suddenly negligible. This accumulating method is a reasonable compromise.

Sample Calculations

As an illustration I take the situation shown in Figure 6, in which the solar wind is radially outward from the Sun, with a speed of 400 km s^{-1}, and the magnetic field is at $45°$ to it. In HSE coordinates, $\theta_{sw} = 90°$, $\phi_{sw} = 180°$, $\theta_B = 90°$, and $\phi_B = 135°$. The magnitude of $\mathbf{B}$ is unimportant.

The O^+ flux (from H_2O) is calculated along the Giotto trajectory for three cases:

TABLE 3. Results of Different Integration Paths

Pitch Angle	v_{gc} km s^{-1}	ϕ_{gc}	O^1 Flux, 10^6 m^{-2} s^{-1} −6 Gm	+3 Gm
143°	285	−128°	0.012	755.0
114°	308	−158°	0.108	8.097
90°	400	180°	0.628	3.022
66°	525	168°	2.051	2.284
37°	666	160°	4.643	2.029
Average	400	180°	1.489	154.2
135°	283	−135°	0.018	295.2

1. for the overall average ion velocity, i.e. with $\beta = 90°$, or the solar wind speed itself (this curve already appears in Figure 5),
2. averaged over 5 pitch angles so distributed that $d\cos\beta$ is uniform (the ions are assumed evenly distributed in solid angle, i.e. in $\sin\beta\, d\beta$),
3. for the injection pitch angle 135°, the case with no pitch angle scattering.

These three curves are plotted in Figure 7. On the inbound flight, the averaged flux (dotted line) is fairly close to the 90° flux (solid line), but on the outbound side, it is closer to the 135° values. This can be understood with the aid of Figure 2: for $\beta = 135°$ the integration path is to the lower left of the observation point, and that is where most of the neutral O is to be found outbound. Inbound beyond 5 Gm there are fewer atoms in this direction resulting in the depressed fluxes for this curve.

Additional insight into this effect is offered by Table 3, in which the results of the integration along each path are listed for distances 6 Gm before and 3 Gm after encounter. Here we can see that the smaller pitch angles correspond to azimuths $\phi_{gc} > 90°$, that is, the ions are moving from the upper left in Figures 2 and 6. Their ion fluxes are the larger ones inbound and the smaller ones outbound.

Both Figure 7 and Table 3 demonstrate that ion fluxes depend very strongly on the path that one takes for the integration. Also the approximation for strong pitch angle scattering in which one takes the solar wind velocity is not valid. To carry out a realistic calculation for comparison with the Giotto ion data, the local conditions for the solar wind and magnetic field must be used. It would also be necessary to know the time and spatial dependence of these quantities sunward of the observations. This is a most difficult proposition, requiring some type of model. Neither the solar wind nor the magnetic field were stable enough during the four days of the Giotto flyby to justify using constant values. An attempt to incorporate realistic v_{sw} and B fields must be left to a future work.

Summary

The Kepler trajectory method of calculating the densities of neutral particles emitted from comet Halley has been used to find the density of oxygen atoms and flux of O$^+$ ions out to 12 Gm from the nucleus. In Paper I this had been done for the dissociation products of the most plenteous parent molecule, H_2O. In that work it was found that the O atoms are distributed in highly elongated ellipsoids rather than spherical shells. Now the contributions from the dissociations of CO and CO_2 have been included. Although these parents are much less abundant than H_2O, their O atoms move with such greater speeds that their ionization scale lengths are larger than that for the slow O from H_2O. The result is that at distances beyond about 3 Gm from the nucleus the densities of these O atoms dominate. Much of the asymmetry in the distribution of O then disappears.

The fluxes of O$^+$ from H_2O along the Giotto trajectory also show large asymmetry between the inbound and outbound legs. This asymmetry is also diminished when the contributions from CO and CO_2 are included.

The flux calculations have been previously done by integrating the neutral density along a solar wind streamline. The justification for this approach has been scrutinized and found to be wanting. The proper integration path should be determined by the pitch angle of the ion at its creation, or if there is strong pitch angle scattering, an average over the fluxes for all pitch angles must be carried out. Integration with the average ion velocity is not the same as averaging over the individual velocities.

For comparison with the energetic ion measurements on board Giotto, the integration path must be selected according to the solar wind velocities and magnetic fields actually measured. However, these quantities were extremely variable, especially the magnetic field. It is also not obvious how one can extrapolate the measurements at Giotto to the vast regions sunward of it where the main contributions to the ion flux are to be found. This is a problem since the results depend very strongly on the integration path, which in turn depends on the local solar wind and magnetic field.

Acknowledgments. The author would like to express his thanks to W.-H. Ip and H.-U. Keller for suggesting the importance of CO and CO_2 at large distances, and to B. Inhester for discussions on integration methods.

References

Balsiger, H., K. Altwegg, F. Bühler, J. Geiss, A. G. Ghielmetti, B. E. Goldstein, R. Goldstein, W. T. Huntress, W.-H. Ip, A. J. Lazarus, A. Meier, M. Neugebauer, U. Rettenmund, H. Rosenbauer, R. Schwenn, R. D. Sharp, E. G. Shelley, E. Ungstrup, and D. T. Young, Ion composition and dynamics at comet Halley, *Nature*, *321*, 330–334, 1986.

Balsiger, H., K. Altwegg, F. Bühler, S. A. Fuselier, J. Geiss, B. E. Goldstein, R. Goldstein, W. T. Huntress, W.-H. Ip, A. J. Lazarus, A. Meier, M. Neugebauer, U. Rettenmund, H. Rosenbauer, R. Schwenn, E. G. Shelley, E. Ungstrup, and D. T. Young, The composition and dynamics of cometary ions in the outer coma of comet P/Halley, *Astron. Astrophys.*, *187*, 163–168, 1987.

Coates, A. J., A. D. Johnstone, B. Wilken, K. Jockers, and K. H. Glassmeier, Velocity space diffusion of pickup ions from the water group at comet Halley, *J. Geophys. Res.*, *94*, 9983–9993, 1989.

Crovisier, J., On the photodissociation of water in cometary atmospheres, *Astron. Astrophys.*, *213*, 459–464, 1989.

Daly, P. W., The use of Kepler trajectories to calculate ion fluxes at multi-gigameter distances from comet Halley, *Astron. Astrophys.*, *226*, 318–334, 1989.

Eberhardt, P., D. Krankowsky, W. Schulte, U. Dolder, P. Lämmerzahl, J. J. Berthelier, J. Woweries, U. Stubbemann, R. R. Hodges, J. H. Hoffman, and H. M. Illiano, The CO and N_2 abundance in comet P/Halley, *Astron. Astrophys.*, *187*, 481–484, 1987.

Feldman, P. D., M. C. Festou, M. F. A'Hearn, C. Arpigny, P. S. Butterworth, C. B. Cosmovici, A. C. Danks, R. Gilmozzi, W. M. Jackson, L. A. McFadden, P. Patriarchi, D. G. Schleicher, G. P. Tozzi, M. K. Wallis, H. A. Weaver, and T. N. Woods, IUE observations of comet P/Halley: Evolution of the ultraviolet spectrum between September 1985 and July 1986, *Astron. Astrophys.*, *187*, 325–328, 1987.

Festou, M. C., The density distribution of neutral compounds in cometary atmospheres I. models and equations, *Astron. Astrophys.*, *95*, 69–79, 1981.

Gloeckler, G., D. Hovestadt, F. M. Ipavich, M. Scholer, B. Klecker, and A. B. Galvin, Cometary pick-up ions observed near Giacobini-Zinner, *Geophys. Res. Lett.*, *13*, 251–254, 1986.

Huebner, W. F., The photochemistry of comets, in *The Photochemistry of Atmospheres*, edited by J. S. Levine, pp. 437–481, Academic Press, Orlando, Florida, 1985.

Hynds, R. J., S. W. H. Cowley, T. R. Sanderson, K.-P. Wenzel, and J. J. van Rooijen, Observations of energetic ions from comet Giacobini-Zinner, *Science, 232*, 361–365, 1986.

Ipavich, F. M., A. B. Galvin, G. Gloeckler, D. Hovestadt, B. Klecker, and M. Scholer, Comet Giacobini-Zinner: in situ observations of energetic heavy ions, *Science, 232*, 366–369, 1986.

Keller, H. U., The interpretations of ultraviolet observations of comets, *Space Sci. Rev., 18*, 641–648, 1976.

Krankowsky, D., P. Lämmerzahl, I. Herrwerth, J. Woweries, P. Eberhardt, U. Dolder, U. Herrmann, W. Schulte, J. J. Berthelier, J. M. Illiano, R. R. Hughes, and J. H. Hoffman, In situ gas and ion measurements at comet Halley, *Nature, 321*, 326–329, 1986.

McKenna-Lawlor, S., E. Kirsch, D. O'Sullivan, A. Thompson, and K.-P. Wenzel, Energetic ions in the environment of comet Halley, *Nature, 321*, 347–349, 1986.

Neugebauer, M., B. E. Goldstein, H. Balsiger, F. M. Neubauer, R. Schwenn, and E. G. Shelley, The density of cometary protons upstream of comet Halley's bow shock, *J. Geophys. Res., 94*, 1261–1269, 1989.

Richardson, I. G., S. W. H. Cowley, R. J. Hynds, P. W. Daly, T. R. Sanderson, and K.-P. Wenzel, Properties of energetic water-group ions in the extended pick-up region surrounding comet Giacobini-Zinner, *Planet. Space Sci., 36*, 1429–1450, 1988.

Sanderson, T. R., K.-P. Wenzel, P. W. Daly, S. W. H. Cowley, R. J. Hynds, E. J. Smith, S. J. Bame, and R. D. Zwickl, The interaction of heavy ions from comet P/Giacobini-Zinner with the solar wind, *Geophys. Res. Lett., 13*, 411–414, 1986.

Somogyi, A. J., K. I. Gringauz, K. Szegő, L. Szabó, G. Kozma, A. P. Remizov, J. Erő, Jr., I. N. Klimenko, I. T. Szücs, M. I. Verigin, J. Windberg, T. E. Cravens, A. Dyachkov, G. Erdős, M. Faragó, T. I. Gombosi, K. Kecskeméty, E. Keppler, T. Kovács, Jr., A. Kondor, Y. I. Logachev, L. Lohonyai, R. Marsden, R. Redl, A. K. Richter, V. G. Stolpovskii, J. Szabó, I. Szentpétery, A. Szepesváry, M. Tátrallyay, A. Varga, G. A. Vladimirova, K.-P. Wenzel, and A. Zarándy, First observations of energetic particles near comet Halley, *Nature, 321*, 285–288, 1986.

Terasawa, T., T. Mukai, W. Miyake, M. Kitayama, and K. Hirao, Detection of cometary pickup ions up to 10^7 km from comet Halley: Suisei observation, *Geophys. Res. Lett., 13*, 837–840, 1986.

NEUTRAL HYDROGEN SHELL STRUCTURE NEAR COMET P/HALLEY DEDUCED FROM VEGA-1 AND GIOTTO ENERGETIC PARTICLE DATA

M.I. Verigin[1], S. McKenna-Lawlor[2], A.K. Richter[3], K. Szego[4], I.S. Veselovsky[5].

(1) Space Research Institute, Moscow, USSR
(2) Space Technology Ireland, St Patrick's College, Maynooth, Ireland
3) Max-Planck Institut fur Aeronomie, Lindau, FRG
(4) Central Research Institute for Physics, Budapest, Hungary
(5) Nuclear Physics Research Institute of Moscow University, Moscow, USSR

Abstract. An existing model based on Vega-1 (Tunde-M) and Giotto (EPONA) energetic particle data, representing neutral gas shells expanding about comet Halley, has been up-dated by incorporating additional information concerning energetic particles recorded by Tunde-M, and neutral gas measurements recorded aboard the Vega-1 and Vega-2 spacecraft, in the original data set. The modified model reproduces reasonably well the positions of the maxima in the intensity profiles of energetic cometary ions observed along the Vega and Giotto trajectories and it is estimated that the velocity of gas in the envisioned neutral shells is ~ 7.3 km/s i.e. close to the velocity (~8 km/s) of the slow hydrogen component of cometary neutrals. Detailed arguments are presented to support the suggestion that, at distances of 2 - 10 x 10^6 km from the comet nucleus, the energetic particles recorded in the quasi-periodic structures identified by the Tunde-M and EPONA instruments were protons.

Introduction

Energetic ion flux measurements, recorded in the energy range from 30 keV to a few hundred keV by the Tunde-M and EPONA experiment aboard Vega-1 and Giotto respectively, during their individual flybys of Halley's Comet (March 1986), reveal the presence of striking quasi-periodic variations in ion flux intensity at cometocentric distances of 2-10 x 10^6 km from the nucleus [Somogyi et al., 1986; McKenna-Lawlor et al., 1989]. The flux variations recorded were characterized by their amplitude (1 to 2 orders of magnitude) and sharpness, see Figure 1. They were separated in time by about 4 hours, and appeared along both the inbound and outbound passes though the comet

A *tentative* explanation of the observations, based on the assumption that the periodic spatial structure of the energetic particle fluxes results from the flight of the spacecraft through a series of expanding neutral gas shells, formed due to injection of neutral gas into the collisional zone from a strong source on the rotating (T ~54 hours) cometary nucleus, has

Cometary Plasma Processes
Geophysical Monograph 61
©1991 American Geophysical Union

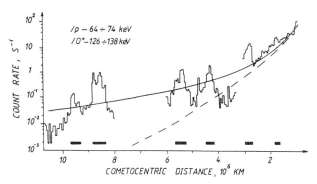

Figure 1. Dependence of energetic ion fluxes on cometocentric distance during the inbound pass of Vega-1 through P/Halley [Kecskemety et al., 1989]. The energy ranges for protons and ions relevant to the channel presented are given above the data. The dashed curve is proportional to the simple Haser dependence of neutral gas number density on cometocentric distance for an ionization time 10^6s and expansion velocity V ~ 1 km/s. The solid curve represents a similar dependence with V_n ~ 1 km/s and V ~ 7.3 km/s for an equal gas production rate. Heavy horizontal bars indicate the locations of those enhanced fluxes used in the least square analysis

already been considered by Kecskemety et al.[1986], Richter et al.[1989], and McKenna-Lawlor et al.[1989]. According to the model developed by these authors, increased neutral gas densities should be observed with a temporal periodicity of $T_{sc} = L/(V_a \pm V_{sc})$, where $L = V_n T$ is the space between the shells, V_{sc} is the spacecraft velocity relative to the nucleus and V_n is the expansion velocity of the shells. Having applied a least squares fitting procedure to the in situ data recorded at the two spacecraft, the expansion velocity of the neutral gas shells was estimated to be approximately 6.2 km/s.

In the present paper, additional information concerning energetic ion flux measurements recorded by Tunde-M [Kecskemety et al. 1989] as well as neutral gas density

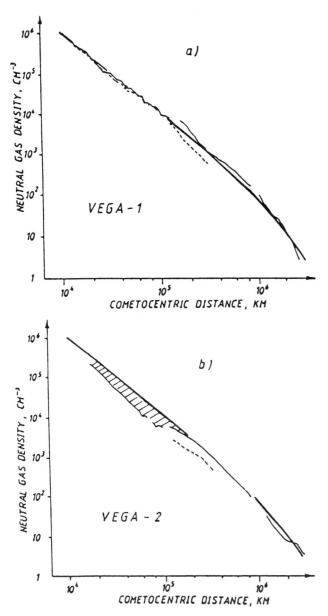

Figure 2. Cometocentric density profiles for neutral particles estimated from the data measured by the RFC instrument (a) onboard Vega-1 (b) onboard Vega-2 during their inbound (broken line) passes through the comet. The thick solid line shows the approximate dependence of $n_n(r) \sim r^2 \exp(-r/\lambda)$ with $\lambda = 2 \times 10^6$ km.

measurements made by the Ram Faraday Cup (RFC) instruments aboard Vega-1 and 2 [Remizov et al. 1986], are incorporated into the previous data set and the expansion velocity of the neutral gas shells recalculated on the basis of this additional information. The possible composition of the quasi-periodic ion fluxes measured at the two spacecraft is then discussed and arguments presented to suggest that the particles recorded were protons.

Data Analysis

The data already considered by Richter et al.[1989], and McKenna-Lawlor et al.[1989] contained (inbound) a data gap at a distance between 6 - 8 x 10^6 km from the nucleus and it was assumed by these authors that this interval contained two unobserved peaks. Recent investigations of the results from Tunde-M reveal, however, that although the instrument operated during this interval in a different mode, and with high background noise, it is nevertheless possible to identify only a single narrow energetic particles peak occurring at ~7.00 UT on March 5th, 1986, approximately in the middle of the previous gap [Kecskemety et al., 1989].

Neutral gas density measurements made by the RFC instruments aboard Vega-1 and Vega-2 are presented in Figure 2a and 2b respectively, following Remizov et al.[1986]. The smooth heavy line in each figure corresponds to the neutral gas density radial dependence calculated by Haser[1957]. This dependence is close to both the Vega-1 (March 6th) and Vega-2 (March 9th) data for cometocentric distances greater than 1-2 x 10^5 km. Closer to the comet however, the neutral gas densities measured by Vega-2 were lower, by approximately a factor of two, than was expected. The gap between the expected and measured densities is shown by cross-hatching in Figure 2b. If the deficit in neutral gas density recorded onboard Vega-2 is interpreted to result from a variation in the gas production rate at the rotating nucleus, then it appears that Vega-2 approached comet Halley during a period of low gas production and, by comparing Vega-1 and Vega-2 data, we can infer that the gas production rate of P/Halley is variable by a factor of at least 2.

Taking now all the data from Tunde-M and from EPONA, presented in Table 1 of Richter et al.[1989] and applying the same best fit procedure employed by Richter et al.[1989] and McKenna-Lawlor et al.[1989] to the data, with the exception that a single, not a double particle peak is assumed to be present within the data gap on March 5th a new representation of the assumed expanding gas shells, as a function of time, can be generated and represented by diagonal lines, see Figure 3. The inbound and outbound spacecraft trajectories of Vega-1 and Giotto are shown superimposed on this spatial structure, with heavy bars indicating the times and positions of the recorded flux enhancements. The position of the peak at 07.00 UT on March 5th is indicated by a triangle and it is seen to be located at a position predicted by the model to be characterized by a maximum in the particle flux. In general, the modified kinematic model is able to predict reasonably the positions of the measured maxima if the expansion velocity of the neutral gas shells is increased from the value of 6.2 km/s, estimated by previous authors, to 7.3 km/s.

The time and location at which Vega-2 recorded the neutral gas density deficit is marked by an asterisk in Figure 3. It corresponds to a minimum in neutral gas density production according to the present updated model. In the previous model, it would have been situated close to the position of a maximum.

Identity of the Recorded Ions

Tunde-M and EPONA utilize silicon surface barrier detectors which do not have the capability to distinguish mass. Full descriptions of these instruments are contained in Somogyi et al.[1986] and in McKenna-Lawlor et al.[1989]. In order to be detected by such an instrument, a freshly created cometary

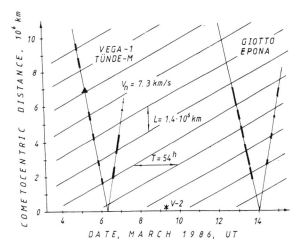

Figure 3. Plot of cometocentric distance vs time. The system of diagonal lines represents the positions of neutral gas shell centres based on a least square solution. The inbound/outbound trajectories of the Vega-1 and Giotto spacecraft appear superimposed on the overall plot. Heavy bars mark the positions of enhanced fluxes of energetic ions recorded by onboard instrumentation.

ion should undergo pitch angle scattering by MHD turbulence and then suitably diffuse in velocity space through the process of second order Fermi acceleration. Figure 4a shows the part of velocity space in the ecliptic plane, solid circle, that can be filled by protons after pitch angle scattering. In order to be registered by Tunde-M, particles should fall within the dashed circle (where the locations, within the observing cone, of the first four channels of the instrument are shown by cross hatching). For successful detection, protons should be accelerated to approximately 25 times their initial energy in the solar wind reference frame (SWRF). Figure 4b shows the corresponding situation for O^+ ions which, for detection, need only be accelerated by a factor of 4. However, under conditions of strong turbulence, protons can be accelerated approximately 16 times more efficiently than can oxygen ions [Ip and Axford, 1986].

The slow component of hydrogen atoms, coming from radical photodissociation of OH (which itself originates from the photodissociation of cometary water molecules), has a velocity of approximately 8 km/s [e.g. Mendis et al., 1985]. There is no intensive source of fast heavy atoms near P/Halley. As a result of CO/CO_2 photodissociation, a minor fraction of C, O fragments can attain velocities of 6 - 10 km/s. However, the production rate of such atoms is some three orders of magnitude less than the production rate of the slow component of hydrogen atoms [Ip and Axford, 1987]. Also, the electron dissociative recombination of CO^+, and CO_2^+ ions can provide a flux of heavy suprathermal atoms but their production rate is even less than that for the C, O fragments [McKenna-Lawlor et al., 1988].

The characteristic ionization time of both the slow hydrogen and oxygen component of cometary neutrals is of the order of 10^6 s. Both components originate from the OH radical within cometocentric distances of $r \sim 2 \times 10^5$ km. Therefore, for the range covered by the data of Tunde-M and

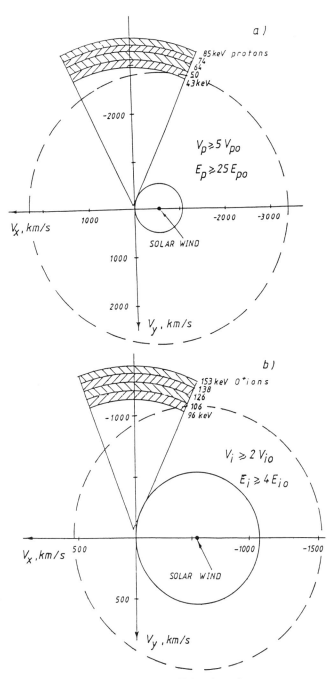

Figure 4. Diagram of TUNDE-M observing geometry in velocity space (ecliptic plane) when ions registered are (a) protons and (b) O^+ ions. The locations of the first four instrument channels on the observing cone are indicated by cross hatching.

EPONA, Haser's [1957] point source consideration is a reasonable assumption and the neutral gas number density can be approximated by the expression,

$$n_n(r) \sim (\exp(-r/v_0\tau_i)/v_0 + \exp(-r/v_H\tau_i)/v_H)/r^2$$

This summed functional dependence with $v_0 = 1$ km/s and $v_H = 7.3$ km/s, is represented by the solid curve in Figure 1. The dashed curve shows the corresponding dependence of the oxygen component alone. Both terms become equal at $r \sim 2 \times 10^6$ km, suggesting that, outside this limit, accelerated protons should mainly be detected while, inside this limit accelerated heavy ions may be observed by energetic particle instruments. This expectation is supported by McKenna-Lawlor et al., [1989] who demonstrated, though comparing data from the Implanted Ion Spectrometer on Giotto and EPONA data recorded at a location close to the inbound bow shock, that the fluxes recorded by EPONA were due to ions of the water group.

Flux variations

The sharp quasi-periodic flux variations of energetic particle fluxes recorded by Tunde-M and EPONA are difficult to explain in terms of expected variations in neutral gas density. The amplitude of different components of the neutral gas variation can be estimated by solving the following system of equations,

$$\frac{dN_{H_2O}}{dt} \approx Q_{H_2O}(t) - \frac{N_{H_2O}}{\tau_{H_2O}}$$

$$\frac{dN_{OH}}{dt} \approx \frac{N_{H_2O}}{\tau_{H_2O}} - \frac{N_{OH}}{\tau_{OH}}$$

where N_{H_2O} and N_{OH} respectively represent the total number of H_2O and OH molecules in cometary space; τ_{H_2O} and τ_{OH} are the photodissociation scale times for these molecules and Q_{H_2O} is the water molecule production rate. Assuming that Q_{H_2O} is subject to temporal variations by a factor of 2 (see Figure 3 and its discussion) and that the nucleus is deemed to be active during a time t (see Figure 5a), then the equations can be easily solved to yield the temporal variation of OH, as well as that of slow hydrogen production rates ($Q_{OH} \sim N_{H_2O}/\tau_{H_2O}$, $Q_H \sim N_{OH}/\tau_{OH}$). The solution concerned is shown in Figure 5b,c (for simplicity, it is taken to be $\tau \sim \tau_{OH} \sim T/2$; $\tau_{H_2O} \sim T/2$; $\tau_{OH} \sim T$). The variation in the density of slow hydrogen is thus expected to be of the order of a few percent ($\delta n/n \sim 4\%$). Similar estimates apply to oxygen atoms which also originate from OH photodissociation.

On the basis of the above, large variations are not expected in the source term $S(v,x)$ of the following equation, which describes the diffusion in velocity space of pick-up ions [see e.g. Gribov et al., 1987; Isenberg, 1987],

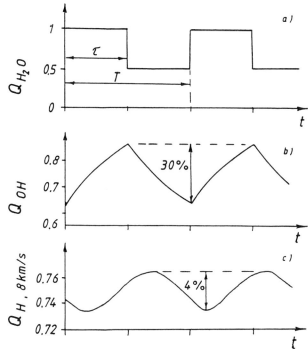

Figure 5. Simplified models of temporal variations in the water production rate based on (a) Vega-1,2 RFC data (a) and (b) estimated temporal variations in OH and (c) slow hydrogen component production rates.

$$\frac{1}{v^2}\frac{\partial}{\partial v}\left[v^2 D(v)\frac{\partial f}{\partial v}\right] + v_{sw}\frac{\partial f}{\partial x} - \frac{1}{3}\frac{\partial v_{sw}}{\partial x}v\frac{\partial f}{\partial v} +$$

$$+ S(v,x) = 0$$

where $f(v,x)$ is the distribution function; v is the velocity and $D(v)$ the velocity diffusion coefficient of pick-up ions and v_{sw} is the solar wind velocity. The analytical solution of this equation presented by Isenberg[1987] shows that the value of $f(v)$, at the local pick-up energy in the SWRF, is proportional to the source term. Thus the amplitude variations of $f(v)$ at the local pickup energy are proportional to the amplitude variations of S: $\delta f/f \sim \delta S/S \sim \delta n/n \sim 4\%$.

The same does not apply to variations of accelerated ions since, in this case, $f(v)$ decreases very rapidly as v increases and the slope of $f(v)$ is a function of the diffusion coefficient D. In consequence, small changes in D and thus in the $f(v)$ slope, could lead to a large change in $f(v)$ itself. In the case where the magnetic field power spectrum density can be approximated by a power law dependence on the wave number $|B_k|^2 \sim A(x)/k^2$, the diffusion coefficient is inversely proportional to the ion velocity $D(x) \sim A(x)/v$ and the asymptotic solution of the diffusion equation reduces to $f(v) \sim \exp(-v/v_0)$ [Isenberg, 1987], where $v_0 \sim A(x)$ is that parameter characterizing the slope of $f(v)$ and so,

$$\frac{\delta f}{f} \sim \frac{v}{v_0}\frac{\delta v}{v_0} \sim \frac{v}{v_0}\frac{\delta A}{A} \sim \frac{v}{v_0}\frac{\delta n}{n}$$

A value for the v_0 parameter can be estimated from experimental spectra. For example from the spectra measured by Tunde-M published in Figure 6 of Kecskemety et al.[1989], we can estimate that $v_0 = 30$ km/s for $v \sim$ 1100km/s if the registered ions are assumed to be oxygen. If the ions are assumed to be protons then $v_0 = 130$ km/s. For protons having an energy that is ≥ 25 times the initial pickup energy the asymptotic solution is reasonable and we can estimate that $\delta f/f \sim 25\ \delta n/n \sim 1$. For heavy ions, the asymptotic solution is not applicable as their energy is quite close to the initial pick-up energy (only ≥ 4 times more) and in this case, variations in their flux are mainly determined by source term variations.

Conclusion

The model of expanding shells of neutral gas about the comet Halley nucleus presented by Richter et al.[1989] and McKenna-Lawlor et al.[1989] has been modified to incorporate additional data from Tunde-M. The model reproduces, reasonably well, the positions of the quasi-periodic maxima in the particle records and shows a minimum in neutral gas density at a location where the neutral gas density measurements from Vega-1 and 2 recorded a deficit. The estimated velocity of gas in these shells is ~ 7.3 km/s, which is very close to that of the slow hydrogen component of cometary neutrals (~ 8 km/s).

Arguments in favour of the interpretation that the energetic particles registered in the quasi periodic flux enhancements at cometocentric distances of $2 - 10 \times 10^6$ km were protons include (a) at $\geq 2 \times 10^6$ km from the nucleus, hydrogen is the main component of cometary neutral gas while there is no significant production of fast component heavy neutrals at the comet (b) small variations in the slope of spectra of 25 times accelerated hydrogen ions can lead to large variations in the ion fluxes observed in the 100 keV range. It is, on the other hand difficult to produce large variations in fluxes of moderately-accelerated heavy ions.

Although the quasi-periodic structures revealed by the energetic particle records may represent intensity variations marking the flight of the spacecraft through expanding shells of neutral hydrogen atoms, more sophisticated quantitative models should be developed to fully elucidate the nature of the phenomena observed.

References

Gribov, E.E., Kecskemety, K., Sagdeev, R.Z., Shapiro, V.D., Shevchenko, V.L., Somogyi, A.J., Szego, K., Erdos, E.G., Gringauz, K.I., Keppler, E., Marsden, R.G., Remizov, A.P., Richter, A.K., Riedler, W., Schwingenschuh, K., Wenzel, K.P.. Stochastic Fermi acceleration of ions in the preshock region of comet P/Halley, *Astron. Astrophys.* <u>187</u>, 293-296, 1987.

Haser, L., Distribution d'intensite dans la tete d'une comete *Bull. Acad. Roy. Belgique, Casse des Sciences,* <u>43</u>, 740-743, 1957.

Ip, W.-H., Axford, W.I..The acceleration of particles in the vicinity of comets, *Planet. Space Sci.,* <u>34</u>, 1061-1065, 1986.

Ip, W.-H., Axford, W.I.. Stochastic acceleration of cometary ions: the effect of composition and source strength distributions, *Symposium on the Diversity and Similarity of Comets*, ESA SP-278, 139-144, 1987.

Isenberg, P.A.. Energy diffusion of pickup ions upstream of comets. *J. Geophys. Res.*, <u>92</u>, 8795-8799, 1987.

Kecskemety, K., Cravens, T.E., Afonin, V.V., Erdos, G., Eroshenko, E.G., Gan, L., Gombosi, T.I, Gringauz,K.I., Keppler, E., Klimenko, I.N., Marsden, R.,Nagy, A.F., Remizov, A.P., Richter, A.K., Riedler,W., Schwingenschuh, K., Somogyi, A.J., Szego, K., Tatrallyay, M., Varga, A., Verigin, M.I., Wenzel,K.P. Energetic pick-up ions outside the comet Halley bow shock. *Exploration of Halley's Comet,* ESA SP-250, <u>1</u>, 109-114, 1986.

Kecskemety, K., Cravens, T.E., Afonin, V.V., Erdos, G., Eroshenko, E.G., Gan, L., Gombosi, T.I., Gringauz, K.I., Keppler, E., Klimenko, I.N., Marsden, R., Nagy, A.F., Remizov, A.P. Richter, A.K., Riedler, W., Schwingenschuh, K., Somogyi, A.J., Szego, K., Tatrallyay, M., Varga, A., Verigin, M.I., Wenzel, K.P.. Pick up ions in the unshocked solar wind at comet Halley. *J. Geophys. Res.*, <u>94</u>, 185-196, 1989.

McKenna-Lawlor, S., Daly, P., Kirsh, E., Wilken, B., O'Sullivan, D., Thomson, A., Kecskemety, K., Somogyi, A., Coates, A.. In-situ energetic particle observations at comet Halley recorded by instrumentation aboard the Giotto and Vega 1 missions. *Ann. Geophys,* <u>7</u>, 121-128, 1989.

Mendis, D.A., Houpis, H.L.F., Marconi, M.L.. The physics of comets, *Fund. Cosmic Phys.,* <u>10</u> , 1-380, 1985.

Remizov, A.P., Verigin, M.I., Gringauz, K.I., Apathy, I., Szemerey, I., Gombosi, I., Richter, A.K.. Measurements of neutral particle density in the vicinity of comet Halley by Plasmag-1 on board Vega-1 and Vega-2, *Exploration of Halley's Comet*, ESA SP-250, <u>1</u>, 387-390, 1986.

Richter, A.K., Daly, P.W., Verigin, M.L., Gringauz, K.I. Erdos, G., Kecskemety, K., Somogyi, A.J., Szego, K., Varga, A., McKenna-Lawlor, S.. Quasi-periodic variations of cometary ion fluxes at large distances from comet Halley. *Ann. Geophys.*, <u>7</u>, 115-120, 1989.

Somogyi, A.J., Gringauz, K.I., Szego, K., Szabo, L., Kosma, D., Remizov, A.P., Ero, J., Klimenko, I.N., Szucs, I.-T., Verigin, M.I., Windberg, J., Cravens, T.E., Dyachkov, A., Eroda, G., Farago, M., Gombosi, T.I., Kecskemety, K., Keppler, E., Kovacs, T., Logachev, Yu.I., Lohonyai., L., Marsden, R., Redl, R., Richter, A.K., Stolpovskii, V.G., Szabo, L., Szentpetery, I., Szepesvary, A., Tatrallyay, M., Varga, A., Vladimirova, G.A.. First observations of energetic particles near comet Halley. *Nature,* <u>321</u>, 285-287, 1986.

Comment on the paper "Neutral Hydrogen Shell Structure near Comet P/Halley Deduced from Vega-1 and Giotto Energetic Particle Data" by M. I. Verigin et al.

M. NEUGEBAUER

Jet Propulsion Laboratory, California Institute of Technology, Pasadena, CA, USA

A. J. COATES

Mullard Space Science Laboratory, University College London, Holmbury St. Mary, Dorking, Surrey, UK

In the previous paper, M. I. Verigin et al. [1991] deduced that the energetic particles detected at large distances from comet Halley by Vega-1 and by Giotto are protons. Their conclusion is based on the spacing of quasiperiodic peaks in particle flux which they relate to enhancements of gas density moving radially outward from the comet at 7.3 km/s. In this Comment, we summarize several arguments against this model and the conclusions drawn from it.

1) The variations in energetic particle flux are, at best, only quasiperiodic. In Figure 3 of Verigin et al., the second enhancement shown for Vega-1 and the second and the last two enhancements shown for Giotto fall at minima of the hypothesized periodic variation.

2) The calculated shell speed depends on the rotation period of the nucleus, and there is still a great deal of controversy about whether or not there was a 54-hour periodicity of gas production near the time of the Vega and Giotto flybys.

3) The factor of two decrease in neutral gas density between the Vega-1 and Vega-2 flybys appears to fit nicely into Verigin et al.'s model, because the Vega-2 encounter corresponded to a minimum in the postulated periodicity. We note, however, that the neutral gas observed inside 10^5 km was moving away from the comet at ~1 km/s, so one would not expect its fluctuations to be in phase with the fluctuations of 8 km/s H atoms detected far from the comet.

4) Verigin et al.'s theoretical expectation that "under conditions of strong turbulence, protons can be accelerated approximately 16 times more efficiently than can oxygen ions" overlooks the greater wave energy available at the O^+ cyclotron frequency than at the proton cyclotron frequency. The diffusion coefficient depends on the ratio of resonant wave energy to ion mass. For the observed wave spectral coefficient of -2 [*Glassmeier et al.*, 1989], the diffusion rates should be the same for the two ion species.

5) Verigin et al.'s argument that "there is no significant production of fast component heavy neutrals at the comet" is contradicted by the observation of water-group pickup ions at $>4 \times 10^6$ km from the comet by the JPA/IIS detector on Giotto [*Coates et al.*, 1989].

6) Verigin et al.'s model requires that pickup protons be accelerated much more efficiently than heavier ions. Neugebauer et al. [1990], however, find that the thicknesses (in velocity space) of the shells of picked-up protons and water-group ions were approximately equal.

7) Figure 1 demonstrates that the variations in the densities of pickup protons and water-group ions are correlated with each other (at the 3.6-σ level). The local maxima and minima in both the proton and water-group ion densities are not only in phase with each other, but also appear to correlate well with the peaks in the energetic particle flux observed by Giotto, although a numerical correlation coefficient has not been calculated.

Although none of the first six arguments above would, by themselves, be fatal to the expanding shell model and the hypothesis that the energetic particles were protons, we believe the last argument is fairly convincing. The expanding shell model is inconsistent with the correlation of densities of species with different outflow speeds. Furthermore, the pickup ions, rather than the neutral species, are the immediate parents of energetic ions. The correlation between pickup proton and water-group ion densities can be understood on the basis of Neugebauer et al.'s [1989] suggestion that the small-scale variations in the density of pickup ions is determined by changes in the angle between the direction to the cometary nucleus and the direction of the guiding-center motion of the ions.

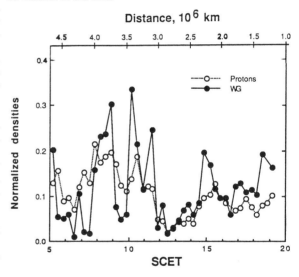

Fig. 1. Densities of pickup protons and water-group ions (WG) observed by Giotto plotted versus time (bottom scale) and distance from the comet (top scale). The densities (units of cm^{-3}) have been normalized by multiplying the proton density by $r^{2.0}$ and the water-group density by $r^{3.7}$, where r is distance from the comet in units of 10^6 km. From Neugebauer et al. [1990].

Acknowledgements. Part of the research described in this paper was carried out by the Jet Propulsion Laboratory of the California Institute of Technology under contract with NASA.

REFERENCES

Coates, A. J., A. D. Johnstone, B. Wilken, K. Jockers, and K.-H. Glassmeier, Velocity space diffusion of pickup ions from the water group at comet Halley, *J. Geophys. Res., 94,* 9983, 1989.

Glassmeier, K.-H., A. J. Coates, M. H. Acuna, M. L. Goldstein, A. D. Johnstone, F. M. Neubauer, and H. Reme, Spectral characteristics of low-frequency plasma turbulence upstream of comet P/Halley, *J. Geophys. Res., 94,* 37, 1989.

Neugebauer, M., B. E. Goldstein, H. Balsiger, F. M. Neubauer, R. Schwenn, and E. G.Shelley, The density of cometary protons upstream of comet Halley's bow shock, *J. Geophys. Res., 94,* 1261, 1989.

Neugebauer, M., A. J. Coates, and F. M. Neubauer, Comparison of picked up protons and water-group ions upstream of comet Halley's bow shock, *J. Geophys. Res.,* accepted, 1990.

Verigin, M. I., S. McKenna-Lawlor, A. K. Richter, K. Szego, and I. S. Veselovsky, Neutral hydrogen shell structure near comet P/Halley deduced from Vega-1 and Giotto energetic particle data, *This volume,* 1991.

Reply to comments by Neugebauer and Coates on
"NEUTRAL HYDROGEN SHELL STRUCTURE NEAR COMET P/HALLEY
DEDUCED FROM VEGA-1 AND GIOTTO ENERGETIC PARTICLE DATA"

M.I. Verigin[1], S. McKenna-Lawlor[2], A.K. Richter[3], K. Szego[4], I.S. Veselovsky[5].

(1) Space Research Institute, Moscow, USSR
(2) Space Technology Ireland, St Patrick's College, Maynooth, Ireland
3) Max-Planck Institut fur Aeronomie, Lindau, FRG
(4) Central Research Institute for Physics, Budapest, Hungary
(5) Nuclear Physics Research Institute of Moscow University, Moscow, USSR

Both Verigin et al [1991](Paper 1) and Neugebauer and Coates [1991](Comments) contain updated information about the problem of interpreting a quasiperiodic variation in energetic particle fluxes recorded on board the Vega-1 and Giotto spacecraft (see references in Paper 1). The present text is a response to Comments.

The reality of the flux variations recorded is accepted (Argument 1 of Comments). However, a complete explanation of the effect meets with severe difficulties and our present understanding of it is far from being complete. At the present time there is no other explanation of the quasiperiodicity other then that based on the assumption that it is a consequence of neutral gas production by a rotating cometary nucleus (Argument 2).

It is not important that the velocity of neutral gas inside $r < 10^5$ km (velocity $\sim$ 1 km/s) is different from that of 'slow' hydrogen atoms (velocity $\sim$ 8 km/s) at a distance of a few million km (Argument 3). These atoms are produced within a region extending outwards only to a few hundred thousand km which, from the point of view of the interpretation of observations made at a distance of a few million km, can be considered to constitute a point of source. Thus, the Vega-2/Plasmag-1 neutral gas records presented in Fig. 2 of Paper 1 can be interpreted to constitute observations of a temporally "empty" reservoir of water group molecules acting like a point source of variable intensity for slow hydrogen atoms.

The remark that, under conditions of strong turbulence, protons are more effectively accelerated than oxygen ions is a conclusion of Ip and Axford, [1986] and not a theoretical expectation derived in Paper 1(Argument 4).

With regard to the production of heavy fast neutrals at the comet (Argument 5), it is not possible, using data from the JPA/IIS detector on Giotto, to conclude that the recorded pick-up ions of the water group originated from fast rather than from slow heavy neutrals. The rapid $\sim r^{-3.7}$ decrease in density of pick-up heavy ions provides evidence that they originate from an exponentially decreasing (due to photoionization), population of slow neutral oxygen atoms, rather than from fast neutrals because, in the latter case, the radial dependence would be close to $-r^{-2}$). Additionally, it is easy to show by the reverse normalization of the proton and water group ion densities presented in Fig. 1 of Comments, that the pick-up proton density in fact exceeded that of the water group ions at $r > 1.5 \cdot 10^6$ km. This observation can be considered to provide strong support for those other arguments in Paper 1 that suggest that the energetic particles detected at large distances from comet Halley aboard Vega-1 and aboard Giotto were mainly protons.

The relative efficiency of acceleration of protons and water group ions is not the essential feature of our model (Argument 6). What is essential is that the protons observed by Tunde-M must have been accelerated by more than 25 times their pick-up energy. Hence they can be highly modulated by rather small changes in the inclination of their spectra whereas, in the case of heavy ions, their energy is only approximately 4 times the original one and it is then very difficult to explain strong variations is their fluxes.

It is not clear that the seventh argument in Comment is fatal to the expanding shells model since the particles recorded on the one hand by the JPA/IIS instrument and on the other by the Tunde-M and EPONA particle telescopes were of significantly different energies. Certainly, the idea of Neugebauer et al [1989] concerning the decisive role of the angle between the direction of the guiding centre of motion and the direction of cometary nucleus, which is reasonably valid for locally picked-up, but as yet not accelerated, ions cannot, of itself, explain the quasiperiodicity. Whether our interpretation can survive will then be, rather, determined downstream on the basis of more sophisticated quantitative models, developed to yet further elucidate the nature of the phenomena observed.

References

Ip, W.-H. and Axford, W.I.. The acceleration of particles in the vicinity of Comets, *Planet and Space Sci.* 34, 1061, 1986

Neugebauer, M., Goldstein, B.E., Balsiger, H., Neubauer, F.M., Schwenn, R. and Shelley, E.G.. The density of cometary protons upstream of comet Halley's bow shock. *J. Geophys. Res.* 94, 1261, 1989.

Neugebauer, M. and Coates, A.J.. Comment on the paper "Neutral hydrogen shell structure near comet P/Halley deduced from Vega-1 and Giotto energetic particle data by Verigin et al". *This volume*, 1991.

Verigin, M.I., McKenna-Lawlor, S., Richter, A.K., Szego, K. and Veselovsky, I.S.. Neutral hydrogen shell structure near comet P/Halley deduced from Vega-1 and Giotto energetic particle data. *This volume*, 1991.

ENERGETIC WATER GROUP ION FLUXES ($E_{H_2O} > 60$ KEV) IN A QUASIPERPENDICULAR AND A QUASIPARALLEL SHOCK FRONT AS OBSERVED DURING THE GIOTTO-HALLEY ENCOUNTER.

E. Kirsch[1], S. McKenna-Lawlor[2], P.W. Daly[1], F.M. Neubauer[3], A. Coates[4], A. Thompson[5], D. O'Sullivan[5], K.-P. Wenzel[6]

[1] Max-Planck-Institut für Aeronomie, Katlenburg-Lindau, FRG

[2] Space Technology, St. Patrick's College, Maynooth, Ireland

[3] Institut für Geophysik, Universität Köln, FRG

[4] Mullard Space Science Lab., University College London, Holmbury St. Mary, Dorking, Surrey, UK

[5] Institute for Advanced Studies, Dublin, Ireland

[6] Space Science Department of ESA/ESTEC, Noordwijk, NL

Abstract. Energetic water group ions ($E_{H_2O} = 1\text{-}86$ keV and $E_{H_2O} = 60$ to >270 keV) recorded by EPA/EPONA and the JPA experiment aboard Giotto, as well as corresponding magnetic field measurements obtained, especially, at the quasiperpendicular inbound and quasiparallel outbound bow shocks of Comet Halley, were used to study particle propagation and acceleration processes in the bowshock environment. It was found that energetic particles were accelerated in association with a strong magnetosonic wave field in the foreshock, in the inbound bowshock, within the near cometosheath, as well as immediately inside and outside the broad outbound bowshock. A relatively hard spectrum was observed in the inbound foreshock ($\gamma = 3.3$). Inside the bowshock itself, the spectrum was somewhat softer ($\gamma = 4.1$), due to an increase in the fluxes in the lowest energy channels. Further acceleration of the particles took place inside the cometosheath ($\gamma = 3.5$). Similar, but somewhat steeper spectra, were observed on the outbound side. First and second order Fermi processes, as well as the transit time damping effect were most likely to have been responsible for particle acceleration at and within the bowshocks of comet Halley. The measurements also showed that more particles escaped from the outbound than from the inbound bowshock (presumably along the magnetic field vector), and that these latter particles produced a somewhat steeper energy spectrum near the bowshock.

1. Introduction

The Energetic Particle Analyser / Energetic Particle ONset Admonitor EPA/EPONA onboard Giotto obtained energetic particle data ($E_{H_2O} > 60$ keV up to about 300 keV total energy) during the crossing of the inbound and outbound bowshocks of comet Halley. For a study of the propagation and acceleration process of such particle measurements are available in four ion energy channels with limited information on the azimuthal propagation direction.

The inbound bowshock has been classified by Neubauer et al. (1986) as a quasiperpendicular shock and the outbound bowshock as a broad quasiparallel shock. Furthermore, an intense magnetosonic wave field

Cometary Plasma Processes
Geophysical Monograph 61
©1991 American Geophysical Union

can be recognized in the magnetic field measurements near the inbound and outbound bowshock with enhanced RMS-noise (Neubauer et al., 1987).

Energetic particle as well as magnetic field (MAG experiment), solar wind and H_2O^+ ion measurements ($E_{H_2O} = 1\text{-}86$ keV, Johnstone Plasma Analyser) were also available for the present study. The inbound foreshock and bowshock have been discussed earlier e.g. by Réme et al. (1986), d'Uston et al. (1987), Fusilier et al. (1987), Anderson et al. (1987), Coates et al. (1987), Neugebauer et al. (1989). Neubauer et al. (1990) investigated the outbound bowshock and classified it as a "draping shock". A further detailed study of comet Halley's bowshock will be published by Coates et al. (1990).

McKenna-Lawlor et al. (1989) concluded from composite particle spectra observed at and inside the inbound bowshock, that the second order Fermi process was mainly responsible for the particle acceleration. The intensity spikes observed by EPA/EPONA inside the inbound bowshock and on the outbound side (14/15 March 1986) have been studied by Kirsch et al. (1989, 1990). Indications of the field line reconnection process and the shock drift mechanism were found also.

2. Experiment description

A detailed experiment description of the EPA/EPONA instrument has been given by McKenna-Lawlor et al. (1987). The four relevant ion energy channels are $E_{1H_2O} = 60\text{-}106$ keV, $E_{2H_2O} = 97\text{-}145$ keV, $E_{3H_2O} = 144\text{-}270$ keV, $E_{4H_2O} = 0.26\text{-}3.5$ MeV. The geometric factor for all energy channels was $8 \cdot 10^{-2}$ cm^2 ster. The count rates presented in the following figures are directly proportional to the particle fluxes. Table 1 shows the look directions of the used sectors in a Comet Solar Ecliptic Coordinate System. EPA/EPONA consists of 3 Telescopes: T1 is directed 45° to the spin axis, and looks backward to the flight direction; T2 and T3 are both mounted at 135° to the spin axis and look in flight direction.

3. Observations

Figs. 1a, b represent the magnetic field measurements of the inbound and outbound bowshock crossings. The bowshock was tra-

GIOTTO – EPONA   1986 Mar 13

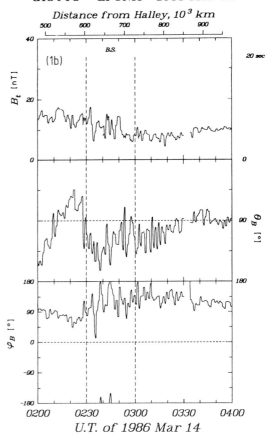

Fig. 1. Magnetic field measurements of the in- and outbound bow-shock. B_t = field strength in nT, Θ_B and φ_B = inclination and azimuthal angle in °

Table 1
Look directions of the sectors and used telescopes

Towards Sun:	sector S2, T3	$\phi = 311°$	$\theta = 29.1°$ southward
Towards East:	sector S3, T1	$\phi = 61.8°$	$\theta = 17.9°$ southward
Towards East:	sector S1, T1	$\phi = 125.6°$	$\theta = 52.4°$ southward
Towards Anti-Sun:	sector S7, T1	$\phi = 150.8°$	$\theta = 03.8°$ northward
Towards West:	sector S8, T3	$\phi = 248.3°$	$\theta = 12.8°$ southward

ϕ = azimuthal angle
θ = inclination angle

versed during the interval ~ 19:24 to 19:38 UT (Fig. 1a) which is characterised by a slight increase of the field magnitude and strong directional changes of the inclination (Θ_B) and azimuthal angle (ϕ_B) of the field vector. Generally ϕ_B showed a continuous directional change from ~ 160° to ~ 360° (18:00 to 19:24 UT) and Θ_B a directional change at 18:20 UT. In addition, a quasiperiodic wave structure of ~ 3 min periodicity can be recognized. On the outbound side (Fig. 1b) a broad bowshock and also a strong magnetosonic wave activity of ~ 4 min quasiperiodicity was observed from 02:30 to 03:40 UT.

Energetic particle measurements ($E_{2H_2O} = 97\text{-}145$ keV) from the inbound side are depicted in Fig. 2a (sectors 1, 3, 7, 2, 8) together with the solar wind velocity and the density of low energy water ions (E = 1-86 keV), see Coates et al., 1990. The sectors S_1 and S_3 detected a separated foreshock with a flux increase from ~ 18:20 UT

onward (minimum at ~ 19:10 UT) whereas S_2 showed a more continuous increase. From the directions Anti-Sun (S_7) and west (S_8) only small particle fluxes were detectable. The solar wind velocity started to decrease from 18:20 UT onward where the foreshock became observable (Fig. 2a) whereas the H_2O^+ ion density increased slowly (18:20 to 19:40 UT).

For comparison the corresponding measurements at the outbound bowshock are shown in Fig. 2b. It must be noted that the sectors S_1 and S_3 are now looking backward to the comet, S_7 still Anti-Sunward, S_2 towards the SUN and S_8 in the direction west (away from the comet). The bowshock extended from 02:30 to 03:00 UT and the intense magnetosonic wave field from 02:30 to 03:40 UT.

The count rates of energetic water ions ($E_2 = 97\text{-}145$ keV) showed a double peak structure. The maxima can be recognized at ~ 02:45 UT and ~ 03:20 UT. The solar wind velocity V_{sw} increased continuously between 02:00 UT and 04:00 UT from ~ 175 km/s to ~ 300 km/s, see also Coates et al., 1990. Superimposed small velocity increases appeared at 02:45 UT and 03:20 UT which seem to be related to the maxima of the energetic ions. The density of the low energy H_2O^+ ions (bottom panel) decreased from ~ 0.5/cm³ to ~ 0.15/cm³ (02:20 to 03:40 UT). At 02:45 UT a small density minimum appeared which may be related to the first maximum in the fluxes of energetic ions. It is assumed here that low energy H_2O^+ ions have been accelerated to > 90 keV. During the second flux maximum (03:20 UT) the H_2O^+ density and the solar wind velocity showed no significant variations.

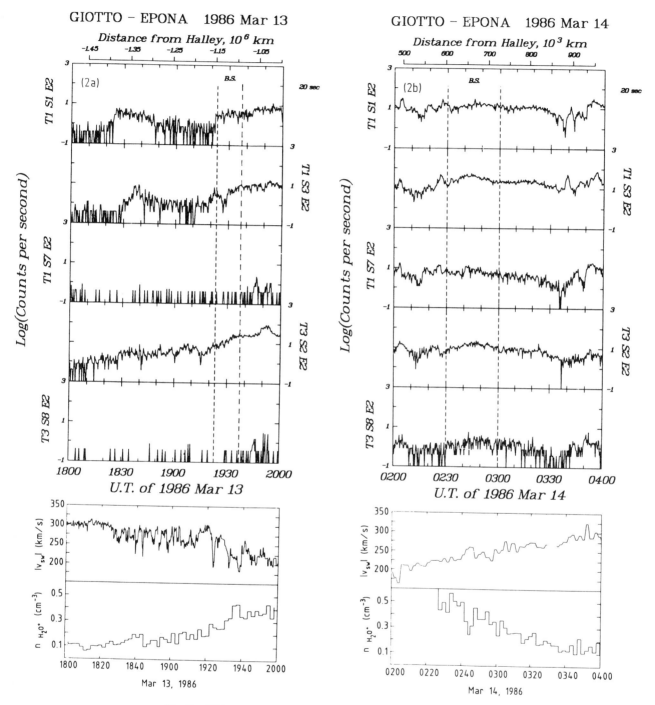

Fig. 2. EPA-sector measurements of the in- and outbound side together with the solar wind velocity and the density of low energy water ions as measured by the JPA instrument (lower 2 panels). The foreshock (Fig. 2a) appeared only in sectors S_1, S_3, S_2, from 18:20–19:10 UT. On the outbound side (Fig. 2b) a double peak structure can be recognized in sectors S_1, S_3, S_2 from 02:30–03:30 UT.

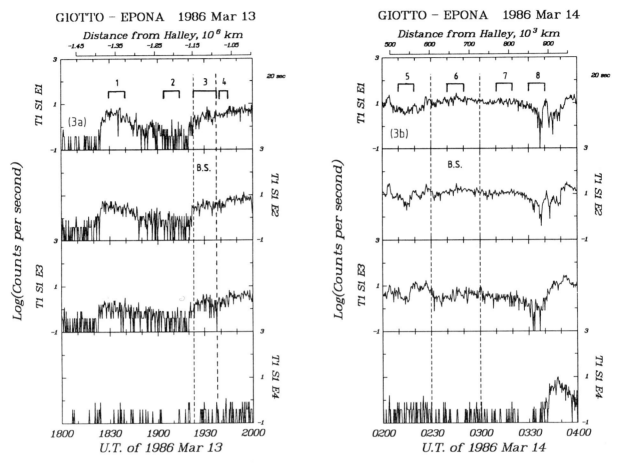

Fig. 3.   Four energy channels of sector 1 from the in- and outbound side. The intervals 1-8, indicated by brackets, are specified in Table 2.

The Figs. 3a, b show all 4 available energy channels for sector 1. They indicate that the particle energies in the foreshock, the inbound bowshock as well as those inside the outbound bowshock (and until the end of the wave field, ∼ 03:40 UT) lie in the range 60-270 keV. Only small count rates were recorded in channel 4 ($E_{H_2O} > 270$ keV). The spike in channels 1-4 observed at 03:40 to 04:00 UT (Fig. 3b) has been discussed earlier by Kirsch et al. (1990).

The magnetic field measurements were used to calculate the pitch angles for the EPA/EPONA sector measurements. A full pitch angle distribution is however not available since the spin axis of Giotto was oriented almost parallel and not perpendicular to the ecliptic plane. Thus the instrument had not a full coverage of the 360° azimuthal angle. The Figs. 4a, b show the pitch angles (angles between the look direction of the sectors and the magnetic field vector) as a function of time. It can be seen that the magnetosonic wave activity of 3-4 min quasiperiodicity (compare Figs. 1a, b) generated fluctuations of the pitch angles from 18:30 to 19:40 UT (foreshock and inbound bowshock) and 02:30 to 03:40 UT (outbound bowshock until the end of the wave activity).

In order to characterise the particle fluxes observed by EPA/EPONA near the inbound and outbound bowshocks the energy spectra (phase space density as function of the particle energy) have been calculated (see Figs. 5a,b) and transformed into the solar wind frame according to the formula (Ipavich, 1974):

$$\frac{dj}{dE} = K \cdot E^{-\gamma} \left[1 - 2\left(\frac{W}{V}\right)\cos\beta + \left(\frac{W}{V}\right)^2 \right]^{-(\gamma+1)} \qquad (1)$$

where $\frac{dj}{dE}$ is the differential flux in a moving frame and

  K = const.
  W = solar wind velocity
  V = particle velocity
  β = angle between solar wind and proton propagation
  γ = spectral index

The dotted lines in Figs. 5a,b are the least square fits of the power law spectra. In Table 2 are compiled the exponents of the calculated power law spectra for the various time intervals which are also indicated in Figs. 3a, b (intervals 1-8). The first spectrum (B) in Figs. 5a, b shows the cosmic ray background spectrum (assuming also mass 16), determined one day before the encounter of Giotto with comet Halley (March 12, 12:00 to 18:00 UT). The background spectrum has to be substracted from all other spectra shown in Figs. 5a, b. The spectral exponents compiled in Table 2 can, however, be used for a relative comparison of the 8 time intervals, since the background is constant.

It can be stated that the outbound spectra are generally somewhat softer than the inbound spectra. Particle acceleration took place in the foreshock and in the cometosheath (intervals 1, 4). The intervals

EPA PITCH ANGLES 1986 Mar 13

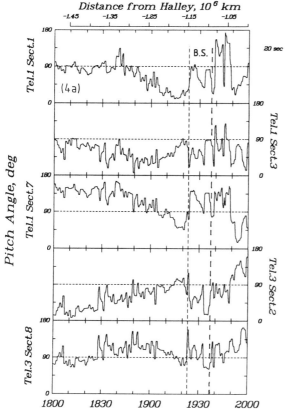

EPA PITCH ANGLES 1986 Mar 14

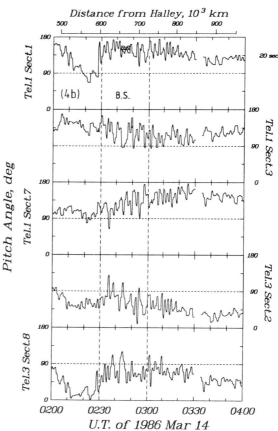

Fig. 4. Pitch angles as function of time. The quasiperiodic wave
structure of the magnetic field vector can be recognized.
0° = particle propagation along the field vector
180° = particle propagation contrary to the field vector.

Table 2

Power law spectra

Inbound
1. Foreshock $\gamma = 3.34$
2. Minimum fore-/bowshock $\gamma = 4.13$
3. Bowshock $\gamma = 4.10$
4. Cometosheath $\gamma = 3.49$

Outbound
5. Cometosheath $\gamma = 4.82$
6. Bowshock $\gamma = 5.87$
7. Wave-field $\gamma = 5.34$
8. End of the wave-field $\gamma = 3.71$

2 and 3 have comparable spectral exponents, but inside the bowshock (interval 3) low energy particles in channels 1 and 2 appeared in addition and caused a steepening of the spectrum. Thus the acceleration process here first increased the low energy particle count rates and later the higher energy channels (intervals 3, 4). On the outbound side the situation is similar. A relatively hard spectrum was found in the cometosheath (interval 5) and softer spectra in the bowshock and the wave field (intervals 6 and 7) because more low energy particles appeared in these intervals. The hardening of the spectrum in inter-

val 8, outside of the wave field, was caused by a decrease of the low energy particle count rate.

4. Discussion

The energetic particle and magnetic field measurements presented in Figs. 1-4 revealed that the particle fluxes appeared at the inbound and outbound bowshock in association with a strong magnetosonic wave field of about 1 hour duration (18:20 to 19:30 UT and 02:30 to 03:40 UT), see also Glassmeier et al. (1987). On the inbound side a separated foreshock could be observed in sectors 1 and 3 (looking backward to the flight direction of Giotto) whereas sector 2 measured a continuous count rate increase until the bowshock was reached. On the outbound side a double peak structure appeared in the particle count rates. The second peak is not separated from the first one detected inside the broad bowshock. The quasiparallel shock of the outbound side obviously allowed the direct escape of energetic particles accelerated inside the cometosheath. The highest count rates appeared in sectors S_1 and S_3 which were directed backward to the comet at that time (Fig. 2b). The pitch angles for S_1 and S_3 varied between 90° and 160° (Fig. 4b).

As a possible acceleration process for energetic particles we shall first consider the pick-up process (see e.g. Ip and Axford, 1989). Fig. 6 shows the sector count rates S_1, S_2 together with $\sin^2\alpha$ (α = angle

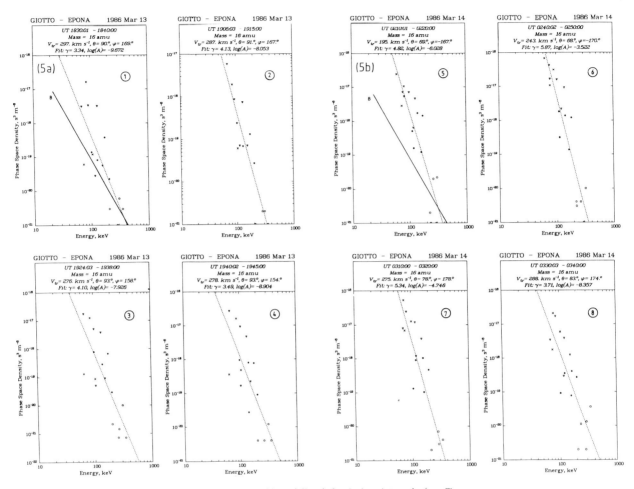

Fig. 5. Energy spectra (dotted lines) for 8 time intervals (see Table 2 and Figs. 3a,b) in the solar wind frame for mass = 16 amu.

$V_t r$ = solar wind velocity in km/s
θ = inclination angle of the solar wind
ϕ = azimuthal angle of the solar wind
γ = exponent of the power law spectrum
B = background spectrum obtained on 12 March 1986 before encounter also for mass = 16 amu

between the solar wind and magnetic field direction, solar wind velocity see Fig. 2a). It can be seen that the foreshock observed with sector S$_1$ was anticorrelated with sin$^2\alpha$. However the small maximum in sin$^2\alpha$ (19:00 to 19:15 UT) and in the count rate of sector S$_2$ at the same time could be caused by the pick-up process followed by additional acceleration up to few hundred keV. As a further acceleration process (e.g. Armstrong et al., 1985) can be invoked the shock drift mechanism which requires a quasiperpendicular shock configuration.

The acceleration of particles takes place according to this mechanism in the $\vec{E} = -\vec{v} \times \vec{B}$ field which is directed parallel to the shock front ($\vec{v}$ = solar wind velocity, $\vec{B}$ = magnetic field vector). In a paper by Fusilier et al. (1987) it has been shown that ϑ_{BN} (angle between the magnetic field vector and the normal to the shockfront) was $\approx 90°$ from 17:40 to 18:25 UT and $\sim 45°$ from 18:25 to 19:15 UT. Thus a quasiperpendicular shock appeared without energetic particles and a more quasiparallel shock when the foreshock has been observed. It is concluded that the shock drift acceleration mechanism cannot explain

the existence of the foreshock. Therefore the first and second order Fermi processes shall be considered (Amata and Formisano, 1985; Ip and Axford, 1986, 1989). Sagdeev et al. (1987) studied the solar wind interaction with comets as a model for the cosmic ray acceleration by shocks. They concluded that the second order Fermi process is the principal acceleration mechanism near comets. The energization of pickup ions by quasilinear energy diffusion in the turbulence upstream of a cometary bow wave was also investigated by Isenberg (1987). He found that the high energy spectra from comet Halley are in reasonable agreement with his model calculations. An interesting combination of the first and second order Fermi processes particularly for the acceleration of heavy ions in the cometary foreshock region, has been studied by Gombosi et al. (1989). According to their model the second order Fermi mechanism accelerates particles up to moderate energies and is then followed by the more efficient diffusive-compressive shock acceleration. They predict a power law spectrum with a spectral index of 5-6 (above 100 keV). It is assumed that the

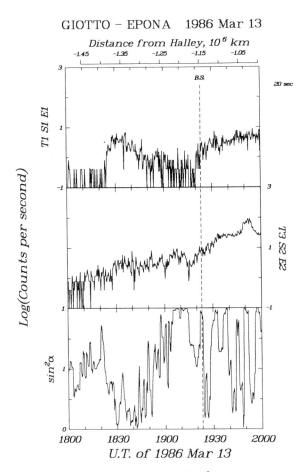

GIOTTO – EPONA   1986 Mar 13

Fig. 6.   Measurements of sectors 1, 2 and $\sin^2\alpha$ (α = angle between solar wind and magnetic field direction).

entire foreshock region where the solar wind speed slows down (compare Figs. 2a) serves as a region of diffusive-compressive acceleration of ions, i.e. a strong shock is realized and energizes the particles. The diffusive compressive acceleration in the foreshock region is however limited by the solar wind convection time.

The foreshock observed by EPA/EPONA in sectors 1, 2, 3 from 18:20 to 19:00 UT (spectral exponent $\gamma = 3.34$) could result from such a mechanism. The energy spectra observed directly in the bowshock (18:24 to 18:38 UT and 02:30 to 03:00 UT) are somewhat steeper (see Table 2) than further inside the cometosheath. This is mainly caused by an enhancement of the count rates in channels 1 and 2 (see Figs. 5a,b). It seems also possible that the accelerated ions can propagate preferentially inside the bowshock from the upstream region toward the Giotto trajectory. Further inside the cometosheath the spectrum becomes harder again (see also McKenna-Lawlor et al., 1989). Glassmeier et al. (1987) found that the interaction region between solar wind and comet Halley was characterized by large amplitude and low frequency magnetic field fluctuations in the upstream region as well as in the cometosheath, both on the in- and outbound pass of Giotto. The compressional magnetic field fluctuation (visible also in Figs. 1a,b, 4a,b) of 3-4 min quasiperiodicity should contribute to the particle acceleration via the transit time damping mechanism (Fisk, 1976) as Glassmeier et al. (1987) and Ip and Axford (1987) suggested. The particles are thereby scattered and accelerated by randomly moving magnetic gradients, which is a Fermi process with

magnetic scattering centers. The particle and magnetic field observations of EPA/EPONA and the Giotto magnetometer on the inbound (foreshock and bowshock) and outbound side (bowshock and magnetosonic wave field) are at least in qualitative agreement with the first and second order Fermi mechanism as discussed by Ip and Axford (1986), Sagdeev et al. (1987), Gombosi et al. (1989), Fisk (1976). The acceleration process starts here with the pick-up ions and increases their energy. From the observed in- and outbound anisotropies it can be concluded that the quasiparallel "draping shock" (Neubauer et al., 1990) allows an easy escape of particles on the outbound side. The discussed transit time damping mechanism lowers the pitch angles (Fisk, 1976) of the energetic particles generally and may cause the escape of particles along the magnetic field vector on the outbound side. These particles appear there superimposed on the locally ionised and accelerated cometary particles in the interplanetary space.

A similar study of the particle energy spectra measured by the International Cometary Explorer (ICE) at and inside the bowshocks of comet Giacobini-Zinner has been published by Richardson et al. (1987, their figures 10 and 11). They found also a harder spectrum inside the bowshock than outside. The first and second order Fermi effect were discussed as particle acceleration processes, but final conclusions could not be drawn from the derived energy spectra.

5. Summary and Conclusions

1. During the inbound pass a foreshock ($E_{H_2O} = 60$-260 keV) has been detected by EPA/EPONA from 18:20 to 19:00 UT and a $\sim 6 \cdot 10^3$ km thick bowshock.
2. Outbound a double peak structure in the particle count rate and a $1.3 \cdot 10^5$ km thick bowshock appeared in association with a quasiparallel shock.
3. The energy spectra calculated from the EPA/EPONA observations were somewhat harder on the inbound than on the outbound side.
4. The solar wind velocity measured by the JPA detector decreased during the foreshock from 18:20 to 20:00 UT (from 300 to 200 km/s) and increased again on the outbound side from 02:00 to 04:00 UT, (double peak structure) whereas the H_2O^+ ion density appeared to be anticorrelated to the solar wind velocity (0.1-0.4 and 0.5-0.1 /cm³, respectively), see Coates et al., 1990.
5. The acceleration of energetic particles in the foreshock, as well as in the in- and outbound bowshock takes place in association with an intense magnetosonic wave field of 3-4 min quasiperiodicity and enhanced RMS noise which has been detected by the Giotto magnetometer.
6. The foreshock (18:20 to 19:00 UT) appeared when $\sin^2\alpha$ was very small and the angle ϑ_{BN} was $\simeq 40°$ (indicating a more quasiparallel shock), i.e. the pickup process and the shock drift acceleration process can be discarded as the reason for the existence of the foreshock.
7. A combination of the first and second order Fermi process and the transit time damping effect are the most likely candidates for the particle acceleration observed by EPA/EPONA at the in- and outbound bowshock.
8. More particles escaped from the outbound than from the inbound cometosheath, indicating a general flux anisotropy, probably caused by the transit time damping mechanism.

Acknowledgments. The authors thank all who contributed to the development of the EPA/EPONA, the MAG and JPA experiment. The experiment EPA/ EPONA has been funded by the Max Planck Gesellschaft zur Förderung der Wissenschaften, by the Bundesministerium für Forschung und Technologie (grant number 010F082) and by the National Board for Science and Technology, Ireland.

References

Amata, E., V. Formisano, Energization of positive ions in the cometary foreshock region, *Planet Space Sci., 23*, 1243-1250, 1985.

Anderson, K.A., C.W. Carlson, D.W. Curtis, R.P. Lin, H. Réme, J.A. Sauvaud, C. d'Uston, A. Korth, A.K. Richter, D.A. Mendis, The upstream region, foreshock and bowshock wave at comet P/Halley from plasma electron measurements, *Astron. Astrophys. 187*, 290-292, 1987.

Armstrong, T.P., M.E. Pesses, R.D. Decker: Shock drift acceleration, In: Collisionless shocks in the heliosphere: Reviews of current research. Geophysical Monograph, Series Vol. 35, p.271-285, 1985. Ed. Tsurutani, B.T. and R.G. Stone.

Coates, A.J., R.P. Lin, B. Wilken, E. Amata, K.A. Anderson, H. Borg, D.A. Bryant, C.W. Carlson, D.A. Curtis, V. Formisano, K. Jockers, A.D. Johnstone, A. Korth, D.A. Mendis, H. Réme, A.K. Richter, H. Rosenbauer, J.A. Sauvaud, W. Stüdemann, M.F. Thomsen, C. d'Uston, J.D. Winningham, Giotto measurements of cometary and solar wind plasma at the Comet Halley bow shock, *Nature 327*, 489-492, 1987.

Coates, A.J., B. Wilken, A.D. Johnstone, K. Jockers, K.-H. Glassmeier, and D.E. Huddleston, Bulk properties and velocity distributions of water group ions at comet Halley: Giotto Measurements, *J. Geophys. Res., 95*, 10,249-10,260, 1990.

D'Uston, C., H. Réme, J.A. Sauvaud, A. Cros. K.A. Anderson, C.W. Carlson, D. Curtis, R.P. Lin, A. Korth, A.K. Richter, A. Mendis, Description of the main boundaries seen by the Giotto electron experiment inside comet P/Halley - solar wind interaction region, *Astron. Astrophys. 187*, 137-140, 1987.

Fisk, L.A, The acceleration of energetic particles in the interplanetary medium by transit time damping, *J. Geophys. Res., 81*, 4633-4640, 1976.

Fuselier, S.A., K.A. Anderson, H. Balsiger, K.H. Glassmeier, B.E. Goldstein, M. Neugebauer, H. Rosenbauer, E.G. Shelley, The Foreshock region upstream from the comet Halley bowshock, Symp. on the Diversity and Similarities of Comets 6-9 April 1989, Brussels, Belgium. ESA SP278, 77-82, 1987.

Glassmeier, K.H., F.M. Neubauer, M.H. Acuna, F. Mariani, Low frequency magnetic field fluctuations in comet P/Halley's magnetosheath: Giotto observations, *Astron. Astrophys., 187*, 65-68, 1987.

Gombosi, T.I., K. Lorenz, J.R. Jokipii, Combined first and second order Fermi acceleration at Comets, *Adv. Space Res., 9*, 337-341, 1989.

Ip, W.-H., W.I. Axford, The acceleration of particles in the vicinity of comets, *Planet. Space Sci., 34*, 1061-1065, 1986.

Ip, W.-H., W.I. Axford, A numerical simulation of charged particle acceleration and pitch angle scattering in the turbulent plasma environment of cometary comas, *20th International Cosmic Ray Conference, Moscow, Aug. 2-15, 1987, Vol. 3*, p.233-236.

Ip, W.-H., W.I. Axford, Cometary plasma physics, In: Physics of Comets in the Space Age, to be published by W.F. Huebner, 1989.

Ipavich, F.M., The Compton Getting effect for low energy particles, *Geophys. Res. Letters 1*, 149-152, 1974.

Isenberg, P.A., Energy diffusion of pickup ions upstream of comets, *J. Geophys. Res. 92*, 8795-8799, 1987.

McKenna-Lawlor, S., E. Kirsch, A. Thompson, D. O'Sullivan, K.-P. Wenzel, The light weight energetic particle detector EPONA and its performance on Giotto, *J. Phys. E. Sci. Instrum., 2*, 732-740, 1987.

McKenna-Lawlor, S., P. Daly, E. Kirsch, B. Wilken, D. O'Sullivan, A. Thompson, K. Kecskemety, A. Somogyi, A. Coates, In situ particle observations at comet Halley recorded by instrumentation aboard the Giotto and Vega 1 mission, *Annales Geophysicae 7*, (2), 121-128, 1989.

Kirsch, E., S. McKenna-Lawlor, P. Daly, A. Korth, F.M. Neubauer, D. O'Sullivan, A. Thompson, K.-P. Wenzel, Evidence for the field line reconnection process in the particle and magnetic field measurements obtained during the Giotto-Halley encounter, *Annales Geophysicae 7*, 107-114, 1989.

Kirsch, E., S. McKenna-Lawlor, P. Daly, W.H. Ip, F.M. Neubauer, A. Thompson, D. O'Sullivan, K.-P. Wenzel, Particle observations by EPA/EPONA during the outbound pass of Giotto from comet Halley and their relationship to large scale magnetic field irregularities, *Annales Geophysicae,8*, (7-8), 455-462, 1990.

Neubauer, F.M., K.H. Glassmeier, M. Pohl, J. Raeder, M.H. Acuna, L.F. Burlaga, N.F. Ness, G. Musmann, F. Mariani, M.K. Wallis, E. Ungstrup, H.U. Schmidt, First results from the Giotto magnetometer experiment at comet Halley, *Nature 321*, 352-355, 1989.

Neubauer, F.M., Giotto magnetic field results on the boundaries of the pile-up region and the magnetic cavity, *Astron. Astrophys. 187*, 73-79, 1987.

Neubauer, F.M., K.H. Glassmeier, M.H. Acuna, F. Mariani, G. Musmann, N.F. Ness, A.J. Coates, Giotto magnetic field observations at the outbound quasiparallel bowshock of comet Halley, *Annales Geophysicae 8*, (7-8), 463-472, 1990.

Neugebauer, M., A.J. Lazarus, H. Balsiger, S.A. Fuselier, F.M. Neubauer, H. Rosenbauer. The velocity distributions of cometary protons picked up by the solar wind, *J. Geophys. Res. 94*, 5227-5239, 1989.

Remé, H., J.A. Sauvaud, C. d'Uston, A. Cross, K.A. Anderson, C.W. Carlson, D.W. Curtis, R.P. Lin, A. Korth, A.K. Richter, D.A. Mendis, General features of comet P/Halley: solar wind interaction from plasma measurements, *Astron. Astrophys.* 33-38, 1987.

Richardson, J.G., S.W.H. Cowley, R.J. Hynds, C. Transquille, T.R. Sanderson, K.-P. Wenzel, P.W. Daly, Observations of energetic water group ions at comet Giacobini-Zinner: Implications for ion acceleration processes, *Planet. Space Sci. 35*, 1323-1345, 1987.

Sagdeev, R.Z., A.A. Galeev, V.D. Shapiro, V.J. Shevchenko, Solar wind interaction with comets as a model for the cosmic ray acceleration by shocks, *20th Intern. Cosmic Ray Conference, Moscow, August 2-15, 1987, Vol.9*, pp.103-120, 1987.